XXV International Conference on Integrable Systems and Quantum Symmetries (ISQS-25)

Journal of Physics: Conference Series Volume 965

Prague, Czech Republic
6 – 10 June 2017

ISBN: 978-1-5108-5852-7
ISSN: 1742-6588

Printed from e-media with permission by:

Curran Associates, Inc.
57 Morehouse Lane
Red Hook, NY 12571

Some format issues inherent in the e-media version may also appear in this print version.

Printed by Curran Associates, Inc. (2018)

Institute of Physics
Dirac House, Temple Back
Bristol BS1 6BE UK

Phone: 44 1 17 929 7481
Fax: 44 1 17 920 0979

techtracking@iop.org

Additional copies of this publication are available from:

Curran Associates, Inc.
57 Morehouse Lane
Red Hook, NY 12571 USA
Phone: 845-758-0400
Fax: 845-758-2633
Email: curran@proceedings.com
Web: www.proceedings.com

Table of contents

Volume 965

The XXV International Conference on Integrable Systems and Quantum Symmetries (ISQS-25)

6–10 June 2017, Prague, Czech Republic

Accepted papers received: 31 January 2018
Published online: 15 February 2018

Preface

Papers

IOP Conf. Series: Journal of Physics: Conf. Series **965** (2018) 011001 doi:10.1088/1742-6596/965/1/011001

XXV International Conference on Integrable Systems and Quantum Symmetries (ISQS-25)

Čestmír Burdík, Ondřej Navrátil and Severin Pošta

Abstract

The XXV International Conference on Integrable Systems and Quantum Symmetries (ISQS-25) organized by the Department of Mathematics, Faculty of Nuclear Sciences and Physical Engineering, Czech Technical University, Prague and the Bogoliubov Laboratory of Theoretical Physics of the Joint Institute for Nuclear Research continues a successful series of conferences held at the Czech Technical University which began in 1992 and is devoted to problems of mathematical physics related to the theory of integrable systems, quantum groups and quantum symmetries. During the last year, the conference gathered around 200 scientists from all over the world. 44 papers of plenary lectures and contributions presented at ISQS-25 are published in the present issue of the Journal of Physics: Conference Series.

IOP Conf. Series: Journal of Physics: Conf. Series **965** (2018) 011002 doi:10.1088/1742-6596/965/1/011002

Peer review statement

All papers published in this volume of *Journal of Physics: Conference Series* have been peer reviewed through processes administered by the proceedings Editors. Reviews were conducted by expert referees to the professional and scientific standards expected of a proceedings journal published by IOP Publishing.

Analytical Solutions of the Schrödinger Equation for the Manning-Rosen plus Hulthén Potential Within SUSY Quantum Mechanics

A. I. Ahmadov[1], Maria Naeem[2], M. V. Qocayeva[3], V. A. Tarverdiyeva[3]

[1] Department of Theoretical Physics, Baku State University, Z. Khalilov st. 23, AZ-1148, Baku, Azerbaijan,
[2] COMSATS Institute of Information Technology Park-Road, Chak-Shahzad, 44000,Islamabad, Pakistan, [3] Institute of Physics, Azerbaijan National Academy of Sciences, H. Javid Avenue,131, AZ-1143, Baku, Azerbaijan

E-mail: ahmadovazar@yahoo.com, maria_naeem91@yahoo.com, mefkureqocayeva@yahoo.com, vefa.tarverdiyeva@mail.ru

Abstract. In this paper, the bound state solution of the modified radial Schrödinger equation is obtained for the Manning-Rosen plus Hulthén potential by implementing the novel improved scheme to surmount the centrifugal term. The energy eigenvalues and corresponding radial wave functions are defined for any $l \neq 0$ angular momentum case via the Nikiforov-Uvarov (NU) and supersymmetric quantum mechanics (SUSYQM) methods. By using these two different methods, equivalent expressions are obtained for the energy eigenvalues, and the expression of radial wave functions transformations to each other is demonstrated. The energy levels are worked out and the corresponding normalized eigenfunctions are represented in terms of the Jacobi polynomials for arbitrary l states. A closed form of the normalization constant of the wave functions is also found. It is shown that, the energy eigenvalues and eigenfunctions are sensitive to n_r radial and l orbital quantum numbers.

1. Introduction

The exact solutions of the stationary Schrödinger equation play an important role in quantum mechanics since they contain all the necessary information regarding the quantum system under consideration. Since the wave function contains all necessary information for full description of a quantum system, therefore, an analytical solution of the Schrödinger, Klein-Fock-Gordon and Dirac equations is of high importance in quantum mechanics [1, 2]. There are few potentials for which the Schrödinger equation can be solved explicitly for all n_r radial and l orbital quantum states. However, analytic solutions are possible only for a few simple quantum systems like the movement in the spherical simmetrical field and the linear harmonic oscillator [1, 3, 4, 5]. In general, many quantum systems can be treated only by approximation methods or numerical solutions.

As known, one of the main objectives in theoretical physics since the early years of quantum mechanics (QM) is to obtain an exact solution of the Schrödinger equation for some special potentials of physical interest.

IOP Conf. Series: Journal of Physics: Conf. Series **965** (2018) 012001 doi:10.1088/1742-6596/965/1/012001

It would be interesting and an other side important to solve the Schrödinger equation for the Manning-Rosen plus Hulthén potential for $l \neq 0$, since it has been extensively used to describe the bound and continuum states of the interacting systems.

The central Manning-Rosen potential is defined by [6, 7]

$$V(r) = \frac{1}{kb^2}\left[\frac{\alpha(\alpha-1)e^{-2r/b}}{(1-e^{-r/b})^2} - \frac{A'e^{-r/b}}{(1-e^{-r/b})}\right], k = 2\mu/\hbar^2. \tag{1}$$

where A' and α are two dimensionless parameters, but parameter b has dimensions of length.

The Hulthén potential is defined by [8, 9]

$$V(r) = -\frac{Ze^2\delta e^{-\delta r}}{(1-e^{-\delta r})} \tag{2}$$

where Z and δ are respectively the atomic number and the screening parameter, determining the range for the Hulthén potential. The Hulthén potential is one of the important short-range potentials in physics.

It should be noted that, the radial Schrödinger equation for the Hulthén potential can be solved analytically for only the states with zero angular momentum [8, 9, 10, 11, 12]. For any l states a number of methods have been employed to evaluate bound-state energies numerically [13, 14, 15, 16, 17, 18, 19, 20, 21, 22, 23].

The combined potential which is considered in this study is obtained by adding Hulthén potential term to Manning-Rosen potential as:

$$V(r) = \frac{1}{k}\left[\frac{\alpha(\alpha-1)e^{-2r/b}}{b^2(1-e^{-r/b})^2} - \frac{A'e^{-r/b}}{b^2(1-e^{-r/b})}\right] - \frac{Ze^2\delta e^{-\delta r}}{(1-e^{-\delta r})}. \tag{3}$$

Unfortunately, for an arbitrary l-states ($l \neq 0$), the Schrödinger equation does not get an exact solution. So far, many methods were developed and applied, for the solution Schrödinger equation, such as supersymmetry (SUSY), factorization, Laplace transform approach and the path integral methods. An other method known as the Nikiforov-Uvarov (NU) method [24] was proposed for solving the Schrödinger equation analytically. Many works show the power and simplicity of NU method in solving central and noncentral potentials. This method is based on solving the second order linear differential equation by reducing to a generalized equation of hypergeometric type which is a second order homogeneous differential equation with polynomials coefficients of degree not exceeding the corresponding order of differentiation.

It is very important to notice that modified radial Schrödinger equation for the Manning-Rosen potential is fully studied in Ref. [25].

Taking into account the facts which are mentioned, in this article we obtain the energy eigenvalues and corresponding eigenfunctions for arbitrary l states by solving the modified radial Schrödinger equation for the Manning-Rosen plus Hulthén potential using Nikiforov-Uvarov (NU) method [24] and the shape invariance concept that was introduced by Gendenshtein [26, 27].

The rest of the present work is organized as follows. Bound-state Solution of the radial Schrödinger equation for Manning-Rosen plus Hulthén potential by NU method within ordinary quantum mechanics is provided in Section 2. In Section 3 we present the Solution of Schrödinger equation for Manning-Rosen plus Hulthén potential within SUSY QM and the numerical results and concluding remarks for energy levels and the corresponding normalized eigenfunctions are presented in Section 4.

2. Bound state Solution of the Radial Schrödinger equation for Manning-Rosen Plus Hulthén potential within ordinary quantum mechanics.

The radial Schrödinger equation for Manning-Rosen plus Hulthén potential is

$$R''(r)+\frac{2}{r}R'(r)+\frac{2\mu}{\hbar^2}\Bigg[E+Ze^2\delta\frac{e^{-\delta r}}{1-e^{-\delta r}}- \\ -\frac{l(l+1)\hbar^2}{2\mu r^2}-\frac{\alpha(\alpha-1)e^{-2r/b}}{b^2(1-e^{-r/b})^2}+\frac{A'e^{-r/b}}{b^2(1-e^{-r/b})}\Bigg]R(r)=0, \tag{4}$$

respectively.

The effective Manning-Rosen plus Hulthén potential is defined in this form:

$$V_{\text{eff}}(r)=-Ze^2\delta\frac{e^{-\delta r}}{1-e^{-\delta r}}+\frac{l(l+1)\hbar^2}{2\mu r^2}+\frac{\alpha(\alpha-1)e^{-2r/b}}{b^2(1-e^{-r/b})^2}-\frac{A'e^{-r/b}}{b^2(1-e^{-r/b})} \tag{5}$$

In order to solve Eq.(4) for $l\neq 0$, we must make an approximation for the centrifugal term. When $\delta r << 1$, $(r/b << 1)$ we use an improved approximation scheme [25, 28, 29] to deal with the centrifugal term,

$$\frac{1}{r^2}\approx\delta^2\left[C_0+\frac{e^{-\delta r}}{(1-e^{-\delta r})^2}\right], \tag{6}$$

where the parameter $C_0=\frac{1}{12}$ (Ref. [30]) is a dimensionless constant. It should be noted that this approximation, is only valid for small δr and it breaks down in the high screening region. After using this approximation radial Schrödinger equation is solvable analytically.

We assume $R(r)=\frac{1}{r}\chi(r)$ and $1/b=\delta$ in Eq.(4) then the modified radial Schrödinger equation becomes

$$\chi''(r)-l(l+1)\delta^2\left(C_0+\frac{e^{-\delta r}}{(1-e^{-\delta r})^2}\right)\chi(r)+ \\ \frac{2\mu}{\hbar^2}\left[E+Ze^2\delta\frac{e^{-\delta r}}{1-e^{-\delta r}}-\frac{\delta^2\alpha(\alpha-1)e^{-2r\delta}}{k^2(1-e^{-r\delta})^2}+\frac{A'\delta^2e^{-r\delta}}{k^2(1-e^{-r\delta})}\right]\chi(r)=0. \tag{7}$$

According to the (6) approximation scheme the modified effective potential $\widetilde{V}_{\text{eff}}(r)$ becomes

$$\widetilde{V}_{\text{eff}}(r)=-Ze^2\delta\frac{e^{-\delta r}}{1-e^{-\delta r}}+\frac{\delta^2\alpha(\alpha-1)e^{-2\delta r}}{(1-e^{-\delta r})^2}-\frac{A'\delta^2e^{-r/b}}{(1-e^{-\delta r})}+ \\ +\frac{l(l+1)\hbar^2}{2\mu}\delta^2\left(C_0+\frac{e^{-\delta r}}{(1-e^{-\delta r})^2}\right) \tag{8}$$

In order to transform Eq.(7), the equation of the generalized hypergeometric-type which is in the form [24]

$$\chi''(s)+\frac{\tilde{\tau}}{\sigma}\chi'(s)+\frac{\tilde{\sigma}}{\sigma^2}\chi(s)=0, \tag{9}$$

we use the following ansatz in order to make the differential equation more compact,

$$-\varepsilon^2=\frac{2\mu}{\hbar^2\delta^2}E,\ \ E<0,\beta^2=\frac{2\mu Ze^2}{\hbar^2\delta},s=e^{-\delta r}. \tag{10}$$

Hence, we obtain

$$\chi''(s) + \frac{\chi'(s)}{s} - \frac{1}{s^2(1-s)^2}\Big[-\varepsilon^2(1-s)^2 - l(l-1)(C_0(1-s)^2+s) + (\beta^2+A')s(1-s) - \alpha(\alpha-1)s^2\Big]\chi(s) = 0, \tag{11}$$

Now, we can successfully apply NU method for defining eigenvalues of energy. Finally for energy eigenvalues we find:

$$E_{nl} = \frac{-h^2\delta^2}{2\mu} \cdot \left[\frac{l(l+1)+1/2+n_r(n_r+1)+\sqrt{\frac{1}{4}+l(l+1)+\alpha(\alpha-1)}(2n_r+1)-\beta^2-A'}{2n_r+1+2\sqrt{\frac{1}{4}+\alpha(\alpha-1)+l(l+1)}}\right]^2 + \frac{h^2\delta^2}{2\mu}l(l+1)C_0, \tag{12}$$

For the radial eigenfunction we find:

$$\chi_{n_r}(s) = C_{n_r}s^{\sqrt{c}}(1-s)^K\frac{\Gamma(n_r+2\sqrt{c}+1)}{n_r!\Gamma(2\sqrt{c}+1)}\underset{21}{F}\left(-n_r, 2\sqrt{c}+2K+n_r, 1+2\sqrt{c}; s\right). \tag{13}$$

The normalization constant C_{n_r} can be found from normalization condition

$$\int_0^\infty |R(r)|^2r^2dr = \int_0^\infty |\chi(r)|^2dr = b\int_0^1 \frac{1}{s}|\chi(s)|^2ds = 1, \tag{14}$$

by using the following integral formula [32]:

$$\int_0^1 (1-s)^{2(\delta+1)}s^{2\lambda-1}\left\{\underset{21}{F}(-n_r, 2(\delta+\lambda+1)+n_r, 2\lambda+1; s)\right\}^2 dz = \frac{(n_r+\delta+1)n_r!\Gamma(n_r+2\delta+2)\Gamma(2\lambda)\Gamma(2\lambda+1)}{(n_r+\delta+\lambda+1)\Gamma(n_r+2\lambda+1)\Gamma(2(\delta+\lambda+1)+n_r)}, \tag{15}$$

for $\delta > \frac{-3}{2}$ and $\lambda > 0$. After simple calculations, we obtain normalization constant as

$$C_{n_r} = \sqrt{\frac{n_r!2\sqrt{c}(n_r+K+\sqrt{c})\Gamma(2(K+\sqrt{c})+n_r)}{b(n_r+K)\Gamma(n_r+2\sqrt{c}+1)\Gamma(n_r+2K)}}. \tag{16}$$

3. The Solution of Schrödinger equation for Manning-Rosen Plus Hulthén potential within SUSY Quantum Mechanics

According to Supersymmetric quantum mechanics, the eigenfunction of ground state $\chi_0(r)$ in Eq.(7) is a form as below [33, 34]

$$\chi_0(r) = Nexp\left(-\frac{\sqrt{2\mu}}{\hbar}\int W(r)dr\right). \tag{17}$$

IOP Conf. Series: Journal of Physics: Conf. Series **965** (2018) 012001 doi:10.1088/1742-6596/965/1/012001

where N and $W(r)$ are normalized constant superpotential, respectively. The connection between the supersymmetric partner potentials $V_1(r)$ and $V_2(r)$ of the superpotential $W(r)$ is as follows [33, 34]:

$$V_1(r) = W^2(r) - \frac{\hbar}{\sqrt{2\mu}} W'(r) + E \,, \quad V_2(r) = W^2(r) + \frac{\hbar}{\sqrt{2\mu}} W'(r) + E. \tag{18}$$

The particular solution of the Riccati equation (18) searches the following form:

$$W(r) = -\frac{\hbar}{\sqrt{2\mu}} \left(A + \frac{Be^{-\delta r}}{1 - e^{-\delta r}} \right), \tag{19}$$

where A and B unknown constants. Since $V_1(r) = \tilde{V}_{eff}(r)$, having inserted the relations Eq.(8) and Eq.(19) into the expression Eq.(18), and from comparison of compatible quantities in the left and right sides of the equation, we find the following relations for A and B constants :

$$A^2 = -\frac{2\mu E}{\hbar^2} + \delta^2 C_0 l(l+1) = \varepsilon^2 \delta^2 + \delta^2 C_0 l(l+1), \tag{20}$$

$$2AB - \delta B = \delta^2 l(l+1) - \beta^2 \delta^2 - A' \delta^2 - \delta^2 \alpha(\alpha - 1), \tag{21}$$

$$B^2 - \delta B = \delta^2 l(l+1) - \delta^2 \alpha(\alpha - 1). \tag{22}$$

Considering extremity conditions to wave functions, we obtain $B > 0$ and $A < 0$.
Solving Eq.(22) yields

$$B = \frac{\delta \pm \sqrt{\delta^2 + 4\delta^2 l(l+1) + \alpha(\alpha+1)}}{2} = \frac{\delta \pm 2\delta \sqrt{\frac{1}{4} + l(l+1) + \alpha(\alpha+1)}}{2}, \tag{23}$$

and considering $B > 0$ from Eqs.(21) and (22), we find

$$2AB - B^2 = \delta^2 l(l+1) - \beta^2 \delta^2 - A' \delta^2, \tag{24}$$

or

$$A = \frac{B}{2} - \frac{\delta^2(\beta^2 + A')}{2B}, \tag{25}$$

From Eq.(20) and (25), we find

$$-\frac{2\mu E_0}{\hbar^2 \delta^2} = \frac{1}{\delta^2} \left(\frac{B}{2} - \frac{\delta^2(\beta^2 + A')}{2B} \right)^2 - C_0 \lambda. \tag{26}$$

Finally, for energy eigenvalue, we obtain

$$E_0 = \frac{\hbar^2 \lambda C_0 \delta^2}{2\mu} - \frac{\hbar^2}{2\mu} \left(\frac{B}{2} - \frac{\delta^2(\beta^2 + A')}{2B} \right)^2, \tag{27}$$

When $r \to \infty$, the chosen superpotential $W(r)$ is $W(r) \to -\frac{\hbar A}{\sqrt{2\mu}}$.
Having inserted the Eq.(19) into Eq.(18), then we can find supersymmetric partner potentials $V_1(r)$ and $V_2(r)$ in the form

$$\begin{aligned} V_1(r) &= W^2(r) - \frac{\hbar}{\sqrt{2\mu}}W'(r) + E = \\ &= \frac{\hbar^2}{2\mu}\left[A^2 + \frac{(2AB - \delta B)e^{-\delta r}}{1 - e^{-\delta r}} + \frac{(B^2 - \delta B)e^{-2\delta r}}{(1 - e^{-\delta r})^2}\right]. \end{aligned} \tag{28}$$

and

$$\begin{aligned} V_2(r) &= W^2(r) + \frac{\hbar}{\sqrt{2\mu}}W'(r) + E = \\ &= \frac{\hbar^2}{2\mu}\left[A^2 + \frac{(2AB + \delta B)e^{-\delta r}}{1 - e^{-\delta r}} + \frac{(B^2 + \delta B)e^{-2\delta r}}{(1 - e^{-\delta r})^2}\right], \end{aligned} \tag{29}$$

By using superpotential $W(r)$ from Eq.(19) we can find $\chi_0(r)$ radial eigenfunction in this form:

$$\begin{aligned} \chi_0(r) = N_0 exp\left(-\frac{\sqrt{2\mu}}{\hbar}\int W(r)dr\right) = N_0 exp\left[\int\left(A + \frac{Be^{-\delta r}}{1 - e^{-\delta r}}\right)dr\right] = \\ N_0 e^{Ar} exp\left[\frac{B}{\delta}\int\frac{d(1 - e^{-\delta r})}{1 - e^{-\delta r}}\right] = N_0 e^{Ar}(1 - e^{-\delta r})^{\frac{B}{\delta}} \end{aligned} \tag{30}$$

here $r \to 0$; $\chi_0(r) \to 0, B > 0$ and $r \to \infty$; $\chi_0(r) \to 0, A < 0$

Two partner potentials $V_1(r)$ and $V_2(r)$ which differ from each other with additive constants and have the same functional form are called invariant potentials [26, 27]. Thus, for the partner potentials $V_1(r)$ and $V_2(r)$ given with Eq.(28) and Eq.(29), the invariant forms are:

$$\begin{aligned} R(B_1) = V_2(B, r) - V_1(B_1, r) = \frac{\hbar^2}{2\mu}\left[A^2 - A_1^2\right] = \\ = \frac{\hbar^2}{2\mu}\left[\left(\frac{B}{2} - \frac{\delta^2(\beta^2 + A')}{2B}\right)^2 - \left(\frac{B + \delta}{2} - \frac{\delta^2(\beta^2 + A')}{2(B + \delta)}\right)^2\right]. \end{aligned} \tag{31}$$

So we have

$$\begin{aligned} R(B_i) = V_2[B + (i-1)\delta, r] - V_1[B + i\delta, r] = \\ -\frac{\hbar^2}{2\mu}\left[\left(\frac{B + i\delta}{2} - \frac{\delta^2(\beta^2 + A')}{2(B + i\delta)}\right)^2 - \left(\frac{B + (i-1)\delta}{2} - \frac{\delta^2(\beta^2 + A')}{2(B + (i-1)\delta)}\right)^2\right]. \end{aligned} \tag{32}$$

where the reminder $R(B_1)$ is independent of r.

If we continue this procedure and make the substitution $B_{n_r} = B_{n_r - 1} + \delta = B + n_r\delta$ at every step until $B_n \geq 0$, the whole discrete spectrum of Hamiltonian $H_-(B)$:

$$E_{n_r l} = E_0 + \sum_{i=0}^{n} R(B_i) =$$

$$= \frac{\hbar^2\lambda}{2\mu}\delta^2 C_0 - \frac{\hbar^2}{2\mu}\left(\frac{B}{2} - \frac{\delta^2(\beta^2+A')}{2B}\right)^2 - \frac{\hbar^2}{2\mu}\left[\left(\frac{B+\delta}{2}\frac{\delta^2(\beta^2+A')}{2(B+\delta)}\right)^2 - \right.$$

$$-\left(\frac{B}{2} - \frac{\delta^2(\beta^2+A')}{2B}\right)^2 + \left(\frac{B+2\delta}{2} - \frac{\delta^2(\beta^2+A')}{2(B+2\delta)}\right)^2 + \ldots.. +$$

$$+\left(\frac{B+(n_r-1)\delta}{2} - \frac{\delta^2(\beta^2+A')}{2(B+(n_r-1)\delta)}\right)^2 - \left(\frac{B+(n_r-2)\delta}{2} - \right.$$

$$\left. - \frac{\delta^2(\beta^2+A')}{2(B+(n_r-2)\delta)}\right)^2 - \left(\frac{B+(n_r-2)\delta}{2} - \frac{\delta^2(\beta^2+A')}{2(B+(n_r-2)\delta)}\right)^2 +$$

$$+\left(\frac{B+n_r\delta}{2} - \frac{\delta^2(\beta^2+A')}{2(B+n_r\delta)}\right)^2 -$$

$$\left. -\left(\frac{B+(n_r-1)\delta}{2} - \frac{\delta^2(\beta^2+A')}{2(B+(n_r-1)\delta)}\right)^2\right] =$$

$$= \frac{\hbar^2\lambda}{2\mu}\delta^2 C_0 - \frac{\hbar^2}{2\mu}\left(\frac{B+n_r\delta}{2} - \frac{\delta^2(\beta^2+A')}{2(B+n_r\delta)}\right)^2, \tag{33}$$

and we obtain

$$E_{n_r l} = \frac{\hbar^2\lambda}{2\mu}\delta^2 C_0 - \frac{\hbar^2}{2\mu}\left(\frac{B+n_r\delta}{2} - \frac{\delta^2(\beta^2+A')}{2(B+n_r\delta)}\right)^2, \tag{34}$$

Finally, for energy eigenvalues we found

$$E_{nl} = \frac{-h^2\delta^2}{2\mu}\cdot$$

$$\left[\frac{l(l+1)+1/2+n(n+1)+\sqrt{\frac{1}{4}+\lambda+\alpha(\alpha-1)}(2n+1)-\beta^2-A'}{2n+1+2\sqrt{\frac{1}{4}+\alpha(\alpha-1)+\lambda}}\right]^2 +$$

$$\frac{h^2\delta^2}{2\mu}l(l+1)C_0, \tag{35}$$

The obtained result of radial Schrödinger equation by using the Eq.(17) of the ground state eigenfunction is exactly same with the result obtained by using NU method.

4. Numerical Results and Conclusion

Solutions of the modified radial Schrödinger equation for the Manning-Rosen plus Hulthén potential are obtained within ordinary quantum mechanics by applying the Nikiforov-Uvarov method and within SUSY QM by applying the shape invariance concept that was introduced by Gendenshtein method in which we have used the improved approximation scheme to the centrifugal potential for arbitrary l states. Both ordinary and SUSY quantum mechanical energy eigenvalues and corresponding eigenfunctions obtained for arbitrary l quantum numbers. Solutions of the modified radial Schrödinger equation for the Manning-Rosen plus Hulthén potential are obtained within ordinary quantum mechanics by applying the Nikiforov-Uvarov method and within SUSY QM by applying the shape invariance concept that was introduced by Gendenshtein method in which we have used the improved approximation scheme to the centrifugal potential for arbitrary l states. Both ordinary and SUSY quantum mechanical energy eigenvalues and corresponding eigenfunctions obtained for arbitrary l quantum numbers. In the Table I, we present numerical results for the energy eigenvalues of the Hulthén, Manning-Rosen and Manning-Rosen plus Hulthén potentials as a function of screening parameter for various state in atomic units which is obtained by using ordinary (NU method) and SUSY QM(shape invariance method) methods. It can be seen from Table I that, the bound states more stable in the case Manning-Rosen plus Hulthén potential the relative Manning-Rosen and Hulthén potentials cases. This opens a new possibilities for determining of the properties of the interactions in atomic system. As a conclusion of the results presented in these tables, the numerical analyses obtained of the analytically solution is very sensitive to the n_r radial and l orbital quantum numbers. The results are sufficiently accurate for practical purpose.

Our detailed analysis show that, if we take $C_0 = 0$, $\alpha = 0$, $A' = 0$ and $l \neq 0$ in Eqs. (12) and (35) leads to the same result obtained in the work [35] and $l = 0$, $\alpha = 0$, and $A' = 0$ to the same result as in [10].

It is shown, that energy eigenvalues and corresponding eigenfunctions are identical for both ordinary and SUSY QM. These results in the special cases are compared to the results [25, 36] and the numerical results for the energy eigenvalues are obtained by using MATHEMATICA package programm.

We firstly, present an alternative method to obtain the energy eigenvalues and corresponding eigenfunctions of the Manning-Rosen plus Hulthén potential within the framework SUSY quantum mechanics for any l states. Here after, an analytical study of the Schrödinger equation have been performed for Manning-Rosen plus Hulthén potential using the improved approximation scheme to the centrifugal term for arbitrary l-states.

The energy eigenvalues of the bound states and corresponding eigenfunctions are analytically found via both of NU and SUSY quantum mechanics. The same expressions were obtained for the energy eigenvalues, and the expression of radial wave functions transformed each other is shown by using these methods. A closed form of the normalization constant of the wave functions is also found.

It is worth to mention that the Manning-Rosen and Hulthén potential is one of the important exponential potential, and it is a subject of interest in many fields of physics and chemistry. The main results of this paper are the explicit and closed form expressions for the energy eigenvalues and the normalized wave functions. The method presented in this paper is a systematic one and in many cases it is one of the most concrete works in this area.

Consequently, studying of analytical solution of the modified radial Schrödinger equation for the Manning-Rosen plus Hulthén potential within the framework ordinary and SUSY QM could provide valuable information on the QM dynamics at atomic and molecule physics and opens new window.

We can conclude that our analytical results of this study are expected to enable new possibilities for pure theoretical and experimental physicist, because the results are exact and more general.

Acknowledgments

One of the authors, A. I. A. thanks the organizers of the ISQS-25, in Prague especially to the Prof. Cestmir Burdik for the opportunity to speak at the conference, financial support and for their exceptional hospitality.

[1] Greiner W 2001 *Quantum Mechanics* **4th. edn.** (Springer, Berlin)
[2] Bagrov V G and Gitman D M 1990 *Exact Solutions of Relativistic Wave Equations* (Kluwer Academic Publishers, Dordrecht)
[3] Schiff L I 1955 *Quantum Mechanics* **3rd edn.** (New York: McGraw-Hill)
[4] Landau L D and Lifshitz E M 1977 *Quantum Mechanics, Non-Relativistic Theory* **3rd edn.** (Oxford: Pergamon)
[5] Dong Shi-Hai 2007 *Factorization Method in Quantum Mechanics* (The Netherlands: Springer)

[6] Manning M F *Phys. Rev.* 1933 **29** 951. doi:10.1103/PhysRev.44.951
[7] Manning M F and Rosen N 1933 *Phys. Rev.* **44** 953. doi:10.1103/PhysRev.44.953
[8] Hulthén L 1942 *Ark. Mat. Astron. Fys.* A **28** 5
[9] Hultheń L 1942 *Ark. Mat. Astron. Fys.* B **29** 1
[10] Flügge S 1994 *Practical Quantum Mechanics* **1** (Springer, Berlin)
[11] Lam C S and Varshni Y P 1971 *Phys. Rev.* A **4** 1875. doi:10.1103/PhysRevA.4.1875
[12] Varshni Y P 1990 *Phys. Rev.* A **41** 4682. doi:10.1103/PhysRevA.41.4682
[13] Lai C S and Lin W C 1980 *Phys. Lett.* A **78** 335. doi:doi.org/10.1016/0375-9601(80)90388-6
[14] Dutt R and Mukherji U 1982 *Phys. Lett.* A **90** 395. doi.org/10.1016/0375-9601(82)90793-9
[15] Patil S H 1984 *J. Phys.* A **17** 575. doi.org/10.1088/0305-4470/17/3/019
[16] Popov V S and Weinberg V M 1985 *Phys. Lett.* A **107** 371 doi.org/10.1016/0375-9601(85)90692-9
[17] Roy B and Roychoudhury R 1987 *J. Phys.* A **20** 3051. doi.org/10.1088/0305-4470/20/10/048
[18] Tang A Z and Chan F T 1987 *Phys. Rev.* A **35** 911. doi:10.1103/PhysRevA.35.911
[19] Lai C H 1987 *J. Math. Phys.* **28** 1801. doi.org/10.1063/1.527439
[20] Matthys P and De Meyer H 1988 *Phys. Rev.* A **38** 1168. doi:10.1103/PhysRevA.38.1168
[21] Laha U, Bhattacharyya C, Roy K, and Talukdar B 1988 *Phys. Rev.* C **38** 558. doi:10.1103/PhysRevC.38.558
[22] Talukdar B, Das U, Bhattacharyya C and Bera P K 1992 *J. Phys.* A **25** 4073. doi.org/10.1088/0305-4470/25/14/021
[23] Filho E D and Ricotta R M 1995 *Mod. Phys. Lett.* A **10** 1613. doi.org/10.1142/S0217732395001733
[24] Nikiforov A F and Uvarov V B 1988 *Special Functions of Mathematical Physics* (Birkhäuser, Basel)
[25] Qiang Wen-Chao and Dong Shi-Hai 2009 *Phys. Scr.* **79** 045004. doi.org/10.1088/0031-8949/79/04/045004
[26] Gendenshtein L E 1983 *JETP Lett.* **38** 356. doi:http://www.jetpletters.ac.ru/ps/1822/index.shtml
[27] Gendenshtein L E and Krive I V 1985 *Sov. Phys. Usp.* **28** 645. doi:dx.doi.org/10.1070/PU1985v028n08ABEH003882
[28] Qiang Wen-Chao and Dong Shi-Hai 2007 *Phys. Lett.* A **363** 169. doi:doi.org/10.1016/j.physleta.2006.10.091
[29] Wei Gao-Feng and Dong Shi-Hai 2008 *Phys. Lett.* A **373** 49. doi:doi.org/10.1016/j.physleta.2008.10.064
[30] Jia C S, Chen T and Cui L G 2009 *Phys. Lett.* A **373** 1621. doi:doi.org/10.1016/j.physleta.2009.03.006
[31] Area I, Godoy E, Ronveaux A and Zarzo A 2003 *J. Comput. Appl. Math.* **157** 93. doi:doi.org/10.1016/S0377-0427(03)00377-7
[32] Abramowitz M and Stegun I A 1964 *Handbook of Mathematical Functions with Formulas, Graphs and Mathematical Tables* (Dover, New York)
[33] Cooper F, Khare A and Sukhatme U 2001 *Supersymmetry in Quantum Mechnics* World Scientific
[34] Cooper F, Khare A and Sukhatme U 1995 *Phys. Rep.* **251** 267. doi:doi.org/10.1016/0370-1573(94)00080-M
[35] Ikhdair S M and Sever R 2007 *J. of. Math. Chem.* **42** 461. doi:10.1007/s10910-006-9115-8
[36] Ahmadov H I, Jafarzade Sh I and Qocayeva M V 2015 *Int. J. Mod. Phys.* A **30** 1550193. doi:dx.doi.org/10.1142/S0217751X15501936

state	δ	Manning-Rosen potential	Hulthén potential	Manning-Rosen plus Hulthén potential
2p	0.025	-0.221994	-0.225573	-0.447567
	0.050	-0.202355	-0.202292	-0.404646
	0.075	-0.183505	-0.180156	-0.363662
	0.10	-0.165446	-0.159167	-0.324613
	0.150	-0.131699	-0.120625	-0.252324
	0.200	-0.101113	-0.0866667	-0.18778
	0.250	-0.0736887	-0.0572917	-0.13098
	0.300	-0.0494253	-0.0325	-0.0819253
	0.350	-0.02833231	-0.0122917	-0.0406147
3p	0.025	-0.216964	-0.0874653	-0.304429
	0.050	-0.192805	-0.0665278	-0.259333
	0.075	-0.169949	-0.0482986	-0.218247
	0.100	-0.148394	-0.0327778	-0.181172
	0.150	-0.109192	-0.00986111	-0.119053
3d	0.025	-0.216964	-0.0874653	-0.304429
	0.050	-0.192805	-0.0665278	-0.259333
	0.075	-0.169949	-0.0482986	-0.218247
	0.100	-0.148394	-0.0327778	-0.181172
	0.150	-0.109192	-0.00986111	-0.119053
4p	0.025	-0.223478	-0.0399479	-0.263426
	0.050	-0.205195	-0.0222917	-0.227486
	0.075	-0.187573	-0.00953125	-0.197105
	0.100	-0.170614	-0.00166667	-0.172281
4d	0.025	-0.223478	-0.0399479	-0.263426
	0.050	-0.205195	-0.0222917	-0.227486
	0.075	-0.187573	-0.00953125	-0.197105
	0.100	-0.170614	-0.00166667	-0.172281

state	δ	Manning-Rosen potential	Hulthén potential	Manning-Rosen plus Hulthén potential
4f	0.025	-0.223478	-0.0399479	-0.263426
	0.050	-0.205195	-0.0222917	-0.227486
	0.075	-0.187573	-0.00953125	-0.197105
	0.100	-0.170614	-0.00166667	-0.172281
5p	0.025	-0.242051	-0.0188542	-0.260905
	0.050	-0.241574	-0.00541667	-0.246991
5d	0.025	-0.242051	-0.0188542	-0.260905
	0.050	-0.241574	-0.00541667	-0.246991
5f	0.025	-0.242051	-0.0188542	-0.260905
	0.050	-0.241574	-0.00541667	-0.246991
5g	0.025	-0.242051	-0.0188542	-0.260905
	0.050	-0.241574	-0.00541667	-0.246991
6p	0.025	-274104	-0.00835069	-0.282454
6d	0.025	-274104	-0.00835069	-0.282454
6g	0.025	-274104	-0.00835069	-0.282454

Table 1. Energy eigenvalues of the Manning-Rosen, Hultheń and Manning-Rosen plus Hulthén potentials as a function of the screening parameter for 2p, 3p, 3d, 4p, 4d, 4f, 5p, 5d, 5f, 5g, 6p, 6d, 6f and 6g states in atomic units ($\hbar = m = e = 1$) and for $Z = 1$.

IOP Conf. Series: Journal of Physics: Conf. Series **965** (2018) 012002 doi:10.1088/1742-6596/965/1/012002

Corrections to Newton's law of gravitation - application to hybrid Bloch brane

C A S Almeida, D F S Veras and D M Dantas

Physics Department - Universidade Federal do Ceará, Fortaleza, CE, BRAZIL

E-mail: carlos@fisica.ufc.br, franklin@fisica.ufc.br, davi@fisica.ufc.br

Abstract. We present in this work, the calculations of corrections in the Newton's law of gravitation due to Kaluza-Klein gravitons in five-dimensional warped thick braneworld scenarios. We consider here a recently proposed model, namely, the hybrid Bloch brane. This model couples two scalar fields to gravity and is engendered from a domain wall-like defect. Also, two other models the so-called asymmetric hybrid brane and compact brane are considered. Such models are deformations of the ϕ^4 and sine-Gordon topological defects, respectively. Therefore we consider the branes engendered by such defects and we also compute the corrections in their cases. In order to attain the mass spectrum and its corresponding eigenfunctions which are the essential quantities for computing the correction to the Newtonian potential, we develop a suitable numerical technique. The calculation of slight deviations in the gravitational potential may be used as a selection tool for braneworld scenarios matching with future experimental measurements in high energy collisions

1. Introduction

Phenomenological application to braneworld models is a very interesting current subject. As can be found in detail in the introduction of reference [1], we have recent works in the braneworld branch explaining the process in the meson B decay and in the neutrinos physics, setting bounds into corrections to Coulomb's law and into electrical conductivity. Selecting parameters to the experimental data of the nucleon-nucleus total cross-section of various chemical elements, and interesting applications to black holes and gravitational waves, are other examples.

In this view, one of most famous scenarios is the Bloch brane [2]. This smooth model consists in a thick brane generated by two interacting scalar fields which solve several issues in the fields localisation present in thin models [3, 4], especially in the context of gauge fields and fermion fields. The Bloch brane allows the existence of internal structures and a rich variety of scalar field deformations. So, there are many other models based on the initial proposal of the Bloch brane, such the degenerate Bloch brane [1] and recently the hybrid Bloch brane [5]. The term "hybrid" comes from the existence of an interval with compact structure [6]. One of the scalar fields that compose the Bloch model is the well-known kink solution, which describes a wide variety of physics phenomena by connecting smoothly two minima. Unlike the kink where the energy density vanishes asymptotically, a compact kink is a linear compact structure, with the energy density that vanishes outside a closed interval [6]. Then, the Bloch brane proposed in the reference [5] is the so-called hybrid, because exhibits a usual thick brane inside an interval and a compact solution outside of this interval. We analyse in this work the gravity localization

IOP Conf. Series: Journal of Physics: Conf. Series **965** (2018) 012002 doi:10.1088/1742-6596/965/1/012002

and a perspective to the application of corrections to Newton's law in the hybrid Bloch brane model.

In next section 2, we made a brief review of the usual Bloch brane of reference [2], next we present the two versions of the hybrid Bloch brane of reference [5]. In the section 3 we introduce the general way to confine gravity in a 5D warped model and comment how to obtain the expression of correction to gravity potential. Finally, in the conclusion, we point out the main points and results of this work.

2. The Hybrid Bloch brane

2.1. The usual Bloch brane

The Bloch brane is built upon a 5D warped geometry with the metric [2]:

$$ds^2 = g_{ab}dx^a dx^b = e^{2A(y)}\eta_{\mu\nu}dx^\mu dx^\nu - dy^2 \ , \tag{1}$$

where the indexes $a, b = 0, 1, 2, 3, 4$, the $A(y)$ is the warp factor and the $\eta_{\mu\nu}$ is the 4D Minkowski metric ($\mu, \nu = 0, 1, 2, 3$).

The action for this model is written in this form [2]:

$$I = \int d^4x dy \sqrt{|g|}\left(-\frac{R}{4} + \mathcal{L}_s(\phi, \chi)\right) \ , \tag{2}$$

where the R represents the scalar curvature and the two scalar fields Lagrangian is given by:

$$\mathcal{L}_s(\phi, \chi) = \frac{1}{2}\left(\partial_a\phi\partial^a\phi + \partial_a\chi\partial^a\chi + 2V(\phi, \chi)\right) \tag{3}$$

For the Bloch brane, the scalar fields ϕ, χ are only dependent on extra dimension coordinate y, so the equation of motions reads [2]:

$$\begin{aligned}
\phi' + 4A'\phi' &= \frac{\partial V}{\partial \phi}, \\
\chi' + 4A'\chi' &= \frac{\partial V}{\partial \chi}, \\
A'^2 &= -\frac{1}{6}\left(\phi'^2 + \chi'^2\right) - \frac{1}{3}V, && (4) \\
A'' &= -\frac{2}{3}\left(\phi'^2 + \chi'^2\right), && (5)
\end{aligned}$$

the primes denote the derivative with relation to y coordinate.

The above equation of motion can be simplified to first order equations if the so-called superpotential W is considered [5]:

$$V(\phi, \chi) = \frac{1}{2}\left[\left(\frac{\partial W}{\partial \phi}\right)^2 + \left(\frac{\partial W}{\partial \chi}\right)^2 - \frac{8}{3}W^2\right] . \tag{6}$$

Thus, the first-order version of equations (5) with the superpotential (6) becomes:

$$\phi' = \frac{\partial W}{\partial \phi}, \ \chi' = \frac{\partial W}{\partial \chi}, \ A' = -\frac{2}{3}W \ . \tag{7}$$

Hence, the Ref. [2] defines the following superpotential to the Bloch Brane as:

$$W(\phi,\chi) = \phi - \frac{1}{3}\phi'^3 - r\phi\chi^2 \, , \tag{8}$$

where the real parameter r, limited to $r \in (0,1)$, regulates the regimen of one field solution to the two fields solution. The choice of superpotential in equation (8) allows the analytical solution to warp factor and fields as follows [2, 5]:

$$\begin{aligned} A(y) &= \frac{1}{9r}\left[(1-3r)\tanh^2(2ry) - 2\ln\cosh(2ry)\right] , \\ \phi(y) &= \tanh(2ry), \\ \chi(y) &= \pm\sqrt{\left(\frac{1}{r}-1\right)}\,\mathrm{sech}(2ry) \, . \end{aligned} \tag{9}$$

Note that the profile of ϕ field is a kink-like solution, while the χ field exhibits a lump-like profile. These solutions provide a thick brane scenario with localised energy given by [2]

$$\rho(y) = \mathrm{e}^{2A(y)}\left[\frac{1}{2}\left(\frac{d\phi(y)}{dy}\right)^2 + \frac{1}{2}\left(\frac{d\chi(y)}{dy}\right)^2 + V\left(\phi(y),\chi(y)\right)\right] . \tag{10}$$

The Bloch brane energy configuration can presents a split according to the r parameter, as can be seen in figure 1. For more details and physical applications to this Bloch brane model, we recommend the work [1].

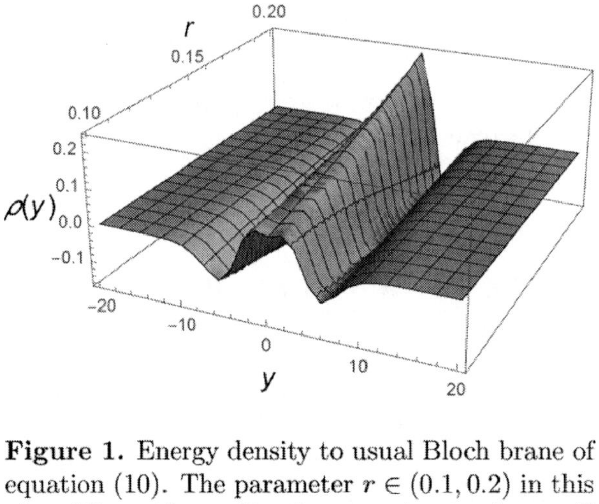

Figure 1. Energy density to usual Bloch brane of equation (10). The parameter $r \in (0.1, 0.2)$ in this plot. We verify a split of energy density for small values of r.

2.2. Symmetric and asymetric hybrid Bloch brane

Other choices to the superpotential $W(\phi,\chi)$ will be analysed in this section. The first case is the one that provides a symmetric profile with respect to an axis where the brane is placed. So, the warp factor and the energy density are even function in this case. This is written by modifying the usual Bloch superpotential in equation (8) as follows [5]:

$$W(\phi,\chi) = \phi - \frac{\phi^{2n+1}}{2n+1} - r\phi\chi^2 \, , \tag{11}$$

where n is a positive integer parameter that, for large values, split the brane in two intervals. We have two domains, one inside interval where the brane is non-compact, and other outside interval where the brane is compact. Due to this behaviour, this model is called Hybrid Bloch Brane [5]. Moreover, for $n = 1$ the usual Bloch brane superpotential (8) is retrieved. The reference [5] shows analytically that for $n \to \infty$ the model achieves, exactly, the thin brane profile in the outside interval due to compact behaviour.

In order to illustrate the differences between the usual structure and the compact structure we compare the ϕ field in the kink-like profile ($n = 1$) to the half-compact profile ($n \to \infty$) in figure 2. Note in the Fig. 2 that while the kink (black full line) achieves the points $\phi(y \to \pm\infty) = \pm 1$ the kink compacton (blue dashed line) achieves $\phi = \pm 1$ for $|y| > 1$.

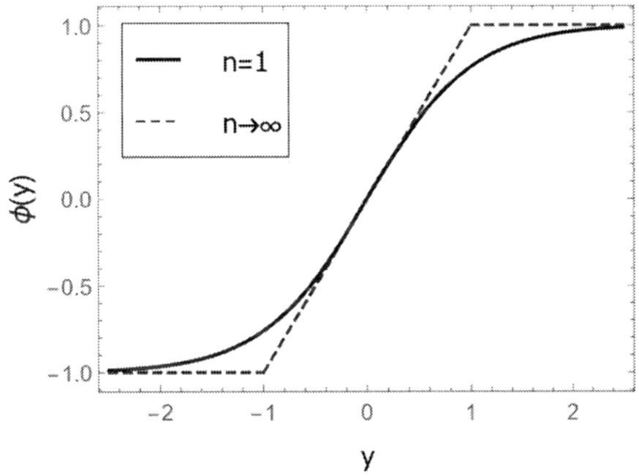

Figure 2. Differences between the kink and the compacton.

Another choice to the superpotential is the one that exhibits an asymmetric profile to the brane, obtained as:

$$W(\phi, \chi) = \phi - \frac{\phi^2}{2} + \frac{\phi^{p+1}}{p+1} - \frac{\phi^{p+2}}{p+2} - r\phi\chi^2 \ . \tag{12}$$

Here the p parameter is an odd integer that also allows the existence of a compact interval to the brane, and also reproduces the usual Bloch brane superpotential (8) when $p = 1$. It is worthwhile to mention the important feature of asymmetric models in the massive modes of gravity. In reference [7] we verify that in these models the resonant state provides a stronger contribution to the correction in the Newtonian potential.

3. Gravity localisation and the correction to Newton's Law

In order to obtain the correction to Newton's law on the hybrid Bloch brane scenario, we need to perform the gravity localisation by introducing a small fluctuation $h_{\mu,\nu}(\vec{x}, y)$ in the metric (1) in the form [8]

$$ds^2 = \mathrm{e}^{2A(y)} \left(\eta_{\mu\nu} + h_{\mu\nu}\right) dx^\mu dx^\nu - dy^2 \ . \tag{13}$$

Now assuming the transverse-traceless gauge and imposing that $h_{5N} = 0$ [8], we get the following equation of motion:

$$h''_{\mu\nu} + 4A'h'_{\mu\nu} = \mathrm{e}^{-2A}\Box h_{\mu\nu} \ , \tag{14}$$

where the $\Box$ is the 4D d'Alembertian. Setting the Kaluza-Klein (KK) decomposition [9]:

$$h_{\mu\nu}(\vec{x}, y) = \sum_{k=0}^{\infty} \tilde{\eta}_{\mu\nu}^{(k)} \varphi_k(y) \ , \tag{15}$$

with the 4D portion

$$\eta_{\mu\nu}\partial^{\mu}\partial^{\nu}\tilde{\eta}_{\mu\nu}^{(k)} = -m_k^2\tilde{\eta}_{\mu\nu}^{(k)} , \tag{16}$$

being m_k the gravitational KK mass, we have the equation of motion in the form

$$\varphi_k''(y) + 4A'\varphi_k'(y) = -m_k^2\varphi_k(y) . \tag{17}$$

Now, we use a conformal transformation in order to transform the equation (17) into a Schrödinger-like equation. Now we use the independent variable transformation $y \to z$ [8]

$$\frac{dz}{dy} = \mathrm{e}^{-A(y)} \to z(y) = \int_y \mathrm{e}^{-A(\tilde{y})}d\tilde{y} . \tag{18}$$

We choose that $z(y = 0) = 0$ in order to cancel the integration constant. Moreover, the dependent variable transformation is requested $\varphi \to \psi$ [8]

$$\varphi_k = \mathrm{e}^{-\frac{3}{2}A}\psi_k , \tag{19}$$

in order to get the following Schrödinger-like equation [9]

$$-\ddot{\psi}_k(y) + U(z)\psi_k(z) = m_k^2\psi_k(z) , \tag{20}$$

where the dot denotes derivative with relation to z coordinate. The analogue quantum potential has the form [8]

$$U(z) = \frac{3}{2}\ddot{A}(z) + \frac{9}{4}\dot{A}^2(z) . \tag{21}$$

The general equation (20) give us all the massive (and massless) eigenstates for a specific model. The zero mode ($m = 0$) needs to be confined (normalized and holding all boundary conditions) in order to reproduce the usual 4D gravity at the brane. Moreover, none massive modes can be normalized, therefore there is not a massive KK state bounded. However, some model can exposes a massive quasi-localised state that have high probability amplitude close to region where the brane is placed. These quasi-localised states are namely resonances.

Independently of the existence of resonances, the massive KK modes perform a slight Yukawa-like correction to the Newton's potential in the form [7, 10]:

$$V(x) = G_N\frac{m_1m_2}{r}\left[1 + \Delta(\vec{x})\right] , \tag{22}$$

where $x = |\vec{x}|$ is a distance in our 4D world, the G_N is the gravitational constant, the m_i are point-like source masses that live in the 4D space and the correction $\Delta(\vec{x})$ is obtained as

$$\Delta(x) = \sum_{k>0}^{\infty}\frac{\mathrm{e}^{-m_kx}}{x}\left.\left(\frac{\hat{\psi}_k(z)}{\hat{\psi}_0(z)}\right)\right|_{z=z^*} , \tag{23}$$

where $\hat{\psi}_k$ represents the normalized massive eigenfunction and $\hat{\psi}_0$ the normalized zero mode (for $k = 0$ this quantity gives exactly 1, the usual gravity, present in the equation (22)). The z^* is the point in the z coordinate where the energy density is maximum. Usually, for symmetric non-splitted models this point match with the origin $z^* = 0$.

The expressions of correction $\Delta(x)$ can be analytically obtained in some thin model, as the Randall-Sundrum (RS) model [3] and the Gherghetta-Shaposhnikov (GS) model [11]. In the RS model $\Delta(x) \approx (cx)^{-2}$ [3], while in the GS model $\Delta(x) \approx (cx)^{-3}$ [11], with c as a parameter associated to the scalar curvature of the model. In another models, this correction is dependent

IOP Conf. Series: Journal of Physics: Conf. Series **965** (2018) 012002 doi:10.1088/1742-6596/965/1/012002

on more parameters, as the case of Resolved Conifold (RC) presented in reference [12]. The RC case is phenomenologically more interesting since that other parameters can be combined in order to satisfy simultaneously the hierarchy problem, the correction of Newton's Law and other issues [7].

In this work, we expose an important result to achieve, namely, the correction of Newton's Law in the Bloch brane scenarios. The mass spectrum presented in figure 3 is fundamental to set the discrete suitable mass value to eigenfunction of the Schödinger-like equation (20). From figure 3 we note that the mass spectrum is approximately linear and decreases for large r parameter. Similar results are also found in reference [8]. After get the numerical results for the eigenfunction and applied it to equation (23), in the hybrid Bloch brane case, we can set experimental bound in the same way that performed in reference [7]. So, we can adjust the models parameters in order to obtain suitable results endorsed by the Newton's Law.

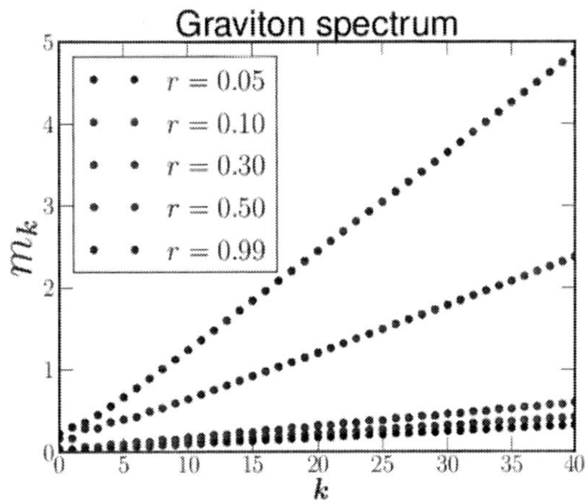

Figure 3. Mass spectrum of the Bloch brane.

4. Conclusions

In this work, we review the Bloch brane model and the recent version of the hybrid Bloch brane. We perform a general method to confine the gravity in a 5D warped model and present the expression to the correction of Newton's Law. Finally, we show the discrete mass spectrum to the Bloch Brane, which has some similarities to hybrid brane model as can be seen in Ref. [8]. For the sake of completeness as further steps, we intend to finish the numerical evaluation of massive eigenfunctions at the z^* point in both symmetric and asymmetric hybrid Bloch brane in order to get the proper plot of gravity modification in these scenarios. We also intend to set experimental bounds in these corrections as presented in the work of Ref. [7].

Acknowledgments

The authors thank the Coordenação de Aperfeiçoamento de Pessoal de Nível Superior (CAPES), the Conselho Nacional de Desenvolvimento Científico e Tecnológico (CNPQ), and Fundação Cearense de Apoio ao Desenvolvimento Científico e Tecnológico (FUNCAP) for financial support.

References

[1] Cruz W T, Dantas DM, Correa R A C and Almeida C A S 2017 *Phys. Lett. B* **772** 592
[2] Bazeia D and Gomes A R 2004 *JHEP* **0405** 012

IOP Conf. Series: Journal of Physics: Conf. Series **965** (2018) 012002 doi:10.1088/1742-6596/965/1/012002

[3] Randall L, Sundrum R 1999 *Phys. Rev. Lett.* **83** 4690
[4] Bajc B and Gabadadze G 2000 *Phys. Lett. B* **474** 282
[5] Bazeia D, Lima E E M and Losano L 2017 *Eur. Phys. J.* C **77** no. 2 127
[6] Bazeia D, Losano L, Marques M A and Menezes R 2014 *Phys. Lett. B* **736** 515
[7] Araujo J B C, Veras D F S, Dantas D M and Almeida C A S *Graviton KK resonant mode in the correction of the Newton's law from 6D braneworlds* hep-th/1610.08124
[8] Veras D F S and Almeida C A S 2017 *Phys. Rev. D* **95** no. 10, 104032
[9] Veras D F S, Cruz W T, Maluf R V and Almeida C A S 2016 *Phys. Lett. B* **754** 201
[10] Csaki C, Erlich J, Hollowood T J and Shirman Y 2000 *Nucl. Phys. B* **581** 309
[11] Gherghetta T and Shaposhnikov M E 2000 *Phys. Rev. Lett.* **85** 240
[12] Dantas D M, Silva J E G and Almeida C A S 2013 *Phys. Lett. B* **725** 425

IOP Conf. Series: Journal of Physics: Conf. Series **965** (2018) 012003 doi:10.1088/1742-6596/965/1/012003

Noncommutative Quantum Mechanics based on Representations of Exotic Galilei Group

R. G. G. Amorim

Faculdade Gama, Universidade de Brasília, 72444-240, Setor Leste (Gama), Brasília, DF, Brazil.

E-mail: `ronniamorim@gmail.com`

S. C. Ulhoa

Instituto de Física, Universidade de Brasília, 70910-900, Brasília, DF, Brazil

E-mail: `sc.ulhoa@gmail.com`

Abstract. Using elements of symmetry, we constructed the Noncommutative Schrödinger Equation from a representation of Exotic Galilei Group. As a consequence, we derive the Ehrenfest theorem using noncommutative coordinates. We also have showed others features of quantum mechanics in such a manifold. As an important result, we find out that a linear potential in the noncommutative Schrödinger equation is completely analogous to the ordinary case. We also worked with harmonic and anharmonic oscillators, giving corrections in the energy for each one.

1. Introduction

The concept of continuous group was introduced by the Norwegian mathematician Sofus Lie [1] while working with differential equations in second half of nineteenth century. Then it has gained great visibility in 1928 with the work of H. Weyl [2], who linked group theory to atomic spectroscopy. Thus it was established, for the first time, a direct relation between quantum theory and group theory. The importance of such a subject increased enormously when Wigner applied the theory of representations of Lie groups to physical systems. Particularly fundamental particles can be described by means of irreducible representations of kinematical groups, such as De Sitter, Galilei and Poincaré groups. Hence fundamental interactions are implemented by gauge symmetries. Later, in order to find an appropriate Hamiltonian structure for atomic systems, P. Dirac also used the theory of representation of Lie groups to investigate such systems [3]. Therefore group theory, which is an important branch of mathematics, became an indispensable tool to understand and formalize symmetries in physics.

The notion of non-commutativity began in physics with the uncertain principle which is a well known relation between coordinates and momenta operators. In a letter addressed to Peierls in 1930, Heisenberg considered that this very feature could be valid also between coordinates operators, in such a way that it could precludes the well known divergences of the self-energy terms calculated for particles. Since then the idea of noncommutative spatial coordinates propagated throughout the scientific community until it reached H. Snyder, a former Ph.D.

IOP Conf. Series: Journal of Physics: Conf. Series **965** (2018) 012003 doi:10.1088/1742-6596/965/1/012003

student of Oppenheimer [4, 5, 6]. Thus, at least formally, Snyder was the first one who stated that spatial coordinates would not commutate with each other at small distances, in his work published in 1947 [7, 8]. He proposed a whole new paradigm in which the spacetime should be understood as a collection of tiny cells of minimum size, where there is no such a concept as the idea of a point. Thus, once the minimum size is reached, in the realm of some high energy phenomenon, the position should be given by the noncommutative coordinate operators. As a direct consequence it would be impossible to precisely measure a particle position.

After Snyder's pioneering, those ideas remained forgotten, mainly because the success achieved by the so called renormalization procedure, which has led to an entire new branch of research in physics, on dealing with ultra-violet divergences of scattering integrals in the context of quantum field theory. Then in the last years the interest of the scientific community on noncommutative geometry has increased due to works on non-abelian theories [9], gravitation [10, 11, 12], standard model [13, 14, 15] and on quantum Hall effect [16]. More recently the discovery that the dynamics of an open string can be associated with noncommutative spaces has contributed to the last revival of noncommutative theories [17].

From the mathematical point of view, the simplest algebra obeyed by the coordinates operators $\widehat{x^\mu}$ is

$$[\widehat{x^\mu}, \widehat{x^\nu}] = i\theta^{\mu\nu},$$

where $\theta^{\mu\nu}$ is a skew-symmetric constant tensor. It worths to point out that the mean values of the position operators do correspond to the actual position observed, thus it is said that such operators are hermitian ones. It is well known in quantum mechanics that a noncommutative relation between two operators lead to a specific uncertain relation, hence the above expression yields

$$\Delta\widehat{x^\mu}\Delta\widehat{x^\nu} \geq \frac{1}{2}|\theta^{\mu\nu}|.$$

Following the ideas introduced by Snyder, it possible to associate the minimum size with a distance of the $\sqrt{|\theta^{\mu\nu}|}$ order of magnitude. Thus the noncommutative effects turn out to be relevant at such scales. Usually the non-commutativity is introduced by means of the Moyal product defined as [7]

$$\begin{aligned} f(x) \star g(x) &\equiv \exp\left(\frac{i}{2}\theta^{\mu\nu}\frac{\partial}{\partial x^\mu}\frac{\partial}{\partial y^\nu}\right) f(x)g(y)|_{y\to x} \\ &= f(x)g(x) + \frac{i}{2}\theta^{\mu\nu}\partial_\mu f \partial_\nu g + \frac{1}{2!}\left(\frac{i}{2}\right)^2 \theta^{\mu_1\nu_1}\theta^{\mu_2\nu_2}(\partial_{\mu_1}\partial_{\mu_2}f)(\partial_{\nu_1}\partial_{\nu_2}g) + \cdots \end{aligned}$$

with a constant $\theta^{\mu\nu}$. Then the usual product is repalced by the Moyal product in the classical lagrangian density.

In this work we study the representation of Galilei group in noncommutative space. Particularly, we analyse the exotic structure of Galilei algebra, that appears in special case of 2+1 dimensions [18, 19]. This exotic structure shows us some physical properties which are not shared by another dimensions, for instance, it allows a natural association between noncommutative coordinates and the second central extension of the algebra. In this sense, the exotic extension of Galilei algebra is responsible for the emergence of noncommutative coordinates [20, 21].

The paper is organized as follows. In Section 2 we construct the noncommutative Galilei algebra in $2+1$ dimensions and write the Schödinger equation in noncommutative plane. As a result, we derive the Ehrenfest Theorem in noncommutative coordinates. In section 3 we study the linear potential in noncommutative plane. Then in section 3 we work with the harmonic oscillator in the noncommutative plane. In section 4 we present the anharmonic oscillator using noncommutative coordinates in two dimensions. We obtain a correction in the energy using perturbation theory. Finally in the last section we present our concluding remarks.

2. Noncommutative Galilei-Lie Algebra

In this section, we study the noncommutative Galilei algebra in 2 + 1 dimension. For this proposal, we construct the operators

$$\widehat{x_i} = x_i - \frac{i}{2}\theta_{ij}\frac{\partial}{\partial x_j}, \tag{1}$$

$$\widehat{p_i} = i\hbar\frac{\partial}{\partial x_i}, \tag{2}$$

$$\widehat{K_i} = m\widehat{x_i} - t\widehat{p_i}, \tag{3}$$

$$\widehat{L} = \widehat{x_i}\widehat{p_j} - \widehat{x_j}\widehat{p_i}, \tag{4}$$

and

$$\widehat{H} = \frac{1}{2m}\sum_{i=1}^{2}\widehat{p_k^2}. \tag{5}$$

From this set of unitary operators we obtain, after some long but simple calculations, the following set of commutation relations,

$$[\widehat{p_i}, \widehat{p_j}] = 0,$$

$$[\widehat{K_i}, \widehat{p_j}] = i\hbar m\delta_{ij},$$

$$[\widehat{L}, \widehat{p_i}] = i\hbar\epsilon_{ij}\widehat{p_j},$$

$$[\widehat{K_i}, \widehat{K_j}] = im^2\theta_{ij},$$

$$[\widehat{p_i}, \widehat{H}] = 0,$$

$$[\widehat{L}, \widehat{H}] = 0.$$

This is the extended exotic Galilei algebra, where $\widehat{L_i}$ stand for rotations, $\widehat{p_i}$ for translations and $\widehat{K_i}$ for boosts. The central extension is given by $[\widehat{K_i}, \widehat{p_j}] = i\hbar m\delta_{ij}$. However, in the plane, the Galilei group admits a second extension, highlighted by the non-commutativity of boost generators. This extension was studied first by Levy-Leblond [18], and this algebra is called the Exotic Galilei Group. In general, the second extension of Galilei algebra is associated to noncommutative coordinates. Here, we show that noncommutativity between coordinates leads to exotic extension. To obtain the physical content, we first notice that $\widehat{p}$ transform under the boost as the physical momentum, i.e.,

$$\exp\left(-i\vec{v}\cdot\frac{\widehat{K}}{\hbar}\right)\widehat{p_j}\exp\left(i\vec{v}\cdot\frac{\widehat{K}}{\hbar}\right) = \widehat{p_j} + mv_j\mathbf{1}.$$

Futhermore, the operators $\widehat{x}$ and $\widehat{p}$ do not commute with each other, that is,

$$[\widehat{x_i}, \widehat{p_j}] = i\hbar\delta_{ij}.$$

Therefore, $\widehat{x}$ and $\widehat{p}$ can be taken to be the physical observables of position and momentum. To be consistent, generators $\widehat{L}$ are interpreted as the angular momentum operator, and $\widehat{H}$ is taken as the Hamiltonian operator. The Casimir invariants of the Lie algebra are given by

$$I_1 = \widehat{H} - \frac{\widehat{P^2}}{2m} \quad \text{and} \quad I_2 = \widehat{L} - \frac{1}{m}\widehat{K}\times\widehat{P},$$

IOP Conf. Series: Journal of Physics: Conf. Series **965** (2018) 012003 doi:10.1088/1742-6596/965/1/012003

where I_1 describes the Hamiltonian of a free particle and I_2 is associated with the spin degrees of freedom. First, we study the scalar representation; i.e. spin zero.

Particularly, $\widehat{H}$ can be identify as generator of temporal translation. In this way, the time evolution of the system state $\psi(x,t) = \langle x|\psi(t)\rangle$ is given by

$$\psi(x,t) = e^{\frac{-i\widehat{H}t}{\hbar}}\psi(x,0). \tag{6}$$

Differentiating Eq.(6) with respect to t, we obtain

$$i\hbar\partial_t\psi(x,t) = \frac{\widehat{p}^2}{2m}\psi(x,t) + V(x_i - \frac{i}{2}\theta_{ij}\partial_{x_j})\psi(x,t), \tag{7}$$

which is the noncommutative Schrödinger equation. Note that Eq.(7) was obtained here from representation theory of symmetry group.

As a important result, we derive the Ehrenfest theorem for the noncommutative coordinates. For this proposal, we take the expansion

$$V\left(x_k - \frac{1}{2\hbar}\theta_{kl}\widehat{p_l}\right) = V(x_k) - \frac{1}{2\hbar}\theta_{kl}\widehat{p_l}\frac{\partial V}{\partial x_k} + o(\theta^2),$$

and to use in the equation $\frac{d}{dt}\langle A\rangle = \frac{i}{\hbar}\langle [H,A]\rangle + \langle\frac{\partial A}{\partial t}\rangle$, we obtain

$$\frac{d}{dt}\langle\widehat{x_k}\rangle = \frac{\langle\widehat{p_k}\rangle}{m} - \frac{\theta_{kl}}{2\hbar}\langle\frac{\partial V}{\partial x_l}\rangle, \tag{8}$$

$$\frac{d}{dt}\langle\widehat{p_k}\rangle = \langle-\frac{\partial V}{\partial x_k}\rangle + \frac{\theta_{jl}}{2\hbar}\langle\widehat{p_l}\frac{\partial^2 V}{\partial x_l \partial x_j}\rangle. \tag{9}$$

Observe that if we take $\theta = 0$, we obtain the same result for commutating coordinates. In sequence, we prove an important theorem valid for linear potential in noncommutative coordinates.

3. Linear Potential

In this section we analyse the linear potential in noncommutative coordinates. The problem of linear potential is applied for example in quantum chromodynamics for studies of gluon condensates (H. G. Dosch). Our focus is in the two dimensional case $V(x,y) = \alpha x + \beta y$, where α and β are complex constants. In this sense, the noncommutative Schrödinger equation with linear interaction can be written by ($\hbar = m = 1$)

$$\frac{\partial^2\psi(x,y)}{\partial x^2} + \frac{\partial^2\psi(x,y)}{\partial y^2} + \alpha x\psi(x,y) + \beta y\psi(x,y) - i\alpha\frac{\theta}{2}\frac{\partial\psi(x,y)}{\partial y} + i\beta\frac{\theta}{2}\frac{\partial\psi(x,y)}{\partial x} = E\psi(x,y). \tag{10}$$

Observe that we utilize the anti-symmetric tensor θ as

$$\theta_{ij} = \begin{bmatrix} 0 & \theta \\ -\theta & 0 \end{bmatrix}.$$

To solve Eq. (10) firstly we note that its imaginary part yields the equation

$$\beta\frac{\partial\psi(x,y)}{\partial x} - \alpha\frac{\partial\psi(x,y)}{\partial y} = 0,$$

IOP Conf. Series: Journal of Physics: Conf. Series **965** (2018) 012003 doi:10.1088/1742-6596/965/1/012003

which implies $\psi = \psi(\alpha x + \beta y)$ and that the solution is real, thus we change variables with the definition $z = (\alpha x + \beta y)$, such that $\frac{\partial\psi}{\partial x} = \alpha\frac{\partial\psi}{\partial z}$, $\frac{\partial\psi}{\partial y} = \beta\frac{\partial\psi}{\partial z}$, $\frac{\partial^2\psi}{\partial x^2} = \alpha^2\frac{\partial^2\psi}{\partial z^2}$, $\frac{\partial^2\psi}{\partial y^2} = \beta^2\frac{\partial^2\psi}{\partial z^2}$. This leads to

$$(\alpha^2 + \beta^2)\frac{\partial^2\psi(z)}{\partial z^2} + z\psi(z) - E\psi(z) = 0. \tag{11}$$

Observe that Eq.(11) is of the general form

$$\frac{d^2u}{dx^2} - xu = 0,$$

that is the Airyt's differential equation. So, the solutions of Eq.(11) are given by

$$\psi(x, y) = \frac{2^{2/3}}{2\pi}Ai(\alpha x + \beta y + E), \tag{12}$$

where $Ai(w)$ é the Airy function. Observe that the solution does not depend on the noncommutative parameter θ, which shows that the Linear Potential does not have noncommutative corrections.

4. Two Dimensional Noncommutative Isotropic Harmonic Oscillator

In this section we will deal with the two dimensional isotropic harmonic oscillator in noncommutative coordinates. The Hamiltonian for just for the 2D harmonic oscillator is given by

$$\widehat{H} = \frac{1}{2m}\widehat{p_x} + \frac{1}{2m}\widehat{p_y} + \frac{1}{2}m(\omega_x^2\widehat{x}^2 + \omega_y^2\widehat{y}^2), \tag{13}$$

If we take $m = \hbar = 1$ then the Hamiltonian reads

$$\widehat{H} = \frac{1}{2m}\widehat{p_x} + \frac{1}{2m}\widehat{p_y} + \frac{1}{2}\omega_x^2(x - \frac{i}{2}\theta_{12}\partial_y)^2 + \frac{1}{2}\omega_y^2(y - \frac{i}{2}\theta_{21}\partial_x)^2, \tag{14}$$

Once we rearrange the terms it yields

$$\widehat{H} = -\frac{1}{2}\left[\left(1 + \frac{\theta_{12}^2\omega_y^2}{4}\right)\partial_x^2 + \left(1 + \frac{\theta_{12}^2\omega_x^2}{4}\right)\partial_y^2\right] + \\ + \frac{1}{2}(\omega_x^2x^2 + \omega_y^2y^2) + \frac{i}{2}\theta_{12}(\omega_y^2 y\partial_x - \omega_x^2 x\partial_y). \tag{15}$$

Thus we would like to analyze the isotropic case which is settled by the condition $\omega_x = \omega_y = \omega$. Performing a change of variables $z = \frac{1}{2}(\omega_x^2x^2 + \omega_y^2y^2)$ which implies, as in the previous section, that we are seeking for real solutions, we obtain

$$\widehat{H}\psi = -\frac{1}{2}\left(1 + \frac{\theta^2\omega^2}{4}\right)\left(2\omega^2\frac{d\psi}{dz} + 2z\omega^2\frac{d^2\psi}{dz^2}\right) + z\psi, \tag{16}$$

we recall that $\theta_{12} = -\theta_{21} = \theta$. Therefore the Schrödinger equation reads

$$-\beta\left(z\frac{d^2\psi}{dz^2} + \frac{d\psi}{dz}\right) + z\psi = E\psi, \tag{17}$$

where $\beta = \omega^2\left(1 + \frac{\theta^2\omega^2}{4}\right)$. If we seek by solutions of the form $\psi = e^{-\xi/\omega}\phi$, where $\xi = \beta z$, then we find

$$\xi\frac{d^2\phi}{d\xi^2} + \left(1 - \frac{2\xi}{\omega}\right)\frac{d\phi}{d\xi} + \frac{1}{\omega}\left(-1 + \frac{\epsilon}{\omega}\right)\phi = 0\,, \tag{18}$$

where $\epsilon = E\omega/\sqrt{\beta}$. Performing another change of variables $\zeta = 2\frac{\xi}{\omega}$ in order to obtain a non-dimensional equation, we get

$$\zeta\frac{d^2\phi}{d\zeta^2} + (1 - \zeta)\frac{d\phi}{d\zeta} + \frac{1}{2}\left(-1 + \frac{\epsilon}{\omega}\right)\phi = 0\,, \tag{19}$$

the above equation is known as Kummer equation and its solution is the confluent hypergeometric function. If we look for polynomial solutions, then we get

$$\psi(z) = c\exp(-z/\beta^{1/2})F(-n;1;2z/\beta^{1/2}), \tag{20}$$

where c is a normalization constant, n is an integer and

$$\epsilon = \omega(2n + 1).$$

Using the properties of the confluent hypergeometric functions, it is possible to write the solution as

$$\psi(x, y) = \left(\frac{\omega}{\beta^{1/4}2^{(n_1+n_2)/2}\sqrt{\pi n_1! n_2!}}\right)\exp\left[-\frac{\omega^2(x^2 + y^2)}{2\beta^{1/2}}\right]H_{n_1}\left(\frac{\omega x}{\beta^{1/4}}\right)H_{n_2}\left(\frac{\omega y}{\beta^{1/4}}\right), \tag{21}$$

where $n = \frac{(n_1+n_2)}{2}$ and $H_n(x)$ are the Hermite polynomials. Thus the energy is given by

$$E_{n_1 n_2} = \left(1 + \frac{\theta^2\omega^2}{4}\right)^{1/2}\omega\,(n_1 + n_2 + 1)\,,$$

we can see that the correction due to the non-commutativity in the coordinates is given by the first factor of the above expression.

5. Two Dimensional Noncommutative Anharmonic Oscillator

In this section we will analyze the anharmonic oscillator in two dimensions under, taking into account the feature of non-commutativity between coordinates. This system consists of small deviations from the normal oscillator. The Hamiltonian that describes it just reads

$$\widehat{H} = \frac{1}{2m}\widehat{p}_x + \frac{1}{2m}\widehat{p}_y + \frac{1}{2}m(\omega_x^2\widehat{x}^2 + \omega_y^2\widehat{y}^2) + \alpha\left(\widehat{x}^3 + \widehat{y}^3\right) + \gamma\left(\widehat{x}^4 + \widehat{y}^4\right), \tag{22}$$

the quantities α and γ are coupling constants, they are supposed to be very small, in such a way that it is possible to apply perturbation theory to calculate the corrections in the energy. Thus given a Hamiltonian of the form $\widehat{H} = \widehat{H}_0 + \Delta\widehat{H}$ where $\widehat{H}_0\Psi_0 = E_0\Psi_0$, it is possible to evaluate the correction by the expression $\Delta E = \left\langle\Psi_0 \mid \Delta\widehat{H} \mid \Psi_0\right\rangle$. If we take $m = \hbar = 1$, then we get

$$\widehat{H}_0 = \frac{1}{2}\widehat{p}_x + \frac{1}{2}\widehat{p}_y + \frac{1}{2}\omega_x^2(x - \frac{i}{2}\theta_{12}\partial_y)^2 + \frac{1}{2}\omega_y^2(y - \frac{i}{2}\theta_{21}\partial_x)^2$$

and

$$\Delta\widehat{H} = \alpha\left[(x - \frac{i}{2}\theta_{12}\partial_y)^3 + (y - \frac{i}{2}\theta_{21}\partial_x)^3\right] + \gamma\left[(x - \frac{i}{2}\theta_{12}\partial_y)^4 + (y - \frac{i}{2}\theta_{21}\partial_x)^4\right].$$

From last section we identify E_0 as $E_{n_1,n_2} = \sqrt{\beta}(n_1+n_2+1)$ and Ψ_0 as the function into expression (21).

In order to calculate the mean values of the quantities appearing in the Hamiltonian, it is necessary to present some integrals. Thus using the following relations

$$\begin{aligned}
\int_{-\infty}^{\infty} e^{-x^2} H_n^2(x)dx &= 2^n n! \sqrt{\pi}, \\
\int_{-\infty}^{\infty} x e^{-x^2} H_n(x)H_m(x)dx &= \sqrt{\pi}2^n n! \left[\frac{1}{2}\delta_{n-1,m} + (n+1)\,\delta_{n+1,m}\right], \\
\int_{-\infty}^{\infty} x^2 e^{-x^2} H_n(x)H_m(x)dx &= \sqrt{\pi}2^n n! \left[\left(n+\frac{1}{2}\right)\delta_{n,m} + (n+2)\,(n+1)\,\delta_{n+2,m} + \frac{1}{4}\delta_{n-2,m}\right],
\end{aligned}$$

we get

$$\begin{aligned}
\int_{-\infty}^{\infty} x^3 e^{-x^2} H_n^2(x)dx &= 0, \\
\int_{-\infty}^{\infty} x^3 e^{-x^2} H_n(x)H_{n-1}(x)dx &= 3\sqrt{\pi}\,2^{n-2} n^2 (n-1)!, \\
\int_{-\infty}^{\infty} x^4 e^{-x^2} H_n^2(x)dx &= 3\sqrt{\pi}\,2^{n-2}(2n^2+2n+1)n!, \\
\int_{-\infty}^{\infty} e^{-x^2/2} H_n(x)\left(\frac{\partial^2}{\partial x^2}\right) e^{-x^2/2} H_n(x)dx &= \sqrt{\pi}2^n \left(n-\frac{1}{2}\right)n!, \\
\int_{-\infty}^{\infty} e^{-x^2/2} H_n(x)\left(\frac{\partial^4}{\partial x^4}\right) e^{-x^2/2} H_n(x)dx &= \frac{3}{2}\sqrt{\pi}2^n \left(3n^2 - 7n + \frac{1}{2}\right)n!.
\end{aligned}$$

Hence we have

$$\begin{aligned}
\left\langle x^4 \right\rangle &= \frac{3\beta}{2\omega^{7/2}}\left(n_1^2 + n_1 + \frac{1}{2}\right), \\
\left\langle y^4 \right\rangle &= \frac{3\beta}{2\omega^{7/2}}\left(n_2^2 + n_2 + \frac{1}{2}\right), \\
\left\langle x^2 \frac{\partial^2}{\partial y^2} \right\rangle &= \left(n_1 + \frac{1}{2}\right)\left(n_2 - \frac{1}{2}\right), \\
\left\langle y^2 \frac{\partial^2}{\partial x^2} \right\rangle &= \left(n_1 - \frac{1}{2}\right)\left(n_2 + \frac{1}{2}\right), \\
\left\langle \frac{\partial^4}{\partial y^4} \right\rangle &= \frac{3\omega^4}{2\beta}\left(3n_2^2 - 7n_2 + \frac{1}{2}\right), \\
\left\langle \frac{\partial^4}{\partial x^4} \right\rangle &= \frac{3\omega^4}{2\beta}\left(3n_1^2 - 7n_1 + \frac{1}{2}\right).
\end{aligned}$$

Therefore the above expressions lead to the following results

$$\begin{aligned}
\left\langle \widehat{x^3} \right\rangle &= 0, \\
\left\langle \widehat{y^3} \right\rangle &= 0, \\
\left\langle \widehat{x^4} \right\rangle &= \frac{3\beta}{2\omega^{7/2}}\left(n_1^2 + n_1 + \frac{1}{2}\right) - \frac{3\theta^2}{2}\left(n_1 + \frac{1}{2}\right)\left(n_2 - \frac{1}{2}\right) + \frac{3\theta^4\omega^4}{32\beta}\left(3n_2^2 - 7n_2 + \frac{1}{2}\right), \\
\left\langle \widehat{y^4} \right\rangle &= \frac{3\beta}{2\omega^{7/2}}\left(n_2^2 + n_2 + \frac{1}{2}\right) - \frac{3\theta^2}{2}\left(n_1 - \frac{1}{2}\right)\left(n_2 + \frac{1}{2}\right) + \frac{3\theta^4\omega^4}{32\beta}\left(3n_1^2 - 7n_1 + \frac{1}{2}\right).
\end{aligned}$$

Then the first order correction of the energy is given by

$$\langle \Delta E \rangle = \gamma \left\{ \frac{3\beta}{2\omega^{7/2}} \left(n_1^2 + n_2^2 + n_1 + n_2 + 1 \right) - \frac{3\theta^2}{2} \left[\left(n_1 + \frac{1}{2} \right) \left(n_2 - \frac{1}{2} \right) + \right.\right.$$
$$\left.\left. + \left(n_1 - \frac{1}{2} \right) \left(n_2 + \frac{1}{2} \right) \right] + \frac{3\theta^4 \omega^4}{32\beta} \left[3 \left(n_2^2 + n_1^2 \right) - 7 \left(n_2 + n_1 \right) + 1 \right] \right\} . \quad (23)$$

We stress out that the corrected energy has, itself, correction in terms of the noncommutative parameter θ. This means that we recover what is known for the anharmonic oscillator plus a new behavior due to the non-commutativity between coordinates.

6. Concluding Remarks

In this paper we have dealt with the noncommutative Galilei group which is also known as Exotic Galilei group. Such a group only has consistency when we work with two spatial dimensions and one temporal dimension (2+1). In this framework we obtain the noncommutative Schrödinger equation from a group structure rather than the well known procedure consisting on to change the normal product by the Moyal product into the Lagrangian. Then as applications we analyze the noncommutative Schrödinger equation or the case of a linear potential, harmonic oscillator and anharmonic oscillator. We find no corrections for the first case. On the other hand we give corrections for the energy in terms of the noncommutative parameter θ for both oscillators (harmonic and anharmonic). We point out that all corrections obtained in this paper goes to the known quantities in the limit $\theta \rightarrow 0$.

[1] P.J. Olver. *Appilcation of Lie Groups to Differential Equation*(Springer-Verlag, N.Y., 1986).
[2] H. Weyl. *The Theory of Group and Quantum Mechanics*(E.P. Dunton and Company Incorporation).
[3] P.A.M. Dirac. Rev. Mod. Phys. **21**, 392 (1949).
[4] W. Pauli. *Scientific Correspondence*, Vol II, p.15, Ed. K. von Meyenn, (Spring-Verlag, Berlin, 1985).
[5] W. Pauli. *Scientific Correspondence*, Vol III, p.380, Ed. K. von Meyenn, (Spring-Verlag, Berlin, 1993).
[6] R. Jackiw. *Physical instances of noncommutating coordinates.* Nucl. Phys. Proc. Suppl. 108 (2002) 30.
[7] H. S. Snyder. *Quantized Spacetime*, Phys. Rev. 71 (1947) 38.
[8] H. S. Snyder. *The Eletromagnetic Field in Quantized Spacetime.* Phys. Rev. 72 (1947) 68.
[9] A. H. chanseddine, G. Felder, J. Frohlich. *Gravity in Noncommutative Geometry.* Commun. Math. Phys. 155 (1993) 205.
[10] W. Kalau, M. Walze. *Gravity Noncommuative Geometry and the Wodzicki Residue.* J. Geom. Phys. 16 (1955) 327.
[11] D. Kastler. *The Dirac Operator and Gravitation* Commun. Math Phys. 166 (1995) 633.
[12] A. H. Chanseddine, A. Connes. *The Spectral Action Principle.* Commun. Math Phys. 186 (1997) 731.
[13] A. Connes, J. Loot. *Particle Models and noncommuative Geometry.* Nucl. Phys. Proc. Suppl. 18B (1991) 29.
[14] J. C. Varilly, J. M. Garcia-Bondia. *Noncommutative Diferencial Geometry and the Standard Model.* J. Geom. Phys. 12 (1993) 223.
[15] C. P Martin, J. C. Varilly, J. M. Garcia-Bondia. *The Standard Model as a Noncommutative Geometry: the low energy regime.* Phys. Rept. 294 (1998) 363.
[16] J. Belissard, A. van Elst, H. Schulz-Baldes. *The Noncommutative Geometry of the Quantum Hall Effect.* J. Math. Phys. 35 (1994) 53.
[17] N. Seiberg, E. Witten, JHEP **9909** (1999) 32, hep-th/9908142.
[18] J. Levy-Leblond. J. Math. Phys. **4**, 776 (1963).
[19] J. Levy-Leblon, H. Bacry. J. Math. Phys. **9**, 1605 (1968).
[20] R. Jackiw, V.P. Nair. Phys. Lett. B **480**, 237 (2000).
[21] G.V. Dunne, R. Jackiw. Nucl. Phys. Proc. Suppl. **33 C**, 114 (1993).

IOP Conf. Series: Journal of Physics: Conf. Series **965** (2018) 012004 doi:10.1088/1742-6596/965/1/012004

A notion of symmetry witness related to Wigner's theorem on symmetry transformations

Paolo Aniello

Dipartimento di Fisica "Ettore Pancini", Università di Napoli "Federico II", and INFN – Sezione di Napoli, C. Universitario di Monte S. Angelo, via Cintia, I-80126 Napoli, Italy

E-mail: `paolo.aniello@na.infn.it`

Abstract. A *symmetry witness* is a subset of the space of selfadjoint trace class operators that allows one to ascertain whether a linear map acting in that space is a symmetry transformation. This notion arises from a certain type of linear preserver problems. Precisely, a symmetry witness is a suitable set which is invariant with respect to an injective linear map in the Banach space of selfadjoint trace class operators where the quantum states live if and only if this map acts as a symmetry transformation. In particular, by a linear version of Wigner's classical theorem, the set of pure states — the rank-one projections — is a symmetry witness. Linearity entails that the usual assumption of preservation of the transition probability between pure states becomes superfluous. This result extends to every set of projections of a fixed (finite) rank, with some suitable constraint on this rank. One then obtains a classification of the sets of projections of a fixed rank that are symmetry witnesses. These symmetry witnesses are projectable. Namely, formulating the mentioned result in terms of quantum states, the sets of 'uniform' density operators of a suitable fixed rank are symmetry witnesses as well.

1. Introduction

According to a celebrated classical theorem of Wigner, a symmetry transformation of a quantum system is implemented by means of a unitary or antiunitary operator, which is uniquely determined up to multiplication by a phase factor [1–9]. Otherwise stated, there is a canonical one-to-one correspondence between the symmetry transformations of a quantum system and the elements of the projective unitary-antiunitary group of the relevant Hilbert space. As a remarkable consequence — in general, projective — unitary group representations play a central role in quantum theory [10].

We stress that in Wigner's seminal work symmetries were regarded as maps on *pure states* only and, as a defining property, the *transition probability* between pairs of states was assumed to be preserved by these maps. *Linearity* — or convex linearity — is simply not contemplated in this approach; it appears in subsequent formulations of Wigner's theorem involving maps defined in different settings (e.g., the Kadison and Jordan-Segal automorphisms [4]). It is clear, however, that the linear structure is essentially already there, because the pure states, realized as rank-one projections, are embedded in the set of all states, which is a convex set in a linear space, the real Banach space of selfadjoint trace class operators. Moreover, the symmetry transformations of pure states extend in a straightforward way to linear maps acting in that space.

It is then natural to wonder whether *assuming in advance* that the maps one deals with be *linear* operators, and under suitable hypotheses, one can achieve theorems 'of the Wigner

IOP Conf. Series: Journal of Physics: Conf. Series **965** (2018) 012004 doi:10.1088/1742-6596/965/1/012004

type' or, more generally, 'pseudo-Wigner' theorems, characterizing certain types of maps. By the latter expression we mean that one ends with maps of the general form $A \mapsto MAN$ or $A \mapsto MA^{\mathsf{T}}N$, where A belongs to some (real or complex) Banach space of bounded linear operators in a separable complex Hilbert space $\mathcal{H}$, M, N are suitable operators in $\mathcal{H}$ and $A \mapsto A^{\mathsf{T}}$ is the transposition map associated with a given orthonormal basis. A class of problems where this kind of maps typically arise are known in the literature under the name of *linear preserver problems*: one requires that a certain *quantity, set* or *relation* is *preserved* by a linear map and the solution consists in providing a suitable characterization of such a map; see [8,11] and references therein. They date back, at least, to the seminal work of Frobenius [12], who considered the matrix determinant as a preserved quantity and obtained a complete solution of the problem, in the form of a pseudo-Wigner theorem (in this case, M, N are arbitrary nonsingular complex square matrices, with $\det(MN) = 1$).

In particular, a theorem of the Wigner type corresponds to the case where M and N are a unitary operator and its adjoint, respectively; i.e., a map that can be called — in general, with a slight abuse of terminology — a 'symmetry transformation' (note that such a result can be reformulated in terms of unitary and antiunitary operators, without using transposition).

Clearly, one can consider preserver problems that are not necessarily linear, and actually an exhaustive survey of such problems, especially in the context of quantum mechanics, can be found in the standard reference book [8]. E.g., considering bijective maps on the set of pure states and regarding the transition probability as a *preserved quantity* leads to Wigner's theorem. On the other hand, one obtains a related linear preserver problem by considering linear maps in the real Banach space of selfadjoint trace class operators and requiring that the set of pure states be this time a *preserved set*, whereas the assumption of preservation of the transition probability between pure states is omitted. More generally, one may require that the set of (orthogonal) projections with a fixed rank be preserved. The problem can be reformulated in terms of certain families of density operators, thus it is relevant from the point of view of physics.

This intriguing problem has been considered, in the finite-dimensional case, in [13], where a partial solution has been found, and then reconsidered in full generality and completely solved in [9], giving rise to a theorem of the Wigner type. In the present contribution, we will discuss this theorem and various other related results. We will also argue that from this circle of ideas an interesting notion of *symmetry witness* emerges in a natural way. In short, a symmetry witness is suitable a set of selfadjoint trace class operators that allows one to detect the symmetry transformations among other linear maps. The set of all pure states of quantum system can be regarded as the *prototype* of this notion. More generally, as it will be clear later on, a symmetry witness arises whenever the solution of a linear preserver problem associated with a suitable *symmetry witness candidate* produces a theorem of the Wigner type.

The paper is organized as follows. In sect. 2, we review Wigner's theorem and some related results. In sect. 3, we recall some pseudo-Wigner theorems in the linear setting. These results suggest that linearity may play a precise role in the derivation of theorems of the Wigner type, and in sect. 4 we show that this is indeed the case by discussing a linear version of Wigner's theorem and other related facts. From this context then the notion of symmetry witness arises in a natural way; see sect. 5. In sect. 6, we provide a complete classification of the set of projections with a fixed rank that are symmetry witnesses; hence, a complete solution of the problem raised in [13]. Finally, in sect. 7, conclusions are drawn.

2. Wigner's theorem and some related results

Let $\mathcal{H}$ be a separable complex Hilbert space, with $\dim(\mathcal{H}) \geq 2$, and let $\mathscr{P}_1(\mathcal{H})$ be the set of rank-one projections in $\mathcal{H}$, regarded as the set of *pure states* of a quantum system. For every pair $P, Q \in \mathscr{P}_1(\mathcal{H})$, one can define the *transition probability*

$$\mathrm{tr}(PQ) \in [0,1]\,. \tag{1}$$

According to Wigner [1], a *symmetry transformation* is a bijective map

$$\Theta\colon \mathscr{P}_1(\mathcal{H}) \to \mathscr{P}_1(\mathcal{H}) \tag{2}$$

that preserves the transition probability; i.e.,

$$\mathrm{tr}(\Theta(P)\,\Theta(Q)) = \mathrm{tr}(PQ)\,, \quad \forall P, Q \in \mathscr{P}_1(\mathcal{H})\,. \tag{3}$$

Symmetry transformations are characterized by the following celebrated theorem [1–9].

Theorem 1 (Wigner). *A bijection $\Theta\colon \mathscr{P}_1(\mathcal{H}) \to \mathscr{P}_1(\mathcal{H})$ is a symmetry transformation if and only if it is of the form*

$$\Theta(P) = UPU^*, \quad \forall P \in \mathscr{P}_1(\mathcal{H})\,, \tag{4}$$

where U is a unitary or antiunitary operator in $\mathcal{H}$, uniquely determined up to multiplication by a phase factor.

Remark 1. Denoting by $A \mapsto A^{\mathsf{T}}$ (A bounded operator in $\mathcal{H}$) the *transposition* map associated with an orthonormal basis, we have:

$$A^{\mathsf{T}} = JA^*J\,, \tag{5}$$

for some *complex conjugation* J in $\mathcal{H}$ (an antiunitary operator such that $J = J^* = J^{-1}$). Hence, if U is an antiunitary operator in $\mathcal{H}$, for every pure state P we can write:

$$UPU^* = UP^*U^* = UJP^{\mathsf{T}}JU^* = WP^{\mathsf{T}}W^*, \tag{6}$$

for some unitary operator $W = UJ$. Therefore, Wigner's theorem can be expressed in terms of unitary operators and a (fixed) transposition.

There is a remarkable generalization of Wigner's theorem, due to Uhlhorn [14], which is based on a somewhat weaker assumption: only the *zero* transition probability is preserved (in both directions). On the other hand, it can be shown by means of examples — see Example 1 below — that this result does not hold for $\dim(\mathcal{H}) = 2$.

Notice that, for $P, Q \in \mathscr{P}_1(\mathcal{H})$,

$$\mathrm{tr}(PQ) = 0 \iff P \perp Q, \;\text{ i.e. } PQ = 0; \tag{7}$$

i.e., Uhlhorn's result can be expressed in terms of maps that preserve the mutual orthogonality of rank-one projections.

Theorem 2 (Uhlhorn). *For $\dim(\mathcal{H}) \geq 3$, let $\Theta\colon \mathscr{P}_1(\mathcal{H}) \to \mathscr{P}_1(\mathcal{H})$ be a bijective map preserving the zero transition probability — or, equivalently, the orthogonality — in both directions:*

$$\mathrm{tr}(PQ) = 0 \iff \mathrm{tr}(\Theta(P)\,\Theta(Q)) = 0, \quad P, Q \in \mathscr{P}_1(\mathcal{H})\,. \tag{8}$$

Then, Θ is of the form

$$\Theta(P) = UPU^*, \quad \forall P \in \mathscr{P}_1(\mathcal{H})\,, \tag{9}$$

where U is a unitary or antiunitary operator in $\mathcal{H}$, uniquely determined up to multiplication by a phase factor.

Example 1. The fact that Uhlhorn's theorem does not hold for $\dim(\mathcal{H}) = 2$ can be illustrated by a simple counterexample. Let us set $\mathcal{H} = \mathbb{C}^2$. In this case, every pure state P can be written (uniquely) in the form

$$P = \frac{1}{2}(I_2 + \boldsymbol{a}\cdot\boldsymbol{\sigma})\,, \quad \boldsymbol{a} \in \mathbb{R}^3, \quad \|\boldsymbol{a}\| = 1\,, \tag{10}$$

where I_2 is the 2×2 identity matrix and $\boldsymbol{\sigma} = (\sigma_1, \sigma_2, \sigma_3)$ are the Pauli matrices; i.e., there is a natural one-to-one correspondence between the elements of $\mathscr{P}_1(\mathcal{H})$ and the points of the unit sphere in $\mathbb{R}^3$ (the 'Bloch sphere'). The transition probability can be computed in terms of this parametrization using the simple formula

$$\mathrm{tr}((I_2 + \boldsymbol{a} \cdot \boldsymbol{\sigma})(I_2 + \boldsymbol{b} \cdot \boldsymbol{\sigma})) = 2(1 + \boldsymbol{a} \cdot \boldsymbol{b}). \tag{11}$$

Moreover, exploiting this correspondence every bijection $\Theta\colon \mathscr{P}_1(\mathcal{H}) \to \mathscr{P}_1(\mathcal{H})$ induces a bijective map

$$\mathbb{R}^2 \supset X \equiv \{(0,0)\} \cup \{(\pi,0)\} \cup \big(]0,\pi[\times[0,2\pi[\big) \ni (\theta,\phi) \mapsto \vartheta(\theta,\phi) \in X, \tag{12}$$

which is determined by the condition that $\vartheta(\theta,\phi) = (\theta',\phi')$, where

$$\frac{1}{2}(I_2 + \sin\theta' \cos\phi' \,\sigma_1 + \sin\theta' \sin\phi' \,\sigma_2 + \cos\theta' \,\sigma_3) = \Theta\Big(\frac{1}{2}(I_2 + \sin\theta \cos\phi \,\sigma_1 + \sin\theta \sin\phi \,\sigma_2 + \cos\theta \,\sigma_3)\Big);$$

conversely, a bijection ϑ of X onto itself determines a bijection Θ of $\mathscr{P}_1(\mathcal{H})$ onto itself.

Now, let us set, in particular,

$$\vartheta(\theta,\phi) := (\theta,\phi), \quad \text{for } \theta \neq \pi/2, \quad \vartheta(\pi/2,\phi) := (\pi/2, \alpha(\phi)), \tag{13}$$

where

$$\alpha(\phi) := \phi^2/\pi, \quad \text{for } 0 \le \phi < \pi, \quad \alpha(\phi) := (\phi - \pi)^2/\pi + \pi, \quad \text{for } \pi \le \phi < 2\pi, \tag{14}$$

and let us denote by Θ_ϑ the associated bijection of $\mathscr{P}_1(\mathcal{H})$ onto itself. By relation (11) it is clear that the antipodal points on the Bloch sphere parametrize mutually orthogonal projections; hence,

$$\mathrm{tr}(PQ) = 0 \iff \mathrm{tr}(\Theta_\vartheta(P)\,\Theta_\vartheta(Q)) = 0, \quad \forall P, Q \in \mathscr{P}_1(\mathcal{H}), \tag{15}$$

because

$$\alpha(\phi + \pi) = \phi^2/\pi + \pi = \alpha(\phi) + \pi, \quad \text{for } \phi \in [0,\pi). \tag{16}$$

On the other hand, by (11) it is also clear that the map Θ_ϑ does not preserve, in general, the transition probability. Thus, it is not a symmetry transformation, and we have indeed a counterexample. Moreover, Θ_ϑ cannot be extended to a *linear* map acting in the space of 2×2 hermitian matrices. The possibility of constructing a counterexample, relying on a bijection of $\mathscr{P}_1(\mathcal{H})$ onto itself that admits a linear extension, is actually excluded by Theorem 7 below.

Another remarkable formulation of Wigner's theorem is expressed in terms of generic — pure or mixed — states. Then, let $\mathscr{S}(\mathcal{H})$ be the convex set of all quantum states (the normalized positive trace class operators in $\mathcal{H}$). A *Kadison automorphism* is a bijective map

$$\Phi\colon \mathscr{S}(\mathcal{H}) \to \mathscr{S}(\mathcal{H}) \tag{17}$$

which preserves the convex structure of $\mathscr{S}(\mathcal{H})$; i.e., such that

$$\Phi(\epsilon\rho + (1-\epsilon)\sigma) = \epsilon\,\Phi(\rho) + (1-\epsilon)\,\Phi(\sigma), \quad \forall \epsilon \in [0,1], \ \forall \rho, \sigma \in \mathscr{S}(\mathcal{H}). \tag{18}$$

The following classical result holds [3–6].

Theorem 3 (Kadison). *Let a bijection $\Phi\colon \mathscr{S}(\mathcal{H}) \to \mathscr{S}(\mathcal{H})$ be a Kadison automorphism. Then, Φ is of the form*

$$\Phi(\rho) = U \rho\, U^*, \quad \forall \rho \in \mathscr{S}(\mathcal{H}), \tag{19}$$

where U is a unitary or antiunitary operator in $\mathcal{H}$, uniquely determined up to multiplication by a phase factor.

IOP Conf. Series: Journal of Physics: Conf. Series **965** (2018) 012004 doi:10.1088/1742-6596/965/1/012004

It is worth noting that in this case there is no 'preserved quantity' in the hypotheses of the theorem, whereas a central role is played by the natural convex structure of $\mathscr{S}(\mathcal{H})$.

Another interesting result is a generalization of Uhlhorn's theorem which is due to Šemrl [15]. Like the original result, it excludes the case where $\dim(\mathcal{H}) = 2$, and it is based on a preserved quantity — $\mathrm{tr}(PQ) = 0$ — or, equivalently, on the preservation of the mutual orthogonality of projections:

$$\mathrm{tr}(PQ) = 0 \iff PQ = 0; \tag{20}$$

but in this case P, Q are not supposed, in general, to be pure states. They are assumed to belong to $\mathscr{P}_k(\mathcal{H})$, the set of (orthogonal) *projections of rank* $k \in \mathbb{N}$, with some constraint on k in the finite-dimensional case.

Theorem 4 (Šemrl). *Let* $\Phi\colon \mathscr{P}_k(\mathcal{H}) \to \mathscr{P}_k(\mathcal{H})$ *— with* $2 \leq 2k < \dim(\mathcal{H})$ *— be a bijection that preserves orthogonality in both directions:*

$$PQ = 0 \iff \Phi(P)\Phi(Q) = 0, \quad P, Q \in \mathscr{P}_k(\mathcal{H}). \tag{21}$$

Then, Φ *is of the canonical form*

$$\Phi(P) = UPU^*, \quad \forall P \in \mathscr{P}_k(\mathcal{H}), \tag{22}$$

where U *is a unitary or antiunitary operator in* $\mathcal{H}$.

Let us observe explicitly that, in principle, a plausible constraint in the hypotheses of the previous theorem would be that, for $\dim(\mathcal{H}) < \infty$, $2k \leq \dim(\mathcal{H}) \neq 2$; but the case where $2k = \dim(\mathcal{H}) \neq 2$ is actually not contemplated. We will further comment about this peculiar point in sect. 6. The physical significance of considering the set $\mathscr{P}_k(\mathcal{H})$, for $k \neq 1$, will also be discussed later on.

3. What about linearity?

In the literature, several pseudo-Wigner theorems arise from preserver problems concerning *linear* maps [8, 11]. Two classical examples, stemming from linear preserver problems where a Banach space norm is regarded as a preserved quantity, are due to Russo [16, 17] and Arazy [18], respectively. We will denote by $\mathscr{B}_1(\mathcal{H})$ the complex Banach space of trace class operators in $\mathcal{H}$ and, more generally, by $\mathscr{B}_p(\mathcal{H})$ the Schatten p-class, $1 \leq p \leq \infty$, with $\mathscr{B}_\infty(\mathcal{H})$ denoting the Banach space of compact operators (endowed with the operator norm).

Theorem 5 (Russo). *Let* $\Phi\colon \mathscr{B}_1(\mathcal{H}) \to \mathscr{B}_1(\mathcal{H})$ *be a surjective linear isometry. Then,* Φ *is of the form*

$$\Phi(A) = UAV \quad \text{or} \quad \Phi(A) = UA^{\mathsf{T}}V, \quad \forall A \in \mathscr{B}_1(\mathcal{H}), \tag{23}$$

for some unitary operators U, V *in* $\mathcal{H}$.

Remark 2. Reasoning as in Remark 1, for every transposition $A \mapsto A^{\mathsf{T}}$ there is a complex conjugation operator J in $\mathcal{H}$ such that $A^{\mathsf{T}} = JA^*J$. Hence, given unitary operators U, V in $\mathcal{H}$, we have that

$$UA^{\mathsf{T}}V = UJA^*JV = WA^*Z, \tag{24}$$

for some *anti*unitary operators $W = UJ$, $Z = JV$ in $\mathcal{H}$.

The previous result admits the following generalization concerning other Schatten classes of operators (with the exception of the Hilbert-Schmidt class $\mathscr{B}_2(\mathcal{H})$).

Theorem 6 (Arazy). *Let* $\Phi\colon \mathscr{B}_p(\mathcal{H}) \to \mathscr{B}_p(\mathcal{H})$ *— with* $1 \leq p \leq \infty$, $p \neq 2$ *— be a linear isometry of the Banach space* $\mathscr{B}_p(\mathcal{H})$ *onto itself. Then,* Φ *is of the form*

$$\Phi(A) = UAV \quad \text{or} \quad \Phi(A) = UA^{\mathsf{T}}V, \tag{25}$$

for all $A \in \mathscr{B}_p(\mathcal{H})$*, where* U, V *are unitary operators in* $\mathcal{H}$.

Remark 3. The fact that $\mathscr{B}_2(\mathcal{H})$ is a Hilbert space entails that it admits further surjective linear isometries (unitary operators) wrt (25). On the other hand, if $\Phi\colon \mathscr{B}_2(\mathcal{H}) \to \mathscr{B}_2(\mathcal{H})$ is a surjective linear isometry such that $\Phi(\mathscr{P}_1(\mathcal{H})) = \mathscr{P}_1(\mathcal{H})$, then $\Phi(A) = UAU^*$ or $\Phi(A) = UA^{\mathsf{T}}U^*$, for some unitary operator U in $\mathcal{H}$ (by Wigner's theorem, because the transition probability can be regarded as a Hilbert-Schmidt product, and by the fact that $\operatorname{span}_{\mathbb{C}}(\mathscr{P}_1(\mathcal{H}))$ is dense in $\mathscr{B}_2(\mathcal{H})$).

Remark 4. Results analogous to Arazy's theorem hold for the complex Banach space $\mathscr{B}(\mathcal{H})$ of bounded operators and for the real Banach space $\mathscr{B}(\mathcal{H})_{\mathsf{s}}$ of selfadjoint bounded operators. Precisely, if $\Phi\colon \mathscr{B}(\mathcal{H}) \to \mathscr{B}(\mathcal{H})$ is a surjective linear isometry, then it is of the form (25), for some unitary operators U, V in $\mathcal{H}$; see Theorem A.9 of [8]. Moreover, if $\Phi\colon \mathscr{B}(\mathcal{H})_{\mathsf{s}} \to \mathscr{B}(\mathcal{H})_{\mathsf{s}}$ is a surjective linear isometry, then $\Phi(A) = s\,UAU^*$ or $\Phi(A) = s\,UA^{\mathsf{T}}U^*$, for some unitary operator U in $\mathcal{H}$ and $s \in \{1, -1\}$; see Theorem 2.6.1 of [8].

In quantum mechanics, one usually relates *linearity* with the Hilbert space associated with a quantum system — i.e., with the linear superposition of 'state vectors' — and with the observables (selfadjoint operators), rather than with the space of states $\mathscr{S}(\mathcal{H})$. Nevertheless, $\mathscr{S}(\mathcal{H})$ is a convex set in $\mathscr{B}_1(\mathcal{H})$, and in several contexts — e.g., the theory of open systems and quantum information — the linear structure of the whole space $\mathscr{B}_1(\mathcal{H})$ is actually exploited. Typical examples of topics where the linear structure of this space is relevant are [19–30]:

- quantum transmission channels;
- quantum dynamical semigroups and the associated master equations;
- entanglement detection.

Moreover, Kadison's theorem and the results outlined in this section suggest that linearity may play a major role in the derivation of theorems of the Wigner type.

It seems then quite natural to reconsider Wigner's theorem and some related results in the linear setting.

4. Symmetries in the linear setting: the role of linearity

From this point onwards, it will be convenient to work in the *real* Banach space $\mathscr{B}_1(\mathcal{H})_{\mathsf{s}}$ of *selfadjoint* trace class operators in $\mathcal{H}$, where the quantum states (density operators) live. We will denote by $\mathscr{F}(\mathcal{H})_{\mathsf{s}} \subset \mathscr{B}_1(\mathcal{H})_{\mathsf{s}}$ the dense linear manifold of all selfadjoint operators in $\mathcal{H}$ of *finite rank*; i.e., $\mathscr{F}(\mathcal{H})_{\mathsf{s}} = \operatorname{span}_{\mathbb{R}}(\mathscr{P}_1(\mathcal{H}))$. Clearly, in the finite-dimensional case $\mathscr{F}(\mathcal{H})_{\mathsf{s}}$, $\mathscr{B}_1(\mathcal{H})_{\mathsf{s}}$ and $\mathscr{B}(\mathcal{H})_{\mathsf{s}}$ — regarded as linear spaces — actually coincide, and we prefer to adopt the 'neutral' notation $\mathscr{B}(\mathcal{H})_{\mathsf{s}}$ whenever that case occurs.

In the linear setting, one can prove the following result [9], which is closely related to the classical form of Wigner's theorem.

Theorem 7 (Linear form of Wigner's theorem). *Let Φ be a densely defined linear operator in $\mathscr{B}_1(\mathcal{H})_{\mathsf{s}}$ — with $\operatorname{dom}(\Phi) = \mathscr{F}(\mathcal{H})_{\mathsf{s}}$ — mapping the set $\mathscr{P}_1(\mathcal{H})$ of pure states onto itself. Then, Φ is closable, and its closure $\widehat{\Phi}$ is a surjective isometry of the canonical form*

$$\widehat{\Phi}(A) = UAU^*, \quad \forall A \in \mathscr{B}_1(\mathcal{H})_{\mathsf{s}}, \tag{26}$$

where U is a unitary or antiunitary operator in $\mathcal{H}$, uniquely determined up to multiplication by a phase factor.

Remark 5. In this setting, the assumption of preservation of the transition probability between pure states is immaterial, and $\mathscr{P}_1(\mathcal{H})$ becomes a *preserved set*. A 'symmetry transformation' will be then simply a map of the form (26), without reference to Wigner's original approach.

Remark 6. The linear manifold $\mathscr{F}(\mathcal{H})_{\mathsf{s}}$ is dense in the Banach space $\mathscr{B}_p(\mathcal{H})_{\mathsf{s}} := \mathscr{B}_p(\mathcal{H}) \cap \mathscr{B}(\mathcal{H})_{\mathsf{s}}$, the real Schatten p-class, for every $1 \le p \le \infty$. Hence, it is easily seen that the statement of Theorem 7 actually holds true with the space of selfadjoint trace class operators $\mathscr{B}_1(\mathcal{H})_{\mathsf{s}}$ replaced with any of the Schatten classes $\mathscr{B}_p(\mathcal{H})_{\mathsf{s}}$.

Also Kadison's theorem, i.e. Theorem 3, can be revisited in the linear setting. In this case, the preserved set is formed by all density operators of finite rank. Precisely, one can prove the following fact [9].

Theorem 8 (Linear form of Kadison's theorem). *Let* Φ *be a densely defined linear operator in* $\mathscr{B}_1(\mathcal{H})_s$ *— with* $\mathrm{dom}(\Phi) = \mathscr{F}(\mathcal{H})_s$ *— mapping the convex set*

$$\mathscr{F}\mathscr{S}(\mathcal{H}) := \mathscr{F}(\mathcal{H})_s \cap \mathscr{S}(\mathcal{H}) \tag{27}$$

of finite rank density operators bijectively onto itself. Then, Φ *is closable, and its closure* $\widehat{\Phi}$ *is a surjective isometry of the canonical form (26); namely, a symmetry transformation.*

Recently, in the finite-dimensional case Sarbicki *et al* have proved the following interesting result [13], which partially generalizes the linear version of Wigner's theorem and is also related to Šemrl's theorem (Theorem 4). Indeed, in this result one deals with linear maps preserving $\mathscr{P}_k(\mathcal{H})$, the set of projections of a fixed rank $k \geq 1$ (the assumption of preservation of the quantity $\mathrm{tr}(PQ) = 0$ becomes superfluous).

Theorem 9 (Sarbicki *et al*). *Let* $n = \dim(\mathcal{H}) < \infty$ *be a prime number and* $\Phi\colon \mathscr{B}(\mathcal{H})_s \to \mathscr{B}(\mathcal{H})_s$ *a linear map, mapping* $\mathscr{P}_k(\mathcal{H})$ *— with* $k < n$ *— bijectively onto itself. Then,* Φ *is a surjective isometry of the form*

$$\Phi(A) = UAU^*, \quad \forall A \in \mathscr{B}(\mathcal{H})_s, \tag{28}$$

where U *is a unitary or antiunitary operator in* $\mathcal{H}$.

Later on, Størmer [31] has proved another interesting theorem of the Wigner type, involving linear maps that act in the complex Banach space $\mathscr{B}(\mathcal{H})$ of bounded operators in $\mathcal{H}$. These maps are assumed to be *positive* (hence, bounded; see, e.g., [32]) and *unital* (the identity is mapped to itself), and again to preserve the set of projections of a given rank $k \geq 1$.

We stress that, whereas the theorems of Wigner and Kadison — as well as their linear counterparts — are formulated in the 'Schrödinger picture' (symmetries act on states), Størmer's result is worked out in the 'Heisenberg picture' (where, instead, observables are the physical objects undergoing symmetry transformations). Restricting to *selfadjoint* operators — i.e., to the *real* Banach space $\mathscr{B}(\mathcal{H})_s$ of (bounded) quantum observables — for a more direct comparison with Theorem 9 and with other results that will be discussed later on, we have the following equivalent version of Størmer's theorem.

Theorem 10 (Størmer). *Let* $\Phi\colon \mathscr{B}(\mathcal{H})_s \to \mathscr{B}(\mathcal{H})_s$ *be a unital positive linear map, mapping* $\mathscr{P}_k(\mathcal{H})$ *— for some* $k < \dim(\mathcal{H})$ *— onto itself. Then,* Φ *is of the form*

$$\Phi(A) = UAU^*, \quad \forall A \in \mathscr{B}(\mathcal{H})_s, \tag{29}$$

where U *is a unitary or antiunitary operator in* $\mathcal{H}$.

Corollary 1. *Let* $\Phi\colon \mathscr{B}(\mathcal{H})_s \to \mathscr{B}(\mathcal{H})_s$ *be a surjective linear map. Then, the following facts are equivalent:*

(i) Φ *is isometric and positive;*
(ii) Φ *is isometric and unital;*
(iii) Φ *is positive and unital, and maps* $\mathscr{P}_k(\mathcal{H})$*, for some* $k < \dim(\mathcal{H})$*, onto itself;*
(iv) Φ *is of the form (29), for some unitary or antiunitary operator* U *in* $\mathcal{H}$.

Proof. The statement follows from Theorem 10 and from the last assertion of Remark 4. □

A comparison between the theorems of Sarbicki *et al* and Størmer suggests that — at least in the finite-dimensional case — the assumptions of positivity and unitality in the latter result may be relaxed. On the other hand, in Theorem 9 the assumption that $n = \dim(\mathcal{H}) < \infty$ be a prime number appears rather singular. We will clarify both these points in sect. 6.

5. Symmetry witnesses

In the light of the results discussed in the previous section, it seems quite natural to replace, in the linear setting, *preserved quantities* — e.g., transition probability, fidelity, various metrics etc.; see [8] and references therein — with suitable *preserved sets* (e.g., $\mathscr{P}_k(\mathcal{H})$).

One can formalize this idea as follows [9].

Definition 1 (Symmetry witnesses)**.** We call a subset $\mathcal{W}$ of $\mathscr{B}_1(\mathcal{H})_{\mathsf{s}}$, invariant with respect to every unitary or antiunitary transformation, and such that its linear span is dense in $\mathscr{B}_1(\mathcal{H})_{\mathsf{s}}$ — i.e., such that

$$\overline{\mathrm{span}_{\mathbb{R}}(\mathcal{W})} = \mathscr{B}_1(\mathcal{H})_{\mathsf{s}} \tag{30}$$

— a *symmetry witness candidate*. A symmetry witness candidate $\mathcal{W}$ is called a *symmetry witness* if every injective linear operator Φ in $\mathscr{B}_1(\mathcal{H})_{\mathsf{s}}$ — densely defined on $\mathrm{dom}(\Phi) = \mathrm{span}_{\mathbb{R}}(\mathcal{W})$, and preserving $\mathcal{W}$: $\Phi(\mathcal{W}) = \mathcal{W}$ — is of the form

$$\Phi(A) = UAU^*, \quad \forall A \in \mathrm{span}_{\mathbb{R}}(\mathcal{W}), \tag{31}$$

for some unitary or antiunitary operator U in $\mathcal{H}$.

Remark 7. Note that every symmetry witness candidate $\mathcal{W}$ induces a linear preserver problem: to characterize the class of all injective linear operators in $\mathscr{B}_1(\mathcal{H})_{\mathsf{s}}$, with domain $\mathrm{span}_{\mathbb{R}}(\mathcal{W})$, that map $\mathcal{W}$ onto itself. Then, $\mathcal{W}$ is a symmetry witness if this class is formed precisely by all the unitary or antiunitary transformations.

Remark 8. Given a symmetry witness $\mathcal{W}$ and an injective linear operator Φ in $\mathscr{B}_1(\mathcal{H})_{\mathsf{s}}$ — such that $\mathrm{dom}(\Phi) = \mathrm{span}_{\mathbb{R}}(\mathcal{W})$ and leaving $\mathcal{W}$ invariant — Φ is closable and its closure $\widehat{\Phi}$ is a surjective isometry of the canonical form (26); i.e., a symmetry transformation. Hence, given a closed injective linear operator in $\mathscr{B}_1(\mathcal{H})_{\mathsf{s}}$ whose domain contains $\mathrm{span}_{\mathbb{R}}(\mathcal{W})$, one can check whether it is a symmetry transformation by verifying whether it preserves the witness $\mathcal{W}$.

Example 2. By the linear version of the theorems of Wigner and Kadison, the set of pure states $\mathscr{P}_1(\mathcal{H})$ and the convex set $\mathscr{FS}(\mathcal{H}) \subset \mathscr{S}(\mathcal{H})$ of finite-rank density operators are symmetry witnesses.

We also want to introduce a notion of *projectable* symmetry witness. To this aim, let us consider the sets

$$\mathscr{B}_1(\mathcal{H})_*^+ := \{A \in \mathscr{B}_1(\mathcal{H})\colon A \geq 0,\ A \neq 0\}, \quad \mathscr{B}_1(\mathcal{H})_*^- := -\mathscr{B}_1(\mathcal{H})_*^+, \tag{32}$$

$$\mathscr{B}_1(\mathcal{H})_*^\pm := \mathscr{B}_1(\mathcal{H})_*^+ \cup \mathscr{B}_1(\mathcal{H})_*^- \subset \mathscr{B}_1(\mathcal{H})_{\mathsf{s}}, \tag{33}$$

and let ϖ be the natural projection map of $\mathscr{B}_1(\mathcal{H})_*^\pm$ onto $\mathscr{S}(\mathcal{H})$:

$$\varpi\colon \mathscr{B}_1(\mathcal{H})_*^\pm \ni A \mapsto \mathrm{tr}(|A|)^{-1}|A| \in \mathscr{S}(\mathcal{H}). \tag{34}$$

We will say that a symmetry witness (candidate) $\mathcal{W}$ is *positive*, *negative* or *signed* if it is contained in $\mathscr{B}_1(\mathcal{H})_*^+$, $\mathscr{B}_1(\mathcal{H})_*^-$ or $\mathscr{B}_1(\mathcal{H})_*^\pm$, respectively.

If $\mathcal{W}$ is a *signed* symmetry witness candidate, it is easy to check that $\check{\mathcal{W}} := \varpi(\mathcal{W}) \subset \mathscr{S}(\mathcal{H})$ is a symmetry witness candidate too, and this fact allows us to formulate the following definition.

Definition 2 (Projectable symmetry witnesses)**.** We call a *signed* symmetry witness *projectable* if its image through the map ϖ is a symmetry witness too.

Example 3. The set $\mathscr{P}_k(\mathcal{H}) \subset \mathscr{B}_1(\mathcal{H})_*^+$, $k < \dim(\mathcal{H})$, and its projection $\mathscr{S}_k(\mathcal{H})_{\mathsf{u}} := \varpi(\mathscr{P}_k(\mathcal{H}))$ are symmetry witness candidates. Indeed, they are invariant with respect to every unitary or antiunitary transformation; moreover, $\mathrm{span}_{\mathbb{R}}(\mathscr{P}_k(\mathcal{H})) = \mathrm{span}_{\mathbb{R}}(\mathscr{S}_k(\mathcal{H})_{\mathsf{u}}) = \mathscr{F}(\mathcal{H})_{\mathsf{s}}$, because every rank-one projection in $\mathcal{H}$ can be written as a linear combination $k+1$ elements of $\mathscr{P}_k(\mathcal{H})$. The set $\mathscr{S}_k(\mathcal{H})_{\mathsf{u}}$ consists of all *uniform* density operators of rank k; namely, of all mixed states whose eigenvalues form a uniform (constant) probability distribution. It is easy to see that if $\mathscr{P}_k(\mathcal{H})$ is a symmetry witness, then it is projectable; i.e., $\mathscr{S}_k(\mathcal{H})_{\mathsf{u}}$ is a symmetry witness too. By Theorem 9, if $n = \dim(\mathcal{H}) < \infty$ is a prime number, then $\mathscr{P}_k(\mathcal{H})$, $k < n$, is a symmetry witness.

IOP Conf. Series: Journal of Physics: Conf. Series **965** (2018) 012004 doi:10.1088/1742-6596/965/1/012004

6. The sets of projections of a fixed rank as symmetry witnesses

As noted in the previous section, a natural linear preserver problem consists in determining the class of all injective linear operators — densely defined on $\mathrm{span}_{\mathbb{R}}(\mathscr{P}_k(\mathcal{H})) = \mathscr{F}(\mathcal{H})_{\mathsf{s}}$, for some $k < \dim(\mathcal{H})$ — that leave the symmetry witness candidate $\mathscr{P}_k(\mathcal{H})$ invariant.

An almost complete solution of this problem is provided by the following result [9].

Theorem 11. *Let Φ be a densely defined linear operator in $\mathscr{B}_1(\mathcal{H})_{\mathsf{s}}$, with* $\mathrm{dom}(\Phi) = \mathscr{F}(\mathcal{H})_{\mathsf{s}}$. *Suppose that, given some $k \in \mathbb{N}$ — with $k = 1$, if $\dim(\mathcal{H}) = 2$; with $k < n/2$ or $n/2 < k < n$, if $3 \le n = \dim(\mathcal{H}) < \infty$; arbitrary, if $\dim(\mathcal{H}) = \infty$ — the following hypotheses are verified:*

(i) $\Phi(\mathscr{P}_k(\mathcal{H})) = \mathscr{P}_k(\mathcal{H})$;

(ii) Φ *is injective.*

(If $\dim(\mathcal{H}) < \infty$ — or $\dim(\mathcal{H}) = \infty$ and $k = 1$ — the second hypothesis is superfluous, because it follows from the first.)

Then, the operator Φ is closable, and its closure $\widehat{\Phi}$ is a surjective isometry of the canonical form

$$\widehat{\Phi}(A) = UAU^*, \quad \forall A \in \mathscr{B}_1(\mathcal{H})_{\mathsf{s}}, \tag{35}$$

where U is a unitary or antiunitary operator in $\mathcal{H}$, uniquely determined up to multiplication by a phase factor.

Remark 9. Theorem 11 can be reformulated in the *complex* Banach space $\mathscr{B}_1(\mathcal{H})$ of all trace class operators. Let $\Phi_{\mathbb{C}}$ be a linear operator in $\mathscr{B}_1(\mathcal{H})$, with $\mathrm{dom}(\Phi_{\mathbb{C}}) = \mathrm{span}_{\mathbb{C}}(\mathscr{P}_k(\mathcal{H})) = \mathscr{F}(\mathcal{H})$ (the linear manifold in $\mathscr{B}_1(\mathcal{H})$ of finite rank operators), and — for some $k \in \mathbb{N}$ satisfying the constraints specified in the statement of Theorem 11 — let $\Phi_{\mathbb{C}}$ map $\mathscr{P}_k(\mathcal{H})$ onto itself. If $\mathcal{H}$ is infinite-dimensional and $k > 1$, let us further assume that $\Phi_{\mathbb{C}}$ be injective. Then, $\Phi_{\mathbb{C}}$ is closable, and its closure $\widehat{\Phi}_{\mathbb{C}}$ is a surjective isometry in $\mathscr{B}_1(\mathcal{H})$ which is either of the form

$$\widehat{\Phi}_{\mathbb{C}}(A) = UAU^*, \quad \forall A \in \mathscr{B}_1(\mathcal{H}), \tag{36}$$

for some unitary operator U, or of the form

$$\widehat{\Phi}_{\mathbb{C}}(A) = UA^*U^*, \quad \forall A \in \mathscr{B}_1(\mathcal{H}), \tag{37}$$

for some *anti*unitary operator U. As observed in Remark 1, we may consider, equivalently, the transformations generated by unitary operators only, possibly composed with a transposition associated with some orthonormal basis in $\mathcal{H}$.

It is worth stressing that — in the case where $3 \le n = \dim(\mathcal{H}) < \infty$ — the constraints set on k in Theorem 11 bring the following consequence: for n *odd*, every value of $k < n$ gives rise to a symmetry witness, whereas, for $n \ge 4$ and *even*, the value $k = n/2$ is not contemplated. As anticipated in sect. 3, this point is related to the exception $2k = \dim(\mathcal{H}) \neq 2$ in Šemrl's theorem (Theorem 4); see the proof [9] of Theorem 11 (which corresponds to Theorem 6 of [9]). The fact that the case where $1 < k = n/2$ is indeed an exception is clarified by the following example.

Example 4. We will now provide a simple counterexample which shows that the statements of Šemrl's theorem and of Theorem 11 actually do not hold in the aforementioned exceptional case. In fact, assuming that $n = \dim(\mathcal{H}) = 2k$, $k \in \mathbb{N}$, let us consider the linear map

$$\Lambda^{[k]}\colon \mathscr{B}(\mathcal{H})_{\mathsf{s}} \equiv \mathscr{B}_1(\mathcal{H})_{\mathsf{s}} \to \mathscr{B}(\mathcal{H})_{\mathsf{s}}, \quad \Lambda^{[k]}(A) := k^{-1}\,\mathrm{tr}(A)\, I - A. \tag{38}$$

Note that $\Lambda^{[k]}$ maps $\mathscr{P}_k(\mathcal{H})$ bijectively onto itself: $\Lambda^{[k]}(P) = I - P = P^{\perp}$, for every $P \in \mathscr{P}_k(\mathcal{H})$, where $P^{\perp} \in \mathscr{P}_k(\mathcal{H})$ is the projection onto the orthogonal complement of the subspace $\mathrm{ran}(P)$. Namely, $\Lambda^{[k]}$ acts in $\mathscr{P}_k(\mathcal{H})$ as the orthocomplementation map; thus, restricted to this set, it

IOP Conf. Series: Journal of Physics: Conf. Series **965** (2018) 012004 doi:10.1088/1742-6596/965/1/012004

preserves orthogonality in both directions. Moreover, $\Lambda^{[k]}$ is injective, trace-preserving and unital; but, for $k > 1$, *not* positive, since it maps every pure state to a non-positive operator of rank $n-1 = 2k-1 \neq 1$. Therefore, if $k > 1$, $\Lambda^{[k]}$ cannot act as a symmetry transformation (not even on $\mathscr{P}_k(\mathcal{H})$, because $\mathrm{span}_{\mathbb{R}}(\mathscr{P}_k(\mathcal{H})) = \mathscr{B}(\mathcal{H})_{\mathsf{s}}$). For $k = 1$ ($\dim(\mathcal{H}) = 2$), we get the well known *reduction map* $\Lambda^{[1]}$, which, represented in terms of 2×2 hermitian matrices associated with an orthonormal basis, is the form

$$M \mapsto \sigma_2\, M^{\mathsf{T}} \sigma_2\,; \tag{39}$$

thus, a symmetry transformation (σ_2 is hermitian and unitary), coherently with Theorem 11.

It is now evident that the somewhat unnatural assumption — in Theorem 9 — that $n = \dim(\mathcal{H}) < \infty$ be a prime number is actually unnecessary; that result holds just because every prime number is odd. On the other hand, excluding all the even values of n one cannot catch the the significant exception $2k = \dim(\mathcal{H}) \neq 2$ versus the case where $2k = \dim(\mathcal{H}) = 2$ (the linear version of Wigner's theorem in dimension two, an 'exception within the exception'). Thus, the second of the two issues raised at the end of sect. 4 is clarified.

Theorem 11, complemented by the previous example, leads us to the following conclusion.

Corollary 2. *The set $\mathscr{P}_k(\mathcal{H})$, of projections of a finite fixed rank k — with $k = 1$, if $\dim(\mathcal{H}) = 2$; with $k < n/2$ or $n/2 < k < n$, if $3 \leq n = \dim(\mathcal{H}) < \infty$; arbitrary, if $\dim(\mathcal{H}) = \infty$ — is a symmetry witness. In the case where $4 \leq n = \dim(\mathcal{H}) < \infty$ and n is even, a linear operator in $\mathscr{B}(\mathcal{H})_{\mathsf{s}}$, mapping $\mathscr{P}_{\frac{n}{2}}(\mathcal{H})$ onto itself, is a trace-preserving and unital bijection, but, in general, not a symmetry transformation; thus, $\mathscr{P}_{\frac{n}{2}}(\mathcal{H})$ is not a symmetry witness.*

Proof. The only assertion to be proven is that, for $4 \leq n = \dim(\mathcal{H}) < \infty$, with n even, every linear operator in $\mathscr{B}(\mathcal{H})_{\mathsf{s}}$ mapping $\mathscr{P}_{\frac{n}{2}}(\mathcal{H})$ onto itself is a trace-preserving and unital bijection. See the proof of Corollary 1 in [9]. □

A this point, we are able to show that Corollary 1 admits — in the finite-dimensional case — the following refinement, so clarifying also the first of the two issues raised at the end of sect. 4.

Corollary 3. *Suppose that $\dim(\mathcal{H}) < \infty$, and let $\Phi\colon \mathscr{B}(\mathcal{H})_{\mathsf{s}} \to \mathscr{B}(\mathcal{H})_{\mathsf{s}}$ be a surjective linear map. Then, the following facts are equivalent:*

(i) Φ is isometric (wrt the operator norm) and positive;

(ii) Φ is isometric and unital;

(iii) for some $k < \dim(\mathcal{H})$, Φ maps $\mathscr{P}_k(\mathcal{H})$ onto itself and, in the case where $2 < 2k = \dim(\mathcal{H})$, is also positive;

(iv) Φ is a symmetry transformation.

Proof. Taking into account Corollary 1 and Theorem 11, the only point that we still need to analyze is the case where $2k = \dim(\mathcal{H})$, $k > 1$, with Φ further assumed to be positive. By Corollary 2, we know that with the given assumptions Φ is unital, as well. Hence, by Corollary 1 the proof is complete. □

Finally, we observe that the previous results can be rephrased in terms of uniform density operators, that are the physically relevant objects.

Corollary 4. *Let Φ be a densely defined linear operator in $\mathscr{B}_1(\mathcal{H})_{\mathsf{s}}$, with $\mathrm{dom}(\Phi) = \mathscr{F}(\mathcal{H})_{\mathsf{s}}$, mapping the set $\mathscr{S}_k(\mathcal{H})_{\mathsf{u}} = \varpi(\mathscr{P}_k(\mathcal{H}))$, of uniform density operators of a fixed rank $k < \dim(\mathcal{H})$, onto itself. Assume, moreover, that*

- *in the case where $\dim(\mathcal{H}) = \infty$ and $k > 1$, Φ is injective;*
- *in the case where $2 < 2k = \dim(\mathcal{H}) < \infty$, Φ is positive.*

Then, Φ is closable, and its closure $\widehat{\Phi}$ is a symmetry transformation. If $k < \dim(\mathcal{H})$ and — for $\dim(\mathcal{H}) \geq 4$ — $2k \neq \dim(\mathcal{H})$, $\mathscr{S}_k(\mathcal{H})_{\mathsf{u}}$ is a symmetry witness; otherwise, it is not.

IOP Conf. Series: Journal of Physics: Conf. Series **965** (2018) 012004 doi:10.1088/1742-6596/965/1/012004

7. Conclusions

We can finally draw the following conclusions:

- A *symmetry witness* is a set $\mathcal{W}$ in the Banach space $\mathscr{B}_1(\mathcal{H})_{\mathsf{s}}$, of selfadjoint trace class operators, that allows one to ascertain whether an injective linear operator Φ acting in $\mathscr{B}_1(\mathcal{H})_{\mathsf{s}}$ is a symmetry transformation, by checking whether it leaves the witness invariant, i.e., whether $\Phi(\mathcal{W}) = \mathcal{W}$; see Definition 1. Therefore, this notion is related to a suitable class of linear preserver problems; see Remark 7.
- By a linear version of Wigner's theorem — Theorem 7 — one can regard the set of pure states $\mathscr{P}_1(\mathcal{H})$ as the *prototype* of a symmetry witness. Further remarkable examples are the convex set $\mathscr{S}(\mathcal{H})$ of all states (see Remark 1 of [9]) and the convex set $\mathscr{F}\mathscr{S}(\mathcal{H}) \subset \mathscr{S}(\mathcal{H})$ of finite-rank density operators (Theorem 8).
- By Theorem 11, the set $\mathscr{P}_k(\mathcal{H})$, $k < \dim(\mathcal{H})$, is a symmetry witness too, with the peculiar exception — when $4 \leq n = \dim(\mathcal{H}) < \infty$, n *even* — of the value $k = n/2$: a linear operator Φ in $\mathscr{B}(\mathcal{H})_{\mathsf{s}} \equiv \mathscr{B}_1(\mathcal{H})_{\mathsf{s}}$, mapping $\mathscr{P}_{\frac{n}{2}}(\mathcal{H})$ onto itself, is a trace-preserving and unital bijection, but, in general, *not* a symmetry transformation (Corollary 2). These symmetry witnesses are *positive*, and *projectable* (Definition 2) onto the symmetry witnesses $\mathscr{S}_k(\mathcal{H})_{\mathsf{u}}$, that have a clear physical interpretation: they consist of those density operators of a fixed rank whose eigenvalues form a *uniform* probability distribution.
- It is worth stressing that the set $\mathscr{S}_k(\mathcal{H})_{\mathsf{u}}$ has a precise role with regard to the *majorization* relation in $\mathscr{S}(\mathcal{H})$ [29, 30]; i.e., it is formed by those density operators that are minimal, with respect to majorization, among all elements of $\mathscr{S}(\mathcal{H})$ with rank not larger than k. Therefore, a (suitably defined) class of quantum entropies — e.g., the von Neumann, the Rényi or the Tsallis entropies — if restricted to the subset of $\mathscr{S}(\mathcal{H})$ of all density operators of rank not larger than k, attain their maximum value precisely on $\mathscr{S}_k(\mathcal{H})_{\mathsf{u}}$ [29,30]. Moreover, considering in particular the finite-dimensional case, we observe that a quantum dynamical semigroup [19, 21, 24–30] that leaves the set $\mathscr{S}_k(\mathcal{H})_{\mathsf{u}}$ invariant, for some $k < \dim(\mathcal{H})$, by Corollary 4 must be a one-parameter (semi-)group of unitary transformations; thus, it must actually describe the dynamics of a *closed* quantum system.
- The results outlined above solve a problem considered in [13]: to find a generalization of the linear version of Wigner's theorem involving projections of a fixed rank. This problem has been first partially solved in [13], in the finite-dimensional case and with the (unnecessary) assumption that $\dim(\mathcal{H})$ be a prime number (Theorem 9). A similar problem has been studied in [31] — see Theorem 10 — with no restriction on the Hilbert space dimension and working in a different setting: the sets of projections with a fixed rank are embedded in the Banach space of bounded operators (the *dual* of the Banach space of trace class operators); moreover, the linear maps acting on bounded operators are supposed to be positive and unital. A direct comparison with the results of [13], and with those discussed in sect. 6, is possible in the finite-dimensional case only, where the distinction between the Schrödinger and the Heisenberg pictures is not sharp. In this case, both positivity and unitality can be dispensed with; the single exception is the case of $\mathscr{P}_{\frac{n}{2}}(\mathcal{H})$, with $n = \dim(\mathcal{H}) \geq 4$ and even. For these values of n, *unitality* is automatically satisfied by a linear map mapping $\mathscr{P}_{\frac{n}{2}}(\mathcal{H})$ onto itself, whereas *positivity* implies that the map is a symmetry transformation.
- As explained in Remark 9, working in the *real* Banach space $\mathscr{B}_1(\mathcal{H})_{\mathsf{s}}$ (alternatively, in $\mathscr{B}(\mathcal{H})_{\mathsf{s}}$) or in the *complex* Banach space $\mathscr{B}_1(\mathcal{H})$ (respectively, in $\mathscr{B}(\mathcal{H})$) is unessential.
- At the beginning of sect. 6, we have observed that Theorem 11 provides an *almost* complete solution of the linear preserver problem associated with the symmetry witness candidate $\mathscr{P}_k(\mathcal{H})$, $k < \dim(\mathcal{H})$. In the exceptional case where $n = \dim(\mathcal{H}) = 2k$, with $k \geq 2$ — according to an interesting *conjecture* [13] — every linear map in $\mathscr{B}(\mathcal{H})_{\mathsf{s}}$ mapping $\mathscr{P}_k(\mathcal{H})$

onto itself would be either a symmetry transformation or a (trace-preserving, unital but) non-positive map of the form $\Lambda_U^{[k]}\colon \mathscr{B}(\mathcal{H})_s \ni A \mapsto \Lambda^{[k]}(UAU^*) \in \mathscr{B}(\mathcal{H})_s$, where U is a unitary or antiunitary operator in $\mathcal{H}$ and $\Lambda^{[k]}$ is the map defined by (38). Proving this conjecture would amount to finding the single missing piece of the puzzle.

References

[1] Wigner E P 1959 *Group Theory and its Application to the Quantum Mechanics of Atomic Spectra* (New York: Academic)
[2] Bargmann V 1964 Note on Wigner's theorem on symmetry operations *J. Math. Phys.* **5** 862–8
[3] Kadison R V 1965 Transformations of states in operator theory and dynamics *Topology* **3** 177–98
[4] Simon B 1976 Quantum dynamics: from automorphism to Hamiltonian *Studies in Mathematical Physics. Essays in Honor of Valentine Bargmann* (*Princeton Series in Physics*) ed E H Lieb *et al* (Princeton, NJ: Princeton University Press) pp 327–49
[5] Cassinelli G, De Vito E, Lahti P J and Levrero A 1997 Symmetry groups in quantum mechanics and the theorem of Wigner on the symmetry transformations *Rev. Math. Phys.* **9** 921–41
[6] Cassinelli G, De Vito E, Lahti P J and Levrero A 2004 *The Theory of Symmetry Actions in Quantum Mechanics with Applications to the Galilei Group* (Berlin: Springer)
[7] Parthasarathy K R 2005 *Mathematical Foundation of Quantum Mechanics* (New Delhi: Hindustan Book Agency)
[8] Molnár L 2007 *Selected Preserver Problems on Algebraic Structures of Linear Operators and on Function Spaces* (Berlin: Springer)
[9] Aniello P and Chruściński D 2017 Symmetry witnesses *J. Phys. A: Math. Theor.* **50** 285302
[10] Varadarajan V S 1985 *Geometry of Quantum Theory* 2nd edn (New York: Springer)
[11] Li C-K and Tsing N-K 1992 Linear preserver problems: a brief introduction and some special techniques *Linear Algebra Appl.* **162–4** 217–35
[12] Frobenius G 1897 Über die Darstellung der endlichen Gruppen durch lineare Substitutionen *Sitzungsber. Königl. Preuss. Akad. Wiss. Berlin* 994–1015
[13] Sarbicki G, Chruściński D and Mozrzymas M 2016 Generalising Wigner's theorem *J. Phys. A: Math. Theor.* **49** 305302
[14] Uhlhorn U 1963 Representation of symmetry transformations in quantum mechanics *Ark. Fys.* **23** 307–40
[15] Šemrl P 2004 Orthogonality preserving transformations on the set of n-dimensional subspaces of a Hilbert space *Illinois J. of Math.* **48** 567–73
[16] Russo B 1969 Trace preserving mappings of matrix algebras *Duke Math. J.* **36** 297–300
[17] Russo B 1969 Isometries of the trace class *Proc. Amer. Math. Soc.* **23** 213
[18] Arazy J 1975 The isometries of C_p *Israel J. Math.* **22** 247–56
[19] Alicki R and Lendi K 1987 *Quantum Dynamical Semigroups and Applications* (Berlin: Springer)
[20] Nielsen M A and Chuang I L 2000 *Quantum Computation and Quantum Information* (Cambridge: Cambridge University Press)
[21] Breuer H-P and Petruccione F 2002 *The Theory of Open Quantum Systems* (Oxford: Oxford University Press)
[22] Bengtsson I and Życzkowski K 2006 *Geometry of Quantum States* (Cambridge: Cambridge University Press)
[23] Lupo C, Pirandola S, Aniello P and Mancini S 2011 On the classical capacity of quantum Gaussian channels *Phys. Scr.* **T143** 014016
[24] Aniello P, Kossakowski A, Marmo G and Ventriglia F 2010 Brownian motion on Lie groups and open quantum systems *J. Phys. A: Math. Theor.* **43** 265301
[25] Chruściński D, Kossakowski A, Aniello P, Marmo G and Ventriglia F 2010 A class of commutative dynamics of open quantum systems *Open Syst. Inf. Dyn.* **17** 255–77
[26] Aniello P 2011 On a certain class of semigroups of operators *Open Syst. Inf. Dyn.* **18** 129–42
[27] Aniello P 2013 Quantum dynamical semigroups, group representations and convolution semigroups *Phys. Scr.* **T153** 014003
[28] Aniello P 2013 Operators versus functions: from quantum dynamical semigroups to tomographic semigroups *J. Phys.: Conf. Series* **474** 012005
[29] Aniello P and Chruściński D 2016 Characterizing the dynamical semigroups that do not decrease a quantum entropy *J. Phys. A: Math. Theor.* **49** 345301
[30] Aniello P 2017 Quantum entropies, Schur concavity and dynamical semigroups *J. Phys.: Conf. Series* **804** 012003
[31] Størmer E 2017 Positive maps which map the set of rank k projections onto itself *Positivity* **21** 509–11
[32] Størmer E 2013 *Positive Linear Maps of Operator Algebras* (Berlin: Springer)

IOP Conf. Series: Journal of Physics: Conf. Series **965** (2018) 012005 doi:10.1088/1742-6596/965/1/012005

Projective Ponzano–Regge spin networks and their symmetries

Vincenzo Aquilanti[1], **Annalisa Marzuoli**[2,3]

[1] Dipartimento di Chimica, Biologia e Biotecnologie, via Elce di Sotto 8, Università di Perugia, 06123 Perugia, I

[2] Dipartimento di Matematica 'F. Casorati', Università degli Studi di Pavia, via Ferrata 1, 27100 Pavia, I

[3] INFN, Sezione di Pavia, via Bassi 6, 27100 Pavia, I

E-mail: annalisa.marzuoli@unipv.it

Abstract. We present a novel hierarchical construction of projective spin networks of the Ponzano-Regge type from an assembling of five quadrangles up to the combinatorial 4-simplex compatible with a geometrical realization in Euclidean 4-space. The key ingredients are the projective Desargues configuration and the incidence structure given by its space-dual, on the one hand, and the Biedenharn–Elliott identity for the $6j$ symbol of SU(2), on the other. The interplay between projective-combinatorial and algebraic features relies on the recoupling theory of angular momenta, an approach to discrete quantum gravity models carried out successfully over the last few decades. The role of Regge symmetry –an intriguing discrete symmetry of the $6j$ which goes beyond the standard tetrahedral symmetry of this symbol– will be also discussed in brief to highlight its role in providing a natural regularization of projective spin networks that somehow mimics the standard regularization through a q-deformation of $SU(2)$.

1. Introduction

The Hilbert system of axioms for an incidence geometry deals with ideal elements such as points, lines and planes, together with the notion of concurrence (incidence) relations [1, 2]. The resulting pre–metric admissible configurations can be also studied as concrete (finite) sets of points, lines and planes in specific geometries, the Euclidean, affine and projective planes or their higher–dimensional counterparts: one then speaks of *realizations* of some given incidence configuration. The simplest abstract incidence structures have been exploited in the (re)coupling theory of $SU(2)$ angular momentum since the earliest times of Fano and Racah [3] and, more in general, combinatorial methods have been widely applied to the study of discretized quantum spacetimes since the seminal contributions of Giorgio Ponzano and Tullio Regge [4] and Roger Penrose [5]. 'Quantum spin networks', a term that nowadays includes quite a wide range of interlaced research fields –from basic features and applications of quantum field theories in low dimensions [6, 7], up to the 'Loop' approaches to the quantization of General Relativity [8, 9]– will be shown to emerge solely from the basic axioms and theorems of finite–dimensional projective geometry suitably combined with quantum angular momentum theory as rooted in the diagrammatic formalism of Yutsis graphs [10] (a self–contained account on projective configurations and diagrammatic–combinatorial tools is collected in two appendices).

The combinatorial–projective structures emerging from the hierarchical construction carried out in section 2 are interpreted as associated with collective modes of elementary 'quantum geometries' on the gravitational side [11, 12, 13, 14], and as related to integrable quantum many-body systems of interest also in atomic and molecular physics [15, 16, 17], on the other. We recognize in particular that –starting from the $6j$ symbol of $SU(2)$ encoded into a quadrangle associated with the projective configuration of four points joined in pairs by six distinct lines– a recollection of five nested quadrangles is projectively nothing but the Desargue configuration $(10)_3$ (ten points and ten lines, with each of its ten points incident to three lines and each of its ten lines incident to three points). We will show how such a configuration can be consistently labeled with $SU(2)$ spin variables as to comply with the algebraic content of the Biedenharn–Elliot (BE) identity treated in section 2. Moreover: *i)* Desargues configuration can be realized in a 3D Euclidean space, a well known fact ensured by Desargues' theorem which holds true in any dimension > 2, see appendix 5.2; *ii)* according to the basic axioms of projective geometry, each theorem must be valid under projective dualities, to be meant as either the usual (point $\leftrightarrow$ line) exchanges in the plane realization or as (point $\leftrightarrow$ plane) and (line $\leftrightarrow$ line) exchanges when viewed in 3D space. On the basis of the above remarks we will remind first how the 'space–dual' of Desargues configuration has a combinatorial content that matches to the realization of an Euclidean 4-simplex (bounded by five tetrahedra) and then we will provide the natural $SU(2)$ labeling of its edges. This result is relevant for all Ponzano–Regge-like quantum gravity models proposed over the years [18], since it would explain the apparent dichotomy in the assignment of spin labels to triads of angular momenta: in Yutsis-type diagrams (and in most models used in Loop quantum gravity) the correspondence is (triad) $\rightarrow$ (3-valent vertex), while in Ponzano–Regge and other approaches [11] it is assumed that (triad) $\rightarrow$ (triangular faces) (*cf.* appendix 5.1). Actually both correspondences are fully consistent in 3D Euclidean space because they are related by Poincaré 2D duality for polyhedra, and the tetrahedron is obviouly self–dual in this sense. However, if we keep on assuming, much in the spirit of Regge Calculus [19], the prominent role of the assignment (edge length) $\rightarrow$ (spin label), then the 4-simplex Desargue spin network automatically complies with this requirement and inherits triangular inequalities on triangular faces just as a consequence of the projective dualization procedure. (Recall that in the Euclidean path integral formulation of Regge–discretized gravity the measure is expressed in terms of edge lengths, the discrete analog of the functional measure over metrics, the latter being the dynamical variables of General Relativity). In the final part of section 2 we will briefly comment on the issues of dimensional reduction and increase in this upgraded projective context.

In section 3 the association between the Desargues configuration and the Biedenharn–Elliott identity for the $6j$ symbol will be established and worked out in details (this algebraic identity has a number of interpretations and implications in many branches of mathematical physics, and we will briefly review some of them in the concluding remarks). In this section it will be recognized how the BE identity complies with the two projective Desargues spin networks viewed, on the one hand, as a collection of five nested quadrangles, and as associated to the 4-simplex on the other. Such an algebraic setting can be further improved by calling into play Regge symmetry of the $6j$ symbol [20], previously discovered for the Wigner $3j$ symbol [21]. The prominent feature of Regge symmetry is its functional, not geometric origin, grounded into the Racah sum rule for the $6j$ symbol, *cf.* [22, 23] for reviews and original references. However, over the years it has been recognized as multi–faceted and still intriguing in many respects, ranging from its interpretation as non trivial automorphism group of a set of not-congruent Euclidean tetrahedra [24], up to its relation with Okamoto symmetry of the Painlevé VI differential equation [25]. We have been working on the aspects related to quadratic operator algebras, a frame that includes general (binary and symmetric) recoupling theory of $SU(2)$ angular momenta, as well as on the semiclassical analysis of three–term recursion relations, *cf.* [12, 13] also for a complete list of

IOP Conf. Series: Journal of Physics: Conf. Series **965** (2018) 012005 doi:10.1088/1742-6596/965/1/012005

references. The definition of Regge symmetry is briefly reviewed at the beginning of the section, and it will be shown that, by suitably ordering and keeping fixed the spin labels assigned to the four edges of a given reference quadrangle, the quantum amplitude of each Desargues spin network is automatically regularized in terms of a positive integer as happens when working *ab initio* within the framework of representation theory of the quantum group $SU(2)_q$ at $q=$ root of unity [26, 27].

In section 4 a glimpse to other interconnections and further developments will be addressed.

2. Desargues spin networks

In this section we are going to present a novel hierarchical construction of projective spin networks, from an assembling of five quadrangles up to the combinatorial 4-simplex (5 vertices, 10 edges, 10 triangular faces, 5 tetrahedra) compatible with a geometrical *realization* in Euclidean 4-space. We proceed in parallel and consistently with 'combinatorial' labeling of configurations, on the one hand, and with labeling provided by angular momentum theory, on the other. The algebraic content of the construction, together with an improved regularization procedure, are addressed in section 3. We refer to appendices 5.1 and 5.2 for a primer on angular momentum (re)coupling theory and for the discussion of the projective features of the configurations and diagrams of this section. Notations and conventions are those of the classic handbook [23].

2.1. Combinatorics of nested quadrangles

The basic building block of the construction is the $6j$ symbol presented through its Yutsis diagram as in Fig.1 or, equivalently, as the complete quadrangle of Fig.15, where $(a, b, c, d, x, y) \in \{0, 1/2, 1, 3/2, \dots\}$ label irreducible representations (*irreps*) of the Lie group $SU(2)$.

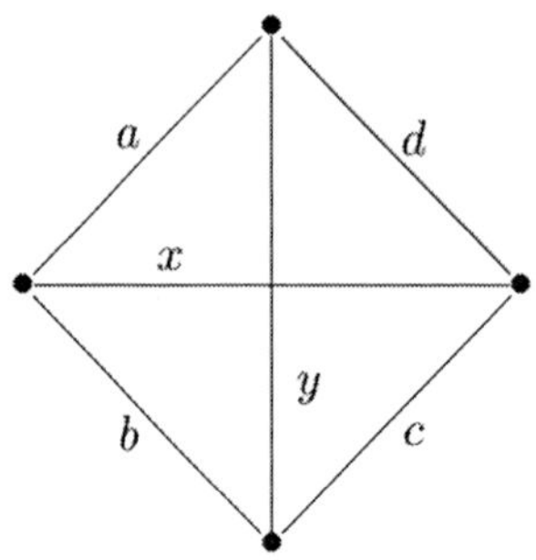

$$\longleftrightarrow \begin{Bmatrix} a & b & x \\ c & d & y \end{Bmatrix}$$

Figure 1. The Yutsis diagram of the $6j$.

The Desargues configuration $(10)_3$ shown in Fig.2 –the symmetric incidence structure made of ten points and ten lines– turns out to be combinatorially consistent with an assembling of five quadrangles, $\mathcal{Q}_1$, $\mathcal{Q}_2$, $\mathcal{Q}_3$, $\mathcal{Q}_4$ and $\mathcal{Q}_5$, in one–to–one correspondence with the coloring (Black, Red, Blue, Purple, Green). The ten bubbles which stand at the intersection of three lines are 3-valent vertices and have a double (unordered) labeling since each of them is shared by two quadrangles. The six edges of each quadrangle inherit colors from (one of) the colors of vertices. The angular momentum (spin) labeling of the five quadrangles –to be associated with the five $6j$ symbols of the BE identity in the following section– is depicted in Fig.3 and is given by the correspondences

$$\mathcal{Q}_1 \leftrightarrow \begin{Bmatrix} a & b & x \\ c & d & p \end{Bmatrix}; \mathcal{Q}_2 \leftrightarrow \begin{Bmatrix} c & d & x \\ e & f & q \end{Bmatrix}; \mathcal{Q}_3 \leftrightarrow \begin{Bmatrix} e & f & x \\ b & a & r \end{Bmatrix}; \mathcal{Q}_4 \leftrightarrow \begin{Bmatrix} p & q & r \\ f & b & c \end{Bmatrix}; \mathcal{Q}_5 \leftrightarrow \begin{Bmatrix} p & q & r \\ e & a & d \end{Bmatrix}. \quad (1)$$

IOP Conf. Series: Journal of Physics: Conf. Series **965** (2018) 012005 doi:10.1088/1742-6596/965/1/012005

It is worth pointing out that now the incidence structure underlying Desargues configuration has to be thought of as *geometrically* realized in a $3D$ projective or Euclidean space (namely the perspective point of Desargues' Theorem must be in general position with respect to the planes of the two triangles, *cf.* Fig.18). In a $3D$ Euclidean space, in particular, the five quadrangles considered so far would appear nested into each other, and not glued to each other to form a polyhedron. However, the duality principle (appendix 5.2) can be called into play to establish the actual existence of the space–dual of Desargues configuration, which will be proven below to be compatible not only with a combinatorial 4-simplex as noted in [2, 28], but also with a spin labeling of the 4-simplex.

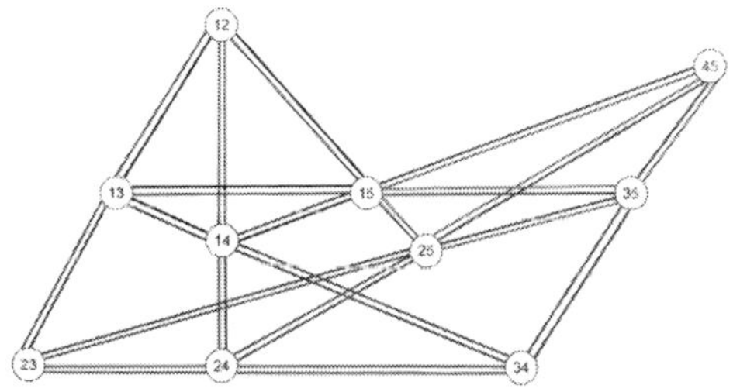

Figure 2. Combinatorial labeling of Desargues configuration $(10)_3$.

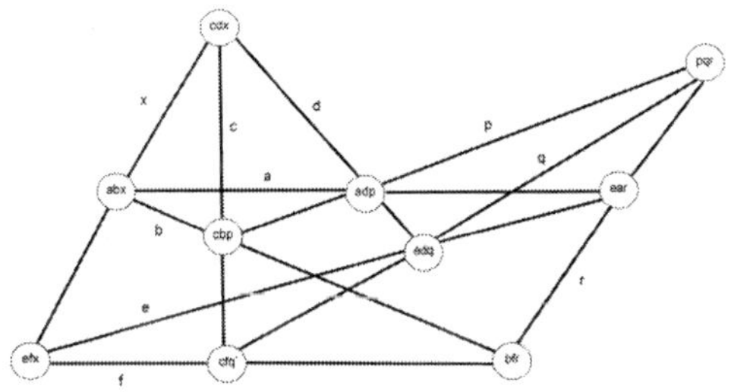

Figure 3. Spin labeling of Desargues configuration $(10)_3$.

2.2. Space–dual of Desargues configuration and the 4-simplex

The notion of duality applies in a straightforward way to the combinatorial labeling used in Fig.2. The points are two-colored, denoted by circles or round brackets, (12), (23), …; quadrangles are listed $1, 2, 3, 4, 5$ as in Eq.(1); lines are enumerated by triples of points, *e.g.* (12) (13) (23). An intermediate step consists in switching the line labeling to the dual one, where for instance the previous line is labeled with the two complementary colors [45] within square brackets, see all the resulting tags in Fig.4.

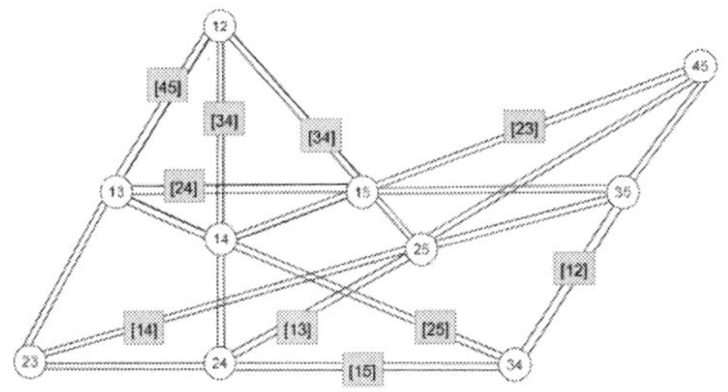

Figure 4. Dual labeling of lines put on top of Desargues configuration of Fig.2.

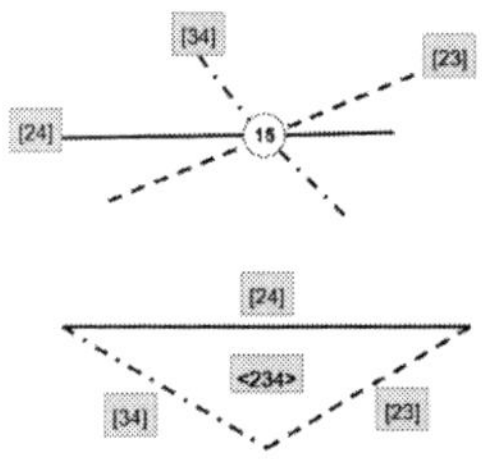

Figure 5. A blown-up 3-valent vertex and its dual triangle.

IOP Conf. Series: Journal of Physics: Conf. Series **965** (2018) 012005 doi:10.1088/1742-6596/965/1/012005

Recall from the final part of appendix 5.2 that in a projective space P^3 there is duality between subspaces of dimension k and $(3-k-1)$ and then the basic correspondences for going through the space–dual of Desargues configuration (theorem) are

point $\longrightarrow$ plane,
line $\longrightarrow$ line,
quadrangle $\longrightarrow$ point,

where the last pairing can be worked out starting from the dualization of labeling of the original quadrangles, *e.g.* $\mathcal{Q}_1 \rightarrow$ vertex $\{2345\}$ (curly brackets) in the space–dual configuration. Moreover, noting that the intersections of pairs of quadrangles represent the vertices of the original configuration of Fig.2, namely

$$\mathcal{Q}_i \cap \mathcal{Q}_j = (ij)\,, \quad i,j = 1,2,3,4,5 \;\;\text{with}\, i \neq j\,, \tag{2}$$

and recalling that any such vertex is actually the intersection of three lines (a *triad* of spin labeling according to the dictionary of Fig.3), we are forced to improve the duality (point $\rightarrow$ plane) as

3-valent vertex $\longrightarrow$ triangle

and to assign the combinatorial labels according to

$$\text{2-color vertex}(ij) \;\rightarrow\; \text{triangle} < klm > \tag{3}$$

with k, l, m complementary to (ij), *cf.* Fig.5.

The statement of the space–dual of Desargues' theorem reads:

> If two trihedra in 3D space are perspective from a triangle, then there exists a line connecting the tips of the two trihedra together with three planes meeting at this line and passing through the sides of the triangle.

The resulting connected ensemble of five vertices, ten lines, ten triangles and five tetrahedra (actually trihedra, each pair of which intersects a fourth triangle) is combinatorially a 4-simplex, denoted $\mathcal{S}_{comb}$ for short. The pictorial representation of this construction is given in Fig.6 where the quadrangles $\mathcal{Q}_1$ and $\mathcal{Q}_5$ –or, respectively, dual vertices $\{2345\}$ and $\{1234\}$– have been employed consistently with the choice made in Fig. 5 (actually any other choice of the selected pair of quadrangles turns out to be combinatorially equivalent to each other).

As for spin labeling of this new configuration, observe preliminary that the dual combinatorial labels on edges established in Fig.4 have been retained also for lines in the dual Desargues configuration on the basis of the duality (line $\rightarrow$ line) in space. Thus, by resorting to the dictionary of Fig.3, and denoting the five tetrahedra of the combinatorial 4-simplex $\mathcal{S}_{comb}$ by $\mathcal{T}_i$, the correspondence with $6j$ symbols arranged into the upgraded $\mathcal{S}_{spin}$, is necessarily given by

$$\mathcal{T}_1 \leftrightarrow \begin{Bmatrix} a & b & x \\ c & d & p \end{Bmatrix};\mathcal{T}_2 \leftrightarrow \begin{Bmatrix} c & d & x \\ e & f & q \end{Bmatrix};\mathcal{T}_3 \leftrightarrow \begin{Bmatrix} e & f & x \\ b & a & r \end{Bmatrix};\mathcal{T}_4 \leftrightarrow \begin{Bmatrix} p & q & r \\ f & b & c \end{Bmatrix};\mathcal{T}_5 \leftrightarrow \begin{Bmatrix} p & q & r \\ e & a & d \end{Bmatrix}. \tag{4}$$

The crucial difference with respect to the correspondence between quadrangles of the original Desargues configuration and $6j$'s given in (1) consists in recognizing that here the triads of the $6j$ have to be assigned to the triangular faces of tetrahedra as a consequence of the appropriate *projective duality* stated in Eq. (3). As mentioned in appendix 5.1 such a presentation of the $6j$ stands at the basis of Ponzano–Regge model for (the semiclassical analysis of) 3D Euclidean gravity, where the spin labels of $SU(2)$ *irreps* are the weights assigned to edges and give rise to the discrete analog of the gravitational path integral measure on the metrics. An exploded view of the space–dual Desargues spin network $\mathcal{S}_{spin}$ is shown in Fig. 7.

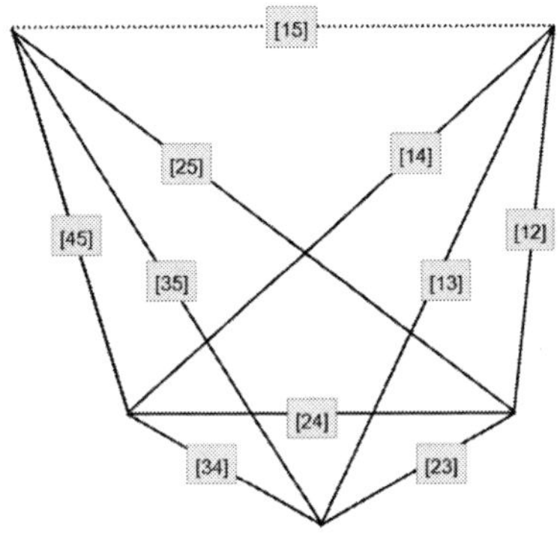

Figure 6. Space–dual of Desargues' theorem. Here the combinatorial labeling is only on lines for simplicity, but the labels of the other substructures are easily reconstructed from it: the two trihedra are incident on the triangle $< 234 > = [23] \cup [24] \cup [34]$; the three planes of each trihedron are spanned by the three (of course not coplanar) lines [25], [35], [45] and [12], [13], [15], respectively; the theorem ensure the existence of line [15] and of three planes (triangles) $< 125 >$, $< 135 >$, $< 145 >$. The two vertices on the top represent the tips of the trihedra, labeled {2345} and {1234} or, by resorting again to complementarity of labeling, denoted 1 and 5, respectively.

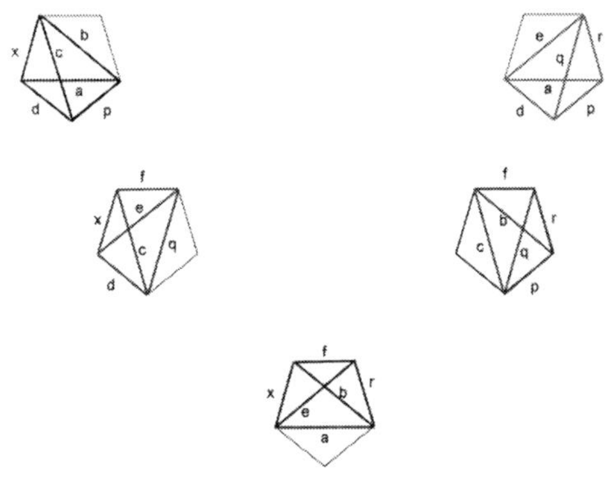

Figure 7. Spin labeling of the tetrahedra of the 4-simplex $\mathcal{S}_{spin}$: ($\mathcal{T}_1$, $\mathcal{T}_2$, $\mathcal{T}_3$, $\mathcal{T}_4$, $\mathcal{T}_5$) $\leftrightarrow$ (Black, Red, Blue, Purple, Green), from top left in anti-clockwise order.

2.3. Geometrical realizations and dimensional reduction and increase

As already pointed out at the end of section 2.1, the hierarchical construction that leads to the Desargues spin network $\mathcal{S}_{spin}$ discussed in section 2.2 (complemented with material of appendices 5.1 and 5.2) relies on the assumption that the underlying incidence structures are realized in a P^n with $n \geq 3$. The projective viewpoint provides, on the one hand, an axiomatic frame for dealing with incidence structures that *emerge* from the simplest configurations of points and lines in the plane, and an effective procedure based on projective *dimensional reductions* to go from the geometric realization of a 4-simplex in Euclidean space down to its low–dimensional cross–sections, on the other. The lines of reasoning underlying these achievements are summarized in the following steps.

- Any two distinct points are incident with just a line and any two lines are incident with at least one point, (Axioms I) and II) of appendix 5.2). The existence of quadrangles (configurations $(4_3\,,6_2)$) is the third axiom of (plane) projective geometry;
- Among the various finite configurations of points and lines, Desargues' (10_3) has a special status: in $n=2$ it must be postulated separately but it is encoded as a theorem for $n \geq 3$, so that in the latter cases no further axiom is necessary;
- In $n = 3$ there exist two projective–dual versions of the theorem: its plane–dual, still denoted (10_3) because the configuration is symmetric under the exchange (points $\leftrightarrow$ lines), and its space–dual, denoted $\mathcal{S}_{comb}$, discussed at length in section 2.2;
- $\mathcal{S}_{comb}$ can be realized through the projection of an Euclidean 4-simplex onto P^3;
- If we slice any such Euclidean realization of $\mathcal{S}_{comb}$ by an arbitrary solid, each line is cut by this portion of space in a point, each plane in a line and each tetrahedron in a plane so that any cross–section of $\mathcal{S}_{comb}$ represents a figure made of ten points, ten lines and five planes, namely Desargues' (10_3) [28].

Obviously the previous remarks touch basic questions about the nature of quantum spacetime, the existence of a fundamental length and a number of questions that arise in connection with discretized, simplicial or spin networks, approaches to quantum gravity, not to mention the issue of dimensional reduction [29]. Our view is inspired by the (quite elementary) frame provided by projective geometry of configurations and by its intriguing relations with angular momentum theory –relations that have been only partially worked out by other authors, sometimes leading to ambiguous conclusions, *cf.* the appendices.
The result we have been proving,

> the Desargues configuration and its space–dual are respectively cross-sections and projections of an Euclidean 4-simplex, and consistent $SU(2)$ spin labeling can be assigned to each of them,

leaves open the crucial question of the existence of a 'quantum of space' in (space or spacetime) dimensions D=3,4. In the light of the previous considerations, one might either *i)* postulate the existence of an Euclidean 4-simplex, assign projective coordinates and go down to a 3D Desargues configuration which admits a spin labeling as in Fig.3, or *ii)* look for an emergent 3D (projective) space from the (re)coupling of angular momenta. The latter option is actually what occurs according to the combined results found by Penrose [5] and by Ponzano and Regge [4] which stand at the basis of all of the spin networks models addressed over the years in quantum gravity *cf.* [6, 18] for brief accounts and [9] for an extended overview on the Loop approach. Then the rationale of our construction relies on *ii)* and exploits the projective nature of the $SU(2)$ spin networks up to the emergency of the 4-simplex $\mathcal{S}_{spin}$.

3. Regge symmetry and improved algebraic setting

As is well known the defining relations for $6j$ symbols of $SU(2)$ are the orthogonality condition,

$$\sum_x (2x+1) \begin{Bmatrix} a & b & x \\ c & d & y \end{Bmatrix} \begin{Bmatrix} c & d & x \\ a & b & y' \end{Bmatrix} = \frac{\delta_{yy'}}{2y'+1} \delta_{(ady)}\, \delta_{(bcy)}, \tag{5}$$

where the symbol $\delta_{(\dots)}$ constrains the entries to belong to a triad, and the Biedenharn–Elliot (BE) identity

$$\sum_x (-1)^{\varphi+x} \begin{Bmatrix} a & b & x \\ c & d & p \end{Bmatrix} \begin{Bmatrix} c & d & x \\ e & f & q \end{Bmatrix} \begin{Bmatrix} e & f & x \\ b & a & r \end{Bmatrix} = \begin{Bmatrix} p & q & r \\ f & b & c \end{Bmatrix} \begin{Bmatrix} p & q & r \\ e & a & d \end{Bmatrix}, \tag{6}$$

where $\varphi = a+b+c+d+e+f+p+q+r$. The content of this algebraic relation, depicted in Fig.8, complies with the assignments of spin labels/triads to lines/vertices –resulting in the association $6j \leftrightarrow$ quadrangles (see Eq.(1))– in the Desargues configuration $(10)_3$ as established in Fig.3.

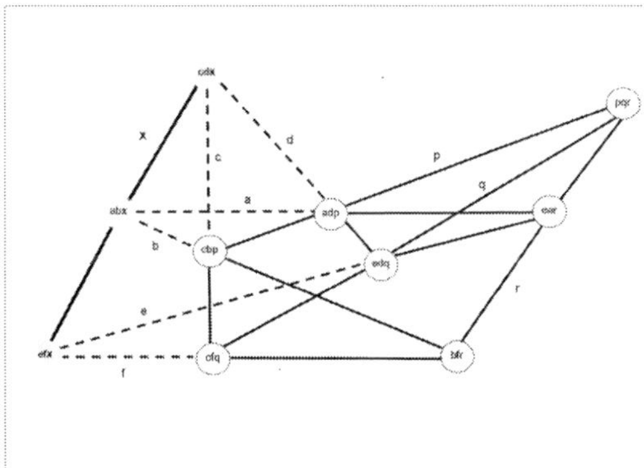

Figure 8. Upon summation over x (bold), three triads of spin labels together with the dashed edges disappear: only the two quadrangles corresponding to the $6j$'s on the right side of the BE identity survive.

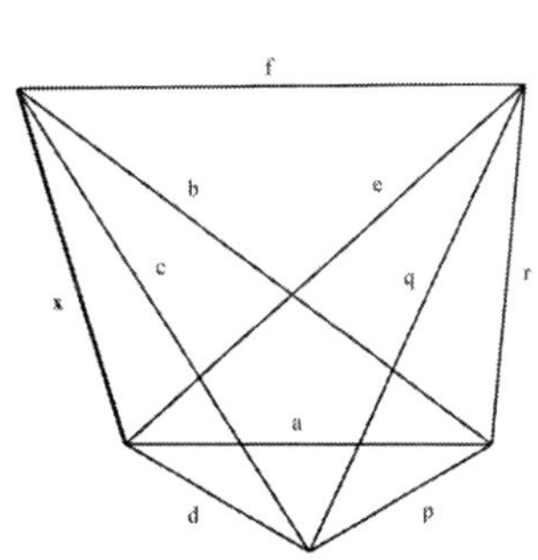

Figure 9. The BE identity in the Desargues space-dual $\mathcal{S}_{spin}$ model: here labelings are identical to those of Fig.7.

On the other hand, *the same spin assignment* turns out to be compatible with the 4-simplex frame $\mathcal{S}_{spin}$ associated with the space–dual of Desargues configuration discussed at length in section 2.2, and the BE identity holds true consistently, *cf.* Fig. 9. In other words, the geometric content of the BE identity in the Ponzano–Regge tetrahedral presentation of the $6j$ (see appendix 5.1) is intrinsically projective, as far as the two arrangements (3 tetrahedra joined along x) and (two tetrahedra joined along one common face) cannot coexist in an Euclidean 3-space but can be realized through $\mathcal{S}_{spin}$ in dimension 4. The latter remark holds true for the coexistence of the arrangements (4 tetrahedra joined together and incident on a common vertex) and (1 tetrahedron) which would correspond to an identity easily derived from (6) by resorting to (5)). In Fig.10 and Fig.11 the purely topological versions of the so-called Pachner moves [30] for simplicial 3-manifolds are depicted.

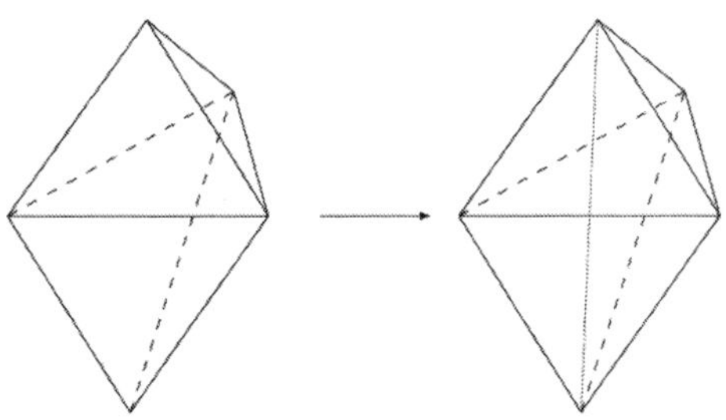

Figure 10. Pachner move 2-3

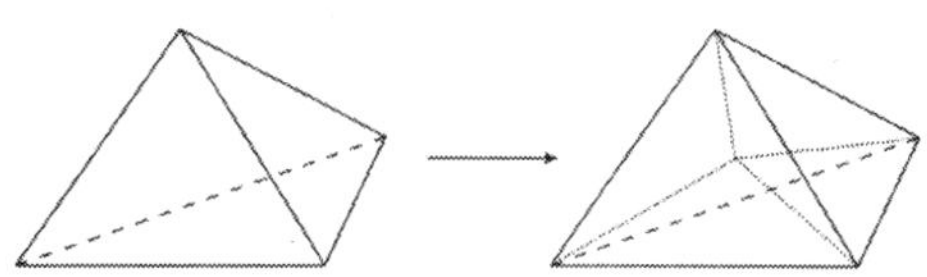

Figure 11. Pachner move 1-4

Looking back at the $6j$ symbol of Fig.1, the four labels (a, b, c, d) –associated with the consecutive sides of a quadrangle, or equivalently of a tetrahedron *cf.* 5.1– are going to be treated as tunable parameters, while $\binom{x}{y}$ is thought of as the pair of variables corresponding

IOP Conf. Series: Journal of Physics: Conf. Series **965** (2018) 012005 doi:10.1088/1742-6596/965/1/012005

to the matrix indexes of the $6j$ when applied to binary coupled states of three $SU(2)$ angular momenta in the space of the total angular momentum [23]. As is well known the 24 'classical' symmetries of the $6j$ can be viewed geometrically as associated to the complete tetrahedral group (isomorphic to S_4) and leave the (values of the) symbol invariant under any permutation of its columns or under interchange of the lower and upper arguments in each of any two columns. The Regge symmetries [20] are expressed as functional relations among parameters and read

$$\begin{Bmatrix} a & b & x \\ c & d & y \end{Bmatrix} = \begin{Bmatrix} s-a & s-b & x \\ s-c & s-d & y \end{Bmatrix} := \begin{Bmatrix} a' & b' & x \\ c' & d' & y \end{Bmatrix}, \tag{7}$$

where $s := (a+b+c+d)/2$ is the semi–perimeter of the quadrangle and in the last equality the new set (a', b', c', d') is defined. The $6j$ associated with the primed entries represents the Regge 'conjugate' of the original quadrangle, and of the Ponzano–Regge tetrahedron as well. Combining the symmetries given in (7) with two more sets obtained by keeping the other pairs of opposite entries fixed, it turns out that the total number of symmetries is 144, which equals the order of the product permutation group $S_3 \times S_4$.

In the $6j$s of Eq.(7) the range of the running entries is $x_{max} - x_{min} = y_{max} - y_{min} = 2\; min(a, b, c, d,', b', c', d')$ on the basis of the choice of a particularly useful ordering for dealing with the discrete, four–parameter $6j$-function, see [12, 13]. This construction highlights in particular 'Regge–invariant' geometric realizations of both quadrangles and tetrahedra, a symmetry which naturally extends to the three-term recursion relations which hold for the $6j$ symbol [23, 31], on the one hand, and for eigenfunctions of the volume operator, on the other [11, 14].
Here we just add a remark about the role Regge symmetry in providing a natural regularization on the representation ring of $SU(2)$, once the set of spins a, b, c, d (and thus the semi-perimeter s) is kept fixed. If we denote a the smallest among the eight (integer and semi-integer) parameters $(a, b, c, d, s-a, s-b, s-, s-d)$, then we assign the label c to the opposite entry, *cf.* the first $6j$ in (7), and d is chosen as the biggest among the two remaining labels. The ordering and the range of c which turn out to be compatible with all of the quadrangular and triangular inequalities are given by

$$a \leq b \leq d \leq s\,;\; d - (b-a) \leq c \leq d + (b-a) \tag{8}$$

and $x_{max} - x_{min} = y_{max} - y_{min} = 2a$. It is easily checked that the semi–perimeter s satisfies

$$s \leq \max\{a, b, c, d\} \;\equiv\; d + (b-a)\,, \tag{9}$$

a condition to be compared with what would happen for the $SU(2)_q$ case with q a root of unity. The following inequality, proven explicitly in [32] and adapted here to our conventions on spin labeling,

$$s \leq \min\{a, b, c, d\} \,+\, \tfrac{(r-2)}{2} \;:=\; a + \kappa\,, \tag{10}$$

involves the integer $r \geq 3$ with $q = e^{2\pi i/r}$. The two inequalities above lead to the compatibility condition

$$\kappa \;\leq\; x_{min} + y_{min} \quad \text{or}\; r \;\leq\; (2x_{min}+1) + (2y_{min}+1)\,, \tag{11}$$

interpreted as associated with a sort of effective q-regularization. The integer r is thus bounded by the sum of the dimensions of the $SU(2)$ *irreps* labeled x_{min}, y_{min} and the geometrical content of Eq.11 relies on the identification of x_{min}, y_{min} with the lower admissible values of the lengths of the diagonals of the quadrangle in Fig.15 or of two opposite edges in the solid Euclidean tetrahedron in Fig.16 of Appendix 5.1. It is worth noting that, owing to the nested structure of the quintuple of quadrangles, it can be easily shown that the regularization conditions above can be fixed on a single reference quadrangle (or even on a single tetrahedron in $\mathcal{S}_{spin}$).

A quite interesting reformulation of the remarks presented so far would be in term of a quaternionic reparametrization of the entries of the $6j$ as given in Eq.7: the semi-perimeter

IOP Conf. Series: Journal of Physics: Conf. Series **965** (2018) 012005 doi:10.1088/1742-6596/965/1/012005

s represent the real part of a quaternion $\mathfrak{Q}$ and three independent linear combinations of (a, b, c, d) give its other real components. Then it turns out that that the Regge–transformed $6j$ is associated with the 'conjugate' $\mathfrak{Q}'$ of quaternion $\mathfrak{Q}$, a circumstance that justifies algebraically the use of term 'Regge–conjugate' for the $6j$ on the right side of Eq.7 (note that octonions should also be called into play as consistent coordinatization of pairs $(\mathfrak{Q}, \mathfrak{Q}')$). More details on such a reparametrization, its implications and a few applications can be found in [12, 13].

4. Outlook and conclusions

Actually quite a lot of issues and interconnections have not been addressed here for lake of space. As for the BE identity, recall that it has been used originally for proving the combinatorial invariance of the Ponzano–Regge state sum under refinement, *cf.* the moves depicted in Fig.10 and Fig.11. In the same paper it can be found the derivation of the three-term recursion relation for the $6j$ symbol that in the semiclassical limit (where all angular momentum variables are much greater than 1 in $\hbar$ units) gives the asymptotic formula for the $6j$. As mentioned already in the introduction, this model has been extensively used in discrete quantum gravity approaches, as well as in topological quantum computing see [33, 34, 35, 36]. Of particular interest in quantum gravity applications is the volume operator of the tetrahedron: on the basis of the paper [11] we have developed recently a complete analysis of the three-term recursion relation for the eigenfuntions of such operator highlightening the proper Regge–invariant frame of Racah quadratic algebra [37, 38, 14], as well as the emergence of a Hamiltonian dynamics at the classical level [12].

In summary, we have emphasized in the present paper the central role played in Regge–regularized Desargues spin networks by the BE identity (together with the orthogonality relation) as the common algebraic counterpart of assemblings of quintuples of projectively–related quadrangles and tetrahedra, as well as of the associated geometric realizations. As for dimensional reduction through projections from 3D to 2D and 1D of Desargues spin networks, we argue that it should be possible to proceed in parallel with semiclassical limiting procedures achieved by taking suitable partial limits of the three–term recursions for the $6j$ symbol and for the volume operator quoted above. The level of persistence of Regge symmetry might represent a guiding principle in this kind of analysis. Finally, as for the 4-simplex – realized through the space-dual of a Deasargues spin network– we have shown in section 2.3 how its 3D cross sections still represents the (original) Desargues spin network, thus opening the possibility of looking at dimensional reduction from 4D to 3D in a new perspective. We argue also that, by properly framing the projective tools within a picture grounded into recursion relationships governed by discrete variables, a Hamiltonian description for the amplitude of the 4-simplex might emerge. Work is in progress in this direction.

5. Appendix: Projective configurations and angular momentum theory

5.1. The triangle of couplings and the quadrangle of recouplings

Triangular inequalities involving three *irreps* labels of $SU(2)$ angular momenta basically arise from the Clebsch–Gordan series, $\mathcal{H}^{j_1} \otimes \mathcal{H}^{j_1} = \oplus_{j=|j_1-j_2|}^{j_1+j_2} \mathcal{H}^{j}$, where $\mathcal{H}^{j}$ is the $(2j+1)$–dimensional Hilbert space of the (square of the) total angular momentum $\mathbf{J}$ $(-j \leq m \leq j)$ with m being the magnetic quantum number associated with J_z, the component of $\mathbf{J}$ along the quantization axis. The two independent systems of sharp angular momenta $\mathbf{J}_1$ and $\mathbf{J}_2$ have dimensions satisfying $(2j_1 + 1)$ $(2j_2 + 1) = (2j + 1)$ and the (orthogonal) transformation between the basis vectors of $\mathcal{H}^{j_1} \otimes \mathcal{H}^{j_1}$ and those of $\mathcal{H}^{j}$ is given by Clebsch–Gordan coefficients or by their symmetrized counterpart, the Wigner $3j$ symbol $\begin{pmatrix} j_1 & j_2 & j \\ m_1 & m_2 & m \end{pmatrix}$ with $m_1 + m_2 = m$. Such an unordered triple of angular momentum variables (spins), say (j, k, l) taking values in $\{0, 1/2, 1, 3/2, \dots\}$, constitutes a *triad* and can be graphically represented as either a trivalent vertex (Fig.12), or as a triangle (Fig.13), or else by three points lying on a same line (Fig. 14). The latter was first noticed by Fano and Racah [3], hence unveiling the intriguing connection between (binary) coupling of angular momenta and (finite) projective configurations of points and lines discussed below.

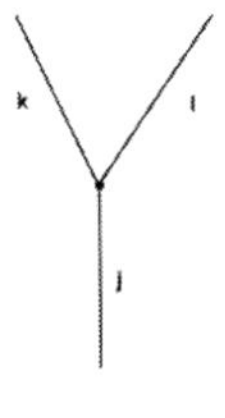

Figure 12. The triad (j, k, l) associated with a trivalent vertex.

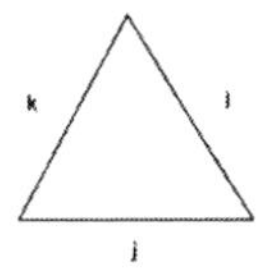

Figure 13. The triad (j, k, l) associated with a triangle.

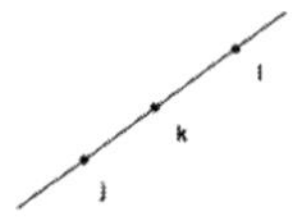

Figure 14. The triad (j, k, l) associated with three points on a same line.

The basic building block of angular momentum theory is the W recoupling coefficient of Racah or, equivalently, the Wigner $6j$ symbol $(a, b, c, d, x, y \in \{0, 1/2, 1, \dots\})$

$$\begin{Bmatrix} a & b & x \\ c & d & y \end{Bmatrix} \text{ with triads } (abx), (bcy), (cdx), (ady) \tag{12}$$

As is well known there exist three alternative graphical presentations of the combinatorial content of this symbol, discussed below focusing on their interpretations in projective geometry.

The $6j$ as a complete quadrangle (Fig.15)
Edmonds [39] showed that the $6j$ –together with other recoupling coefficients such as the $9j$ and the $12j$ symbols– can be associated with trivalent graphs obtained by joining the free ends of binary coupling trees whose labeling are consistent with that of the trivalent vertex of Fig.12. On this basis Yutsis and collaborators [10] elaborated a general diagrammatical scheme able to work out complicated calculations –involving generalized Clebsch–Gordan coefficients and $3nj$ recoupling coefficients– by resorting to combinatorial operations on such labeled graphs. An upgraded version of these methods is collected in the handbook by Varshalovich and collaborators [23], the notation of which is used throughly in this paper. In particular, the $6j$ is depicted as a complete graph on four vertices as in Fig.1, each vertex being associated with a triad of spin labels. Notably, when each edge is extended to a line, the resulting configuration can be perceived projectively as a *complete quadrangle*, denoted $(4_3, 6_2)$.

The $6j$ as a tetrahedron (Fig.16)
The second graphical presentation dates back to Wigner [40] and Ponzano and Regge [4] and has been especially exploited in connection with the semiclassical limit of the $6j$ symbol representing an Euclidean tetrahedron in the classically allowed region where its square volume is positive. Here spin labels are associated with the edge lengths and the triads with the triangular faces as in Fig.13. The relationship between the two presentations –quadrangle and tetrahedron– is easily recognized in polyhedral geometry: if the quadrangle of Fig.15 is realized in Euclidean 3-space as a tetrahedron (with the condition on the volume given above), or, equivalently, if it represents the projection on a plane of a solid tetrahedron, then under topological duality of 2D simplicial complexes, (vertex $\leftrightarrow$ triangle) and (edge $\leftrightarrow$ edge), the two diagrammatical presentations are turned into each other. (Obviously this correspondence involves the triangulation of the boundary of the tetrahedron while in the study of 3D simplicial dissections of manifolds most of the results and applications rely on the duality (0-simplex $\leftrightarrow$ 3-simplex) and (1-simplex $\leftrightarrow$ 2-simplex)).

The $6j$ as a complete quadrilateral (Fig.17)
This presentation, derived from the assignment (triad $\leftrightarrow$ three points on a line) of Fig.14, has

IOP Conf. Series: Journal of Physics: Conf. Series **965** (2018) 012005 doi:10.1088/1742-6596/965/1/012005

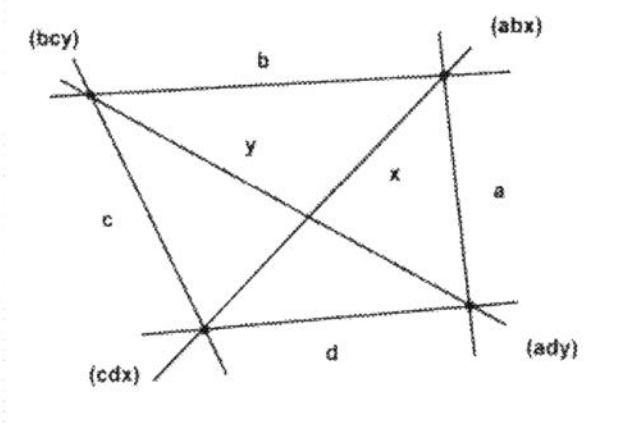

Figure 15. The configuration $(4_3\,,6_2)$: four points in a plane are joined in pairs by six distinctlines.

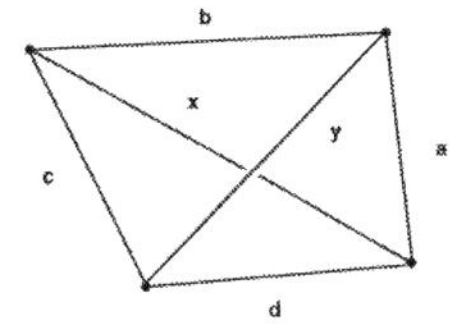

Figure 16. The tetrahedron (the boundary of a 3-simplex) has four vertices, six edges and four triangular faces.

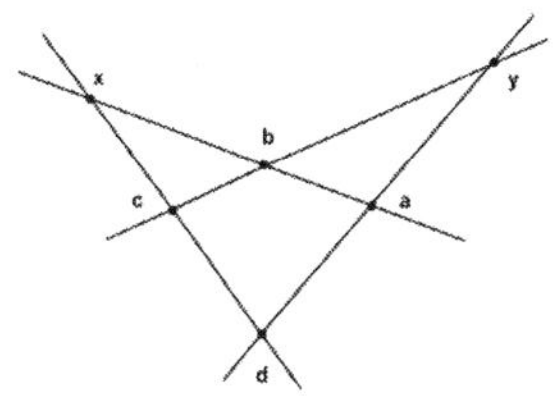

Figure 17. The configuration $(6_2\,,4_3)$: four lines in a plane meet by pairs in six distinct points.

been the basic tool to disclose the projective content of angular momentum recoupling theory since Fano and Racah. We refer the reader interested in more details to the papers of Robinson [41], Judd [42], Labarthe [43] and to the book of Biedenharn and Louck [22]. The quadrangle of Fig.15 is related to the quadrilateral by the standard projective duality in the plane, (point $\leftrightarrow$ line).

It is worth to remark that it is actually the configuration (7_3) of seven points and seven lines (known as the Fano plane) that historically has attracted much attention, and has been also the object of recent investigations by the authors in connection with applications to atomic and molecular physics, *cf.* [44] and references therein.

5.2. Assembling of projective configurations: from quadrangles to Desargues configuration

The three presentations of the $6j$ symbol (12) –two of which are truly projective configurations as recognized above– have been used alternately in different approaches to quantum spin networks, even though from the combinatorial viewpoint they are related by simple (projective or Poincaré) duality transformations as pointed out above. Taken for granted that each of them matches with the standard algebraic setting (in particular with the three-term recursion relation for the $6j$ [23]), we are implementing a bottom–up approach to construct spin networks from the basic axioms of projective geometry (as stated in [1]) and showing how assembling quadrangles – rather than quadilaterals– turns out to be the most economical and consistent way of looking at emergent geometric realizations of interest also in discretized quantum gravity models, *cf.* the final remarks at the end of section 2.3. The first two incidence axioms are

I) Any two distinct points are incident with just a line.

II) Any two lines are incident with at least one point.

A configuration in the plane is denoted (p_γ, ℓ_π), where p is the number of points, ℓ the number of lines, γ the number of lines per point, and π the number of points per line. The equality $p\gamma = \ell\pi$ is easily verified, and configurations for which $p = \ell$ and $\gamma = \pi$, so that $(p_\gamma, p_\gamma) \equiv (p_\gamma)$, are called symmetric (for instance (3_2) is a triangle and (n_2) a polygon with n sides). The (standard) projective dual to a configuration (p_γ, ℓ_π) is a configuration (ℓ_π, p_γ) in which the roles of points and lines are interchanged, and symmetric configurations are self-dual in the plane by definition. The complete quadrangle $(4_3, 6_2)$ of Fig.15 and quadilateral $(6_2, 4_3)$ of Fig.17 are dual to each other, as already noted. The existence of the quadrangle bears on the axiom

III) There exist four points, no three of which are collinear.

Actually this is the simplest configuration that allows for the assignment of coordinates in the plane, looking at the complete quadrangle $(4_3\,,6_2)$ of Fig. 15 with vertices labeled by the pairs $(0,0)$ $(0,1)$ $(1,0)$ $(1,1)$ taking values in the smallest finite number field $\mathbb{F}_2$ [45].

Focusing on symmetric configurations, it should be clear that the the symbolic expression (p_γ) does not determine uniquely a projective configuration up to incidence isomorphisms. In particular, among the ten different (10_3) configurations, the Desargues configuration plays a prominent role in projective geometry essentially because it encodes Desargue's theorem. The statement is illustrated in Fig.18 Actually Desargues' theorem holds true in any projective

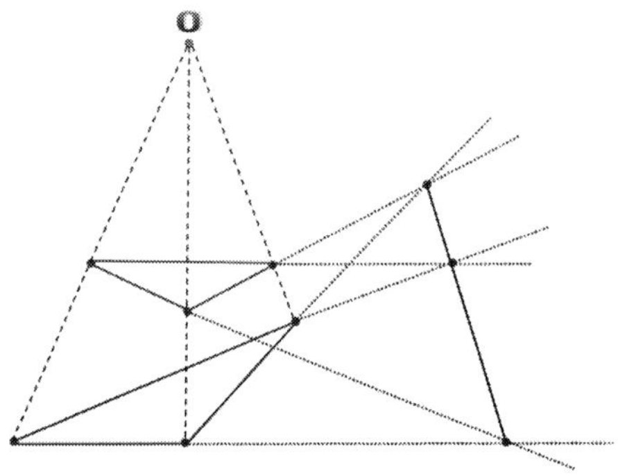

Figure 18. Illustration of Desargues' theorem: if two triangles (red and blue) are in perspective from a center O, then the extensions to lines of their sides meet in a line, the perspective line. In the (plane) dual of this theorem the existence of the perspective line with three points is assumed and the existence of the center of perspective O is proven.

geometry P^n for $n \geq 3$ while its realization in the plane P^2 must be separately postulated (*cf.* [1]). The principle that characterizes projective geometry and is crucial for the hierarchical construction of spin networks is *duality* (the notion of plane duality, or duality in P^2, has been already quoted above). In general, suppose that in a proposition –or a figure, or a configuration– in P^n one interchanges P^n and P^{n-r-1} $(0 \leq r \leq n)$ and switches 'is contained'/ 'contains', 'collinear'/ 'concurrent' and similar. Then the duality principle can be stated as follows
Duality principle. If a proposition (about a configuration) is true in P^n then it is also true in *any of the dual configurations* of the original one lying in P^{n-r-1} for $0 \leq r \leq n$.

References

[1] Coxeter H S M 1974 *Projective Geometry* (Berlin: Springer–Verlag)
[2] Hilbert D and Cohn–Vossen 1952 *Geometry and the Imagination* (Chelsea: Amer. Math. Soc.)
[3] Fano U and Racah G 1959 *Irreducible Tensorial Sets* (New York: Academic Press)
[4] Ponzano G and Regge T 1968 *Semiclassical Limit of Racah Coefficients* in *Spectroscopic and Group Theoretical Methods in Physics* ed F Bloch et al (Amsterdam: North–Holland) pp 1-58
[5] Penrose R 1971 *Angular Momentum: an Approach to Combinatorial Space-time* in *Quantum Theory and beyond* Bastin T ed (Cambridge: Cambridge University Press) pp 151-180
[6] Carfora M, Marzuoli A and Rasetti M 2009 *J. Phys. Chem.* A **113** 15376
[7] Carfora M and Marzuoli A 2012 *Quantum Triangulations: Moduli Spaces, Strings, and Quantum Computing* (Lect. Notes in Physics **845**) (Berlin: Springer–Verlag)
[8] Rovelli C and Vidotto F 2017 *Covariant Loop Quantum Gravity: An Elementary Introduction To Quantum Gravity And Spinfoam Theory* (Cambridge: Cambridge University Press)
[9] Ashtekar A and Pullin J 2017 *Loop Quantum Gravity: The First 30 Years* (Singapore: World Scientific)
[10] Yutsis A, Levinson I, Vanagas V 1962 *Mathematical Apparatus of the Theory of Angular Momentum* (Israel Program of for Scientific Translation)
[11] Carbone G, Carfora M and Marzuoli A 2002 *Class. Quantum Grav.* **19** 3761
[12] Aquilanti V, Marinelli D and Marzuoli A 2013 *J. Phys. A: Math. Theor.* **46** 175303

[13] Marinelli D 2013 *Single and Collective Dynamics of Discretized Geometries* PhD Thesis (Pavia, Italy: University Press) ISBN 978-88-95767-73-4
[14] Aquilanti V, Marinelli D and Marzuoli A 2014 *J. Phys.: A: Conf. Series* **46** 175303
[15] Ragni M, Bitencourt A C P, Da S Ferreira C, Aquilanti V, Anderson R W and Littlejohn R G 2010 *Int. J. Quantum Chem.* **110** 731
[16] Aquilanti V, Haggard H M, Hedeman A, Jeevanjee N, Littlejohn R G and Liang Y 2012 *J. Phys. A: Math. Theor.* **45** 065209
[17] Bitencourt A C P, Marzuoli A, Ragni M, Anderson R W and Aquilanti V 2012 *Lect. Notes Comput. Science* **7333** Part I ed B Murgante et al (Berlin Heidelberg: Springer–Verlag) p 723
[18] Regge T and Williams R M *J. Math. Phys.* **41** 3964
[19] Regge T 1961 *Nuovo Cim.* **19** 558
[20] Regge T 1959 *Nuovo Cim.* **11** 116
[21] Regge T 1958 *Nuovo Cim.* **10** 544
[22] Biedenharn L C and Louck J D 1981 *The Racah–Wigner Algebra in Quantum Theory* (Encyclopedia of Mathematics and its Applications Vol 9) ed G-C Rota (Reading, MA: Addison–Wesley)
[23] Varshalovich D A, Moskalev A N and Khersonskii V K 1988 *Quantum Theory of Angular Momentum* (Singapore: World Scientific)
[24] Roberts J 1999 *Geom. Topology* **3** 21
[25] Boalch P P 2007 *Commun. Math. Phys.* **276** 117
[26] Biedenharn L C and Lohe M A 1995 *Quantum Group Symmetry and q-Tensor Algebra* (Singapore: World Scientific)
[27] Koekoek R, Lesky P A and Swarttouw R F 2010 *Hypergeometric Orthogonal Polynomials and Their q-Analogues* (Heidelberg: Springer)
[28] Barnes J 2012 *Gems of Geometry* (Heidelberg: Springer)
[29] Carlip S 2012 *Spontaneous Dimensional Reduction? Preprint* 1207.4503 [gr-qc]
[30] Pachner U 1991 *Eur. J. Combin.* **12** 129
[31] Ragni M, Littlejohn R G, Bitencourt A C P, Aquilanti V and Anderson R W 2013 *Lect. Notes Comput. Science* **7972** (Berlin Heidelberg: Springer–Verlag) p 60
[32] Khavkine I 2015 *Int. J. Geom. Meth. Mod. Phys.* **12** 1550117
[33] Marzuoli A and Rasetti M 2005 *Ann. Phys.* **318** 345
[34] Marzuoli A and Rasetti M 2002 *Phys. Lett.* A **306** 79
[35] Garnerone S, Marzuoli A and Rasetti M 2009 *Adv. Theor. Math. Phys.* **13** 1601
[36] Kádár Z, Marzuoli A and Rasetti M 2009 *Int. J. Quantum Inf.* **7** 195
[37] Sklyanin E K 1982 *Funct. Anal.* **16** 263
[38] Granovskii Ya I, Lutzenko I M and Zhedanov A S 1992 *Ann. Phys.* **217** 1
[39] Edmonds A 1960 *Angular Momentum in Quantum Mechanics* (Princeton NJ: Princeton Univ. Press)
[40] Wigner E 1959 *Group Theory and its Application to the Quantum Mechanics of Atomic Spectra* (New York: Academic Press)
[41] Robinson deB G 1970 *J. Math. Phys.* **11** 3428
[42] Judd B 1983 *Found. Phys.* **13** 51
[43] Labarthe J-J 2000 *J. Phys.* A **13** 763
[44] Santos R F, Arruda M S, Bitencourt A C P, Ragni M, Prudente F V, Coletti C, Marzuoli A and Aquilanti V 2017 *J. Mol. Spectrosc.* **337** 153
[45] Hall M 1983 *Trans. Amer. Math. Soc.* **45** 229

IOP Conf. Series: Journal of Physics: Conf. Series **965** (2018) 012006 doi:10.1088/1742-6596/965/1/012006

Hom-associative Ore extensions

P Bäck, J Richter and S Silvestrov

Division of Applied Mathematics, The School of Education, Culture and Communication, Mälardalen University, Box 883, SE-721 23 Västerås, Sweden

E-mail: per.back@mdh.se, johan.richter@mdh.se and sergei.silvestrov@mdh.se

Abstract. We introduce hom-associative Ore extensions as non-associative, non-unital Ore extensions with a hom-associative multiplication, as well as give some necessary and sufficient conditions when such exist. Within this framework, we also construct a family of hom-associative Weyl algebras as generalizations of the classical analogue, and prove that they are simple.

1. Introduction

Hom-Lie algebras and related hom-algebra structures have recently become a subject of growing interest and extensive investigations, in part due to the prospect of providing a general framework in which one can produce many types of natural deformations of (Lie) algebras, in particular q-deformations which are of interest both in mathematics and in physics. One of the main initial motivations for this development came from mathematical physics works on q-deformations of infinite-dimensional algebras, primarily the q-deformed Heisenberg algebras (q-deformed Weyl algebras), oscillator algebras, and the Virasoro algebra [1–6, 8, 10, 11, 15–17].

Quasi-Lie algebras, subclasses of quasi-hom-Lie algebras, and hom-Lie algebras as well as their general colored (graded) counterparts were introduced between 2003 and 2005 in [9, 12–14, 24]. Further on, between 2006 and 2008, Makhlouf and Silvestrov introduced the notions of hom-associative algebras, hom-(co, bi)algebras and hom-Hopf algebras and also studied their properties [18–20]. Hom-associative algebras are a generalization of associative algebras with the associativity axiom extended by a linear twisting map. Moreover are they hom-Lie admissible, meaning that the commutator multiplication in any hom-associative algebra yields a hom-Lie algebra [18]. In hom-associative algebras, the associativity is replaced by hom-associativity; hom-coassociativity for a hom-coalgebra can be considered in a similar way.

One of the main tools in these important developments and in many constructions of examples and classes of hom-algebra structures in physics and in mathematics are based on twisted derivations or σ-derivations, which are generalized derivations twisting the Leibniz rule by means of a linear map. These types of twisted derivation maps are central for the associative Ore extension algebras or rings, introduced in algebra in the 1930s, generalizing crossed product (semidirect product) algebras or rings incorporating both actions and twisted derivations.

Non-associative Ore extensions were introduced in the unital case in [21] (see also [22] for an extension to monoid Ore extensions). In the present article, we generalize this construction to the non-unital case, and investigate when the Ore extensions are hom-associative. We do not make use of any previous results about non-associative Ore extensions, but our construction of simple hom-associative Weyl algebras in Example 1 has some similarities to the simple non-associative Weyl algebras in [21].

IOP Conf. Series: Journal of Physics: Conf. Series **965** (2018) 012006 doi:10.1088/1742-6596/965/1/012006

2. Preliminaries

In this section, we present some definitions and review some results from the theory of hom-associative algebras and that of non-associative Ore extensions.

First, by a non-unital ring, we mean a ring which is not necessarily unital. In the same way, a non-associative ring is a ring not necessarily associative. $\mathbb{N}$ will denote the set of non-negative integers, and $\mathbb{N}_{>0}$ the set of positive integers.

2.1. Hom-associative algebras

Here we define what we mean for an algebraic structure to be *hom-associative*, and review a couple of results concerning the construction of them.

Definition 1 (Hom-associative algebra). A *hom-associative algebra* over an associative, commutative, and unital ring R, is a triple $(M, \cdot, \alpha)$ consisting of an R-module M, a binary operation $\cdot\colon M \times M \to M$ linear in both arguments, and a linear map $\alpha\colon M \to M$ satisfying, for all $a, b, c \in M$,

$$\alpha(a) \cdot (b \cdot c) = (a \cdot b) \cdot \alpha(c).$$

Since α twists the associativity, we will refer to it as the *twisting map*.

Definition 2 (Hom-associative ring). A *hom-associative ring* is a hom-associative algebra over the ring of integers.

Definition 3 (Weakly unital algebra). Let $\mathcal{A}$ be a hom-associative algebra. If for all $a \in \mathcal{A}$, $e \cdot a = a \cdot e = \alpha(a)$ for some $e \in \mathcal{A}$, we say that $\mathcal{A}$ is *weakly unital* with *weak unit* e.

Any associative algebra can be extended to a non-trivial hom-associative algebra, which the following proposition demonstrates:

Proposition 1 ([25]). Let A be an associative algebra, α an algebra endomorphism on A and define $*\colon A \times A \to A$ by $a * b := \alpha(a \cdot b)$ for all $a, b \in A$. Then $(A, *, \alpha)$ is a hom-associative algebra.

Proof. Linearity follows immediately, while for all $a, b, c \in A$, we have

$$\alpha(a) * (b * c) = \alpha(a) * (\alpha(b \cdot c)) = \alpha(\alpha(a) \cdot \alpha(b \cdot c)) = \alpha(\alpha(a \cdot b \cdot c)),$$
$$(a * b) * \alpha(c) = \alpha(a \cdot b) * \alpha(c) = \alpha(\alpha(a \cdot b) \cdot \alpha(c)) = \alpha(\alpha(a \cdot b \cdot c)),$$

which proves that $(A, *, \alpha)$ is hom-associative. □

Note that we are abusing the notation in Definition 1 a bit here; A in $(A, *, \alpha)$ does really denote the algebra and not only its module structure. From now on, we will always refer to this construction when writing $*$.

Corollary 1 ([23]). If A is an associative, unital algebra, then $(A, *, \alpha)$ is a weakly unital hom-associative algebra with weak unit 1.

Proof. $1 * x = \alpha(1 \cdot x) = \alpha(x) = \alpha(x \cdot 1) = x * 1$. □

2.2. Non-associative, non-unital Ore extensions

In what follows, we define *non-associative, non-unital Ore extensions*, together with some new terminology.

Definition 4 (Left R-additivity). If R is a non-associative, non-unital ring, we say that a map $\beta\colon R \to R$ is *left R-additive* if for all $r, s, t \in R$, $r \cdot \beta(s + t) = r \cdot (\beta(s) + \beta(t))$.

IOP Conf. Series: Journal of Physics: Conf. Series **965** (2018) 012006 doi:10.1088/1742-6596/965/1/012006

Given a non-associative, non-unital ring R with left R-additive maps $\delta\colon R \to R$ and $\sigma\colon R \to R$, by a *non-associative, non-unital Ore extension* of R, $R[X;\sigma,\delta]$, we mean the set of formal sums $\sum_{i\in\mathbb{N}} a_iX^i$, $a_i \in R$, with finitely many a_i non-zero, endowed with the addition

$$\sum_{i\in\mathbb{N}} a_iX^i + \sum_{i\in\mathbb{N}} b_iX^i = \sum_{i\in\mathbb{N}}(a_i+b_i)X^i, \quad a_i, b_i \in R,$$

where two polynomials are equal if and only if their corresponding coefficients are equal, and for all $a, b \in R$ and $m, n \in \mathbb{N}$, a multiplication

$$aX^m \cdot bX^n = \sum_{i\in\mathbb{N}} \left(a \cdot \pi_i^m(b)\right) X^{i+n}. \tag{1}$$

Here π_i^m denotes the sum of all $\binom{m}{i}$ possible compositions of i copies of σ and $m-i$ copies of δ in arbitrary order. Then, for example $\pi_0^0 = \mathrm{id}_R$ and $\pi_1^3 = \sigma\circ\delta\circ\delta + \delta\circ\sigma\circ\delta + \delta\circ\delta\circ\sigma$. We also extend the definition of π_i^m by setting $\pi_i^m \equiv 0$ whenever $i < 0$, or $i > m$. Imposing distributivity of the multiplication over addition makes $R[X;\sigma,\delta]$ a ring.

Note that when $m = n = 0$, $aX^0 \cdot bX^0 = \sum_{i\in\mathbb{N}} \left(a\cdot\pi_i^0(b)\right) X^i = (a\cdot b)X^0$, so $R \cong RX^0$ by the isomorphism $r \mapsto rX^0$ for any $r \in R$. Since RX^0 is a subring of $R[X;\sigma,\delta]$, we can view R as a subring of $R[X;\sigma,\delta]$, making sense of expressions like $a \cdot bX^0$.

Remark 1. Unless R contains a unit, there is no element X in the ring $R[X;\sigma,\delta]$.

The left-distributivity of the multiplication over addition forces δ and σ to be left R-additive: for any $r, s, t \in R$, $rX\cdot(s+t) = rX\cdot s + rX\cdot t$, so by expanding the left- and right-hand side and then comparing coefficients proves the statement;

$$\begin{aligned} rX\cdot(s+t) &= r\cdot\sigma(s+t)X + r\cdot\delta(s+t),\\ rX\cdot s + rX\cdot t &= r\cdot\sigma(s)X + r\cdot\delta(s) + r\cdot\sigma(t)X + r\cdot\delta(t). \end{aligned}$$

3. Hom-associative Ore extensions of non-associative rings

The following section is devoted to the question what non-associative, non-unital Ore extensions $R[X;\sigma,\delta]$ of non-associative, non-unital rings R are hom-associative?

Proposition 2 (Hom-associative Ore extension)**.** Let $R[X;\sigma,\delta]$ be a non-associative, non-unital Ore extension of a non-associative, non-unital ring R. Furthermore, let $\alpha_{i,j}(a) \in R$ be dependent on $a \in R$ and $i, j \in \mathbb{N}_{>0}$, and put for an additive map $\alpha\colon R[X;\sigma,\delta] \to R[X;\sigma,\delta]$,

$$\alpha\left(aX^m\right) = \sum_{i\in\mathbb{N}} \alpha_{i+1,m+1}(a)X^i, \quad \forall a\in R, \forall m \in \mathbb{N}. \tag{2}$$

Then $R[X;\sigma,\delta]$ is hom-associative with the twisting map α if and only if for all $a, b, c \in R$ and $k, m, n, p \in \mathbb{N}$,

$$\sum_{j\in\mathbb{N}}\sum_{i\in\mathbb{N}} \alpha_{i+1,m+1}(a)\cdot\pi_{k-j}^i\left(b\cdot\pi_{j-p}^n(c)\right) = \sum_{j\in\mathbb{N}}\sum_{i\in\mathbb{N}} \left(a\cdot\pi_i^m(b)\right)\cdot\pi_{k-j}^{i+n}\left(\alpha_{j+1,p+1}(c)\right). \tag{3}$$

Proof. For any $a, b, c \in R$ and $m, n, p \in \mathbb{N}$,

$$\alpha\left(aX^m\right)\cdot\left(bX^n\cdot cX^p\right) = \alpha\left(aX^m\right)\cdot\left(\sum_{q\in\mathbb{N}}\left(b\cdot\pi_q^n(c)\right)X^{q+p}\right)$$

$$
\begin{aligned}
&= \sum_{q\in\mathbb{N}} \alpha\left(aX^m\right)\cdot\left(\left(b\cdot\pi_q^n(c)\right)X^{q+p}\right)\\
&= \sum_{q\in\mathbb{N}}\sum_{i\in\mathbb{N}} \alpha_{i+1,m+1}(a)X^i\cdot\left(\left(b\cdot\pi_q^n(c)\right)X^{q+p}\right)\\
&= \sum_{q\in\mathbb{N}}\sum_{i\in\mathbb{N}}\sum_{l\in\mathbb{N}} \alpha_{i+1,m+1}(a)\cdot\pi_l^i\left(b\cdot\pi_q^n(c)\right)X^{l+q+p}\\
&= \sum_{l\in\mathbb{N}}\sum_{j\in\mathbb{N}}\sum_{i\in\mathbb{N}} \alpha_{i+1,m+1}(a)\cdot\pi_l^i\left(b\cdot\pi_{j-p}^n(c)\right)X^{l+j}\\
&= \sum_{k\in\mathbb{N}}\sum_{j\in\mathbb{N}}\sum_{i\in\mathbb{N}} \alpha_{i+1,m+1}(a)\cdot\pi_{k-j}^i\left(b\cdot\pi_{j-p}^n(c)\right)X^k,
\end{aligned}
$$

$$
\begin{aligned}
\left(aX^m\cdot bX^n\right)\cdot\alpha\left(cX^p\right) &= \left(\sum_{i\in\mathbb{N}}\left(a\cdot\pi_i^m(b)\right)X^{i+n}\right)\cdot\alpha\left(cX^p\right)\\
&= \sum_{i\in\mathbb{N}}\left(a\cdot\pi_i^m(b)\right)X^{i+n}\cdot\alpha\left(cX^p\right)\\
&= \sum_{i\in\mathbb{N}}\left(a\cdot\pi_i^m(b)\right)X^{i+n}\cdot\sum_{j\in\mathbb{N}}\alpha_{j+1,p+1}(c)X^j\\
&= \sum_{i\in\mathbb{N}}\sum_{j\in\mathbb{N}}\left(a\cdot\pi_i^m(b)\right)X^{i+n}\cdot\alpha_{j+1,p+1}(c)X^j\\
&= \sum_{i\in\mathbb{N}}\sum_{j\in\mathbb{N}}\sum_{l\in\mathbb{N}}\left(a\cdot\pi_i^m(b)\right)\cdot\pi_l^{i+n}\left(\alpha_{j+1,p+1}(c)\right)X^{l+j}\\
&= \sum_{k\in\mathbb{N}}\sum_{j\in\mathbb{N}}\sum_{i\in\mathbb{N}}\left(a\cdot\pi_i^m(b)\right)\cdot\pi_{k-j}^{i+n}\left(\alpha_{j+1,p+1}(c)\right)X^k.
\end{aligned}
$$

Comparing coefficients completes the proof. □

Corollary 2. Let $R[X;\sigma,\delta]$ be a hom-associative, non-unital Ore extension of a non-associative, non-unital ring R, with the twisting map defined by (2). Then the following assertions hold for all $a,b,c\in R$ and $k,p\in\mathbb{N}$:

$$
\sum_{i=k-p}^{I_{0,a}} \alpha_{i+1,1}(a)\cdot\pi_{k-p}^i(b\cdot c) = (a\cdot b)\cdot\alpha_{k+1,p+1}(c), \tag{4}
$$

$$
\begin{aligned}
\sum_{i=k-p-1}^{I_{0,a}} &\alpha_{i+1,1}(a)\cdot\left(\pi_{k-p-1}^i(b\cdot\sigma(c))\right)\\
&+\sum_{i=k-p}^{I_{0,a}} a_{i+1,1}\left(\pi_{k-p}^i(b\cdot\delta(c))\right) = (a\cdot b)\cdot\left(\delta(\alpha_{k+1,p+1}(c))+\sigma(\alpha_{k,p+1}(c))\right)\\
&\qquad = (a\cdot b)\cdot\left(\alpha_{k+1,p+1}(\delta(c))+\alpha_{k,p+1}(\sigma(c))\right),
\end{aligned}
\tag{5}
$$

$$
\begin{aligned}
\sum_{i=k-p}^{I_{1,a}} \alpha_{i+1,2}(a)\cdot\pi_{k-p}^i(b\cdot c) &= (a\cdot\sigma(b))\cdot\left(\delta(\alpha_{k+1,p+1}(c))+\sigma(\alpha_{k,p+1}(c))\right)\\
&\quad + (a\cdot\delta(b))\cdot\alpha_{k+1,p+1}(c),
\end{aligned}
\tag{6}
$$

where $\alpha_{0,p+1}(\cdot) := 0$, and $I_{p,a}$ is the smallest natural number, depending on p and a, such that $\alpha_{i+1,p}(a) = 0$ for all $i > I_{p,a}$.

Proof. We get (4), the first equality in (5), and (6) immediatly from the cases $m = n = 0$, $m = 0, n = 1$, and $m = 1, n = 0$ in (3), respectively. The second equality in (5) follows from comparison with (4). □

Remark 2. In case $k < p$, or $k > I_{0,a}$ in (4), $(a \cdot b) \cdot \alpha_{k+1,p+1}(c) = 0$. The statement is analogous for (5) and (6).

There are two immediate consequences of (5):

Corollary 3. Let $R[X;\sigma,\delta]$ be a hom-associative, non-unital Ore extension of a non-associative, non-unital ring R, with the twisting map defined by (2). Then the following assertions hold for all $a, b, c \in R$ and $j, p \in \mathbb{N}$:

$$(a \cdot b) \cdot \sigma(\alpha_{I+1,p+1}(c)) = (a \cdot b) \cdot \alpha_{I+1,p+1}(\sigma(c)), \quad I = \max(I_{p,c}, I_{p,\delta(c)}), \tag{7}$$

$$(a \cdot b) \cdot \delta(\alpha_{1,p+1}(c)) = (a \cdot b) \cdot \alpha_{1,p+1}(\delta(c)) = \begin{cases} (a \cdot b) \cdot \alpha_{j+1,j+1}(\delta(c)) & \text{if } p = 0, \\ 0 & \text{if } p \neq 0. \end{cases} \tag{8}$$

Proof. Put $k = \max(I_{p,c}, I_{p,\delta(c)})$ and $k = 0$ in (5), respectively. □

Definition 5 (Homogeneous map)**.** Let $R[X;\sigma,\delta]$ be a non-associative, non-unital Ore extension of a non-associative, non-unital ring R. Then we say that a map $\beta\colon R[X;\sigma,\delta] \to R[X;\sigma,\delta]$ is *homogeneous* if for all $a \in R$ and $m \in \mathbb{N}$, $\beta(aX^m) = \beta(a)X^m$. If $\alpha\colon R \to R$ is any (additive) map, we may *extend it homogeneously* to $R[X;\sigma,\delta]$ by $\alpha(aX^m) := \alpha(a)X^m$ (imposing additivity).

Lemma 1 (Homogeneously extended ring endomorphism)**.** Let $R[X;\sigma,\delta]$ be a non-associative, non-unital Ore extension of a non-associative, non-unital ring R. If α is an endomorphism on R, then the homogeneously extended map on $R[X;\sigma,\delta]$ is an endomorphism if and only if

$$\alpha(a) \cdot \pi_i^m(\alpha(b)) = \alpha(a) \cdot \alpha(\pi_i^m(b)), \quad \text{for all } i, m \in \mathbb{N} \text{ and } a, b \in R. \tag{9}$$

Proof. Additivity follows from the definition, while for any monomials aX^m and bX^n,

$$\alpha(aX^m \cdot bX^n) = \alpha\left(\sum_{i \in \mathbb{N}} a \cdot \pi_i^m(b)X^{i+n}\right) = \sum_{i \in \mathbb{N}} \alpha(a) \cdot \alpha(\pi_i^m(b))X^{i+n},$$

$$\alpha(aX^m) \cdot \alpha(bX^n) = \alpha(a)X^m \cdot \alpha(b)X^n = \sum_{i \in \mathbb{N}} \alpha(a) \cdot \pi_i^m(\alpha(b))X^{i+n}.$$

Comparing coefficients between the two completes the proof. □

Corollary 4 (Homogeneously extended unital ring endomorphism)**.** Let $R[X;\sigma,\delta]$ be a non-associative, unital Ore extension of the non-associative, unital ring R. If α is an endomorphism on R, and there exists an $a \in R$ such that $\alpha(a) = 1$, then the homogeneously extended map on $R[X;\sigma,\delta]$ is an endomorphism if and only if α commutes with δ and σ.

Proof. This follows from Lemma 1 by choosing a so that $\alpha(a) = 1$: if α commutes with δ and σ, then $\pi_i^m(\alpha(b)) = \alpha(\pi_i^m(b))$. On the other hand, if $\pi_i^m(\alpha(b)) = \alpha(\pi_i^m(b))$, then by choosing $m = 1$ and $i = 0$, $\pi_0^1(\alpha(b)) = \delta(\alpha(b))$, and $\alpha(\pi_0^1(b)) = \alpha(\delta(b))$. If choosing $m = 1$ and $i = 1$, $\pi_1^1(\alpha(b)) = \sigma(\alpha(b))$, $\alpha(\pi_1^1(b)) = \alpha(\sigma(b))$. □

4. Hom-associative Ore extensions of hom-associative rings

In this section, we will continue our previous investigation, but narrowed down to hom-associative Ore extensions of hom-associative rings.

Corollary 5. Let $R[X;\sigma,\delta]$ be a non-associative, non-unital Ore extension of a hom-associative, non-unital ring R, and extend the twisting map $\alpha\colon R\to R$ homogeneously to $R[X;\sigma,\delta]$. Then $R[X;\sigma,\delta]$ is hom-associative if and only if for all $a,b,c\in R$ and $l,m,n\in\mathbb{N}$,

$$\sum_{i\in\mathbb{N}}\alpha(a)\cdot\pi_i^m\left(b\cdot\pi_{l-i}^n(c)\right)=\sum_{i\in\mathbb{N}}\left(a\cdot\pi_i^m(b)\right)\cdot\pi_l^{i+n}\left(\alpha(c)\right). \tag{10}$$

Proof. A homogeneous α corresponds to $\alpha_{i+1,m+1}(a)=\alpha(a)\cdot\delta_{i,m}$ and $\alpha_{j+1,p+1}(c)=\alpha(c)\cdot\delta_{j,p}$ in Proposition 2, where $\delta_{i,m}$ is the Kronecker delta. Then the left-hand side reads

$$\begin{aligned}\sum_{j\in\mathbb{N}}\sum_{i\in\mathbb{N}}\alpha_{i+1,m+1}(a)\cdot\pi_{k-j}^i\left(b\cdot\pi_{j-p}^n(c)\right)&=\sum_{j\in\mathbb{N}}\sum_{i\in\mathbb{N}}\alpha(a)\cdot\delta_{i,m}\cdot\pi_{k-j}^i\left(b\cdot\pi_{j-p}^n(c)\right)\\&=\sum_{j\in\mathbb{N}}\alpha(a)\cdot\pi_{k-j}^m\left(b\cdot\pi_{j-p}^n(c)\right)\\&=\sum_{i\in\mathbb{N}}\alpha(a)\cdot\pi_i^m\left(b\cdot\pi_{k-p-i}^n(c)\right)\\&=\sum_{i\in\mathbb{N}}\alpha(a)\cdot\pi_i^m\left(b\cdot\pi_{l-i}^n(c)\right),\end{aligned}$$

and the right-hand side

$$\begin{aligned}\sum_{j\in\mathbb{N}}\sum_{i\in\mathbb{N}}\left(a\cdot\pi_i^m(b)\right)\cdot\pi_{k-j}^{i+n}\left(\alpha_{j+1,p+1}(c)\right)&=\sum_{j\in\mathbb{N}}\sum_{i\in\mathbb{N}}\left(a\cdot\pi_i^m(b)\right)\cdot\pi_{k-j}^{i+n}\left(\alpha(c)\cdot\delta_{j,p}\right)\\&=\sum_{i\in\mathbb{N}}\left(a\cdot\pi_i^m(b)\right)\cdot\pi_{k-p}^{i+n}\left(\alpha(c)\right)\\&=\sum_{i\in\mathbb{N}}\left(a\cdot\pi_i^m(b)\right)\cdot\pi_l^{i+n}\left(\alpha(c)\right),\end{aligned}$$

which completes the proof. □

Remark 3. If $\alpha=\mathrm{id}_R$, i.e. if R is associative, then (10) becomes

$$a\cdot\sum_{i\in\mathbb{N}}\pi_i^m\left(b\cdot\pi_{l-i}^n(c)\right)=a\cdot\sum_{i\in\mathbb{N}}\pi_i^m(b)\cdot\pi_l^{i+n}\left(c\right), \tag{11}$$

which is the necessary and sufficient condition for an Ore extension of an associative ring to be associative. The unital case, corresponding to $a=1$, has been proved in e.g. [26] using a counting argument and in [27] in a more direct way.

Corollary 6. Let $R[X;\sigma,\delta]$ be a hom-associative, non-unital Ore extension of a hom-associative, non-unital ring R, with the twisting map $\alpha\colon R\to R$ extended homogeneously to $R[X;\sigma,\delta]$. Then, for all $a,b,c\in R$,

$$(a\cdot b)\cdot\delta(\alpha(c))=(a\cdot b)\cdot\alpha(\delta(c)), \tag{12}$$

$$(a\cdot b)\cdot\sigma(\alpha(c))=(a\cdot b)\cdot\alpha(\sigma(c)), \tag{13}$$

$$\alpha(a)\cdot\delta(b\cdot c)=\alpha(a)\cdot\left(\delta(b)\cdot c+\sigma(b)\cdot\delta(c)\right), \tag{14}$$

$$\alpha(a)\cdot\sigma(b\cdot c)=\alpha(a)\cdot\left(\sigma(b)\cdot\sigma(c)\right). \tag{15}$$

Proof. Using the same technique as in the proof of Corollary 5, this follows from Corollary 2 with a homogeneous α. □

For the two last equations, it is worth noting the resemblance to the associative case; if R is an associative, unital ring, then it is necessary that σ be an endomorphism and δ a σ-derivation, i.e. an additive map satisfying $\delta(b \cdot c) = \delta(b) \cdot c + \sigma(b) \cdot \delta(c)$ for all $b, c \in R$, for $R[X; \sigma, \delta]$ to be an associative, unital Ore extension.

Corollary 7. Assume $\alpha\colon R \to R$ is the twisting map of a hom-associative ring R, and extend the map homogeneously to $R[X; \sigma, \delta]$. Assume further that α commutes with δ and σ. Then $R[X; \sigma, \delta]$ is hom-associative if and only if for all $a, b, c \in R$ and $l, m, n \in \mathbb{N}$,

$$\alpha(a) \cdot \sum_{i\in\mathbb{N}} \pi_i^m \left(b \cdot \pi_{l-i}^n(c)\right) = \alpha(a) \cdot \sum_{i\in\mathbb{N}} \left(\pi_i^m(b) \cdot \pi_l^{i+n}(c)\right). \tag{16}$$

Proof. Using Corollary 5, we know that $R[X; \sigma, \delta]$ is hom-associative if and only if for all $a, b, c \in R$ and $l, m, n \in \mathbb{N}$,

$$\sum_{i\in\mathbb{N}} \alpha(a) \cdot \pi_i^m \left(b \cdot \pi_{l-i}^n(c)\right) = \sum_{i\in\mathbb{N}} \left(a \cdot \pi_i^m(b)\right) \cdot \pi_l^{i+n}\left(\alpha(c)\right).$$

However, since α commutes with both δ and σ, and R is hom-associative, the right-hand side can be rewritten as

$$\begin{aligned}\sum_{i\in\mathbb{N}} \left(a \cdot \pi_i^m(b)\right) \cdot \pi_l^{i+n}\left(\alpha(c)\right) &= \sum_{i\in\mathbb{N}} \left(a \cdot \pi_i^m(b)\right) \cdot \alpha\left(\pi_l^{i+n}(c)\right) \\ &= \sum_{i\in\mathbb{N}} \alpha(a) \cdot \left(\pi_i^m(b) \cdot \pi_l^{i+n}(c)\right).\end{aligned}$$

As a last step, we use left-distributivity to pull out $\alpha(a)$ from the sums. □

We refer to Remark 3 for the resemblance with the associative case. With this in mind, one may further ask oneself whether it is possible to construct non-trivial hom-associative Ore extensions, starting from associative rings? The answer is affirmative, and the remaining part of the article will be devoted to show that.

Proposition 3. Let $R[X; \sigma, \delta]$ be an associative, non-unital Ore extension of an associative, non-unital ring R, and $\alpha\colon R \to R$ a ring endomorphism that commutes with δ and σ. Then $(R[X; \sigma, \delta], *, \alpha)$ is a hom-associative, non-unital Ore extension with α extended homogeneously to a ring endomorphism on $R[X; \sigma, \delta]$.

Proof. Since α is an endomorphism on R that commutes with δ and σ, (9) holds, so by Lemma 1, the homogeneously extended map α on $R[X; \sigma, \delta]$ is an endomorphism. Referring to Proposition 1, $(R[X; \sigma, \delta], *, \alpha)$ is thus a hom-associative ring. Furthermore, we see that $*$ is the multiplication (1) of a non-associative, non-unital Ore extension, since for all $a, b \in R$ and $m, n \in \mathbb{N}$,

$$\begin{aligned}aX^m * bX^n = \alpha\left(\sum_{i\in\mathbb{N}} \left(a \cdot \pi_i^m(b)\right) X^{i+n}\right) &= \sum_{i\in\mathbb{N}} \alpha\left(a \cdot \pi_i^m(b)\right) X^{i+n} \\ &= \sum_{i\in\mathbb{N}} \left(a * \pi_i^m(b)\right) X^{i+n}.\end{aligned}$$

□

Remark 4. Note in particular that if $R[X;\sigma,\delta]$ is unital, then $(R[X;\sigma,\delta],*,\alpha)$ is weakly unital with weak unit 1 due to Corollary 1.

Proposition 4 (Hom-associative σ-derivation)**.** Let A be an associative algebra, α and σ algebra endomorphisms, and δ a σ-derivation on A. Assume α commutes with δ and σ. Then σ is an algebra endomorphism and δ a σ-derivation on $(A,*,\alpha)$.

Proof. Linearity follows immediately, while for any $a,b\in A$,

$$\begin{aligned}\sigma(a*b)&=\sigma(\alpha(a\cdot b))=\alpha(\sigma(a\cdot b))=\alpha(\sigma(a)\cdot\sigma(b))=\sigma(a)*\sigma(b),\\ \delta(a*b)&=\delta(\alpha(a\cdot b))=\alpha(\delta(a\cdot b))=\alpha(\sigma(a)\cdot\delta(b)+\delta(a)\cdot b)\\ &=\alpha(\sigma(a)\cdot\delta(b))+\alpha(\delta(a)\cdot b)=\sigma(a)*\delta(b)+\delta(a)*b,\end{aligned}$$

which completes the proof. □

Example 1 (Hom-associative Weyl algebras)**.** Consider the first Weyl algebra exhibited as an associative, unital Ore extension, $K[Y][X;\mathrm{id}_{K[Y]},\delta]=:A$, where K is a field and $\delta=\frac{\mathrm{d}}{\mathrm{d}Y}$, i.e. the standard derivation on $K[Y]$. Clearly any algebra endomorphism α on $K[Y]$ commutes with $\mathrm{id}_{K[Y]}$, but what about δ? Since $\alpha(\delta(Y))=\alpha(1)=1$, we need to have $\delta(\alpha(Y))=1$ which implies $\alpha(Y)=Y+k$, for some $k\in K$. On the other hand, if α is an algebra endomorphism such that $\alpha(Y)=Y+k$ for any $k\in K$, then for any monomial aY^m,

$$\begin{aligned}\alpha(\delta(aY^m))&=am\alpha(Y^{m-1})=am\alpha^{m-1}(Y)=am(Y+k)^{m-1},\\ \delta(\alpha(aY^m))&=a\delta(\alpha^m(Y))=a\delta((Y+k)^m)=am(Y+k)^{m-1}.\end{aligned}$$

Hence any algebra endomorphism α on $K[Y]$ that satisfies $\alpha(Y)=Y+k$ for any $k\in K$ will commute with δ (and any algebra endomorphism that commutes with δ will be on this form). Since α commutes with δ and σ, we know from Corollary 4 that α extends to a ring endomorphism on A as well by $\alpha(aX^m)=\alpha(a)X^m$. Linearity over K follows from the definition, so in fact α extends to an algebra endomorphism on A. Appealing to Proposition 3 and Remark 4, we thus have a family of hom-associative, weakly-unital Ore extensions $(A,*,\alpha_k)$ with weak unit 1, where $k\in K$ and α_k is the K-algebra endomorphism defined by $\alpha_k\left(p(Y)X^m\right)=p(Y+k)X^m$ for all polynomials $p(Y)\in K[Y]$ and $m\in\mathbb{N}$. Since Proposition 4 assures δ to be a K-linear σ-derivation on any member $(A,*,\alpha_k)$ as well, we call these *hom-associative Weyl algebras*, including the associative Weyl algebra in the member corresponding to $k=0$.

A well-known fact about the associative Weyl algebras are that they are simple, meaning that any non-zero ideal of such an algebra is necessarily the algebra itself. This fact is also true in the case of the non-associative Weyl algebras introduced in [21], and it turns out that the hom-associative Weyl algebras as described above have this property as well.

Proposition 5. The hom-associative Weyl algebras are simple.

Proof. The main part of the proof follows the line of reasoning for the associative case; let $(A,*,\alpha_k)$ be any hom-associative Weyl algebra, and I any non-zero ideal of it. Let $p=\sum_{i\in\mathbb{N}}p_i(Y)X^i\in I$ be an arbitrary non-zero polynomial with $p_i(Y)\in K[Y]$, and put $m:=\max_i(\deg(p_i(Y)))$. Then

$$\begin{aligned}X*p&=\sum_{i\in\mathbb{N}}X*p_i(Y)X^i=\sum_{i\in\mathbb{N}}1*\left(p_i'(Y)X^i+p_i(Y)X^{i+1}\right)\\ &=\sum_{i\in\mathbb{N}}\alpha_k\left(p_i'(Y)X^i+p_i(Y)X^{i+1}\right)=\sum_{i\in\mathbb{N}}\left(\alpha_k\left(p_i'(Y)X^i\right)+\alpha_k\left(p_i(Y)X^{i+1}\right)\right),\\ p*X&=\sum_{i\in\mathbb{N}}p_i(Y)X^i*X=\sum_{i\in\mathbb{N}}p_i(Y)*X^{i+1}=\sum_{i\in\mathbb{N}}\alpha_k\left(p_i(Y)X^{i+1}\right),\end{aligned}$$

where $'$ denotes differentiation with respect to Y. By using the commutator $[\cdot,\cdot]$,

$$[X,p] = \sum_{i\in\mathbb{N}} \alpha_k \left(p_i'(Y)X^i\right) = \sum_{i\in\mathbb{N}} p_i'(Y+k)X^i.$$

Since $\max_i(\deg(p_i'(Y+k)) = m-1$, by applying the commutator to the resulting polynomial with X m times, we get a polynomial $\sum_{j\in\mathbb{N}} a_jX^j$ of degree n, where $a_n \in K$ is non-zero. Then

$$\sum_{j\in\mathbb{N}} a_jX^j * Y = \sum_{j\in\mathbb{N}}\sum_{i\in\mathbb{N}} a_j * \pi_i^j(Y)X^i = \sum_{j\in\mathbb{N}} \left(a_j * X^{j-1} + a_j * YX^j\right),$$

$$Y * \sum_{j\in\mathbb{N}} a_jX^j = \alpha_k\left(Y\sum_{j\in\mathbb{N}} a_jX^j\right) = \alpha_k\left(\sum_{j\in\mathbb{N}} a_jYX^j\right) = \sum_{j\in\mathbb{N}} \alpha_k\left(a_jYX^j\right)$$
$$= \sum_{j\in\mathbb{N}} a_j * YX^j.$$

Therefore $\deg\left(\left[\sum_{j\in\mathbb{N}} a_jX^j, Y\right]\right) = n-1$, where $\deg(\cdot)$ now denotes the degree of a polynomial in X. By applying the commutator to the resulting polynomial with Y n times, we get $a_n * 1 \in I$;

$$a_n * 1 = \alpha_k(a_n) = a_n \in I \implies a_n^{-1} * (a_n * 1) = a_n^{-1} * a_n = \alpha_k(1) = 1 \in I.$$

Take any polynomial $q = \sum_{i\in\mathbb{N}} q_i(Y)X^i$ in $(A, *, \alpha_k)$. Then

$$1 * \sum_{i\in\mathbb{N}} q_i(Y-k)X^i = \sum_{i\in\mathbb{N}} q_i(Y)X^i = q \in I$$

and therefore $I = (A, *, \alpha_k)$. □

Acknowledgments
We would like to thank Lars Hellström for discussions leading to some of the results presented in the article.

References
[1] Aizawa N and Sato H 1991 q-deformation of the Virasoro algebra with central extension *Phys. Lett.* B **256** 185–90
[2] Chaichian M, Ellinas D and Popowicz Z 1990 Quantum conformal algebra with central extension *Phys. Lett.* B **248** 95–9
[3] Chaichian M, Isaev A P, Lukierski J, Popowic Z and Prešnajder P 1991 q-deformations of Virasoro algebra and conformal dimensions *Phys. Lett.* B **262** 32–8
[4] Chaichian M, Kulish P and Lukierski J 1990 q-deformed Jacobi identity, q-oscillators and q-deformed infinite-dimensional algebras *Phys. Lett.* B **237** 401–6
[5] Chaichian M, Popowicz Z and Prešnajder P 1990 q-Virasoro algebra and its relation to the q-deformed KdV system *Phys. Lett.* B **249** 63–5
[6] Curtright T L and Zachos C K 1990 Deforming maps for quantum algebras *Phys. Lett.* B **243** 237–44
[7] Curtright T L, Fairlie D B and Zachos C K 2008 Ternary Virasoro-Witt algebra *Phys. Lett.* B **666** 386–90
[8] Daskaloyannis C 1992 Generalized deformed Virasoro algebras, *Modern Phys. Lett.* A **7** 809–16
[9] Hartwig J T, Larsson D and Silvestrov S D 2006 Deformations of Lie algebras using σ-derivations *J. of Algebra* **295** 314–61 (*Preprint* arXiv:math/0408064)
[10] Hu N 1999 q-Witt algebras, q-Lie algebras, q-holomorph structure and representations *Algebra Colloq.* **6** 51–70
[11] Kassel C 1992 Cyclic homology of differential operators, the Virasoro algebra and a q-analogue *Comm. Math. Phys.* **146** 343–56

[12] Larsson D and Silvestrov S D 2005 Quasi-Hom-Lie algebras, Central Extensions and 2-cocycle-like identities *J. Algebra* **288** 321–44 (*Preprint* arXiv:math/0408061)
[13] Larsson D and Silvestrov S D 2005 Quasi-Lie algebras *Noncommutative Geometry and Representation Theory in Mathematical Physics (Contemp. Math. vol 391)* ed J Fuchs, J Mickelsson, G. Rozenblioum, A. Stolin, A. Westerberg (Providence, RI: Amer. Math. Soc.) pp 241–8
[14] Larsson D and Silvestrov S D 2005 Graded quasi-Lie agebras *Czechoslovak J. Phys.* **55** 1473–78
[15] Liu K Q 1991 Quantum central extensions *C. R. Math. Rep. Acad. Sci. Canada* **13** 135–40
[16] Liu K Q 1992 Characterizations of the Quantum Witt Algebra *Lett. Math. Phys.* **24** 257–65
[17] Liu K Q 1992 *The Quantum Witt Algebra and Quantization of Some Modules over Witt Algebra* (Edmonton, Alberta: University of Alberta) (*PhD Thesis*)
[18] Makhlouf A and Silvestrov S D 2008 Hom-algebra structures *J. Gen. Lie Theory Appl.* **2** 51–64 (*Preprint* arXiv:math/0609501)
[19] Makhlouf A and Silvestrov S D 2010 Hom-Algebras and Hom-Coalgebras *J. Algebra Appl.* **9** 553–89 (*Preprint* arXiv:0811.0400)
[20] Makhlouf A and Silvestrov S D 2009 Hom-Lie admissible Hom-coalgebras and Hom-Hopf algebras *Generalized Lie Theory in Mathematics, Physics and Beyond* ed S. Silvestrov, E. Paal, V. Abramov, A. Stolin (Berlin: Springer) chapter 17 pp 189–206 (*Preprint* arXiv:0709.2413)
[21] Nystedt P, Öinert J and Richter J 2015 Non-associative Ore extensions *Preprint* arXiv:1509.01436
[22] Nystedt P, Öinert J and Richter J 2017 Monoid Ore extensions *Preprint* arXiv:1705.02778
[23] Fregier Y and Gohr A 2009 On unitality conditions for Hom-associative algebras *Preprint* arXiv:0904.4874
[24] Sigurdsson G and Silvestrov S 2006 Graded quasi-Lie algebras of Witt type *Czech J Phys.* **56** 1287–91
[25] Yau D 2009 Hom-algebras and Homology *J. Lie Theory* **19** 409–21
[26] Nystedt P 2013 A combinatorial proof of associativity of Ore extensions *Discrete Math.* **313** 2748–50
[27] Richter J 2014 A note on "A combinatorial proof of associativity of Ore extensions" *Discrete Math.* **315–316** 156–7

IOP Conf. Series: Journal of Physics: Conf. Series **965** (2018) 012007 doi:10.1088/1742-6596/965/1/012007

Notion of Random Domino Automaton revisited

Arpan Bagchi

Institute of Geophysics, Polish Academy of Sciences, ul. Ks. Janusza 64, 01-452 Warszawa, Poland

E-mail: arpan@igf.edu.pl

Abstract.

Inspired by a need of effective simulations of a system of Random Domino Automaton type defined for Bethe lattice, new variables are introduced. Main results obtained for 1-dimensional system — including a set of equations describing stationary state and relation to Motzkin numbers — are investigated in this new notion.

1. Introduction

Random Domino Automaton (RDA) was introduced as a straightforward totally discrete stochastic dynamical framework serving as a toy model of earthquakes [9, 10]. It may be considered as an extension of 1 dimensional Drossel and Schwabl forest fire model [14, 13, 17]. However, contrary to the Drossel-Schwabl model where parameters are constants, RDA allows some parameters to be functions of cluster's size (forests' size) and a good mathematical structure of the equations of the model is preserved. As a consequence, RDA generates bigger variety of distributions depending on respective parameters — form exponential distributions to inverse-power ones [10] and also quasi-periodic for a finite version [5]. Moreover, it is remarkable that the inverse problem of finding parameters (that define the dynamics of the system) from the given distribution can be solved [10, 4]. That last property is useful for explaining statistical properties of occurrence of earthquakes [6]. The extension also leads to a significant link to the Motzkin number recurrence [3, 7].

The published results are related to 1-dimensional RDA. Construction of RDA type systems on more complex geometries leads to more complicated equations, mainly due to loss of one-to-one correspondence between a size of a cluster and a size of its boundary. For 1-dimensional systems there are always just two end cells of a cluster, and the cluster may be enlarged only through them. For 2-dimensional systems the perimeter depends on a shape of a cluster. Thus, in order to extend the geometry of RDA preserving its good mathematical structure, one may choose Bethe lattice — an infinite loop free graph with fixed coordination number K, introduced by Hans Bethe in 1935 [12, 19, 15, 11]. In Bethe lattice there is a fixed relation between a cluster size and the size of its perimeter. This property allows to derive a set of equations using mean-field approximation in a stationary state of RDA type system on Bethe lattice [2].

Bethe lattice is infinite and homogenous everywhere, so any of its point can be regarded as an origin. However, when considering a finite part of a Bethe Lattice — for example in simulations — one may notice that majority of cells within that part are placed close to its boundary — see FIG. 1, where the boundary is created by fixing a certain radius. It follows, one can find many clusters which cross the boundary and extend outside. When an avalanche happen, i.e. a cluster is removed from the system (or equivalently all cells of a clusters become empty) one can not estimate how many cells are actually responsible for the avalanche by observing that finite part only. The definition of RDA (and other similar systems) on Bethe lattice in natural way leads to the difficulty described above. To avoid this difficulty, a new variable S_i (for $i = 1, 2, \ldots$) is introduced as follows [8]

$$S_i = \sum_{j=i}^{\infty} n_j,$$

where n_j is the number of clusters of size j (i.e. with exactly j occupied cells) and (S_i) is the number of clusters of size $\geq i$.

In the FIG. 1 it is visible, that the finite part contains only 10 cells in which 8 are occupied and 2 are empty. There are five clusters and they are extended through the boundary. So it is not possible to find out information about the distribution of clusters (n_i), because there is no cluster which lay just inside the finite part. By the definition of new variables one can say all 5 clusters contribute to S_1 and at least one cluster contribute to S_4.

In the present paper, this new variable is investigated for 1-dimensional RDA. For example, FIG. 2 presents a finite section of line where two clusters cross left and right boundaries of fixed finite part of the line. Thus it is impossible to know which n_i they contribute to, but it is clear that they both contribute to S_2 and at least one to S_3. Main results obtained for 1-dimensional RDA are rewritten in the new notion. Study of the RDA on Bethe lattice — using both "old" and "new" notions — is a matter of a forthcoming paper.

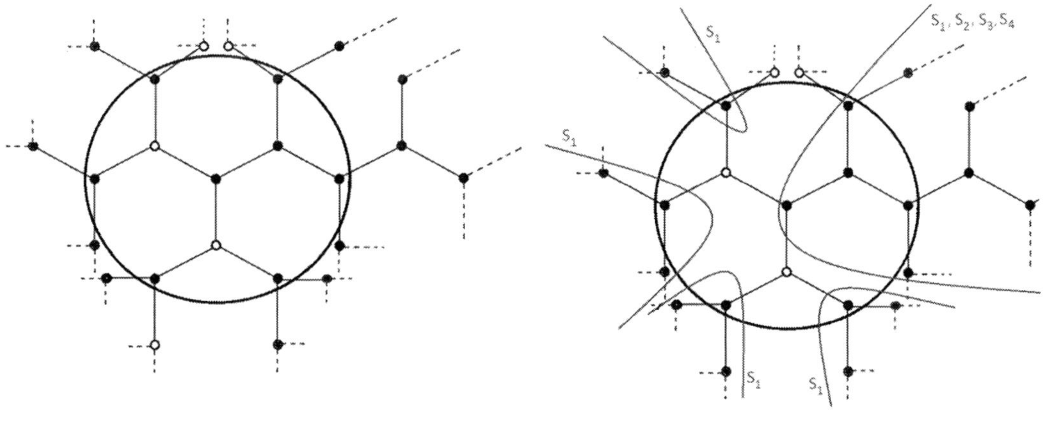

Figure 1. New variable in Bethe lattice with coordination number 3

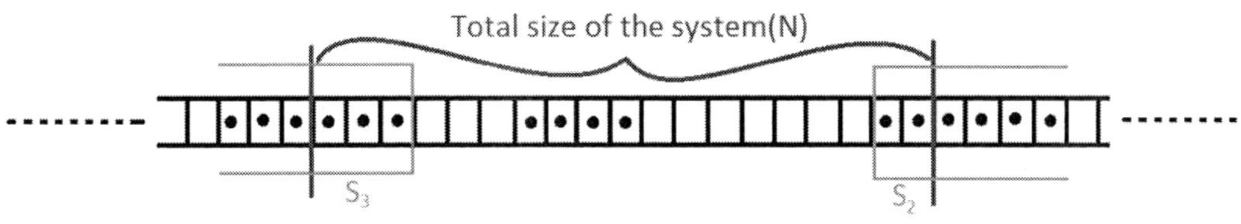

Figure 2. New variable in line

2. Evolution rules of extended RDA

We shortly recall evolution rules of RDA. Assume that any cell may be in one of two states: empty or occupied by a ball (portion energy). The state of automaton in the discrete time t is characterized by instantaneous density ρ_t equal to number of occupied cells in a given time t divided by the size of the system N.

Rules of evolution include parameters c and μ and are as follows. In each time step t one cell is chosen (we assume that each has the same probability).

The probability that the chosen cell is empty is equal to $(1-\rho_t)$. Then the empty cell becomes occupied with the probability c, or it remains empty (the ball is rebounded) with probability $(1-c)$ and the state of automaton is unchanged.

If the chosen cell is already occupied by energy (the respective probability is ρ_t), then either the energy will be rebounded with the probability $(1-\mu)$ or it triggers an avalanche with the probability μ. The avalanche size is equal to the number of cells changing their state.

Then the update procedure repeats in next time step.

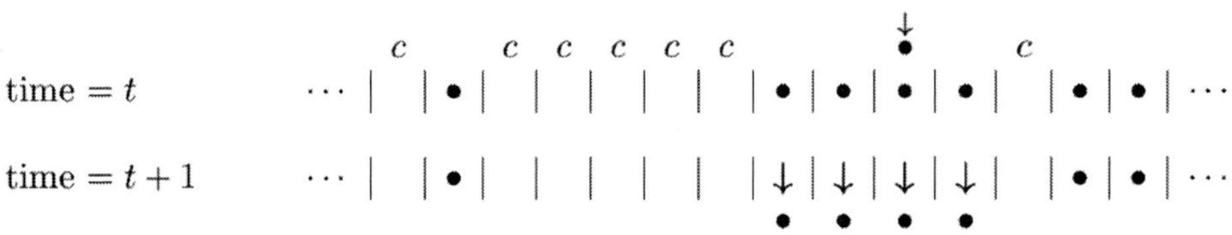

3. Equations for RDA

3.1. *Equations for RDA with the variable* n_i

The stationary state of the system may be described by the distribution of clusters n_i. The number of all clusters n and and the density ρ are

$$n = \sum_{i\geq 1} n_i, \qquad \rho = \frac{1}{N}\sum_{i\geq 1} i n_i. \tag{1}$$

The set of equations for the distribution of clusters n_i and the distribution of empty clusters n_i^0 for $i = 1, 2, \ldots$ in stationary state is [10]

$$n_1 = \frac{1}{\frac{\mu_1}{c} + 2}\left((1-\rho)N - 2n + n_1^0\right), \tag{2}$$

$$n_2 = \frac{2n - 2n_1^0}{\left(4 + \frac{2}{cn}\sum_{i\geq 1}\mu_i i n_i\right)}, \tag{3}$$

$$n_i = \frac{1}{\frac{\mu_i}{c} i + 2}\left[2n_{i-1}\left(1 - \frac{n_1^0}{n}\right) + \frac{n_1^0}{n^2}\sum_{k=1}^{i-2} n_k n_{i-1-k}\right], \text{for } i \geq 3 \tag{4}$$

and

$$n_1^0 = \frac{2n}{\left(3 + \frac{2}{cn}\sum_{i\geq 1}\mu_i i n_i\right)}, \tag{5}$$

$$n_2^0 = \frac{2n - 2n_1^0}{\left(4 + \frac{2}{con}\sum_{i\geq 1}\mu_i i n_i\right)}, \tag{6}$$

$$n_k^0 = \frac{2(n - \sum_{i=1}^{k-1} n_i^0) + \sum_{j=1}^{k-2}\sum_{l=1}^{k-1-j} \frac{\mu_j}{c} j n_j \frac{n_l^0}{n}\frac{n_{k-j-l}^0}{n}}{2 + k + \frac{2}{cn}\sum_{i\geq 1}\mu_i i n_i}, \quad \text{for } k \geq 3. \tag{7}$$

The balance equation for the total number of clusters n and for the density ρ reads [10]

$$(1-\rho)N - 2n = \sum_{i\geq 1} \frac{\mu_i}{c} n_i i, \tag{8}$$

$$(1-\rho)N = \sum_{i\geq 1} \frac{\mu_i}{c} n_i i^2. \tag{9}$$

*3.2. **Equations for RDA with the variable** S_i*

Some properties of S_i:

1. By the basic definition of S_i, S_1 can be written as the total number of clusters.

$$S_1 = n = \sum_{i\geq 1} n_i. \tag{10}$$

2. S_i is decreasing function i.e.

$$S_1 \geq S_2 \geq S_3 \geq \ldots\ldots \tag{11}$$

3. n_i can be witten as the difference between S_i and S_{i+1}

$$n_i = S_i - S_{i+1}. \tag{12}$$

4. Density can be written interms of S_i as follows.

$$\rho = \frac{1}{N}\sum_{i=1}^{\infty} i n_i = \frac{1}{N}\sum_{i=1}^{\infty} S_i. \tag{13}$$

5. 2nd order moment of n_i is

$$\frac{1}{N}\sum_{i=1}^{\infty} i^2 n_i = \frac{1}{N}\left(2\sum_{i=1}^{\infty} i S_i - \sum_{i=1}^{\infty} S_i\right). \tag{14}$$

3.3. Equations for extended RDA with the variable S_i

The system of the equations ((2)-(4)) can be reconstructed in terms of S_i as follows.

$$\sum_{i\geq 1}\frac{\mu_i}{c}in_i+2S_1 = (1-\rho)N, \tag{15}$$

$$\sum_{i\geq 2}\frac{\mu_i}{c}in_i+2S_2 = 2S_1-n_1^0, \tag{16}$$

$$\sum_{i\geq 3}\frac{\mu_i}{c}in_i+2S_3 = 2S_2-\left(\frac{2S_2}{S_1}-1\right)n_1^0, \tag{17}$$

$$\sum_{i\geq j}\frac{\mu_i}{c}in_i+2S_j = (2-\frac{n_1^0}{S_1})S_{j-1}+\frac{n_1^0}{S_1^2}\sum_{k=1}^{j-2}\left[S_k-S_{k+1}\right]S_{j-1-k} \quad \text{for } i\geq 4. \tag{18}$$

In general it is very difficult to solve the equation (18) because for the 1st term. Here we will investigate the solution for special choice of rebound parameter.

4. Special case: $\mu_i=\frac{\delta}{i}$, where $\theta=$ constant

4.1. Solution for n_i with special setting of rebound parameter

Choosing this setting of rebound parameters, equations (2)-(4) gives

$$n_1 = \frac{1}{\theta+2}\left((1-\rho)N-2n+n_1^0\right), \tag{19}$$

$$n_2 = \frac{2}{\theta+2}\left(1-\frac{n_1^0}{n}\right)n_1, \tag{20}$$

$$n_i = \frac{1}{\theta+2}\left[2n_{i-1}\left(1-\frac{n_1^0}{n}\right)+\frac{n_1^0}{n^2}\sum_{k=1}^{i-2}n_kn_{i-1-k}\right] \tag{21}$$

for $i\geq 3$, where $\theta=\frac{\delta}{c}, \theta\in[0,\infty)$.

We define new variables M_i for $i=0,1,\ldots,$ by

$$M_i=\frac{\beta}{\alpha^{i+1}}n_{i+1}, \tag{22}$$

where

$$\alpha = \frac{2(1-\frac{n_1^0}{n})}{\theta+2}, \tag{23}$$

$$\beta = \frac{\frac{n_1^0}{n}}{2(1-\frac{n_1^0}{n})n}. \tag{24}$$

Then, the equation (21) can be rewritten in the form

$$M_{m+2}=M_{m+1}+\sum_{k=0}^{m}M_kM_{m-k}. \tag{25}$$

which is valid for $m \geq 0$ $(m = i-2)$. Initial data M_0 and M_1 are easily obtained from equations (19)-(20), when it is transformed according the rule of equation (22), namely

$$M_0 = M_1 = \frac{\frac{n_1^0}{n^2}[(1-\rho)N - 2n + n_1^0]}{4(1-\frac{n_1^0}{n})^2} = \frac{\frac{n_1^0}{n}(\theta + \frac{n_1^0}{n})}{4(1-\frac{n_1^0}{n})^2}. \quad \text{(Using(8))} \tag{26}$$

Notice, M_0 and M_1 are equal.

The above equation (25) has the form of Motzkin numbers recurrence [18, 1, 16, 3]. From the Motzkin numbers: $1, 1, 2, 4, 9, 21, 51, 127, 323, 835, 2188, \ldots$, etc. we can find the condition $M_0 = M_1 = 1$ which gives,

$$\frac{n_1^0}{n^2}[(1-\rho)N - 2n + n_1^0] = 4(1-\frac{n_1^0}{n})^2 = 1. \tag{27}$$

The solution of the equation (25) is

$$M_m = \frac{1}{2}\sum_{j=0}^{\lfloor\frac{m+2}{2}\rfloor} \frac{(2M_0 - \frac{1}{2})^j}{(m-j+2)2^{m-j}} \binom{2(m-j)+1}{m-j+1}\binom{m-j+2}{j}. \tag{28}$$

Thus formula (22) gives the explicit solution of the equation (19)-(21) for the distribution of n_i for any values of θ.

By the above choice of μ_i (8) and (9) give the values of ρ, n and n_1^0 [10] as follows.

$$\rho = \frac{1}{\theta+1}, \tag{29}$$

$$n = \frac{N\theta}{(\theta+1)(\theta+2)}, \tag{30}$$

$$n_1^0 = \frac{2n}{2\theta+3}. \tag{31}$$

4.2. Solution for S_i with special setting of rebound parameter

By this chosen setting of rebound parameter, the set of equations (15)-(16) transform into the following set of equations.

$$(\theta+2)S_1 = (1-\rho)N, \tag{32}$$

$$(\theta+2)S_2 = 2S_1 - n_1^0, \tag{33}$$

$$(\theta+2)S_3 = 2S_2 - \left(\frac{2S_2}{S_1} - 1\right)n_1^0, \tag{34}$$

$$(\theta+2)S_j = (2 - \frac{n_1^0}{S_1})S_{j-1} + \frac{n_1^0}{S_1^2}\sum_{k=1}^{j-2}\left[S_k - S_{k+1}\right]S_{j-1-k} \quad \text{for } i \geq 4 \tag{35}$$

To solve (35) we define new variables M_i^* for $i = 0, 1, \ldots,$ by

$$M_i^* = \frac{B}{A^{i+1}}S_{i+1}, \tag{36}$$

IOP Conf. Series: Journal of Physics: Conf. Series **965** (2018) 012007 doi:10.1088/1742-6596/965/1/012007

where

$$A = \frac{2 - \frac{n_1^0}{S_1}}{\theta + 2}, \tag{37}$$

$$B = \frac{\frac{n_1^0}{S_1}}{S_1(\theta + 2)}. \tag{38}$$

Then, the equation (35) can be rewritten in the form

$$M^*_{m+2} = M^*_{m+1} + \frac{1}{A}\sum_{k=0}^{m} M^*_k M^*_{m-k} - \sum_{k=0}^{m} M^*_{k+1} M^*_{m-k}. \tag{39}$$

This recurrence is equivalent to Motzkin number recurrence [3]. Initial data M_0^* and M_1^* are easily obtained from equations (32)-(33).

$$M_0^* = M_1^* = \frac{\frac{n_1^0}{S_1}}{2 - \frac{n_1^0}{S_1}}. \tag{40}$$

Notice, M_0^* and M_1^* are equal.

We will solve (39) by using generating function [3]. Let us consider the generating function

$$D(z) = \sum_{m\geq 0} M^*_m z^m. \tag{41}$$

By this generating function the recurrence (39) gives

$$z\left(\frac{z}{A} - 1\right)D^2(z) + \left(z + zM_0^* - 1\right)D(z) + \left((M_1^* - M_0^*)z + M_0^*\right) = 0.$$

Since $M_0^* = M_1^*$ it gives,

$$z\left(\frac{z}{A} - 1\right)D^2(z) + \left(z + zM_0^* - 1\right)D(z) + M_0^* = 0. \tag{42}$$

Let D(z) be the solution of (42). Then,

$$D(z) = \frac{-(-1 + z + zM_0^*) - \sqrt{1 - 2z(p + qz)}}{2zM_0^*\left(\frac{z}{A} - 1\right)}, \tag{43}$$

where,

$$p = (1 - M_0^*), \tag{44}$$

$$q = \frac{1}{2}\left(\frac{4M_0^*}{A} - (1 + M_0^*)^2\right). \tag{45}$$

We choose negative sign in order D(z) to be a power series of z.

We are interested to find out the coefficient of z^m of $D(z)$.

$$D(z) = \left[1 - \frac{q}{2M_0^*}z - \frac{1}{2M_0^*}\sum_{n\geq 2}\frac{(-1)}{n2^{n-2}}\binom{2n-3}{n-1}z^n(p+qz)^n\right]\left(1-\frac{z}{A}\right)^{-1}$$

$$= \left[1 - \frac{q}{2M_0^*}z - \frac{1}{2M_0^*}\sum_{n\geq 2}\frac{p^n}{n2^{n-2}}\binom{2n-3}{n-1}z^{n-1}\sum_{j=0}^{n}\binom{n}{j}\left(\frac{q}{p}\right)^j z^j\right] \times \left[\sum_{n\geq 0}\left(\frac{z}{A}\right)^n\right]. \tag{46}$$

So, coefficient of z^m is

$$M_m^* = \frac{M_0^*}{A^m} - \frac{q}{2A^{m-1}} - \frac{1}{2}\left[\sum_{x=2}^{\lfloor\frac{m+1}{2}\rfloor}\sum_{y=0}^{x}\frac{p^x}{x2^{x-2}}\binom{2x-3}{x-1}\binom{x}{y}\left(\frac{q}{p}\right)^y\frac{1}{A^{m+1-x-y}}\right.$$
$$\left.+\sum_{x=\lfloor\frac{m+3}{2}\rfloor}^{m+1}\sum_{y=0}^{m+1-x}\frac{p^x}{x2^{x-2}}\binom{2x-3}{x-1}\binom{x}{y}\left(\frac{q}{p}\right)^y\frac{1}{A^{m+1-x-y}}\right], \tag{47}$$

where

$$p = (1 - M_0^*),$$
$$q = \frac{1}{2}\left(\frac{4M_0^*}{A} - (1+M_0^*)^2\right),$$
$$A = \frac{2 - \frac{n_1^0}{S_1}}{\theta+2}.$$

By the above formula of ρ, n and n_1^0 ((29)-(31)) we can calculate the values of S_i, for $i = 1, 2, \ldots$ as follows.

$$S_i = \begin{cases} \dfrac{N\theta}{\left(\theta+1\right)\left(\theta+2\right)}, & i=1 \\ \dfrac{4N\theta}{\left(\theta+2\right)^2\left(2\theta+3\right)}, & i=2 \\ \dfrac{2^{(2i-1)}N\theta\left(\theta+1\right)^{(i-1)}}{\left(\theta+2\right)^i\left(2\theta+3\right)^{i-1}}M_{i-1}^* & i\geq 3 \end{cases} \tag{48}$$

Now we will find out the distribution of n_i by (22) and (36) and finally we will show that both distributions are same.

From the equation (22) we get,

$$n_i = \frac{\alpha^i}{\beta}M_{i-1}$$
$$= \begin{cases} \dfrac{\left(\theta+\frac{n_1^0}{n}\right)n}{\left(\theta+2\right)}, & i=1 \\ \dfrac{\left(2(1-\frac{n_1^0}{n})\right)^{i+1}}{2^{\frac{n_1^0}{n^2}}\left(\theta+2\right)^i}\left[\sum_{j=0}^{\lfloor\frac{i+2}{2}\rfloor}\dfrac{(2M_0-\frac{1}{2})^j}{(i-j+1)2^{i-j-1}}\binom{2(i-j)-1}{i-j}\binom{i-j+1}{j}\right]. & i\geq 2 \end{cases} \tag{49}$$

Again from the equation (12) and (36) we get

$$
\begin{aligned}
n_i &= S_i - S_{i-1} \\
&= \frac{A^i}{B}\Big[M^*_{i-1} - M^*_i\Big] \\
&= \begin{cases} \frac{A}{B}\Big(1-A\Big)M_0^*, & i=1 \\ \frac{A^{i+1}}{2B}\Bigg[\displaystyle\sum_{k=\lfloor\frac{i+2}{2}\rfloor}^{i+1} \frac{p^k}{k2^{k-2}}\binom{2k-3}{k-1}\binom{k}{i+1-k}\Big(\frac{q}{p}\Big)^{i+1-k}\Bigg]. & i\geq 2 \end{cases} \\
&= \begin{cases} \frac{\Big(\theta+\frac{n_1^0}{S_1}\Big)n}{\Big(\theta+2\Big)}, & i=1 \\ \frac{\Big(2-\frac{n_1^0}{S_1}\Big)^{i+1}}{2\frac{n_1^0}{S_1^2}\Big(\theta+2\Big)^i}\Bigg[\displaystyle\sum_{k=\lfloor\frac{i+2}{2}\rfloor}^{i+1} \frac{p^k}{k2^{k-2}}\binom{2k-3}{k-1}\binom{k}{i+1-k}\Big(\frac{q}{p}\Big)^{i+1-k}\Bigg]. & i\geq 2 \end{cases}
\end{aligned}
\tag{50}
$$

We need to show that (49) and (50) are equivalent. If we substitute $k = i+1-j$ and $S_1 = n$ in the equation (50) then we get

$$
n_i = \frac{\Big(2-\frac{n_1^0}{n}\Big)^{i+1}}{2\frac{n_1^0}{n^2}\Big(\theta+2\Big)^i}\Bigg[\sum_{j=0}^{\lfloor\frac{i+1}{2}\rfloor} \frac{p^{i+1}}{(i-j+1)2^{i-j-1}}\binom{2(i-j)-1}{i-j}\binom{i-j+1}{j}\Big(\frac{q}{p^2}\Big)^j\Bigg]. \qquad \text{for} \quad i\geq 2 \tag{51}
$$

Now if we proof the following hypothesis then we can tell that (49) and (50) are equivalent.

Hypothesis:

$$
2^{i+1}\Big(1-\frac{n_1^0}{n}\Big)^{i+1}(2M_0-\frac{1}{2})^j = p^{i+1}\Big(2-\frac{n_1^0}{n}\Big)^{i+1}\Big(\frac{q}{p^2}\Big)^j. \qquad \forall\ j \tag{52}
$$

$$
\Rightarrow \left(\frac{2(1-\frac{n_1^0}{n})}{2-\frac{n_1^0}{n}}\right)^{i+1}(2C_0-\frac{1}{2})^j = p^{i+1}\Big(\frac{q}{p^2}\Big)^j. \qquad \forall\ j
$$

We need to prove

$$
p = \frac{2(1-\frac{n_1^0}{n})}{2-\frac{n_1^0}{n}}. \tag{53}
$$

and

$$
(2M_0-\frac{1}{2}) = \frac{q}{p^2} \Rightarrow \frac{\frac{n_1^0}{n}\Big(\theta+\frac{n_1^0}{n}\Big)}{2\Big(1-\frac{n_1^0}{n}\Big)} - \frac{1}{2} = \frac{1}{2}\left(\frac{4M_0^*\Big(\frac{1}{A}-1\Big)}{\Big(1-M_0^*\Big)^2} - 1\right). \tag{54}
$$

By the values of p,M_0^* and A form the equation (44), (37) and (40) we can easily prove (53) and (54).

5. Conclusions

We pointed out a difficulty in analysis of finite part of the Random Domino Automaton related to clusters crossing boundary of the system. The problem is present for any geometry, but essential for Bethe lattice — a natural candidate for extension RDA beyond dimension 1.

Following the remark [8] we introduce a new variable $S_i = \sum_{j \geq i} n_j$ and repeated main results obtained for 1-dimensional Random Domino Automaton using the new notion. Obtained formulas are in the form convenient to comparisons with numerical simulations. Moreover, we solve the system by choosing special rebound parameters by generating function method which produce a counterpart of Motzkin number recurrence.

Obtained results will be useful for extension and investigation of Random Domino Automaton type system defined on Bethe lattice.

Acknowledgments

This work was partially supported by Institute of Geophysics, Polish Academy of Science , project 500-10-28E and partially supported within statutory activities No. 3841/E-41/S/2017 of the Ministry of Science and Higher Education of Poland.

References

[1] Aigner M 2007 A Course in Enumeration (Berlin: Springer)
[2] Bagchi A, Białecki M Random Domino Automaton on Bethe lattice (in preparation)
[3] Białecki M 2012 Motzkin numbers out of random domino automaton *Phys. Lett A* **376** 3098
[4] Białecki M 2013 From statistics of avalanches to microscopic dynamics parameters in a toy model of earthquakes *Acta Geophys.* **61** 1677
[5] Białecki M 2015 Properties of a finite stochastic cellular automaton toy model of earthquakes *Acta Geophys.* **63** 923
[6] Białecki M 2015 On mechanistic explanation of the shape of the universal curve of earthquake recurrence time distributions *Acta Geophys.* **63** 1205
[7] Białecki M 2016 Solvable structures of a simple model of earthquakes *J Phys. Conf. Series IOP Publishing* **670** 012010
[8] Białecki M Private communication
[9] Białecki M and Czechowski Z 2010 On a simple stochastic cellular automaton with avalanches: simulation and analytical results. In Synchronization and Triggering: from Fracture to Earthquake Processes (GeoPlanet: Earth and Planetary Sciences) ed V De Rubeis et al (Berlin: Springer) chapter 5 pp 63-75 DOI: 10.1007/978-3-642-12300-9_5
[10] Białecki M and Czechowski Z 2013 On one-to-one dependence of rebound parameters on statistics of clusters: exponential and inverse-power distributions out of Random Domino Automaton *J. Phys. Soc. Jpn.* **82** 014003
[11] De Bacco C, Majumdar S N and Sollich P 2015 The average number of distinct sites visited by a random walker on random graphs *J Phys. A* **48** 205004
[12] Dhar D and Majumdar S N 1990 Abelian sandpile model on the Bethe lattice. *J Phys. A* **23** 4333
[13] Drossel B, Clar S and Schwabl F, 1993. Exact results for the one-dimensional self-organized critical forest-fire model *Phys. Rev. Lett.* **71** 3739
[14] Drossel B and Schwabl F 1992 Self-organized critical forest-fire model *Phys. Rev. Lett.* **69** 1629
[15] Eckstein M, Kollar M, Byczuk K and Vollhardt D 2005. Hopping on the Bethe lattice: Exact results for densities of states and dynamical mean-field theory *Phys. Rev. B* **71** 235119
[16] Flajolet P and Sedgewick R 2008 Analytic Combinatorics *Cambridge University Press*
[17] Malamud B D , Morein G and Turcotte D L 1998 Forest fires: an example of self-organized critical behavior *Science* **281** 1840
[18] Sloane N J A 1995 The encyclopedia of integer sequences *Academic Press*
[19] Zapperi S, Lauritsen K B and Stanley H E 1995 Self-organized branching processes: mean-field theory for avalanches *Phys. Rev. Lett.* **75** 4071

IOP Conf. Series: Journal of Physics: Conf. Series **965** (2018) 012008 doi:10.1088/1742-6596/965/1/012008

Generalized generating functional for mixed-representation Green's functions: Coherent-state path integral representation

M Blasone[1], P Jizba[2] and L Smaldone[1]

[1] Dipartimento di Fisica, Università di Salerno, Via Giovanni Paolo II, 132 84084 Fisciano, Italy & INFN Sezione di Napoli, Gruppo collegato di Salerno, Italy

[2] FNSPE, Czech Technical University in Prague, Břehová 7, 115 19 Praha 1, Czech Republic

E-mail: blasone@sa.infn.it

E-mail: petr.jizba@fjfi.cvut.cz

E-mail: luca.smaldone@studenti.unina2.it

Abstract. The aim of this paper is to study mixed-representation Green's functions from the point of view of coherent-state path integrals. This is achieved by using the machinery of generalized generating functionals of Green's functions, previously introduced by the present authors in the context of standard phase-pace path integrals. The obtained results are illustrated in the context of the linear harmonic oscillator.

1. Introduction

Coherent states (CS's) were first introduced by E. Schrödinger in 1926 [1], to describe a dispersionless wave packet of the linear harmonic oscillator (LHO). J. von Neumann [2] used similar ideas in order to investigate the measurement of coordinates and momenta in quantum mechanics (QM). With the advent of laser physics in early 60's this subject firmly established itself through a number of pioneering works [3, 4, 5]. An excellent review, complemented by many original papers, can be found in Ref. [6].

A path integral (PI) formulation of QM in terms of CS's were first proposed by J. R. Klauder in Ref. [3]. As shown in Ref. [7], CS PI's are better mathematically behaved than the more conventional phase-space (PS) PI's [8, 9]. For these reasons, CS PI's found a wide applicability, for example, in quantum field theory (QFT) [9, 10, 11, 12] and theoretical chemistry [13, 14].

The present authors proved in Ref. [15] the usefulness of CS PI's in connection with the study of canonical transformations in the context of quantum field theoretical functional integral (FI). In particular, it was found that the CS PI's are instrumental in helping to exhibit features related to the existence of unitarily inequivalent representations of canonical commutation relations. In addition, in Ref. [16], it was introduced the concept of *generalized generating functional of Green's functions* (GGF), in order to evaluate *mixed representations Green's functions*, i.e. Green's functions constructed on different vacua, connected with each other via unitary canonical transformations. There, it was found that the latter is intimately connected with the Schwinger closed-time-path formalism [17]. In Ref. [18] we applied this method to evaluate the generating functional of Green's functions on the *flavor vacuum*, for a two-flavor bosonic system [19].

In this paper, GGF's are used to provide a new viewpoint on CS PI's that were otherwise treated only within the standard PS PI's. More specifically, we will see that the CS evolution kernel can be regarded as a generalized generating functional, where the defining unitary transformation belongs to the quotient group $\mathcal{W}_1/U(1)$, where $\mathcal{W}_1$ is the one-dimensional Weyl–Heisenberg group [20].

The paper is organized as follows: in Section 2 we briefly remind some essentials of canonical coherent states. To this end we will employ the notation of Ref. [12]. In Section 3 we review key results of Ref. [16] and, in particular, the concept of GGF. In Section 4 we show how to connect CS PI's with GGF's. We apply this result to evaluate some correlation functions of the position operator. Various remarks and generalizations are addressed in the concluding Section 5.

2. Coherent state PI's

A canonical (or Glauber) CS in the Heisenberg picture is defined as [6, 12]

$$|p\,q,t\rangle \;\equiv\; \exp\left[-iq\hat{P}(t)\right]\,\exp\left[ip\hat{Q}(t)\right]|0\rangle\,, \tag{1}$$

where $|0\rangle$ satisfies the equation

$$\left(\hat{Q}(t)+i\hat{P}(t)\right)|0\rangle \;=\; 0\,, \tag{2}$$

and hence it can be identified with the ground state of the LHO, with Hamiltonian

$$\hat{H}\left(\hat{P}(t),\hat{Q}(t)\right) \;=\; \frac{\hat{P}^2(t)}{2}+\frac{\hat{Q}^2(t)}{2}\,. \tag{3}$$

In analogy with QFT it is common to call $|0\rangle$ the *vacuum* state. The unitary operator

$$\hat{G}_{pq}(t) \;=\; \hat{G}^1_p(t)\,\hat{G}^2_q(t)\,, \tag{4}$$

(with $\hat{G}^1_p(t)=e^{-ip\hat{Q}(t)}$ and $\hat{G}^2_q(t) \;=\; e^{iq\hat{P}(t)}$) belongs to the quotient group $\mathcal{W}_1/U(1)$.

If we now employ the Baker–Campbell–Hausdorff formula we obtain:

$$\langle p\,q,t|\hat{Q}(t)|p\,q,t\rangle \;=\; q\,, \tag{5}$$

$$\langle p\,q,t|\hat{P}(t)|p\,q,t\rangle \;=\; p\,. \tag{6}$$

One can easily check that the resolution of unity can be written in terms of CS's as

$$\mathbb{1} \;=\; \int\frac{\mathrm{d}p\,\mathrm{d}q}{2\pi}\,|p\,q,t\rangle\langle p\,q,t|\,. \tag{7}$$

This form can be now used to derive a CS PI form for the evolution kernel, namely (see, e.g. Ref. [6])

$$\langle q_f\,p_f,t_f|q_i\,p_i,t_i\rangle \;=\; \int\mathcal{D}p\,\mathcal{D}q\,\exp\left(iS_{cs}\left[q,p\right]\right)\,, \tag{8}$$

where $q(t_f),p(t_f)=q_f,p_f\,,\;q(t_i),p(t_i)=q_i,p_i$, and

$$S_{cs}[q,p] \;=\; \int_{t_i}^{t_f}\mathrm{d}t\,\left[p(t)\dot{q}(t)\;-\;H\left(p(t),q(t)\right)\right]\,, \tag{9}$$

is the classical action. Note that, as shown in Eqs. (5)-(6), p and q can be specified at the equal time. So, in a sense, CS's provide a PS representation of the QM. The precise meaning of Eq. (8) together with a precise definition of the measure can be found, e.g., in Ref. [12].

For our purposes is also interesting to note that the Feynman–Matthews–Salam formula, i.e.

$$\langle q_f\, p_f, t_f | T\left[\hat{Q}(t_n)\ldots\hat{Q}(t_1)\right] | q_i\, p_i, t_i\rangle \;=\; \int \mathcal{D}p\, \mathcal{D}q\, q(t_n)\ldots q(t_1)\, \exp\left(iS_{cs}\left[q,p\right]\right)\,, \tag{10}$$

holds without changes. Here $T[\ldots]$ denotes the time-ordered symbol and the boundary conditions are the same specified above. The PI on the RHS of Eq. (10) can be evaluated with the Schwinger trick, i.e. by introducing

$$\langle q_f\, p_f, t_f | q_i\, p_i, t_i\rangle_J \;=\; \int \mathcal{D}q\, \mathcal{D}p\; \exp\left(iS_{cs\,J}\left[p,q\right]\right)\,, \tag{11}$$

where

$$S_{cs\,J}\left[p,q\right] \;=\; \int_{t_i}^{t_f} \mathrm{d}t\left[p(t)\dot{q}(t) \;-\; H(q(t),p(t)) \;+\; J(t)q(t)\right]\,, \tag{12}$$

and $J(t)$ is the Schwinger-type current (i.e., it has a compact support). With this we can write

$$\langle q_f\, p_f, t_f | T\left[\hat{Q}(t_n)\ldots\hat{Q}(t_1)\right] | q_i\, p_i, t_i\rangle \;=\; \left\{(-i)^n \frac{\delta\langle q_f\, p_f, t_f|q_i\, p_i, t_i\rangle_J}{\delta J(t_1)\ldots\delta J(t_n)}\right\}_{J=0}\,, \tag{13}$$

or, in other words

$$\begin{aligned}&\langle q_f\, p_f, t_f | q_i\, p_i, t_i\rangle_J \\ &\quad= \sum_{n=0}^{\infty} \frac{i^n}{n!}\int\cdots\int \mathrm{d}t_1\ldots\mathrm{d}t_n\, \langle q_f\, p_f, t_f | T\left[\hat{Q}(t_n)\ldots\hat{Q}(t_1)\right]|q_i\, p_i, t_i\rangle\, J(t_1)\ldots J(t_n)\,.\end{aligned} \tag{14}$$

We say that $\langle q_f\, p_f, t_f|q_i\, p_i, t_i\rangle_J$ is the *generating functional* of the correlation functions (10).

We now ask the question: how are Eqs. (8) and (11) related to the standard phase-space PI?

3. Generalized generating functional of Green's functions

Let us denote the PS evolution kernel in presence of currents as[1]

$$\langle Q_f, t_f | Q_i, t_i\rangle_{J_Q,J_P} \;=\; \int_{Q(t_i)\,=\,Q_i}^{Q(t_f)\,=\,Q_f} \mathcal{D}Q\,\mathcal{D}P\;\exp\left(iS_{ps\,J}\left[P,Q\right]\right)\,, \tag{15}$$

where the action functional has the form

$$S_{ps\,J}\left[P,Q\right] \;=\; \int_{t_i}^{t_f}\mathrm{d}t\left[P(t)\dot{Q}(t) \;-\; H(Q(t),P(t)) \;+\; J_Q(t)Q(t) \;+\; J_P(t)P(t)\right]\,, \tag{16}$$

and $J_Q(t)$ and $J_P(t)$ are again Schwinger-type currents.

We now perform a canonical transformation of the operators $\hat{Q}(t)$ and $\hat{P}(t)$:

$$\hat{Q}(t;\alpha) \;\equiv\; \hat{G}_\alpha^\dagger(t)\,\hat{Q}(t)\,\hat{G}_\alpha(t)\,, \tag{17}$$

$$\hat{P}(t;\alpha) \;\equiv\; \hat{G}_\alpha^\dagger(t)\,\hat{P}(t)\,\hat{G}_\alpha(t)\,, \tag{18}$$

[1] We denote by Q, P the eigenvalues of the operators $\hat{Q}, \hat{P}$, to distinguish them from q, p which are just expectation values (see Eqs.(5),(6)).

where

$$\hat{G}_\alpha(t) \;=\; \exp\left[i\alpha\hat{K}\left(\hat{P}(t),\hat{Q}(t)\right)\right], \tag{19}$$

is a unitary continuous transformation which leaves invariant the canonical commutation relations

$$\left[\hat{Q}(t;\alpha),\hat{P}(t;\alpha)\right] \;=\; i\,. \tag{20}$$

At this stage we introduce

$$|0(\alpha,t)\rangle \;=\; \hat{G}^{-1}_\alpha(t)|0\rangle\,, \tag{21}$$

and we call $|0(\alpha,t)\rangle$ the *transformed vacuum state*, satisfying

$$\left(\hat{Q}(t;\alpha) \;+\; i\hat{P}(t;\alpha)\right)|0(\alpha,t)\rangle \;=\; 0\,. \tag{22}$$

It can be proved [16] that the correlation function

$$i\mathcal{G}_{\beta 0}\left(t'-t\right) \;=\; \langle 0(\beta,t)|T\left[\hat{Q}(t')\hat{Q}(t)\right]|0(\beta,t)\rangle\,, \tag{23}$$

is indeed a Green's function. The aim is now to derive a simple formula, similar to Eq. (13), which will allow to evaluate Eq. (23). To this end, we introduce the q–ordering as

$$\mathcal{O}^q\left[\exp\left(iK\left(\hat{P}(t),\hat{Q}(t)\right)\right)\right] \;\equiv\; \sum_{k,l=0}^{\infty} K_{kl}\,\hat{Q}^k(t)\hat{P}^l(t)\,. \tag{24}$$

The latter orders the operator in such a way that all $\hat{Q}(t;\alpha)$'s are on the left and $\hat{P}(t;\alpha)$'s are on the right. Then we define the "classical" q–ordering as

$$\mathcal{O}^q_{cl}\left[\exp\left(iK\left(\hat{P}(t),\hat{Q}(t)\right)\right)\right] \;=\; \frac{\langle Q,t|\mathcal{O}^q\left[\exp\left(iK\left(\hat{P}(t),\hat{Q}(t)\right)\right)\right]|P,t\rangle}{\langle Q,t|P,t\rangle}\,.$$

By following the usual route (see, e.g., Ref [9]) we arrive at

$$\langle 0(\gamma,t_+)|T\left[\hat{Q}(t_n)\ldots\hat{Q}(t_1)\right]|0(\beta,t_-)\rangle \;=\; \langle 0|T\left[\hat{G}_\gamma(t_+)\hat{Q}(t_n)\ldots\hat{Q}(t_1)\hat{G}^\dagger_\beta(t_-)\right]|0\rangle$$

$$= \lim_{\substack{t_f\to+\infty\\ t_i\to-\infty}} \frac{\int_{Q(t_i)=Q_i}^{Q(t_f)=Q_f}\mathcal{D}Q\,\mathcal{D}P\,\mathcal{O}^q_{cl}\left[G_\gamma(t_+)\right]\mathcal{O}^q_{cl}\left[G_{-\beta}(t_-)\right]\,Q(t_n)\ldots Q(t_1)\;e^{iS(P,Q)}}{\int_{Q(t_i)=Q_i}^{Q(t_f)=Q_f}\mathcal{D}Q\,\mathcal{D}P\;e^{iS(P,Q)}}\,. \tag{25}$$

Following Ref. [16], we introduce[2].

$$\mathcal{Z}^{+\,-}_{\gamma\,\beta\,0}[J_Q] = \left\{e^{if(\gamma)K\left(\frac{\delta}{\delta J_P(t_+)},\frac{\delta}{\delta J_Q(t_+)}\right)}\,\mathcal{Z}[J_Q,J_P]\,e^{if(-\beta)K\left(\frac{\overleftarrow{\delta}}{\delta J_P(t_-)},\frac{\overleftarrow{\delta}}{\delta J_Q(t_-)}\right)}\right\}_{J_P=0}, \tag{26}$$

where

$$\mathcal{Z}[J_Q,J_P] \;=\; \lim_{\substack{t_f\to+\infty\\ t_i\to-\infty}}\frac{\langle Q_f,t_f|Q_i,t_i\rangle_{J_Q,J_P}}{\langle Q_f,t_f|Q_i,t_i\rangle_{J_Q,J_P=0}}\,, \tag{27}$$

[2] This is not the most general definition but fully sufficient for our purposes. For more technical details and ensuing discussion see Ref. [16].

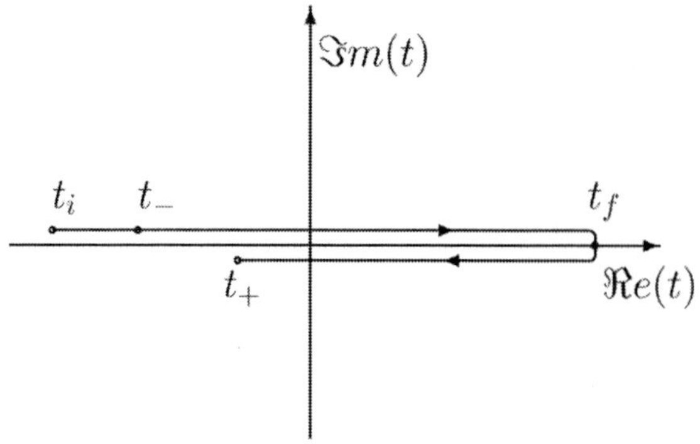

Figure 1. The complex time contour $\mathcal{C}$ in the Schwinger CTP formalism. Dots on the forward and the backward branches of the contour denote discrete time points.

and $f(\alpha)$ would be determined by imposing $\hat{G}_\alpha(t)$ to through the q–ordering as

$$\mathcal{O}^q_{cl}\left[\hat{G}_\alpha(t)\right] = \mathcal{O}^q_{cl}\left[e^{i\alpha K(\hat{P}(t),\hat{Q}(t))}\right] = e^{if(\alpha)\,K(Q(t),P(t))}\,. \tag{28}$$

Eq. (26) defines a *generalized generating functional of Green's functions.* In fact, it can be shown that [16]:

$$\langle 0(\gamma,t_+)|T\left[\hat{Q}(t_n)\ldots\hat{Q}(t_1)\right]|0(\beta,t_-)\rangle = \left\{(-i)^n\frac{\delta^n}{\delta J_Q(t_n)\ldots\delta J_Q(t_1)}\mathcal{Z}^{+-}_{\gamma\beta 0}[J_Q]\right\}_{J_Q=0}, \tag{29}$$

where $t_+ > \{t_i\}_{i=1,\ldots,n} > t_-$. At the end one can consider the limit $t_+ \to t_- \to t$:

$$\langle 0(\gamma,t)|T\left[\hat{Q}(t_n)\ldots\hat{Q}(t_1)\right]|0(\beta,t)\rangle = \lim_{t_+\to t_-=t}\left\{(-i)^n\frac{\delta^n}{\delta J_Q(t_n)\ldots\delta J_Q(t_1)}\mathcal{Z}^{+-}_{\gamma\beta 0}[J_Q]\right\}_{J_Q=0}. \tag{30}$$

This procedure is formally identical to the so-called Schwinger closed-time-path integral [17], namely

$$\langle 0(\beta,t)|T\left[\hat{Q}(t_n)\ldots\hat{Q}(t_1)\right]|0(\beta,t)\rangle = \langle 0|T_{\mathcal{C}}\left[\hat{G}_\beta(t_+)\hat{Q}(t_n)\ldots\hat{Q}(t_1)\hat{G}^\dagger_\beta(t_-)\right]|0\rangle\Big|_{t_+\to t_-=t}$$

$$= \lim_{\substack{t_f\to+\infty\\ t_i\to-\infty}}\frac{\int_{\mathcal{C}\;Q(t_i)=Q_i}^{Q(t_f)=Q_f}\mathcal{D}Q\,\mathcal{D}P\,\mathcal{O}^q_{cl}\left[G_\beta(t_+)\right]\mathcal{O}^q_{cl}\left[G_{-\beta}(t_-)\right]Q(t_n)\ldots Q(t_1)\;e^{iS(P,Q)}\Big|_{t_+\to t_-=t}}{\int_{Q(t_i)=Q_i}^{Q(t_f)=Q_f}\mathcal{D}Q(\alpha)\,\mathcal{D}P(\alpha)\;e^{iS(P,Q)}}, \tag{31}$$

where the contour $\mathcal{C}$ is shown in Figure 1.

Finally, the Green's function (23) can be now easily evaluated:

$$\mathcal{G}_{\beta 0}(t_2-t_1) = \lim_{t_+\to t_-=t}\left\{(-i)^2\frac{\delta^2}{\delta J_Q(t_2)\delta J_Q(t_1)}\mathcal{Z}^{+-}_{\beta\beta 0}[J_Q]\right\}_{J_Q=0}. \tag{32}$$

4. Coherent states PIs as generalized generating functionals

Let us perform the canonical transformation:

$$\hat{P}(t;p\,q) \;=\; \hat{G}^{-1}_{pq}(t)\,\hat{P}(t)\,\hat{G}_{pq}(t)\,, \tag{33}$$

$$\hat{Q}(t;p\,q) \;=\; \hat{G}^{-1}_{pq}(t)\,\hat{Q}(t)\,\hat{G}_{pq}(t)\,, \tag{34}$$

where $\hat{G}_{pq}(t)$ was defined in Eq. (4). From Eqs. (1) and (21), it is thus evident that

$$\langle q_f\,p_f, t_f|q_i\,p_i, t_i\rangle \;=\; \langle 0(q_f\,p_f, t_f)|0(q_i\,p_i, t_i)\rangle\,. \tag{35}$$

In the present case, Eq. (26) reduces to

$$\mathcal{Z}_{f\,i}[J_Q] = \left\{ e^{q_f \frac{\delta}{\delta J_P(t_f)}}\, e^{-p_f \frac{\delta}{\delta J_Q(t_f)}}\, \mathcal{Z}[J_P, J_Q] e^{p_i \frac{\overleftarrow{\delta}}{\delta J_Q(t_i)}}\, e^{-q_i \frac{\overleftarrow{\delta}}{\delta J_P(t_i)}} \right\}_{J_P=0}\,, \tag{36}$$

where, in order to simplify the notation, we introduced $\mathcal{Z}_{f\,i}[J_Q] \equiv \mathcal{Z}^{+-}_{p_f q_f\, p_i q_i\, 0}[J_Q]$. Hence, from Eqs. (30) and (35) follows that

$$\langle q_f\,p_f, t_f|q_i\,p_i, t_i\rangle \;=\; \mathcal{Z}_{fi}[0]\,. \tag{37}$$

That is the first result we were looking for. On the LHS we have the CS kernel Eq. (8) that can be evaluated as CS PI. On the RHS we have a GGF, which can be evaluated as a PS PI (see Eq. (25)). Moreover, when applying Eq. (29) to the present case, we get

$$\langle q_f\,p_f, t_f|T\left[\hat{Q}(t_n)\ldots\hat{Q}(t_1)\right]|q_i\,p_i, t_i\rangle \;=\; \left\{(-i)^n \frac{\delta^n}{\delta J_Q(t_n)\ldots\delta J_Q(t_1)}\mathcal{Z}_{fi}[J]\right\}_{J_Q=0}\,. \tag{38}$$

This means that

$$\mathcal{Z}_{fi}[J] = \sum_{n=0}^{\infty}\frac{i^n}{n!}\int\cdots\int \mathrm{d}t_1\ldots\mathrm{d}t_n\,\langle q_f\,p_f, t_f|T\left[\hat{Q}(t_n)\ldots\hat{Q}(t_1)\right]|q_i\,p_i, t_i\rangle\, J_Q(t_1)\ldots J_Q(t_n)\,. \tag{39}$$

Because the currents are arbitrary, we can compare Eq. (14) with Eq. (39), deducing that

$$\langle q_f\,p_f, t_f|q_i\,p_i, t_i\rangle_J \;=\; \mathcal{Z}_{fi}[J]\,, \tag{40}$$

which completes our result.

Let us now consider, for example, the n–point correlation functions

$$i\mathcal{G}^n_{qp}(t_1,\ldots,t_n) \;=\; \langle q\,p, t|T\left[\hat{Q}(t_n)\ldots\hat{Q}(t_1)\right]|q\,p, t\rangle\,. \tag{41}$$

These can be evaluated (see Eq. (30)) as

$$i\mathcal{G}^n_{qp}(t_1,\ldots,t_n) \;=\; \lim_{t_f\to t_i=t}(-i)^n\left\{\frac{\delta^n}{\delta J_Q(t_n)\ldots J_Q(t_1)}\mathcal{Z}_{fi}[J_Q]\right\}_{J_Q=0}\,. \tag{42}$$

We treat the explicit case of a LHO (see Eq. (3)). The generating functional Eq. (27) then reads [9]:

$$\begin{aligned}\mathcal{Z}[J_Q, J_P] \;=\;& e^{-\frac{i}{2}\int_{-\infty}^{+\infty}\mathrm{d}\tau\int_{-\infty}^{+\infty}\mathrm{d}\tau' J_Q(\tau)\mathcal{G}_0(\tau-\tau')J_Q(\tau')} \\ &\times\, e^{-i\int_{-\infty}^{+\infty}\mathrm{d}\tau\int_{-\infty}^{+\infty}\mathrm{d}\tau' J_Q(\tau)\partial_{\tau'}\mathcal{G}_0(\tau-\tau')J_P(\tau') - \frac{i}{2}\int_{-\infty}^{+\infty}\mathrm{d}\tau\int_{-\infty}^{+\infty}\mathrm{d}\tau' J_P(\tau)\mathcal{G}_0(\tau-\tau')J_P(\tau')}\,,\end{aligned} \tag{43}$$

where $\mathcal{G}_0(\tau - \tau')$ is the Feynman propagator:

$$i\mathcal{G}_0\left(t'-t\right) \equiv i\mathcal{G}_{00}\left(t'-t\right) = \langle 0|T\left[\hat{Q}(t')\hat{Q}(t)\right]|0\rangle . \tag{44}$$

By using the well known formula

$$\exp\left(\alpha\frac{\delta}{\delta J(x)}\right) F[J] = F[J+\alpha_x], \tag{45}$$

with $\alpha_x = \alpha\,\delta(x-y)$, we arrive at

$$\mathcal{Z}_{fi}[J_Q] = e^{-\frac{i}{2}\int_{-\infty}^{+\infty}\mathrm{d}\tau\int_{-\infty}^{+\infty}\mathrm{d}\tau' J_Q(\tau)\mathcal{G}_0(\tau-\tau')J_Q(\tau')}\, e^{-i\int_{-\infty}^{+\infty}\mathrm{d}\tau J_Q(\tau)\mathcal{F}(\tau;t_f,t_i)}\,\mathcal{Z}_{fi}[0], \tag{46}$$

where

$$\mathcal{F}(\tau;t_f,t_i) \equiv p_i\mathcal{G}_0(\tau-t_i) - p_f\mathcal{G}_0(\tau-t_f) + q_f\partial_{t_f}\mathcal{G}_0(\tau-t_f) - q_i\partial_{t_i}\mathcal{G}_0(\tau-t_i), \tag{47}$$

and $\mathcal{Z}_{fi}[0]$ is nothing but the coherent state kernel of the LHO:

$$\begin{aligned}\mathcal{Z}_{fi}[0] &= e^{-\frac{1}{4}(q_i^2+q_f^2)+\frac{1}{2}q_f q_i e^{-i(t_f-t_i)}}\, e^{-\frac{1}{4}(p_i^2+p_f^2)+\frac{1}{2}p_f p_i e^{-i(t_f-t_i)}} \\ &\times\, e^{\frac{i}{2}\left(p_i q_f e^{-i(t_f-t_i)} - p_f q_i e^{-i(t_f-t_i)}\right)-\frac{i}{2}(p_i q_i - p_f q_f)} .\end{aligned} \tag{48}$$

In particular, performing the limit $t_f \to t_i$, we get the well known result [12]:

$$\lim_{t_f\to t_i}\mathcal{Z}_{fi}[0] = \langle q_f\,p_f|q_i\,p_i\rangle = e^{-\frac{1}{4}\left[(q_i-q_f)^2+(p_i-p_f)^2\right]}\, e^{\frac{i}{2}(p_f+p_i)(q_f-q_i)} . \tag{49}$$

Passing to one-point correlation functions we get, by using Eq. (42),

$$\begin{aligned}\langle q_f\,p_f,t_f|\hat{Q}(t)|q_i\,p_i,t_i\rangle &= \frac{1}{2}\left\{i\left[p_i e^{-i(t-t_i)} - p_f e^{i(t-t_f)}\right] + \left[q_i e^{-i(t-t_i)} + q_f e^{i(t-t_f)}\right]\right\} \\ &\times\, \langle q_f\,p_f,t_f|q_i\,p_i,t_i\rangle .\end{aligned} \tag{50}$$

Performing the limit $t_f \to t_i = t$, we get the integral kernel of the operator $\hat{Q}$

$$\langle q_f\,p_f,t|\hat{Q}(t)|q_i\,p_i,t\rangle = \frac{1}{2}\left[i\,(p_i-p_f) + q_i + q_f\right]\langle q_f\,p_f|q_i\,p_i\rangle . \tag{51}$$

This coincides with the result in Ref. [12]. Moreover, one can now easily derive Eq. (5) by considering diagonal elements.

Passing to two-point functions, we get

$$\langle q_f\,p_f,t_f|T\left[\hat{Q}(t_2)\hat{Q}(t_1)\right]|q_i\,p_i,t_i\rangle = \left[i\mathcal{G}_0(t_2-t_1) + \mathcal{F}(t_2;t_f,t_i)\mathcal{F}(t_1;t_f,t_i)\right]\mathcal{Z}_{fi}[0] . \tag{52}$$

In particular, if $t_f \to t_i = t = t_1$, $t_2 = t'$, $p_f = p_i = p$ and $q_f = q_i = q$ we get the Green's function (32):

$$i\mathcal{G}_{p\,q\,0}\left(t'-t\right) = i\mathcal{G}_0(t_2-t_1) + q\left[p\sin\left(t'-t\right) + q\cos\left(t'-t\right)\right] . \tag{53}$$

Note that, if $q = p = 0$ we can recover the standard Feynman propagator Eq. (44) and

$$\mathcal{H}\left(t'-t;p,q\right) = q\left[p\sin\left(t'-t\right) + q\cos\left(t'-t\right)\right] , \tag{54}$$

is a solution of the homogeneous equation

$$\left(\partial_t^2 + 1\right)\mathcal{H}\left(t'-t;p,q\right) = \left(\partial_{t'}^2 + 1\right)\mathcal{H}\left(t'-t;p,q\right) = 0 . \tag{55}$$

Therefore, the result of Eq. (53) differs from the Feynman propagator Eq. (44) just by the choice of boundary conditions.

IOP Conf. Series: Journal of Physics: Conf. Series **965** (2018) 012008 doi:10.1088/1742-6596/965/1/012008

5. Conclusions

In this paper we have proved that coherent-state path integrals can be viewed as GGF's (see Eqs. (37) and (40)). As an illustrative toy model system we considered the LHO and performed the ensuing computations of the 1-point and 2-point mixed-representation correlation functions. All results obtained proved to be in agreement with known results from the literature.

Our present analysis focused only on canonical (Glauber) coherent states. It would be clearly interesting to extend our study to other coherent states such as Perelomov group-related CS's [20] and Barut–Girardello CS's [21]. Particularly, Perelomov's CS's are an attractive option in view of their relevance in QFT [22, 23].

Finally, the issue of boundary conditions, touched upon in Section 4, will be discussed in more detail in our future work.

Acknowledgments

P.J. was supported by the Czech Science Foundation Grant No. 17-33812L

References

[1] Schrödinger E 1926 *Naturwissenschaften* **14** 664
[2] von Neumann J 1932 *Mathematatische Grundlagen der Quantenmechanik* (Berlin: Springer)
[3] Klauder J R 1960 *Ann. Phys.* **11** 123
[4] Glauber R J 1963 *Phys. Rev.* **130** 2529
[5] Glauber R J 1963 *Phys. Rev.* **131** 2766
[6] Klauder J R and Skagerstam B *Coherent States-Applications in Physics and in Mathematical Physics* (New York: World Scientific)
[7] Schweber S S (1962) *J. Math. Phys.* **3** 831
[8] Schulman L S 1985 *Techniques and Applications of Path Integration* (New York: John Wiley and Sons)
[9] Kleinert H 2009 *Path Integrals in Quantum Mechanics, Statistics, Polymer Physics, and Financial Markets* (London: World Scientific)
[10] De Facio B, Hammer C L and Shrauner J E 1978 *Phys. Rev.* D **18** 373
[11] Faddeev L D and Slavnov A A 1980 *Gauge Fields-Introduction to Quantum Theory* (New York: The Benjamin Cummings Publishing Company)
[12] Klauder J R 2011 *A Modern Approach to Functional Integration* (London: Birkhäuser)
[13] Heller E J 1977 *Journ. Chem. Phys.* **66** 5777
[14] Weissman Y 1982 *Journ. Chem. Phys.* **76** 4067
[15] Blasone M, Jizba P and Smaldone L 2017 *Ann. Phys.* **383** 207–238; *J. Phys: Conf. Ser.* **804** 012006
[16] Blasone M, Jizba P and Smaldone L 2017 eprint arXiv:1709.00461 [hep-th]
[17] Schwinger J 1961 J. Math. Phys. **2** 407
[18] Blasone M, Jizba P and Smaldone L 2017 *J. Phys: Conf. Ser.* **880** 012051
[19] Blasone M and Vitiello 1995 *Ann. Phys.* **244**, 283 ; Blasone M, Capolupo A, Romei O and Vitiello G 2001 Phys. Rev. D **63** 125015
[20] Perelomov A 1986 *Generalized Coherent states and Their Applications* (Berlin: Springer-Verlag)
[21] Barut A O and Girardello L 1971 *Commun. Math. Phys.* **21** 41
[22] Blasone M and Jizba P 2012 *J. Phys.* A *Math. Theor.* **45** 244009
[23] Blasone M, Jizba P and Vitiello G 2011 *Quantum Field Theory and its Macroscopic Manifestations* (London: World Scientific & ICP)

IOP Conf. Series: Journal of Physics: Conf. Series **965** (2018) 012009 doi:10.1088/1742-6596/965/1/012009

Hurwitz Algebras and the Octonion Algebra

Čestmir Burdik[1,2,†], **Sultan Catto**[3,4,††],

[1]Department of Mathematics, Faculty of Nuclear Sciences and Physical Engineering, Czech Technical University, Prague, Trojanova 13, CZ-120 00 Czech Republic
[2]Bogoliubov Laboratory of Theoretical Physics, JINR, 141980 Dubna, Russia
[3]Physics Department, The Graduate School, City University of New York, New York, NY 10016-4309
[4]Theoretical Physics Group, Rockefeller University, 1230 York Avenue, New York, NY 10021-6399

[†]Grant No. SG15/215/OHK4/3T/14 Czech Technical University in Prague
[††] Work supported in DOE contracts No. DE-AC-0276 ER 03074 and 03075; and NSF Grant No. DMS-8917754.

Abstract.
We explore some consequences of a theory of internal symmetries for elementary particles constructed on exceptional quantum mechanical spaces based on Jordan algebra formulation that admit exceptional groups as gauge groups.

1. Introduction

Soon after the discovery of Quantum Mechanics which brought miraculous light and established law and order in the hitherto mysterious world of atomic physics, physicists were confronted with the puzzling phenomena of the nuclear frontier. This happened in the early thirties with the emergence of the neutron, the riddle of beta decay and Heisenberg's concept of isotopic spin to describe the symmetrical proton-neutron doublet. Emboldened with the striking success of radical ideas which had replaced classical mechanics by quantum mechanics, some of the founding fathers of Quantum Theory were already looking beyond its boundaries to an even more revolutionary form of mechanics, hoping that this new structure would be more appropriate to the description of the nuclear world. Now to go from classical to Quantum Mechanics we represent observables by hermitian matrices (which can be infinite) which obey an associative but non-commutative algebra. The elements of these matrices are in general complex numbers. Thus we drop both the real numbers and the commutativity of the algebra of observables in the transition from classical to quantum theory. What further algebraic properties can we change if we want to construct an "ultra mechanics"? Clearly we can drop associativity and replace complex numbers by the two remaining normed algebras of quaternions and octonions.

Now, in a reformulation of Quantum Mechanics, Jordan introduced an algebra of observables in which the product (anticommutator) leads again to an observable. This is the Jordan product, which is symmetrical but not associative. In this formulation states are not represented by column matrices in Hilbert space but by projection operators which are again Hermitian matrices but are also idempotent.

IOP Conf. Series: Journal of Physics: Conf. Series **965** (2018) 012009 doi:10.1088/1742-6596/965/1/012009

Thus we can trade associativity for commutativity and still remain within the Quantum Mechanical framework.

In 1934, Jordan, von Neumann and Wigner[?] put the algebra of observables on an axiomatic basis and found solutions which are not complex matrices. In fact they are 3×3 octonionic hermitian matrices. Hence these observables correspond to a new mechanics with finite degrees of freedom. Such observables were called exceptional observables by the authors who hoped they would have some relevance to nuclear physics.

Thanks to a wealth of new experimental discoveries Nuclear Theory developed in the late thirties along the lines of conventional quantum mechanics. The key ideas were the Fermi theory of the beta decay using Pauli's neutrinos and Yukawa's idea of new bosons (mesons) mediating strong interactions. Hence the exceptional observables associated with exceptional Jordan algebras dropped out of physics to be captured, studied and developed further by mathematicians.

Meanwhile the theory of Lie groups as formulated by Cartan was taking new shapes. It was realized that the three infinite series of semi simple classical groups, namely the orthogonal groups (B_r and C_r), the unitary groups (A_r) and the symplectic groups (C_r) were respectively associated with real, complex and quaternionic numbers. In fact they arise as automorphism groups of Jordan algebras over these three Hurwitz algebras. On the other hand, it was already known to Cartan that the exceptional group G_2 is the automorphism group of the non-associative octonion algebra. We had to wait until 1949 before Chevalley and Schafer[2] realized that the exceptional group F_4 is the automorphism group of the Jordan exceptional algebra of 3×3 hermitian octonionic matrices. The same year Jordan[3] showed that the homogeneous space $F_4/SO(9)$ coincides with the projective Moufang plane where a point is represented by a pair of octonions.

These discoveries initiated a flurry of activity by Borel, Rozenfeld, Freudenthal, Springer and others, culminating in the work of Tits[4] who gave a unified treatment of exceptional groups and some of their subgroups that fit in Freudenthal's magic square. There followed a unification of the new projective geometries which generalize the Moufang plane and have non-Desarguesian properties. Since there is a correspondence between projective geometries and ray representations of physical states in Hilbert spaces, (Weyl, von Neumann, Varadarajan) the new geometries associated with exceptional groups provided new exceptional Quantum Mechanical spaces extending the Jordan-von Neumann-Wigner (JNW) construction and combining exceptional with ordinary observables in well determined patterns.

During the decades (from the fifties to the seventies) in which the new geometries were developed, a parallel explosion of modern particle physics took place. Besides the electric charge and isotopic spin of the pre-war years, the space of internal quantum numbers expanded to include strangeness, which combined with isospin within an $SU(3)$ group, and charm which extended $SU(3)$ to $SU(4)$. These groups of the strong interactions had to be broken spontaneously. A field theory of strong interactions based on elementary quarks necessitated yet a new and a fundamental exact $SU(3)$ group, namely the color group $SU(3)^c$ (Greenberg, Nambu, Gell Mann)[5]. On the other hand a unification of weak and electromagnetic interactions had to be based on $SU(2) \times U(1)$ (the Weinberg-Salam group) which acted differently on left handed fermions (leptons and quarks), and right handed fermions. Thus a unified theory of fundamental interactions (excluding gravitation) had to be reformulated in a large but finite charge space in which the invariance group would have at least rank five. A local, spontaneously broken gauge field theory based on such a group would be a good candidate for a fundamental theory of elementary particles.

At this point we are naturally led to the speculative identification of the phenomenological finite space of quantum numbers with the exceptional quantum mechanical spaces associated with octonions. The simplest is the space of JNW exceptional observables with the associated invariance group F_4, and the others are the geometrical spaces related to the groups E_6, E_7 and E_8. If this were so we would not only get a natural explanation of the color group $SU^c(3)$, but we would also be in a position to predict the nature of the "flavor groups" like $SU(3)$ of Gell-Mann and Neeman. Finally the new non-Desarguesian geometries would be realized in nature as the Hilbert spaces of generalized charges. So far this view has not been refuted by experiment. Hence we can take advantage of this possibility and explore the consequences of a theory of internal symmetries for elementary particles based on exceptional quantum mechanical spaces that admit exceptional groups as gauge groups.

Before this view was proposed, exceptional groups have made only sporadic appearances on the stage of physics. Racah introduced the group G_2 in connection with the level degeneracies of rare earth atoms. G_2 has also been proposed as a possibility for the symmetry of strong interactions before the supremacy of $SU(3)$ was established. Gamba pointed out the $SU(3)$ structure of the 3×3 Jordan algebras without mentioning the groups F_4 and E_6 with which they are associated.

2. Speculation concerning the origin of internal symmetries of quarks and leptons

There are some unique exceptional geometries which are connected with octonions and correspond to a finite number of degrees of freedom that cannot be extended. This is the new Hilbert space that we want tentatively to identify with the Hilbert space of internal symmetries carrying color and flavor quantum numbers.

One of the fundamental questions in particle physics is the understanding of the quark substructure of hadrons. We would like to understand the emergence of the hierarchy of interactions, the mass spectrum of quarks and leptons and the wider mass spread of the fundamental gauge bosons through the spontaneous breaking of a local symmetry based on a unifying group or supergroup associated with a gauge field theory.

The concept of a field theory that unifies both the various kinds of interactions and the great variety of symmetries of fundamental processes is philosophically and aesthetically more satisfying than the old division of labor, concepts and methods in the main areas of physics; geometry for gravitation, field theory for electromagnetism, effective field theory for weak interactions and S-matrix methods (based on analyticity and unitarity) for strong interactions. The only trouble with the new natural philosophy is that it deals largely with an unobservable substructure (quarks, color quantum numbers, superheavy gauge bosons, essentially undetectable gravitinos, etc.) which makes physicists rather uncomfortable and reminds them of their imponderable past (the caloric fluid, the aether) ridiculed by later generations.

The confrontation of the new theory with experiment is by its very nature indirect, in contrast to the foundations of the twentieth century physics (special relativity and quantum mechanics) that had their roots in the principle of dealing only with the observables. We are now back to a view more akin to the one held by Einstein in his old age, spinning out geometrical structures far removed from space-time geometry based on equivalence principle, trying to unify gravity with other fields that seemed to have non-geometrical properties. It is ironical that Heisenberg also in his old age rejected his positivist S-matrix approach in favor of underlying (non-linear but non-renormalizable) field theory for the unification of electromagnetism and strong interactions.

In defense of the new philosophy it might be said that other past substructures that were regarded imponderable or too abstract by contemporary physicists did prevail in the end and were accepted as physical reality. Atoms, fields (both electromagnetic and gravitational), antimatter, the neutrino, the particle-wave dual nature of matter and radiation, Hilbert space, the complex wave function, etc., might provide a few examples.

In this note we shall move to a still more abstract plane, in fact the octonionic planes that may provide the geometrical foundation for the existence of internal symmetries like color and flavor. These octonionic geometries allow us to construct new finite Hilbert spaces with unique properties. The non-Desargues'ian geometric property makes them non-embeddeble in higher spaces, hence essentially finite. It also leads to peculiarities in the superposition principle in the color sector of the Hilbert space. This theory of the charge space, if correct, suggests a new geometric picture for the substructure of the material world, that of an octonionic geometry attached at each point of Einstein's Riemannian manifold for space time with local symmetry that leaves the properties of charge space invariant.

There are deep connections between the lattices of certain groups that arise in superstrings and

IOP Conf. Series: Journal of Physics: Conf. Series **965** (2018) 012009 doi:10.1088/1742-6596/965/1/012009

discrete elements of Jordan algebras associated with the magic square. Such correspondence suggests a profound role that division algebras and exceptional Jordan algebras are likely to play in superstring theories and their compactification. It also points out to the existence of a larger theory connected with a chiral 27 dimensional lattice that generalizes Conway's lattice and unifies all superstring theories. There is the possibility that the presently known superstring theories may be special cases of a unique theory associated with the ultra-hyperbolic group E_∞ that admits Conway's lattice as a root system. We shall deal with this issue extensively in another publication.

Exceptional Lie algebras are related to the exceptional Jordan algebra of 3×3 hermitian matrices with octonionic entries. This relation yields descriptions of certain real forms of the complex Lie algebras F_4, E_6 and E_7 which can be interpreted as octonionic versions of Lie algebras of, respectively, antihermitian 3×3 matrices, traceless 3×3 matrices and symplectic 6×6 matrices. Color can combine with flavors within a semi-simple Lie groups if the color $SU(3)$ group is identified with an octonion automorphism. The groups F_4, E_6, E_7 and E_8 and some of their subgroups fitting in Freudenthal's[8] Magic Square can be obtained in a unified way by combining the subalgebras of the octonion algebra, namely the four Hurwitz algebras with quadratic norm with the subalgebras of the exceptional Jordan algebra by way of Tits' construction[4]. For a general review we refer the reader to Gürsey and Tze's book.[9] Also recently Barton and Sudbery[10],[11] gave an alternative construction of the Magic Square showing split composition algebras lead to analogues of matrix Lie algebras $SU(3)$, $Sl(3)$ and $Sp(6)$.

3. Hurwitz algebras and the octonion algebra

Let $\alpha=(a_1,a_2)$, $\beta=(b_1,b_2)$ such that

$$\sqrt{a_1\bar{a}_1}=N(a_1)\geq 0,\quad N(ab)=N(a)N(b) \tag{1}$$

$$\bar{\alpha}=(\bar{a}_1,-a_2), \tag{2}$$

and associator

$$(ab)c-a(bc)=0 \tag{3}$$

Then with the Cayley-Dickson process we define a new normed algebra

$$\alpha\beta=(a_1,a_2)(b_1,b_2)=(a_1b_1-b_2\bar{a}_2,\bar{a}_1b_2+b_1a_2) \tag{4}$$

Thus

$$\alpha\bar{\alpha}=(a_1,a_2)(\bar{a}_1,-a_2)=(a_1\bar{a}_1+a_2\bar{a}_2,-\bar{a}_1a_2+\bar{a}_1a_2)=(a_1\bar{a}_1+a_2\bar{a}_2,0) \tag{5}$$

The unit is $e_0=(1,0)$, so that

$$\alpha\bar{\alpha}=(a_1\bar{a}_1+a_2\bar{a}_2)e_0 \tag{6}$$

and the inverse is

$$\alpha^{-1}=(\alpha\bar{\alpha})^{-1}\bar{\alpha}, \tag{7}$$

so that $\alpha^{-1}\alpha=\alpha\bar{\alpha}^{-1}=e_0$.

(a) Complex numbers:

Take $(0,1)=i$

$$\alpha=a_1+ia_2=a_1(1,0)+a_2(0,1), \tag{8}$$

$$\bar{\alpha}=a_1-ia_2. \tag{9}$$

Here the field A so that $a_1,a_2\in A$ is commutative and associative. So is the field $\text{Æ}=(a_1,a_2)$.

(b) Quaternions:

Writing

$$(i, 0) = j_1, \quad (0, 1) = j_2, \quad (0, -i) = j_3 \tag{10}$$

so that

$$j_1 j_2 = (i.0 - 1.0, -i.1 + 0.1) = (0, -i) = j_3, \tag{11}$$

similarly

$$j_2 j_1 = (0, 1)(i, 0) = (0, i) = -j_3. \tag{12}$$

Quaternions are associative but not commutative.

(c) Octonions

Writing

$$(j_i, 0) = e_i, \quad (0, 1) = e_7, \quad (0, j_i) = e_{i+3}, \quad (i = 1, 2, 3) \tag{13}$$

we get, for example

$$e_1 e_7 = (j_1, 0)(0, 1) = (0, \bar{j}_1.1) = (0, -j_1) = -e_4, \tag{14}$$

$$e_7 e_1 = (0, 1)(j_1, 0) = (0, j_1) = e_4 \tag{15}$$

and

$$(j_1, 0)(j_2, 0) = (j_1 j_2, 0) = j_3. \tag{16}$$

Thus we can write

$$e_a e_b = -\delta_{ab} + f_{abc} e_c \tag{17}$$

with the involution

$$\bar{e}_a = -e_a. \tag{18}$$

The tensor f_{abc} takes values $1, 0, -1$ and is antisymmetrical in all three indices a, b, c, taking the value 1 for $(a, b, c) = (1, 2, 3)$ and its cyclic permutations with $e_1, e_2, e_4, e_6, e_5, e_7$ arranged clockwise on a circle shown in our previous paper[12],

These values correspond to the successive columns of the seven dimensional finite projective plane with the standard arrangement

$$\begin{matrix} 1 & 2 & 4 & 3 & 6 & 5 & 7 \\ 2 & 4 & 3 & 6 & 5 & 7 & 1 \\ 3 & 6 & 5 & 7 & 1 & 2 & 4 \end{matrix} \tag{19}$$

In geometrical language each column represent lines with 3 points on them so that we have seven lines (seven triangles on the circle) and 7 points. The same number always occurs in 3 different columns so that 3 lines pass through each point.

The corresponding algebra is non associative as the associator

$$[e_a, e_b, e_c] = 2\widetilde{f}_{abcd} e_d \tag{20}$$

where

$$\widetilde{f}_{abcd} = \frac{1}{3!} \epsilon_{abcdklm} f_{klm} \tag{21}$$

is the dual tensor for f_{abc} of the multiplication table, ϵ being the completely anti symmetrical Levi-Civita symbol in 7 dimensions. Due to alternativity the octonions form a power associative algebra.

Defining the conjugates

$$\bar{e}_\alpha = -e_\alpha, \quad \bar{e}_0 = e_0 \tag{22}$$

the octonionic conjugate $\bar{x}$ is written as

$$\bar{x} = x_0 - x_\alpha e_\alpha = x_i \bar{e}_i \tag{23}$$

IOP Conf. Series: Journal of Physics: Conf. Series **965** (2018) 012009 doi:10.1088/1742-6596/965/1/012009

The scalar part of the octonion ($Sc(x)$) and its vector part ($Vec(x)$) are

$$Sc(x) = \frac{1}{2}(x + \bar{x}) = x_0 \tag{24}$$

and

$$Vec(x) = \frac{1}{2}(x - \bar{x}) = x_\alpha e_\alpha \tag{25}$$

Conjugate of the product of two octonions x and y is

$$(\overline{xy}) = \bar{y}\bar{x} \tag{26}$$

and their scalar product is defined by

$$< x, y >= x_i y_i = \frac{1}{2}(x\bar{y} + y\bar{x}) = \frac{1}{2}(\bar{x}y + \bar{y}x) \tag{27}$$

which, in terms of octonion units gives

$$< e_i, e_j >= \frac{1}{2}(\bar{e}_i e_j + \bar{e}_j e_i) = \frac{1}{2}(e_i\bar{e}_j + e_j\bar{e}_i) = \delta_{ij} \tag{28}$$

The norm $N(x)$ of an octonion is

$$N(x) = \bar{x}x = x\bar{x} = x_i x_i \tag{29}$$

and is zero if $x = 0$, and is always positive otherwise. For a nonzero octonion x, its inverse is given by

$$x^{-1} = \frac{\bar{x}}{N(x)} \tag{30}$$

with

$$(xy)^{-1} = y^{-1}x^{-1} \tag{31}$$

The norm defined above satisfies

$$N(xy) = N(x)N(y) \tag{32}$$

In analogous way to the quaternionic case in R^4,we now make the following definitions of vectorial products of octonion units, namely two new antisymmetric tensors:

$$e_{ij} = \frac{1}{2}(\bar{e}_i e_j - \bar{e}_j e_i) \tag{33}$$

and

$$e'_{ij} = \frac{1}{2}(e_i\bar{e}_j - e_j\bar{e}_i) \tag{34}$$

Componentwise they read as

$$e_{\alpha\beta} = e'_{\alpha\beta} = -\epsilon_{\alpha\beta\gamma}e_\gamma \tag{35}$$

and

$$e_{0\alpha} = -e'_{0\alpha} = e_\alpha \tag{36}$$

These octonionic tensors naturally enter into covariant formulation of various cross-products in R^8. We can now write

$$\bar{e}_i e_j = \frac{1}{2}(\bar{e}_i e_j + \bar{e}_j e_i) + \frac{1}{2}(\bar{e}_i e_j - \bar{e}_j e_i) = \delta_{ij} + e_{ij} \tag{37}$$

$$e_i\bar{e}_j = \frac{1}{2}(e_i\bar{e}_j + e_j\bar{e}_i) + \frac{1}{2}(e_i\bar{e}_j - e_j\bar{e}_i) = \delta_{ij} + e'_{ij} \tag{38}$$

Using the definition of commutator of two octonions

$$[x, y] = 0 \tag{39}$$

the product rule implies

$$[e_\alpha, e_\beta] = 2\epsilon_{\alpha\beta\gamma}e_\gamma \tag{40}$$

and

$$[e_0, e_\alpha] = 0 \tag{41}$$

so that

$$[x, y] = 2x_\alpha y_\beta \epsilon_{\alpha\beta\gamma}e_\gamma \tag{42}$$

We see the relation between the commutator and e_{ij} and e'_{ij} through the triality relation

$$[e_i, e_j] + e_{ij} + e'_{ij} = 0 \tag{43}$$

We now define the associator $[x, y, z]$ of three octonions by

$$[x, y, z] = (xy)z - x(yz) \tag{44}$$

which is completely antisymmetric:

$$[x, y, z] = [y, z, x] = [z, x, y] \tag{45}$$

and

$$[x, y, z] = -[y, x, z] = -[x, z, y] = -[z, y, x] \tag{46}$$

It is also purely vectorial since

$$\overline{[x, y, z]} = -[x, y, z] \tag{47}$$

Since e_0 commutes and associates with other octonion units, only the purely vectorial parts of x, y, z contribute to the associator $[x, y, z]$.

We now define a completely antisymmetric 4-index object $\psi_{\alpha\beta\mu\nu}$ related to the associator as

$$[e_\alpha, e_\beta, e_\mu] = 2\psi_{\alpha\beta\mu\nu}e_\nu \tag{48}$$

By means of the associator and the product rules, we find

$$\psi_{\alpha\beta\mu\nu} = \frac{1}{2}(\delta_{\beta[\mu}\delta_{\alpha]\nu} + \epsilon_{\beta\gamma[\mu}\epsilon_{\alpha]\gamma\nu}) \tag{49}$$

Explicit calculation of the values of $\psi_{\alpha\beta\mu\nu}$ show it is dual to $\epsilon_{\lambda\sigma\rho}$ in R^7. $\psi_{\alpha\beta\mu\nu}$ take value 1 for the following combinations:

$$(\alpha\beta\mu\nu) = (1346), (2635), (4567), (3751), (6172), (5214), (7423) \tag{50}$$

Duality property between $\epsilon_{\lambda\sigma\rho}$ and $\psi_{\alpha\beta\mu\nu}$ in R^7 is best seen in the following construction:

$$\left\{\begin{matrix} 2 & 4 & 3 & 6 & 5 & 7 & 1 \\ 5 & 7 & 1 & 2 & 4 & 3 & 6 \\ 7 & 1 & 2 & 4 & 3 & 6 & 5 \end{matrix}\right\} = \epsilon_{\lambda\sigma\rho}$$
$$\left\{\begin{matrix} 1 & 2 & 4 & 3 & 6 & 5 & 7 \\ 3 & 6 & 5 & 7 & 1 & 2 & 4 \\ 4 & 3 & 6 & 5 & 7 & 1 & 2 \\ 6 & 5 & 7 & 1 & 2 & 4 & 3 \end{matrix}\right\} = \psi_{\alpha\beta\mu\nu} \tag{51}$$

IOP Conf. Series: Journal of Physics: Conf. Series **965** (2018) 012009 doi:10.1088/1742-6596/965/1/012009

This table is read as

$$\frac{1}{2}[e_2, e_5] = e_7, \quad \frac{1}{2}[e_4, e_7] = e_1, \dots \tag{52}$$

etc., for the triads,and

$$\frac{1}{2}[e_1, e_3, e_4] = -e_6, \quad \frac{1}{2}[e_2, e_6, e_3] = -e_5, \dots \tag{53}$$

etc., for the associators.

We see that there are seven associative planes and their transverse non-associative planes. While an associative plane is closed under commutation, a non-associative plane closes under the associator.

From the definition of the involution for octonions we get

$$w = w_0 + e_a w_a \qquad (a = 1, \dots, 7) \tag{54}$$

$$\bar{w} = w_0 - e_a w_a \tag{55}$$

$$|w|^2 = [N(w)]^2 = w\bar{w} = \bar{w}w = w_0^2 + w_1^2 + \dots + w_7^2 \tag{56}$$

which is positive definite if w_α $(\alpha = 0, \dots, 7)$ are real.

The inverse octonion is

$$w^{-1} = |w|^{-2}\bar{w}. \tag{57}$$

It is also convenient to introduce the scalar part

$$s(w) = \frac{1}{2}(w + \bar{w}) = \frac{1}{2}t(w) = w_0 \tag{58}$$

where $t(w)$ is the trace part. We have the norm property

$$|ab| = |a|\,|b|. \tag{59}$$

Because of non associativity the octonion units can not be represented by matrices.

The following trilinear identities

$$a(ab) = a^2 b, \quad a(ba) = (ab)a, \quad ba^2 = (ba)a \tag{60}$$

which follow from alternativity so that

$$[a, a, b] = [a, b, a] = [b, a, a] = 0 \tag{61}$$

These lead to

$$a(\bar{a}x) = (a\bar{a})x, \quad a(x\bar{a}) = (ax)\bar{a}, \quad (xa)\bar{a} = x(a\bar{a}) \tag{62}$$

so that axa and axa^{-1} have an unambiguos meaning.

The following quadrilinear Moufang identities also hold

$$(xax)b = x[a(xb)] \tag{63}$$

$$b(xax) = [(bx)a]x. \tag{64}$$

$$(xa)(bx) = x(ab)x \tag{65}$$

We also have

$$[\bar{x}a, \bar{x}, \bar{(b)}] = -x[x, a, b] = x[x, \bar{a}, b]. \tag{66}$$

If we continue the Cayley-Dickson process we use alternativity, hence power associativity and the norm property.

We shall also use the split octonionic units by introducing an imaginary unit i that commute with e_a. We define

$$u_0 = \frac{1}{2}(1 + ie_7), \quad u_j = \frac{1}{2}(e_j + ie_{j+3}) \qquad (j = 1, 2, 3) \tag{67}$$

$$u_0^* = \frac{1}{2}(1 - ie_7), \quad u_j^* = \frac{1}{2}(e_j - ie_{j+3}). \tag{68}$$

A real octonion can be written as

$$w = \psi_0^* u_0 + \psi_j^* u_j + \psi_0 u_0^* + \psi_j u_j^* = \psi^* + \psi \tag{69}$$

where

$$\psi = \psi_0 u_0^* + \psi_j u_j^* \tag{70}$$

and

$$\psi_0 = w_0 + iw_7, \quad \psi_j = w_j + iw_{j+3} \tag{71}$$

A complex octonion has the form

$$\xi = \chi + \zeta^* = \chi_0 u_0^* + \chi_j u_j^* + \zeta_0^* u_0 + \zeta_j u_j. \tag{72}$$

Complex octonions, unlike real octonions can have zero norm so that they do not form a division algebra. Nevertheless complex octonions are important in constructing new exceptional quantum mechanical spaces.

The split units obey the following multiplication table:

$$u_i u_j = \epsilon_{ijk} u_k^* \qquad (i, j = 1, 2, 3) \tag{73}$$

$$u_i u_j^* = -\delta_{ij} u_0 \tag{74}$$

$$u_0 u_i = u_i u_0^* = u_i \tag{75}$$

$$u_i u_0 = u_0^* u_i = 0 \tag{76}$$

$$u_o u_0^* = 0 \tag{77}$$

$$u_0^2 = u_0 \tag{78}$$

and the complex conjugate equations obtained by changing the sign of i. We can anticipate from this table the $SU(3)$ structure of the split units.

Now, we put together the compactified multiplication table for the split octonion units:

	u_0	u_0^*	u_k	u_k^*
u_0	u_0	0	u_k	0
u_0^*	0	u_0^*	0	u_k^*
u_j	0	u_j	$\epsilon_{jki} u_i^*$	$-\delta_{jk} u_0$
u_j^*	u_j^*	0	$-\delta_{jk} u_0^*$	$\epsilon_{jki} u_i$

It is worth noting that u_i and u_j^* behave like fermionic annihilation and creation operators:

$$\{u_i, u_j\} = \{u_i^*, u_j^*\} = 0, \quad \{u_i, u_k^*\} = -\delta_{ij} \tag{79}$$

This fermionic Heisenberg algebra shows the three split units u_i to be Grassmann numbers. Being non-associative, these split units give rise to an exceptional Grassmann algebra.

Operators u_i, unlike ordinary fermion operators, are not associative. We also have $\frac{1}{2}[u_i, u_j] = \epsilon_{ijk}\, u_k^*$. The Jacobi identity does not hold since $[u_i, [u_j, u_k]] = -ie_7 \neq 0$, where e_7, anticommute with u_i and u_i^*.

We note that, like the imaginary units e_α, the split units cannot be represented by matrices. Unlike the octonion algebra, the split octonion algebra contains zero divisors and is therefore not a division algebra.

The associators of split octonion units are given below:

$$[u_i, u_j, u_k] = \epsilon_{ijk}(u_0^* - u_0) \tag{80}$$

$$[u_i^*, u_j^*, u_k^*] = \epsilon_{ijk}(u_0 - u_0^*) \tag{81}$$

$$[u_i, u_j, u_0] = -\epsilon_{ijk} u_k^* \tag{82}$$

$$[u_i, u_j, u_0^*] = \epsilon_{ijk} u_k^* \tag{83}$$

$$[u_i, u_j, u_k^*] = \delta_{jk} u_i - \delta_{ik} u_j \tag{84}$$

$$[u_i, u_j^*, u_k^*] = \delta_{ki} u_j^* - \delta_{ij} u_k^* \tag{85}$$

$$[u_i^*, u_j^*, u_0] = \epsilon_{ijk} u_k \tag{86}$$

$$[u_i^*, u_j^*, u_0^*] = -\epsilon_{ijk} u_k \tag{87}$$

$$[u_i, u_j^*, u_0] = 0 \tag{88}$$

$$[u_i, u_j^*, u_0^*] = 0 \tag{89}$$

Defining hermitian conjugation as both complex and octonionic conjugation we write

$$u_i^\dagger = \bar{u}_i^* = -u_i^*, \quad u_0^\dagger = \bar{u}_0^* = u_0 \tag{90}$$

We also make new definitions:

$$u_{\mu\nu} = \frac{1}{2}(u_\mu^\dagger u_\nu - u_\nu^\dagger u_\mu) \tag{91}$$

and

$$u'_{\mu\nu} = \frac{1}{2}(u_\mu u_\nu^\dagger - u_\nu u_\mu^\dagger) \tag{92}$$

and see that the left handed product

$$u'_{\mu\nu} = 0 \tag{93}$$

and, in the component form the right handed product $u'_{\mu\nu}$ survives only as $u_{0i} = \frac{1}{2}e_i$, i.e.

$$u_{ij} = 0 \tag{94}$$

with

$$u_{0i} = \frac{1}{2}(u_i + u_i^*) = \frac{1}{2}e_i \tag{95}$$

thereby reducing the octonions to purely vectorial quaternions.

Now, through $w = \mathcal{R}e\ \psi$ we can associate with each real octonion a special complex octonion ψ so that

$$\psi = \begin{pmatrix}\psi_0\\ \psi_1\\ \psi_2\\ \psi_3\end{pmatrix} \begin{pmatrix}u_0^* & u_1^* & u_2^* & u_3^*\end{pmatrix} = \Psi^T u^* \tag{96}$$

where

$$u = \begin{pmatrix} u_0 \\ u_1 \\ u_2 \\ u_3 \end{pmatrix}, \qquad \Psi = \begin{pmatrix} \psi_0 \\ \psi_1 \\ \psi_2 \\ \psi_3 \end{pmatrix} \tag{97}$$

and T denotes transposition. The multiplication table gives

$$uu^T = \begin{pmatrix} u_0^* \\ u_1^* \\ u_2^* \\ u_3^* \end{pmatrix} \begin{pmatrix} u_0^* & -u_1 & -u_2 & -u_3 \end{pmatrix} = \begin{pmatrix} 0 & -u_1 & -u_2 & -u_3 \\ u_1 & 0 & -u_3^* & u_2^* \\ u_2 & u_3^* & 0 & -u_1^* \\ u_3 & -u_2^* & u_1^* & 0 \end{pmatrix} \tag{98}$$

$$uu^\dagger = u\bar{u}^{*T} = \begin{pmatrix} u_0^* \\ u_1^* \\ u_2^* \\ u_3^* \end{pmatrix} \begin{pmatrix} u_0 & -u_1^* & -u_2^* & -u_3^* \end{pmatrix} = \begin{pmatrix} u_0 & 0 & 0 & 0 \\ 0 & u_0 & 0 & 0 \\ 0 & 0 & u_0 & 0 \\ 0 & 0 & 0 & u_0 \end{pmatrix} \tag{99}$$

$$\psi^2 = (\psi_0 u_0^* + \psi_j u_j^*)(\psi_0 u_0^* + \psi_k u_k^*) \tag{100}$$
$$= \psi_0^2 u_o^* + \psi_j \psi_k \epsilon_{jkl} u_l + \psi_0 \psi_k u_k^* \tag{101}$$
$$= \psi_0^2 u_0^* + \psi_0 \psi_k u_k^* = \psi_0(\psi_0 u_0^* + \psi_k u_k^*) \tag{102}$$

so that

$$\psi^2 = \psi_0 \psi \tag{103}$$

or

$$\frac{\psi^2}{\psi_0^2} = \frac{\psi}{\psi_0}. \tag{104}$$

Thus $\frac{\psi}{\psi_0}$ is idempotent.

Hence a real octonion is characterized by a complex multiple of an idempotent complex octonion.

4. Construction of octonionic Hilbert spaces
(a) Jordan algebra formulation of Quantum Mechanics

Consider finite n-dimensional quantum mechanical spaces.

In the usual formulation of quantum mechanics an observable is represented by a $n \times n$ hermitian matrix over complex numbers

$$\Omega = \Omega^\dagger = \Omega^{*T}. \tag{105}$$

A state is a ket vector with n components $|\alpha >= a$ so that

$$\alpha|\alpha >= a^\dagger a = 1. \tag{106}$$

Another state b is orthogonal to it if

$$< \beta|\alpha >= b^\dagger a = 0 \tag{107}$$

A complete set of states will obey

$$< \alpha_i|\alpha_j >= \delta_{ij} \tag{108}$$

Thus for the state α we can write $\alpha = \alpha_i|\alpha_i >$. We have

$$\Omega = \sum_{i=1}^{n} w_i \, |\alpha_i >< \alpha_i| \tag{109}$$

IOP Conf. Series: Journal of Physics: Conf. Series **965** (2018) 012009 doi:10.1088/1742-6596/965/1/012009

so that

$$\Omega\ |\alpha_i>= w_i|\alpha_i> \tag{110}$$

and α_i are the real eigenvalues of w_i. They are the quantities that are measured for the observable Ω.

The hermitian matrices

$$P_i = |\alpha_i><\alpha_i| = a_i a_i^\dagger \tag{111}$$

are in one-to-one correspondence with the kets $|\alpha_i>$.

The P_i are idempotent hermitian operators since

$$P_i^2 = P_i, \tag{112}$$

$$Tr\ P_i = 1, \tag{113}$$

and

$$\Omega = \sum w_i P_i. \tag{114}$$

We define

$$(\Omega_1, \Omega_2) = Tr(\omega_1 \Omega_2). \tag{115}$$

The transition amplitude between states α and β is

$$<\alpha|\beta>= (\alpha_i^* <\alpha_i|)(\beta_i|\beta_i>) = \alpha_i^* \beta_i \tag{116}$$

and the transition probability is

$$\Pi_{\alpha\beta} = |<\alpha|\beta>|^2 = (\alpha_i^* \beta_i)(\alpha_j \beta_j^*) =<\alpha|\beta><\beta|\alpha> \tag{117}$$

We consider the density matrices

$$P_\alpha = |\alpha><\alpha|, \quad P_\beta = |\beta><\beta| \tag{118}$$

so that

$$P_\alpha P_\beta = |\alpha><\alpha|\beta><\beta| =<\alpha|\beta><|\alpha><\beta| \tag{119}$$

$$(P_\alpha, P_\beta) = Tr\ P_\alpha P_\beta =<\alpha|\beta> Tr\ |\alpha><\beta| =<\alpha|\beta><\beta|\alpha| \tag{120}$$

Thus

$$(P_\alpha, P_\beta) = \Pi_{\alpha\beta} = Tr\ P_\alpha . P_\beta \tag{121}$$

where

$$P_\alpha . P_\beta = \frac{1}{2}(P_\alpha P_\beta + P_\beta P_\alpha) \tag{122}$$

is the Jordan product of P_α and P_β under which hermitian matrices form a closed algebra.

The observable $\Omega = \sum w_i P_i$ is a density matrix in general.

The matrix elements of the observable Ω are given by

$$x <\alpha_i|\Omega|\alpha_j>= \Omega_{ij} \tag{123}$$

and

$$|\Omega_{ij}|^2 = Tr\ (\Omega P_i \Omega P_j) = Tr\ (\Omega^{\frac{1}{2}} P_i \Omega^{\frac{1}{2}}).(\Omega^{\frac{1}{2}} P_j \Omega^{\frac{1}{2}}) \tag{124}$$

The transition probability $|<\alpha|\beta>|^2$ has a simple geometrical meaning. Let

$$d_{\alpha\beta}^2 = \frac{1}{2} Tr\ (P_\alpha - P_\beta)^2 \tag{125}$$

where P_α and P_β are idempotents. Then

$$d^2_{\alpha\beta} = \frac{1}{2}(1 + 1 - 2Tr\ P\alpha P_\beta) = 1 - \Pi_{\alpha\beta} \tag{126}$$

or

$$\Pi_{\alpha\beta} = 1 - d^2_{\alpha\beta} \tag{127}$$

Now in a projective geometry the normalized idempotent P_α represents a point with n homogeneous and $(n-1)$ inhomogeneous coordinates.

The state $\lambda|\alpha>$ is equivalent to the state $|\alpha>$ in a ray representation. The corresponding projection operator is $|\lambda|^2 P_\alpha$ which is equivalent to P_α as we can take $|\lambda| = 1$ by normalization. This corresponds to homogeneous coordinates α_i and $\lambda\alpha_i$ representing the same point P_α.

Thus a point is in a one-to-one correspondence with a state or an idempotent hermitian matrix (projection operator).

The transition probabilities are invariant under the unitary transformation $(U \in U(n))$.

$$|\alpha> \to U|\alpha>, \quad |\beta> \to U|\beta> \tag{128}$$

or

$$P_\alpha \to UP_\alpha U^{-1}, \quad P_\beta \to UP_\beta U^{-1} \qquad (U \in SU(n)) \tag{129}$$

The invariant distance $d_{\alpha\beta}$ between P_α and P_β is defined as

$$Tr\ P_\alpha P_\beta = d\alpha\beta^2. \tag{130}$$

When $\alpha = \beta$ their relative distance vanishes while the transition probability becomes unity. Thus

$$\Pi_{\alpha\beta} = 1 - d^2_{\alpha\beta} \tag{131}$$

is the complement of the square of the relative distance.

A point is left invariant by the subgroup of $U(n)$ which leaves the idempotent invariant. Hence it is $U(n-1) \times U(1)$. This is the stability group of $U(n)$.

Note that

$$|\Omega_{ij}|^2 = Tr\ (\Omega P_i \Omega P_j) \tag{132}$$

can also be expressed by means of the Jordan product alone. The Jordan product is commutative and non associative. It is not alternative but power associative. We have

$$(A, B, C) = (A.B).C) - A.(B.C) = -(C, B, A) \tag{133}$$

so that

$$(A, B, A) = 0 \tag{134}$$

Writing

$$\begin{aligned} (A, B, A^2) &= (AB + BA)A^2 + A^2(AB + BA) \qquad (135) \\ &\quad -A(BA^2 + A^2B) - (BA^2 + A^2B)A \\ &= ABA^2 + BA^3 + A^3B + A^2BA \\ &\quad -ABA^2 - A^3B - BA^3 - A^2BA = 0 \qquad (136) \end{aligned}$$

and

$$\begin{aligned}(A, B, A^{-1}) &= (A.B).A^{-1} - A.(B.A^{-1}) \\ &= \frac{1}{4}[(AB + BA)A^{-1} + A^{-1}(AB + BA) \\ &\quad -A(BA^{-1} + A^{-1}B) - (BA^{-1} + A^{-1}B)A] \\ &= \frac{1}{4}[ABA^{-1} + B + B + A^{-1}BA \\ &\quad -ABA^{-1} - B - B - A^{-1}BA = 0 \end{aligned} \tag{137}$$

We have shown

$$(A, B, A^2) = 0 \tag{138}$$

and for a special Jordan algebra of matrices

$$(A, B, A^{-1}) = 0 \tag{139}$$

as well.

We also have

$$\begin{aligned}(P, A, A) &= \frac{1}{4}[(PA + A.P)A + A(PA + AP) - 2PA^2 - 2A^2P] \\ &= \frac{1}{4}[PA^2 + APA + APA + A^2P - 2PA^2 - 2A^2P] \\ &= \frac{1}{4}(2APA - A^2P - PA^2) \\ &= \frac{1}{2}APA - \frac{1}{4}(A^2P + PA^2) \\ &= \frac{1}{2}APA - \frac{1}{2}A^2.P \end{aligned} \tag{140}$$

Hence

$$\frac{1}{2}APA = \frac{1}{2}A^2.P + (P, A, A) \tag{141}$$

can be expressed by means of the Jordan product

$$\begin{aligned}APA &= A^2P + 2(P, A, A) \\ &= A^2.P + 2(P.A).A - 2P.A^2 \end{aligned} \tag{142}$$

simply

$$APA = 2(P.A).A - A^2.P, \tag{143}$$

or

$$\Omega^{\frac{1}{2}} = 2(P_i.\Omega^{\frac{1}{2}}).\Omega^{\frac{1}{2}} - \Omega.P_i \tag{144}$$

Inverse matrix A^{-1} satisfies

$$A^{-1}.A = I, \quad A^{-1}.A^2 = A \tag{145}$$

where I is the identity matrix.

We now express the unitary transformation in infinitesimal form as $U = 1 + iA$, so that

$$P'_\alpha = UP_\alpha U^{-1} = (1 + iA)P_\alpha(1 - iA) = P_\alpha + i[A, P_\alpha] \tag{146}$$

IOP Conf. Series: Journal of Physics: Conf. Series **965** (2018) 012009 doi:10.1088/1742-6596/965/1/012009

Let

$$-4iA = [H_1, H_2] \tag{147}$$

where H_1 and H_2 are hermitian. Then

$$P'_\alpha = P_\alpha - \frac{1}{4}[[H_1, H_2], P_\alpha]. \tag{148}$$

Now

$$-\frac{1}{4}[[H_1, H_2], P_\alpha] = (H_1.P_\alpha).H_2 - H_1.(P_\alpha.H_2) = (H_1, P_\alpha, H_2) \tag{149}$$

for matrices, so that

$$P'_\alpha = P_\alpha + (H_1, P_\alpha, H_2) \tag{150}$$

By iteration we can express the finite unitary transformation as

$$TP_\alpha = E_{H_1,H_2}P\alpha = P_\alpha + (H_1, P\alpha, H_2) + \frac{1}{2!}(H_1, (H_1, P_\alpha, H_2)H_2 + \cdots \tag{151}$$

which is an expression involving multiple associators. For the usual matrices we have

$$\begin{aligned} TP_\alpha &= e^{-\frac{1}{4}[h_1,H_2]}p_\alpha e^{\frac{1}{4}[h_1,H_2]} \\ &= P_\alpha - \frac{1}{4}[[H_1, H_2], P_\alpha] \\ &= \frac{1}{2!}(-\frac{1}{4}^2)[[H_1, H_2], [[H_1, H_2], P\alpha]] + \cdots \end{aligned} \tag{152}$$

If $|\alpha_i >$ is an orthogonal set $U|\alpha_i >= |\alpha'_i >$ is another.
If P_i form a complete set $\sum P_i$, then $UP_iU^{-1} = P'_i$ are also complete.
Thus, states, transition probabilities, unitary transformations in the algebraic formulation of quantum mechanics involves only hermitian matrices and their Jordan product.

(b) Exceptional observables:

We start from the axioms of Jordan algebras.

$$A.B = B.A \tag{153}$$

gives

$$[A, B, C] = [C, B, A] \tag{154}$$

giving $[A, B, A] = 0$. We do require

$$[A, B, A^2] = 0 \tag{155}$$

proven above.

This system is also satisfied by 3×3 octonionic hermitian matrices (Jordan, von Neumann, Wigner)

$$J = \begin{pmatrix} \alpha & c & \bar{b} \\ \bar{c} & \beta & a \\ b & \bar{a} & \gamma \end{pmatrix} \tag{156}$$

which gives rise to a power associative algebra. Entries a, b, c are real octonions.

Now, these matrices are closed under the Jordan product. What fails is the identity

$$-\frac{1}{4}[[H_1, H_2], J] = (H_1, J, H_2) = (H_1.J).H_2 - H_1.(J.H_2) \tag{157}$$

Instead we have

$$-\frac{1}{4}[[H_1,H_2],J] = (H_1,J,H_2)-\frac{1}{4}[J,H_1,H_2]+\frac{1}{4}[J,H_2,H_1] -\frac{1}{4}[H_1,J,H_2]+\frac{1}{4}[H_2,J,H_1] -\frac{1}{4}[H_1,H_2,J]+\frac{1}{4}[H_2,H_1,J] \quad (158)$$

note that the first term on the left is not a hermitian matrix, and the first term on the right, namely $(H_1,J,H_2)=J'$ is a hermitian matrix.

These extra terms vanish for an ordinary matrix algebra as they refer to associators with respect to the usual matrix product

$$[H_1,J,H_2]=(H_1J)H_2-H_1(JH_2) \quad (159)$$

etc. They vanish for ordinary matrices but not for the octonionic matrices. It is with the Jordan associators that we obtain an automorphism. Because the left hand side is in general not hermitian.

We define the scalar product

$$(J,J)=\frac{1}{2}Tr\ J^2, \quad (160)$$

also the Freudenthal product by

$$J\times J=(J^2-JTr\ J)-\frac{1}{2}ITr(J^2-JTr\ J) \quad (161)$$

with

$$Det\ J=\frac{1}{3}Tr\ J.(J\times J)=\frac{2}{3}(J,J\times J) \quad (162)$$

If $DetJ\neq 0$, then $J\times J$ has been defined so that

$$J^{-1}=(DetJ)^{-1}(J\times J) \quad (163)$$

and

$$J\times J=J^{-1}DetJ \quad (164)$$

We have

$$J^{-1}.J=I \quad (165)$$

and

$$J^{-1}.J^2=J \quad (166)$$

so that J^{-1} is the inverse matrix.

Also from $J\times J=J^{-1}DetJ$ we find

$$J.(J\times J)=(J.J^{-1})DetJ=IDetJ \quad (167)$$

Now we have for any two non singular JNW matrices J and K

$$(J,K,J^2)=0 \quad (168)$$

also

$$(J,K,\lambda J)=0 \quad (169)$$

and

$$(J,K,\kappa J)=0 \quad (170)$$

Thus

$$(J, K, J \times J) = 0 \tag{171}$$

Because we have a 3×3 matrix we find

$$(J, K, J^{-1}) = 0 \tag{172}$$

This ensures power associativity for non singular matrices. For $K = J$ we have

$$(J, J, J^2) = 0 \tag{173}$$

or

$$J^2.J^2 = J.J^3 \tag{174}$$

$$(J, J, J^{-1}) = 0 \tag{175}$$

or

$$J^2.J^{-1} = J \tag{176}$$

Also

$$(J, J^n, J^2) = 0 \tag{177}$$

gives

$$(J.J^n).J^2 = (J.J^{n+2} \tag{178}$$

$$(J, J^n, J^{-1}) = 0 \tag{179}$$

$$(J.J^n).J^{-1} = J.(J^{(n-1)}) \tag{180}$$

Note that we have the identity

$$(J \times J).J = IDetJ = J^3 = J^2TrJ + \frac{1}{2}J(TrJ)^2 - TrJ^2 \tag{181}$$

where $J^3 = J^2.J$.

Explicitly we find

$$(J, J) = \frac{1}{2}(\alpha^2 + \beta^2 + \gamma^2) + |a|^2 + |b|^2 + |c|^2 \tag{182}$$

where $|a|^2 = a\bar{a}$, etc.

$$\begin{aligned} DetJ &= \alpha\beta\gamma - \alpha|a|^2 - \beta|b|^2\gamma|c|^2 + (ab)c + \bar{c}(\bar{b}\bar{a}) \\ &= \alpha\beta\gamma - \alpha|a|^2 - \beta|b|^2 = \gamma|c|^2 + t[(abc)] \end{aligned} \tag{183}$$

This expression is unambiguous since we have

$$(ab)c + \bar{c}(\bar{b}\bar{a}) = a(bc) + (\bar{c}\bar{b})\bar{a}. \tag{184}$$

The proof is simple. The above equation is equivalent to

$$[a, b, c] = [\bar{c}, \bar{b}, \bar{a}]. \tag{185}$$

Because

$$\bar{c} = -c + 2c_0, \quad \bar{b} = -b + 2b_0, \quad \bar{a} = -a + 2a_0, \tag{186}$$

the right hand side takes the form

$$[\bar{c}, \bar{b}, \bar{a}] = -[c, b, a] \tag{187}$$

Using alternativity we find

$$-[c,b,a] = [a,b,c] \tag{188}$$

so that

$$t[(ab)c] = t[a(bc)] = t(abc) \tag{189}$$

We also have

$$J \times J = \begin{pmatrix} \beta\gamma - a\bar{a} & \bar{b}\bar{a} - \gamma c & ca - \beta\bar{b} \\ ab - \gamma\bar{c} & \gamma\alpha - b\bar{b} & \bar{c}\bar{b} - \alpha a \\ \bar{a}\bar{c} - \beta b & bc - \alpha\bar{a} & \alpha\beta - c\bar{c} \end{pmatrix} \tag{190}$$

Now in ordinary quantum mechanics a 3×3 projection operator has the form

$$P = UEU^{-1} \tag{191}$$

where E is diagonal, idempotent with trace unity. For instance

$$E = E_3 = \begin{pmatrix} 0 & & \\ & 0 & \\ & & 1 \end{pmatrix} \tag{192}$$

It satisfies

$$E_3 \times E_3 = 0 \tag{193}$$

Similarly

$$E_1 = \begin{pmatrix} 1 & & \\ & 0 & \\ & & 0 \end{pmatrix}, \tag{194}$$

$$E_2 = \begin{pmatrix} 0 & & \\ & 1 & \\ & & 0 \end{pmatrix}, \tag{195}$$

$$E_1 + E_2 + E_3 = I \tag{196}$$

For E_i we have $E_i \times E_i = 0$, so that

$$P \times P = 0 \tag{197}$$

with

$$TrP = I, \quad DetP = 0 \tag{198}$$

Hence, in the octonionic case we can define a projection operator by

$$TrP = I, \quad P \times P = 0 \tag{199}$$

and with

$$P = \begin{pmatrix} \alpha & c & \bar{b} \\ \bar{c} & \beta & a \\ b & \bar{a} & \gamma \end{pmatrix}, \quad \alpha + \beta + \gamma = 1 \tag{200}$$

we find

$$\beta\gamma = a\bar{a}, \quad \gamma\alpha = b\bar{b}, \quad \alpha\beta = c\bar{c}, \quad bc = \alpha\bar{a}, \quad ca = \beta\bar{b}, \quad ab = \gamma\bar{c} \tag{201}$$

This is satisfied by

$$P_3 = \lambda_3 \, \bar{\lambda}_3^{\,T} \tag{202}$$

where

$$\lambda_3 = \begin{pmatrix} \gamma^{-\frac{1}{2}}\bar{b} \\ \gamma^{-\frac{1}{2}}a \\ \gamma^{\frac{1}{2}} \end{pmatrix} \tag{203}$$

so that

$$P_3 = \begin{pmatrix} \gamma^{-1}b\bar{b} & \gamma^{-1}\bar{b}\bar{a} & \bar{b} \\ \gamma^{-1}ab & \gamma^{-1}a\bar{a} & a \\ b & \bar{a} & \gamma \end{pmatrix} \tag{204}$$

giving

$$\alpha = \gamma^{-1}b\bar{b}, \quad \beta = \gamma^{-1}a\bar{a}, \quad c = \gamma^{-1}\bar{b}\bar{a} \tag{205}$$

with the constraint

$$\gamma^2 - \gamma + |a|^2 + |b|^2 = 0 \tag{206}$$

or

$$\begin{aligned} \gamma = \gamma(a,b) &= \frac{1}{2} \pm \frac{1}{2}\sqrt{1 - 4(|a|^2 + |b|^2)} \\ &= \frac{1}{2} \pm \sqrt{\frac{1}{4} - (|a|^2 + |b|^2)} \end{aligned} \tag{207}$$

Reality of γ requires

$$0 \leq (|a|^2 + |b|^2) < \frac{1}{4} \tag{208}$$

then $\gamma > 0$.

Thus an idempotent JNW matrix is determined by two octonions a, b with

$$\sqrt{|a|^2 + |b|^2} < \frac{1}{2} \tag{209}$$

and corresponds to a point on a generalized sphere. We can represent a physical state by such a point. We have

$$\gamma(0,0) = 1, \quad P_3 = E_3. \tag{210}$$

Now we turn to an infinitesimal transformation that leaves J, TrJ, TrJ^2 and $DetJ$ invariant. Let

$$\delta J = (H_1, J, H_2) = L_{H_1,H_2}J \tag{211}$$

Then we find

$$\delta TrJ = 0 \tag{212}$$

$$\delta(J,J) = (\delta J, J) = 0 \tag{213}$$

and

$$\delta DetJ = (\delta J, J \times J) + (J, \delta J \times J) + (J, J \times \delta J) = 0 \tag{214}$$

or

$$(\delta J, J \times J) + 2(J, \delta J \times J) = 0 \tag{215}$$

We also find that if

$$J \to J' = J + \delta J, \quad K' = K + \delta K \tag{216}$$

then

$$(J', K', J' \times J') = 0 \tag{217}$$

or

$$(J+\delta J, K, J\times J)+(J, K+\delta K, J\times J)+2(J, K, J+\delta J)=0 \tag{218}$$

so that $J\to J'$ is an automorphism of the algebra.

It has H_1, H_2 with $TrH_1 = 0,\ TrH_2 = 0$ so that the automorphism group is defined by 26 parameters. It is the group F_4.

The projection operator transforms as

$$P'=\lambda'\lambda'^{\dagger}=\lambda\lambda^{\dagger}+(H_1,\lambda\lambda^{\dagger},H_2)+\frac{1}{2!}(H_1,(H_1,\lambda\lambda^{\dagger},H_2),H_2)+\cdots \tag{219}$$

In the non octonionic case this can be written

$$\lambda'\lambda'^{\dagger}=e^{-\frac{1}{4}[H_1,H_2]}\lambda\lambda^{\dagger}e^{\frac{1}{4}[H_1,H_2]} \tag{220}$$

which is equivalent to

$$\lambda'=e^{-\frac{1}{4}[H_1,H_2]}\lambda=U\lambda \tag{221}$$

so that λ transforms linearly under the group and therefore forms a representation of the automorphism group. In the octonionic case such a small representation does not exist as

$$U(\lambda\lambda^{\dagger})U^{\dagger}\neq(U\lambda)(\lambda^{\dagger}U^{\dagger}) \tag{222}$$

so that there is no ket representation of the state, only a projection operator representation.

Also note that in this case, besides the states P_1, P_2, P_3 we can also define observables H_1, H_2 associated with the "charges" carried by the states.

Another remark contains the transformation property of c. The c transforms like

$$(\frac{1}{2}\pm\sqrt{\frac{1}{4}-(|a|^2+|b|^2)})^{-1}\bar{b}\bar{a} \tag{223}$$

If we choose a subgroup of F_4 which preserve $|a|$ and $|b|$, it belongs to $SO(8)$. In that case

$$t[ab(\bar{b}\bar{a})] \tag{224}$$

is invariant, and similarly

$$t[a(\bar{a}\bar{c})c] \tag{225}$$

and

$$t[(\bar{c}\bar{b})bc] \tag{226}$$

are invariant.

(c) Octonionic form of G_2

The norm group of the octonions is the group that preserves

$$[N(w)]^2=w_0^2+w_1^{@}+\cdots+w_7^2 \tag{227}$$

Hence it is $SO(8)$.

The norm group for traceless octonions is $SO(7)$. A subgroup of $SO(7)$ is the automorhism group

$$u=\begin{pmatrix}u_1\\u_2\\u_3\\u_0\end{pmatrix}\to e^{i\lambda_3\alpha_7}u,\quad e^{i\lambda_8\beta_7}u \tag{228}$$

leave u_0 invariant and preserve the commutation relations.

Let S be the discrete symmetry that gives a cyclic permutation to e_a Then $S^7 = 1$.

Consider $S\ u$ with $7 \to 1$.

$$S\ u \to e^{i\lambda_3 \alpha_1} S\ u, \tag{229}$$

$$S\ u \to e^{i\lambda_8 \beta_1} S\ u \tag{230}$$

thus is also an automorphism.

Hence automorphism group has $7 \times 2 = 14$ parameters.

$$w \to awa \in SO(8) \tag{231}$$

7 parameters in $SO(8)/SO(7)$

$$w \to bwb^{-1} \in SO(7) \tag{232}$$

in $SO(7)/G_2$.

$$(cd)^{-1}[d(cwc_{-1})d^{-1}](cd) \tag{233}$$

is a 14 parameter subgroup of $SO(7) \in G_2$.

Finally, the infinitesimal form is

$$w' = w + \frac{1}{4}[[\widetilde{\alpha}, \widetilde{\beta}], \widetilde{w}] - \frac{3}{4}[\widetilde{\alpha}, \widetilde{\beta}, \widetilde{w}]. \tag{234}$$

We also note that G_2 is also the automorphism group of the anti-symmetrical algebra defined by

$$e_a \wedge e_b = f_{abc} e_c \tag{235}$$

which is not a Lie algebra.

Invariance principles of exceptional quantum mechanics based on F_4 and its generalization to the case of complex Jordan algebra E_6 were worked out before.[14] It is clear that the fundamental symmetries of elementary particles seem to point out to field theories based on local internal symmetries connected with structures that can be constructed by using octonions (see some of the references at the end of the paper). Octonionic observables form a Jordan algebra whose automorhism group is an exceptional group F_4 or E_6 and we are led into a gauge field theory of quarks and leptons based on exceptional groups.

It is also possible to consider super Jordan algebras for generalized Jordan algebras involving both bosonic and fermionic observables. In this case the automorphism group is a supergroup. If the generalized Jordan algebra is octonionic then its automorphism is given by an exceptional supergroup, it could be F_4 for example[15] which has 40 parameters and its Lie subgroup is $SO(7) \times SU(2)$ which admits $SU(2) \times U(1) \times SU(3)^c$ as a subgroup. Instead of leptoquarks and diquarks[16], the super F_4 would introduce spin $\frac{3}{2}$ gauge particles alongside with gauge bosons, so it would be similar to supergravity when used as a local supersymmetry group.

There remain many unsolved problems. The outstanding ones are concerned with the building of Fock space (tensor products with exceptional structures), the understanding of color confinement and realization of the ultimate synthesis with supergravity. At this point, for many reasons it seems impossible to imbed a successful grand unified theory based on E_6 with extended supergravity. A way out may be the reformulation of supergravity in superspace consisting of space-time manifold and Grassmann manifold. Grassmann numbers are obtained by using direct products of quaternions. Exceptional Grassmann numbers are based on octonions, hence it is possible that the invariance properties of such exceptional superspace will involve new supersymmetries that generalize the exceptional groups.

5. Acknowledgments
The development of ideas presented here is due in large part to stimulating discussions with our colleagues and friends including Vladimir Akulov, Robert L. Blumenblatt, Vladimir Dobrev, Yasemin Gürcan, Yuan K. Ha, Francesco Iachello, Amish Khalfan, Alexander Kheyfits, Levent Kurt, David Tepper and Francesco Toppan.

Bibliography

[1] Jordan P, von Neumann J and Wigner E P 1934 *Ann. Math.* **35**, 29.
[2] Chevalley C and Schafer R D 1950 *Proc. Natl. Acad. Sci. U.S.* **36** 137
[3] Jordan P 1933 *Nachr. Ges. Wiss. Göttingen* 209
[4] Tits J 1966 *Nederl. Akad. Wetensch. Proc. Ser. A* **69**, 223
[5] Greenberg 1964 *Phys. Rev. Lett.* **13**, 598
[6] Nambu Y 1966 in *"Preludes in Theoretical Physics" eds. A. de Shalit, H. Fesbach and L. van Hove, p.133 (Amsterdam)*
[7] Fritzsch H, Gell-Mann M and Leutwyler H 1973 *Phys. Lett.* **47B**,
[8] Freudenthal H 1965 *Adv. in Math. I* 145
[9] Gürsey F and Tze C H 1966 *"On the Role of Division, Jordan and Related Algebras in Particle Physics," World Scientific*
[10] Barton C H 2000 *Magic Squares of Lie Algebras," Ph.D. Thesis, University of York*
[11] Barton C H and Sudbery A 2002 *Mat. RA/0001803, and RA/0203010*
[12] Burdik C, Catto S, Gürcan Y, Khalfan A and Kurt L 2017 *Physics of Particles and Nuclei Letters* **14** 390
[13] Günaydin M and Gürsey F 1973 *J. Math. Phys.* **14**, 1651
[14] Catto S , Choun Y S and Kurt L 2013 *"Invariance Properties of the Exceptional Quantum Mechanics (F_4) and its Generalization to Complex Jordan Algebras (E_6)" in Lie Theory and Its Applications in Physics, Ed. V. Dobrev (Springer), Japan* 469
[15] Gürsey F and Marchildon L 1977 *J. Math. Phys.* **19**, 942
[16] Catto S 1993 *"Colored Supersymmetry of Mesons and Baryons Based on Octonionic Algebras," in "Symmetries in Science VI: From the Rotation Group to Quantum Algebras." Ed. B. Gruber, Plenum Publishers* 129
[17] Gürsey F 1977 "Non-associative Algebras in Quantum Mechanics and Particle Physics." In Proceedings of the Charlotsville Conference on Non Associative Algebras, U. Virginia
[18] Jacobson N 1951 *Trans. Amer. Math. Soc. 70* 509
[19] Albert A A 1934 *Ann. Math.* **35**, 65
[20] Mostow G D 1973 *"Strong Rigidity of Locally Symmetric Spaces." (Princeton U. Press)*
[21] Moufang R 1933 *Abh. Math. Sem. Univ. Hamburg 9* 207
[22] Günaydin M, Piron C and Ruegg H 1978 *Comm. Math. Phys. 61* 69
[23] Günaydin M and Gürsey F 1974 *Phys. Rev. D9* 3387 (1974)
[24] G.B. Seligman, *"Modular Lie Algebras." Springer-Verlag, Berlin, Heidelberg, New York, 1967.*
[25] Schafer R D 1966 *"An Introduction to Non-associative Algebras." (Academic Press, New York*
[26] Gürsey F , Ramond P and Sikivie P 1976 *Phys. Lett. 60B* 177
[27] Faulkner J R 1970 *Memoirs of the Am. Math. Soc. No.104 (Providence, RI).*
[28] Kac V 1977 it Comm. Math. Phys. **53**, 31.
[29] Pati J and Salam A 1973 *Phys. Rev.* **D8**, 1240

Infinitesimal deformations of Poisson bi-vectors using the Kontsevich graph calculus

Ricardo Buring[1], **Arthemy V Kiselev**[2] **and Nina Rutten**[2]
[1] Institut für Mathematik, Johannes Gutenberg–Universität, Staudingerweg 9, D-55128 Mainz, Germany
[2] Johann Bernoulli Institute for Mathematics and Computer Science, University of Groningen, P.O. Box 407, 9700 AK Groningen, The Netherlands
E-mail: rburing@uni-mainz.de, A.V.Kiselev@rug.nl

Abstract. Let $\mathcal{P}$ be a Poisson structure on a finite-dimensional affine real manifold. Can $\mathcal{P}$ be deformed in such a way that it stays Poisson? The language of Kontsevich graphs provides a universal approach – with respect to all affine Poisson manifolds – to finding a class of solutions to this deformation problem. For that reasoning, several types of graphs are needed. In this paper we outline the algorithms to generate those graphs. The graphs that encode deformations are classified by the number of internal vertices k; for $k \leqslant 4$ we present all solutions of the deformation problem. For $k \geqslant 5$, first reproducing the pentagon-wheel picture suggested at $k = 6$ by Kontsevich and Willwacher, we construct the heptagon-wheel cocycle that yields a new unique solution without 2-loops and tadpoles at $k = 8$.

Introduction. This paper contains a set of algorithms to generate the Kontsevich graphs that encode polydifferential operators – in particular multi-vectors – on Poisson manifolds. We report a result of implementing such algorithms in the problem of finding symmetries of Poisson structures. Namely, continuing the line of reasoning from [1, 2], we find all the solutions of this deformation problem that are expressed by the Kontsevich graphs with at most four internal vertices. Next, we present one six-vertex solution (based on the previous work by Kontsevich [10] and Willwacher [13]). Finally, we find a heptagon-wheel eight-vertex graph which, after the orientation of its edges, gives a new universal Kontsevich flow. We refer to [8, 9] for motivations, to [2, 4] for an exposition of basic theory, and to [6] and [5] for more details about the pentagon-wheel (5+1)-vertex and heptagon-wheel (7+1)-vertex solutions respectively. Let us remark that all the algorithms outlined here can be used without modification in the course of constructing all k-vertex Kontsevich graph solutions with higher $k \geqslant 5$ in the deformation problem under study.

Basic concept. We work with real vector spaces generated by finite graphs of the following two types: (1) k-vertex non-oriented graphs, without multiple edges nor tadpoles, endowed with a wedge ordering of edges, e.g., $E = e_1 \wedge \cdots \wedge e_{2k-2}$; (2) oriented graphs on k internal vertices and n sinks such that every internal vertex is a tail of two edges with a given ordering Left $\prec$ Right. Every connected component of a non-oriented graph γ is fully encoded by an ordering E

IOP Conf. Series: Journal of Physics: Conf. Series **965** (2018) 012010 doi:10.1088/1742-6596/965/1/012010

on the set of adjacency relations for its vertices.[1] Every such oriented graph is given by the list of ordered pairs of directed edges. An edge swap $e_i \wedge e_j = -e_j \wedge e_i$ and the reversal Left $\leftrightarrows$ Right of those edges' order in the tail vertex implies the change of sign in front of the graph at hand.[2]

Example 1. The sum γ_5 of two 6-vertex 10-edge graphs,

$$\gamma_5 = \mathsf{12}^{(I)} \wedge \mathsf{23}^{(II)} \wedge \mathsf{34}^{(III)} \wedge \mathsf{45}^{(IV)} \wedge \mathsf{51}^{(V)} \wedge \mathsf{16}^{(VI)} \wedge \mathsf{26}^{(VII)} \wedge \mathsf{36}^{(VIII)} \wedge \mathsf{46}^{(IX)} \wedge \mathsf{56}^{(X)}$$
$$+\tfrac{5}{2}\cdot \mathsf{12}^{(I)} \wedge \mathsf{23}^{(II)} \wedge \mathsf{34}^{(III)} \wedge \mathsf{41}^{(IV)} \wedge \mathsf{45}^{(V)} \wedge \mathsf{15}^{(VI)} \wedge \mathsf{56}^{(VII)} \wedge \mathsf{36}^{(VIII)} \wedge \mathsf{26}^{(IX)} \wedge \mathsf{13}^{(X)},$$

is drawn in Theorem 7 on p. 10 below.

Example 2. The sum $\mathcal{Q}_{1:\frac{6}{2}}$ of three oriented 8-edge graphs on $k = 4$ internal vertices and $n = 2$ sinks (enumerated using 0 and 1, see the notation on p. 4),

$$\mathcal{Q}_{1:\frac{6}{2}} = 2\ 4\ 1 \quad 0\ 1\ \ 2\ 4\ \ 2\ 5\ \ 2\ 3 - 3(2\ 4\ 1 \quad 0\ 3\ \ 1\ 4\ \ 2\ 5\ \ 2\ 3 + 2\ 4\ 1 \quad 0\ 3\ \ 4\ 5\ \ 1\ 2\ \ 2\ 4)$$

is obtained from the non-oriented tetrahedron graph $\gamma_3 = \mathsf{12}^{(I)} \wedge \mathsf{13}^{(II)} \wedge \mathsf{14}^{(III)} \wedge \mathsf{23}^{(IV)} \wedge \mathsf{24}^{(V)} \wedge \mathsf{34}^{(VI)}$ on four vertices and six edges by taking all the admissible edge orientations (see Theorem 4 and Remark 1).

I.1. Let γ_1 and γ_2 be connected non-oriented graphs. The definition of insertion $\gamma_1 \circ_i \gamma_2$ of the entire graph γ_1 into vertices of γ_2 and the construction of Lie bracket $[\cdot,\cdot]$ of graphs and differential d in the non-oriented graph complex, referring to a sign convention, are as follows (cf. [8] and [7, 11, 12]); these definitions apply to sums of graphs by linearity.

Definition 1. The insertion $\gamma_1 \circ_i \gamma_2$ of a k_1-vertex graph γ_1 with ordered set of edges $E(\gamma_1)$ into a graph γ_2 with $\#E(\gamma_2)$ edges on k_2 vertices is the sum of k_2 graphs on $k_1 + k_2 - 1$ vertices and $\#E(\gamma_1) + \#E(\gamma_2)$ edges. Topologically, the sum $\gamma_1 \circ_i \gamma_2 = \sum(\gamma_1 \to v \text{ in } \gamma_2)$ consists of all the graphs in which a vertex v from γ is replaced by the entire graph γ_1 and the edges touching v in γ_2 are re-attached to the vertices of γ_1 in all possible ways.[3] By convention, in every new term the edge ordering is $E(\gamma_1) \wedge E(\gamma_2)$.

To simplify sums of graphs, first eliminate the zero graphs. Now suppose that in a sum, two non-oriented graphs, say α and β, are isomorphic (topologically, i.e. regardless of the respective vertex labellings and edge orderings $E(\alpha)$ and $E(\beta)$). By using that isomorphism, which establishes a 1–1 correspondence between the edges, extract the sign from the equation $E(\alpha) = \pm E(\beta)$. If "+", then $\alpha = \beta$; else $\alpha = -\beta$. Collecting similar terms is now elementary.

Lemma 1. The bi-linear graded skew-symmetric operation,

$$[\gamma_1, \gamma_2] = \gamma_1 \circ_i \gamma_2 - (-)^{\#E(\gamma_1)\cdot\#E(\gamma_2)} \gamma_2 \circ_i \gamma_1,$$

[1] The edges are antipermutable so that a graph which equals minus itself – under a symmetry that induces a parity-odd permutation of edges – is proclaimed to be equal to zero. In particular (view •–•–•), every graph possessing a symmetry which swaps an odd number of edge pairs is a zero graph. For example, the 4-wheel $\mathsf{12} \wedge \mathsf{13} \wedge \mathsf{14} \wedge \mathsf{15} \wedge \mathsf{23} \wedge \mathsf{25} \wedge \mathsf{34} \wedge \mathsf{45} = I \wedge \cdots \wedge VIII$ or the 2ℓ-wheel at any $\ell > 1$ is such; here, the reflection symmetry is $I \rightleftarrows III$, $V \rightleftarrows VII$, and $VI \rightleftarrows VIII$.

[2] An oriented graph equals minus itself, hence it is a zero graph if there is a permutation of labels for its internal vertices such that the adjacency tables for the two vertex labellings coincide but the two realisations of the same graph differ by the ordering of outgoing edges at an odd number of internal vertices (see Example 3 below).

[3] Let the enumeration of vertices in every such term in the sum start running over the enumerated vertices in γ_2 until v is reached. Now the enumeration counts the vertices in the graph γ_1 and then it resumes with the remaining vertices (if any) that go after v in γ_2.

is a Lie bracket on the vector space $\mathfrak{G}$ of non-oriented graphs.[4]

Lemma 2. The operator d(graph) = [•–•, graph] is a differential: $\mathrm{d}^2 = 0$.

In effect, the mapping d blows up every vertex v in its argument in such a way that whenever the number of adjacent vertices $N(v) \geqslant 2$ is sufficient, each end of the inserted edge •–• is connected with the rest of the graph by at least one edge.

Summarising, the real vector space $\mathfrak{G}$ of non-oriented graphs is a differential graded Lie algebra (dgLa) with Lie bracket $[\cdot,\cdot]$ and differential d = [•–•, ·]. The graphs γ_5 and γ_3 from Examples 1 and 2 are d-cocycles. Neither is exact, hence marking a nontrivial cohomology class in the non-oriented graph complex.

Theorem 3 ([7, Th. 5.5]). *At every $\ell \in \mathbb{N}$ in the connected graph complex there is a* d*-cocycle on $2\ell+1$ vertices and $4\ell+2$ edges. Such cocycle contains the $(2\ell+1)$-wheel in which, by definition, the axis vertex is connected with every other vertex by a spoke so that each of those 2ℓ vertices is adjacent to the axis and two neighbours; the cocycle marked by the $(2\ell+1)$-wheel graph can contain other $(2\ell+1, 4\ell+2)$-graphs (see Example* 1 *and* [5]*).*

I.2. The *oriented* graphs under study are built over n sinks from k wedges $\xleftarrow{i_\alpha} \bullet \xrightarrow{j_\alpha}$ (here $\xleftarrow{i_\alpha}\prec\xrightarrow{j_\alpha}$) so that every edge is decorated with its own summation index which runs from 1 to the dimension of a given affine Poisson manifold $(\mathcal{N}, \mathcal{P})$. Each edge $\xrightarrow{i}$ encodes the derivation $\partial/\partial x^i$ of the arrowhead object with respect to a local coordinate x^i on $\mathcal{N}$. By placing an αth copy $P^{i_\alpha j_\alpha}(\boldsymbol{x})$ of the Poisson bi-vector $\mathcal{P}$ in the wedge top ($1 \leqslant \alpha \leqslant k$), by taking the product of contents of the $n+k$ vertices (and evaluating all objects at a point $\boldsymbol{x} \in \mathcal{N}$), and summing over all indices, we realise a polydifferential operator in n arguments; the operator coefficients are differential-polynomial in $\mathcal{P}$. Totally skew-symmetric operators of differential order one in each argument are well-defined n-vectors on the affine manifolds $\mathcal{N}$.

The space of multi-vectors G encoded by oriented graphs is equipped with a graded Lie algebra structure, namely the Schouten bracket $[\![\cdot,\cdot]\!]$. Its realisation in terms of oriented graphs is shown in [2, Remark 4]. Recall that by definition the bi-vectors $\mathcal{P}$ at hand are Poisson by satisfying the Jacobi identity $[\![\mathcal{P},\mathcal{P}]\!] = 0$. The Poisson differential $\partial_\mathcal{P} = [\![\mathcal{P},\cdot]\!]$ now endows the space of multi-vectors on $\mathcal{N}$ with the differential graded Lie algebra (dgLa) structure. The cohomology groups produced by the two dgLa structures introduced so far are correlated by the edge orientation mapping $\vec{\mathsf{Or}}$.

Theorem 4 ([8] and [12, App. K]). *Let $\gamma \in \ker \mathrm{d}$ be a cocycle on k vertices and $2k-2$ edges in the non-oriented graph complex. Denote by $\{\Gamma\} \subset G$ the subspace spanned by all those bi-vector graphs Γ which are obtained from (each connected component in) γ by adding to it two edges to the new sink vertices and then by taking the sum of graphs with all the admissible orientations of the old $2k-2$ edges (so that a set of Kontsevich graphs built of k wedges is produced). Then in that subspace $\{\Gamma\}$ there is a sum of graphs that encodes a nonzero Poisson cocycle $Q(\mathcal{P}) \in \ker \partial_\mathcal{P}$.*

Consequently, to find some cocycle $Q(\mathcal{P})$ in the Poisson complex on any affine Poisson manifold it suffices to find a cocycle in the non-oriented graph complex and then consider the sum of graphs which are produced by the orientation mapping $\vec{\mathsf{Or}}$. On the other hand, to list all the $\partial_\mathcal{P}$-cocycles $Q(\mathcal{P})$ encoded by the bi-vector graphs made of k wedges ←•→, one must generate all the relevant oriented graphs and solve the equation $\partial_\mathcal{P}(Q) \doteq 0$ via $[\![\mathcal{P},\mathcal{P}]\!] = 0$, that is, solve

[4] The postulated precedence or antecedence of the wedge product of edges from γ_1 with respect to the edges from γ_2 in every graph within $\gamma_1 \circ_i \gamma_2$ produce the operations $\circ_i$ which coincide with or, respectively, differ from Definition 1 by the sign factor $(-)^{\#E(\gamma_1)\cdot\#E(\gamma_2)}$. The same applies to the Lie bracket of graphs $[\gamma_1,\gamma_2]$ if the operation $\gamma_1 \circ_i \gamma_2$ is the insertion of γ_2 into γ_1 (as in [11]). Anyway, the notion of d-cocycles which we presently recall is well defined and insensitive to such sign ambiguity.

IOP Conf. Series: Journal of Physics: Conf. Series **965** (2018) 012010 doi:10.1088/1742-6596/965/1/012010

graphically the factorisation problem $[\![\mathcal{P}, Q(\mathcal{P})]\!] = \lozenge(\mathcal{P}, [\![\mathcal{P}, \mathcal{P}]\!])$ in which the cocycle condition in the left-hand side holds by virtue of the Jacobi identity in the right. Such construction of some and classification (at a fixed $k > 0$) of all universal infinitesimal symmetries of Poisson brackets are the problems which we explore in this paper.

Remark 1. To the best of our knowledge [10], in a bi-vector graph $Q(\mathcal{P}) = \mathsf{O\vec{r}}(\gamma)$, at every internal vertex which is the tail of two oriented edges towards other internal vertices, the edge ordering Left $\prec$ Right is inherited from a chosen wedge product $E(\gamma)$ of edges in the non-oriented graph γ. How are the new edges towards the sinks ordered, either between themselves at a vertex or with respect to two other oriented edges, coming from γ and issued from different vertices in $Q(\mathcal{P})$? Our findings in [6] will help us to verify the order preservation claim and assess answers to this question.

1. The Kontsevich graph calculus

Definition 2. Let us consider a class of oriented graphs on $n+k$ vertices labelled $0, \ldots, n+k-1$ such that the consecutively ordered vertices $0, \ldots, n-1$ are sinks, and each of the internal vertices $n, \ldots, n+k-1$ is a source for two edges. For every internal vertex, the two outgoing edges are ordered using $L \prec R$: the preceding edge is labelled L (Left) and the other is R (Right). An oriented graph on n sinks and k internal vertices is a *Kontsevich graph* of type (n, k).

For the purpose of defining a graph normal form, we now consider a Kontsevich graph Γ together with a *sign* $s \in \{0, \pm 1\}$, denoted by concatenation of the symbols: $s\Gamma$.

Notation (Encoding of the Kontsevich graphs)**.** The format to store a signed graph $s\Gamma$ for a Kontsevich graph Γ is the integer number $n > 0$, the integer $k \geqslant 0$, the sign s, followed by the (possibly empty, when $k = 0$) list of k ordered pairs of targets for edges issued from the internal vertices $n, \ldots, n+k-1$, respectively. The full format is then (n, k, s; list of ordered pairs).

Definition 3 (Normal form of a Kontsevich graph)**.** The list of targets in the encoding of a graph Γ can be considered as a $2k$-digit integer written in base-$(n+k)$ notation. By running over the entire group $S_k \times (\mathbb{Z}_2)^k$, and by this over all the different re-labellings of Γ, we obtain many different integers written in base-$(n+k)$. The *absolute value* $|\Gamma|$ of Γ is the re-labelling of Γ such that its list of targets is *minimal* as a nonnegative base-$(n+k)$ integer. For a signed graph $s\Gamma$, the *normal form* is the signed graph $t|\Gamma|$ which represents the same polydifferential operator as $s\Gamma$. Here we let $t = 0$ if the graph is zero (see Example 3 below).

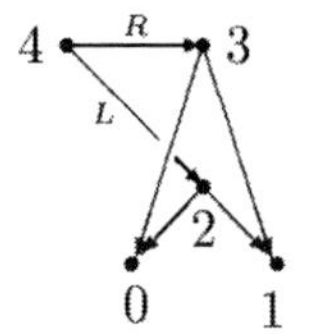

Example 3 (Zero Kontsevich graph)**.** Consider the graph with the encoding `2 3 1  0 1 0 1 2 3`. The swap of vertices $2 \rightleftarrows 3$ is a symmetry of this graph, yet it also swaps the ordered edges $(4 \to 2) \prec (4 \to 3)$, producing a minus sign. Equal to minus itself, this Kontsevich graph is zero.

Notation. Every Kontsevich graph Γ on n sinks (or every sum Γ of such graphs) yields the sum $\mathrm{Alt}\,\Gamma$ of Kontsevich graphs which is totally skew-symmetric with respect to the n sinks content $s_1, \ldots, s_n$. Indeed, let

$$(\mathrm{Alt}\,\Gamma)(s_1, \ldots, s_n) = \sum\nolimits_{\sigma \in S_n} (-)^{\sigma}\, \Gamma(s_{\sigma(1)}, \ldots, s_{\sigma(n)}). \tag{1}$$

Due to skew-symmetrisation, the sum of graphs $\mathrm{Alt}\,\Gamma$ can contain zero graphs or repetitions.

Example 4 (The Jacobiator)**.** The left-hand side of the Jacobi identity is a skew sum of Kontsevich graphs (e.g. it is obtained by skew-symmetrizing the first term)

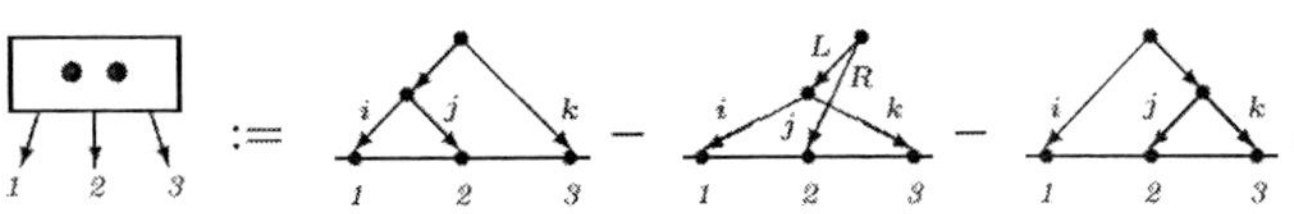

(2)

The default ordering of edges is the one which we see.

IOP Conf. Series: Journal of Physics: Conf. Series **965** (2018) 012010 doi:10.1088/1742-6596/965/1/012010

Definition 4 (Leibniz graph). A *Leibniz graph* is a graph whose vertices are either sinks, or the sources for two arrows, or the Jacobiator (which is a source for three arrows). There must be at least one Jacobiator vertex. The three arrows originating from a Jacobiator vertex must land on three distinct vertices. Each edge falling on a Jacobiator works by the Leibniz rule on the two internal vertices in it.

Example 5. The Jacobiator itself is a Leibniz graph (on one tri-valent internal vertex).

Definition 5 (Normal form of a Leibniz graph with one Jacobiator). Let Γ be a Leibniz graph with one Jacobiator vertex Jac. From (2) we see that expansion of Jac into a sum of three Kontsevich graphs means adding one new edge $w \to v$ (namely joining the internal vertices w and v within the Jacobiator). Now, from Γ construct three Kontsevich graphs by expanding Jac using (2) and letting the edges which fall on Jac in Γ be directed only to v in every new graph. Next, for each Kontsevich graph find the relabelling τ which brings it to its normal form and re-express the edge $w \to v$ using τ. Finally, out of the three normal forms of three graphs pick the minimal one. By definition, the *normal form* of the Leibniz graph Γ is the pair: normal form of Kontsevich graph, that edge $\tau(w) \to \tau(v)$.

We say that a sum of Leibniz graphs is a *skew Leibniz graph* $\operatorname{Alt}\Gamma$ if it is produced from a given Leibniz graph Γ by alternation using formula (1).

Definition 6 (Normal form of a skew Leibniz graph with one Jacobiator). Likewise, the normal form of a skew Leibniz graph $\operatorname{Alt}\Gamma$ is the minimum of the normal forms of Leibniz graphs (specifically, of the graph but not edge encodings) which are obtained from Γ by running over the group of permutations of the sinks content.

Lemma 5 ([3]). In order to show that a sum S of weighted skew-symmetric Kontsevich graphs vanishes for all Poisson structures $\mathcal{P}$, it suffices to express S as a sum of skew Leibniz graphs: $S = \Diamond\big(\mathcal{P}, \operatorname{Jac}(\mathcal{P})\big)$.

1.1. Formulation of the problem

Let $\mathcal{P} \mapsto \mathcal{P} + \varepsilon\mathcal{Q}(\mathcal{P}) + \bar{o}(\varepsilon)$ be a deformation of bi-vectors that preserves their property to be Poisson at least infinitesimally on all affine manifolds: $[\![\mathcal{P} + \varepsilon\mathcal{Q} + \bar{o}(\varepsilon), \mathcal{P} + \varepsilon\mathcal{Q} + \bar{o}(\varepsilon)]\!] = \bar{o}(\varepsilon)$. Expanding and equating the first order terms, we obtain the equation $[\![\mathcal{P}, \mathcal{Q}(\mathcal{P})]\!] \doteq 0$ via $[\![\mathcal{P}, \mathcal{P}]\!] = 0$. The language of Kontsevich graphs allows one to convert this infinite analytic problem within a given set-up $\big(\mathcal{N}^n, \mathcal{P}\big)$ in dimension n into a set of finite combinatorial problems whose solutions are universal for all Poisson geometries in all dimensions $n < \infty$.

Our first task in this paper is to find the space of flows $\dot{\mathcal{P}} = \mathcal{Q}(\mathcal{P})$ which are encoded by the Kontsevich graphs on a fixed number of internal vertices k, for $1 \leqslant k \leqslant 4$. Specifically, we solve the graph equation

$$[\![\mathcal{P}, \mathcal{Q}(\mathcal{P})]\!] = \Diamond\big(\mathcal{P}, \operatorname{Jac}(\mathcal{P})\big) \tag{3}$$

for the Kontsevich bi-vector graphs $\mathcal{Q}(\mathcal{P})$ and Leibniz graphs $\Diamond$. We then factor out the Poisson-trivial and improper solutions, that is, we quotient out all bi-vector graphs that can be written in the form $\mathcal{Q}(\mathcal{P}) = [\![\mathcal{P}, X]\!] + \nabla\big(\mathcal{P}, \operatorname{Jac}(\mathcal{P})\big)$, where X is a Kontsevich one-vector graph and ∇ is a Leibniz bi-vector graph. (The bi-vectors $[\![\mathcal{P}, X]\!]$ make $[\![\mathcal{P}, \mathcal{Q}(\mathcal{P})]\!]$ vanish since $[\![\mathcal{P}, \cdot]\!]$ is a differential. The improper graphs $\nabla(\mathcal{P}, \operatorname{Jac}(\mathcal{P}))$ vanish identically at all Poisson bi-vectors $\mathcal{P}$ on every affine manifold.

Before solving factorisation problem (3) with respect to the operator $\Diamond$, we must generate – e.g., iteratively as described below – an ansatz for expansion of the right-hand side using skew Leibniz graphs with undetermined coefficients.

IOP Conf. Series: Journal of Physics: Conf. Series **965** (2018) 012010 doi:10.1088/1742-6596/965/1/012010

1.2. How to generate Leibniz graphs iteratively

The first step is to construct a layer of skew Leibniz graphs, that is, all skew Leibniz graphs which produce at least one graph in the input (in the course of expansion of skew Leibniz graphs using formula (1) and then in the course of expansion of every Leibniz graph at hand to a sum of Kontsevich graphs). For a given Kontsevich graph in the input S, one such Leibniz graph can be constructed by contracting an edge between two internal vertices so that the new vertex with three outgoing edges becomes the Jacobiator vertex. Note that these Leibniz graphs, which are designed to reproduce S, may also produce extra Kontsevich graphs that are not given in the input. Clearly, if the set of Kontsevich graphs in S coincides with the set of such graphs obtained by expansion of all the Leibniz graphs in the ansatz at hand, then we are done: the extra graphs, not present in S, are known to all cancel. Yet it could very well be that it is not possible to express S using only the Leibniz graphs from the set accumulated so far. Then we proceed by constructing the next layer of skew Leibniz graphs that reproduce at least one of the extra Kontsevich graphs (which were not present in S but which are produced by the graphs in the previously constructed layer(s) of Leibniz graphs). In this way we proceed iteratively until no new Leibniz graphs are found; of course, the overall number of skew Leibniz graphs on a fixed number of internal vertices and sinks is bounded from above so that the algorithm always terminates. Note that the Leibniz graphs obtained in this way are the only ones that can in principle be involved in the vanishing mechanism for S.

Notation. Let v be a graph vertex. Denote by $N(v)$ the set of neighbours of v, by $H(v)$ the (possibly empty) set of arrowheads of oriented edges issued from the vertex v, and by $T(v)$ the (possibly empty) set of tails for oriented edges pointing at v. For example, $\#N(\bullet) = 2$, $\#H(\bullet) = 2$, and $T(\bullet) = \varnothing$ for the top $\bullet$ of the wedge graph $\leftarrow \bullet \rightarrow$.

Algorithm Consider a skew-symmetric sum S_0 of oriented Kontsevich graphs with real coefficients. Let $S_{\text{total}} := S_0$ and create an empty table L. We now describe the ith iteration of the algorithm ($i \geqslant 1$).

Loop ↻ Run over all Kontsevich graphs Γ in S_{i-1}: for each internal vertex v in a graph Γ, run over all vertices $w \in T(v)$ in the set of tails of oriented edges pointing at v such that $v \notin T(w)$ and $H(v) \cap H(w) = \varnothing$ for the sets of targets of oriented edges issued from v and w. Replace the edge $w \to v$ connecting w to v by Jacobiator (2), that is, by a single vertex Jac with three outgoing edges and such that $T(\text{Jac}) = \big(T(v) \setminus w\big) \cup T(w)$ and $H(\text{Jac}) = H(v) \cup \big(H(w) \setminus v\big) =: \{a, b, c\}$.

Because we shall always expand the skew Leibniz graphs in what follows, we do not actually contract the edge $w \to v$ (to obtain a Leibniz graph explicitly) in this algorithm but instead we continue working with the original Kontsevich graphs containing the distinct vertices v and w.

For every edge that points at w, redirect it to v. Sum over the three cyclic permutations that provide three possible ways to attach the three outgoing edges for v and w (excluding $w \to v$) – now seen as the outgoing edges of the Jacobiator – to the target vertices a, b, and c depending on w and v. Skew-symmetrise[5] each of these three graphs with respect to the sinks by applying formula (1).

For every marked edge $w \to v$ indicating the internal edge in the Jacobiator vertex in a graph, replace each sum of the Kontsevich graphs which is skew with respect to the sink content by using the normal form of the respective skew Leibniz graph, see Definition 6. If this skew Leibniz graph is not contained in L, apply the Leibniz rule(s) for all the derivations acting on the Jacobiator vertex Jac. Otherwise speaking, sum over all possible ways to attach the incoming edges of the target v in the marked edge $w \to v$ to its source w and target. To each Kontsevich

[5] This algorithm can be modified so that it works for an input which is not skew, namely, by replacing skew Leibniz graphs by ordinary Leibniz graphs (that is, by skipping the skew-symmetrisation). For example, this strategy has been used in [3, 4] to show that the Kontsevich star product $\star$ mod $\bar{o}(\hbar^4)$ is associative: although the associator $(f \star g) \star h - f \star (g \star h) = \lozenge(\mathcal{P}, \text{Jac}(\mathcal{P}))$ is not skew, it does vanish for every Poisson structure $\mathcal{P}$.

graph resulting from a skew Leibniz graph at hand assign the same undetermined coefficient, and add all these weighted Kontsevich graphs to the sum S_i. Further, add a row to the table L, that new row containing the normal form of this skew Leibniz graph (with its coefficient that has been made common to the Kontsevich graphs).

By now, the new sum of Kontsevich graphs S_i is fully composed. Having thus finished the current iteration over all graphs Γ in the set S_{i-1}, redefine the algebraic sum of weighted graphs S_{total} by subtracting from it the newly formed sum S_i. Collect similar terms in S_{total} and reduce this sum of Kontsevich graphs modulo their skew-symmetry under swaps $L \rightleftarrows R$ of the edge ordering in every internal vertex, so that all zero graphs (see Example 3) also get eliminated. ⟲ **end loop**

Increment i by 1 and repeat the iteration until the set of weighted (and skew) Leibniz graphs L stabilizes. Finally, solve – with respect to the coefficients of skew Leibniz graphs – the linear algebraic system obtained from the graph equation $S_{\text{total}} = 0$ for the sum of Kontsevich graphs which has been produced from its initial value S_0 by running the iterations of the above algorithm.

Example 6. For the skew sum of Kontsevich graphs in the right-hand side of (2), the algorithm would produce just one skew Leibniz graph: namely, the Jacobiator itself.

Example 7 (The 3-wheel)**.** For the Kontsevich tetrahedral flow $\dot{\mathcal{P}} = \mathcal{Q}_{1:6/2}(\mathcal{P})$ on the spaces of Poisson bi-vectors $\mathcal{P}$, see [8, 9] and [1, 2], building a sufficient set of skew Leibniz graphs in the r.-h. s. of factorisation problem (3) requires two iterations of the above algorithm: 11 Leibniz graphs are produced at the first step and 50 more are added by the second step, making 61 in total. One of the two known solutions of this factorisation problem [2] then consists of 8 skew Leibniz graphs (expanding to 27 Leibniz graphs). In turn, as soon as all the Leibniz rules acting on the Jacobiators are processed and every Jacobiator vertex is expanded via (2), the right-hand side $\Diamond(\mathcal{P}, \text{Jac}(\mathcal{P}))$ equals the sum of 39 Kontsevich graphs which are assembled into the 9 totally skew-symmetric terms in the left-hand side $[\![\mathcal{P}, \mathcal{Q}_{1:6/2}]\!]$.

Example 8 (The 5-wheel)**.** Consider the factorisation problem $[\![\mathcal{P}, \mathsf{O\vec{r}}(\gamma_5)]\!] = \Diamond(\mathcal{P}, \text{Jac}(\mathcal{P}))$ for the pentagon-wheel deformation $\dot{\mathcal{P}} = \mathsf{O\vec{r}}(\gamma_5)(\mathcal{P})$ of Poisson bi-vectors $\mathcal{P}$, see [10, 13] and [6]. The ninety skew Kontsevich graphs encoding the bi-vector $\mathsf{O\vec{r}}(\gamma_5)$ are obtained by taking all the admissible orientations of two $(5+1)$-vertex graphs γ_5, one of which is the pentagon wheel with five spokes, the other graph complementing the former to a cocycle in the non-oriented graph complex. By running the iterations of the above algorithm for self-expanding construction of the Leibniz tri-vector graphs in this factorisation problem, we achieve stabilisation of the number of such graphs after the seventh iteration, see Table 1 below.

Table 1. The number of skew Leibniz graphs produced iteratively for $[\![\mathcal{P}, \mathsf{O\vec{r}}(\gamma_5)]\!]$.

No. iteration i	1	2	3	4	5	6	7	8
# of graphs	1609	14937	41122	54279	56409	56594	56600	56600
of them new	all	+13328	+26185	+13157	+2130	+185	+6	none

2. Generating the Kontsevich multi-vector graphs

Let us return to problem (3): it is the ansatz for bi-vector Kontsevich graphs $\mathcal{Q}(\mathcal{P})$ with k internal vertices, as well as the Kontsevich 1-vectors $\mathcal{X}$ with $k-1$ internal vertices (to detect trivial terms $\mathcal{Q}(\mathcal{P}) = [\![\mathcal{P}, \mathcal{X}]\!]$) which must be generated at a given k. (At $1 \leqslant k \leqslant 4$, one can still expand with respect to *all* the Leibniz graphs in the r.-h.s. of (3), not employing the iterative algorithm from §1.2. So, a generator of the Kontsevich (and Leibniz) tri-vectors will also be described presently.)

The Kontsevich graphs corresponding to n-vectors are those graphs with n sinks (each containing the respective argument of n-vector) in which exactly one arrow comes into each sink, so that the order of the differential operator encoded by an n-vector graph equals one w.r.t. each argument, and which are totally skew-symmetric in their n arguments. Let us explain how one can economically obtain the set of one-vectors and skew-symmetric bi- and tri-vectors with k internal vertices in three steps (including graphs with eyes $\bullet \rightleftarrows \bullet$ but excluding graphs with tadpoles). This approach can easily be extended to the construction of n-vectors with any $n \geqslant 1$.

2.1. *One-vectors*
Each one-vector under study is encoded by a Kontsevich graph with one sink. Since the sink has one incoming arrow, there is an internal vertex as the tail of this incoming arrow. The target of another edge issued from this internal vertex can be any internal vertex other then itself.

Step 1. Generate all Kontsevich graphs on $k-1$ internal vertices and one sink (i.e. graphs including those with eyes yet excluding those with tadpoles, and not necessarily of differential order one with respect to the sink content).
Step 2. For every such graph with $k-1$ internal vertices, add the new sink and make it a target of the old sink, which itself becomes the kth internal vertex. Now run over the $k-1$ internal vertices excluding the old sink and – via the Leibniz rule – make every such internal vertex the second target of the old sink.

2.2. *Bi-vectors*
There are two cases in the construction of bi-vectors encoded by the Kontsevich graphs. At all $k \geqslant 1$ the first variant is referred to those graphs with an internal vertex that has both sinks as targets.

Variant 1: *Step 1.* Generate all k-vertex graphs on $k-1$ internal vertices and one sink.
Variant 1: *Step 2.* For every such graph, add two new sinks and proclaim them as targets of the old sink.

Note that the obtained graphs are skew-symmetric.

The second variant produces those graphs which contain two internal vertices such that one has the first sink as target and the other has the second sink as target. The second target of either such internal vertex can be any internal vertex other then itself. Note that for $k = 1$ only the first variant applies.

Variant 2: *Step 1.* Generate all k-vertex Kontsevich graphs on $k-2$ internal vertices and two sinks. These sinks now become the $(k-1)$th and kth internal vertices.
Variant 2: *Step 2.* For every such graph, add two new sinks, make the first new sink a target of the first old sink and make the second new sink a target of the second old sink. Now run over the $k-1$ internal vertices excluding the first old sink, each time proclaiming an internal vertex the second target of the first old sink. Simultaneously, run over the $k-1$ internal vertices excluding the second old sink and likewise, declare an internal vertex to be the second target of the second old sink.
Variant 2: *Step 3.* Skew-symmetrise each graph with respect to the content of two sinks using (1).

2.3. *Tri-vectors*
For $k \geqslant 3$, there exist two variants of tri-vectors. The first variant at all $k \geqslant 2$ yields those Kontsevich graphs with two internal vertices such that one has two of the three sinks as its

IOP Conf. Series: Journal of Physics: Conf. Series **965** (2018) 012010 doi:10.1088/1742-6596/965/1/012010

targets while another internal vertex has the third sink as one of its targets. The second target of this last vertex can be any internal vertex other then itself. The second variant contains those graphs with three internal vertices such that the first one has the first sink as a target, the second one has the second sink as a target, and the third one has the third sink as a target. For each of these three internal vertices with a sink as target, the second target can be any internal vertex other then itself.

Variant 1: *Step 1.* Generate all k-vertex Kontsevich graphs on $k-2$ internal vertices and two sinks.
Variant 1: *Step 2.* For every such graph, add three new sinks, make the first two new sinks the targets of the first old sink and make the third new sink a target of the second old sink. Now run over the $k-1$ internal vertices excluding the second old sink and every time declare an internal vertex the second target of the second old sink.
Variant 1: *Step 3.* Skew-symmetrise all graphs at hand by applying formula (1) to each of them.

Note that for $k=1$ there are no tri-vectors encoded by Kontsevich graphs and also note that for $k=2$ only the first variant applies.

Variant 2: *Step 1.* Generate all Kontsevich graphs on $k-3$ internal vertices and three sinks.
Variant 2: *Step 2.* For every such graph, add three new sinks, make the first new sink a target of the first old sink, make the second new sink a target of the second old sink and make the third new sink a target of the third old sink. Now run over the $k-1$ internal vertices excluding the first old sink and declare every such internal vertex the second target of the first old sink. Independently, run over the $k-1$ internal vertices excluding the second old sink and declare each internal vertex to be the second target of the second old sink. Likewise, run over the $k-1$ internal vertices excluding the third old sink and declare each internal vertex to be the second target of the third old sink.
Variant 2: *Step 3* Skew-symmetrise all the graphs at hand using (1).

2.4. *Non-iterative generator of the Leibniz n-vector graphs*

The following algorithm generates all Leibniz graphs with a prescribed number of internal vertices and sinks. Note that not only multi-vectors, but also all graphs of arbitrary differential order with respect to the sinks can be generated this way.

Step 1: Generate all Kontsevich graphs of prescribed type on $k-1$ internal vertices and n sinks, e.g., all n-vectors.
Step 2: Run through the set of these Kontsevich graphs and in each of them, run through the set of its internal vertices v. For every vertex v do the following: re-enumerate the internal vertices so that this vertex is enumerated by $k-1$. This vertex already targets two vertices, i and j, where $i<j<k-1$. Proclaim the last, $(k-1)$th vertex to be the placeholder of the Jacobiator (see (2)), so we must still add the third arrow. Let a new index ℓ run up to $i-1$ starting at n if only the n-vectors are produced.[6] For every admissible value of ℓ, generate a new graph where the ℓth vertex is proclaimed the third target of the Jacobiator vertex $k-1$. (Restricting ℓ by $<i$, we reduce the number of possible repetitions in the set of Leibniz graphs. Indeed, for every triple $\ell<i<j$, the same Leibniz graph in which the Jacobiator stands on those three vertices would be produced from the three Kontsevich graphs: namely, those in which the $(k-1)$th vertex targets at the ℓth and ith, at the ℓth and jth, and at the ith and jth vertices. In these three cases it is the jth, ith, and ℓth vertex, respectively, which would be appointed by the algorithm as the Jacobiator's third target.)

We use this algorithm to generate the Leibniz tri- and bi-vector graphs: to establish Theorem 6, we list all possible terms in the right-hand side of factorisation problem (3) at $k\leqslant 4$

[6] If we want to generate not only n-vectors but all graphs of arbitrary differential orders, then we let ℓ start at 0 (so that the sinks are included).

and then we filter out the improper bi-vectors in the found solutions $\mathcal{Q}(\mathcal{P})$.

Remark 2. There are at least 265,495 Leibniz graphs on 3 sinks and 6 internal vertices of which one is the Jacobiator vertex. When compared with Table 1 on p. 7, this estimate suggests why at large $k \gtrsim 5$, the breadth-first-search iterative algorithm from §1.2 generates a smaller number of the Leibniz tri-vector graphs, namely, only the ones which can in principle be involved in the factorisation under study.

3. Main result

Theorem 6 ($k \leqslant 4$). *The few-vertex solutions of problem* (3) *are these* (*note that disconnected Kontsevich graphs in* $\mathcal{Q}(\mathcal{P})$ *are allowed*!):

- $k = 1$: *The dilation* $\dot{\mathcal{P}} = \mathcal{P}$ *is a unique, nontrivial and proper solution.*
- $k = 2$: *No solutions exist* (*in particular, neither trivial nor improper*).
- $k = 3$: *There are no solutions*: *neither Poisson-trivial nor Leibniz bi-vectors.*
- $k = 4$: *A unique nontrivial and proper solution is the Kontsevich tetrahedral flow* $\mathcal{Q}_{1:\frac{6}{2}}(\mathcal{P})$ *from Example* 2 (*see* [8, 9] *and* [1, 2]). *There is a one-dimensional space of Poisson trivial* (*still proper*) *solutions* $[\![\mathcal{P}, X]\!]$; *the Kontsevich* 1*-vector* X *on three internal vertices is drawn in* [2, App. F]. *Intersecting with the former by* $\{0\}$, *there is a three-dimensional space of improper* (*still Poisson-nontrivial*) *solutions of the form* $\nabla\big(\mathcal{P}, \mathrm{Jac}(\mathcal{P})\big)$.

None of the solutions $\mathcal{Q}(\mathcal{P})$ *known so far contains any* 2*-cycles* (*or "eyes"* $\bullet \rightleftarrows \bullet$).[7]

We now report a classification of Poisson bi-vector symmetries $\dot{\mathcal{P}} = \mathcal{Q}(\mathcal{P})$ which are given by those Kontsevich graphs $\mathcal{Q} = \mathsf{Or}(\gamma)$ on k internal vertices that can be obtained at $5 \leqslant k \leqslant 9$ by orienting k-vertex connected graphs γ without double edges. By construction, this extra assumption keeps only those Kontsevich graphs which may not contain eyes.

We first find such graphs γ that satisfy $\mathrm{d}(\gamma) = 0$, then we exclude the coboundaries $\gamma = \mathrm{d}(\gamma')$ for some graphs γ' on $k - 1$ vertices and $2k - 3$ edges.

Theorem 7 ($5 \leqslant k \leqslant 9$). *Consider the vector space of non-oriented connected graphs on* k *vertices and* $2k - 2$ *edges, without tadpoles and without multiple edges. All nontrivial* d*-cocycles for* $5 \leqslant k \leqslant 9$ *are exhausted by the following ones:*

- $k = 5, 7, 9$: *No solutions.*
- $k = 6$: *A unique solution*[8] *is given by the Kontsevich–Willwacher pentagon-wheel cocycle* (*see Example* 1). *The established factorisation* $[\![\mathcal{P}, \mathsf{Or}(\gamma_5)]\!] = \Diamond\big(\mathcal{P}, \mathrm{Jac}(\mathcal{P})\big)$ *will be addressed in a separate paper* (*see* [6]).

$\gamma_5 =$ 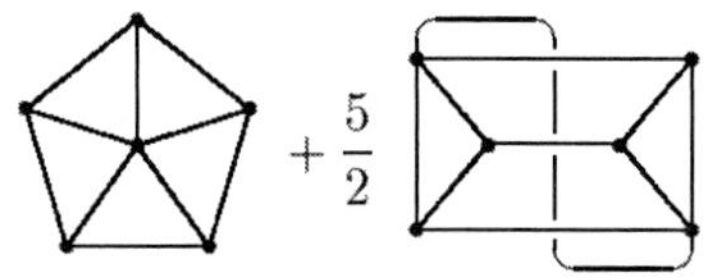

- $k = 8$: *The only solution* γ_7 *consists of the heptagon-wheel and* 45 *other graphs* (*see Table* 2, *in which the coefficient of heptagon graph is* **1** *in bold, and* [5]).

Remark 3. The wheel graphs are built of triangles. The differential d cannot produce any triangle since multiple edges are not allowed. Therefore, all wheel cocycles are nontrivial. Note also that every wheel graph with 2ℓ spokes is invariant under a mirror reflection with respect to a diagonal consisting of two edges attached to the centre. Hence there exists an edge permutation that swaps $2\ell - 1$ pairs of edges. By footnote 1 such graph equals zero.

Acknowledgments

A. V. Kiselev thanks the Organising committee of the international conference ISQS'25 on integrable systems and quantum symmetries (6–10 June 2017 in ČVUT Prague, Czech Republic)

[7] Finding solutions $\mathcal{Q}(\mathcal{P})$ with tadpoles or extra sinks –with fixed arguments– is a separate problem.

[8] There are only 12 admissible graphs to build cocycles from; of these 12, as many as 6 are zero graphs. This count shows to what extent the number of graphs decreases if one restricts to only the flows $\mathcal{Q} = \mathsf{Or}(\gamma)$ obtained from cocycles $\gamma \in \ker(\mathrm{d})$ in the non-oriented graph complex.

Table 2. The heptagon-wheel graph cocycle γ_7.

Graph encoding	Coeff.	Graph encoding	Coeff.
16 17 18 23 25 28 34 38 46 48 57 58 68 78	1	12 13 18 25 26 37 38 45 46 47 56 57 68 78	−7
12 14 18 23 27 35 37 46 48 57 58 67 68 78	−21/8	12 14 16 23 25 36 37 45 48 57 58 67 68 78	77/8
13 14 18 23 25 28 37 46 48 56 57 67 68 78	−77/4	13 16 17 24 25 26 35 37 45 48 58 67 68 78	−7
12 13 15 24 27 35 36 46 48 57 58 67 68 78	−35/8	14 15 17 23 26 28 37 38 46 48 56 57 68 78	49/4
12 13 18 24 26 37 38 46 47 56 57 58 68 78	49/8	12 16 18 27 28 34 36 38 46 47 56 57 58 78	−147/8
14 17 18 23 25 26 35 37 46 48 56 58 67 78	77/8	12 15 16 27 28 35 36 38 45 46 47 57 68 78	−21/8
12 13 18 26 27 35 38 45 46 47 56 57 68 78	−105/8	12 14 18 23 27 35 36 45 46 57 58 67 68 78	−35/8
12 14 18 23 27 36 38 46 48 56 57 58 67 78	7/8	14 15 16 23 26 28 37 38 46 48 57 58 67 78	−49/4
12 14 15 23 27 35 36 46 48 57 58 67 68 78	35/8	12 15 18 23 28 34 37 46 48 56 57 67 68 78	105/8
12 13 14 27 28 36 38 46 47 56 57 58 68 78	−49/8	12 14 17 23 26 37 38 46 48 56 57 58 68 78	−49/8
12 13 18 25 27 34 36 47 48 56 58 67 68 78	35/4	12 16 18 25 27 35 36 37 45 46 48 57 68 78	49/16
12 13 14 25 26 36 38 45 47 57 58 67 68 78	−119/16	12 13 18 25 27 35 36 46 47 48 56 57 68 78	7
12 13 15 24 28 36 38 47 48 56 57 67 68 78	49/8	12 14 18 25 28 34 36 38 47 57 58 67 68 78	−7
12 13 14 23 28 37 46 48 56 57 58 67 68 78	77/4	12 16 18 25 27 35 36 37 45 46 48 58 67 78	−77/16
12 15 17 25 26 35 36 38 45 47 48 67 68 78	−49/8	12 14 18 23 27 35 38 46 47 57 58 67 68 78	77/4
13 15 18 24 26 28 37 38 46 47 56 57 68 78	−49/4	12 14 15 23 27 36 38 46 48 57 58 67 68 78	35/2
13 14 18 25 26 28 36 38 47 48 56 57 67 78	−49/4	12 13 18 25 27 34 36 46 48 57 58 67 68 78	−105/8
12 14 18 23 28 35 37 46 48 56 57 67 68 78	−7	12 15 16 25 27 35 36 38 46 47 48 57 68 78	−7
12 14 18 23 28 36 38 46 47 56 57 58 67 78	−7	12 13 16 25 28 34 37 47 48 57 58 67 68 78	−147/16
12 15 16 25 27 35 36 38 46 47 48 58 67 78	49/8	12 13 17 25 26 35 37 45 46 48 58 67 68 78	−77/4
12 14 18 23 28 36 37 46 47 56 57 58 68 78	49/8	12 14 17 23 27 35 38 46 48 57 58 67 68 78	−49/8
12 13 15 26 27 35 36 45 47 48 58 67 68 78	−7	12 13 15 26 28 35 37 45 46 47 58 67 68 78	−7/4
12 13 18 24 28 35 38 46 47 57 58 67 68 78	7	12 14 18 23 26 36 38 47 48 56 57 58 67 78	−7

for a warm atmosphere during the meeting. The authors are grateful to M. Kontsevich and T. Willwacher for helpful discussion. We also thank Center for Information Technology of the University of Groningen for providing access to Peregrine high performance computing cluster. This research was supported in part by JBI RUG project 106552 (Groningen, The Netherlands). A part of this research was done while R. Buring and A. V. Kiselev were visiting at the IHÉS (Bures-sur-Yvette, France) and A. V. Kiselev was visiting at the MPIM (Bonn, Germany).

Appendix A. How the orientation mapping $\vec{\mathsf{Or}}$ is calculated

The algorithm lists all ways in which a given non-oriented graph can be oriented in such a way that it becomes a Kontsevich graph on two sinks. It consists of two steps:

(i) choosing the source(s) of the two arrows pointing at the first and second sink, respectively;

(ii) orienting the edges between the internal vertices in all admissible ways, so that only Kontsevich graphs are obtained.

Step 1. Enumerate the k vertices of a given non-oriented, connected graph using $2, \ldots, k+1$. They become the internal vertices of the oriented graph. Now add the two sinks to the non-oriented graph, the sinks enumerated using 0 and 1. Let a and b be a non-strictly ordered ($a \leqslant b$) pair of internal vertices in the graph. Extend the graph by oriented edges $a \to 0$ and $b \to 1$ from vertices a and b to the sinks 0 and 1, respectively.

Remark 4. The choice of such a base pair, that is, the vertex or vertices from which two arrows are issued to the sinks, is an external input in the orientation procedure. Let us agree that if, at any step of the algorithm, a contradiction is achieved so that a graph at hand cannot be of Kontsevich type, the oriented graph draft is discarded; one proceeds with the next options in that loop, or if the former loop is finished, with the next level-up loops, or – having returned to the choice of base vertices – with the next base. In other words, we do not exclude in principle a possibility to have no admissible orientations for a particular choice of the base for a given non-oriented graph.

Notation. Let v be an internal vertex. Recalling from p. 6 the notation for the set $N(v)$ of neighbours of v, the (initially empty) set $H(v)$ of arrowheads of oriented edges issued from the vertex v, and the (initially empty) set $T(v)$ of tails for oriented edges pointing at v, we now put by definition $F(v) := N(v) \setminus (H(v) \cup T(v))$. In other words, $F(v)$ is the subset of neighbours connected with v by a non-oriented edge.

Step 2.1. Inambiguous orientation of (some) edges. Here we use that every internal vertex of a Kontsevich graph should be the tail of exactly two outgoing arrows. We run over the set of all internal vertices v. For every vertex such that the number of elements $\#H(v) = 2$, proclaim $T(v) := N(v) \setminus H(v)$, whence $F(v) = \varnothing$. If for a vertex v we have that $\#H(v) = 1$ and $\#F(v) = 1$, then include $F(v) \hookrightarrow H(v)$, that is, convert a unique non-oriented edge touching v into an outgoing edge issued from this vertex. If $\#H(v) = 0$ and $\#F(v) = 2$, also include $F(v) \hookrightarrow H(v)$, effectively making both non-oriented edges outgoing from v.

Repeat the three parts of Step 2.1 while any of the sets $F(v)$, $T(v)$, or $S(v)$ is modified for at least one internal vertex v unless a contradiction is revealed. Summarising, Step 2.1 amounts to finding the edge orientations which are implied by the choice of the base pair a, b and by all the orientations of edges fixed earlier.

Step 2.2. Fixing the orientation of (some) remaining edges. Choose an internal vertex v such that $H(v) < 2$ and such that $H(v) \neq \varnothing$ or $T(v) \neq \varnothing$, that is, choose a vertex that is not yet equipped with two outgoing edges and that is attached to an oriented edge. If $\#H(v) = 1$, then run over the non-empty set $F(v)$: for every vertex w in $F(v)$, include $\{w\} \hookrightarrow H(v)$ and start over at Step 2.1. Otherwise, i.e. if $H(v) = \varnothing$, run over all ordered pairs (u, v) of vertices in the set $F(v)$: for every such pair, make $H(v) := \{u, w\}$ and start over at Step 2.1.

By realising Steps 1 and 2 we accumulate the sum of fully oriented Kontsevich graphs.

References

[1] Bouisaghouane A and Kiselev A V 2017 Do the Kontsevich tetrahedral flows preserve or destroy the space of Poisson bi-vectors? *J. Phys.: Conf. Ser.* **804** 012008 (*Preprint* `arXiv:1609.06677` [q-alg])
[2] Bouisaghouane A, Buring R and Kiselev A 2017 The Kontsevich tetrahedral flow revisited *J. Geom. Phys.* **119** 272–85 (*Preprint* `arXiv:1608.01710` [q-alg])
[3] Buring R and Kiselev A V 2017 On the Kontsevich $\star$-product associativity mechanism *PEPAN Letters* **14**:2 403–7 (*Preprint* `arXiv:1602.09036` [q-alg])
[4] Buring R and Kiselev A V 2017 Software modules and computer-assisted proof schemes in the Kontsevich deformation quantization *Preprint* IHÉS/M/17/05
[5] Buring R, Kiselev A V and Rutten N J 2017 The heptagon-wheel cocycle in the Kontsevich graph complex *J. Nonlin. Math. Phys.*
[6] Buring R, Kiselev A V and Rutten N J 2017 The Kontsevich–Willwacher pentagon-wheel symmetry of Poisson structures
[7] Dolgushev V A, Rogers C L and Willwacher T H 2015 Kontsevich's graph complex, GRT, and the deformation complex of the sheaf of polyvector fields *Ann. Math.* **182**:3 855–943 (*Preprint* `arXiv:1211.4230` [math.KT])
[8] Kontsevich M 1997 Formality conjecture *Deformation theory and symplectic geometry* June 17–21 1996 Ascona *Math. Phys. Stud.* vol 20 ed D Sternheimer et al (Dordrecht: Kluwer Acad. Publ.) pp 139–56
[9] Kontsevich M 2017 Derived Grothendieck–Teichmüller group and graph complexes [after T. Willwacher] *Séminaire Bourbaki* 1126
[10] Kontsevich M 2017 Private communication
[11] Khoroshkin A, Willwacher T and Živković M 2017 Differentials on graph complexes *Adv. Math.* **307** 1184–214 (*Preprint* `arXiv:1411.2369` [q-alg])
[12] Willwacher T 2015 M. Kontsevich's graph complex and the Grothendieck–Teichmüller Lie algebra *Invent. Math.* **200**:3 671–760 (*Preprint* `arXiv:1009.1654` [q-alg])
[13] Willwacher T 2017 Private communication

IOP Conf. Series: Journal of Physics: Conf. Series **965** (2018) 012011 doi:10.1088/1742-6596/965/1/012011

On the vibron-polaron damping in quasi 1D macromolecular chains

D. Cevizovic[1], A. Chizhov[2,3], A. Reshetnyak[4], Z. Ivic[1], S. Galovic[1], S. Petkovic[1]

[1]University of Belgrade, "Vinča" Institute of Nuclear sciences, Laboratory for Theoretical and Condensed Matter Physics, P.O. BOX 522, 11001, Belgrade, Serbia
[2]Joint Institute for Nuclear Research, Bogoliubov Laboratory of Theoretical Physics, Dubna, 141980, Russia
[3]Dubna State University, Dubna, 141980, Russia
[4]Institute of Strength Physics and Materials Science SB RAS, Tomsk, 634055, Russia

E-mail: cevizd@vinca.rs

Abstract. The properties of the intramolecular vibrational excitation (vibron) in a quasi 1D macromolecular structure are studied. It is supposed that due to the vibron interaction with optical phonon modes, a vibron might form partially dressed small polaron states. The properties of these states are investigated in dependence on the basic system parameters and temperature of a thermal bath. We also investigate the process of damping of the polaron amplitude as a function of temperature and vibron-phonon coupling strength. Two different regimes of the polaron damping are found and discussed.

1. Introduction

The knowledge of exciton properties and especially physics of exciton transport in quasi 1D structures is very important in many fields of material sciences. Here, it is pertinent to mention the problem of charge and energy transport in biological macromolecules (MCs) and the problem of the quantum information transport in quantum circuits, especially in realistic conditions, where a thermal bath is an important source of quantum information losses. For example, the energy and charge transport in biological macromolecules (such as protein macromolecules, RNA, and DNA) plays an important role in the processes which take place in functioning of living cells. One of the most important processes is the photosynthesis, in which a photon quantum is absorbed by special cell structures (light-harvesting antennas) as an electronic excitation. After that these excitations are transported to the reaction centres, where they are converted to the chemical energy. It is remarkable fact that during its migration almost every absorbed photon is successfully transferred to the reaction centre despite its short life-time (the characteristic life-time of an electron excitation is approximately 1 ns).

There are two main subsystems that we are interested in, namely a vibronic excitation and phonons. Due to the presence of intersite coupling, a vibronic excitation can migrate between two adjacent sites. The phonon subsystem affects to the excitation migration on different ways. First, due to the presence of the phonon-excitation interaction, the excitation can be self-trapped and form a new quasiparticle dressed by the cloud of virtual phonons, i.e. a polaron quasiparticle. Under some circumstances, such a quasiparticle can be more stable than the bare excitation, and in principle such a model might explain the transfer of the excitation over long distances along a macromolecular spine [1, 2, 3, 4]. Second, the phonon subsystem is a source of dissipative processes in relation to the excitation (polaron) subsystem. Phonons prevent the formation of the polaron quasiparticle and reduce its life-time. As a consequence, the transfer properties of polarons in MCs are determined by the interplay of dynamic and dissipative processes [3, 5, 6].

In this paper, we investigate the process of relaxation of exciton states on the lattice vibrations in 1D macromolecular chains. We suppose that, due to the presence of exciton-phonon coupling, an exciton quasiparticle is self-trapped and forms a non-adiabatic polaron state. It is then assumed that the dissipation of such a state is the consequence of the presence of the residual polaron-phonon interaction (all other sources of dissipation, such as polaron-polaron interaction, are disregarded). The process of the exciton self-trapping is described in the Sec. 2. Main properties of the formed polaron state are presented in dependence on the values of the basic energy parameters of macromolecular systems. In the Sec. 3 we consider the dependence of the polaron amplitude damping on the system parameters and the thermal bath temperature.

2. Physical foundation of the problem

The framework of our investigation is the Holstein molecular crystal model (MCM) [7]. We start from the Hamiltonian which includes the exciton (vibron) interaction with the optical phonon subsystem.

$$H = E\sum_n a_n^\dagger a_n - J\sum_n a^\dagger\left(a_{n-1}+a_{n+1}\right) + \sum_q \hbar\omega_q b_q^\dagger b_q + \frac{1}{\sqrt{N}}\sum_{n,q} F_q e^{iqnR_0} a_n^\dagger a_n\left(b_q + b_{-q}^\dagger\right). \quad (1)$$

In the above Hamiltonian E is the exciton excitation energy, J is the resonant transfer integral (intersite coupling strength), which is responsible for quantum jumps of the particle between neighbouring sites, R_0 is the parameter of the macromolecular chain, a_n ($a_n^\dagger$) are the operators of annihilation (creation) of an exciton on the n-th structure element of the macromolecular chain, and b_q ($b_q^\dagger$) are the phonon annihilation (creation) operators. The relation for the exciton-phonon coupling constant F_q depends on the nature of the phonon subsystem. For the optical phonons, this parameter has the form $F_q = \chi\sqrt{\hbar/(2M\omega_q)}$ with χ and M being by the interaction parameter and the mass of the chain structure element respectively.

The standard procedure to pass into the polaron picture (dressed vibron point of view), is based on the Lang-Firsov (LF) unitary transformation [4, 8]. Such approach is applicable in the case where the conditions for a formation of the nonadiabatic polaron are fulfilled (i.e. for the strong-coupling exciton–phonon case). However, in the case of a vibron excitation in the biological macromolecules these conditions are not satisfied (and therefore vibron-phonon interaction can belong to intermediate or weak coupling cases!) [9]. Consequently, in this case, the variational approaches are much more promising ones [10, 11, 12, 13, 14]. Here, we use the variational approach based on the modified Lang-Firsov unitary transformation (MLFUT): $U = \exp(-\sum_n a_n^\dagger a_n S_n)$, where $S_n = 1/\sqrt{N}\sum_q f_q e^{-iqnR_0}(b_{-q} - b_q^\dagger)$, and $f_q = f_{-q}^*$ being by the parameters that determine the degree of exciton dressing by a cloud of phonons. They can be obtained with use of the variational approach, i.e. by the minimization of the polaron ground state energy. Using MLFUT, we obtain a transformed Hamiltonian with new operators that correspond to dressed exciton (polaron) operators and the new phonons (which describe oscillations of structure elements around new equilibrium positions):

$$\begin{aligned}\tilde{H} &= \mathcal{E}\sum_n a_n^\dagger a_n - J\sum_n a_n^\dagger\left(a_{n-1}\theta_{n-1} + a_{n+1}\theta_{n+1}\right) + \sum_n \hbar\omega_q b_q^\dagger b_q + \\ &+ \frac{1}{\sqrt{N}}\sum_{n,q}(F_q - \hbar\omega_q f_q^*)e^{iqnR_0} a_n^\dagger a_n\left(b_q + b_{-q}^\dagger\right) + \tilde{H}_R, \end{aligned} \quad (2)$$

where $\theta_{n\pm1} = \exp(S_{n\pm1} - S_n)$ is dressing operator, and $\mathcal{E} = E + N^{-1}\sum_q \hbar\omega_q|f_q|^2 - N^{-1}\sum_q F_q(f_q + f^*_{-q})$ is the energy shift of dressed excitons. The fourth term in (2) describes the remaining interaction between the polaron and new phonon, which is responsible for the phonon-induced polaron dissipation processes (i.e. for polaron amplitude damping). The rest part of the Hamiltonian reads as $\tilde{H}_R = N^{-1}\sum_{q,n,m\neq n}\left\{\hbar\omega_q|f_q|^2 - F_q(f_q + f_q^*)\right\}e^{iq(n-m)R_0}a_n a_m a_n^\dagger a_m^\dagger$ corresponds to the multi-polaron processes.

In order to account for the influence of thermal fluctuations of structure elements on polaron properties, we make averaging of our Hamiltonian over the new phonon subsystem to get $\mathcal{H} = \mathcal{H}_0 + \mathcal{H}_{rest}$. Here, $\mathcal{H}_0 = \mathcal{H}_{vib} + \tilde{H}_{ph}$ stands for the effective mean-field Hamiltonian consists of effective vibron ($\mathcal{H}_{vib} = \left\langle \tilde{H} - \tilde{H}_{ph}\right\rangle_{ph}$) and the phonon ($\tilde{H}_{ph}$) parts. Further, $\mathcal{H}_{rest} = \tilde{H} - \tilde{H}_{ph} - \left\langle \tilde{H} - \tilde{H}_{ph}\right\rangle_{ph}$ is the rest part of the interaction between the dressed exciton and new phonon. The symbol $\langle\ \rangle_{ph}$ denotes the averaging over new-phonon ensemble. Due to the fact that relaxation processes for the phonons which forms the thermal bath are rather fast (with the typical time constants in the biological macromolecular chains of order $10^{-14} \div 10^{-11}$ s) in comparison with the characteristic time-scale of the transfer processes, they are populated according to an equilibrium statistical distribution. The explicit form of effective vibron Hamiltonian is

$$\mathcal{H}_{vib} = \mathcal{E}\sum_n a_n^\dagger a_n - J\mathrm{e}^{-W(T)}\sum_n\left\{a_n^\dagger a_{n-1} + a_n^\dagger a_{n+1}\right\} \tag{3}$$

and the rest part of the Hamiltonian has the form

$$\begin{aligned}\mathcal{H}_{rest} &= -J\sum_n\left\{a_n^\dagger a_{n-1}\left(\theta_{n-1} - \langle\theta_{n-1}\rangle_{ph}\right) + a_n^\dagger a_{n+1}\left(\theta_{n+1} - \langle\theta_{n+1}\rangle_{ph}\right)\right\} + \\ &+\frac{1}{\sqrt{N}}\sum_{n,q}\left(F_q - \hbar\omega_q f_q^*\right)\mathrm{e}^{iqnR_0}a_n^\dagger a_n(b_q + b_{-q}^\dagger),\end{aligned} \tag{4}$$

where $\langle\theta_{n\pm1}\rangle_{ph} = \mathrm{e}^{-W(T)}$, $W(T) = N^{-1}\sum_q|f_q|^2(2\nu_q+1)(1-\cos(qR_0))$, is dressing fraction and $\nu_q = \left(\mathrm{e}^{\hbar\omega_q/k_BT} - 1\right)^{-1}$ describes the equilibrium population of the q-th phonon mode. The first term in (4) represents the fluctuations of the vibron-phonon interaction energy around its mean value. We assume that these fluctuations are small, and they can be disregarded. The second part is the main part of the polaron-phonon interaction which is responsible for the attenuation of the polaron amplitude.

We then transform our Hamiltonian from the coordinate to the wave vector k representation using $a_n = 1/\sqrt{N}\sum_k \mathrm{e}^{iknR_0}a_k$, and $a_n^\dagger = 1/\sqrt{N}\sum_k \mathrm{e}^{-iknR_0}a_k^\dagger$ can be given as

$$\mathcal{H} = \mathcal{H}_{vib} + \tilde{H}_{ph} + \mathcal{H}_{rest} = \sum_k \mathcal{E}_k a_k^\dagger a_k + \sum_q \hbar\omega_q b_q^\dagger b_q + \mathcal{H}_{rest}, \tag{5}$$

where

$$\mathcal{H}_{rest} = \frac{1}{\sqrt{N}}\sum_{k,q}\left(F_q - \hbar\omega_q f_q^*\right)a_{k+q}^\dagger a_k(b_q + b_{-q}^\dagger), \tag{6}$$

and

$$\mathcal{E}_k = \frac{1}{N}\sum_q \hbar\omega_q|f_q|^2 - \frac{1}{N}\sum_q F_q(f_q + f_{-q}^*) - 2J\mathrm{e}^{-W(T)}\cos(kR_0) \tag{7}$$

is the polaron energy band measured from the vibron excitation energy. At this place, we will mention that the nature of self-trapped exciton state can be characterized by the mutual ratio of three basic energy parameters: vibron energy band width $2J$, characteristic phonon energy $\hbar\omega_C$, and polaron binding energy $E_b = N^{-1}\sum_q |F_q|^2/(\hbar\omega_q)$. In the process of vibron self-trapping in biological macromolecules, optical phonon modes are of particular significance [4, 15, 16]. For that reason, we restricted our calculations to the case of exciton interaction with non-dispersive optical phonon modes ($\omega_q = \omega_0 = \omega_C$). Standard small-polaron theory corresponds to such systems where $E_b \gg \hbar\omega_C \gg 2J$, i.e. to the systems with narrow energy band. In such structures, quasiparticle and phonon cloud forms new entity, dressed quasiparticle, and formed deformation instantaneously follows the motion of quasiparticle.

Further, in order to facilitate the analytic treatment, we assume that all phonon modes equally influence to exciton dressing. In this case, we introduce only one variational parameter δ ($0 < \delta < 1$), which can be defined as $f_q = \delta \cdot F_q^*/(\hbar\omega_q)$. This parameter measures the degree of the narrowing of the polaron band. Its value determines whether the vibron motion takes place in the coherent fashion (when δ takes low values) by means of the band mechanism, or it is achieved incoherently by the random jumps between neighbouring sites (for high values of δ). Although it is technically less demanding, such an approach provides the results that are quite satisfying, especially for system parameter values that are of our interests [17].

Finally, the calculations are performed in terms of two system parameters, namely the adiabatic parameter ($B = 2J/(\hbar\omega_C)$) and the coupling constant ($S = E_b/(\hbar\omega_C)$). In this case, we have $W(\tau) = \delta^2 S \coth(1/2\tau)$, where $\tau = k_B T/\hbar\omega_0$ is the normalized temperature. Under these assumptions, the ground state polaron energy ($\mathcal{E}_{k=0}$) normalized to the characteristic phonon energy $\hbar\omega_0$ takes the form

$$\mathcal{E}_{gs} = S\delta(\delta - 2) - B\mathrm{e}^{-S\delta^2 \coth(1/2\tau)}. \tag{8}$$

An optimal polaron state should be determined by the minimization of the polaron ground state energy, from the condition $\partial\mathcal{E}_{gs}/\partial\delta = 0$. Consequently, the variational parameter can be found by solving of the following transcendental equation

$$\delta - 1 + \delta B \coth(1/2\tau)\mathrm{e}^{-S\delta^2 \coth(1/2\tau)} = 0. \tag{9}$$

The dependence of the exciton dressing and its ground-state energy upon the basic system parameters and temperature are presented in Figs. 1 and 2.

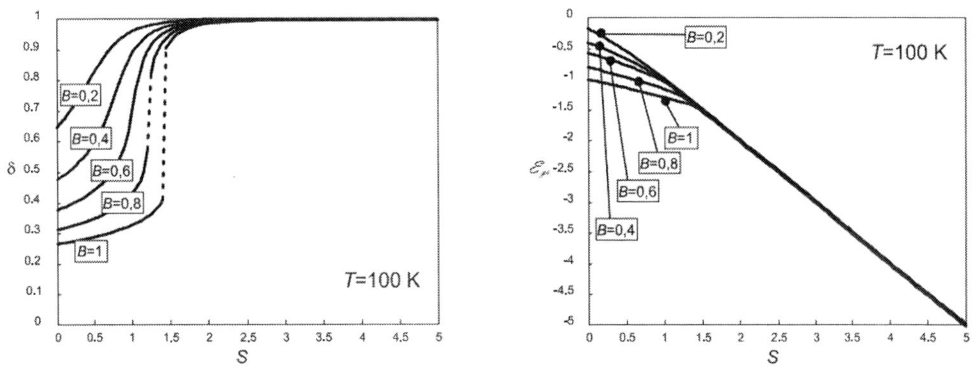

Figure 1. The dependence of the vibron dressing (left) and the ground state energy (right) on the coupling constant for various vales of the adiabatic parameter at temperature $T = 100$ K. Critical values of the adiabatic parameter $B_C = 0.68$ and the coupling constant $S_C = 1.3$ belong to moderately non-adiabatic and strong coupling limits.

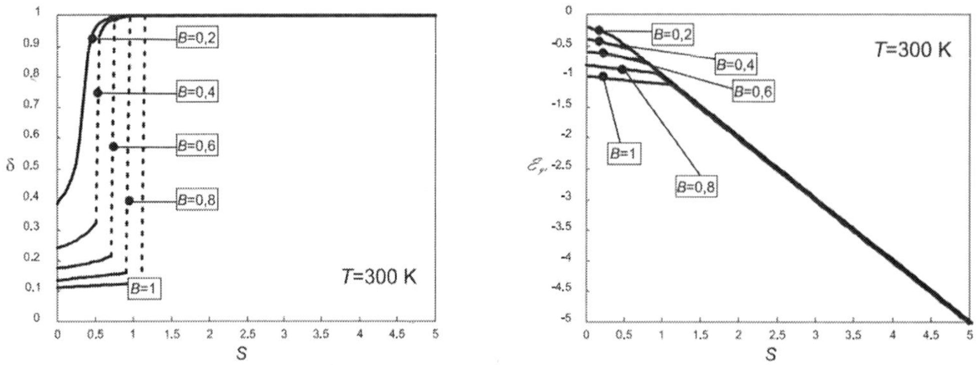

Figure 2. The dependence of the vibron dressing (left) and the ground state energy (right) on the coupling constant for various vales of the adiabatic parameter at temperature $T = 300$ K. Critical values of the adiabatic parameter $B_C = 0.22$ and the coupling constant $S_C = 0.4$ belong to non-adiabatic and weak coupling limits.

Obtained results appear to be consistent with the predictions of our earlier work concerned with the vibron self-trapping in the alpha-helix and two parallel chain structures [14, 18]. Briefly, we found that the system parameter space is divided on two characteristic areas: in the first area vibron dressing is a continuous function of the coupling constant S for fixed values of T and B. Such a behavior corresponds to the standard small-polaron quasiparticle. In the second area the vibron excitation abruptly changes its dressing: being initially a slightly dressed (practically free) quasiparticle, after an abrupt change its nature corresponds to the strongly dressed, practically immobile quasiparticle. The border between these two areas (determined by the critical values B_C and S_C) depends on the temperature of the bath: with increasing of the temperature, its moves towards higher values of S and B. It means that for the low temperatures the so called critical values belong to non-adiabatic ($B \ll 1$) and weak ($S \ll 1$) limits. However, for the room temperatures, the critical values may still belong to the non-adiabatic and intermediate-coupling! This result is interesting for understanding of the nature of the vibron dressing mechanism which is usually modelled in the framework of the adiabatic (soliton) polaron picture [1, 2, 3, 19]. On the other hand, it is consistent with some numerical investigations provided by Hamm and Tsironis [20].

3. Finding of the polaron amplitude damping

Since the characteristic time-scale of the exciton transfer processes is much larger than the time of phonon relaxation, in order to describe exciton migration processes it is necessary to use the non-equilibrium statistical distribution for exciton subsystem, i.e. the density matrix of excitons must be described by non-equilibrium density operator. On the other hand, as it was mentioned, the phonon subsystem is regarded as a subsystem in the equilibrium state and it is described by the equilibrium statistical operator. Consequently, in order to examine the polaron damping, we follow the well-known approach developed within the theory of the open quantum systems [21, 22], and we derive a set of evolution equations for non-equilibrium statistical average values $\langle a_k \rangle = \mathrm{Tr}\,\langle \rho(t) a_k \rangle$ of polaron operators a_k. Here $\rho(t) = \rho_e(t) \otimes \rho_{ph}$ is the time dependent non-equilibrium statistical operator (whereas $\rho_e(t)$ is the non-equilibrium time-dependent exciton statistical operator, while ρ_{ph} is phonon statistical operator). Evolution of the mean value of operator A is determined by the relation $i\hbar d\langle A \rangle/dt = \langle [A, H] \rangle$. For the Hamiltonian (5) we obtain

$$i\hbar\frac{d}{dt}\langle a_k\rangle = \mathcal{E}_k\langle a_k\rangle + \sum_q \phi_q\left\{\langle a_{k-q}b_q\rangle + \left\langle a_{k-q}b^\dagger_{-q}\right\rangle\right\}, \tag{10}$$

where $\phi_q = (F_q - \hbar\omega_q f_q^*)/\sqrt{N}$ (in the case of δ variational approach this parameter becomes $\phi_q = F_q(1-\delta)/\sqrt{N}$). In the equation above, there appear two new amplitudes, namely $\langle a_{k-q}b_q\rangle$ and $\left\langle a_{k-q}b^\dagger_{-q}\right\rangle$. The equation of motion for these amplitudes contain other new amplitudes: $\langle a_k b_{q_1} b_{q_2}\rangle$, $\left\langle a_k b^\dagger_{q_1} b_{q_2}\right\rangle$, $\left\langle a_{k_1} a^\dagger_{k_2} a_{k_3}\right\rangle$, $\left\langle a_k b_{q_1} b^\dagger_{q_2}\right\rangle$, and $\left\langle a_k b^\dagger_{q_1} b^\dagger_{q_2}\right\rangle$. The equation of motion for these amplitudes contain amplitudes that correspond to multiparticle processes which probability is lower in order of magnitude than the probability of single-particle processes. For this reason, it is possible to cut-off the above mentioned amplitudes as follows: $\langle a_k b_{q_1} b_{q_2}\rangle = \langle a_k\rangle\langle b_{q_1} b_{q_2}\rangle = 0$, $\left\langle a_k b^\dagger_{q_1} b^\dagger_{q_2}\right\rangle = \langle a_k\rangle\left\langle b^\dagger_{q_1} b^\dagger_{q_2}\right\rangle = 0$, $\left\langle a_k b^\dagger_{q_1} b_{q_2}\right\rangle = \langle a_k\rangle\left\langle b^\dagger_{q_1} b_{q_2}\right\rangle = \langle a_k\rangle \nu_{q_1}\delta_{q_1,q_2}$, $\left\langle a_k b_{q_1} b^\dagger_{q_2}\right\rangle = \langle a_k\rangle\left\langle b_{q_1} b^\dagger_{q_2}\right\rangle = \langle a_k\rangle(1+\nu_{q_1})\delta_{q_1,q_2}$. One should have in mind that the phonon subsystem is in the equilibrium state. Finally, we neglect all the multipolaron processes $\left\langle a_{k_1} a^\dagger_{k_1} a_{k_2}\right\rangle \approx 0$. Under these assumptions the equations of motion for $\langle a_{k-q}b_q\rangle$, and $\left\langle a_{k-q}b^\dagger_{-q}\right\rangle$ take the form:

$$i\hbar\frac{d}{dt}\langle a_{k-q}b_q\rangle = (\mathcal{E}_{k-q} + \hbar\omega_q)\langle a_{k-q}b_q\rangle + \phi_{-q}(\nu_q+1)\langle a_k\rangle, \tag{11}$$

$$i\hbar\frac{d}{dt}\left\langle a_{k-q}b^\dagger_{-q}\right\rangle = (\mathcal{E}_{k-q} - \hbar\omega_q)\left\langle a_{k-q}b^\dagger_{-q}\right\rangle + \phi_{-q}\nu_q\langle a_k\rangle. \tag{12}$$

After integration of Eqs. (11) and (12), one can obtain

$$\langle a_{k-q}b_q\rangle = -\langle a_k\rangle\left\{\mathcal{P}\left(\frac{1}{\mathcal{E}_{k-q}+\hbar\omega_q}\right) + i\pi\delta(\mathcal{E}_{k-q}+\hbar\omega_q)\right\}\phi_{-q}(\nu_q+1), \tag{13}$$

$$\left\langle a_{k-q}b^\dagger_{-q}\right\rangle = -\langle a_k\rangle\left\{\mathcal{P}\left(\frac{1}{\mathcal{E}_{k-q}-\hbar\omega_q}\right) - i\pi\delta(\mathcal{E}_{k-q}-\hbar\omega_q)\right\}\phi_{-q}\nu_q, \tag{14}$$

where $\mathcal{P}(1/[\mathcal{E}_k \pm \hbar\omega_q])$ are the principal values of corresponding integrals. After substitution of Eqs. (13), (14) into Eq. (10), we obtain a Schrödinger-like equation with damping

$$i\hbar\frac{d}{dt}\langle a_k\rangle = (\mathcal{E}_k - \Delta_k)\langle a_k\rangle - i\gamma_k\langle a_k\rangle \tag{15}$$

Such an equation describes the non-equilibrium dynamics of polaron, and it determines the polaron energy shift and polaron amplitude damping caused by the polaron interaction with the environment (i.e. with the thermal oscillations of the macromolecular chain). Here

$$\Delta_k = \sum_q |\phi_q|^2\nu_q\mathcal{P}\left(\frac{1}{\mathcal{E}_{k+q}-\hbar\omega_q}\right) + \sum_q |\phi_q|^2(1+\nu_q)\mathcal{P}\left(\frac{1}{\mathcal{E}_{k+q}+\hbar\omega_q}\right) \tag{16}$$

is the shift of the polaron energy, and

$$\gamma_k = \pi\left\{\sum_q |\phi_q|^2\nu_q\delta\left(\mathcal{E}_{k+q}-\hbar\omega_q\right) + \sum_q |\phi_q|^2(1+\nu_q)\delta\left(\mathcal{E}_{k+q}+\hbar\omega_q\right)\right\} \tag{17}$$

is the damping factor of the polaron amplitude which determines the characteristic relaxation time $\tau_k \sim 1/\gamma_k$. As can be remarked, the polaron amplitude damping is caused by the emission and absorption of real phonons, because the presence of the delta-functions in γ_k shows that the energy is conserved during such processes. On the other hand, the shift of the polaron energy Δ_k arises as a consequence of the emission and absorption of virtual phonons, because it is meaningful only in the case when $\mathcal{E}_{k+q} \neq \pm\hbar\omega_q$.

In the case when the vibron interacts with non-dispersive optical phonon modes the normalized energy of damped system looks as

$$\Delta_k = S(1-\delta)^2 I_\Delta, \tag{18}$$

where integral I_Δ has the form

$$I_\Delta = \frac{1}{N}\sum_q \frac{\coth(1/2\tau)\left[S\delta(\delta-2) - B\mathrm{e}^{-S\delta^2\coth(1/2\tau)}\cos{(k-q)R_0}\right]-1}{\left[S\delta(\delta-2) - B\mathrm{e}^{-S\delta^2\coth(1/2\tau)}\cos{(k-q)R_0}\right]^2-1},$$

and the damping factor becomes

$$\gamma_k = \pi E_b(1-\delta)^2 I_\gamma, \tag{19}$$

where

$$I_\gamma = \frac{\mathrm{e}^{1/\tau}}{\mathrm{e}^{1/\tau}-1}\frac{1}{N}\sum_q \delta(\bar{\mathcal{E}}_{k-q}+1) + \frac{1}{\mathrm{e}^{1/\tau}-1}\frac{1}{N}\sum_q \delta(\bar{\mathcal{E}}_{k-q}-1).$$

The dependence of the damping factor on the system parameters and temperature is shown in Fig. 3.

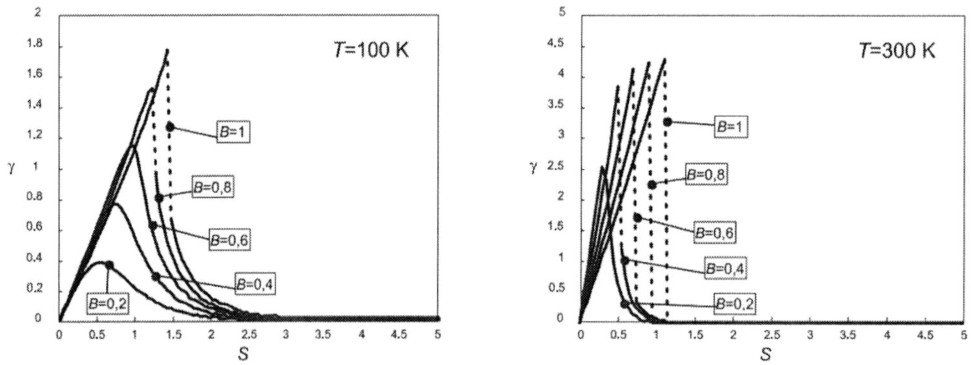

Figure 3. The dependence of the vibron damping on the coupling constant for various vales of the adiabatic parameter at temperatures $T = 100$ K (left) and $T = 300$ K (right).

From the above results, one can remark that the damping of the vibron amplitudes induced by the residual polaron-phonon interaction is significant in the area of the system parameter space where low or partial vibron dressing takes place (i.e., where $\delta < 1$). In the system parameter area where the exciton is slightly dressed, the amplitude damping rises with the increasing of S. But in the area where the exciton becomes heavy dressed, γ sharply decreases. Such a behavior is much more expressed for high B (i.e., when the system becomes more adiabatic). Additionally, this tendency increases with the rise of B, and for the system parameters where the dressing abruptly changes, the damping becomes also discontinuous. The maximal value of γ increases with increasing of B and T.

Obtained results suggest that the increasing of system temperature can drive the biological MCs to the system parameter area where the polaron crossover is possible. In this area of the parameter space, the systems are quite sensitive to small variations of temperature and the values of other system parameters. Amplitude damping of slightly dressed vibrons (caused by residual vibron-phonon interaction) in (moderately) non-adiabatic region strongly increases with the increasing of the coupling constant and temperature of the thermal bath. In the nearly adiabatic and adiabatic region such increasing is noticeably lower. At the same time, the residual vibron-phonon interaction practically has no influence on the amplitude damping of heavy dressed vibrons. The mean lifetime of nearly free vibrons is very short (as compared to the mean lifetime in the low temperature limit), and consequently the width of vibron spectral lines of the absorption spectra should be high. These results could explain the absence of observation of weak shift of the anomalous peak in the absorption spectra of crystalline acetanilide, which was predicted by the use of the partially dressed SP picture.

Acknowledgments

This work was supported by the Serbian Ministry of Education and Science under Contract Nos. III-45010, III-45005 and by the Project within the Cooperation Agreement between the JINR, Dubna, Russian Federation and Ministry of Education and Science of Republic of Serbia. One of the authors (A.Ch.) thanks the Blokhintsev-Votruba Program (JINR, Dubna) for the financial support of participation in the conference ISQS25.

References

[1] Davydov A S 1973 *J. Theor. Biol.* **38** 559
[2] Davydov A S 1979 *Phys. Scr.* **20** 387
[3] Davydov A S 1981 *Phys.* D **3** 1
[4] Alexander D M and Krumhansl J A 1986 *Phys. Rev.* B **33** 7172
[5] Petrov E G, Goychuk I A and May V 1996 *Phys. Rev.* B **54** 476
[6] Agranovich V M and Galin M D 1978 *Transfer of Energy of Electronic Excitation in Condensed Media* (Moskva: Nauka)
[7] Holstein T 1959 *Annals of Physics* **8** 325
[8] Lang I G and Firsov Yu A 1962 *Zh. Eksp. Teor. Fiz.* **43** 1843
[9] Pouthier V 2010 *J. Chem. Phys.* **132** 035106
[10] Yarkony D and Silbey R 1976 *J. Chem. Phys.* **65** 1042
[11] Brown D W and Ivić Z 1989 *Phys. Rev.* B **40** 9876
[12] Pržulj Ž, Čevizović D, Zeković S and Ivić Z 2008 *Chem. Phys. Lett.* **462** 213
[13] Čevizović D, Galović S and Ivić Z 2011 *Phys. Rev.* E **84** 011920
[14] Cevizović D, Galović S, Reshetnyak A and Ivić Z 2013 *Chin.Phys.* B **22** 060501
[15] Barthes M 1989 *J. Mol. Liq.* **41** 143
[16] Scott A C 1992 *Phys.Rep.* **217** 1
[17] Cevizovic D, Galovic S, Reshetnyak A and Ivic Z 2012 *J. Phys. Conf. Ser.* **393** 012033
[18] Cevizovic D, Ivic Z, Galovic S, Reshetnyak A and Chizhov A 2016 *Phys.* B **490** 9
[19] Brizhik L, Eremko A, Piette B and Zakrzewski W 2004 *Phys. Rev.* E **70** 031914
[20] Hamm P and Tsironis G 2007 *Eur. Phys. J. S. T.* **147** 303
[21] Valasek K, Zubarev D N and Kuzemskii A L 1970 *Teor. Mat. Fiz.* **5** 281 (in Russian)
[22] Sadovskii M V 1999 *Lekcii po statisticheskoi fizike* (Ekaterinburg: Institut elektrofiziki UrO RAN) (In Russian)

IOP Conf. Series: Journal of Physics: Conf. Series **965** (2018) 012012 doi:10.1088/1742-6596/965/1/012012

On a Microscopic Representation of Space-Time VII – On Spin

Rolf Dahm

Permanent address: beratung für Informationssysteme und Systemintegration, Gärtnergasse 1, D-55116 Mainz, Germany

E-mail: dahm@bf-is.de

Abstract. We recall some basic aspects of line and line Complex representations, of symplectic symmetry emerging in bilinear point transformations as well as of Lie transfer of lines to spheres. Here, we identify SU(2) spin in terms of (classical) projective geometry and obtain spinorial representations from lines, i.e. we find a natural non-local geometrical description associated to spin. We discuss the construction of a Lagrangean in terms of line/Complex invariants. We discuss the edges of the fundamental tetrahedron which allows to associate the most real form SU(4) with its various related real forms covering SO(n,m), $n + m = 6$.

1. Introduction

At this stage (see [5] and references), we've established the ties from our original Lie group based approach of hadronic (and quantum) degrees of freedom in terms of SU(4) and SU∗(4), respectively, to (classical and advanced) projective geometry of real 3-space. Now, in parts VII and VIII of this series, we want to focus more detailed on two of the included major aspects or building blocks – spin and relativity. So departing from a description common to both parts and based on elementary line geometry, we are going to present a more detailed treatment of each spin and relativity related to classical line and projective geometry. This, however, at the same time almost immediately and inherently comprises a necessary branch of the two topics here and in upcoming papers.

It is the task of this paper to recall briefly some (very) elementary geometrical properties and features, and to relate this approach to spin and associated algebraical concepts and physical background. As such, we'll rearrange coordinates and relate this rep[1] to nowadays common spin (and 'quantum') aspects and some of their reps. Moreover, we use this approach to discuss further aspects as mentioned in the abstract so that inherently we obtain a relation between the various reps and, moreover, between their respective interpretations and related concepts. Here, the crucial point with respect to spin discussions is the application of Lie's transfer principle [7] and a special complexification of the real (inhomogeneous) coordinates. On the other hand, this is also the branching point mentioned above so that, based on the same Ansatz but on real coordinates, in part VIII we'll concentrate on the original real (and the line geometric) aspects, and there we'll continue this second branch to discuss special linear Complexe and relativity. We feel that the next few pages grant to state the relation of the usually disconnected 'quantum'

[1] As in the papers before, we use the shorthand notation 'rep' to denote both representations or realizations as long as the respective context is unambiguous. 'PG' subsequently denotes 'projective geometry'.

IOP Conf. Series: Journal of Physics: Conf. Series **965** (2018) 012012 doi:10.1088/1742-6596/965/1/012012

world to classical geometrical concepts, however, the wealth of possibilities comprised in Lie's mechanism [7], e.g. with respect to higher order (or class) objects or using further choices of the transfer functions f_i, are still far from being applied or even exhausted. The 'degenerate cases' mentioned in Lie's paper are related to very special choices and cannot serve to comprise the full power of this Ansatz based on Plücker's algebraic concept of a general (higher dimensional) description of ordinary space and Lie's extension thereof (see [7], especially page 151 above/end of §2 with respect to Lie's understanding of Plücker's original algebraical Ansatz [9]).

So although at this time we'll include general aspects only, we feel it is the time to split our original series on a microscopic rep of space-time at this point into a spin (using complex numbers and elliptic geometry) and a relativistic (hyperbolic) part. On one hand, it is worth to investigate Lie transfer in more detail and to relate Clifford's and Study's work on elliptic geometry as well as especially Study's further work on (different) line geometries and on kinematics (see e.g. [11] and later publications). To our opinion, in conjunction with (quaternary) invariant and form theory, this should be sufficient to exhaust and comprise what is nowadays called 'quantum theory'. On the other hand, it is worth to separate a discussion focussing on real aspects and hyperbolic geometry in order to keep in parallel the track to relativity - special relativity using (special) linear Complexe as well as the full geometry of regular linear Complexe in P^5 and/or higher order Complexe and nonlinear transformations. Stepping back to line and Complex geometry as the basic and underlying framework of Lie's approach [8], we feel to escape the various restrictions of early use of (linear) rep theory of groups and algebras as well as concepts of differential geometry, and we think it's sufficient when asking for integral and/or nonlinear representations only afterwards.

As such, in the upcoming sections, we use Plücker's line rep (in inhomogeneous coordinates) and feel free to rearrange and reinterpret this rep in a special form so that we can write the coordinates and transformations in matrix notation. This helps to identify elements and relate them to quantum reps, e.g. in Lagrangean or current algebra notation, and to ongoing discussions. Please note, too, that the next section is useful both for spin and relativity, especially after switching to homogeneous coordinates and departing from affine to non-euclidean geometry, and that it includes nothing but elementary PG. Afterwards, in section 3, we branch from real projective geometry by Lie's transfer principle [7] to the derivation of (complex) spin reps and the known algebraic formalism. Throughout this section, we feel free to recall some historic contexts and include them in the text, not to write some science-history related paper[2] but to put some known facts in an understandable context right from the original development. The last section is devoted to a brief outlook in order to sketch the wealth of opening possibilities which appear by Lie transfer and sphere geometry.

2. Line representation and coordinates

The reason to cite some (very elementary) issues from [10] is mainly to connect the reasoning of the 19th century view of projective geometry to the dynamic picture used nowadays in terms of 'time' and 'velocity' coordinates which emerge only after we tag points in the respective

[2] According to current discussions and the attribution of achievements, we feel the urgent necessity to reconsider the real historical development, however, personally, we neither have the competency nor the time to evaluate and honor those achievements scientifically and relate them in detail to the original authors. So we focus here on few spin-related aspects only which to our opinion need (among other authors) to be attributed mainly to S. Lie and especially E. Study already back in the late 19th and first decade of 20th century instead of E. Cartan in 1938 [1] (or even 1913, as he claimed there, see his introduction). However, we cannot rise a fundamental and historically complete priority discussion here.

IOP Conf. Series: Journal of Physics: Conf. Series **965** (2018) 012012 doi:10.1088/1742-6596/965/1/012012

geometry and 'observe' them moving around[3]. So in choosing [10] the (euclidean) rep

$$x = r\,z + \rho\,, \quad y = s\,z + \sigma\,, \tag{1}$$

we obtain first of all a (cartesian) line rep of 3-space which is 'globally' defined by two linear functions in terms of the third cartesian coordinate z. In each of the planes (x, z), i.e. $y = 0$, and (y, z), i.e. $x = 0$, we find linear line projections, too, which we may map to separate line reps as well. So far, we do not have tagged special points on the line like the origin as a reference or a unit point or two (fundamental) points to declare anharmonic ratios or three points to declare a coordinate (or measure) on the line itself. Moreover, in the same picture, we do *not yet* apply the vector calculus of real 3-space common today by pointing with a vector to the line and defining a 'time' parameter and a velocity on the line (or at least its direction).

2.1. Coordinates

So the following arguments are concerned with the (real) parameters r, ρ, s and σ from eq. (1), i.e. what enters these equations are position and orientation (i.e. 'the geometry') of the line itself and as a 'whole' object in the 3-dim euclidean[4] coordinate system x, y, z. Now Plücker founds his definition of (4-dim) line coordinates and their necessary extension[5] by a fifth line coordinate $\eta = r\sigma - s\rho$ on this basis, and in [10] he shows how to define and use sets of lines in terms of Complexe, Congruences and Configurations as foundation of a space description, his so-called 'new geometry', in distinction to standard (projective) geometry which uses the point as base element and planes to complete the point-based framework with respect to duality (see e.g. his contemporary summary [9] or later text books on PG also by other authors).

Suppressing PG details here, we want to prepare our discussion of both parts VII and VIII, i.e. spin and relativity, by rearranging the (euclidean) coordinates x, y, still as a function of z only, with respect to the line projection onto the (x, y)-plane at $z = 0$. Our main concern and driving force having been inclusion and understanding of special Lorentz transformations, however, it turned out that if we transform (in the point picture) along the z-axis, and if we require as usual the x- and y-coordinates to remain invariant, we have remarkable freedom in that this transform concerns only $z \sim \frac{x_3}{x_0}$ as (euclidean) normal to (x, y). Moreover, weakening the invariance requirement(s), we can rearrange the variables in the projection plane[6] (x, y) as needed to investigate invariances of products, quotients, etc., too.

So if we arrange e.g. the ratio of the two (cartesian) coordinate projections x and y of the line into a new coordinate ζ by $\zeta = \frac{x}{y}$, using eq. (1) we obtain the relation $\frac{x}{y} = \zeta = \frac{r\,z+\rho}{s\,z+\sigma}$. Now we can interpret the rhs as projective transformation[7] and, of course, map the parameters to a

[3] We want to emphasize the difference of the 'pictures' because there are alternative reps of points e.g. by considering points in terms of (inter-)sections of lines in 3-space. This leads almost immediately to a discussion of Congruences when including (moving) observers into the physical description, see also our preparatory discussions in [4], [5]. We have to remark as well the *dependent* relation of velocities to the respective time coordinates of the individual lines, or axes, or of the (local) coordinate systems which in its simplest form of (linear) motion with uniform velocity read as $s = v \cdot t$ in those coordinates. Note also, that Plücker (as well as Lie in his approach) uses (inhomogeneous) euclidean coordinates x, y, z and not fourfold/quaternary point coordinates, neither homogeneous nor (affine) Hesse coordinates by introducing the additional t-coordinate common at that time to denote the fourth (affine) point coordinate. This reduces the anharmonic ratio using point reps to a simple quotient of point distances or 'coordinate' projections.

[4] We are going to extend this approach by homogeneous coordinates mainly in part VIII and postpone the more general discussion, also with respect to anharmonic ratios and projective transformations of the lines, to upcoming parts.

[5] This ensures to preserve the grade during transformations whereas a sixth coordinate is necessary to linearize the transformations of the line coordinates.

[6] We discuss more details related to relativity in part VIII.

[7] The three (cartesian) projections of the original line onto each of the planes (x, z), (y, z) and (x, y) in the general case are, of course, related to the line by perspective.

(real) matrix formalism by

$$\frac{x}{y} = \zeta \longrightarrow \begin{pmatrix} x \\ y \end{pmatrix} = \begin{pmatrix} r & \rho \\ s & \sigma \end{pmatrix} \begin{pmatrix} z \\ 1 \end{pmatrix} . \quad (2)$$

2.2. Homogeneous Coordinates

Moreover, by formally[8] introducing homogeneous coordinates for $\zeta \longrightarrow \frac{\zeta_1}{\zeta_2}$ and $z \longrightarrow \frac{x_3}{x_0}$, we can rewrite eq. (2) according to

$$\begin{pmatrix} \zeta_1 \\ \zeta_2 \end{pmatrix} = \begin{pmatrix} r & \rho \\ s & \sigma \end{pmatrix} \begin{pmatrix} x_3 \\ x_0 \end{pmatrix} \quad (3)$$

and apply the usual 2×2 matrix formalism.

Now, dependent on whether we focus on the rhs of eqns. (2) and (3), or on the lhs of these equations, we steer towards a discussion of spin for the rhs or (special) relativity for the lhs as we'll discuss soon. Note, however, that throughout our approach $\frac{x}{y}$ remains 'invariant' if we replace the cartesian coordinates x and y homogeneously by $x \longrightarrow \frac{x_1}{x_0}$ and $y \longrightarrow \frac{x_2}{x_0}$, i.e. $\frac{x}{y} = \frac{x_1}{x_2}$, following the affine interpretation by tagging a certain plane within PG.

Plücker's fifth line coordinate $\eta = r\sigma - s\rho$ then turns out to represent the determinant of the transformation matrix above, i.e. via the line rep in eq. (1), we have associated (formally) a projective transformation of the (euclidean) z-axis to a (projective) transformation of the coordinate ζ. Accordingly, we can relate this 'new' coordinate axis of ζ to z by associated projective transformations $\zeta = f(z)$. More general, the projective relation of z and ζ allows to apply the full framework of point transformations on first order (projective) elements and to investigate binary forms (and the related invariant theory) in terms of mappings of point sets on lines to each other. In addition, our approach suggests the discussion of involutions on projective elements of first kind, so that we may include complex numbers via von Staudt's or Lüroth's geometric interpretation of complex numbers naturally. By considering this approach from a more general viewpoint, we have now set up the very foundation of the complete framework of PG, here in terms of point geometry on two lines coordinated by z and ζ.

3. Lie Transfer and Spin

Following Lie's approach here, we are going to discuss the rhs of eq. (2) only. Also, we require the matrix determinant (or Plücker coordinate η) here as nonzero, i.e. the original line doesn't hit the z-axis, and the line of the respective projections in the (x, y)-plane is skew versus the z-axis. In other words, $(0, 0)$ being the (cartesian) projection of the z-axis in the (x, y)-plane (and all its parallel planes, i.e. planes hitting the same 'absolute line' within the spatial affine picture), the non-vanishing determinant ensures considering the case of two skew lines, and moreover eases the introduction of homogeneous coordinates by avoiding singular/special cases. Some aspects related to the general case, including $\eta = 0$, will be discussed in part VIII, also in the context of using homogeneous coordinates.

3.1. Lie's Mapping – aka 'Lie Transfer'

Lie restricted the general cases of his mapping at first by considering lines instead of general curves ([7], #5 and #21), and further by choosing special constants in that the second Complex in the degenerate case ([7], #21) hits the conic section in the absolute plane of the *affine* geometry. Note, that it is this linear extension of euclidean geometry[9] which introduces the

[8] Here, we may only use the affine (linear) coordinate extension by an absolute plane $x_0 = 0$ (better by $\pm\sqrt{x_0^2}$) in the denominator.

[9] ...which also forces the switch to homogeneous coordinates to describe absolute elements. We refer to Study's remarks [12] and some of his further work cited here. See also our few remarks in section 3.6.

isotropy discussion (and as such the 'spinor' rep), whereas with respect to relativistic (or non-euclidean) transformations, leaving a general second order (i.e. quadratic) surface invariant, we have to switch to 3-space instead of (linear) planar discussions. There will remain the connecting link that 2nd order surfaces can be generated by two families of lines, i.e. in each point of the surface, we can find appropriate members of the two generating families, however, in order to handle spin and (special) relativity from the viewpoint of the degeneracies discussed here and in the upcoming part VIII with respect to special linear Complexe, it is obvious that one should switch to Complexe in general as geometrical base elements as we have worked out and proposed throughout this paper sequence (see [5] and references there). A strongly supporting, direct argument for switching to line and Complex geometry of P^5 would be to refer back to Lie's original investigations [7] and by noting that the general theory underlying Lie's approach and reasoning is Complex geometry. Indeed, Lie obtained this (bilinear) approach given in eq. (4) only by further specializing his originally more general mapping to a line Complex hitting the absolute circle of affine geometry[10], or isotropic lines, respectively. For now, we think that Lie's Ansatz in general establishes sphere geometry in a separate space besides the line (and the represented Complex) geometry of ordinary 3-space, whereas concerning 'spin', here we treat only simple/linear and (multiply) degenerate cases of Lie's general mapping which 'overlap' or relate in certain aspects due to the (linear) cases and/or the degeneracies considered.

3.2. Matrix Reps and Spin

Nevertheless, we depart from Lie's restricted (bilinear) Ansatz (see [7], p. 146, or [7], §8, #24 (1)),

$$-Zz = x - (X + iY)\,, \quad (X - iY)z = y - Z \tag{4}$$

to relate the (cartesian) coordinates of the spaces r and R, which we rearrange into

$$x = -Zz + (X + iY)\,, \quad y = (X - iY)z + Z\,, \tag{5}$$

and according to our choice[11] of the line representation in eq. (1) and the matrix definition in (2), this allows to identify the transformation matrix directly by

$$\begin{pmatrix} r & \rho \\ s & \sigma \end{pmatrix} \sim \begin{pmatrix} -Z & X + iY \\ X - iY & +Z \end{pmatrix}. \tag{6}$$

Note here, that on the lhs, we originally describe a *linear* transformation of *cartesian* coordinate projections which yields the context for discussion of [3] later on. So the original line coordinates of space r are represented (or 'replaced') by (complexified point) coordinate combinations of the space R, where the rhs of eq. (6), however, may be re-expressed in nowadays common notation according to

$$\begin{pmatrix} -Z & X + iY \\ X - iY & +Z \end{pmatrix} = X\sigma_1 - Y\sigma_2 - Z\sigma_3 \tag{7}$$

when introducing standard 'Pauli matrices' σ_i. So the important point here – derived from the background of Complex geometry – is the identification of real (point) coordinates of R in the

[10] It is obvious from the Cayley-Klein construction of a metric that this construction is relevant to metrical properties, too. However, we do not want to discuss general lengths, 'masses' or 'charges' here as this approach is, first of all, connected to absolute elements of an affine Ansatz only.

[11] For detailed analytic consideration and calculations one should recall that at the time of Plücker's and Lie's writing, the authors were used to coordinate systems with different handedness, especially to a different sign of the y-coordinate, which sometimes leads to signs especially in linear equations involving y or σ.

Pauli/spin basis by eq. (7) with (inhomogeneous) line coordinates[12] of real 3-space, and it is the su(2) (Lie) algebra which maps the geometries and the respective coordinate interpretation. However, for the general environment and the context of our current discussion, it is necessary to mention and cite three more important aspects given by Lie in his paper [7].

3.3. Geometry in R-space

Casting the equations above in a form to identify coordinates in R-space, Lie shows ([7], p. 168) that those (still inhomogeneous) coordinates fulfil the equation $R^2 + S^2 + 1 = 0$, or when cast in differential form[13], based on [7], §8, eq. (3), by[14] $R = \frac{\mathrm{d}X}{\mathrm{d}Z}$, $S = \frac{\mathrm{d}Y}{\mathrm{d}Z}$, which yields $\mathrm{d}X^2 + \mathrm{d}Y^2 + \mathrm{d}Z^2 = 0$ and suggests the considerations of isotropic elements. This led Lie to the statement that the line Complex in R-space is being built of imaginary lines of zero length[15]. We want to address those aspects later and separately.

Here, to connect to 'Cartan's' spin rep, we want to recall Lie's equations ([7], p. 168) of the line coordinates R and S,

$$R = \frac{1}{2}\left(z - \frac{1}{z}\right) , \quad S = \frac{1}{2i}\left(z + \frac{1}{z}\right) , \tag{8}$$

and the quadratic equation $R^2 + S^2 + 1 = 0$ from above. We can now start from either 'side' – using the quadratic equation and introducing homogeneous coordinates, or recalling the inhomogeneous character of z and introducing homogeneous quaternary coordinates of 3-space in (8). By the second approach, we obtain from eq. (8)

$$R = \left(\frac{z^2 - 1}{2z}\right) = \left(\frac{x_3^2 - x_0^2}{2x_0x_3}\right) , \quad S = \left(\frac{z^2 + 1}{2iz}\right) = \left(\frac{x_3^2 + x_0^2}{2ix_0x_3}\right) . \tag{9}$$

By introducing primed 'homogeneous' variables $R =: \frac{R'}{T'}$, $S =: \frac{S'}{T'}$, we can extract the identifications

$$R' = x_0^2 - x_3^2 \, , \; S' = i\left(x_0^2 + x_3^2\right) \, , \; T' = -2x_0x_3 \, , \tag{10}$$

where we've used the standard relation $z = \frac{x_3}{x_0}$ from affine geometry with respect to the z-axis in all places. The equation $R^2 + S^2 + 1 = 0$ converts to $R'^2 + S'^2 + T'^2 = 0$, thus featuring (finite) line coordinates and 4th order surfaces in the original (point) coordinates. Eq. (10) yields Cartan's coordinate and 'spinor' 'definition' in [1], #52.

3.4. Lie Transfer and 'Spinors'

Now regarding eqns. (9) and (10), it is easy to trace back Cartan's miraculous 'definition' of 'a spinor' emerging in [1], #52 by recalling the relation of inhomogeneous and homogeneous coordinates of (real) 3-space, the definition of the absolute element of affine geometry. What

[12] This is especially noteworthy facing the current discussions on non-local aspects of quantum theories if we depart from real parametrizations of Pauli matrices. Note in this contexts, that already special linear Complexe are *sets* of lines in 3-space, and that using Complexe, we also have to take into account regular and higher order Complexe, i.e. the geometry of P^5 as well as quadratic manifolds. Here, we are discussing linear mappings and degenerate cases only, so these are only the very beginnings of a potentially relevant description of observations in nature.

[13] We cite the expression as given, not discussing partial derivatives. Please note, that this replacement of coordinates by derivatives is attached to Plücker's original line rep and to a *linear* choice of the coordinates. So this Ansatz yields the very foundation of Lie's contact transformations based on line elements (or five-parameter space reps (x, y, z, p, q)) later on. However, due to the intrinsically linear rep, we may also switch directly to the point rep on lines in terms of binary forms [3] and the related symbolism.

[14] We want to mention the relation to three homogeneous (planar) coordinates only.

[15] Isotropic lines.

enters in addition, are the absolute circle, by $x_1^2 + x_2^2 + x_3^2 = 0$, $x_0 = 0$ in *homogeneous* spatial coordinates, and an identification $x_0^2 - x_3^2$ here with $\xi_0^2 - \xi_1^2$ in Cartan's notation. Accordingly, his 'spinor' coordinates ξ_0 and ξ_1 map to two homogeneous point coordinates x_0 and x_3 of real 3-space, whereas by eqns. (9) or (10) the three linear coordinates x_i of Cartan map to Lie's 'line coordinates' R and S, or R', S' and T' from above and from the 'inverse' Lie mapping, obeying $R^2 + S^2 + 1 = 0$ or $R'^2 + S'^2 + T'^2 = 0$, respectively.

There is, however, a much easier way to establish the 'spinor' interpretation versus Lie's mapping by starting from [1], #55, eq. (1), and by recalling $z = \frac{x_3}{x_0} \sim \frac{\xi_1}{\xi_0}$ from above, i.e.

$$\xi_0 x_3 + \xi_1(x_1 - ix_2) = 0\,,\ \xi_0(x_1 + ix_2) - \xi_1 x_3 = 0\,, \tag{11}$$

Here, we can at first change the order of the equations and divide by x_0, thus switching back to inhomogeneous, cartesian coordinates X, Y and Z. We obtain

$$-\xi_1 Z + \xi_0(X + iY) = 0\,,\ \xi_0 Z + \xi_1(X - iY) = 0\,, \quad \text{or} \tag{12}$$

$$0 = -Z\xi_1 + (X + iY)\xi_0\,,\ 0 = (X - iY)\xi_1 + Z\xi_0\,. \tag{13}$$

We may put these equations in matrix form, too, by extracting a 'spinor' $(\xi_1, \xi_0)^T$ to the right, and celebrating the su(2) (Lie) algebra in terms of Pauli matrices. What is more important, however, is the comparison with eq. (5) if we introduce an inhomogeneous coordinate $\xi := \frac{\xi_1}{\xi_0}$ (which, as we see above, is Lie's original z). Dividing eq. (13) by ξ_0, we obtain

$$0 = -Zz + (X + iY)\,,\ 0 = (X - iY)z + Z\,, \tag{14}$$

which recovers a special case of Lie transfer, i.e. instead of Lie's general coordinate projections x and y, Cartan's definition is obviously fixed to the z-axis at $(0, 0)$ in the (x, y)-plane, which from Lie's approach is related to vanishing η (or determinant).

More important is the missing dependence of x and y at all, or according to (14) $x = 0$ and $y = 0$. So in addition to the restrictions and the resulting degeneracies/singularities applied by Lie, there is *a priori no suitable coordinate dependence* besides the 'spinor' itself describing a line in terms of two homogeneous coordinates ξ_i, or a simple 1-dim projective transformation of the quotient. The quotient may be reversed by the exchange $\xi_1 \longleftrightarrow \xi_0$, in 'spinor' notation induced by σ_1 or $i\sigma_2$, which correspond to reciprocal transformations[16] $z \longleftrightarrow \pm\frac{1}{z}$. Due to the remarkable situation that (in the usual point picture) special relativity doesn't transform x and y when we choose the transformation along z, Cartan's approach – being independent of x and y – survives and reflects those transformations. One should, however, place question marks to general space-dependent SU(2) spinors like $\psi(x)$ or $\psi(x^\mu)$ founding on Cartan's definition.

3.5. Further Linear Aspects

The second aspect covers a more general case. Here, Lie's equations (4) mapping the two spaces r and R, relate both spaces in a manner that points in r map to isotropic lines in R, and points in R map to the linear Complex $r + s = 0$ in r. In upcoming papers, we hope to have the possibility to discuss more geometrical details, especially with respect to further details and geometrical relations mentioned by Lie in [7]. For now, we want to close this subsection by reference to Lie's remarks in §9, p. 171 with respect to the second unnumbered equation (see also eq. (15)) holding

[16] By a suitable setup of the system, arranging a (metric) radius along $\hat{z}$, we have $r \longleftrightarrow \pm\frac{1}{r}$, and planar duality as well. Conjugation of eq. (14) in R-space is invariant if, at the same time, we perform $z \longleftrightarrow -\frac{1}{z}$, etc. Note also sign behaviour of R and S in eq. (8) in the $\pm$-cases. Such methods can be applied throughout Cartan's 'spinor' formalism, however, it should at least be founded on Lie's approach. The more general treatment has to include and respect polar systems.

IOP Conf. Series: Journal of Physics: Conf. Series **965** (2018) 012012 doi:10.1088/1742-6596/965/1/012012

a quadratic relation. Tracing this back with our matrix rep, it is obvious that the variable H in R corresponds to the *unit matrix* σ_0, and it thus relates the unit matrix of the R matrix rep with the linear Complex $r + \sigma$ of r-space and it's special rôle. In Lie's original work [7], p. 171, the spheres in R-space (i.e. using Plücker's inhomogeneous coordinates[17] to describe Complex lines) fulfil

$$[X - (s + \rho)]^2 + [Y - i(s - \rho)]^2 + [Z - (\sigma - r)]^2 = (\sigma + r)^2 . \tag{15}$$

So the coordinate set[18] $X' = s + \rho$, $Y' = i(s - \rho)$ and $Z' = \sigma - r$ describes the center of the sphere in R-space in terms of line coordinates (or a shift from an original center $(0, 0, 0)$) whereas $\sigma + r$ takes the rôle of an oriented coordinate or even a 'radius' $H = r + \sigma$ in terms of line coordinates of the line Complex in r-space. This enhances the mapping $r \longleftrightarrow -Z$, $\rho \longleftrightarrow X + iY$, $s \longleftrightarrow X - iY$ and $\sigma \longleftrightarrow +Z$ in eq. (6) into (see [7], §9, eq. (3))

$$r = \frac{1}{2}(\pm H' - Z') \,, \rho = \frac{1}{2}(X' + iY') \,, s = \frac{1}{2}(X' - iY') \,, \sigma = \frac{1}{2}(\pm H' + Z') \tag{16}$$

which now corresponds to the expression

$$\frac{1}{2} \begin{pmatrix} -Z' \pm H' & X' + iY' \\ X' - iY' & +Z' \pm H' \end{pmatrix} \sim X'\sigma_1 - Y'\sigma_2 - Z'\sigma_3 \pm H'\sigma_0 . \tag{17}$$

Recalling Plücker's fifth coordinate $\eta = r\sigma - s\rho$, eq. (16) yields $r\sigma = H'^2 - Z'^2$ and $s\rho = X'^2 + Y'^2$, so $\eta = H'^2 - Z'^2 - X'^2 - Y'^2$. This prepares the basis of the 2-spinor formalism and all its applications and analytical forms, in addition, we thus map the quest for 'relativistic' invariance to 2×2-matrix notation (17) or complexified quaternions, and invariance of the determinant η.

Especially, this gives some insight on how (due to the quadratic nature of the manifold) the unit matrix serves [7] in this rep: A line given in r-space determines a sphere in R. However, a sphere in R given by (X, Y, Z, H^2) maps to two lines in r, both being reciprocal polars with respect to the linear line Complex $\pm H = r + \sigma = 0$ in r. H in R-space plays the rôle of a sphere radius, so choosing $H = 0$ amounts to 'point spheres'[19] mapping uniquely to the Complex $r + \sigma = 0$ of r-space.

Finally, and pointing towards a third important aspect, it is worth mentioning [7] article #28 which relates the contact of Lie's spheres to intersection of lines. This emphasizes the relevance of contact interactions in both pictures when constructing and investigating symmetry and invariance principles.

For us, the essence of the reasoning so far – remembering all the various restrictions applied up to this point – justifies according to our current use of the Pauli spin (or the quaternion) algebra in various physical contexts to reverse the viewpoint: If we are to interpret Pauli matrices with *real* coefficients and including the unit matrix to represent a 'radius', we have to recall their origin from transformations in R-space and their identification with some (inhomogeneous) line coordinates which themselves describe, according to their direct relation with the underlying linear line Complex, automatically an extended (i.e. 'nonlocal') object – a set of lines – in real 3-space r.

[17] Generalizing the coordinates via affine homogeneous coordinate extensions, note that ρ and σ are related to the 'new' affine coordinate t or x_0 and as such to the absolute plane in r-space. Recall also, that transformations of the inhomogeneous Plücker coordinates r, s, ρ, σ do not preserve grade, i.e. (16) pretends a linear relationship due to the projective rep.
[18] Note, that Y' is imaginary, or iY' is real, and remember polar relationships in affine geometry, too.
[19] German: Punktkugeln [7], p. 171

IOP Conf. Series: Journal of Physics: Conf. Series **965** (2018) 012012 doi:10.1088/1742-6596/965/1/012012

3.6. Some Aspects of Study's work

Driven by some discussions throughout and following this year's conferences with respect to Cartan's spin and to unitary symmetries in QFT, especially when based on one or more su(2) algebras, we've searched through literature. So before performing some algebra related to sect. 3.3, it is necessary and important to recall few historical but obviously forgotten aspects which allows to rearrange some aspects back into their original context.

Especially when facing the discussion of spin and unitary symmetries, typically attributed to Pauli, Cartan and su(2) (and certain su(n)) algebras, the origins can be easily traced back at least to Hesse's transfer principle (1866) and Meyer's generalization (1883) (for references and an outline see [6], §51) of rational curves of a point $\lambda = \frac{\lambda_1}{\lambda_2}$ on a line. For $n = 2$, we thus obtain (planar) conic sections parametrized by three homogeneous coordinates $\rho x_0 = \lambda_1^2$, $\rho x_1 = \lambda_1\lambda_2$ and $\rho x_2 = \lambda_2^2$, or appropriate linear combinations. Choosing $n = 3$ for later use, we obtain 3rd order curves parametrized by the four coordinates $\rho x_0 = \lambda_1^3$, $\rho x_1 = \lambda_1^2\lambda_2$, $\rho x_2 = \lambda_1\lambda_2^2$ and $\rho x_3 = \lambda_2^3$. Now linear transformations of the line, alternatively re-expressed by matrix transformations of the vector/'spinor' $(\lambda_1, \lambda_2)^T$ with non-vanishing determinant, i.e. by a 3-dim group[20], lead to collineations in $\mathbb{R}^n$ which exchange points on the (invariant) curve, for $n = 2$ the conic section. For further details see [6] or [3].

In the context of line and Complex geometry – in a certain sense continuing the heritage of Plücker and Lie – one *must* consider and honor Study's work (see [11], or [12] and references) which has been definitely known to Cartan. Ad hoc, without enough time to work through and synchronize notation, we found references and work in Münchner and Leipziger Berichte (where Cartan published in german, too), and Mathematische Annalen, besides Study's remarks on Lie's transfer principle in [13]. The main task to make the mappings unique will focus on introducing 'orientation' so that e.g. (projective) lines correspond to two (oriented) 'Speere' which – on the other hand – introduces and requires metric (and sometimes euclidean) concepts and objects, see e.g. [11].

Here, mentioning [12], part II, p. 151, eqns. (51,l) and (51,r), Study explains the above 'spinor' notation, and as such Cartan's 'definition', in 'elliptic' context and SU(2)×SU(2) (or quaternionic) transformations, whereas in eq. (52), he resolves the 'spinor' components linearly in terms of his 'elliptic' line coordinates.

From our point of view, this emphasizes once more the use of Complex geometry. However, Study's part II and his reasoning with respect to 'elliptic' geometry has to be regarded with time and care in detail, because in eq. (1) on p. 119 he assumes the absolute polar system as $(xy) = x_0y_0 + x_1y_1 + x_2y_2 + x_3y_3$. This allows him to introduce real combinations of (homogeneous) line coordinates in his eq. (3). However, his eq. (4), re-expressing the Plücker condition of the line coordinates $X_{\alpha\beta}$ and using six line coordinate combinations X_i, can be cast into the Complex expression

$$X_1^2 - X_2^2 + X_3^2 - X_4^2 + X_5^2 - X_6^2 = 0 \tag{18}$$

while exhibiting SO(3,3) symmetry for a point interpretation of X_i, which directly relates to the fundamental form of Plücker's line geometry (see also [6], §23, especially p. 98).

However, if we switch Study's absolute polar system to $(xy) = -x_0y_0 + x_1y_1 + x_2y_2 + x_3y_3$ by introducing an i in the first (point) coordinate $\{\cdot\}_0$ so that the Plücker coordinates X_{0k} change to iX_{0k}, the coordinates (up to overall i's) in Study's eq. (3) are Klein's coordinates z ([6], §23).

Anyhow, citing Klein's work on six Complexe in involution and their threefold ±-handedness ([6], §23, p. 99), or facing Study's discussion of two spheres (or left/right quaternions) above, it is evident that these properties represent PG in terms of line and Complex geometry which reflects in certain known symmetry groups[21] like SU(2), SU(2)×SU(2) or even SU(4) of the

[20] German: dreigliedrige Gruppe

[21] U(1) is almost always evident by planar projections.

(spinorial) formulation of quantum theories.

3.7. Basic Spin Commutators

Now, having established the above matrix form of Lie transfer, we want to insert a brief section related to practical application and identification.

This time, we start from the matrix reps above in that we define points in R-space right from the beginning and ask for the rôle of the commutator as natural (antisymmetric) 'product' of two such objects[22]. As such, we define two 'points' $\mathcal{X} \sim \{X, Y, Z\}$ and $\mathcal{X}' \sim \{X', Y', Z'\}$, keeping the notation of the inhomogeneous coordinates from above, but switching to positive signs in eqns. (7) and (17), respectively, by

$$\mathcal{X} := X\sigma_1 + Y\sigma_2 + Z\sigma_3 = \begin{pmatrix} Z & X - iY \\ X + iY & -Z \end{pmatrix}, \quad \mathcal{X}' := \begin{pmatrix} Z' & X' - iY' \\ X' + iY' & -Z' \end{pmatrix}.$$

Defining $\mathcal{C}$ mainly as commutator by $\mathcal{C}(\mathcal{X}, \mathcal{X}') := \frac{1}{2}[\mathcal{X}, \mathcal{X}']$, straightforward calculation yields

$$\mathcal{C} = \begin{pmatrix} i(XY' - YX') & ZX' - XZ' + i(YZ' - ZY') \\ XZ' - ZX' + i(YZ' - ZY') & -i(XY' - YX') \end{pmatrix}. \tag{19}$$

In order to get more insight, by using the two points $\mathcal{X}$ and $\mathcal{X}'$, we construct line coordinates in R-space, i.e. we define and use $P_{12} = XY' - YX'$, $P_{23} = YZ' - ZY'$ and $P_{31} = ZX' - XZ'$. So eq. (19) reads as

$$\mathcal{C} = \begin{pmatrix} iP_{12} & P_{31} + iP_{23} \\ -P_{31} + iP_{23} & -iP_{12} \end{pmatrix} = i \begin{pmatrix} P_{12} & P_{23} - iP_{31} \\ P_{23} + iP_{31} & -P_{12} \end{pmatrix}, \tag{20}$$

where the rhs can be re-expressed in terms of Pauli spin matrices as $\mathcal{C} = i(P_{23}\sigma_1 + P_{31}\sigma_2 + P_{12}\sigma_3)$. Although we are used to having the i accompanying the commutator of (Pauli) spin matrices, it is easy to transform eq. (20) into a real equation by introducing quaternions, $q_k = -i\sigma_k$. Eq. (20) changes into the equation

$$\mathcal{C} = \frac{1}{2}(\mathcal{X}\mathcal{X}' - \mathcal{X}'\mathcal{X}) = i(P_{23}\sigma_1 + P_{31}\sigma_2 + P_{12}\sigma_3). \tag{21}$$

By multiplying both sides with -1, i.e. $(-i)^2$ on the lhs, we obtain $-\mathcal{C} =: \underline{\mathcal{C}} = \frac{1}{2}(\underline{X}\,\underline{X}' - \underline{X}'\underline{X}) = \underline{P}$ where each underlining now denotes a real quaternion, respectively, e.g. $\underline{X} = Xq_1 + Yq_2 + Zq_3$, $\underline{P} = P_{23}q_1 + P_{31}q_2 + P_{12}q_3$), etc. If we transfer the spin notation above of $\underline{\mathcal{C}} = -\mathcal{C}(\mathcal{X}, \mathcal{X}')$ to real quaternion calculus, eq. (21) reads as $\underline{\mathcal{C}}(\underline{X}, \underline{X}') = \underline{P}$, i.e. *the product (or commutator)* $\underline{\mathcal{C}}$ *of two real 'point' quaternions* $\underline{X}$ *and* $\underline{X}'$ *yields a real quaternion holding the three spatial Plücker coordinates.* We may, in addition, recognize the striking index ordering in eqns. (20) and (21) as well as ask for further commutators of those objects.

3.8. Advanced Spin Commutators

If we denote the components P_i of the 'Plücker' quaternion $\underline{P}$ above through $\underline{P} = P_i q_i$, by performing some algebra we obtain $\underline{\mathcal{C}}(\underline{X}, \underline{P}) = (P_{kj}x_j)q_k =: \underline{U}$. Whereas in QFT, we would now have to consider higher and increasing order of 'vector' products, reducable by quadratic (mass/momentum) constraints and angular momentum algebra, here we can stress the origin of $\underline{P}$.

[22] There is some parallel background from within classical polar theory of second order surfaces, however, the calculations here are self-contained so we'll discuss those issues in projective geometry at a later time.

By switching to the general context in terms of homogeneous coordinates, we can use the relation of linear Complexe and null systems. As such, in terms of homogeneous coordinates it is $P_{\alpha\beta}$ which relates x_β to the dual plane coordinatized by u_α. Whereas in the polar case, incidence $u_\alpha x_\alpha = 0$ of point and plane is special, treating null systems, the incidence relation is always satisfied, i.e. point and plane intersect because of antisymmetry of $P_{\alpha\beta}$ in $u_\alpha x_\alpha = P_{\alpha\beta} x_\beta x_\alpha = 0$.

Recalling this background from PG, by considering the real quaternion $\underline{U}$ above, its three components realize (almost) the spatial part of a null system in terms of the associated planar coordinates u_i.

It is obvious that we have to enhance the mechanism because at this stage we've lost contact to the 0-components of the complete null system, and to the 0-coordinates of points and planes in homogeneous rep. What remains open here, is the question whether this quaternionic calculus is sufficient to represent inhomogeneous calculations in mechanics and electrodynamics, i.e. comparison to experimental (classical) data.

Possible enhancements are obvious because the su(2) rep above is related to one line only. By recalling the two skew (and thus independent) lines above which we can identify either as opposite edges of a (fundamental) tetrahedron or as the two lines of a Congruency [4], we can almost automatically enhance the scheme by another independent su(2) (or quaternion) of the second (skew) line, so that the general construction scheme comprises SU(2)×SU(2) (or SL(1,$\mathbb{H}$)×SL(1,$\mathbb{H}$)) or SU(4) (or SL(2,$\mathbb{H}$)≅SU∗(4)), so we've closed the loop to our original starting point SU(4) or Dirac algebra in terms of SU∗(4) (see [4], [5] and refs there). SO(6) and the various real forms SO(n,m), $m+n = 6$, can be related either to their unitary covering groups, or by starting from SO(3,3) or SO(6) in line geometry by coordinate complexification, either of lines or points (thus, however, changing the respective absolute element or polar system, and in consequence the related underlying geometry). We postpone further discussions here.

Last not least, while realizing PG aspects by quaternions and their commutators, it is important to note that (in general) $\underline{\mathcal{C}}(\underline{X}'', \underline{P}) \sim [\underline{X}'', \underline{P}] \sim [\underline{X}'', [\underline{X}, \underline{X}']]$. This discussion can be extended to Jacobi identities, triple systems, curvature, and further geometric relations.

3.9. Lagrangean Construction

So far, after having related 'spinors' and SU(2) spin reps by means of Lie transfer to line and Complex geometry, we have to focus on appropriate Lagrangean constructions in order to reflect Complex invariants and dynamics, and provide the Lagrangean (or Hamiltonian) apparatus.

Here, we'll firstly focus on the photon rep of the QED Lagrangean, i.e. we'll postpone the construction of matter fields and the discussion of Dirac or Clifford algebra for a moment and focus on the construction principle using Complexe. This reflects the fact[23] that the photon rep almost automatically corresponds to a special linear Complex [4], [5] while obeying (and respecting) special relativity. Relativistic aspects will be treated in more detail in part VIII.

Above, we've shown that the 'spinor' discussion is a subset of Complex geometry right from Lie's paper. So to discuss invariance and construct invariant elements in Complex geometry, we have lots of geometrical possibilities, some of which we've already mentioned elsewhere. On the ground of Klein's Erlanger program, being in the regime of projective transformations, we may stress directly (quadratic) tetrahedral Complexe leaving anharmonic ratios of line intersections with a (fundamental) tetrahedron invariant. With respect to second order surfaces to be discussed in part VIII as absolute elements of non-euclidean geometry, we may as well consider polar or tangential Complexe. And possible degeneracy or Clebsch's paper on Complex symbolism [2] suggests to consider assemblies or powers of individual (special) linear Complexe as well in the basic construction process.

[23] For details, see [5] sec. III with respect to relativistic symmetry and to symplectic geometry, thus stabilizing a certain linear Complex.

Now the first major step is to identify electromagnetism by means of a special linear Complex. The Plücker condition $P := p_{01}p_{23} + p_{02}p_{31} + p_{03}p_{12} = 0$ suggests to identify the field strengths $\vec{E}$ and $\vec{B}$, i.e. we have $\vec{E} \cdot \vec{B} = 0$, so initially we can identify $\vec{E}_i$ and $\vec{B}_i$ as six Plücker coordinates. Having argued in [5] with Klein's paper of 1869 and 1871 (see also [6]), here we introduce some linear combinations y of Plücker's line coordinates $p \sim x$ according to the notation [6], p. 98, and we obtain (up to a factor 4)

$$\vec{E}^2 - \vec{B}^2 = \vec{E}^2 + i\vec{B}^2 =: \Omega\,, \tag{22}$$

which Klein defines as an invariant Ω of the linear Complex and states $\Omega = 0$ as condition for a special linear Complex. Recall that this decomposition historically gave rise to various discussions of SU(2)×SU(2) reps, as well as various complexifications thereof, mostly celebrating vector/'spinor' calculus.

Re-expressed in contemporary notation, we can compare e.g. with

$$-F_{\mu\nu}F^{\mu\nu} \sim \vec{E}^2 - \vec{B}^2 = \Omega\,, \quad G_{\mu\nu}F^{\mu\nu} \sim \vec{E} \cdot \vec{B} = P\,,\, G_{\mu\nu} = \frac{1}{2}\epsilon_{\alpha\beta\mu\nu}F^{\alpha\beta}$$

being the hodge dual which we've related earlier to conjugation and polar theory of Complexe.

So working with simple invariants of linear special Complexe, only, comprises known QED Lagrangean formulations. From the decomposition of quadratic Complexe or investigating pencils of linear Complexe, we know that we have to treat regular Complexe as well, so besides using the above quadratic expressions to enforce projective invariance or absolute elements, we expect the individual as well as the simultaneous invariance theory of general linear Complexe (special as well as regular) to comprise an appropriate basis of Lagrangean construction. Standard unitary symmetries are included as we've shown already above.

4. Summary and Outlook

Essentially, we have shown that su(2) and the Pauli rep according to Lie's setup is an artifact emerging from line geometry, and a degenerated case of Complex geometry. Lie transfer puts classical PG and sphere geometry side by side, and we've found especially real quaternion calculus formally realizing known geometry based on null systems, as soon as one understands the commutator as a generalized product. However, this needs deeper and more complete research. The general framework superseding this type of algebra is Complex geometry, and we have connected previous work by closing the circle to twofold quaternions and certain unitary symmetries we've started with. The outline and development, however, suggests to consider typical spin applications or SU(2) symmetry in physics from the viewpoint of line or Complex geometry as well.

References

[1] Cartan E 1938 *Leçons sur La Théorie des Spineurs* (Paris: Hermann & Cie Éditeurs)
[2] Clebsch A 1869 *Math. Ann.* **2** 1
[3] Clebsch A 1872 *Theorie der binären algebraischen Formen* (Leipzig: B. G. Teubner)
[4] Dahm R 2017 *Phys. Atom. Nuclei* **80** 512
[5] Dahm R 2017 *Journal of Physics : Conference Series* **804** 012013
[6] Klein F 1926 *Vorlesungen über höhere Geometrie* (Die Grundlehren der mathematischen Wissenschaft XXII) (Berlin: Springer)
[7] Lie S 1872 *Math. Ann.* **5** 145; *Math. Ann.* **5** 209
[8] Lie S 1888/93 *Theorie der Transformationsgruppen* (Leipzig: B. G. Teubner)
[9] Plücker J 1846 *System der Geometrie des Raumes* (Düsseldorf: Schaub'sche Buchhandlung, W. H. Scheller)
[10] Plücker J 1868/1869 *Neue Geometrie des Raumes* ed A Clebsch and F Klein (Leipzig: B. G. Teubner)
[11] Study E 1903 *Geometrie der Dynamen* (Leipzig: B. G. Teubner)
[12] Study E 1907 *Amer. J. Math.* **29** 101
[13] Study E 1917 *Jahresber. der Deutschen Mathematiker-Vereinigung* **25** 96

IOP Conf. Series: Journal of Physics: Conf. Series **965** (2018) 012013 doi:10.1088/1742-6596/965/1/012013

Chiral trace relations in Ω-deformed $\mathcal{N} = 2$ theories

A Fachechi, G Macorini and M Beccaria

Dipartimento di Matematica e Fisica Ennio De Giorgi & INFN, Via Arnesano, 73100 Lecce, Italy.

E-mail: alberto.fachechi@le.infn.it

Abstract. We study chiral trace relations in $\mathcal{N} = 2$ supersymmetric theories. Applying localization formulae for chiral observables, we derive closed chiral trace relations relating the vacuum expectation values of chiral ring elements. In this setting, we discuss how the Ω-background breaks the polynomial nature of such relations. These results are interpreted in the light of AGT duality, thus making contact with the integrable structure of conformal field theories on Riemann surfaces.

1. Introduction

Instanton calculus in gauge theories with extended supersymmetry is a useful tool to investigate non-perturbative effects. As is well-known [1], the exact solution of the Wilsonian effective theory of $SU(2)$ $\mathcal{N} = 2$ model without matter hypermultiplet (given in term of an holomorphic function called the prepotential) is fully encoded in an hyperelliptic curve capturing both perturbative and non-perturbative contributions. The low-energy effective theory is given implicitely in terms of contour integrals of the Seiberg-Witten differential along non-contractible cycles encircling the cuts of this curve. This result was then extended to generic mass content and gauge groups [2]. SW solution shed also light on the relation between $\mathcal{N} = 2$ supersymmetric theories and integrable systems, with the Seiberg-Witten curve playing the role of the spectral curve of its integrable counterpart [3, 4, 5, 6].

Despite this suggestive picture and the presence of several undirect checks of SW solution, a direct test of non-perturbative contributions was difficult since instanton moduli space is not compact and requires a systematical regularization. Investigations in this sense (see for instance [7]) culminated in the work by Nekrasov introducing the so-called Ω-deformation [8], see also [9]. It consists in a 2-parametric Poincaré-breaking regularization which allows to use explicit localization formulae to compute the partition function order by order in the instanton number. The regularized partition function plays a central role in the AGT duality [10] relating $\mathcal{N} = 2$ supersymmetric theories and Liouville CFT on 2d Riemann surfaces.[1] In more quantitative terms, AGT relation links the instanton partition function of the deformed gauge theory with the conformal blocks of correlation functions with a suitable operator insertion in the CFT.

Recently, the interest in supersymmetric gauge theories has been extended to specific observables in the chiral ring [17], *i.e.* trace operators $\mathrm{Tr}\,\varphi^n$ involving powers of the scalar field

[1] The original AGT duality is formulated only for $\mathcal{N} = 2$ superconformal $SU(2)$ quivers, see also [11]. Similar correspondence also holds for asymptotically free theories [12]. Finally, there exist AGT relations for bigger quivers, see for example [13, 14, 15, 16].

in the vector multiplet. Chiral observables play an important role in integrable properties of supersymmetric gauge theories [18, 19] and in the physics of surface operators [20, 21, 22, 23, 24]. For further details on the topic of the present paper, we refer to our original work [25].

As it is well-known, vacuum expectation values of chiral trace operators $\mathfrak{t}_n = \langle \mathrm{Tr}\, \varphi^n \rangle$ are not independent at the classical level [17]. This is a consequence of gauge group matrix properties, so it is possible to find simple relations involving them. In the quantum theory, such relations get modified non-perturbatively with contributions involving the instanton counting parameter q. The next natural question is how such identities are further modified when the Ω-background is switched on. The aim of this paper is to discuss chiral trace relations in the pure $\mathcal{N} = 2$ deformed supersymmetric theory for various gauge groups.

2. Chiral trace relations in undeformed Seiberg-Witten theories

Chiral trace relations are easily found in undeformed theories with resolvent techniques. In pure $SU(N)$ gauge theories, the resolvent reads [17]

$$\left\langle \mathrm{Tr}\, \frac{1}{z-\varphi} \right\rangle = \frac{P'(z)}{\sqrt{P^2(z) - 4\Lambda^{2N}}}. \tag{1}$$

Here, $P(z) = z^N - \sum_{l=2}^{N} \frac{\boldsymbol{u}_l}{l} z^{N-l}$ with the $\boldsymbol{u}_l$ parametrizing the $(N-1)$-dimensional quantum moduli space, while Λ is the dynamically generated scale related to the instanton counting parameter as $q \sim \Lambda^{2N}$. Chiral trace relations are obtained expanding (1) at large z. For the $SU(2)$ gauge group, the moduli space is parametrized by the quantity $\mathfrak{t}_2 \equiv \boldsymbol{u}$. From (1), we get the trace relations

$$\mathfrak{t}_4 = 4q + \frac{\boldsymbol{u}^2}{2}, \qquad \mathfrak{t}_6 = 6q\boldsymbol{u} + \frac{\boldsymbol{u}^3}{4}, \qquad \mathfrak{t}_8 = 12q^2 + 6q\boldsymbol{u}^2 + \frac{\boldsymbol{u}^4}{8}, \quad \dots \tag{2}$$

while odd traces identically vanish. Increasing the rank of the gauge group, chiral trace relations are less trivial, since more independent quantities appear, but their structure is totally similar. For the $SU(3)$ gauge group, they are $\mathfrak{t}_2 = \boldsymbol{u}_2$ and $\mathfrak{t}_3 = \boldsymbol{u}_3$. Then:

$$\begin{aligned} &\mathfrak{t}_4 = \frac{\mathfrak{t}_2^2}{2}, && \mathfrak{t}_5 = \frac{5\mathfrak{t}_2\mathfrak{t}_3}{6}, && \mathfrak{t}_6 = -6q + \frac{\mathfrak{t}_2^3}{4} + \frac{\mathfrak{t}_3^2}{3}, \\ &\mathfrak{t}_7 = \frac{7}{12}\mathfrak{t}_2^2\mathfrak{t}_3, && \mathfrak{t}_8 = -8q\mathfrak{t}_2 + \frac{\mathfrak{t}_2^4}{8} + \frac{4}{9}\mathfrak{t}_2\mathfrak{t}_3^2, && \dots \end{aligned} \tag{3}$$

Finally, for the $SU(4)$ gauge group, the independent quantities are $\mathfrak{t}_2 = \boldsymbol{u}_2$, $\mathfrak{t}_3 = \boldsymbol{u}_3$ and $\mathfrak{t}_4 = \boldsymbol{u}_4 + \frac{\boldsymbol{u}_2^2}{2}$. The other relations read

$$\begin{aligned} &\mathfrak{t}_5 = \frac{5\mathfrak{t}_2\mathfrak{t}_3}{6}, && \mathfrak{t}_6 = -\frac{\mathfrak{t}_2^3}{8} + \frac{\mathfrak{t}_3^2}{3} + \frac{3\mathfrak{t}_2\mathfrak{t}_4}{4}, \\ &\mathfrak{t}_7 = \frac{7\mathfrak{t}_2^2\mathfrak{t}_3}{24} + \frac{7\mathfrak{t}_3\mathfrak{t}_4}{12}, && \mathfrak{t}_8 = 8q - \frac{\mathfrak{t}_2^4}{16} + \frac{4\mathfrak{t}_2\mathfrak{t}_3^2}{9} + \frac{\mathfrak{t}_2^2\mathfrak{t}_4}{4} + \frac{\mathfrak{t}_4^2}{4}, \\ &\dots \end{aligned} \tag{4}$$

In pure Seiberg-Witten theories, chiral trace relations are clearly coherent with the natural gradation $[\varphi] = 1$ (which also implies that $[\mathfrak{t}_n] = n$) and $[q] = 2N$. The latter condition means that quantum contributions to trace relations start at the order $n = 2N$.

3. Chiral trace relations in Ω-deformed theories

Chiral trace relations obviously change when Ω-background is switched on. One of the most important reasons is that the original Poincaré symmetry is here downsized by the presence of the gravitational background. This has remarkable effects on chiral operators, since factorizations of the form $\langle \mathrm{Tr}\,\varphi^n\,\mathrm{Tr}\,\varphi^m\rangle \to \langle \mathrm{Tr}\,\varphi^n\rangle\langle \mathrm{Tr}\,\varphi^m\rangle$ no longer holds [26]. As a consequence, chiral trace relations are much more intricated in deformed theories. Moreover, resolvent techniques are not available in this setup, so localization formulae [27, 28] have to be adopted. Applying them, we get the vacuum expectation values of chiral trace operators in power series in the instanton counting parameter, with coefficients being functions of the scalar field vacuum expectation values $a_1, \ldots, a_N$, the deformation parameters $\epsilon_{1,2}$ and eventually the masses of matter fields. The problem is therefore to express higher trace in terms of the independent ones in closed form. This is of very simple solution in the undeformed limit, since it only requires the inversion $\mathbf{t}_i(a_1,\ldots,a_N)$ in favor of $a_i(\mathbf{t}_1,\ldots,\mathbf{t}_N)$ with $i = 1,\ldots,N$. For instance, computing $\langle \mathrm{Tr}\,\varphi^2\rangle$ in the $SU(2)$ gauge theory and then sending the deformation parameters to zero, we easily get

$$\boldsymbol{u} = 2a^2 + \frac{q}{a^2} + \frac{5q^2}{16a^6} + \frac{9q^3}{32a^{10}} + \frac{1469q^4}{4096a^{14}} + \frac{4471q^5}{8192a^{18}} + \mathcal{O}(q^6), \tag{5}$$

Inverting this equality, we express the scalar field vev in power series of the moduli $\boldsymbol{u}$ in the form

$$a = \sqrt{\frac{\boldsymbol{u}}{2}} - \frac{q}{\sqrt{2}\left(\sqrt{\boldsymbol{u}}\right)^3} - \frac{15q^2}{4\sqrt{2}\left(\sqrt{\boldsymbol{u}}\right)^7} - \frac{105q^3}{4\sqrt{2}\left(\sqrt{\boldsymbol{u}}\right)^{11}} - \frac{15015q^4}{64\sqrt{2}\left(\sqrt{\boldsymbol{u}}\right)^{15}} - \frac{153153q^5}{64\sqrt{2}\left(\sqrt{\boldsymbol{u}}\right)^{19}} + O\left(\boldsymbol{u}^{-10}\right). \tag{6}$$

Repeating the computation for higher order traces, we find

$$\begin{aligned} \mathbf{t}_4 &= 2a^4 + 6q + \frac{9q^2}{8a^4} + \frac{7q^3}{8a^8} + \frac{2145q^4}{2048a^{12}} + \frac{1575q^5}{1024a^{16}} + \mathcal{O}(q^6), \\ \mathbf{t}_6 &= 2a^6 + 15a^2q + \frac{135q^2}{16a^2} + \frac{125q^3}{32a^6} + \frac{16335q^4}{4096a^{10}} + \frac{44343q^5}{8192a^{14}} + \mathcal{O}(q^6), \\ &\ldots \end{aligned} \tag{7}$$

It is not hard to verify that, inserting the expansion (6) into higher order traces, we recover the relations (10). This method can be in principle used also in the general Ω-background. In this case, we get the more intricated quantity

$$\begin{aligned} \mathbf{t}_2 = {} & 2a^2 + \frac{q}{a^2} + \frac{5q^2}{16a^6} + \frac{9q^3}{32a^{10}} + \frac{1469q^4}{4096a^{14}} + \ldots \\ & + \epsilon_1\epsilon_2\left(-\frac{q^2}{8a^8} - \frac{q^3}{2a^{12}} - \frac{1647q^4}{1024a^{16}} + \ldots\right) \\ & + (\epsilon_1+\epsilon_2)^2\left(\frac{q}{4a^4} + \frac{21q^2}{32a^8} + \frac{55q^3}{32a^{12}} + \frac{18445q^4}{4096a^{16}} + \ldots\right) \\ & + (\epsilon_1\epsilon_2)^2\left(\frac{11q^2}{256a^{10}} + \frac{351q^3}{512a^{14}} + \frac{171201q^4}{32768a^{18}} + \ldots\right) + \ldots \end{aligned} \tag{8}$$

We can now try to invert this equation and express again a as a function of $\langle \mathrm{Tr}\,\varphi^2\rangle = \boldsymbol{u}$. If we replace this expansion in higher order traces, say for example $\mathbf{t}_4$, we get the involved expression

$$\begin{aligned} \mathbf{t}_4 = {} & \frac{\boldsymbol{u}^2}{2} + \frac{\left(8\boldsymbol{u} - 4\left(\epsilon_1^2 + 3\epsilon_1\epsilon_2 + \epsilon_2^2\right)\right)}{2\boldsymbol{u} - (\epsilon_1+\epsilon_2)^2}q \\ & + \frac{8\epsilon_1\epsilon_2\left(36\boldsymbol{u}^2 + 9\epsilon_1^4 + 50\epsilon_1^3\epsilon_2 + 86\epsilon_1^2\epsilon_2^2 + 50\epsilon_1\epsilon_2^3 + 9\epsilon_2^4 - 4\boldsymbol{u}\left(9\epsilon_1^2 + 13\epsilon_1\epsilon_2 + 9\epsilon_2^2\right)\right)}{\left(-2\boldsymbol{u} + (\epsilon_1+\epsilon_2)^2\right)^3\left(4\boldsymbol{u}^2 + \left(2\epsilon_1^2 + 5\epsilon_1\epsilon_2 + 2\epsilon_2^2\right)^2 - 2\boldsymbol{u}\left(5\epsilon_1^2 + 8\epsilon_1\epsilon_2 + 5\epsilon_2^2\right)\right)}q^2 + \mathcal{O}(q^3). \end{aligned} \tag{9}$$

Clearly, this method would not give closed trace relations. Moreover, even in the pure case similar expressions do not truncate at any order in the instanton number, as opposite to the undeformed case in which quantum corrections are polynomial in q.

These problems could discourage in proceeding further, but we can ascribe it *a posteriori* to the unnecessary requirement that chiral trace relations will depend polynomially on $\boldsymbol{u}$ in a general Ω-background. In fact, relaxing this assumption and allowing for the presence of derivatives of $\boldsymbol{u}$ with respect to the instanton counting parameter, we are able to derive closed trace relations! Although it can appear somewhat unexpected, this new feature has a nice AGT justification, see Section 4. The procedure we used to generate closed relation is simple. For each $\mathbf{t}_n$, we express it as combinations of independent traces and their derivatives, then we fix the coefficients in order to get agreement with localization formulae. Denoting $\boldsymbol{u}' = q\partial_q \boldsymbol{u}$, we get for the $SU(2)$ pure case

$$\begin{aligned}
\mathbf{t}_3 &= 0, \\
\mathbf{t}_4 &= \frac{1}{2}\boldsymbol{u}^2 - \epsilon_1\epsilon_2\boldsymbol{u}' + 4q, \\
\mathbf{t}_5 &= 10\,(\epsilon_1+\epsilon_2)\,q, \\
\mathbf{t}_6 &= \frac{1}{4}\boldsymbol{u}^3 - \frac{3}{2}\epsilon_1\epsilon_2\boldsymbol{u}\boldsymbol{u}' + \epsilon_1^2\epsilon_2^2\boldsymbol{u}'' + 6q\boldsymbol{u} + 6q\left(3\epsilon_1^2 + 4\epsilon_1\epsilon_2 + 3\epsilon_2^2\right), \\
\mathbf{t}_7 &= (\epsilon_1+\epsilon_2)\left(21q\boldsymbol{u} + 7q\left(4\epsilon_1^2 + 3\epsilon_1\epsilon_2 + 4\epsilon_2^2\right)\right), \\
\mathbf{t}_8 &= \frac{1}{8}\boldsymbol{u}^4 + 6q\boldsymbol{u}^2 + 12q^2 - \epsilon_1^3\epsilon_2^3\boldsymbol{u}''' + 2\epsilon_1^2\epsilon_2^2\boldsymbol{u}\boldsymbol{u}'' + \frac{3}{2}\epsilon_1^2\epsilon_2^2\boldsymbol{u}'^2 - 12\epsilon_1\epsilon_2 q\boldsymbol{u}' \\
&\quad - \frac{3}{2}\epsilon_1\epsilon_2\boldsymbol{u}^2\boldsymbol{u}' + (52\epsilon_1^2 + 72\epsilon_1\epsilon_2 + 52\epsilon_2^2)q\boldsymbol{u} + 8q(5\epsilon_1^4 + 11\epsilon_1^3\epsilon_2 + 15\epsilon_1^2\epsilon_2^2 + 11\epsilon_1\epsilon_2^3 + 5\epsilon_2^4), \\
&\ldots
\end{aligned} \tag{10}$$

For the $SU(3)$ gauge group, the modification coming from Ω-background reads

$$\begin{aligned}
\mathbf{t}_4 &= \frac{1}{2}\mathbf{t}_2^2 - \epsilon_1\epsilon_2\mathbf{t}_2', \\
\mathbf{t}_5 &= \frac{5}{6}\mathbf{t}_2\mathbf{t}_3 - \frac{5}{3}\epsilon_1\epsilon_2\mathbf{t}_3', \\
\mathbf{t}_6 &= -6q + \frac{\mathbf{t}_{3,3}}{3} + \frac{1}{4}\mathbf{t}_2^3 - \frac{3}{2}\epsilon_1\epsilon_2\mathbf{t}_2\mathbf{t}_2' + \epsilon_1^2\epsilon_2^2\mathbf{t}_2'', \\
\mathbf{t}_7 &= \frac{7\mathbf{t}_2^2\mathbf{t}_3}{12} - 21q(\epsilon_1+\epsilon_2) - \frac{7}{6}\epsilon_1\epsilon_2\mathbf{t}_3\mathbf{t}_2' - \frac{7}{3}\epsilon_1\epsilon_2\mathbf{t}_2\mathbf{t}_3' + \frac{7}{3}\epsilon_1^2\epsilon_2^2\mathbf{t}_3'', \\
&\ldots
\end{aligned} \tag{11}$$

while for the $SU(4)$ gauge group, we find

$$\begin{aligned}
\mathbf{t}_5 &= \frac{5\mathbf{t}_2\mathbf{t}_3}{6} - \frac{5}{3}\epsilon_1\epsilon_2\mathbf{t}_3', \\
\mathbf{t}_6 &= \frac{3\mathbf{t}_{2,4}}{4} - \frac{\mathbf{t}_2^3}{8} + \frac{\mathbf{t}_{3,3}}{3} + \frac{3}{4}\epsilon_1\epsilon_2\mathbf{t}_2\mathbf{t}_2' - \frac{1}{2}\epsilon_1^2\epsilon_2^2\mathbf{t}_2'', \\
\mathbf{t}_7 &= \frac{7\mathbf{t}_2^2\mathbf{t}_3}{24} + \frac{7\mathbf{t}_{3,4}}{12} - \frac{7}{12}\epsilon_1\epsilon_2\mathbf{t}_3\mathbf{t}_2' - \frac{7}{6}\epsilon_1\epsilon_2\mathbf{t}_2\mathbf{t}_3'' + \frac{7}{6}\epsilon_1^2\epsilon_2^2\mathbf{t}_3'', \\
&\ldots
\end{aligned} \tag{12}$$

We checked these expressions up to $\sim q^8$ instanton order.

To conclude this section, we would like to do some comments. First, gravitational corrections also hold in the so-called Nekrasov-Shatashvili limit $\epsilon_2 \to 0$, but in this case the differential structure of chiral ring is lost. Moreover, in this case odd traces no longer vanish (unless in the limit $\epsilon_1 = -\epsilon_2$). Finally, a relevant feature of the gravitational background is the appearence of new universal terms (*i.e.* $-\epsilon_1\epsilon_2\boldsymbol{u}'$ for $n = 4$, $-\frac{3}{2}\epsilon_1\epsilon_2\boldsymbol{u}\boldsymbol{u}'$ and $\epsilon_1^2\epsilon_2^2\boldsymbol{u}''$ for $n = 6$ *etc.*). All these features have a clear AGT explanation.

4. AGT interpretation

AGT duality is a correspondence relating deformed $\mathcal{N} = 2$ supersymmetric theories and conformal field theories on 2d Riemann surfaces. The topology of this surface is uniquely determined by the flavor and gauge groups of the 4d field theory. Quantitatively, the duality states the equivalence between the instanton partition function of the gauge theory and the conformal block of the correlation function with a suitable choice of the external operator insertions. In doing this, matter field masses and the vacuum expectation value of the scalar field are respectively mapped to the conformal dimensions of the external and intermediate operators in the conformal block, while the deformation parameters determine the central charge.

AGT duality provides an interesting point of view about the chiral ring structure. As is well-known [29], two-dimensional CFTs present a suggestive integrable structure which is proved by the existence of an infinite tower of local integrals of motion. The basic observation in understanding the duality at this level is the strong evidence that the chiral ring of the gauge theory retains the structure of integrals of motion, so chiral trace relations should emerge in a very natural way in this setup. This is clear in particular from the results of [30, 31] for the $SU(2)$ $\mathcal{N} = 2$ theory with $N_f = 4$ fundamentals hypermultiplets. As stated by AGT duality, this theory is related to a Liouville CFT on the 4-punctured sphere, with the instanton partition function being identified with the 4-point conformal block. A convenient way [32] to link the two AGT sides consists in considering the symmetry algebra to a Vir $\oplus$ Heis, with commutation relations

$$\begin{aligned} [L_m, L_n] &= (m-n)L_{m+n} + \frac{c}{12}m(m^2-1)\delta_{m+n,0}, \\ [a_m, a_n] &= \frac{m}{2}\delta_{m+n,0}, \qquad [L_m, a_n] = 0. \end{aligned} \tag{13}$$

The first local integrals of motion are in this case

$$\begin{aligned} I_2 &= L_0 - \frac{c}{24} + 2\sum_{k=1}^{\infty} a_{-k}a_k, \\ I_3 &= \sum_{k\in\mathbb{Z}/\{0\}} a_{-k}L_k + 2iQ\sum_{k=1}^{\infty} ka_{-k}a_k + \frac{1}{3}\sum_{i+j+k=0} a_ia_ja_k, \\ I_4 &= 2\sum_{k=1}^{\infty} L_{-k}L_k + L_0^2 - \frac{c+2}{12}L_0 + \ldots, \\ &\ldots \end{aligned} \tag{14}$$

where Q parametrizes the central charge as $c = 1 + 6Q^2$.[2] The authors of [30, 31] found a simple relation between the vacuum expectation values of chiral trace operators and the CFT correlation function with charge insertions:

$$G_k^{N_f=4}(\alpha_i|z) = \langle V_{\alpha_1}(\infty)V_{\alpha_2}(1)I_kV_{\alpha_3}(q)V_{\alpha_4}(0)\rangle, \tag{15}$$

[2] Note that $Q = b + b^{-1}$ where b is related to the deformation parameter as $b = \sqrt{\frac{\epsilon_2}{\epsilon_1}}$.

where V_α are primary fields in this extended algebra with conformal dimension $\Delta_\alpha = \alpha(Q-\alpha)$. Guided by these results, in [25] we searched for similar relations in the $SU(2)$ $\mathcal{N}=2^*$ and pure gauge theories. In the former case, AGT correspondence states that the dual quantity in the CFT side is the 1-point conformal block on the torus

$$Z_{\text{inst}} = \left[q^{-\frac{1}{24}}\eta(q)\right]^{2\Delta_m - 1} \text{Tr}_\Delta\left(\mathcal{O}_{\Delta_m} q^{L_0-\Delta}\right), \tag{16}$$

where the trace is taken on the descendants of the Virasoro primary in the intermediate channel and Δ_m is the conformal dimension of the external operator. Starting from Matone's relation [27, 33] and with simple calculations, it is easy to prove that[3]

$$\boldsymbol{u} = (4m^2 - \epsilon_1^2 - \epsilon_2^2)\frac{\text{E}_2 - 1}{24} - \frac{\epsilon_1\epsilon_2}{12} - 2\epsilon_1\epsilon_2 \frac{\text{Tr}_\Delta\left(\mathcal{O}_{\Delta_m}(L_0 - \frac{c}{24})q^{L_0-\frac{c}{24}}\right)}{\text{Tr}_\Delta\left(\mathcal{O}_{\Delta_m} q^{L_0-\frac{c}{24}}\right)}, \tag{17}$$

where the last term is nothing but the insertion of the Virasoro part of the integral of motion I_2 in the instanton partition function. This feature is very similar to the $N_f = 4$ case, so we defined the CFT quantities

$$G_k^{\mathcal{N}=2^*} = \text{Tr}_\Delta\left(\mathcal{O}_{\Delta_m} I_k^{\text{Vir}} q^{L_0-\frac{c}{24}}\right), \tag{18}$$

where I_k^{Vir} is the Virasoro part of integrals (14). We argued that they are related to the chiral trace vacuum expectation values.

The pure theory is something different from the previous cases. In fact, the duality presented in [10] only holds for $\mathcal{N}=2$ superconformal quivers of $SU(2)$. However, asymptotically free theories such as the pure $SU(2)$ case also admit a similar statement. To this aim, Gaiotto introduced in [12] a peculiar state lying in the Verma module of the highest weight of conformal dimension $\Delta = \frac{Q^2}{4} - \frac{a^2}{\epsilon_1\epsilon_2}$. This state is defined as a formal power series in Λ,

$$|\Lambda, \Delta\rangle = \sum_{k=0}^{\infty} \Lambda^{2k} v_k, \tag{19}$$

with each component satisfies the following constraints:

$$L_1 v_k = v_{k-1}, \quad L_2 v_k = 0. \tag{20}$$

The starting point of the sequence is the condition $v_0 = |\Delta\rangle$, which is of course the highest weight state in the Verma module. The AGT statement for the pure $SU(2)$ theory is that the norm of such a state exactly reproduces the Nekrasov instanton partition function, so Λ^4 acquires the meaning of instanton counting parameter:

$$Z_{\text{inst}} = \langle\Lambda, \Delta|\Lambda, \Delta\rangle = \sum_{k=0}^{\infty} \Lambda^{4k} ||v_k||^2. \tag{21}$$

[3] Here, E_2 is the Eisenstein quasi-modular form

$$\text{E}_2(q) = 1 - 24\sum_{n=1}^{\infty} \sigma_1(n) q^n,$$

with $\sigma_1(n)$ is the sum of divisors on n.

With the same approach as before, we can built new quantities in the CFT side by inserting the local integrals of motion in the brackets:

$$G_k^{\text{pure}} = \langle \Lambda, \Delta | I_k^{\text{Vir}} | \Lambda, \Delta \rangle, \quad G^{\text{pure}} = \langle \Lambda, \Delta | \Lambda, \Delta \rangle. \tag{22}$$

Again, we claim that these quantities are related to the chiral trace vacuum expectation values in the gauge theory. However, a systematical check of this conjecture requires a basis choice in the algebra of integrals of motion and a precise matching between the two sides, which unfortunately is still missing in general.

4.1. Leading terms and universality

Despite the absence of the dictionary between the chiral ring elements and correlation functions with the insertion of integrals of motion, it is possible to extract some predictions. The basic observation is that in all cases chiral trace relations share some common terms:

$$\begin{aligned}
\mathfrak{t}_4 &= \frac{1}{2}\boldsymbol{u}^2 - \epsilon_1\epsilon_2\boldsymbol{u}' + \dots, \\
\mathfrak{t}_6 &= \frac{1}{4}\boldsymbol{u}^3 - \frac{3}{2}\epsilon_1\epsilon_2\boldsymbol{u}\boldsymbol{u}' + \epsilon_1^2\epsilon_2^2\boldsymbol{u}'' + \dots, \\
\mathfrak{t}_8 &= \frac{1}{8}\boldsymbol{u}^4 - \frac{3}{2}\epsilon_1\epsilon_2\boldsymbol{u}^2\boldsymbol{u}' + \frac{3}{2}\epsilon_1^2\epsilon_2^2(\boldsymbol{u}')^2 + 2\epsilon_1^2\epsilon_2^2\boldsymbol{u}\boldsymbol{u}'' - \epsilon_1^3\epsilon_2^3\boldsymbol{u}''' + \dots, \\
&\dots
\end{aligned} \tag{23}$$

These leading terms are special monomials of $\boldsymbol{u}$ and its derivatives with coefficients depending only the deformation parameters ϵ_i. They are indeed universal, as they not depend on the particular model. In order to understand their origin, the key point is that even integrals of motion present powers of the Virasoro generators L_0^n, which act in CFT correlation function through the differential operators $(q\partial_q)^n$. It is therefore natural to guess that leading terms are associated to such special terms. In mathematical terms, this is nothing but the requirement that

$$G_{2n} \underset{\text{leading}}{\sim} (q\partial_q)^n G. \tag{24}$$

We can therefore write

$$\mathfrak{t}_{2n} \underset{\text{leading}}{\sim} 2\frac{(-\epsilon_1\epsilon_2 q\partial_q)^n G}{G}. \tag{25}$$

From Matone's relation

$$\boldsymbol{u} = -2\epsilon_1\epsilon_2 q\partial_q \log Z, \tag{26}$$

where Z now is the total partition function $Z = Z_{\text{cl}}Z_{\text{1-loop}}Z_{\text{inst}}$. The classical part is $Z_{\text{cl}} = \exp\left(-(\epsilon_1\epsilon_2)^{-1}a^2\log q\right)$, while the 1-loop contribution is q-independent. Then (apart for contributions which are not relevant in this discussion) we can identify G with the whole partition function Z, or in other words

$$G = \exp\left(-\frac{U}{2\epsilon_1\epsilon_2}\right), \tag{27}$$

where $\boldsymbol{u} = q\partial_q U$. Using this representation and (25), we can easily determine the terms in (23), even without mentioning which theory we are considering. Indeed, the results reported in this section depends only on the structure of the local integrals of motion of the conformal field theory. Eq. (24) holds both for G_k defined through (18) in the $\mathcal{N} = 2^*$ model and through (22) for the pure theory. Therefore, we conclude the leading terms are universal, in perfect agreement with the explicit form of chiral trace relations.

IOP Conf. Series: Journal of Physics: Conf. Series **965** (2018) 012013 doi:10.1088/1742-6596/965/1/012013

References

[1] Seiberg N and Witten E 1994 *Nucl. Phys.* **B426** 19–52 (*Preprint* hep-th/9407087)
[2] Seiberg N and Witten E 1994 *Nucl. Phys.* **B431** 484–550 (*Preprint* hep-th/9408099)
[3] Donagi R and Witten E 1996 *Nucl. Phys.* **B460** 299–334 (*Preprint* hep-th/9510101)
[4] Martinec E J 1996 *Phys. Lett.* **B367** 91–6 (*Preprint* hep-th/9510204)
[5] Gorsky A, Krichever I, Marshakov A, Mironov A and Morozov A 1995 *Phys. Lett.* **B355** 466–74 (*Preprint* hep-th/9505035)
[6] D'Hoker E and Phong D H 1998 *Nucl. Phys.* **B513** 405–44 (*Preprint* hep-th/9709053)
[7] Moore G W, Nekrasov N and Shatashvili S 2000 *Commun. Math. Phys.* **209** 97–121 (*Preprint* hep-th/9712241)
[8] Nekrasov N A 2003 *Adv. Theor. Math. Phys.* **7** 831–64 (*Preprint* hep-th/0206161)
[9] Nekrasov N and Okounkov A 2006 *Prog. Math.* **244** 525–96 (*Preprint* hep-th/0306238)
[10] Alday L F, Gaiotto D and Tachikawa Y 2010 *Lett. Math. Phys.* **91** 167–97 (*Preprint* arxiv:0906.3219)
[11] Gaiotto D 2012 *J. High Energy Phys.* JHEP08(2012)034 (*Preprint* arxiv:0904.2715)
[12] Gaiotto D 2013 *J. Phys.: Conf. Series* **462** 012014 (*Preprint* arxiv:0908.0307)
[13] Wyllard N 2009 *J. High Energy Phys.* JHEP11(2009)002 (*Preprint* arxiv:0907.2189)
[14] Schiappa R and Wyllard N 2010 *J. Math. Phys.* **51** 082304 (*Preprint* arxiv:0911.5337)
[15] Mironov A and Morozov A 2010 *Nucl. Phys.* **B825** 1–37 (*Preprint* arxiv:0908.2569)
[16] Kanno S, Matsuo Y and Shiba S 2010 *Phys. Rev.* **D82** 066009 (*Preprint* arxiv:1007.0601)
[17] Cachazo F, Douglas M R, Seiberg N and Witten E 2002 *J. High Energy Phys.* **12** 071 (*Preprint* hep-th/0211170)
[18] Minahan J A and Zarembo K 2003 *J. High Energy Phys.* **03** 013 (*Preprint* hep-th/0212208)
[19] Nekrasov N A and Shatashvili S L 2009 *Proc. 16th Int. Cong. on Mathematical Physics (Prague)* ed P Exner (Singapore: World Scientific) pp 265–89 (*Preprint* 0908.4052)
[20] Gukov S and Witten E 2006 *Preprint* hep-th/0612073
[21] Gaiotto D, Gukov S and Seiberg N 2013 *J. High Energy Phys.* JHEP09(2013)070 (*Preprint* arxiv:1307.2578)
[22] Cordova C, Gaiotto D and Shao S H 2017 *J. High Energy Phys.* JHEP05(2017)140 (*Preprint* arxiv:1704.01955)
[23] Ashok S K, Billo M, Dell'Aquila E, Frau M, John R R and Lerda A 2017 *J. High Energy Phys.* JHEP07(2017)068 (*Preprint* arxiv:1702.02833)
[24] Ashok S K, Billo M, Dell'Aquila E, Frau M, Gupta V, John R R and Lerda A 2017 *Preprint* arxiv:1707.08922
[25] Beccaria M, Fachechi A and Macorini G 2017 *J. High Energy Phys.* JHEP05(2017)023 (*Preprint* arxiv:1702.01254)
[26] Fucito F, Morales J F, Poghossian R and Tanzini A 2006 *J. High Energy Phys.* **01** 031 (*Preprint* hep-th/0510173)
[27] Flume R, Fucito F, Morales J F and Poghossian R 2004 *J. High Energy Phys.* JHEP04(2004)008 (*Preprint* hep-th/0403057)
[28] Billo M, Frau M, Fucito F, Giacone L, Lerda A, Morales J F and Pacifici D R 2012 *J. High Energy Phys.* JHEP08(2012)166 (*Preprint* arxiv:1206.3914)
[29] Bazhanov V V, Lukyanov S L and Zamolodchikov A B 1996 *Commun. Math. Phys.* **177** 381–98 (*Preprint* hep-th/9412229)
[30] Fucito F, Morales J F and Poghossian R 2015 *J. High Energy Phys.* JHEP11(2015)064 (*Preprint* arxiv:1507.05426)
[31] Fucito F, Morale J F and Poghossian R 2017 *J. Geom. Phys.* **118** 169–80 (*Preprint* arxiv:1603.02586)
[32] Alba V A, Fateev V A, Litvinov A V and Tarnopolskiy G M 2011 *Lett. Math. Phys.* **98** 33–64 (*Preprint* arxiv:1012.1312)
[33] Matone M 1995 *Phys. Lett.* **B357** 342–8 (*Preprint* hep-th/9506102)

IOP Conf. Series: Journal of Physics: Conf. Series **965** (2018) 012014 doi:10.1088/1742-6596/965/1/012014

On Quantum Cosmology in Teleparallel Gravity

A. S. Fernandes
Instituto de Física, Universidade de Brasília, 70910-900, Brasília, DF, Brazil
E-mail: alexandre.fernandes@catolica.edu.br

S. C. Ulhoa
Instituto de Física, Universidade de Brasília, 70910-900, Brasília, DF, Brazil
E-mail: sc.ulhoa@gmail.com

R. G. G. Amorim
Faculdade Gama, Universidade de Brasília, 72444-240, Setor Leste (Gama), Brasília, DF, Brazil.
E-mail: ronniamorim@gmail.com

Abstract. A quantum cosmology in teleparallel gravity is presented in this article. Teleparallel gravity is used to perform such an analysis once in General Relativity (GR) the concept of gravitational energy is misleading preventing the establishment of a concise quantum cosmology. The Wheeler-DeWitt like equation is obtained using the Weyl quantization and the teleparallel expression of energy.

1. Introduction

The modern observational cosmology inaugurated at the Mount Wilson Observatory gave a great impetus to understanding the Universe [1]. The Standard Cosmological Model, alongside the Cosmological Principle and field equations of GR, describes all knowledge about large structures with good approximation. The Hubble Law shows how fast galaxies move away from each other at a relatively small distances. Thus it could be used to test new cosmological theories. The Cosmological Principle states that the Universe is isotropic (above 100 Mpc) and homogeneous (there is no center) in addition its dynamics is given by the Einstein field equations, $R_{\mu\nu} - \frac{1}{2}g_{\mu\nu}R = 8\pi T_{\mu\nu}$. In such a way it is possible to trace a complete time evolution of the Universe. If the time is set backwards we see that everything started in a warm and dense state with domination of the radiation energy. The metric that admits the Cosmological Principle and the dynamics given by the GR is that of Friedman-Lemaître-Robertson-Walker (FLRW) [2, 3, 4, 4, 6, 7, 8]:

$$ds^2 = -dt^2 + a^2(t)\left[\frac{dr^2}{1-\kappa r^2} + r^2 d\Omega^2\right], \qquad d\Omega^2 \equiv d\theta + \sin^2\theta\, d\phi^2 \tag{1}$$

where k assumes values of -1 (negative or closed spatial curvature), 0 (null or flat spatial curvature) or $+1$ (positive or open spatial curvature).

IOP Conf. Series: Journal of Physics: Conf. Series **965** (2018) 012014 doi:10.1088/1742-6596/965/1/012014

Due to the rapid improvement in observational cosmology a quantum theory of cosmology is increasingly needed in order to describe new effects. Particularly quantum effects should be relevant at singular states as it is expected in the beginning of the Universe. Such a theory should have an operator and time evolving structure. Since the seminal paper of Rosenfeld in 1930 [9, 10], a lot of research has been done in this subject. It should be noticed that in order to proper treat a quantum theory of gravitation it is necessary to apply any quantization procedure to a physical quantity. On one hand the Hamiltonian is the most recognizable quantity with such a feature, on the other hand the existence of a gravitational energy is perhaps one of the most controversial and oldest problems in General Relativity. Thus a Hamiltonian formulation of GR was presented by Arnowitt, Deser and Misner in 1962 [11] and used by Wheeler and DeWitt in 1967 to settle an Einstein-Schrödinger equation [12] which happened to be known as Wheeler-DeWitt equation. Such an equation has been focus of extensive research in the last 50 years. However the so called problem of time still persists [13] which prevents a dynamical interpretation of one possible solution.

In this article we address the problem of the establishment of a quantum theory of gravitation applied to cosmology using teleparallel gravity and the Weyl quantization procedure. Such a procedure allows the mapping of any function into a operator structure. Thus it is not necessary to write the functions in phase space which is the standard procedure to quantize fields. The Weyl's prescription only require two parameters. Teleparallel gravity is an alternative theory of gravitation and entirely equivalent to General Relativity. Einstein was one of the first to use teleparallel gravity to investigate a form to unify gravity and electromagnetism [14]. In 1961 Møller showed that it is impossible to cancel the gravitational field by coordinate transformation, this means that the pseudo-tensor approach should be avoided. Rescuing the initial idea of Einstein, Møller demonstrated that it was possible to obtain a gravitational energy density [15]. In 1963 Pellegrini and Plebanski found a Lagrangian formulation of the teleparallel gravity based on Møller [16]. Already in 1994 Maluf described the Hamiltonian formulation for Teleparallelism Equivalent to GR (TEGR) [17]. Several applications of the TERG have been made as a description of the gravitational angular momentum [18], neutron stars [19], quantum Schwarzschild spacetime [20], gravitational waves [21] among others.

This article is divided as follows: in section 2 we briefly recall the ideas of teleparallel gravity using natural unities with $c = G = 1$. We show how to obtain the energy-momentum tensor of gravitational and matter fields. In section 3 we present the Weyl quantization procedure and apply it to the total energy of FLRW metric. As a consequence we establish two quantum cosmological equations. One quantizing the total energy density and the other using the total energy itself which is function only of the scale factor and its derivative. Finally in the last section we present our conclusions.

2. Teleparallelism Equivalent to General Relativity

Teleparallel Gravity is dynamically equivalent to GR since the Lagrangian density of both theories differ only by a total divergence which means that the field equations are equivalent. Teleparallel gravity is constructed in terms of the tetrad field, $e^a{}_\mu$, which has two symmetries [17, 22, 23]. The latin indices $(a = (0), (i))$ represent the $SO(3,1)$ group while the greek indices stand for diffeomorphisms. Thus the tetrad field projects one symmetry into another. General Relativity is described in terms of the metric tensor that relates to tetrad field by $g_{\mu\nu} = e^a{}_\mu e_{a\nu}$.

Let's consider a geometry endowed with two connections, one $(\omega_{\mu ab})$ acting on Lorentz indices and another one $(\Gamma^\lambda{}_{\mu\nu})$ acting on transformations of coordinates. If we require that this space is metric compatible then $\nabla_\mu e_{a\nu} = 0$. As a consequence $\Gamma^\lambda{}_{\mu\nu} = e^{a\lambda} e^b{}_\nu \omega_{\mu ab} + e^{a\lambda}\partial_\mu e_{a\nu}$. The curvature of this connection is $R^\lambda{}_{\gamma\mu\nu}(e,\omega) = e_a{}^\lambda e^b{}_\gamma(\partial_\mu\omega_{\nu ab} - \partial_\nu\omega_{\mu ab} + \omega_{\mu ac}\omega_{\nu cb} - \omega_{\nu ac}\omega_{\mu cb})$ and the torsion associated to such a connection is

$$T^{\lambda}{}_{\mu\nu} = \Gamma^{\lambda}{}_{\mu\nu} - \Gamma^{\lambda}{}_{\nu\mu}\,, \tag{2}$$

which is explicitly given by

$$T^{a}{}_{\mu\nu}(e,\omega) = \partial_{\mu}e^{a}{}_{\nu} - \partial_{\nu}e^{a}{}_{\mu} + \omega_{\mu a\nu} - \omega_{\nu a\mu}\,. \tag{3}$$

The connection $\omega_{\mu ab}$ is known as the spin connection since it is used to couple gravity to a fermion field [24]. It is related to the Levi-Civita connection by means the mathematical identity

$$\omega_{\mu ab} = {}^{0}\omega_{\mu ab} + K_{\mu ab} \tag{4}$$

where ${}^{0}\omega_{\mu ab} = -\frac{1}{2}e^{c}{}_{\mu}(\Omega_{abc} - \Omega_{bac} - \Omega_{cab})$, with $\Omega_{abc} = e_{a\nu}(e_{b}{}^{\mu}\partial_{\mu}e_{c}{}^{\nu} - e_{c}{}^{\mu}\partial_{\mu}e_{b}{}^{\nu})$, is the Levi-Civita connection which is a torsionless connection and $K_{\mu ab} = -\frac{1}{2}e_{a}{}^{\lambda}e_{b}{}^{\nu}(T_{\lambda\mu\nu} + T_{\nu\lambda\mu} + T_{\mu\lambda\nu})$ is the contorsion tensor.

The so called teleparallel condition $\omega_{\mu ab} = 0$ establishes a geometric gravitational theory with a vanishing curvature and with the presence a torsion tensor. In other words the non-zero torsion is responsible for gravitational effects [22]. The teleparallel condition allows one to relate the scalar curvature of the Levi-Civita connection to a scalar constructed out of the torsion tensor. Such a relations reads

$$e\,R(e) \equiv -e\left(\frac{1}{4}T^{abc}T_{abc} + \frac{1}{2}T^{abc}T_{bac} - T^{a}{}_{a}\right) + 2\,\partial_{\mu}\left(e\,T^{\mu}\right) \tag{5}$$

where $T^{a}{}_{a} = T^{b}{}_{b}{}^{a}$ and

$$T^{a}{}_{\mu\nu}(e) = \partial_{\mu}e^{a}{}_{\nu} - \partial_{\nu}e^{a}{}_{\mu}\,. \tag{6}$$

Hence the teleparallel Lagrangian density is written as

$$\mathcal{L}(e_{a\mu}) = -k\,e\left(\frac{1}{4}T^{abc}T_{abc} + \frac{1}{2}T^{abc}T_{bac} - T^{a}T_{a}\right) - \mathcal{L}_{M}\,, \tag{7}$$

with $k = \frac{1}{16\pi}$ and $\mathcal{L}_M$ is lagrangian density for matter fields. It should be noticed that the total divergence in (5) was neglected since it gives no dynamical contributions. In addition one should take into account that the above Lagrangian density is equivalent to the Hilbert-Einstein Lagrangian density. As a consequence GR is dynamically equivalent to teleparallel gravity. Expression (7) can be rewritten as $\mathcal{L}(e_{a\mu}) = -k\,e\,\Sigma^{abc}T_{abc} - \mathcal{L}_M$ where Σ^{abc} is defined as [17]

$$\Sigma^{abc} = \frac{1}{4}\left(T^{abc} + T^{bac} - T^{cab}\right) + \frac{1}{2}\left(\eta^{ac}T^{b} - \eta^{ab}T^{b}\right). \tag{8}$$

The field equations are calculated from functional variation of the Lagrangian density $\mathcal{L}$ with respect to tetrads $e^{a\mu}$. They read

$$\partial_{\lambda}(e\,\Sigma^{a\mu\lambda}) - e\left(\Sigma^{b\lambda\mu}T_{b\lambda}{}^{a} - \frac{1}{4}e^{a\mu}T_{bcd}\Sigma^{bcd}\right) = \frac{1}{4k}e\,T^{a\mu}\,. \tag{9}$$

Due to the equivalence between teleparallel gravity and General Relativity the following important relationship is established

$$\begin{aligned} \partial_{\lambda}(e\Sigma^{a\mu\lambda}) - e\left(\Sigma^{b\lambda\mu}T_{b\lambda}{}^{a} - \frac{1}{4}e^{a\mu}T_{bcd}\Sigma^{bcd}\right) &= \frac{1}{2}e\left[R^{a\mu}(e) - \frac{1}{2}e^{a\mu}R(e)\right], \\ R_{a\mu}(e) - \frac{1}{2}e_{a\mu}R(e) &= \frac{1}{2k}T_{a\mu}\,, \end{aligned} \tag{10}$$

where $T_{a\mu}$ is the energy-momentum tensor of matter fields. Therefore the Riemann geometry which is set in terms of a metric tensor and the curvature tensor is equivalent to the Weitzenböck geometry which is established in terms of the tetrad field and the torsion in equation (6). On the other hand given a metric tensor there are infinite possible tetrads. This apparent paradox is solved by considering the physical interpretation of the tetrad field. Once the zero component of the tetrad field is always tangent along of a world line of an observer, it is interpreted as $e_{(0)}{}^{\mu} = u^{\mu}(\tau) = dx^{\mu}/d\tau$. Generalizing this idea the tetrad field can be established by the choice of the reference frame [22, 25].

Equation (9) can be expressed as

$$\partial_{\nu}\left(e\,\Sigma^{a\lambda\nu}\right) = \frac{1}{4}\,e\,e^{a}{}_{\mu}(t^{\lambda}{}_{\mu} + T^{\lambda\mu}), \tag{11}$$

where $t^{\lambda\mu} = k(4\,\Sigma^{bc\lambda}T_{bc}{}^{\mu} - g^{\lambda\mu}\Sigma^{bcd}T_{bcd})$ is interpreted as the momentum-energy tensor of the gravitational field. Since the tensor $\Sigma^{a\lambda\nu}$ is antisymmetric in the last two indices, then $\partial_{\mu}\partial_{\nu}\left(e\Sigma^{a\lambda\nu}\right) \equiv 0$. Thereby,

$$\partial_{\lambda}\left[e\,e^{a}{}_{\mu}(t^{\lambda}{}_{\mu} + T^{\lambda\mu})\right] = 0. \tag{12}$$

In other words, this relation is a law of local conservation for the total energy-momentum tensor [22]. This equation enables one to write the following continuing equation

$$\frac{d}{dt}\int_{V} d^3x\,e\,e^{a}{}_{\mu}\,(t^{0\mu} + T^{0\mu}) = -\oint_{S} dS_j[e\,e^{a}{}_{\mu}(t^{j\mu} + T_{j\mu})]. \tag{13}$$

As mentioned above the tensors $t^{\lambda\mu}$ and $T^{\lambda\mu}$ are gravitational momentum-energy tensor and matter momentum-energy tensor, respectively. Thus the total energy-momentum vector is

$$P^{a} = -\int_{V} d^3x\,\partial_j\Pi^{aj} = -\oint_{S} dS_j\,\Pi^{aj}, \tag{14}$$

with

$$\Pi^{aj} = -4\,\kappa\,e\,\Sigma^{a0j}. \tag{15}$$

This equation represents the total energy-momentum vector because $\Pi^{a\mu}$ is related to $t^{a\mu}$ and $T^{a\mu}$ by the field equations. It is interesting to notice that the above expression do represent the gravitational energy-momentum for vanishing matter fields. We also point out that the expression (14) is invariant under coordinates transformations, but it is a vector under Lorentz transformation. Both features are mandatory rather than only desirable for any definition of energy-momentum vector. They are present in Special Relativity and there is no reason to abandoning them once dealing with a gravitational theory.

3. Quantum cosmology in teleparallel gravity

Since 1925 the quantization procedures have long been known in the literature [20, 26, 27, 28]. In the 1920s another method of quantization was presented by Hermann Weyl [30]. From the mathematical point of view, such a procedure establishes a clear correspondence between classical functions and quantum operators. The Weyl quantization formalism will be used here in quantum cosmology. However before that let's briefly recall the most recognizable method in Quantum Mechanics.

3.1. Canonical quantization

Canonical quantization is represented by a correspondence between the classical observables, which are real values of functions $f(p, q)$ (with $(p, q) = (p_i, q_i) \in \Re^n \times \Re^n$) in phase space to self-adjoint operators Q_f in the Hilbert space $L^2(\Re^n)$. By the theorem of Stone and von Neumann [27, 29] the operators $H[f(q_j)]\psi \equiv \hat{q}_i\psi = q_j\psi$ and $H[f(p_j)]\psi \equiv \hat{p}_i\psi = -\frac{ih}{2\pi}\frac{\partial\psi}{\partial q_j}$ are the only ones acting in Hilbert space that satisfies the irreducibility condition and the commutation relationships $[Q_{p_j}, Q_{p_k}] = [Q_{q_j}, Q_{q_k}] = 0$ and $[Q_{p_k}, Q_{p_j}] = \frac{ih}{2\pi}\delta_{jk}I$. A canonically classic system is composed by momentum and position coordinates, q_i and p_i respectively, with $i = 1, 2, \ldots, n$ where n is the number of degrees of freedom. The quantum states of system is evolved by Schrödinger equation $i\hbar\frac{\partial\Psi(t)}{\partial t} = \hat{H}\Psi(t)$.

3.2. Weyl quantization

Contrary to canonical quantization which requires the functions to be defined in the phase space, the Weyl's quantization allows the quantization of functions in curved spaces. In fact the method require at least two independent parameters and it is invertible and univocal. However it doesn't defines the quantum equation that the operators should obey. It is maps a function f of a classical system of n variables denoted by z_i into quantum operators $W[f(z_i)]$. Thus Weyl's map $W : f \to W[f(z_i)]$ is [20]

$$W[f(z_i)] := \frac{1}{(2\pi)^n}\int d^nk\, d^nz\, f(z_i)\, exp\left(ik^l(z_l - \hat{z}_l)\right). \tag{16}$$

The space formed by $W[z_i] \equiv \hat{z}_i$, with $i = 1, 2, \ldots, n$ where n is the number of degrees of freedom, is non-commutative. In others words, it exchanges local coordinates z_i by hermitian operators $\hat{z}_i$ obeying the following commutation relation

$$[\hat{z}_i, \hat{z}_j] = i\alpha_{ij}, \tag{17}$$

where α_{ij} is a anti-symmetric quantity.

As an example this method, let's consider the function $f(qp) = qp$, in this specific case, we have $[\hat{q}, \hat{p}] = i\hbar$. If we apply equation (16) then $f(q, p) \to W[f(q, p)]$ turns out

$$W[f(q, p)] := \frac{1}{(2\pi\hbar)^2}\int dk\, d\eta\, f(q, p)\, exp\left\{\frac{i}{\hbar}\left[k(q - \hat{q}) + \eta(p - \hat{p})\right]\right\}.$$

Using the Baker-Campbell-Hausdorff formula [31] it yields

$$W[f(q, p)] = \frac{1}{(2\pi\hbar)^2}\int dk\, d\eta\, dq\, dp\, qp\, exp\left[\frac{i}{\hbar}k(q - \hat{q})\right] exp\left[\frac{i}{\hbar}\eta(p - \hat{p})\right] exp(-k\eta/2).$$

The last equation, after simplification, becomes

$$W[f(q, p)] = \frac{1}{(2\pi\hbar)^2}\int dk\, d\eta\, qp\left\{\frac{1}{(2\pi\hbar)}\int dk\, exp\left[\frac{i}{\hbar}k(q - \hat{q} - \hbar/i\eta)\right]\right\} exp\left(\frac{i}{\hbar}\eta(p - \hat{p})\right).$$

Identifying delta function and using the property $\int f(x)\delta(x - a)\, dx = f(a)$, we obtain

$$\begin{aligned}
W[f(q, p)] &= \frac{1}{(2\pi\hbar)}\int dq\, dp\, d\eta\, qp\, \delta(q - \hat{q} - \hbar\eta/2i)\, exp\left[\frac{i}{\hbar}\eta(p - \hat{p})\right] \\
&= \frac{1}{(2\pi\hbar)}\int dp\, d\eta\, \hat{q}p\, exp\left[\frac{i}{\hbar}\eta(p - \hat{p})\right] + (\hbar/2i)\frac{1}{(2\pi\hbar)}dp\, d\eta\, \eta\, p\, exp\left[\frac{i}{\hbar}\eta(p - \hat{p})\right] \\
&= \hat{q}\int dp\, p\, \delta(p - \hat{p}) + \frac{1}{(2\pi\hbar)}\int dp\, d\eta\, (\hbar/2i)\, \eta\, p\, exp\left[\frac{i}{\hbar}\eta(p - \hat{p})\right].
\end{aligned}$$

Using the property $\int f(x)\,\frac{d\delta(x-a)}{dx}\,dx = -f'(a)$, and identifying the delta function again, it yields

$$W\,[f(q,\,p)] = \frac{1}{2}(\hat{q}\,\hat{p} + \hat{p}\,\hat{q})\,,$$

as expected.

3.3. *Quantum cosmology*

In order to construct a quantum cosmology we will calculate the energy of FLRW metric in (1) and, using the teleparallel expression (15), we'll quantize it by means the Weyl's method. The tetrad adapted to a stationary reference frame that yields the FLRW metric is given by

$$e^{a}{}_{\mu} = \begin{pmatrix} 1 & 0 & 0 & 0 \\ 0 & \frac{a}{\sqrt{1-kr^2}}\,sin\theta\,cos\phi & a\,r\,cos\theta\,cos\phi & -a\,r\,sin\theta\,\sin\phi \\ 0 & \frac{a}{\sqrt{1-kr^2}}\,sin\theta\,sin\phi & a\,r\,cos\theta\,sin\phi & a\,r\,sin\theta\,cos\phi \\ 0 & \frac{a}{\sqrt{1-kr^2}}\,cos\theta & -a\,rsin\phi & 0 \end{pmatrix}. \tag{18}$$

The relevant component of Σ^{abc} to calculate the energy is [22]

$$\Sigma^{001} = \frac{1}{4}\,(2g^{11}\,T_{001}) - \frac{1}{2}\,g^{11}\,(T_{001} + g^{22}\,T_{212} + g^{33}\,T_{313})\,, \tag{19}$$

where we used the relation $T^{\beta} = T^{\alpha}{}_{\alpha}{}^{\beta} = g^{\alpha\lambda}g^{\beta\gamma}T_{\lambda\alpha\gamma}$. Then after some algebraic manipulations we arrive at

$$\Sigma^{001} = \frac{\sqrt{1-k\,r^2} - (1-k\,r^2)}{a^2\,r}\,. \tag{20}$$

The total energy density of the Universe is

$$\Pi^{(0)1} = 4\,\kappa\,a\,r\,\sin\theta\,\left(\sqrt{1-k\,r^2} - 1\right) \tag{21}$$

where $\kappa \equiv \frac{1}{16\pi}$, k spatial curvature and $a = a(t)$ is the scale factor.

Let's firstly quantize the energy density, thus $\Pi^{(0)1} = \epsilon(r,\,\tau) = A\,r\,\tau\left(\sqrt{1-k\,r^2} - 1\right)$ where $A = 4\,a\,\kappa$ and $\tau = \sin\theta$. Then the Weyl procedure yields $\hat{\mathcal{H}} = W[\epsilon(r,\,\tau)]$, it reads

$$\hat{\mathcal{H}} = \left[\frac{A}{2}\,(\hat{r}\,\hat{\tau} - \hat{\tau}\,\hat{r})\right]\left(-k\,\hat{r}^2 + 1\right)^{\frac{1}{2}} - \left[\frac{A}{2}\,(\hat{r}\,\hat{\tau} - \hat{\tau}\,\hat{r})\right]. \tag{22}$$

If we use $\hat{r} = r$ and $\hat{\tau} = -i\alpha\frac{\partial}{\partial r}$, then $[\hat{r},\,\hat{\tau}] = i\alpha$. This implies that the Hamiltonian density can be expanded in terms of the non-commutative parameter α which is supposed to be very small. As a consequence we overcome the problem of dealing with a squared operator. Hence

$$\hat{\mathcal{H}} = 3\,a\,\kappa\,k\,i\,\alpha\,r^2 + 2\,a\,\kappa\,k\,r^3\,i\,\alpha\,\frac{\partial}{\partial r}\,. \tag{23}$$

We assume that the dynamics is given by a Schrödinger like equation $\hat{\mathcal{H}}\Psi = i\hbar\frac{\partial\Psi}{\partial t}$, therefore

$$3\,a\,\kappa\,k\,\alpha\,r^2\,\Psi + 2\,a\,\kappa\,k\,r^3\,\alpha\,\frac{\partial\Psi}{\partial r} = \hbar\,\frac{\partial\Psi}{\partial t}\,. \tag{24}$$

It should be noted that in the above equation the scale factor acts as a parameter. It is possible to solve the spatial equation which reads

$$3\,\kappa\,k\,\alpha\,r^2\,\psi + 2\,\kappa\,k\,r^3\,\alpha\,\frac{\partial\psi}{\partial r} = \xi\psi\,,$$

but the temporal part,

$$\hbar \frac{\partial \phi}{\partial t} = a\,\xi \phi\,,$$

requires an explicit form of the scale factor. Here ξ is the constant of separation of variables $\Psi(r,t) = \psi(r)\,\phi(t)$. Possibly the use of an altered temporal scale can solve the problem. Another attempt would be using a semi-classical approach in which the scale factor comes up from the Einstein equation.

The next possibility is to quantize the energy $P^{(0)} = E$ itself. Integrating the energy density over a spherical 3D surface with radius r_0 we get

$$E = -\lim_{r\to r_0} 8\,\pi\,\kappa\,a\,r\left[\sqrt{1-k\,r^2} - 1\right]. \tag{25}$$

It should be noted that r_0 is arbitrary. It specifies the region where we want to calculate the energy. Thus we can choose it to be the radius of the dynamical horizon of FLRW metric, $r_0 = \frac{1}{\sqrt{H^2+\frac{k}{a^2}}}$ with $H = \frac{\dot a}{a}$. In this way E will represent the energy of the observable Universe. In order to obtain an expression that describes the beginning of the Universe, we expand the energy as

$$E\,(a,\,\dot a) \approx \frac{2\,a\,k\,\sqrt{k} - 2\,a^2\,k\,\sqrt{k} - a^3\,k\,\sqrt{k} + \frac{1}{2}\,a^4\,(\dot a)^2\,\sqrt{k} + a^6\,k\,\sqrt{k}}{2\,k^2}\,. \tag{26}$$

Using the Weyl's method the Hamiltonian $\hat H = W\,[E]$ reads

$$2k^2\,\hat H \approx 2\,\hat a\,k\,\sqrt{k} - 2\,\hat a^2\,k\,\sqrt{k} - \hat a^3\,k\,\sqrt{k} + \frac{1}{2}\,\hat a^4\,\sqrt{k}\,(\hat{\dot a})^2 + \hat a^6\,k\,\sqrt{k} - 3\,\omega^2\,\hat a^2\,\sqrt{k} - i\,2\,\omega\,\hat a\,\sqrt{k}\,\hat{\dot a}\,, \tag{27}$$

where $[\hat a,\,\hat{\dot a}] = i\,\omega$ with $\hat a = a$, $\hat{\dot a} = -i\,\omega\frac{\partial}{\partial a}$. Therefore following the Wheeler-DeWitt prescription $\mathcal{H}\Psi = 0$ we get

$$\left(2\,a\,k\,\sqrt{k} - 2\,a^2\,k\,\sqrt{k} - a^3\,k\,\sqrt{k} - \frac{1}{2}\,\omega^2\,a^4\,\sqrt{k}\,\frac{\partial^2}{\partial a^2} + a^6\,k\,\sqrt{k} - 3\,\omega^2\,a^2\,\sqrt{k} - 2\,\omega^2\,a\,\sqrt{k}\,\frac{\partial}{\partial a}\right)\Psi = 0\,. \tag{28}$$

Such an equation has the same dependence of the WDW equation and has a regime of validity for a small Universe.

4. Conclusion

In this article we have derived two quantum equations for cosmology. One of them was obtained by quantizing the energy density and the another one quantizing the energy itself. The quantization of energy leads to a Wheeler-DeWitt like equation, which is valid for a small Universe. However a similar equation should hold for the Universe's present time since the non-commutative parameter should be very small. Since the publication of the RG the vision about the world has changed. With the coming of Friedman [2], Lemaître [3, 4], Robertson [5, 6, 7], Walker [8] and Hubble [1] the cosmological knowledge was raised to a very high plateau. With the coming of Quantum Physics the picture of the Universe is almost complete. What was missing? A union between GR and Quantum Physics. However we believe that GR is not suitable theory to achieve such an aim, since it is not enough to deal with the problem of gravitational energy. On the other hand in the framework of teleparallel gravity it is possible to obtain a well defined expression for energy. Using the Weyl quantization, which is a procedure with a natural correspondence between the classical and quantum operators, the quantum gravity is established [20] and here applied to an Universe governed by the cosmological principle.

[1] Hubble E 1929 *Proc. Nati. Acad. Sci.* **15** 168
[2] Friedman A 1991 *Gen. Relativ. Gravit.* **31** 1991
[3] Lemaître A G 1927 *Annales de la Société Scientifique de Bruxelles* **47** 483 (1927)
[4] Lemaître A G 1931 *Mon. Not. R. Astron. Soc.* **92** 49
[5] Robertson H P 1935 *Astrohpys. J.* **82** 284
[6] H. P. Robertson 1936 *Astrohpys. J.* **83** 187
[7] H. P. Robertson 1936 *Astrohpys. J.* **83** 257
[8] Walker A G 1937 *Proceedings of the London Mathematical Society* **s2-42** 90
[9] Rosenfeld L 1930 *Ann. Phys.* **397** 113
[10] DeWitt B S 1970 *Gen. Relativ. Gravit.* **1** 181
[11] Arnowitt R, Deser S and Misner C W 1962 The dynamics of General Relativity *Gravitation: an introduction to current research* ed L. Witten (New York: John Wiley & Sons)
[12] DeWitt B S 1967 *Phys. Rev.* **160** 1113
[13] Anderson E 2010 The problem of time in quantum gravity *Preprint* gr-qc/1009.2157v3
[14] Einstein A 1930 *Math. Annal.* **102** 687
[15] Møller C 1961 *Ann. Phys.* **12** 118
[16] Pellegrini C and Plebanski J 1963 *Mat. Fys. Skr.* **2** 4
[17] Maluf J W 1994 *J. Math. Phys.* **35** 335
[18] Maluf J W, Ulhoa S C, Faria F F and da Rocha-Neto J F 2009 *Class. Quantum Grav.* **23** 6245
[19] Ulhoa S C and Rocha P M M 2013 *Braz. J. Phys.* **43** 162
[20] Ulhoa S C and Amorim R G G 2014 *Advances in High Energy Physics* **2014** 1
[21] Maluf J W, Ulhoa S C and da Rocha-Neto J F 2013 *Gen. Relativ. Gravit.* **45** 1163
[22] Maluf J W 2013 *Anna. Phys.* **525** 339
[23] Aldrovandi R and Pereira J G 2013 *Teleparallel gravity: an introduction* (New York: Springer)
[24] Maluf J W 2003 *Phys. Rev. D* **67** 108501
[25] Maluf J W, Veiga M V O and da Rocha-Neto J F 2007 *Gen. Relativ. Gravit.* **39** 227
[26] Berezin F A 1974 *Math. USSR Izvestiya* **5** 1109
[27] Ali S T e EngliŝM 2005 *Rev. Math. Phys.* **4**, 391
[28] Giulini D 2003 Lect. Notes Phys. 631, 17 *Preprint* quant-ph/0304202v1
[29] von Neumann J 1955 *Mathematical foundations of quantum mechanics* (Princeton: Princeton University Press)
[30] Weyl H 1930 *The theory of groups and quantum mechanics* (New York: Dover)
[31] Gilmore R 1974 *J. Math. Phys.* **15**, 2090

IOP Conf. Series: Journal of Physics: Conf. Series **965** (2018) 012015 doi:10.1088/1742-6596/965/1/012015

Quantised relativistic membranes and non-perturbative checks of gauge/gravity duality

Veselin G. Filev [1] and Denjoe O'Connor[2]

[1] Institute of Mathematics and Informatics,
Bulgarian Academy of Sciences, Acad. G. Bonchev Str., 1113 Sofia, Bulgaria
[2] School of Theoretical Physics, Dublin Institute for Advanced Studies,
10 Burlington Road, Dublin 4, Ireland.

E-mail: denjoe@stp.dias.ie, vfilev@math.bas.bg

Abstract. We test the background geometry of the BFSS model using a D4–brane probe. This proves a sensitive test of the geometry and we find excellent agreement with the D4–brane predictions based on the solution of a membrane corresponding to the D4–brane propagating on this background.

1. Introduction

From the perspective of a quantised eleven dimensional supermembrane it is perhaps natural to argue that there ought to be a dual gravitational background to the matrix regulated supermembrane proposed by Banks et al [1] as a non-perturbative formulation of M-theory. However, it is important to test this hypothesis as far as possible. In recent years several non-perturbative tests of this hypothesis have been performed [2, 3, 18, 5, 6, 7, 8, 9] at the level of comparing the expectation value of the microscopic Hamiltonian with the energy prediction coming from the black hole background. These are increasingly sensitive tests and the agreement found between theory and simulation in the most recent, and currently most precise test, by Berkowitz et al [9] while impressive is open to question. They observe that if they fit their data "by a single powerlaw $E_0(T) = aT^p$, then $E_0(T) = (3.13 \pm 0.03)T^{2.02\pm0.03} \simeq \pi T^2$ describes our continuum large-N data very well ($\chi^2/DOF = 7.7/5$) in the whole temperature range." While their precision tests are still rather convincing it is important to have other tests that check the geometry in more detail. This can be done, just as one does in practice in the everyday world, by using probes which are sensitive to the geometry yet do not in turn change it significantly.[1] This corresponds to adding $M5$-membrane density to the BFSS matrix model – the Berkooz-Douglas (BD) matrix model. From a string theory point of view the BD model describes the low energy regime of the D0/D4–brane intersection. It is a flavoured $1+0$ dimensional gauge theory with $\mathcal{N} = 8$ supersymmetry and is the dimensional reduction of $4-d$ $\mathcal{N} = 2$ supersymmetric Yang–Mills theory. This makes the BD model particularly interesting in the context of holographic studies of flavour dynamics, since the $\mathcal{N} = 16$ supersymmetric Yang–Mills theory which it probes has a well know gravitation dual background.

[1] A similar study using probe D0-branes has been performed in ref. [10].

IOP Conf. Series: Journal of Physics: Conf. Series **965** (2018) 012015 doi:10.1088/1742-6596/965/1/012015

The proposal that there is a gravitational dual to gauged matter systems represents a dramatic insight into certain non-perturbative phenomena. In the original formulation of Maldacena [11], the duality relates string theory in the $AdS_5 \times S^5$ background space-time to the large N limit of $3+1$ dimensional $\mathcal{N}=4$ Supersymmetric Yang-Mill theory living on the asymptotic boundary of the AdS$_5$ space-time. This idea has inspired numerous extensions of the duality with ever increasing phenomenological relevance, currently ranging from heavy ion collisions to condensed matter physics. In this paper we are interested in holographic flavour dynamics–the generalisation of the AdS/CFT correspondence to flavoured gauge theories.

The first such generalisation was proposed by Karch and Katz [12] , who introduced a probe D7–brane to the $AdS_5 \times S^5$ supergravity background. On the field theory side this corresponds to introducing an $\mathcal{N}=2$ fundamental hypermultiplet in the quenched approximation. The classical dynamics of the probe brane is governed by an effective Dirac-Born-Infeld action. Remarkably the AdS/CFT dictionary relates the classical properties of the brane to quantum vacuum expectation values in the dual flavoured gauge theory. One such quantity is the fundamental condensate of the theory, which is encoded in the classical profile of the probe brane near the asymptotic boundary. In refs. [13] and [14] the finite temperature set-up has been considered. The authors uncovered a first order meson melting phase transition corresponding to a topology change transition of the possible D7–brane embeddings. In ref. [15] these studies have been extended to the general Dp/Dq–brane system and certain universal properties of the corresponding holographic gauge theories have been uncovered. In fact the BD matrix model is dual to the D0/D4–brane system, which falls into the same universality class as the phenomenologically relevant D3/D7–brane system.

Another attractive feature of the BD matrix model is that it is $1+0$ dimensional supersymmetric Yang–Mills theory, which makes it super renormalizable and thus accessible with lattice simulations. Therefore, the BD model can serve as a bridge between lattice gauge theory and holography allowing a direct precision test of the gauge/gravity duality with flavours. This was our main motivation to study the D0/D4 system and the BD matrix model [16].

Our studies provide a highly non-trivial test of the AdS/CFT correspondence with matter. The results provide substantial evidence for the validity of the holographic approach to flavour dynamics. They also test the predicted D0-brane geometry in a highly non-trivial way since as the fundamental mass is varied, the D4-brane probes the radial dependence of this dual geometry. The agreement we find with the predictions from the D4-brane probe embedded in the dual D0 black hole geometry is remarkable. Although it is not a mathematical proof, we believe that the remarkable agreement between theory and simulation, which we uncovered is due to the cancellation of α' corrections in the black hole embedding.

2. Holographic flavours in one dimension

In this section we review the description of the D0/D4–brane system in the quenched approximation adapting the general discussion of references [15] and [17].

2.1. D0-brane background

In the near horizon limit the D0-brane background is given by the metric:

$$\begin{aligned} ds^2 &= -H^{-\frac{1}{2}}\, f\, dt^2 + H^{\frac{1}{2}}\left(\frac{du^2}{f} + u^2\, d\Omega_8^2\right) , \\ e^{\Phi} &= H^{\frac{3}{4}} , \qquad C_0 = H^{-1} , \end{aligned} \tag{1}$$

where $H=(L/u)^7$, $f(u)=1-(u_0/u)^7$ is the blackening factor, Φ is the dilaton field and C_0 is the only component of the RR one form coupled to the D0-branes . Here u_0 is the radius of the horizon and the length scale L can be expressed in terms of string theory units as:

$$L^7 = 60\,\pi^3\, g_s\, N_c\, \alpha'^{7/2}\ , \tag{2}$$

where N_c is the number of D0–branes corresponding to the rank of the gauge group of the dual field theory[2]. According to the general gauge/gravity duality [17], the Yang-Mills coupling of the corresponding dual gauge theory is given by:

$$g^2_{\rm YM} = g_s\,(2\pi)^{-2}\,\alpha'^{-3/2}\ . \tag{3}$$

The Yang-Mills coupling is dimensionful and the corresponding dimensionless effective coupling runs with the energy scale according to:

$$g^2_{\rm eff} = \lambda\, U^{-3}\ , \tag{4}$$

where $\lambda = g^2_{\rm YM}\,N_c$ is the t'Hooft coupling. The supergravity background can be trusted if both the curvature and the dilaton are small, which leads to the restriction [17]:

$$1 \ll g_{\rm eff} \ll N_c^{\frac{4}{7}}\ . \tag{5}$$

and the theory is strongly coupled in this regime. From equations (1) and (4) it follows that the upper bound in equation (5) can be violated at low energies (small radial distances) when the dilaton blows, however at finite temperature and fixed 'tHooft coupling, $g_{\rm eff}$ peaks at the black hole horizon and the bound $\lambda/T^3 \ll N_c^{8/7}$ is satisfied in the large N limit. At high energies (large radial distances) the curvature of the background grows, while the effective coupling decreases. As a result the lower bound in (5) is violated at energies higher than approximately $\lambda^{1/3}$ and hence α' corrections are increasingly important at large radial distances.

Finally, the Hawking temperature of the background is given by:

$$T = \frac{7}{4\,\pi\,L}\left(\frac{u_0}{L}\right)^{\frac{5}{2}} \tag{6}$$

and is identified with the temperature of the dual gauge theory.

2.2. Flavour D4-branes

To introduce matter in the fundamental representation we consider the addition of N_f D4-branes to the D0-brane background. In the probe approximation $N_f \ll N_c$, the dynamics of the D4-branes is governed by the Dirac-Born-Infeld, which in the absence of a background B-field is given by:

$$S_{\rm DBI} = -N_f\,T_4 \int d^4\xi\, e^{-\Phi}\sqrt{-\det||G_{\alpha,\beta} + (2\pi\alpha')F_{\alpha,\beta}||}\ , \tag{7}$$

where $G_{\alpha,\beta}$ is the induced metric and $F_{\alpha,\beta}$ is the $U(1)$ gauge field of the D4-brane, which we will set to zero. The D4-brane tension is given by:

$$T_4 = \frac{\mu_4}{g_s} = \frac{1}{(2\pi)^4\,\alpha'^{5/2}\,g_s}\ . \tag{8}$$

The D4-brane embedding that we consider extends along the radial and time directions and wraps a three sphere, S^3, in the directions transverse to the D0-brane. To parametrise it let us split the unit S^8 in the metric (1) into:

$$d\Omega_8^2 = d\theta^2 + \cos^2\theta\, d\Omega_3^2 + \sin^2\theta\, d\Omega_4^2\ . \tag{9}$$

[2] Note that we will abbreviate N_c to N when the context is clear.

Our embedding now extends along t and Ω_3 and has a non-trivial profile in the (u, θ) plane, which we parametrise as $(u, \theta(u))$. Next we Wick rotate the action (7) and periodically identify time with period $\beta = 1/T$. Using equation (1) we obtain:

$$S_{\rm DBI}^E = \frac{N_f\,\beta}{8\,\pi^2\,\alpha'^{5/2}\,g_s}\int du\,u^3\cos^3\theta(u)\,\sqrt{1+u^2\,f(u)\,\theta'(u)^2}\;. \tag{10}$$

In the limit of zero temperature ($u_0 \to 0$) the regular solution to the equation of motion for $\theta(u)$ is given by $u\sin\theta = m$, where the constant m is proportional to the bare mass of the flavours [12], [15]. At finite temperature the separation $L(u) = u\sin\theta(u)$ has a non-trivial profile reflecting the non-vanishing condensate of the theory. To analyse this case it is convenient to define dimensionless radial coordinate $\tilde{u} = u/u_0$. At large $\tilde{u}$ the general solution $\theta(\tilde{u})$ has the expansion:

$$\sin\theta = \frac{\tilde{m}}{\tilde{u}} + \frac{\tilde{c}}{u^3} + \dots . \tag{11}$$

Holography relates the dimensionless constants $\tilde{m}, \tilde{c}$ to the bare mass and condensate of the theory via [15][3]:

$$\begin{aligned} m_q &= \frac{u_0\,\tilde{m}}{2\pi\alpha'} = \left(\frac{120\,\pi^2}{49}\right)^{1/5}\left(\frac{T}{\lambda^{1/3}}\right)^{2/5}\lambda^{1/3}\,\tilde{m}\;, \\ \langle\mathcal{O}_m\rangle &= -\frac{N_f\,u_0^3}{2\,\pi\,g_s\,\alpha^{3/2}}\,\tilde{c} = \left(\frac{2^4\,15^3\,\pi^6}{7^6}\right)^{1/5} N_f\,N_c\left(\frac{T}{\lambda^{1/3}}\right)^{6/5}(-2\,\tilde{c})\;. \end{aligned} \tag{12}$$

Note that equation (11) implies that the D7-branes are described by a one parameter family of embeddings (parametrised by $\tilde{m}$). In the case of the D3/D7 system this is natural due to the scaling symmetry (everything depends on the dimensionless ratio m_q/T), but is this consistent with the D0/D4 system, which has a dimensionful 'tHooft coupling? Indeed, the D0/D4 system has a 'tHooft coupling λ of dimension three suggesting that there are two independent dimensionless parameters $m_q/\lambda^{1/3}$ and $T/\lambda^{1/3}$. However, while the holographic set-up allows rescaling of the radial coordinate by u_0 and the description of D7-brane embeddings by the single parameter $\tilde{m}$, as can be seen from the first equation (12) we have $\tilde{m} \sim (m_q/\lambda^{1/3})(T/\lambda^{1/3})^{-2/5}$. As a result the condensate in the second equation in (12) is indeed a function of the two dimensionless parameters $(T/\lambda^{1/3},\, m_q/\lambda^{1/3})$ which is consistent with dimensional analysis and is in contrast to the D3/D7 system, where the condensate depends on the dimensionless parameter m_q/T.

3. Testing the correspondence

In this section we compare the result of the lattice simulations of the model to the predictions of gauge gravity duality. Our main focus is the fundamental condensate of the theory. As definition of the condensate we use the derivative of the free energy of the theory with respect to the bare mass parameter m_q. Note that on the lattice we use a dimensionless mass parameter m and temperature $\tilde{T}$ given by:

$$m = \frac{m_q}{\lambda^{1/3}}\;, \qquad \tilde{T} = \frac{T}{\lambda^{1/3}}\;, \tag{13}$$

where λ is the dimensionful 'tHooft coupling. Note also that in 1+0 dimensions the fundamental condensate $\langle\mathcal{O}_m\rangle$ is dimensionless. The AdS/CFT dictionary equations (12) can be easily

[3] Note that our expressions differ slightly from the ones presented in [15] due to the different choice of radial variable.

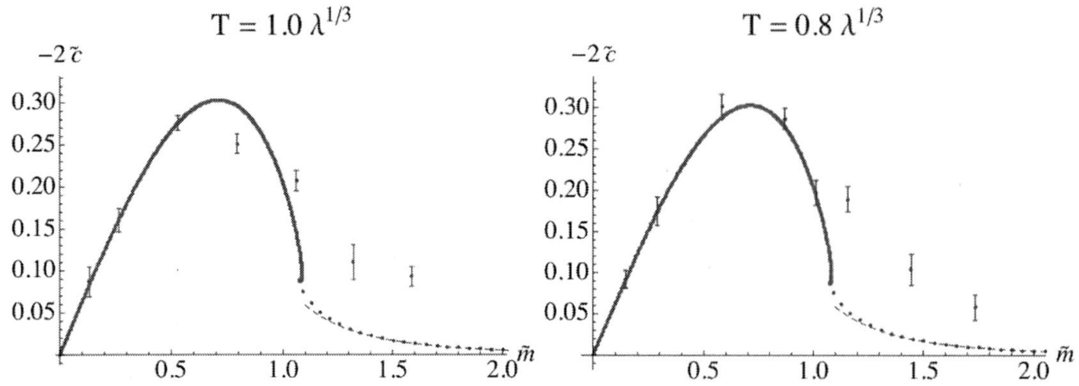

Figure 1. Plots of the condensate versus bare mass parameter curve for $N = 10$, $\Lambda = 16$ and two different temperatures . *left:* For temperature $T = 1.0\,\lambda^{1/3}$ the curve shows excellent agreement at small masses, but deviates quickly from the theoretical curve at greater masses. *right:* The curve at temperature $T = 0.8\,\lambda^{1/3}$ exhibits excellent agreement throughout the whole range of masses corresponding to the black hole phase (blue error bars). Similarly to the higher temperature curve there is a significant deviation from the theoretical curve for the Minkowski phase (red error bars).

rewritten in terms of the lattice dimensionless quantities. One gets:

$$\begin{aligned} m &= \left(\frac{120\,\pi^2}{49}\right)^{1/5} \tilde{T}^{2/5}\,\tilde{m}\ , \\ \langle \mathcal{O}_m \rangle &= \left(\frac{2^4\,15^3\,\pi^6}{7^6}\right)^{1/5} N_f\,N_c\,\tilde{T}^{6/5}\,(-2\,\tilde{c})\ . \end{aligned} \tag{14}$$

Equations (14) can now be used to scale the dependence $\langle \mathcal{O}_m \rangle$ versus m obtained from computer simulations and compare to the dependence $-2\,\tilde{c}$ versus $\tilde{m}$ obtained from holography. The resulting plots for two temperatures are presented in figure 1. The plots are for matrix size $N = 10$ and lattice spacing $\Lambda = 16$.

The left plot corresponds to temperature $T = 1.0\,\lambda^{1/3}$. One can observe excellent agreement between the gauge gravity duality and lattice simulations at small masses ($\tilde{m} < 1$). However, for greater masses there is a significant deviation from the theoretical curve. The right plot corresponds to temperature $T = 0.8\,\lambda^{1/3}$. The excellent agreement between gauge/gravity predictions and lattice simulations extends for the whole range of masses within the deconfined (black hole) phase (blue error bars). In the deconfined (Minkowski) phase there are still significant deviations from the theoretical curve. These results, we argue [16], indicate that the α' corrections to the supergravity background affect the black hole and Minkowski D4-brane embeddings differently. All black hole embeddings reach the horizon and as a result experience similar curvature effects for different values of the mass parameter therefore, the α' corrections largely cancel when one takes a derivative with respect to the mass to calculate the condensate. In contrast, Minkowski embeddings close at different radial distances above the horizon depending on the mass parameter. As a result the effect of the α' corrections depends strongly on the mass and contributes to the calculation of the condensate. The overall better agreement of the lower temperature curve to the theoretical predictions is another signature that the observed deviations at large masses are due to α' corrections as opposed to lattice effects, although at sufficiently high masses ($|m^a| \lesssim 1/a$) lattice effect also become significant.

4. Conclusion

In this paper we review the precision test of holographic flavour dynamics discussed in detail in [16]. We focused on the study of a one-dimensional flavoured Yang-Mills theory holographically dual to the D0/D4–brane intersection, also known as the Berkooz–Douglas matrix model. We considered a lattice discretisation of the model and the super-renormalizability of the model ensures that in the continuum limit, supersymmetry is broken only by the effect of finite temperature, which enabled us to simulate it on a computer.

Our results for the condensate versus bare mass curve (which is universal for different temperatures) show an excellent agreement with holography in the regime of small bare masses and at lower temperature this agreement extends to the whole range of masses in the deconfined phase. We believe that this agreement can be explained by a cancellation of the α' corrections to the condensate for black hole embeddings (deconfined phase). This allows a direct comparison between computer simulations and AdS/CFT predictions at relatively high temperatures compared to similar studies of the pure BFSS matrix model. Furthermore, this remarkable agreement (in the black hole phase) is obtained without any parameter fitting in contrast to the analogous studies of the BFSS matrix model [18], where the authors performed a fit to estimate the α' corrections to the internal energy. We believe that it is the cancellation mechanism described in section 3, which allows this highly non-trivial test of the gauge/gravity correspondence. The improved agreement as the temperature is reduced is also in accord with this interpretation.

Acknowledgements: We wish to thank Yuhma Asano for useful comments and discussions. Part of the simulations were performed within the ICHEC "Discovery" project "dsphy003c". The support from Action MP1405 QSPACE of the COST foundation is gratefully acknowledged.

References

[1] T. Banks, W. Fischler, S. H. Shenker and L. Susskind, "M theory as a matrix model: A Conjecture," Phys. Rev. D **55**, 5112 (1997) [hep-th/9610043].

[2] K. N. Anagnostopoulos, M. Hanada, J. Nishimura and S. Takeuchi, Phys. Rev. Lett. **100** (2008) 021601 doi:10.1103/PhysRevLett.100.021601 [arXiv:0707.4454 [hep-th]].

[3] S. Catterall and T. Wiseman, Phys. Rev. D **78** (2008) 041502 doi:10.1103/PhysRevD.78.041502 [arXiv:0803.4273 [hep-th]].

[4] M. Hanada, Y. Hyakutake, J. Nishimura and S. Takeuchi, Phys. Rev. Lett. **102** (2009) 191602 doi:10.1103/PhysRevLett.102.191602 [arXiv:0811.3102 [hep-th]].

[5] M. Hanada, Y. Hyakutake, G. Ishiki and J. Nishimura, Science **344** (2014) 882 doi:10.1126/science.1250122 [arXiv:1311.5607 [hep-th]].

[6] D. Kadoh and S. Kamata, arXiv:1503.08499 [hep-lat].

[7] V. G. Filev and D. O'Connor, JHEP **1605** (2016) 167 doi:10.1007/JHEP05(2016)167 [arXiv:1506.01366 [hep-th]].

[8] M. Hanada, Y. Hyakutake, G. Ishiki and J. Nishimura, Phys. Rev. D **94** (2016) no.8, 086010 doi:10.1103/PhysRevD.94.086010 [arXiv:1603.00538 [hep-th]].

[9] E. Berkowitz, E. Rinaldi, M. Hanada, G. Ishiki, S. Shimasaki and P. Vranas, Phys. Rev. D **94** (2016) no.9, 094501 doi:10.1103/PhysRevD.94.094501 [arXiv:1606.04951 [hep-lat]].

[10] E. Rinaldi, E. Berkowitz, M. Hanada, J. Maltz and P. Vranas, arXiv:1709.01932 [hep-th].

[11] J. M. Maldacena, Int. J. Theor. Phys. **38**, 1113 (1999) [Adv. Theor. Math. Phys. **2**, 231 (1998)] doi:10.1023/A:1026654312961 [hep-th/9711200].

[12] A. Karch and E. Katz, JHEP **0206**, 043 (2002) doi:10.1088/1126-6708/2002/06/043 [hep-th/0205236].

[13] J. Babington, J. Erdmenger, N. J. Evans, Z. Guralnik and I. Kirsch, Phys. Rev. D **69**, 066007 (2004) doi:10.1103/PhysRevD.69.066007 [hep-th/0306018].

[14] C. Hoyos-Badajoz, K. Landsteiner and S. Montero, JHEP **0704**, 031 (2007) doi:10.1088/1126-6708/2007/04/031 [hep-th/0612169].

[15] D. Mateos, R. C. Myers and R. M. Thomson, JHEP **0705**, 067 (2007) doi:10.1088/1126-6708/2007/05/067 [hep-th/0701132].

[16] V. G. Filev and D. O'Connor, JHEP **1605** (2016) 122 doi:10.1007/JHEP05(2016)122 [arXiv:1512.02536 [hep-th]].

[17] N. Itzhaki, J. M. Maldacena, J. Sonnenschein and S. Yankielowicz, Phys. Rev. D **58**, 046004 (1998) doi:10.1103/PhysRevD.58.046004 [hep-th/9802042].

[18] M. Hanada, Y. Hyakutake, J. Nishimura and S. Takeuchi, Phys. Rev. Lett. **102**, 191602 (2009) [arXiv:0811.3102 [hep-th]].

Parametric bases for elliptic boundary value problem

A.A. Gusev[1], S.I. Vinitsky[1,2], O. Chuluunbaatar[1,3], V.L. Derbov[4], A. Góźdź[5], P. M. Krassovitskiy[6]

[1]Joint Institute for Nuclear Research, Dubna, Russia
[2]RUDN University, Moscow, Russia, 6 Miklukho-Maklaya st, Moscow, 117198
[3]Institute of Mathematics, National University of Mongolia, Ulaanbaatar, Mongolia
[4]N.G. Chernyshevsky Saratov National Research State University, Saratov, Russia
[5]Institute of Physics, Maria Curie Skłodowska University, Lublin, Poland
[6]Institute of Nuclear Physics, Almaty, Kazakhstan

E-mail: gooseff@jinr.ru

Abstract. We consider the calculation schemes in the framework of Kantorovich method that consist in the reduction of a 3D elliptic boundary-value problem (BVP) to a set of second-order ordinary differential equations (ODEs) using the parametric basis functions. These functions are solution of the 2D parametric BVP. The coefficients in the ODEs are the parametric eigenvalues and the potential matrix elements expressed by the integrals of the eigenfunctions multiplied by their first derivatives with respect to the parameter. We calculate the parametric basis functions numerically in the general case using the high-accuracy finite element method. The efficiency of the proposed calculation schemes and algorithms is demonstrated by the example of the BVP describing the bound states of helium atom.

1. Introduction

The adiabatic representation is widely applied for solving multichannel scattering and bound-state problems for systems of a few quantum particles in molecular, atomic and nuclear physics [1]. Such problems are described by elliptic boundary value problems (BVPs) in a multidimensional domain of the configuration space, solved using the Kantorovich method[2], i.e., the reduction to a set of self-adjoint ordinary differential equations (ODEs) using the basis of surface functions of an auxiliary BVP depending on the independent variable of the ODEs parametrically.

We propose new calculation schemes and algorithms for solving the parametric self-adjoint elliptic BVP with the Dirichlet and/or Neumann type boundary conditions in a 2D finite domain, using high-accuracy finite element method (FEM) with triangular Lagrange elements [3]. The corresponding algorithm and programs calculate with given accuracy the parametric eigenvalues, the surface eigenfunctions with their parametric derivatives, and the potential matrix elements, expressed as integrals of the products of surface eigenfunctions and/or their first derivatives with respect to the parameter, which is a key problem in the adiabatic representation. These parametric eigenvalues (potential curves) and the potential matrix elements can be used for reducing the 3D BVP to bound-state and multi-channel scattering problems for sets of coupled second-order ODEs with the appropriate asymptotic boundary conditions [4].

In the present paper we demonstrate the efficiency of the proposed approach by benchmark calculations of helium atom bound states.

2. The statement of the problem

Let us consider the BVP for the parametric self-adjoint 2D partial differential equation in the domain Ω_x, $x{=}(x_1,x_2)$ with the piecewise continuous boundary $S{=}\partial\Omega_x$,

$$(D(x;z){-}\varepsilon_i(z))\,\Phi_i(x;z){=}0, \tag{1}$$
$$D{\equiv}D(x;z){=}{-}\frac{1}{g_0(x)}\left(\sum_{ij=1}^{2}\frac{\partial}{\partial x_i}g_{ij}(x)\frac{\partial}{\partial x_j}\right){+}U(x;z),$$

with the mixed Dirichlet/Neumann boundary conditions

$$(I):\Phi(x;z)|_S=0,\quad (II):\left.\frac{\partial\Phi(x;z)}{\partial n_D}\right|_S=0,\quad \frac{\partial\Phi(x;z)}{\partial n_D}=\sum_{ij=1}^{2}(\hat{n},\hat{e}_i)g_{ij}(x)\frac{\partial\Phi(x;z)}{\partial x_j}. \tag{2}$$

Here $\frac{\partial\Phi_m(z)}{\partial n_D}$ is the derivative along the conormal direction, $\hat{n}$ is the outer normal to the boundary $S=\partial\Omega$ of Ω, $\hat{e}_i$ is the unit vector of $x=\sum_{i=1}^{2}\hat{e}_i x_i$, and $(\hat{n},\hat{e}_i)$ is the scalar product in $\mathcal{R}^2$, $z\in\Omega_z=[z_{\min},z_{\max}]$ is a parameter. We assume that the functions $g_0(x)>0$, $g_{ij}(x)>0$, and $\partial_{x_k}g_{ij}(x)$, $U(x;z)$, $\partial_z U(x;z)$ and $\partial_z\Phi_i(x;z)$ are continuous and bounded for $x\in\Omega_x$; $g_{12}(x)=g_{21}(x)$, $g_{11}(x)g_{22}(x)-g_{12}^2(x)>0$. We also assume that the elliptic BVP (1)–(2) has only the discrete spectrum, so that $\varepsilon(z):\varepsilon_1(z)<\ldots<\varepsilon_{j_{\max}}(z)<\ldots$ is the desired set of real eigenvalues. The eigenfunctions satisfy the orthonormality conditions

$$\langle\Phi_i|\Phi_j\rangle=\int_\Omega g_0(x)\Phi_i(x;z)\Phi_j(x;z)dx{=}\delta_{ij},\quad dx=dx_1dx_2. \tag{3}$$

The FEM solution of the BVP (1)–(3) is reduced to the determination of stationary points of the variational functional [3, 5]

$$\Xi(\Phi_m,\varepsilon_m(z))\equiv\int_\Omega dx g_0(x)\Phi_m(x;z)\,(D-\varepsilon_m(z))\,\Phi(x;z)=\Pi(\Phi_m,\varepsilon_m(z)), \tag{4}$$

where $\Pi\equiv\Pi(\Phi_m,\varepsilon_m(z))$, $\Phi_m\equiv\Phi_m(x;z)$ is the symmetric quadratic functional

$$\Pi{=}\int_\Omega dx\left[\sum_{ij=1}^{2}g_{ij}(x)\frac{\partial\Phi_m}{\partial x_i}\frac{\partial\Phi_m}{\partial x_j}{+}g_0(x)\Phi_m(U(x;z){-}\varepsilon_m(z))\Phi_m\right].$$

In the FEM, the function $\Phi(x;z)\in\mathcal{F}_z^h\sim\mathcal{H}^1(\Omega_x)$ is approximated by the finite sum [3]

$$\Phi^h(x;z)=\sum_{l=1}^{N}\Phi_l^h(z)N_l^p(x), \tag{5}$$

where the piecewise polynomial functions $N_l^p(x)$ in the domain Ω are constructed by joining the shape functions $\varphi_l^p(x)$ in the triangular subdomains Δ_q:

$$N_l^p(x)=\{\varphi_l^p(x),A_l\in\Delta_q;0,A_l\notin\Delta_q\}$$

and possess the following properties: i)the functions $N_l^p(x)$ are continuous in the domain Ω; ii)the functions $N_l^p(x)$ equal 1 at one of the points A_l and zero at the rest points; iii) $N_l^p(x_{1l'},x_{2l'})=\delta_{ll'}$ in the entire domain Ω. Here l takes the values $l=\overline{1,N}$.

The functions $N_l^p(x)$ form a basis in the space of polynomials of the p-th order. The vector function $\boldsymbol{\Phi}^h = \{\Phi_l^h(z)\}_{l=1}^N$ has a generalized first-order partial derivative and belongs to the Sobolev space $\mathcal{H}^1(\Omega_x)$ [6]. After substituting the expansion (5) into the variational functional and minimizing it [5, 6], we arive at the generalized eigenvalue problem

$$\mathbf{A}^p \boldsymbol{\Phi}^h = \varepsilon^h \mathbf{B}^p \boldsymbol{\Phi}^h. \tag{6}$$

Here $\mathbf{A}^p$ is the stiffness matrix; $\mathbf{B}^p$ is the positive definite mass matrix; $\boldsymbol{\Phi}^h$ is the vector approximating the solution on the finite-element grid; and $\varepsilon^h \equiv \varepsilon^h(z)$ is the corresponding eigenvalue. The matrices $\mathbf{A}^p$ and $\mathbf{B}^p$ have the form $\mathbf{A}^p{=}\{a_{ll'}^p\}_{ll'=1}^N$, $\mathbf{B}^p{=}\{b_{ll'}^p\}_{ll'=1}^N$, where the matrix elements $a_{ll'}^p$ and $b_{ll'}^p$ are calculated for triangular elements as

$$\begin{aligned} a_{ll'}^p &= \sum_{i,j=1}^{2} \int_{\Delta_q} g_{ij} \frac{\partial \varphi_l^p(x;z)}{\partial x_i} \frac{\partial \varphi_{l'}^p(x;z)}{\partial x_j}\, dx + \int_{\Delta_q} g_0(x) \varphi_l^p(x;z) \varphi_{l'}^p(x;z) U(x;z)\, dx, \\ b_{ll'}^p &= \int_{\Delta_q} g_0(x) \varphi_l^p(x;z) \varphi_{l'}^p(x;z) dx. \end{aligned} \tag{7}$$

3. The Algorithm for Calculating the Parametric Derivatives of Eigenfunctions

Differentiating Eqs. (1)–(3) with respect to the parameter z, we find that $\partial_z \Phi_i(x;z)$ is a solution of the following boundary-value problem with the mixed boundary conditions

$$(D(x;z) - \varepsilon_i(z)) \frac{\partial \Phi_i(x;z)}{\partial z} = - \left[\frac{\partial}{\partial z} (U(x;z) - \varepsilon_i(z)) \right] \Phi_i(x;z), \tag{8}$$

$$\left.\frac{\partial \Phi(x;z)}{\partial z}\right|_S = 0 \quad \text{or} \quad \left.\frac{\partial^2 \Phi(x;z)}{\partial n_D \partial z}\right|_S = 0.$$

The parametric BVP (8) has a unique solution, if and only if it satisfies the conditions

$$\frac{\partial \varepsilon_i(z)}{\partial z} = \int_\Omega dx g_0(x) \Phi_i(x;z) \frac{\partial U(x;z)}{\partial z} \Phi_i(x;z), \quad \int_\Omega dx g_0(x) \Phi_i(x;z) \frac{\partial \Phi_i(x;z)}{\partial z} = 0. \tag{9}$$

Below we present the efficient numerical method that allows the calculation of $\partial_z \Phi_i(x;z)$ with the same accuracy as the one achieved for the eigenfunctions of the BVP (1)–(3) and the use of it for computing the effective potentials $H_{ij}(z){=}H_{ji}(z)$ and $Q_{ij}(z){=}{-}Q_{ji}(z)$ defined as

$$H_{ij}(z) = \int_\Omega dx g_0(x) \frac{\partial \Phi_i(x;z)}{\partial z} \frac{\partial \Phi_j(x;z)}{\partial z}, \quad Q_{ij}(z) = - \int_\Omega dx g_0(x) \Phi_i(x;z) \frac{\partial \Phi_j(x;z)}{\partial z}. \tag{10}$$

The boundary-value problem (8)–(9) is reduced to the linear system of inhomogeneous algebraic equations with respect to the unknown $\partial \boldsymbol{\Phi}^h / \partial z$:

$$\mathbf{L} \frac{\partial \boldsymbol{\Phi}^h}{\partial z} \equiv (\mathbf{A}^p - \varepsilon^h \mathbf{B}^p) \frac{\partial \boldsymbol{\Phi}^h}{\partial z} = b, \quad b = - \left(\frac{\partial \mathbf{A}^p}{\partial z} - \frac{\partial \varepsilon^h}{\partial z} \mathbf{B}^p \right) \boldsymbol{\Phi}^h. \tag{11}$$

The normalization condition (3), the condition of orthogonality between the function and its parametric derivative and the additional conditions (9) for the solution of (11) read as

$$\left(\boldsymbol{\Phi}^h\right)^T \mathbf{B}^p \boldsymbol{\Phi}^h = 1, \quad \left(\frac{\partial \boldsymbol{\Phi}^h}{\partial z}\right)^T \mathbf{B}^p \boldsymbol{\Phi}^h = 0, \quad \frac{\partial \varepsilon^h}{\partial z} = \left(\boldsymbol{\Phi}^h\right)^T \frac{\partial \mathbf{A}^p}{\partial z} \boldsymbol{\Phi}^h. \tag{12}$$

Then the potential matrix elements $H_{ij}^h(z)$ and $Q_{ij}^h(z)$ (10) can be calculated using the formulas

$$H_{ij}^h(z) = \left(\frac{\partial \boldsymbol{\Phi}_i^h}{\partial z}\right)^T \mathbf{B}^p \frac{\partial \boldsymbol{\Phi}_j^h}{\partial z}, \quad Q_{ij}^h(z) = -\left(\boldsymbol{\Phi}_i^h\right)^T \mathbf{B}^p \frac{\partial \boldsymbol{\Phi}_j^h}{\partial z}. \tag{13}$$

Since ε^h is an eigenvalue of (6), the matrix $\mathbf{L}$ in Eq. (11) is degenerate. In this case, the algorithm for solving Eq. (11) can be written in three steps as follows:

Step k1. Calculate the solutions $\mathbf{v}$ and $\mathbf{w}$ of the auxiliary inhomogeneous systems of algebraic equations

$$\bar{\mathbf{L}}\mathbf{v} = \bar{\mathbf{b}}, \quad \bar{\mathbf{L}}\mathbf{w} = \mathbf{d}, \tag{14}$$

with the non-degenerate matrix $\bar{\mathbf{L}}$ and the right-hand sides $\bar{\mathbf{b}}$ and $\mathbf{d}$

$$\bar{L}_{ss'} = \begin{cases} L_{ss'}, & (s-S)(s'-S) \neq 0, \\ \delta_{ss'}, & (s-S)(s'-S) = 0, \end{cases} \tag{15}$$

$$\bar{b}_s = \begin{cases} b_s, & s \neq S, \\ 0, & s = S, \end{cases} \quad d_s = \begin{cases} L_{sS}, & s \neq S, \\ 0, & s = S, \end{cases} \tag{16}$$

where S is the number of the element of the vector $\mathbf{B}^p\boldsymbol{\Phi}^h$ having the greatest absolute value.

Step k2. Evaluate the coefficient γ

$$\gamma = -\frac{\gamma_1}{(\mathbf{D}_S - \gamma_2)}, \quad \gamma_1 = \mathbf{v}^T\mathbf{B}^p\boldsymbol{\Phi}^h, \quad \gamma_2 = \mathbf{w}^T\mathbf{B}^p\boldsymbol{\Phi}^h, \quad \mathbf{D}_S = (\mathbf{B}^p\boldsymbol{\Phi}^h)_S. \tag{17}$$

Step k3. Evaluate the vector $\partial_z\boldsymbol{\Phi}^h$

$$\frac{\partial \Phi_s^h}{\partial z} = \begin{cases} v_s - \gamma w_s, & s \neq S, \\ \gamma, & s = S. \end{cases} \tag{18}$$

From the above consideration, it is evident that the computed derivative has the same accuracy as the calculated eigenfunction itself. Let $D(x;z)$ in Eq. (1) be a continuous and bounded positive definite operator on the space $\mathcal{H}^1$ with the energy norm, $\varepsilon_i(z)$, $\Phi_i(x,z) \in \mathcal{H}^2$ being the exact solutions of Eqs. (1)–(3), and $\varepsilon_i^h(z)$, $\Phi_i^h(x;z) \in \mathcal{H}^1$ being the corresponding numerical solutions. Then the following estimates are valid [6]

$$\left|\varepsilon_i(z) - \varepsilon_i^h(z)\right| \leq c_1 h^{2p}, \quad \left\|\Phi_i(x;z) - \Phi_i^h(x;z)\right\|_0 \leq c_2 h^{p+1}, \tag{19}$$

$$\|\Phi_i(x;z)\|_0^2 = \int_{\Omega_x} dx g_0(x)\Phi_i(x;z)\Phi_i(x;z), \tag{20}$$

where h is the largest distance between any two points in Δ_q, p is the order of the finite elements, i is the number of the corresponding solutions, and the constants c_1 and c_2 are independent of the step h. The following theorem can be formulated.

Theorem Let $D(x;z)$ in Eq. (1) be a continuous and bounded positive definite operator on the space $\mathcal{H}^1$ with the energy norm. Also let $\partial_z U(x;z)$ be continuous and bounded for each value of the parameter z. Then for the exact values of the solutions $\partial_z\varepsilon_i(z)$, $\partial_z\Phi_i(x;z) \in \mathcal{H}^2$, $H_{ij}(z)$, $Q_{ij}(z)$ from (8)–(10) and the corresponding numerical values $\partial_z\varepsilon_i^h(z)$, $\partial_z\Phi_i^h(x;z) \in \mathcal{H}^1$, $H_{ij}^h(z)$, $Q_{ij}^h(z)$ from (11)–(13), the following estimates are valid:

$$\left|\frac{\partial\varepsilon_i(z)}{\partial z} - \frac{\partial\varepsilon_i^h(z)}{\partial z}\right| \leq c_3 h^{2p}, \quad \left\|\frac{\partial\Phi_i(x;z)}{\partial z} - \frac{\partial\Phi_i^h(x;z)}{\partial z}\right\|_0 \leq c_4 h^{p+1}, \tag{21}$$

$$\left|Q_{ij}(z) - Q_{ij}^h(z)\right| \leq c_5 h^{2p}, \quad \left|H_{ij}(z) - H_{ij}^h(z)\right| \leq c_6 h^{2p},$$

where h is the largest distance between any two points of the finite element Δ_q, p is the order of finite elements, i, j are the numbers of the corresponding solutions, and the constants c_3, c_4, c_5, and c_6 are independent of the step h.

The proof is straightforward following the scheme in accordance with [6].

4. Benchmark calculations of helium atom bound states

In the hyperspheroidal coordinates, $0 \le \rho < \infty$, $1 \le \xi < \infty$, $-1 \le \eta \le 1$, related to the perimetrical ones r_1, r_2 and r_{12} as

$$r_{12} = \sqrt{2}\rho/\sqrt{\xi^2+\eta^2}, \quad r_1 = \rho(\xi+\eta)/(\sqrt{2}\sqrt{\xi^2+\eta^2}), \quad r_2 = \rho(\xi-\eta)/(\sqrt{2}\sqrt{\xi^2+\eta^2}), \tag{22}$$

the equation for the solution $\Psi(\rho,\xi,\eta) = \sqrt{\xi^2+\eta^2}\Phi(\rho,\xi,\eta)$ describing S-states of the Helium atom reads as [7]

$$\left[-\frac{1}{\rho^5}\frac{\partial}{\partial\rho}\rho^5\frac{\partial}{\partial\rho} - \frac{3}{\rho^2} + \frac{1}{\rho^2}\frac{(\xi^2+\eta^2)^2}{\xi^2-\eta^2}h(\xi,\eta;\rho) - 2E_m\right]\Phi_m(\rho,\xi,\eta) = 0, \tag{23}$$

$$h(\xi,\eta;\rho) = -\frac{\partial}{\partial\xi}(\xi^2-1)\frac{\partial}{\partial\xi} - \frac{\partial}{\partial\eta}(1-\eta^2)\frac{\partial}{\partial\eta} + \frac{\sqrt{2}\rho\,(\xi^2-\eta^2-8\xi)}{\sqrt{\xi^2+\eta^2}^3}. \tag{24}$$

The function $\Phi_m(\rho,\xi,\eta)$ satisfies the boundary conditions(BCs)

$$\lim_{\rho\to0}\rho^5\frac{\partial\Phi_m(\rho,\xi,\eta)}{\partial\rho} = 0, \quad \lim_{\rho\to\infty}\rho^5\Phi_m(\rho,\xi,\eta) = 0, \tag{25}$$

$$\lim_{\xi\to1}(\xi^2-1)\frac{\partial\Phi_m(\rho,\xi,\eta)}{\partial\xi} = 0, \quad \lim_{\xi\to\infty}\Phi_m(\rho,\xi,\eta) = 0, \quad \lim_{\eta\to\pm1}(1-\eta^2)\frac{\partial\Phi_m(\rho,\xi,\eta)}{\partial\eta} = 0, \tag{26}$$

and is normalized by the condition

$$8\pi^2\int_0^\infty dRR^5\int_1^\infty d\xi\int_{-1}^1 d\eta\frac{\xi^2-\eta^2}{(\xi^2+\eta^2)^2}\Phi_m(R,\xi,\eta)\Phi_{m'}(R,\xi,\eta) = \delta_{mm'}. \tag{27}$$

The parametric functions $\phi_i \equiv \phi_i(\xi,\eta;\rho)$ and the eigenvalues $\varepsilon_i(\rho)$ are eigensolutions of the 2D BVP having purely discrete spectrum. The functions satisfy the BCs (26) and the normalization condition

$$\left[h(\xi,\eta;\rho) - \varepsilon_i(\rho)\frac{\xi^2-\eta^2}{(\xi^2+\eta^2)^2}\right]\phi_i = 0, \ \langle\phi_i|\phi_j\rangle = \int_1^\infty d\xi\int_{-1}^1 d\eta\frac{\xi^2-\eta^2}{(\xi^2+\eta^2)^2}\phi_i(\xi,\eta;\rho)\phi_j(\xi,\eta;\rho) = \delta_{ij}. \tag{28}$$

The 2D BVP (28) was solved in terms of scaled variable and parametric surface functions

$$\xi = \xi(\lambda) = \frac{1+\lambda}{1-\lambda}, \quad 0 \le \lambda < 1, \quad \phi_i(\xi,\eta;\rho) = \frac{p_i(\xi,\eta;\rho)}{\xi+1} \equiv \frac{p_i(\lambda,\eta;\rho)}{\xi(\lambda)+1}. \tag{29}$$

As an example consider the calculation of the potential curves $E_j(\rho) = (\varepsilon_j(\rho)-3)/\rho^2$ and the matrix elements

$$Q_{ij}(\rho) = -\left\langle\phi_i(\xi,\eta;\rho)\left|\frac{\partial\phi_j(\xi,\eta;\rho)}{\partial\rho}\right.\right\rangle, \quad H_{ij}(\rho) = \left\langle\frac{\partial\phi_i(\xi,\eta;\rho)}{\partial\rho}\left|\frac{\partial\phi_j(\xi,\eta;\rho)}{\partial\rho}\right.\right\rangle.$$

The calculations were implemented by means of the program POTHEA 2.0 using the server with 2x4 kernels i7k (i7-3770K 4.5 GHz, 32 GB RAM, GPU GTX680), and the Intel Fortran compiler 17.0. The computing time per one point ρ for the considered examples with 10^{-12} accuracy with triangular Lagrange elements of 5th order on the uniform 2D grids $\lambda = \{0(L)1\}$, $\eta = \{0(L)1\}$ at $L = 10, 20, 40$ is 0.38, 5.08, and 41.21 seconds, respectively.

Let us find the solution of the BVP (23)–(27) using the Kantorovich expansion [2]

$$\Phi_m(\rho,\xi,\eta) = \sum_{j=1}^{j_{\max}} \phi_j(\xi,\eta;\rho)\chi_{jm}(\rho) \tag{30}$$

where $\phi_j(\xi,\eta;\rho)$ are eigenfunctions of the parametric 2D BVP with purely discrete spectrum $E_j(\rho) = (\varepsilon_j(\rho) - 3)/\rho^2, j = 1, 2, \ldots$. Substituting the expansion (30) into the 3D BVP (23)–(26), we arrive at the 1D BVP for a finite set of $j_{\max}$ coupled second-order ODEs for $\boldsymbol{\chi}_m(\rho) = \{\chi_{1m}(\rho), \ldots, \chi_{Nm}(\rho)\}^T$

$$\left(-\frac{1}{\rho^5}\mathbf{I}\frac{d}{d\rho}\rho^5\frac{d}{d\rho} + \mathbf{V}(\rho) + \mathbf{Q}(\rho)\frac{d}{d\rho} + \frac{1}{\rho^5}\frac{d\rho^5\mathbf{Q}(\rho)}{d\rho} - 2E_m\mathbf{I}\right)\boldsymbol{\chi}_m(\rho) = 0, \tag{31}$$

with the boundary and normalization conditions for the discrete spectrum of E: $E_1 < E_2 < \ldots < E_m < \ldots$

$$\lim_{\rho\to 0}\rho^5\left(\frac{d\boldsymbol{\chi}_m(\rho)}{d\rho} - \mathbf{Q}(\rho)\boldsymbol{\chi}_m(\rho)\right) = 0,\ \lim_{\rho\to\infty}\rho^5\boldsymbol{\chi}_m(\rho) = 0,\ 8\pi^2\int_0^\infty d\rho\rho^5(\boldsymbol{\chi}_m(\rho))^T\boldsymbol{\chi}_{m'}(\rho) = \delta_{mm'}. \tag{32}$$

The solution of this BVP with the help of KANTBP program [8] using Lagrange elements of the 5th order on the non-uniform grids $\rho = \{0(50)5(75)20\}$ using calculated $E_j(\rho)$, $V_{ij}(\rho) = H_{ij}(\rho) + E_j(\rho)\delta_{ij}$, $Q_{ij}(\rho)$, $i, j = 1, \ldots, 12$ yields the upper estimate of the helium atom energy in the ground and the first exited state $E_1 = -2.903\,724\,30$ a.u. and $E_2 = -2.145\,973\,22$ a.u. with 8 significant digits similar to the results of POTHEA[9].

5. Conclusion

The proposed calculation schemes, algorithms and software implementing the high-accuracy FEM for solving the elliptic boundary value problems can be applied to the analysis of few-body scattering dynamics and quantum tunneling and diffraction models[10, 11, 12, 13]. Detailed description of the implementation of the FEM algorithm and program will be given elswhere.

The authors thank Prof. F.M. Penkov for collaboration. The work was supported partially by grant MES RK 0333/GF4 and grant of Plenipotentiary Representative of the Government of the Republic of Kazakhstan in the framework of collaboration program JINR-RK N 337, 2017. The work was partially funded by the RFBR (grants Nos. 16-01-00080 and 17-51-44003 Mong), the Bogoliubov-Infeld program and the RUDN University Program 5-100.

References

[1] Greene C H, Giannakeas P and Pérez-Ríos J 2017 *Rev. Mod. Phys.* **89** 035006
[2] Kantorovich L V and Krylov V I 1964 *Approximate methods of higher analysis* (New York, Wiley)
[3] Ciarlet P 1978 *The Finite Element Method for Elliptic Problems* (Amsterdam, North-Holland Publ.)
[4] Kadomtsev M B, Vinitsky S I and Yukajlovic F R 1987 *Phys.Rev. A* **36** 4652
[5] Bathe K J 1982 *Finite Element Procedures in Engineering Analysis*, (New York, Englewood Cliffs, Prentice Hall)
[6] Strang G. and Fix G J 1973 *An Analysis of the Finite Element Method* (New York, Englewood Cliffs, Prentice Hall)
[7] Vinitsky S I, Gusev A A, Chuluunbaatar O et al 2016 *Proc. SPIE* **9917** 99172Z
[8] Chuluunbaatar O, Gusev A A, Vinitsky S I and Abrashkevich A G 2008 *Comput. Phys. Commun.* **179** 685
[9] Gusev A A, Chuluunbaatar O, Vinitsky S I and Abrashkevich A G 2014 *Comput. Phys. Commun.* **185** 2636
[10] Gusev A A, Vinitsky S I, Chuluunbaatar O et al 2014 *Phys. Atom. Nucl.* **77** 389
[11] Krassovitskiy P M and Penkov F M 2014 *J. Phys. B* **47** 225210
[12] Krassovitskiy P M and Penkov F M 2017 *Bull. Russian Acad. Sci.: Phys.* **81** 730
[13] Gusev A A, Vinitsky S I, Chuluunbaatar O et al 2017 *Acta Phys. Polonica B Proc. Sup.* **10** 269

IOP Conf. Series: Journal of Physics: Conf. Series **965** (2018) 012017 doi:10.1088/1742-6596/965/1/012017

Integrability properties of minimal surfaces in hyperbolic space

Yifei He, Changyu Huang, Martin Kruczenski
Department of Physics, Purdue University, 525 Northwestern Avenue, W. Lafayette, IN 47907-2036, USA
E-mail: he163@purdue.edu, cyhuang@purdue.edu, markru@purdue.edu

Abstract. We present a method, based on integrability, to find minimal area surfaces in hyperbolic space ending on a given contour at the boundary. The problem has physical interest since the AdS/CFT correspondence relates the area of the minimal surface to the expectation value of the Wilson loop defined by the boundary contour. We give particular solutions using the Mathieu functions and more general results using a numerical method.

1. Introduction
According to the AdS/CFT correspondence [1, 2, 3], Wilson loops in $\mathcal{N}=4$ SYM can be calculated holographically from the area of the minimal surface ending on the boundary of AdS_5 [4, 5] at a contour defined by the Wilson loop. Much work has been done to calculate Wilson loops of different shapes. In Euclidean signature, examples include the circular Wilson loop [6, 7], wavy lines [8, 9] and solutions constructed using Riemann-theta function [10, 11]. In Minkowski signature, the most interesting example is the light-like cusp [12] due to its relation to scattering amplitudes [13, 14, 15]. In this paper, we study the integrability structure of a general smooth Wilson loop on the boundary of Euclidean AdS_3 using the Pohlmeyer reduction [16]. We construct analytical solutions in terms of Mathieu functions and also provide a numerical recipe for studying a Wilson loop of a general shape.

The paper is organized as follows: in section 2, we review the general setup for studying minimal surfaces in Euclidean AdS_3. Section 3 describes the Schrödinger equation defined on the boundary contour. In section 4, we construct analytical solutions using Mathieu functions. In section 5 we briefly describe a method to find the contour and area associated with a given analytic function $f(z)$ appearing in the Pohlmeyer reduction. Finally, we provide a numerical method for solving a general boundary contour in section 6 where we also give some results we found by implementing such numerical method.

2. General setup
Euclidean AdS_3 is embedded as a hyperboloid in $\mathbb{R}^{1,3}$:

$$-X_0^2+X_1^2+X_2^2+X_3^2=-1. \tag{1}$$

The Poincaré coordinates are given by

$$Z=\frac{1}{X_0-X_3}, \quad X=\frac{X_1+iX_2}{X_0-X_3}, \quad \bar{X}=\frac{X_1-iX_2}{X_0-X_3}, \tag{2}$$

with the metric

$$ds^2 = \frac{dZ^2 + dXd\bar{X}}{Z^2}. \tag{3}$$

The string worldsheet can be conformally mapped to the unit disk ($|z| \leq 1$) on the complex plane $z = \sigma + i\tau = re^{i\theta}$ so the Poincaré coordinates are parametrized as $X(r,\theta)$ and $Z(r,\theta)$. Near the boundary, i.e., $r \to 1$, one has

$$Z(r=1,\theta) = 0, \quad X(r=1,\theta) = X(s(\theta)). \tag{4}$$

$X(s)$ is a given contour defined on the boundary of Euclidean AdS_3 parametrized by s, which is related to the conformal angle θ by an unknown reparametrization $s(\theta)$.

The string action is

$$S = \frac{1}{2}\int d\sigma d\tau(\partial X_\mu \bar{\partial} X^\mu + \Lambda(X_\mu X^\mu + 1)), \tag{5}$$

where Λ is a Lagrange multiplier. The classical string solution satisfies the equation of motion

$$\partial\bar{\partial}X - \Lambda X = 0, \quad \Lambda = \bar{\partial}X \cdot \partial X. \tag{6}$$

This should be supplemented by the Virasoro constraints

$$\bar{\partial}X \cdot \bar{\partial}X = 0 = \partial X \cdot \partial X. \tag{7}$$

Writing $\mathbb{X} = X_0 + X_i\sigma^i$, the embedding condition, the equation of motion and the Virasoro constraints become

$$\det \mathbb{X} = 1, \quad \partial\bar{\partial}\mathbb{X} = \Lambda\mathbb{X}, \quad \det(\partial\mathbb{X}) = \det(\bar{\partial}\mathbb{X}) = 0, \tag{8}$$

where we have the reality condition $\mathbb{X}^\dagger = \mathbb{X}$. This condition can be solved by writing

$$\mathbb{X} = \mathbb{A}\mathbb{A}^\dagger, \tag{9}$$

with

$$\det \mathbb{A} = 1, \quad \mathbb{A} \in SL(2,\mathbb{C}). \tag{10}$$

The matrix $\mathbb{A}$ is the solution to the linear problem

$$\partial\mathbb{A} = \mathbb{A}J, \quad \bar{\partial}\mathbb{A} = \mathbb{A}\bar{J}, \tag{11}$$

where

$$J = \begin{pmatrix} -\frac{1}{2}\partial\alpha & fe^{-\alpha} \\ \lambda e^{\alpha} & \frac{1}{2}\partial\alpha \end{pmatrix}, \quad \bar{J} = \begin{pmatrix} \frac{1}{2}\bar{\partial}\alpha & \frac{1}{\lambda}e^{\alpha} \\ -\bar{f}e^{-\alpha} & -\frac{1}{2}\bar{\partial}\alpha \end{pmatrix}. \tag{12}$$

The consistency condition requires f to be holomorphic and , $\bar{f}$ anti-holomorphic, and $\alpha(z,\bar{z})$ to satisfy the generalized cosh-Gordon equation:

$$\partial\bar{\partial}\alpha = e^{2\alpha} + f\bar{f}e^{-2\alpha}. \tag{13}$$

If $|\lambda| = 1$ in J and $\bar{J}$, one obtains a family of minimal surfaces ending on different boundary contours with the same area. This reflects the integrability of the problem and has been studied recently in [17, 18].

The metric induced on the minimal surface is

$$ds^2 = 4e^{2\alpha}dzd\bar{z} \tag{14}$$

so the area is calculated using

$$A = 4\int e^{2\alpha}d\sigma d\tau. \tag{15}$$

After regularization, the finite area is

$$A_f = -2\pi - \int f\bar{f}e^{2\alpha}d\sigma d\tau \tag{16}$$

3. Linear problem on the boundary

Near the boundary, $\alpha(z,\bar{z})$ has the expansion

$$\alpha(\xi,\theta) \simeq -\ln\xi + \beta_2(\theta)(1+\xi)\xi^2 + O(\xi^4), \tag{17}$$

where $\xi = 1 - r^2$ and $\beta_2(\theta)$ can be defined as

$$\beta_2(\theta) = \frac{1}{6}e^{2i\theta}(\partial^2\alpha - (\partial\alpha)^2)\big|_{r\to 1}. \tag{18}$$

All the higher order coefficients in (17) are fixed by $\beta_2(\theta)$ and $f(\theta) = f(e^{i\theta})$.

The solution to the linear problem (11) can be written in terms of two linearly independent row vectors as

$$\mathbb{A} = \begin{pmatrix} \psi_1 & \psi_2 \\ \tilde{\psi}_1 & \tilde{\psi}_2 \end{pmatrix}. \tag{19}$$

Then the linear equations (11) are reduced to

$$\partial\psi = \psi J, \quad \bar{\partial}\psi = \psi\bar{J}, \tag{20}$$

and the same equations for $\tilde{\psi}$.

Taking this linear problem to the boundary, one gets a Schrödinger-like equation[19]:

$$-\partial_\theta^2\chi(\theta) + V^\lambda(\theta)\chi(\theta) = 0 \tag{21}$$

where the potential is given by the Schwarzian derivative of the contour $X^\lambda(\theta)$:

$$V^\lambda(\theta) = -\frac{1}{2}\{X^\lambda(\theta),\theta\} = -\frac{1}{4} + 6\beta_2(\theta) + \lambda f(\theta)e^{2i\theta} - \frac{1}{\lambda}\bar{f}(\theta)e^{-2i\theta}, \tag{22}$$

and the contour $X^\lambda(\theta)$ can be constructed with two anti-periodic linearly independent solutions as

$$X^\lambda(\theta) = \frac{\tilde{\chi}^\lambda}{\chi^\lambda}. \tag{23}$$

Therefore, knowing the original boundary contour in terms of the conformal angle $X(\theta)$, one can calculate the Schwarzian derivative $\{X(\theta),\theta\}$. The λ-deformed contours can be easily obtained by solving eq.(21) with the deformed potential according to eq.(22). Furthermore, one can extract $f(\theta)$ from the imaginary part of $\{X(\theta),\theta\}$, analytically continue it inside the unit disk, and plug it into eq.(13) to solve for $\alpha(z,\bar{z})$. With f, $\bar{f}$ and α, the area can be calculated using eq.(16). It is worth mentioning that in the case where $f(z)$ has no zeros inside the unit disk, the area can be written in terms of the Schwarzian derivative

$$A_f = -2\pi \pm \frac{i}{2}\oint d\theta \frac{\mathrm{Re}\{X(\theta),\theta\} - \{\omega,\theta\}}{\partial_\theta \ln\omega}, \tag{24}$$

with

$$\omega(\theta) = i\int^\theta d\theta' \sqrt{f(\theta')e^{2i\theta'}}. \tag{25}$$

This area formula is manifestly global conformal invariant and does not require solving eq. (13).

4. Mathieu solutions

From the previous section, we see that the minimal surfaces ending on a boundary contour can be studied by considering a Schrödinger equation with periodic potential and the contour can be constructed from two anti-periodic linearly independent solutions. Given a boundary contour $X(s)$, one can rewrite the Schrödinger equation using the parameter s

$$-\partial_s^2\chi + V(\lambda, s)\chi = 0, \tag{26}$$

with $V(\lambda = 1, s) = -\frac{1}{2}\{X(s), s\}$. The prototypical Schrödinger equation with periodic potential is the Mathieu equation

$$\partial_u^2\chi(u) + (a - 2q\cos 2u)\chi(u) = 0. \tag{27}$$

Here q is a parameter and a is the energy eigenvalue. Writing the potential as

$$V(\lambda, s) = -\frac{1}{4} + 6\beta_2 - \lambda f_0 e^{i(n+2)s} + \frac{f_0}{\lambda} e^{-i(n+2)s}, \tag{28}$$

with $\lambda = e^{i\phi}$, the Schrödinger equation (26) becomes the Mathieu equation if we make the following identifications [20]:

$$a = \frac{1 - 24\beta_2}{(n+2)^2}, \quad q = \frac{4if_0}{(n+2)^2}, \tag{29}$$

with the change of variable

$$u(s) = \frac{(n+2)s + \phi}{2} + \frac{\pi}{4}. \tag{30}$$

We look for Floquet solutions of Mathieu equation with characteristic exponent ν

$$\begin{aligned} \chi_\nu(u + \pi) &= e^{i\pi\nu}\chi_\nu(u), \\ \chi_\nu(u) &= e^{i\nu u}p_\nu(q, u). \end{aligned} \tag{31}$$

To guarantee the anti-periodicity of the solution, we take $\nu = \frac{2k+1}{n+2}$, with $k \in \mathbb{Z}$. The eigenvalue $a_\nu(q)$ can be determined given ν and the constant β_2 has the form

$$\beta_2 = \frac{1}{24}(1 - (n+2)^2 a_\nu(q)). \tag{32}$$

The boundary contour can be found from

$$X(u) = \frac{\chi_\nu(u)}{\chi_\nu(-u)} = e^{2i\nu u}\frac{p_\nu(q, u)}{p_\nu(q, -u)}, \tag{33}$$

and the area is given by

$$A_f = -2\pi + \frac{\pi}{n+2} - (n+2)\pi a_\nu(q). \tag{34}$$

It should be noted that the case $n = 0$ has already been studied by J. Toledo in [21] using a different method. The present construction allows for a large class of new solutions where the contours are given in terms of Mathieu functions, the area by the eigenvalues of the Mathieu equation, and the shape of the surface by a numerical solution to the linear problem (11). An example is given in fig.1.

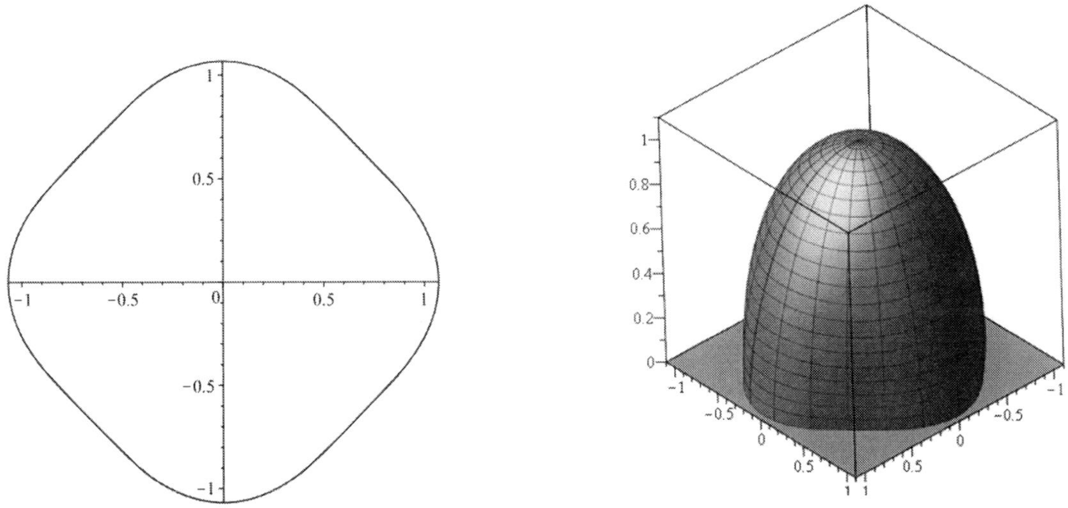

(a) Boundary contour (b) Surface

Figure 1: Example of Mathieu surface with $n = 2$. The area is $\mathcal{A}_f = -6.660397$

5. Numerical method for solving the cosh-Gordon equation

In the next section we describe a numerical method to solve the problem of finding the reparameterization $s(\theta)$ necessary to obtain the minimal area surface ending on a given contour. As a preliminary step it is necessary to implement a method to solve the cosh-Gordon equation (13) for a given function $f(z)$. This is useful in its own right since we can propose an arbitrary $f(z)$ and obtain the contour on which the surface ends together with the area of the surface. The equation can be solved by a simple collocation method as described in [22]. As an example, in figure 2, we plot the contour associated with the function

$$f(z) = 2(z - 0.3)(z - \frac{i}{2})(z + 0.7)(z + \frac{i}{2}) \tag{35}$$

We just chose a polynomial function with four zeros inside the unit disk at arbitrary positions.

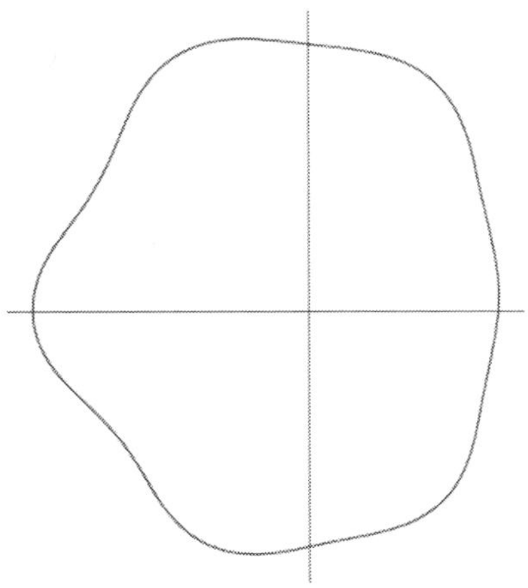

Figure 2: Contour associated with the function $f(z) = 2(z - 0.3)(z - \frac{i}{2})(z + 0.7)(z + \frac{i}{2})$. The area of the corresponding minimal surface is $\mathcal{A}_f = -6.99563$.

6. Numerical method for finding $s(\theta)$

For a general given contour $X(s)$, the problem remains of how to find the conformal reparametrization $s(\theta)$. Using the property of the Schwarzian derivative

$$\{F,\theta\} = \{s,\theta\} + (\partial_\theta s)^2\{F,s\}, \tag{36}$$

eq. (22) becomes

$$\begin{aligned} \{s,\theta\} + (\partial_\theta s)^2 \mathrm{Re}\{X(s),s\} &= \frac{1}{2} - 12\beta_2(\theta), \\ (\partial_\theta s)^2 \mathrm{Im}\{X(s),s\} &= -4\mathrm{Im}(e^{2i\theta} f(\theta)). \end{aligned} \tag{37}$$

For a general contour, one can propose an arbitrary reparametrization $s(\theta)$, extract a $f(\theta)$ from the second equation in (37), analytically continue inside the unit disk to solve for $\alpha(z,\bar{z})$ and calculate $\beta_2(\theta)$ according to eq. (18), which we call $\tilde{\beta}_2(\theta)$. One can then compare this $\tilde{\beta}_2$ with the β_2 calculated from the first equation in (37). If the $s(\theta)$ is the conformal reparametrization, then we should have $\tilde{\beta}_2 = \beta_2$. We therefore perform a minimization procedure on the difference between $\tilde{\beta}_2$ and β_2 to search for the correct $s(\theta)$. For details of the numerics, see [23]. The procedure is summarized in Fig.3.

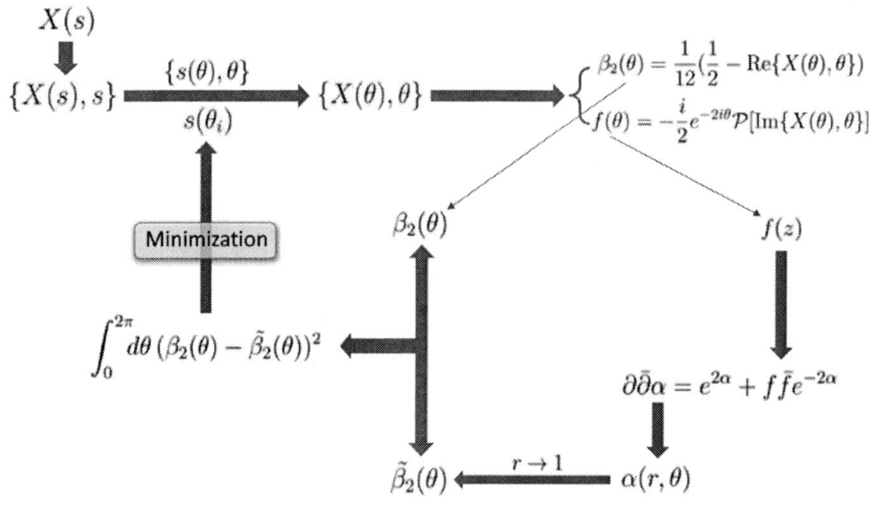

Figure 3: The procedure for finding the reparametrization $s(\theta)$.

This method was used in [9] to analytically study the perturbations from the circular contour. We perform the numerical implementation which does not depend on a perturbative expansion, to study various contours and find their conformal reparametrization $s(\theta)$ and area beyond the regime where the perturbative expansion converges. In Fig. 4, we plot the calculated area compared to the perturbative calculation for symmetric contours

$$X(s) = e^{is + a\sin(ps)}. \tag{38}$$

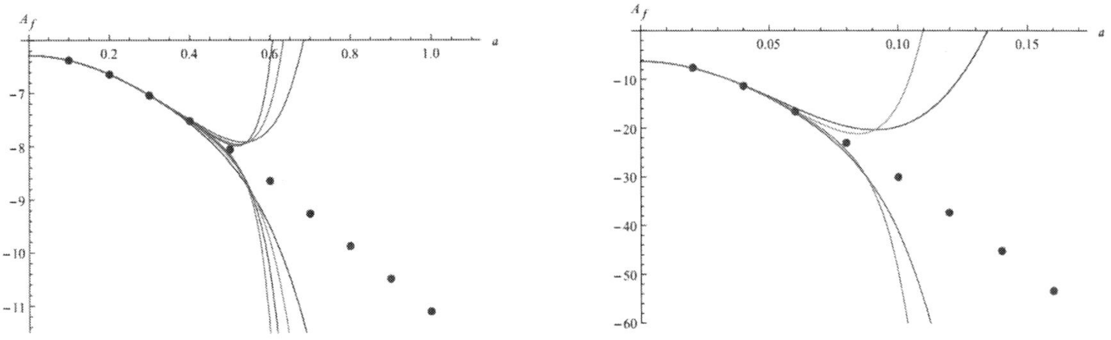

(a) $p = 2$ (b) $p = 13$

Figure 4: The areas of the symmetric contours with $p = 2$ and $p = 13$. We plot a few partial sums (continuous curves) of the perturbative calculations as well as the results from our calculation (dots), which goes beyond the range of a where the perturbative series converges.

Once the reparametrization is found, it is easy to perform the λ-deformation of the contour by solving eq. (21). In Figs. 5 and 6, we plot the λ-deformation for symmetric contours with $p = 2$ and $p = 13$.

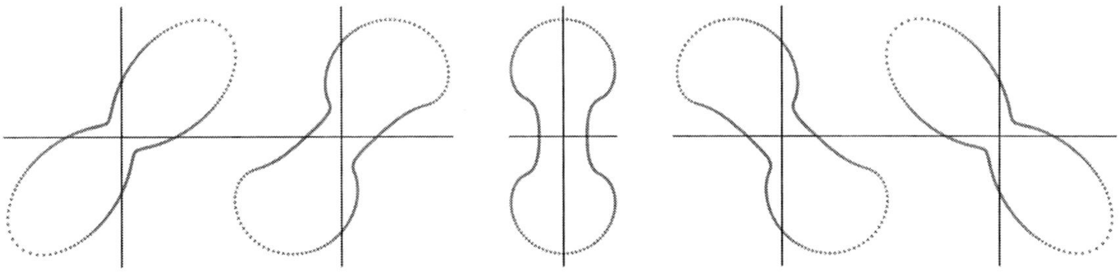

Figure 5: The λ-deformation of symmetric contour with $p = 2$ and $a = 1$. From the left to the right we take $\lambda = 1, e^{i\frac{\pi}{4}}, e^{i\frac{\pi}{2}}, e^{i\frac{3\pi}{4}}, e^{i\pi}$.

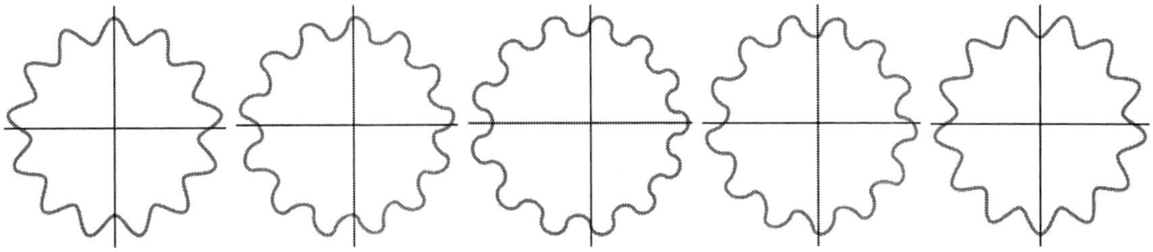

Figure 6: The λ-deformation of symmetric contour with $p = 13$ and $a = 0.12$. From the left to the right we take $\lambda = 1, e^{i\frac{\pi}{4}}, e^{i\frac{\pi}{2}}, e^{i\frac{3\pi}{4}}, e^{i\pi}$.

7. Acknowledgments

We are very grateful to A. Dekel, P. Vieira for useful comments and discussions on this topic. This work was supported in part by DOE through grant DE-SC0007884.

IOP Conf. Series: Journal of Physics: Conf. Series **965** (2018) 012017 doi:10.1088/1742-6596/965/1/012017

References

[1] Maldacena J M 1999 *Int. J. Theor. Phys.* **38** 1113
[2] Gubser S S, Klebanov I R and Polyakov A M 1998 *Phys. Lett.* B **428** 105
[3] Witten E 1998 *Adv. Theor. Math. Phys.* **2** 253
[4] Maldacena J M 1998 *Phys. Rev. Lett.* **80** 4859
[5] Rey S J and Yee J T 2001 *Eur. Phys. J.* C **22** 379
[6] Drukker N, Gross D J and Ooguri H 1999 *Phys. Rev.* D **60** 125006
[7] Berenstein D E, Corrado R, Fischler W and Maldacena J M 1999 *Phys. Rev.* D **59** 105023
[8] Semenoff G W and Young D 2005 *Int. J. Mod. Phys.* A **20** 2833
[9] Dekel A 2015 *J. High Energy Phys.* **1503** 085
[10] Ishizeki R, Kruczenski M and Ziama S 2012 *Phys. Rev.* D **85** 106004
[11] Kruczenski M and Ziama S 2014 *J. High Energy Phys.* **1405** 037
[12] Kruczenski M 2002 *J. High Energy Phys.* **0212** 024
[13] Alday L F and Maldacena J M 2007 *J. High Energy Phys.* **0706** 064
[14] Alday L F and Maldacena J M 2009 *J. High Energy Phys.* **0911** 082
[15] Alday L F, Maldacena J M, Sever A and Vieira P 2010 *J. Phys.* A **43** 485401
[16] Pohlmeyer K 1976 *Commun. Math. Phys.* **46** 207
[17] Klose T, Loebbert F and Munkler H 2016 *Phys. Rev.* D **94** 066006
[18] Klose T, Loebbert F and Munkler H 2017 *Nucl. Phys.* B **916** 320
[19] Kruczenski M 2014 *J. High Energy Phys.* **1411** 065
[20] Huang C, He Y and Kruczenski M *J. High Energy Phys.* **1608** 088
[21] Toledo J 2014 Smooth Wilson loops from the continuum limit of null polygons *Preprint* 1410.5896
[22] Trefethen L N 2000 *Spectral Methods in MATLAB* (Oxford: Oxford University)
[23] He Y and Kruczenski M 2017 Minimal area surfaces in AdS_3 through integrability *Preprint* 1705.10037

IOP Conf. Series: Journal of Physics: Conf. Series **965** (2018) 012018 doi:10.1088/1742-6596/965/1/012018

On superintegrable monopole systems

Md Fazlul Hoque, Ian Marquette and Yao-Zhong Zhang
School of Mathematics and Physics, The University of Queensland, Brisbane, QLD-4072, Australia

E-mail: m.hoque@uq.edu.au, i.marquette@uq.edu.au, yzz@maths.uq.edu.au

Abstract. Superintegrable systems with monopole interactions in flat and curved spaces have attracted much attention. For example, models in spaces with a Taub-NUT metric are well-known to admit the Kepler-type symmetries and provide non-trivial generalizations of the usual Kepler problems. In this paper, we overview new families of superintegrable Kepler, MIC-harmonic oscillator and deformed Kepler systems interacting with Yang-Coulomb monopoles in the flat and curved Taub-NUT spaces. We present their higher-order, algebraically independent integrals of motion via the direct and constructive approaches which prove the superintegrability of the models. The integrals form symmetry polynomial algebras of the systems with structure constants involving Casimir operators of certain Lie algebras. Such algebraic approaches provide a deeper understanding to the degeneracies of the energy spectra and connection between wave functions and differential equations and geometry.

1. Introduction

Classical and quantum superintegrable Hamiltonian systems with non-scalar potentials are significant research topics in mathematical physics. Well-known examples are the Kepler problems with additional magnetic monopole interactions known as MICZ-Kepler monopole systems [1, 2].

The family of Taub-NUT metrics provide interesting models for investigating classical and quantum conserved quantities on curved spaces. The geodesic of the generalized Taub-NUT metrics properly explains the motion of well-separated monopole-monopole interactions which provide non-trivial generalizations of the Kepler and harmonic oscillator systems [3, 4, 5, 6, 7, 8, 9, 10, 11]. The 5D Kepler systems interacting with $su(2)$ monopoles or Yang-Coulomb monopoles have been investigated in [12, 13, 14, 15, 16, 17, 18, 19, 20, 21, 22]. Recently, there is a lot of international research interest in generalizing such monopole models to higher dimensions via different approaches [11, 23, 24, 25, 26, 27].

Quadratic algebras are one important class of symmetry algebras for certain superintegrable systems [28]. The quadratic algebra $Q(3)$ with three generators given by 2nd-order integrals of motion of 3D superintegrable models and its realizations in term of deformed oscillators have been presented in [29] by Daskaloyannis. We extended this work to N-dimensional superintegrable quantum systems with non-central scalar potentials [30, 31]. It was shown that the symmetry algebras of the N-dimensional superintegrable models have the intrinsic form $Q(3)\oplus L_1\oplus L_2\oplus\ldots$, where $L_1, L_2, \ldots$ are certain Lie algebras [32]. We also applied the recurrence approach to the N-dimensional superintegrable systems and constructed the corresponding higher-rank polynomial algebras [33]. Recently, we applied the direct and recurrence approaches to superintegrable models with monopole interactions in both the flat and curved Taub-NUT spaces [34, 35, 36].

In this paper, we review some of the main results obtained in the three papers [34, 35, 36] and show the monopole systems have the similar algebra structure $P(3) \oplus \mathcal{L}$ as their symmetry algebra, where $P(3)$ is a polynomial algebra involving three generators and $\mathcal{L} = \mathcal{L}_1 \oplus \mathcal{L}_2, \ldots$ is a direct sum of certain Lie algebras $\mathcal{L}_1, \mathcal{L}_2 \ldots$. Such algebraic structure allowed us to give purely algebraic derivations for the complete energy spectra of the models.

2. Superintegrable monopole models

In this section, we present the new families of Hamiltonian models interacting with monopoles in the flat and curved Taub-NUT spaces which were introduced in our recent papers [34, 35, 36].

Consider the generalized Taub-NUT metric in 3D [3]

$$ds^2 = f(r)d\mathbf{r}^2 + g(r)(d\psi + \mathbf{A}.d\mathbf{r})^2, \tag{1}$$

where $r = \sqrt{x_0^2 + x_1^2 + x_2^2}$, $f(r)$ and $g(r)$ are certain functions of r, ψ is the additional angular variable, $d\mathbf{r}^2 = dx_0^2 + dx_1^2 + dx_2^3$ is the 3D line element and $\mathbf{A}$ is the vector potential for the magnetic field. The following two monopole systems in the 3D Taub-NUT space were introduced in [34, 36]

$$H_K = \frac{1}{2}\left[\frac{1}{f(r)}\left\{\mathbf{p}^2 + \frac{c_0}{2r} + c_4\right\} + \frac{Q^2}{g(r)}\right] + \frac{1}{4f(r)}\left\{\frac{c_2}{r(r+x_0)} + \frac{c_3}{r(r-x_0)}\right\}, \tag{2}$$

$$H_{MIC} = \frac{1}{2}\left[\frac{1}{f(r)}\left\{\mathbf{p}^2 + \frac{c_0 r^2}{2} + c_4\right\} + \frac{Q^2}{g(r)}\right], \tag{3}$$

where $p_i = -i(\partial_i - iA_iQ)$, $Q = -i\partial_\psi$ and $A_1 = 0$, $A_2 = \frac{x_3}{r(r+x_0)}$, $A_3 = \frac{-x_1}{r(r+x_0)}$ are the components of $\mathbf{A}$ which are identified with the monopole interactions; $f(r) = \frac{a_1}{r} + b_1$, $g(r) = \frac{r(a_1+b_1 r)}{1+c_1 r+d_1 r^2}$ for model H_k, $f(r) = a_1 r^2 + b_1$, $g(r) = \frac{r^2(a_1 r^2+b_1)}{1+c_1 r^2+d_1 r^4}$ for model H_{MIC}, and a_1, b_1, c_0, c_1, c_2, c_3, c_4, d_1 are constants. The operators p_i, Q obey the commutation relations

$$[p_i, p_j] = i\epsilon_{ijk}B_kQ, \quad [p_i, Q] = 0, \quad \mathbf{B} = \frac{\mathbf{r}}{r^3}. \tag{4}$$

The systems (2) and (3) are generalized Hartmann systems [37] in the curved Taub-NUT space with abelian monopole interactions. System (2) is referred to as Kepler monopole system which contains the Kaluza-Klein [4, 27] and MICZ monopole system [1, 2] as special cases. System (3) is referred to as MIC-harmonic oscillator monopole system.

We also consider the monopole model in the 5D flat Euclidean space [35],

$$H_{YC} = \frac{1}{2}(\pi_j)^2 + \frac{\hbar^2}{2r^2}\hat{T}^2 - \frac{c_0}{r} + \frac{c_1}{r(r+x_0)} + \frac{c_2}{r(r-x_0)}, \tag{5}$$

where $\pi_j = -i\hbar\frac{\partial}{\partial x_j} - \hbar A_j^a T_a$, $j = 0, 1, 2, 3, 4$, c_1, c_2 are positive real constants and $\{T_a, a = 1, 2, 3\}$ are the $su(2)$ generators which satisfy $[T_a, T_b] = i\epsilon_{abc}T_c$. $\hat{T}^2 = T^aT^a$ is the Casimir operator of $su(2)$ and A_j^a are the components of the monopole potential $\mathbf{A}^a$ of the form

$$A_j^a = \frac{2i}{r(r+x_0)}\tau_{jk}^a x_k. \tag{6}$$

Here τ^a are the 5×5 matrices

$$\tau^1 = \frac{1}{2}\begin{pmatrix} 0 & 0 & 0 \\ 0 & 0 & -i\sigma^1 \\ 0 & i\sigma^1 & 0 \end{pmatrix}, \quad \tau^2 = \frac{1}{2}\begin{pmatrix} 0 & 0 & 0 \\ 0 & 0 & i\sigma^3 \\ 0 & -i\sigma^3 & 0 \end{pmatrix}, \quad \tau^3 = \frac{1}{2}\begin{pmatrix} 0 & 0 & 0 \\ 0 & \sigma^2 & 0 \\ 0 & 0 & \sigma^2 \end{pmatrix}, \tag{7}$$

which satisfy $[\tau^a, \tau^b] = i\epsilon_{abc}\tau^c$ and σ^i are the Pauli matrices. The system (5) is referred to as the 5D Kepler system in Yang-Coulomb monopole field with non-central potentials.

3. Integrals of motion and algebra structures

In this section, we highlight the common feature of the algebra structures generated by the integrals of motion of the superintegrable monopole models (2), (3) and (5).

Consider a superintegrable system H which allows two additional second-order integrals of motion A and B and other first-order integrals of motion $\{F_i\}$. Suppose that A, B and F_i form the quadratic algebra structure $Q(3) \oplus \mathcal{L}$ [29, 34, 35],

$$\left.\begin{aligned} &[A, B] = C, \\ &[A, C] = \gamma\{A, B\} + \epsilon B + \zeta(H, G_i), \\ &[B, C] = -\gamma B^2 + d(H, G_i)A + z(H, G_i) \end{aligned}\right\} \tag{8}$$

$$[F_i, F_j] = \sum_k C_{ij}^k F_k, \tag{9}$$

where γ, ϵ are constants, ζ, d, z are functions of H, G_i with $\{G_i\}$ being the central elements of $Q(3) \oplus \mathcal{L}$, $\mathcal{L} = \mathcal{L}_1 \oplus \cdots \oplus \mathcal{L}_n$ is a semi-simple Lie algebra and C_{ij}^k are the structure constants of $\mathcal{L}$. Thus the structure of the full symmetry algebra for the Hamiltonian H is a direct sum $Q(3) \oplus \mathcal{L}$ of the quadratic algebra $Q(3)$ and the semi-simple Lie algebra $\mathcal{L}$.

Let K be the Casimir operator of this quadratic algebra. Then K commutes with all generators of the algebra, that is, $[K, A] = [K, B] = 0 = [K, C]$, and it has the form [29, 34, 35]

$$K = C^2 - \gamma\{A, B^2\} + (\gamma^2 - \epsilon)B^2 - 2\zeta(H, G_i)B + 2z(H, G_i)A. \tag{10}$$

The Casimir operator K can also be realized as a polynomial of the central elements H and G_i,

$$K = \sum_k C_{ij} G_i^j H^k. \tag{11}$$

The algebraically independent integrals of motion of model (2) in parabolic coordinates are given by [34]

$$A = L^2 + \frac{c_3\xi}{4\eta} + \frac{c_2\eta}{4\xi}, \quad B = M_3 + \frac{\xi - \eta}{\xi + \eta}\left\{\frac{c_3\xi^2 - c_2\eta^2}{2\xi^2\eta - 2\xi\eta^2} + \frac{c_0}{4}\right\}, \tag{12}$$

where $\mathbf{L} = \mathbf{r} \times \mathbf{p} - \frac{\mathbf{r}}{r}Q$, $M_3 = \frac{1}{2}\{(p_1L_2 - p_2L_1) - (L_1p_2 - L_2p_1)\} - \frac{z}{r}(a_1H - \frac{c_1}{2}Q^2)$, and for the model (5) they are [35]

$$A = \sum_{j<k} L_{jk}^2 + \frac{2rc_1}{(r + x_0)} + \frac{2rc_2}{(r - x_0)}, \quad B = M_k + \frac{c_1(r - x_0)}{\hbar^2 r(r + x_0)} + \frac{c_2(r + x_0)}{\hbar^2 r(r - x_0)}, \tag{13}$$

where $L_{jk} = (x_j\pi_k - x_k\pi_j) - r^2\hbar F^a_{jk}T_a$, $j,k = 0,1,2,3,4$, and $M_k = \frac{1}{2}(\pi_j L_{jk} + L_{jk}\pi_j) + c_0\frac{x_k}{r}$ and $F^a_{jk} = \partial_j A^a_k - \partial_k A^a_j + \epsilon_{abc}A^b_j A^c_k$ is the Yang-Mills field tensor. F^a_{jk} and $\pi_j = -i\hbar\frac{\partial}{\partial x_j} - \hbar A^a_j T_a$ obey the commutation relations

$$[\pi_j, x_k] = -i\hbar\delta_{jk}, \quad [\pi_j, \pi_k] = i\hbar^2 F^a_{jk}T_a. \tag{14}$$

By direct computations and using various commutation identities, it was shown [34, 35] that the integrals of motion $\{A, B, C\}$ close to form the quadratic algebra $Q(3)$ (8). For system (2), Q and $L_3 = \sqrt{\xi\eta}\cos\phi p_2 - \sqrt{\xi\eta}\sin\phi p_1 - \frac{\xi+\eta}{\xi-\eta}Q$ are central elements and they form the two dimensional Abelian algebra,

$$[L_3, L_3] = 0 = [Q, Q]. \tag{15}$$

On the other hand, model (5) has the first order integrals of motion L_{jk}, $j,k = 1,2,3,4$ which generate the $so(4)$ Lie algebra. Let $\hat{L}^2 = \sum_{j<k} L^2_{jk}$, $j,k = 1,2,3,4$. Then $\hat{L}^2$ is the Casimir operator of $so(4)$ and is also a central element of $Q(3)$. In the following table, we list the values of the structure constants in $Q(3)$ and K for the models H_K and H_{YC}, respectively,

	$\mathbf{H_K}$	$\mathbf{H_{YC}}$
γ	2	2
ϵ	$c_2 + c_3$	8
d	$8b_1 H - 4d_1 Q^2 - 4c_4$	$8H$
ς	$-4a_1 HQL_3 + 2c_1 Q^3 L_3 + c_0 QL_3 + a_1(c_2 - c_3)H - \frac{1}{2}c_1(c_2 - c_3)Q^2 - \frac{1}{4}c_0(c_2 - c_3)$	$-2c_0(c_1 - c_2) - 4c_0\hat{T}^2$
z	$2a^2H^2 - 2(ac_1 + 2b)HQ^2 - 4bHL_3^2 + \frac{1}{2}(c_1^2 + 4d)Q^4 + 2dQ^2L_3^2 + (4b - ac_0)H + \frac{1}{2}(c_0c_1 + 4c_4 - 4d)Q^2 + 2c_4L_3^2 + \frac{1}{8}(c_0^2 - 16c_4)$	$-4\hat{L}^2H + (16 - 4c_1 - 4c_2)H + 2c_o^2$

Table 1. Coefficients in $Q(3)$ and K.

Thus both models H_K and H_{YC} have the quadratic algebra $Q(3) \oplus \mathcal{L}$ as their symmetry algebra, where $\mathcal{L}$ is $so(4)$ for H_{YC} and $\mathcal{L} = 0$ for H_K. Moreover, it was shown that the quadratic algebra $Q(3)$ for H_K has structure constants involving Q, L_3, while $Q(3) \equiv Q(3; L^{so(4)}, T^{su(2)})$ for H_{YC} with structure constants involving Casimir operators $\hat{L}^2$ of $so(4)$ and $\hat{T}^2$ of $su(2)$.

We now turn our attention to model (3). The integrals of motion for this system were constructed via the recurrence approach [36]. This approach was previously restricted to systems with scalar potentials. We first extended the application of this recurrence method to superintegrable monopole systems in [36]. We here highlight the main results of [36].

The integrals of motion for the Schrödinger Stäckel equivalent $H'_{MIC}\Psi = E'\Psi$ of $H_{MIC}\Psi = E\Psi$, where $E' = c_4 + c_1\nu_2^2 - 2bE$ and $\Psi(r,\theta,\phi,\psi) = \chi(r,\theta)e^{i(\nu_1\phi+\nu_2\psi)}$, were constructed from the suitable combinations of the ladder and shift operators [36],

$$D_1 = K^+_{l-\nu_2+\frac{1}{2},n} J^+_{l-\nu_1+1} J^+_{l-\nu_1} B, \quad D_2 = B J^-_{l-\nu_1} J^-_{l-\nu_1-1} K^-_{l-\nu_2+\frac{1}{2},n}, \tag{16}$$

where $K^\pm_{l-\nu_2+\frac{1}{2},n}$ are the ladder operators, $J^\pm_{l-\nu_1}$ are the shift operators and $B = \sqrt{\mathbf{L}^2 + \frac{1}{4}}$ is a well-defined operator [7].

The action of the combinations of operators D_1D_2 and D_2D_1 can be obtained on the wave functions. D_1 and D_2 form an algebraically independent set of differential operators and they close to form the higher-rank polynomial algebra $P(3)\oplus\mathcal{L}$ (see details in Ref. [36]),

$$\begin{aligned}&[D_1,H'_{MIC}]=0=[D_2,H'_{MIC}],\quad [B,D_1]=2D_1,\quad [B,D_2]=-2D_2,\\ &D_1D_2=f_1(H'_{MIC},B,L_3,Q),\quad D_2D_1=f_2(H'_{MIC},B,L_3,Q),\end{aligned}\tag{17}$$

where f_1 and f_2 are polynomials in the generators H'_{MIC}, B, L_3 and Q.

4. Unirreps and spectra

In this section, we focus on the finite dimensional unitary representations of the quadratic algebras and apply the results to derive the energy spectra for systems (2), (3) and (5).

The quadratic algebra $Q(3)$ can be realized in terms of the deformed oscillator algebra

$$[\aleph,b^\dagger]=b^\dagger,\quad [\aleph,b]=-b,\quad bb^\dagger=\Phi(\aleph+1),\quad b^\dagger b=\Phi(\aleph),\tag{18}$$

with the structure function $\Phi(x)$ given by [34, 35]

$$\begin{aligned}\Phi(x;u,H)\ &=-3072\gamma^6K\{2(x+u)-1\}^2+48d\gamma^8\{2(x+u)-3\}^2\{2(x+u)-1\}^4\\ &\times\{2(x+u)+1\}^2+12288\gamma^4\zeta^2+128\gamma^6(2\gamma z-d\epsilon)\{2(x+u)-1\}^2\{12(x+u)^2\\ &-12(x+u)-1\}-256\gamma^4(2d\epsilon\gamma^2+12\epsilon\gamma z-4\gamma^3z-3d\epsilon^2)\{2(x+u)-1\}^2.\end{aligned}\tag{19}$$

The corresponding values of the structure constants are given in Table 1 for models (2) and (5), respectively.

On the other hand, for model (3), the symmetry algebra (17) was already in the form of the deformed oscillator algebra (18) by letting $\aleph=\frac{B}{2}$, $b^\dagger=D_1$ and $b=D_2$. The structure function is given by [36]

$$\begin{aligned}\Phi(x;u,H'_{MIC})\ &=\frac{(2x+u-2)}{256}[2(2x+u)-3]^2[2(2x+u)-2L_3-1][2(2x+u)-1]\\ &\times[2(2x+u)-2L_3-3][2(2x+u)-5][2(2x+u)-2L_3-4Q-3]\\ &\times[H'_{MIC}-2\varepsilon\{(2x+u)-Q-1\}][2(2x+u)-4Q-1][H'_{MIC}+2\varepsilon\{(2x+u)\\ &-Q-1\}][2(2x+u)-4Q-3][2(2x+u)-2L_3-4Q-1].\end{aligned}\tag{20}$$

Using appropriate Fock space we can evaluate the structure functions (19) and (20) and obtain, for model (2), (5) and the Stäckel equivalent of model (3), respectively,

$$\begin{aligned}\Phi(x;u,E_K)\ &=\ -3145728(c_4-2bE_K+dq_2^2)[x+u-\tau_{-+}][x+u-\tau_{+-}][x+u-\tau_{--}]\\ &\times[x+u-\tau_{++}]\left[x+u-\left(\frac{1}{2}-\beta(E_K)\right)\right]\left[x+u-\left(\frac{1}{2}+\beta(E_k)\right)\right],\end{aligned}\tag{21}$$

where $\tau_{\pm\pm}=1\pm m_1\pm m_2$, $m_1^2=c_2+(q_1-q_2)^2$, $m_2^2=c_3+(q_1+q_2)^2$, $\beta(E_K)=\frac{4aE_K-2c_1q_2^2-c_0}{4\sqrt{c_4-2bE_K+dq_2^2}}$;

$$\begin{aligned}\Phi(x;u,E_{YC})\ &=[x+u-\mu_{++}][x+u-\mu_{+-}][x+u-\mu_{-+}][x+u-\mu_{--}]\\ &\times[x+u-\eta(E_{YC})][x+u-\eta(E_{YC})],\end{aligned}\tag{22}$$

where $\mu_{\pm\pm}=\frac{1}{2}(1\pm n_1\pm n_2)$, $n_1^2=1+2c_1+\hbar^2l_4(l_4+2)+2\hbar^2T(T+1)$, $n_2^2=1+2c_2+\hbar^2l_4(l_4+2)-2\hbar^2T(T+1)$, $\eta(E_{YC})=\frac{1}{2}+\frac{c_0}{\sqrt{-2E_{YC}}}$;

$$\begin{aligned}\Phi(x;u,E'_{MIC})\ &=\frac{(2x+u-2)}{256}[2(2x+u)-3]^2[2(2x+u)-2q_2-1][2(2x+u)-1]\\ &\times[2(2x+u)-2q_2-3][2(2x+u)-5][2(2x+u)-2q_2-4q_1-3]\\ &\times[E'_{MIC}-2\varepsilon\{(2x+u)-q_1-1\}][2(2x+u)-4q_1-1][E'_{MIC}+2\varepsilon\{(2x+u)\\ &-q_1-1\}][2(2x+u)-4q_1-3][2(2x+u)-2q_2-4q_1-1].\end{aligned}\tag{23}$$

For finite dimensional unitary representations, we have the following constraints on the structure functions (21), (22) and (23),

$$\Phi(p+1;u,E)=0,\quad \Phi(0;u,E)=0,\quad \Phi(x)>0,\quad \forall x>0, \tag{24}$$

where p is a positive integer. This gives $(p+1)$-dimensional unitary representations and their solution gives the corresponding energy E_K [34], E_{YC} [35], E_{MIC} [36], for $\epsilon_1=\pm1$, $\epsilon_2=\pm1$,

$$2\beta(E_K)=2+2p+\epsilon_1 m_1+\epsilon_2 m_2; \tag{25}$$

$$E_{YC}=-\frac{c_0^2}{2(p+1+\frac{n_1+n_2}{2})^2}; \tag{26}$$

$$E'_{MIC}=-\varepsilon(4p-2q_1+2\epsilon_2 q_2+3). \tag{27}$$

By the coupling constant metamorphosis $E'\leftrightarrow c_4+c_1\nu_2^2-2bE$ and $\varepsilon^2\leftrightarrow\frac{c_0}{2}-2aE+d\nu_2^2$, we obtain the energy of the original Hamiltonian (3) from (27)

$$\frac{2bE_{MIC}-c_1q_2^2-c_4}{\sqrt{\frac{c_0}{2}-2aE_{MIC}+dq_2^2}}=4p-2q_1+2\varepsilon_2 q_2+3. \tag{28}$$

The structure functions are positive for $\epsilon_1=1$, $\epsilon_2=1$ and $m_1,m_2>0$. The energies have degeneracy of $p+1$ only when these other quantum numbers would be fixed. The total number of degeneracies may be calculated by taking into account the further constraints on these quantum numbers.

Acknowledgements The research of FH was supported by International Postgraduate Research Scholarship and Australian Postgraduate Award. IM was supported by the Australian Research Council, Discovery Project DP 160101376, YZZ was partially supported by the Australian Research Council, Discovery Project DP 140101492.

5. Conclusion

In this paper, we have reviewed families of superintegrable Hamiltonian systems with Yang-Coulomb monopole interactions in the flat and curved Taub-NUT spaces [34, 35, 36] and highlighted their unified algebra structures of the form $P(3)\oplus\mathcal{L}$. We have applied the direct and recurrence approaches to the superintegrable Hamiltonian systems interacting with monopoles and have given the algebraic derivations of their energy spectra. Our work represents the first application of the recurrence approach to models beyond scalar potentials such as the superintegrable monopole systems [36]. The direct and recurrence approaches could be extended to other N-dimensional MICZ-Kepler monopole type problems (e.g. [25, 38]).

References

[1] McIntosh H V, Cisneros A 1970, *Journ. Math. Phys.* **11**, 896
[2] Zwanziger D 1968, *Phys. Rev.* **176**, 1480
[3] Gross D J, Perry M J 1983, *Nucl. Phys.* **B226**, 29
[4] Gibbons G W, Manton N S 1986, *Nucl. Phys.* **B274**, 183
[5] Feher Gy L, Horvathy P A 1987, *Phys. Lett.* **B183**, 182
[6] Manton N S 1985, *Phys. Lett.* **B154**, 397
[7] Labelle S, Mayrand M, Vinet L 1991, *Journ. Math. Phys.* **32**, 1516
[8] Grosche C, Pogosyan G S, Sissakian A N 1997, *Ann. Phys.* **6**, 144
[9] Iwai T, Uwano Y 1986, *Journ. Math. Phys.* **27**, 1523
[10] Iwai T, Katayama N 1994, *Journ. Phys. A: Math. Gen.* **27**, 3179
[11] Marquette I 2011, *Journ. Phys. A: Math. Theor.* **44**, 235203
[12] Mardoyan L G, Sissakian A N, Ter-Antonyan V M 1996, *JINR Dubna Preprint E2* **96**, 24

[13] Mardoyan L G, Sissakian A N, Ter-Antonyan V M 1998, *Phys. Atom. Nucl.* **61**, 1746
[14] Mardoyan L G, Sissakian A N, Ter-Antonyan V M 1999, *Mod. Phys. Lett.* **A14**, 1303
[15] Pletyukhov M V, Tolkachev E A 1999, *Journ. Phys.* **A32**, L249
[16] Mardoyan L G, Sissakian A N, Ter-Antonyan V M 2000, *Theor. Math. Phys.* **123**, 451
[17] Pletyukhov M V, Tolkachev E A 2000, *Journ. Math. Phys.* **41**, 187
[18] Kalnins E G, Miller W Jr, Pogosyan G S 2000, *Journ. Math. Phys.* **41**, 2629
[19] Nersessian A, Pogosyan G 2001, *Phys. Rev.* **A63**, 020103
[20] Mardoyan L G 2002, *Phys. At. Nucl.* **65**, 1063
[21] Karayan Kh H, Mardoyan L G, Ter-Antonyan V M 2002, *Phys. Part. Nucl.* **33**, 202
[22] Bellucci S, Toppan J, Yeghikyan V 2010, *Journ. Phys.* **A43**, 045205
[23] Mardoyan L G 2003, *Journ. Math. Phys.* **44**, 4981
[24] Mardoyan L G, Petrosyan M G 2007, *Phys. At. Nucl.* **70**, 572
[25] Meng G 2007, *Journ. Math. Phys.* **48**, 032105
[26] Ranjan G P 2008, *Mod. Phys. Lett.* **A23**, 895
[27] Marquette I 2010, *Journ. Math. Phys.* **51**, 102105
[28] Granovskii Ya, Zhedanov A S, Lutzenko I M 1991, *Journ. Phys. A: Math. Theor.* **24**, 3887
[29] Daskaloyannis C 2001, *Journ. Math. Phys.* **42**, 1100
[30] Hoque M F, Marquette I, Zhang Y Z 2015, *Journ. Phys. A: Math. Theor.* **48**, 185201
[31] Hoque M F, Marquette I, Zhang Y Z 2015, *Journ. Phys. A: Math. Theor.* **48**, 445207
[32] Hoque M F, Marquette I, Zhang Y Z 2016, *Journ. Phys.: Conf. Seri.* **670**, 012024
[33] Hoque M F, Marquette I, Zhang Y Z 2016, *Journ. Phys. A: Math. Theor.* **49**, 125201
[34] Hoque M F, Marquette I, Zhang Y Z 2016, *Journ. Math. Phys.* **57**, 092104
[35] Hoque M F, Marquette I, Zhang Y Z 2017, *Ann. Phys.* **380**, 121
[36] Hoque M F, Marquette I, Zhang Y Z 2016, "Recurrence approach and higher order polynomial algebras for superintegrable monopole systems", *arXiv:* 1605.06213
[37] Hartmann H 1972, *Theor. Chim. Acta.* **24**, 201
[38] Krivonos S, Nersessian A, Ohanyan V 2007, *Phys. Rev.* **D75**, 085002

IOP Conf. Series: Journal of Physics: Conf. Series **965** (2018) 012019 doi:10.1088/1742-6596/965/1/012019

A conditionally integrable Schrödinger potential of a bi-confluent Heun class

T A Ishkhanyan[1,2,3], A M Manukyan[1,4] and A M Ishkhanyan[1,4]

[1] Institute for Physical Research, NAS of Armenia, Ashtarak, 0203Armenia
[2] Moscow Institute of Physics and Technology, Dolgoprudny, 141700 Russia
[3] LICB, UMR CNRS 6303-Université de Bourgogne, Dijon, 21078 France
[4] Russian-Armenian University, H. Emin 123, Yerevan, 0051Armenia

Abstract. We present a bi-confluent Heun potential for the Schrödinger equation involving inverse fractional powers and a repulsive centrifugal-barrier term the strength of which is fixed to a constant. This is an infinite potential well defined on the positive half-axis. Each of the fundamental solutions for this conditionally integrable potential is written as an irreducible linear combination of two Hermite functions of a shifted and scaled argument. We present the general solution of the problem, derive the exact energy spectrum equation and construct a highly accurate approximation for the bound-state energy levels.

1. Introduction

We introduce a generalization of the recently presented exactly solvable inverse square root potential for the one-dimensional stationary Schrödinger equation [1]. This is a conditionally integrable singular potential which belongs to a bi-confluent Heun family of potentials involving negative half-integer powers of the space coordinate [2,3]. The term "conditionally integrable" is conventionally used to indicate that the parameters involved in the potential are not varied independently or a potential parameter is fixed to a constant. Well known examples of such potentials discussed by many authors in the past are the two Stillinger confluent hypergeometric potentials [4] and the Dutt-Khare-Varshni ordinary hypergeometric potential [5] (for other examples see, e.g., [6-12]).

The potential we introduce is given as

$$V = V_0 + \frac{5\hbar^2/(32m)}{x^2} + \frac{V_1}{x^{3/2}} - \frac{16m^2V_1^3/\hbar^4}{\sqrt{x}}. \tag{1}$$

The shape of the potential is shown in figure 1. For $V_1 > 0$ this is a potential well defined on the positive half-axis $x > 0$ so that it sustains bound states. Because of the long-range component $\sim x^{-1/2}$ the number of the bound states is infinite. The potential is a member of the bi-confluent Heun family with $m_1 = -1$ (see [2,3]), which generalizes the second Stillinger [4] and the first Exton [7] potentials as well as the potential by López-Ortega [11].

We present the general solution of the one-dimensional Schrödinger equation for a particle of mass m and energy E:

$$\frac{d^2\psi}{dx^2} + \frac{2m}{\hbar^2}\left(E - V(x)\right)\psi = 0, \tag{2}$$

where $\hbar$ is the reduced Planck's constant, derive the exact equation for the energy spectrum and construct a highly accurate approximation for the bound-state energy levels.

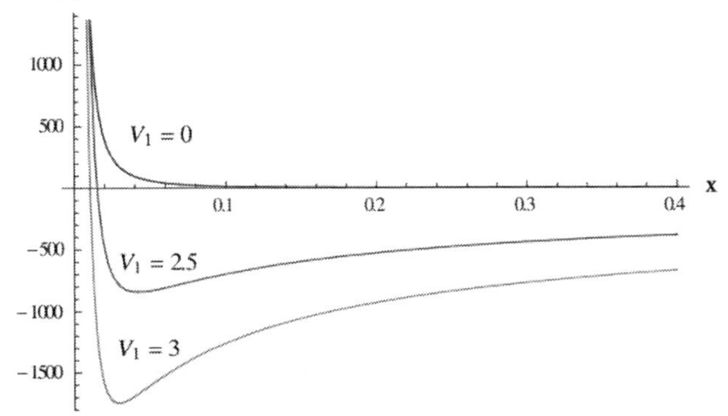

Figure 1. Potential (2) for $V_1 \geq 0$ ($m = \hbar = 1$, $V_0 = 0$).

A peculiarity of the solution is that none of the two independent fundamental solutions composing the general solution of the problem is written through the widely used one-term ansatz involving a single hypergeometric function. Rather, both fundamental solutions are written as irreducible linear combinations of two Hermite functions of a shifted and scaled argument.

We show that for bound states only one of the two fundamental solutions survives. This leads to an energy spectrum equation involving two Hermite functions of an energy-dependent auxiliary parameter. Applying an accurate approximation for these Hermite functions, we derive a highly accurate approximation for the energy levels.

2. The general solution

The solution of the Schrödinger equation for potential (1) is derived by reducing equation (2) to the bi-confluent Heun equation [13-15], which is one of the four confluent forms of the general Heun equation [16]. This is a second-order linear differential equation which has one regular singularity conventionally located at the origin and an irregular singularity of rank 2 conventionally located at the infinity. In an auxiliary five-parametric representation the bi-confluent Heun equation is written as

$$\frac{d^2u}{dz^2} + \left(\frac{\gamma}{z} + \delta + \varepsilon z\right)\frac{du}{dz} + \frac{\alpha z - q}{z}u = 0. \tag{3}$$

This equation is widely encountered in various domains of current physics research such as quantum mechanics, quantum gravity, general relativity, solid state physics, etc. (see [13-15]). For example, in particle physics the equation appears when studying the quarkonium and related models describing the confinement of quarks and antiquarks [17,18].

The reduction of the Schrödinger equation to the bi-confluent Heun equation is presented in Refs. [2,3] (see also [19]). The next step is expanding the solution of the bi-confluent Heun equation (3) in terms of the Hermite functions as described in [3]. Then, by requiring the series to terminate on the third term and allowing the potential parameters to be dependent, we arrive at the following potential:

$$V = V_0 + \frac{5\hbar^2}{32mx^2} + \frac{V_1}{x^{3/2}} + \frac{V_2}{x} + \frac{8mV_1\left(-2mV_1^2 + \hbar^2 V_2\right)}{\hbar^4\sqrt{x}}. \tag{4}$$

Choosing here $V_2 = 2mV_1^2/\hbar^2$ cancels the last term thus reproducing the López-Ortega potential [1] with an additional centrifugal-barrier term $5\hbar^2 x^{-2}/(32m)$, while by putting $V_2 = 0$ we cancel the coulombic term and get potential (1).

Omitting the details in order to avoid the text overlap (since the calculations are merely technical and exactly follow the lines of [3]), the resultant three-term Hermite-function solution of the bi-confluent Heun equation (3) for the given set of the potential parameters leads to the following general solution of the Schrödinger equation. The two independent fundamental solutions composing the general solution for real positive $V_1 > 0$ and $E < V_0$ are written as

$$\psi_F(x) = x^{-1/4} e^{-y^2/2} \left(H_{a+1/2}(y) + \left(\sqrt{2a} + A\sqrt[6]{2a}\right)\left(1 + \frac{\varepsilon\hbar^2\sqrt{x}}{4mV_1}\right) H_{a-1/2}(y) \right), \tag{5}$$

where H is the Hermite function of the shifted and scaled argument

$$y = \sqrt{-\varepsilon x} - \sqrt{2a} \tag{6}$$

and the involved parameters a and ε are given as

$$a = -\frac{2^{11} m^6 V_1^6}{\hbar^{12}\varepsilon^3}, \quad \varepsilon = \pm\sqrt{\frac{8m(-E+V_0)}{\hbar^2}}. \tag{7}$$

The parameter A involved in the fundamental solutions (5) is equal to 1 if the minus sign for ε is chosen in the second equation of (7) and $A = e^{4i\pi/3}$ for the plus sign of ε. The general solution of the Schrödinger equation is thus written as

$$\psi(x) = c_1 \psi_F^- \Big|_{A=1} + c_2 \psi_F^+ \Big|_{A=e^{4i\pi/3}}, \tag{8}$$

where $c_{1,2}$ are arbitrary constants. We note that this solution can alternatively be written in terms of the confluent hypergeometric functions.

3. Bound states and energy spectrum

The energy spectrum is derived by demanding the wave function to vanish at the origin and at infinity (see the discussion in [20]). An observation now is that the second fundamental solution diverges at infinity for any set of $V_1 > 0$ and energy $E < V_0$. This is readily understood if one recalls that the Hermite function behaves at infinity as a power of its argument: $H_\nu(y) \sim 2^\nu y^\nu$, so that the asymptote of the fundamental solution (5) at infinity is governed by the pre-factor $e^{-y^2/2}$. Since $-y^2 \sim \varepsilon x$, it is understood that the second fundamental solution ψ_F^+, for which $\varepsilon > 0$, exponentially diverges, while the first fundamental solution ψ_F^- with $\varepsilon < 0$ always vanishes at $x \to +\infty$. Hence, the requirement for the wave function to vanish at infinity is satisfied only if $c_2 = 0$. The requirement for the wave function to vanish at the origin then immediately leads to the following equation:

$$H_{a+1/2}\left(-\sqrt{2a}\right) + \left(\sqrt{2a} + \sqrt[6]{2a}\right) H_{a-1/2}\left(-\sqrt{2a}\right) = 0. \tag{9}$$

This is the exact equation for the energy spectrum, which is constructed through relations (7). The equation possesses a countable infinite set of positive roots a_n. Except the smallest root $a_0 = 1/2$, each of these roots stands for a bound state energy. The root $a_0 = 1/2$ should be disregarded because for this a the wave function becomes zero. The next root, which is close to $3/2$, produces the first bound state with a single maximum. We prescribe to this root the number 1.

An accurate approximation for the spectrum equation (9) is achieved by noting that the arguments and indexes of the involved Hermite functions $H_\nu(z)$ belong to the left transient layer for which $z \approx -\sqrt{2\nu}$ [21]. We then apply the Airy-function approximation (which is further reduced to trigonometric functions) for the Hermite functions suggested in [22] to derive the approximation

$$F \equiv 1 + \left(\sqrt{2a} + \sqrt[6]{2a}\right) \frac{H_{a-1/2}\left(-\sqrt{2a}\right)}{H_{a+1/2}\left(-\sqrt{2a}\right)} \approx \frac{\sin(\pi a - \pi/3) - 6B_0 2^{-1/3} \sin(\pi a + \pi/3)}{\sin(\pi a - \pi/3) + 6B_0 a^{1/3} \sin(\pi a + \pi/3)}, \tag{10}$$

where

$$B_0 = \frac{\Gamma(1/3)}{6\sqrt[3]{3}\,\Gamma(2/3)}. \tag{11}$$

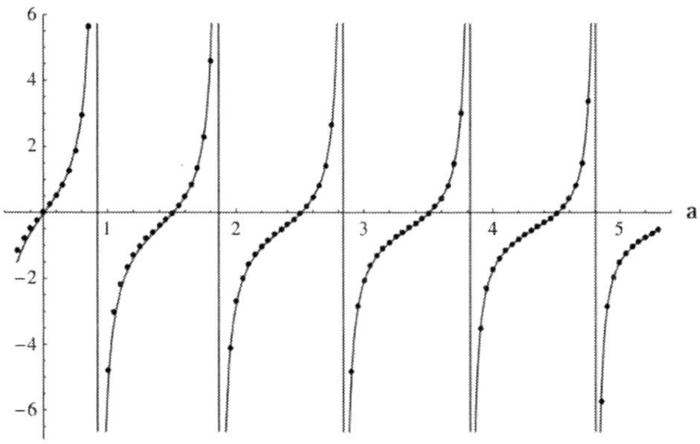

Figure 2. Filled circles – exact numerical result, solid curves – approximation (10).

The derived approximation is compared with the exact function F in figure 2. As seen, we have a highly accurate result. Hence, as it follows from equation (9), the spectrum equation is well approximated by the simple trigonometric equation

$$\sin(\pi a - \pi/3) - 6B_0 2^{-1/3} \sin(\pi a + \pi/3) = 0. \tag{12}$$

It is immediately seen from figure 2 that the roots of this equation (and thus the roots of the exact spectrum equation (9)) are close to positive half-integers. This result is readily deduced if we note that $6B_0 2^{-1/3} \approx 1$ so that the approximate spectrum equation is finally reduced to the very simple equation $-\sqrt{3}\cos(a\pi) = 0$. Thus, $a_n = n + 1/2$, $n \in \mathbb{N}$. According to equation (7), the energy spectrum is then given by the compact formula

$$E_n \approx -\frac{32m^3 V_1^4 / \hbar^6}{(2n+1)^{2/3}}, \quad n = 1, 2, 3, \ldots. \tag{13}$$

It is understood that this result stands for the semiclassical limit for large $n \to \infty$, the Maslov correction index [22,23] being 1/2. However, this is a fairly good approximation for all levels. The relative error is less than 2×10^{-3} even for the ground state (see the inset of figure 3). Having this result, it is not difficult to calculate the next approximations, e.g., by expanding equation (10) in terms of powers of $(2n+1)^{-2/3}$. The second approximation then provides relative error less than 10^{-5} for all energy levels. The first three non-normalized wave functions are shown in figure 4.

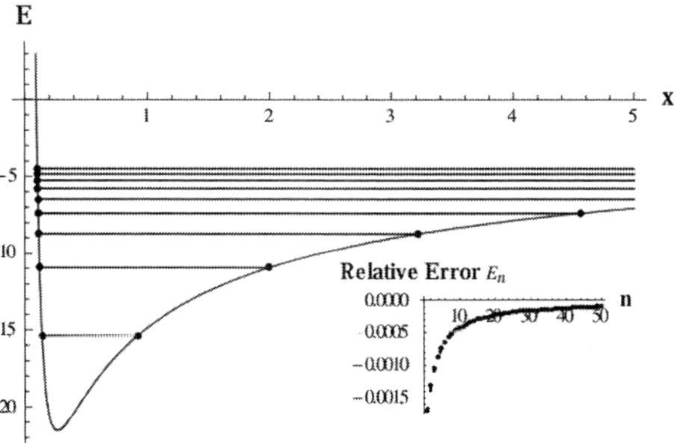

Figure 3. Energy spectrum of potential (1) for $V_1 = 1$ ($m = \hbar = 1$, $V_0 = 0$).

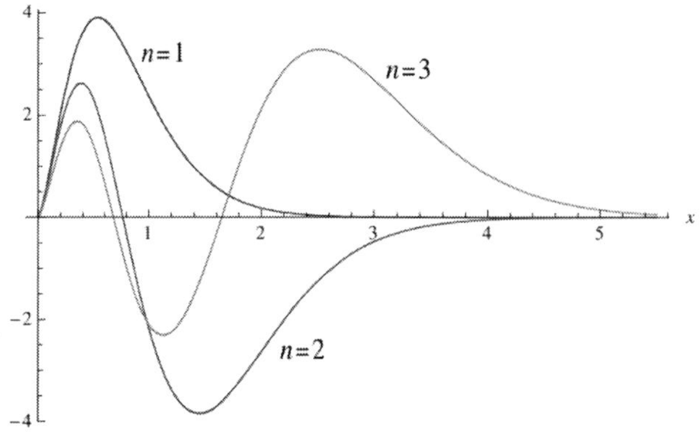

Figure 4. The first three non-normalized wave functions for $V_1 = 1$ ($m = \hbar = 1$, $V_0 = 0$).

4. Discussion

Discussing the reducibility of the Schrödinger equation to the bi-confluent Heun equation, Lemieux and Bose have presented five families of potentials [2] of which four involve power-law terms and the fifth one is a generalization of the Morse potential which involves exponentials [24]. These are rich families including, as particular cases, the three exactly solvable classical confluent hypergeometric potentials (harmonic oscillator, Coulomb and Morse) and many of the known conditionally integrable potentials discussed in the past.

The potential we have introduced belongs to the first Lemieux-Bose family which involves negative integer and half-integer power-law terms. This is a remarkable family the known members of which include, for instance, the second Stillinger [4] and the first Exton [7] potentials as well as the inverse square root potential [1], the potential by López-Ortega [11] and the recent generalization of the latter two potentials [12].

Because of its wide appearance, the solutions of the bi-confluent Heun equation have been studied by many authors [13-15]. Among the recent developments, of particular interest for this research are the expansions of the solutions in terms of special functions other than simple powers [25,26], in particular, the expansion in terms of the Hermite functions of a shifted and scaled argument [3]. The latter expansion indicates that there exists a large set (presumably, infinite) of exactly or conditionally integrable potentials the solution for which is written as a linear combination with constant coefficients of a finite number of Hermite functions. The first potentials of this set are the classical harmonic oscillator potential and the two potentials by Stillinger [4] the solution for which involves a single Hermite function. The next come the potentials for which the fundamental solutions present irreducible combinations of two Hermite functions. These are the two Exton potentials [7] the first of which involves the inverse square root potential [1] and its conditionally integrable generalization [12] which includes the López-Ortega supersymmetric pair of potentials [11].

The potential we have presented is just the next in the list, i.e., the solution of the Schrödinger equation for this potential is a linear combination with constant coefficients of three Hermite functions. Because these are contiguous functions, using the known recurrence relations, the combination is reduced to a combination with non-constant coefficients of two Hermite functions. We have presented the general solution of the Schrödinger equation for this potential and have derived the exact energy spectrum equation for the bound states. Using an accurate approximation for the involved Hermite functions, we have approximated this equation by a simple transcendental equation involving trigonometric functions. We have seen that the solution of the latter equation provides a highly accurate approximation for all bound-state energy levels: the relative error is less than 2×10^{-4} even for the ground state.

Acknowledgments
This research has been conducted within the scope of the International Associated Laboratory IRMAS (CNRS-France & SCS-Armenia). The work has been supported by the Armenian State Committee of Science (Grant No. 15T-1C323) and the Armenian National Science and Education Fund (ANSEF Grant No. PS-4558). T.A. Ishkhanyan acknowledges the support from SPIE through a 2017 Optics and Photonics Education Scholarship and thanks the French Embassy in Armenia for a doctoral grant.

References
[1] Ishkhanyan A M 2015 *EPL* **112** 10006
[2] Lemieux A and Bose A K 1969 *Ann. Inst. Henri Poincaré* A **10** 259–70
[3] Ishkhanyan T A and Ishkhanyan A M 2017 *Ann. Phys.* **383** 79–91
[4] Stillinger F H 1979 *J. Math. Phys.* **20** 1891–95
[5] Dutt R, Khare A and Varshni Y P 1995 *J. Phys.* A **28** L107–13
[6] Ginocchio J N 1984 *Ann. Phys.* **152** 203–19
[7] Exton H 1995 *J. Phys.* A **28** 6739–41
[8] Grosche C 1996 *J. Phys.* A **29** 365–83
[9] Junker G and Roy P 1997 *Phys. Lett.* **232** 155–61
[10] Lévai G, Konya B, Papp Z 1998 *J. Math. Phys.* **39** 5811–23
[11] López-Ortega A 2015 *Phys. Scr.* **90** 085202
[12] Ishkhanyan A M 2016 *Phys. Lett.* A **380** 3786–90
[13] Ronveaux A 1995 *Heun's Differential Equations* (London: Oxford University Press)
[14] Slavyanov S Yu and Lay S.Yu 2000 *Special Functions* (Oxford: Oxford University Press)
[15] Olver F W J, Lozier D W, Boisvert R F and Clark C W 2010 *NIST Handbook of Mathematical Functions* (New York: Cambridge University Press).
[16] Heun K 1889 *Math. Ann.* **33** 161–79
[17] Hamzavi M and Rajabi A A 2013 *Ann. Phys.* **334** 316-20
[18] Karwowski J and Witek H A 2014 *Theor. Chem. Accounts* **133** 1494
[19] Ishkhanyan A and Krainov V 2016 *Eur. Phys. J. Plus* **131** 342
[20] Znojil M 2000 *Phys. Rev.* A **61** 066101
[21] Szegö G 1975 *Orthogonal Polynomials* 4th ed (Providence: Amer. Math. Soc.)
[22] Ishkhanyan A M and Krainov V P 2017 *JETP Lett.* **105** 43–6
[23] Quigg C and Rosner J L 1979 *Phys. Rep.* **56** 167–235
[24] Morse P M 1929 *Phys. Rev.* **34** 57–64
[25] Leroy C and Ishkhanyan A M 2015 *Integral Transforms and Special Functions* **26**, 451–9
[26] Ishkhanyan T A, Pashayan-Leroy Y, Gevorgyan M R, Leroy C and Ishkhanyan A M 2016 *J. Contemp. Phys. (Arm. Ac. Sci.)* **51** 229–36

IOP Conf. Series: Journal of Physics: Conf. Series **965** (2018) 012020 doi:10.1088/1742-6596/965/1/012020

Quantization of noncompact coverings and its physical applications

Petr Ivankov
Researcher, *Dinamika*, *Moscow State University*, Moscow
E-mail: monster.ivankov@gmail.com

Abstract. A rigorous algebraic definition of noncommutative coverings is developed. In the case of commutative algebras this definition is equivalent to the classical definition of topological coverings of locally compact spaces. The theory has following nontrivial applications:

- Coverings of continuous trace algebras,
- Coverings of noncommutative tori,
- Coverings of the quantum $SU(2)$ group,
- Coverings of foliations,
- Coverings of isospectral deformations of Spin – manifolds.

The theory supplies the rigorous definition of noncommutative Wilson lines.

1. Motivation. Preliminaries

Gelfand-Naĭmark theorem [2] states the correspondence between locally compact Hausdorff topological spaces and commutative C^*-algebras.

Theorem 1.1. *[2] (Gelfand-Naĭmark). Let A be a commutative C^*-algebra and let $\mathcal{X}$ be the spectrum of A. There is the natural $*$-isomorphism $\gamma : A \to C_0(\mathcal{X})$.*

So any (noncommutative) C^*-algebra may be regarded as a generalized (noncommutative) locally compact Hausdorff topological space. But $*$-homomorphisms are not good analogs of continuous maps, because there is no a $*$-homomorphism $C_0(\mathcal{X}) \to C_0\left(\widetilde{\mathcal{X}}\right)$ which corresponds to a continuous map from a non-compact topological space $\widetilde{\mathcal{X}}$ to a compact one $\mathcal{X}$. However there is the natural *-homomorphism $C_0(\mathcal{X}) \to C_b\left(\widetilde{\mathcal{X}}\right)$. Following theorem gives a pure algebraic description of finite-fold coverings of compact spaces.

Theorem 1.2. *[9] Suppose $\mathcal{X}$ and $\mathcal{Y}$ are compact Hausdorff connected spaces and $p : \mathcal{Y} \to \mathcal{X}$ is a continuous surjection. If $C(\mathcal{Y})$ is a projective finitely generated Hilbert module over $C(\mathcal{X})$ with respect to the action*

$$(f\xi)(y) = f(y)\xi(p(y)),\ f \in C(\mathcal{Y}),\ \xi \in C(\mathcal{X}),$$

then p is a finite-fold covering.

IOP Conf. Series: Journal of Physics: Conf. Series **965** (2018) 012020 doi:10.1088/1742-6596/965/1/012020

This article contains pure algebraic generalizations of following topological objects:

- Coverings of noncompact spaces,
- Infinite coverings.

The words "set", "family" and "collection" are synonyms.

Following table contains special symbols.

Symbol	Meaning
A_+	Cone of positive elements of C^*- algebra, i.e. $A_+ = \{a \in A \mid a \geq 0\}$
A^G	Algebra of G - invariants, i.e. $A^G = \{a \in A \mid ga = a, \forall g \in G\}$
$\mathrm{Aut}(A)$	Group of * - automorphisms of C^*- algebra A
$B(\mathcal{H})$	Algebra of bounded operators on a Hilbert space $\mathcal{H}$
$\mathbb{C}$ (resp. $\mathbb{R}$)	Field of complex (resp. real) numbers
$C(\mathcal{X})$	C^*- algebra of continuous complex valued functions on a compact space $\mathcal{X}$
$C_0(\mathcal{X})$	C^*- algebra of continuous complex valued functions on a locally compact topological space $\mathcal{X}$ equal to 0 at infinity
$C_c(\mathcal{X})$	Algebra of continuous complex valued functions on a topological space $\mathcal{X}$ with compact support
$C_b(\mathcal{X})$	C^*- algebra of bounded continuous complex valued functions on a locally compact topological space $\mathcal{X}$
$G\left(\widetilde{\mathcal{X}} \mid \mathcal{X}\right)$	Group of covering transformations of covering $\widetilde{\mathcal{X}} \to \mathcal{X}$ [10]
δ_{ij}	Delta symbol. If $i = j$ then $\delta_{ij} = 1$. If $i \neq j$ then $\delta_{ij} = 0$
$\mathcal{H}$	Hilbert space
$\mathcal{K} = \mathcal{K}(\mathcal{H})$	C^*- algebra of compact operators on the separable Hilbert space $\mathcal{H}$
$\varinjlim$	Direct limit
$\varprojlim$	Inverse limit
$M(A)$	A multiplier algebra of C^*-algebra A
$\mathbb{M}_n(A)$	The $n \times n$ matrix algebra over C^*- algebra A
$\mathbb{N}$	A set of positive integer numbers
$\mathbb{N}^0$	A set of nonnegative integer numbers
$U(A) \subset A$	Group of unitary operators of algebra A
$\mathbb{Z}$	Ring of integers
$\mathbb{Z}_n$	Ring of integers modulo n
$\overline{k} \in \mathbb{Z}_n$	An element in $\mathbb{Z}_n$ represented by $k \in \mathbb{Z}$
$X \backslash A$	Difference of sets $X \backslash A = \{x \in X \mid x \notin A\}$
$\|X\|$	Cardinal number of a finite set X
$[x]$	The range projection of element x of a von Neumann algebra.
$f\|_{A'}$	Restriction of a map $f : A \to B$ to $A' \subset A$, i.e. $f\|_{A'} : A' \to B$

IOP Conf. Series: Journal of Physics: Conf. Series **965** (2018) 012020 doi:10.1088/1742-6596/965/1/012020

2. Prototype. Inverse limits of coverings in topology

This subsection is concerned with a topological construction of the inverse limit in the category of coverings.

Definition 2.1. [10] Let $\widetilde{\pi} : \widetilde{\mathcal{X}} \to \mathcal{X}$ be a continuous map. An open subset $\mathcal{U} \subset \mathcal{X}$ is said to be *evenly covered* by $\widetilde{\pi}$ if $\widetilde{\pi}^{-1}(\mathcal{U})$ is the disjoint union of open subsets of $\widetilde{\mathcal{X}}$ each of which is mapped homeomorphically onto $\mathcal{U}$ by $\widetilde{\pi}$. A continuous map $\widetilde{\pi} : \widetilde{\mathcal{X}} \to \mathcal{X}$ is called a *covering projection* if each point $x \in \mathcal{X}$ has an open neighborhood evenly covered by $\widetilde{\pi}$. $\widetilde{\mathcal{X}}$ is called the *covering space* and $\mathcal{X}$ the *base space* of the covering.

Definition 2.2. [10] A fibration $p : \widetilde{\mathcal{X}} \to \mathcal{X}$ with unique path lifting is said to be *regular* if, given any closed path ω in $\mathcal{X}$, either every lifting of ω is closed or none is closed.

Definition 2.3. [10] A topological space $\mathcal{X}$ is said to be *locally path-connected* if the path components of open sets are open.

Denote by π_1 the functor of fundamental group [10].

Theorem 2.4. *[10] Let $p : \widetilde{\mathcal{X}} \to \mathcal{X}$ be a fibration with unique path lifting and assume that a nonempty $\widetilde{\mathcal{X}}$ is a locally path-connected space. Then p is regular if and only if for some $\widetilde{x}_0 \in \widetilde{\mathcal{X}}$, $\pi_1(p)\,\pi_1\left(\widetilde{\mathcal{X}}, \widetilde{x}_0\right)$ is a normal subgroup of $\pi_1(\mathcal{X}, p(\widetilde{x}_0))$.*

Definition 2.5. [10] Let $p : \widetilde{\mathcal{X}} \to \mathcal{X}$ be a covering. A self-equivalence is a homeomorphism $f : \widetilde{\mathcal{X}} \to \widetilde{\mathcal{X}}$ such that $p \circ f = p$. This group of such homeomorphisms is said to be the *group of covering transformations* of p or the *covering group*. Denote by $G\left(\widetilde{\mathcal{X}} \mid \mathcal{X}\right)$ this group.

Proposition 2.6. *[10] If $p : \widetilde{\mathcal{X}} \to \mathcal{X}$ is a regular covering and $\widetilde{\mathcal{X}}$ is connected and locally path connected, then $\mathcal{X}$ is homeomorphic to space of orbits of $G\left(\widetilde{\mathcal{X}} \mid \mathcal{X}\right)$, i.e. $\mathcal{X} \approx \widetilde{\mathcal{X}} / G\left(\widetilde{\mathcal{X}} \mid \mathcal{X}\right)$. So p is a principal bundle.*

Corollary 2.7. *[10] Let $p : \widetilde{\mathcal{X}} \to \mathcal{X}$ be a fibration with a unique path lifting. If $\widetilde{\mathcal{X}}$ is connected and locally path-connected and $\widetilde{x}_0 \in \widetilde{\mathcal{X}}$ then p is regular if and only if $G\left(\widetilde{\mathcal{X}} \mid \mathcal{X}\right)$ transitively acts on each fiber of p, in which case*

$$\psi : G\left(\widetilde{\mathcal{X}} \mid \mathcal{X}\right) \approx \pi_1(\mathcal{X}, p(\widetilde{x}_0)) / \pi_1(p)\,\pi_1\left(\widetilde{\mathcal{X}}, \widetilde{x}_0\right).$$

Remark 2.8. Above results are copied from [10]. Below the *covering projection* word is replaced with *covering*.

Definition 2.9. [7] *A compactification* of a space $\mathcal{X}$ is a compact Hausdorff space $\mathcal{Y}$ containing $\mathcal{X}$ as a subspace and the closure $\overline{\mathcal{X}}$ of $\mathcal{X}$ is $\mathcal{Y}$, i.e $\overline{\mathcal{X}} = \mathcal{Y}$.

Remark 2.10. In the following text we denote compactifications by $\mathcal{X} \hookrightarrow \mathcal{Y}$.

The algebraic construction requires a following definition.

Definition 2.11. [5] A covering $\pi : \widetilde{\mathcal{X}} \to \mathcal{X}$ is said to be a *covering with compactification* if there are compactifications $\mathcal{X} \hookrightarrow \mathcal{Y}$ and $\widetilde{\mathcal{X}} \hookrightarrow \widetilde{\mathcal{Y}}$ (cf. Definition 2.9) such that:

- There is a covering $\overline{\pi} : \widetilde{\mathcal{Y}} \to \mathcal{Y}$,
- The covering π is the restriction of $\overline{\pi}$, i.e. $\pi = \overline{\pi}|_{\widetilde{\mathcal{X}}}$ and $\overline{\pi}\left(\widetilde{\mathcal{X}}\right) = \mathcal{X}$.

Definition 2.12. The sequence of regular finite-fold coverings with compactification

$$\mathcal{X} = \mathcal{X}_0 \leftarrow ... \leftarrow \mathcal{X}_n \leftarrow ...$$

is said to be a *(topological) finite covering sequence* if following conditions hold:

- The space $\mathcal{X}_n$ is a second-countable [7] locally compact connected Hausdorff space for any $n \in \mathbb{N}^0$,
- If $k < l < m$ are any nonnegative integer numbers then there is the natural exact sequence

$$\{e\} \to G(\mathcal{X}_m \mid \mathcal{X}_l) \to G(\mathcal{X}_m \mid \mathcal{X}_k) \to G(\mathcal{X}_l \mid \mathcal{X}_k) \to \{e\}.$$

For any finite covering sequence we will use a following notation

$$\mathfrak{S} = \{\mathcal{X} = \mathcal{X}_0 \leftarrow ... \leftarrow \mathcal{X}_n \leftarrow ...\} = \{\mathcal{X}_0 \leftarrow ... \leftarrow \mathcal{X}_n \leftarrow ...\}, \quad \mathfrak{S} \in \mathfrak{FinTop}.$$

Definition 2.13. Let $\{\mathcal{X} = \mathcal{X}_0 \leftarrow ... \leftarrow \mathcal{X}_n \leftarrow ...\} \in \mathfrak{FinTop}$, and let $\widehat{\mathcal{X}} = \varprojlim \mathcal{X}_n$ be the inverse limit in the category of topological spaces and continuous maps (See [10]). If $\widehat{\pi}_0 : \widehat{\mathcal{X}} \to \mathcal{X}_0$ is the natural continuous map then homeomorphism g of the space $\widehat{\mathcal{X}}$ is said to be a *covering transformation* if the following condition holds

$$\widehat{\pi}_0 = \widehat{\pi}_0 \circ g.$$

The group $\widehat{G}$ of covering homeomorphisms is said to be the *group of covering transformations* of $\mathfrak{S}$. Denote by $G\left(\widehat{\mathcal{X}} \mid \mathcal{X}\right) \stackrel{\text{def}}{=} \widehat{G}$.

Definition 2.14. Let $\mathfrak{S} = \{\mathcal{X}_0 \leftarrow ... \leftarrow \mathcal{X}_n \leftarrow ...\}$ be a finite covering sequence. The pair $\left(\mathcal{Y}, \{\pi^{\mathcal{Y}}_n\}_{n \in \mathbb{N}}\right)$ of a (discrete) set $\mathcal{Y}$ with and surjective maps $\pi^{\mathcal{Y}}_n : \mathcal{Y} \to \mathcal{X}_n$ is said to be a *coherent system* if for any $n \in \mathbb{N}^0$ following diagram

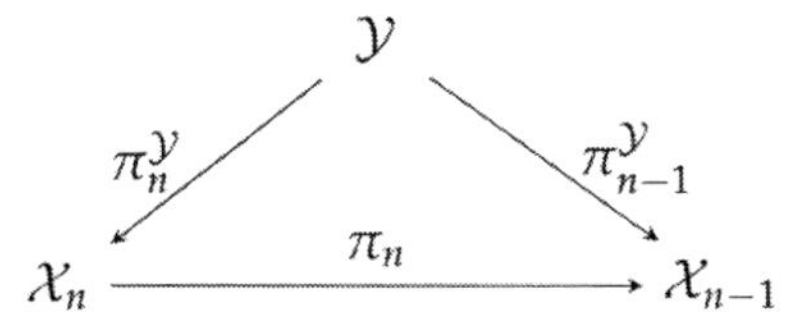

is commutative.

Definition 2.15. Let $\mathfrak{S} = \{\mathcal{X}_0 \leftarrow ... \leftarrow \mathcal{X}_n \leftarrow ...\}$ be a topological finite covering sequence. A coherent system $(\mathcal{Y}, \{\pi^{\mathcal{Y}}_n\})$ is said to be a *connected covering* of $\mathfrak{S}$ if $\mathcal{Y}$ is a connected topological space and $\pi^{\mathcal{Y}}_n$ is a regular covering for any $n \in \mathbb{N}$. We will use following notation $(\mathcal{Y}, \{\pi^{\mathcal{Y}}_n\}) \downarrow \mathfrak{S}$ or simply $\mathcal{Y} \downarrow \mathfrak{S}$.

Definition 2.16. Let $(\mathcal{Y}, \{\pi^{\mathcal{Y}}_n\})$ be a coherent system of $\mathfrak{S}$ and $y \in \mathcal{Y}$. A subset $\mathcal{V} \subset \mathcal{Y}$ is said to be *special* if $\pi^{\mathcal{Y}}_0(\mathcal{V})$ is evenly covered by $\mathcal{X}_1 \to \mathcal{X}_0$ and for any $n \in \mathbb{N}^0$ following conditions hold:

- $\pi^{\mathcal{Y}}_n(\mathcal{V}) \subset \mathcal{X}_n$ is an open connected set,
- The restriction $\pi^{\mathcal{Y}}_n|_{\mathcal{V}} : \mathcal{V} \to \pi^{\mathcal{Y}}_n(\mathcal{V})$ is a bijection.

Remark 2.17. If $(\mathcal{Y}, \{\pi^{\mathcal{Y}}_n\})$ is a covering of $\mathfrak{S}$ then the topology of $\mathcal{Y}$ is generated by special sets.

Definition 2.18. Let us consider the situation of the Definition 2.15. A *morphism* from $\left(\mathcal{Y}', \left\{\pi_n^{\mathcal{Y}'}\right\}\right) \downarrow \mathfrak{S}$ to $\left(\mathcal{Y}'', \left\{\pi_n^{\mathcal{Y}''}\right\}\right) \downarrow \mathfrak{S}$ is a covering $f : \mathcal{Y}' \to \mathcal{Y}''$ such that

$$\pi_n^{\mathcal{Y}''} \circ f = \pi_n^{\mathcal{Y}'}$$

for any $n \in \mathbb{N}$.

2.19. There is a category with objects and morphisms described by Definitions 2.15, 2.18. Denote by $\downarrow \mathfrak{S}$ this category.

Lemma 2.20. *[5] There is the final object of the category $\downarrow \mathfrak{S}$ described in 2.19.*

Remark 2.21. In [5] it is proven that if $\widetilde{\mathcal{X}}$ is the final object of the category $\downarrow \mathfrak{S}$ then following condition holds

$$\overline{\mathcal{X}} = \bigsqcup_{g \in \overline{\widehat{G}/G}} g\widetilde{\mathcal{X}} \tag{2.1}$$

where

- $\overline{\mathcal{X}}$ is a topological space which coincides with $\varprojlim \mathcal{X}_n$ as a set and topology of $\overline{\mathcal{X}}$ is generated by special sets,
- $\widehat{G} = \varprojlim G_n$,
- $G = G\left(\widetilde{\mathcal{X}}, \mathcal{X}\right)$.
- $\overline{\widehat{G}/G}$ is a set of representatives of $\widehat{G}/G$.

Definition 2.22. The final object $\left(\widetilde{\mathcal{X}}, \left\{\pi_n^{\widetilde{\mathcal{X}}}\right\}\right)$ of the category $\downarrow \mathfrak{S}$ is said to be the *(topological) inverse limit* of $\downarrow \mathfrak{S}$. The notation $\left(\widetilde{\mathcal{X}}, \left\{\pi_n^{\widetilde{\mathcal{X}}}\right\}\right) = \varprojlim \downarrow \mathfrak{S}$ or simply $\widetilde{\mathcal{X}} = \varprojlim \downarrow \mathfrak{S}$ will be used. The space $\overline{\mathcal{X}}$ from the proof of the Lemma 2.20 is said to be the *disconnected inverse limit* of $\mathfrak{S}$.

3. Noncommutative finite-fold coverings

Definition 3.1. If A is a C^*- algebra then an action of a group G is said to be *involutive* if $ga^* = (ga)^*$ for any $a \in A$ and $g \in G$. The action is said to be *non-degenerated* if for any nontrivial $g \in G$ there is $a \in A$ such that $ga \neq a$.

Definition 3.2. Let $A \hookrightarrow \widetilde{A}$ be an injective *-homomorphism of unital C^*-algebras. Suppose that there is a non-degenerated involutive action $G \times \widetilde{A} \to \widetilde{A}$ of a finite group G, such that $A = \widetilde{A}^G \stackrel{\text{def}}{=} \left\{a \in \widetilde{A} \mid a = ga;\ \forall g \in G\right\}$. There is an A-valued product on $\widetilde{A}$ given by

$$\langle a, b \rangle_{\widetilde{A}} = \sum_{g \in G} g\left(a^* b\right) \tag{3.1}$$

and $\widetilde{A}$ is an A-Hilbert module. We say that a triple $\left(A, \widetilde{A}, G\right)$ is an *unital noncommutative finite-fold covering* if $\widetilde{A}$ is a finitely generated projective A-Hilbert module.

Remark 3.3. Above definition is motivated by the Theorem 1.2.

Definition 3.4. Let A, $\widetilde{A}$ be C^*-algebras such that following conditions hold:

(a) There are unital C^*-algebras $B, \widetilde{B}$ and inclusions $A \subset B$, $\widetilde{A} \subset \widetilde{B}$ such that A (resp. $\widetilde{A}$) is an essential ideal of B (resp. $\widetilde{B}$),

(b) There is an unital noncommutative finite-fold covering $\left(B, \widetilde{B}, G\right)$,

(c)
$$\widetilde{A} = \left\{ a \in \widetilde{B} \mid \left\langle \widetilde{B}, a \right\rangle_{\widetilde{B}} \in A \right\}. \tag{3.2}$$

The triple $\left(A, \widetilde{A}, G\right)$ is said to be a *noncommutative finite-fold covering*. The group G is said to be the *covering transformation group* (of $\left(A, \widetilde{A}, G\right)$) and we use the following notation

$$G\left(\widetilde{A} \mid A\right) \stackrel{\text{def}}{=} G. \tag{3.3}$$

Lemma 3.5. *[5] Let us consider the situation of the Definition 3.4. Following conditions hold:*

(i) From (3.2) it turns out that $\widetilde{A}$ is a closed two sided ideal of $\widetilde{B}$,

(ii) The action of G on $\widetilde{B}$ is such that $G\widetilde{A} = \widetilde{A}$, i.e. there is the natural action of G on $\widetilde{A}$,

(iii)
$$A \cong \widetilde{A}^G = \left\{ a \in \widetilde{A} \mid a = ga; \ \forall g \in G \right\}. \tag{3.4}$$

Remark 3.6. The Definition 3.5 is motivated by the Theorem 5.2.

Definition 3.7. The injective *-homomorphism $A \hookrightarrow \widetilde{A}$, which follows from (3.4) is said to be a *noncommutative finite-fold covering*.

Definition 3.8. Let $\left(A, \widetilde{A}, G\right)$ be a noncommutative finite-fold covering. The algebra $\widetilde{A}$ is a Hilbert A-module with an A-valued product given by

$$\langle a, b \rangle_{\widetilde{A}} = \sum_{g \in G} g(a^* b); \ a, b \in \widetilde{A}. \tag{3.5}$$

We say that this structure of Hilbert A-module is *induced by the covering* $\left(A, \widetilde{A}, G\right)$. Henceforth we shall consider $\widetilde{A}$ as a right A-module, so we will write $\widetilde{A}_A$.

4. Noncommutative infinite coverings

This section contains a noncommutative generalization of infinite coverings.

Definition 4.1. Let
$$\mathfrak{S} = \left\{ A = A_0 \xrightarrow{\pi_1} A_1 \xrightarrow{\pi_2} \dots \xrightarrow{\pi_n} A_n \xrightarrow{\pi^{n+1}} \dots \right\}$$

be a sequence of C^*-algebras and noncommutative finite-fold coverings such that:

(a) Any composition $\pi_{n_1} \circ \dots \circ \pi_{n_0+1} \circ \pi_{n_0} : A_{n_0} \to A_{n_1}$ corresponds to the noncommutative covering $\left(A_{n_0}, A_{n_1}, G\left(A_{n_1} \mid A_{n_0}\right)\right)$;

(b) If $k < l < m$ then $G\left(A_m \mid A_k\right) A_l = A_l$ (Action of $G\left(A_m \mid A_k\right)$ on A_l means that $G\left(A_m \mid A_k\right)$ acts on A_m, so $G\left(A_m \mid A_k\right)$ acts on A_l since A_l a subalgebra of A_m);

(c) If $k < l < m$ are nonegative integers then there is the natural exact sequence of covering transformation groups

$$\{e\} \to G\left(A_m \mid A_l\right) \xrightarrow{\iota} G\left(A_m \mid A_k\right) \xrightarrow{\pi} G\left(A_l \mid A_k\right) \to \{e\}$$

where the existence of the homomorphism $G\left(A_m \mid A_k\right) \xrightarrow{\pi} G\left(A_l \mid A_k\right)$ follows from (b).

The sequence $\mathfrak{S}$ is said to be an *(algebraical) finite covering sequence.* For any finite covering sequence we will use the notation $\mathfrak{S} \in \mathfrak{FinAlg}$.

Definition 4.2. Let $\widehat{A} = \varinjlim A_n$ be the C^*-inductive limit [8], and suppose that $\widehat{G} = \varprojlim G(A_n \mid A)$ is the projective limit of groups [10]. There is the natural action of $\widehat{G}$ on $\widehat{A}$. A non-degenerate faithful representation $\widehat{A} \to B(\mathcal{H})$ is said to be *equivariant* if there is an action of $\widehat{G}$ on $\mathcal{H}$ such that for any $\xi \in \mathcal{H}$ and $g \in \widehat{G}$ following condition holds

$$(ga)\xi = g\left(a\left(g^{-1}\xi\right)\right). \tag{4.1}$$

Definition 4.3. Let $\pi : \widehat{A} \to B(\mathcal{H})$ be an equivariant representation. A positive element $\overline{a} \in B(\mathcal{H})_+$ is said to be *special* (with respect to $\mathfrak{S}$) if following conditions hold:

(a) For any $n \in \mathbb{N}^0$ the following series

$$a_n = \sum_{g\in\ker\left(\widehat{G}\to G(A_n \mid A)\right)} g\overline{a}$$

is strongly convergent and the sum lies in A_n, i.e. $a_n \in A_n$;

(b) If $f_\varepsilon : \mathbb{R} \to \mathbb{R}$ is given by

$$f_\varepsilon(x) = \left\{ \begin{array}{cc} 0 & x \le \varepsilon \\ x - \varepsilon & x > \varepsilon \end{array} \right. \tag{4.2}$$

then for any $n \in \mathbb{N}^0$ and for any $z \in A$ following series

$$\begin{aligned} b_n &= \sum_{g\in\ker\left(\widehat{G}\to G(A_n \mid A)\right)} g\left(z\overline{a}z^*\right), \\ c_n &= \sum_{g\in\ker\left(\widehat{G}\to G(A_n \mid A)\right)} g\left(z\overline{a}z^*\right)^2, \\ d_n &= \sum_{g\in\ker\left(\widehat{G}\to G(A_n \mid A)\right)} gf_\varepsilon\left(z\overline{a}z^*\right) \end{aligned}$$

are strongly convergent and the sums lie in A_n, i.e. $b_n,\ c_n,\ d_n \in A_n$;

(c) For any $\varepsilon > 0$ there is $N \in \mathbb{N}$ (which depends on $\overline{a}$ and z) such that for any $n \ge N$ a following condition holds

$$\left\|b_n^2 - c_n\right\| < \varepsilon. \tag{4.3}$$

An element $\overline{a}' \in B(\mathcal{H})$ is said to be *weakly special* if

$$\overline{a}' = x\overline{a}y; \text{ where } x, y \in \widehat{A}, \text{ and } \overline{a} \in B(\mathcal{H}) \text{ is special.}$$

Lemma 4.4. *[5] If $\overline{a} \in B(\mathcal{H})_+$ is a special element and $G_n = \ker\left(\widehat{G} \to G(A_n \mid A)\right)$ then from*

$$a_n = \sum_{g\in G_n} g\overline{a},$$

it follows that $\overline{a} = \lim_{n\to\infty} a_n$ in the sense of the strong convergence. Moreover one has $\overline{a} = \inf_{n\in\mathbb{N}} a_n$.

Corollary 4.5. *[5] Any weakly special element lies in the enveloping von Neumann algebra $\widehat{A}''$ of $\widehat{A} = \varinjlim A_n$. If $\overline{A}_\pi \subset B(\mathcal{H})$ is the C^*-norm completion of an algebra generated by weakly special elements then $\overline{A}_\pi \subset \widehat{A}''$.*

Lemma 4.6. *[5] If $\overline{a} \in B(\mathcal{H})$ is special, (resp. $\overline{a}' \in B(\mathcal{H})$ weakly special) then for any $g \in \widehat{G}$ the element $g\overline{a}$ is special, (resp. $g\overline{a}'$ is weakly special).*

Corollary 4.7. *[5] If $\overline{A}_\pi \subset B(\mathcal{H})$ is the C^*-norm completion of algebra generated by weakly special elements, then there is a natural action of $\widehat{G}$ on $\overline{A}_\pi$.*

Definition 4.8. Let $\mathfrak{S} = \left\{A = A_0 \xrightarrow{\pi^1} A_1 \xrightarrow{\pi^2} ... \xrightarrow{\pi^n} A_n \xrightarrow{\pi^{n+1}} ...\right\}$ be an algebraical finite covering sequence. Let $\pi : \widehat{A} \to B(\mathcal{H})$ be an equivariant representation. Let $\overline{A}_\pi \subset B(\mathcal{H})$ be the C^*-norm completion of algebra generated by weakly special elements. We say that $\overline{A}_\pi$ is the *disconnected inverse noncommutative limit* of $\downarrow \mathfrak{S}$ (*with respect to* π). The triple $\left(A, \overline{A}_\pi, G\left(\overline{A}_\pi \mid A\right) \stackrel{\text{def}}{=} \widehat{G}\right)$ is said to be the *disconnected infinite noncommutative covering* of $\mathfrak{S}$ (*with respect to* π). If π is the universal representation then "with respect to π" is dropped and we will write $\left(A, \overline{A}, G\left(\overline{A} \mid A\right)\right)$.

Definition 4.9. A maximal irreducible subalgebra $\widetilde{A}_\pi \subset \overline{A}_\pi$ is said to be a *connected component* of $\mathfrak{S}$ (*with respect to* π). The maximal subgroup $G_\pi \subset G\left(\overline{A}_\pi \mid A\right)$ among subgroups $G \subset G\left(\overline{A}_\pi \mid A\right)$ such that $G\widetilde{A}_\pi = \widetilde{A}_\pi$ is said to be the *$\widetilde{A}_\pi$-invariant group* of $\mathfrak{S}$. If π is the universal representation then "with respect to π" is dropped.

Remark 4.10. From the Definition 4.9 it follows that $G_\pi \subset G\left(\overline{A}_\pi \mid A\right)$ is a normal subgroup.

Definition 4.11. Let

$$\mathfrak{S} = \left\{A = A_0 \xrightarrow{\pi^1} A_1 \xrightarrow{\pi^2} ... \xrightarrow{\pi^n} A_n \xrightarrow{\pi^{n+1}} ...\right\} \in \mathfrak{FinAlg},$$

and let $\left(A, \overline{A}_\pi, G\left(\overline{A}_\pi \mid A\right)\right)$ be a disconnected infinite noncommutative covering of $\mathfrak{S}$ with respect to an equivariant representation $\pi : \varinjlim A_n \to B(\mathcal{H})$. Let $\widetilde{A}_\pi \subset \overline{A}_\pi$ be a connected component of $\mathfrak{S}$ with respect to π, and let $G_\pi \subset G\left(\overline{A}_\pi \mid A\right)$ be the $\widetilde{A}_\pi$ - invariant group of $\mathfrak{S}$. Let $h_n : G\left(\overline{A}_\pi \mid A\right) \to G\left(A_n \mid A\right)$ be the natural surjective homomorphism. The representation $\pi : \varinjlim A_n \to B(\mathcal{H})$ is said to be *good* if it satisfies to following conditions:

(a) The natural *-homomorphism $\varinjlim A_n \to M\left(\widetilde{A}_\pi\right)$ is injective,

(b) If $J \subset G\left(\overline{A}_\pi \mid A\right)$ is a set of representatives of $G\left(\overline{A}_\pi \mid A\right)/G_\pi$, then the algebraic direct sum

$$\bigoplus_{g \in J} g\widetilde{A}_\pi$$

is a dense subalgebra of $\overline{A}_\pi$,

(c) For any $n \in \mathbb{N}$ the restriction $h_n|_{G_\pi}$ is an epimorphism, i. e. $h_n\left(G_\pi\right) = G\left(A_n \mid A\right)$.

If π is the universal representation we say that $\mathfrak{S}$ is *good*.

Definition 4.12. Let $\mathfrak{S} = \{A = A_0 \to A_1 \to ... \to A_n \to ...\} \in \mathfrak{FinAlg}$ be an algebraical finite covering sequence. Let $\pi : \widehat{A} \to B(\mathcal{H})$ be a good representation. A connected component $\widetilde{A}_\pi \subset \overline{A}_\pi$ is said to be the *inverse noncommutative limit of $\downarrow \mathfrak{S}$ (with respect to π)*. The $\widetilde{A}_\pi$-invariant group G_π is said to be the *covering transformation group of $\mathfrak{S}$* (*with respect to* π). The triple $\left(A, \widetilde{A}_\pi, G_\pi\right)$ is said to be the *infinite noncommutative covering* of $\mathfrak{S}$ (*with respect to* π). We will use the following notation

$$\varprojlim_\pi \downarrow \mathfrak{S} \stackrel{\text{def}}{=} \widetilde{A}_\pi,$$

$$G\left(\widetilde{A}_\pi \mid A\right) \stackrel{\text{def}}{=} G_\pi.$$

If π is the universal representation then "with respect to π" is dropped and we will write $\left(A, \widetilde{A}, G\right)$, $\varprojlim \downarrow \mathfrak{S} \stackrel{\text{def}}{=} \widetilde{A}$ and $G\left(\widetilde{A} \mid A\right) \stackrel{\text{def}}{=} G$.

5. Examples

5.1. *Quantization of topological coverings*

Following theorems state the equivalence between algebraical and topological definitions of coverings.

Theorem 5.2. *[5] If $\mathcal{X}$, $\widetilde{\mathcal{X}}$ are locally compact spaces, and $\pi : \widetilde{\mathcal{X}} \to \mathcal{X}$ is a surjective continuous map, then following conditions are equivalent:*

(i) The map $\pi : \widetilde{\mathcal{X}} \to \mathcal{X}$ is a finite-fold covering with a compactification,

(ii) There is a natural noncommutative finite-fold covering $\left(C_0\left(\mathcal{X}\right), C_0\left(\widetilde{\mathcal{X}}\right), G\right)$.

Theorem 5.3. *[5] If $\mathfrak{S}_\mathcal{X} = \{\mathcal{X} = \mathcal{X}_0 \leftarrow ... \leftarrow \mathcal{X}_n \leftarrow ...\} \in \mathfrak{FinTop}$ and*

$$\mathfrak{S}_{C_0(\mathcal{X})} = \{C_0(\mathcal{X}) = C_0(\mathcal{X}_0) \to ... \to C_0(\mathcal{X}_n) \to ...\} \in \mathfrak{FinAlg}$$

is an algebraical finite covering sequence then following conditions hold:

(i) $\mathfrak{S}_{C_0(\mathcal{X})}$ is good.

(ii) There are isomorphisms:

- $\varprojlim \downarrow \mathfrak{S}_{C_0(\mathcal{X})} \approx C_0\left(\varprojlim \downarrow \mathfrak{S}_\mathcal{X}\right)$;
- $G\left(\varprojlim \downarrow \mathfrak{S}_{C_0(\mathcal{X})} \mid C_0\left(\mathcal{X}\right)\right) \approx G\left(\varprojlim \downarrow \mathfrak{S}_\mathcal{X} \mid \mathcal{X}\right)$.

5.4. *Coverings of noncommutative C^*-algebras*

There are known finite-fold and infinite coverings of the following C^*-algebras:

- Continuous trace algebras,
- Noncommutative torus,
- Quantum $SU_q(2)$ group,
- Coverings of foliations,
- Isospectral deformations of spectral triples.

These coverings are described in [5,6].

6. Coverings and fundamental groups in physics

The math definition of fundamental group is presented in [10]. Below there is a physical explanation. The *fundamental group* $\pi_1(\mathcal{X})$, which counts how many ways a loop can be mapped into a path-connected space $\mathcal{X}$. More precisely, we define $\pi_1(\mathcal{X}, x)$ to be the set of all homotopy classes of parameterized loop mappings that begin and end at a base-point x. For a path-connected space we can add a path from x to any other point and back as part of the loop, so that $\pi_1(\mathcal{X}, x)$ is independent of x and is written as $\pi_1(\mathcal{X})$. There is a connection between fundamental group and coverings, i.e. a fundamental group is a group of self-equivalences of coverings. The rigorous math foundation of the above statement is described in [10]. This connection is implicitly described in [3] (Section 16.3) where it is noted that there are two equivalent descriptions of Wilson lines. First description is based on representation of the fundamental group, second one uses the group of self-equivalences of coverings. The noncommutative geometry has no points, so it has not paths. It turns out that applications of fundamental group is impossible in noncommutative physics. However we can use group of self-equivalences of noncommutative coverings. The physical explanation of this fact is noted in [3]. The heuristic physical description of noncommutative coverings is already described in [1] where authors calculate Wilson lines on noncommutative tori. However [1] does not contain the rigorous math foundation. The rigorous foundation of noncommutative Wilson lines is described in [4].

References

[1] Alekseev A and Bytsko A 2000 Wilson lines on noncommutative tori *Preprint* hep-th/0002101
[2] Arveson W 1976 *An Invitation to C*-Algebras* (Berlin: Springer)
[3] Green M, Schwarz J, Witten E 1988. *Superstring Theory: Volume 2, Loop Amplitudes, Anomalies and Phenomenology* (Cambridge Monographs on Mathematical Physics)
[4] Ivankov P 2014 Noncommutative Generalization of Wilson Lines *Preprint* math.OA/1408.4101
[5] Ivankov P 2017 Quantization of noncompact coverings *Preprint* math.OA/1702.07918
[6] Ivankov P 2017 Coverings of foliation algebras *Preprint* math.OA/1708.02128
[7] Munkres J 2000 *Topology* (Prentice Hall, Incorporated)
[8] Murphy G 1990 *C*-Algebras and Operator Theory* (Academic Press)
[9] Pavlov A and Troitskii E 2011 *Russian J. of Math. Phys.* **18** 338
[10] Spanier E 1966 *Algebraic Topology* (New York McGraw-Hill)

Classical and quantum superfield invariants in $\mathcal{N} = (1,1), 6D$ SYM theory

Evgeny Ivanov

Bogoliubov Laboratory of Theoretical Physics, JINR, 141980 Dubna, Moscow Region, Russia

E-mail: eivanov@theor.jinr.ru

Abstract. We give an overview of recent results on the classical and quantum superfield invariants of $\mathcal{N} = (1,1), 6D$ supersymmetric Yang-Mills theory in the off-shell $\mathcal{N} = (1,0)$ and on-shell $\mathcal{N} = (1,1)$, $6D$ harmonic superspaces.

1. Introduction

For the last few years, maximally extended supersymmetric gauge theories (with 16 supersymmetries) are under intensive study. These can be represented by the following chain

$$\mathcal{N} = 4,\ 4D \implies \mathcal{N} = (1,1),\ 6D \implies \mathcal{N} = (1,0),\ 10D\,.$$

Among them, the $\mathcal{N} = 4$, $4D$ SYM (Super Yang-Mills) theory is most known. It is UV finite and perhaps completely integrable at the quantum level [1]. On the other hand, $\mathcal{N} = (1,1)$, $6D$ SYM is not renormalizable by formal counting (the coupling constant is dimensionful) but is also expected to possess various unique properties. In particular, it respects the so-called "dual conformal symmetry" like its $4D$ counterpart [2]. It provides the effective theory descriptions of some particular low energy sectors of string theory, such as D5-brane dynamics. The full effective action of D5-brane is expected to be of non-abelian Born-Infeld type, generalizing the microscopic $\mathcal{N} = (1,1)$ SYM action [3]. $\mathcal{N} = (1,1)$ SYM is anomaly free [4], as distinct from $\mathcal{N} = (1,0)$ SYM. The $\mathcal{N} = (1,1)$ and $\mathcal{N} = (1,0)$ SYM theories are analogs of $\mathcal{N} = 8$ supergravity which is also formally non-renormalizable, so their study in the quantum domain can shed more light on the quantum properties of the latter theory.

Recent perturbative explicit calculations in $\mathcal{N} = (1,1)$ SYM (treated as a low-energy limit of type II superstrings) show a lot of cancelations of the UV divergencies which cannot be expected in advance. The theory is UV-finite up to 2 loops, while at 3 loops only a single-trace counterterm of canonical dim 10 is required. The allowed double-trace counterterms do not appear [5] - [7]. To explain this, one needs new non-renormalization theorems. As usual, the maximally supersymmetric off-shell formulations are needed to clarify these issues.

However, maximum that one can achieve in $6D$ is off-shell $\mathcal{N} = (1,0)$ supersymmetry [1]. The most natural off-shell formulation of $\mathcal{N} = (1,0)$ SYM is achieved in harmonic $\mathcal{N} = (1,0), 6D$ superspace (HSS) [9, 10] as a generalization of $\mathcal{N} = 2, 4D$ HSS [11, 12]. In HSS, the $\mathcal{N} = (1,1)$

[1] The maximal off-shell supersymmetry with 16 supercharges is claimed to be attainable in the "pure spinor" superfield formalism [8], but here we limit our attention to the standard superspaces.

SYM theory action can be schematically presented as a sum [$\mathcal{N} = (1,1)$ SYM] = [$\mathcal{N} = (1,0)$ SYM + $6D$ hypermultiplet], with the second hidden on-shell $\mathcal{N} = (0,1)$ supersymmetry.

In order to know the putative structure of the effective action and candidate counterterms for $\mathcal{N} = (1,1)$ SYM theory, one needs to learn how to construct higher-dimension $\mathcal{N} = (1,1)$ invariants in terms of $\mathcal{N} = (1,0)$ superfields. In the "brute-force" method one starts with the appropriate $\mathcal{N} = (1,0)$ SYM invariant and then completes it to $\mathcal{N} = (1,1)$ invariant by adding the proper hypermultiplet terms. In practice, it is very cumbersome technically, though the situation is somewhat simplified by the fact that for finding all admissible superfield counterterms it is enough to stay *on shell*[2].

In [13] there was developed a new approach to constructing higher-dimension $\mathcal{N} = (1,1)$ invariants, based on the concept of on-shell $\mathcal{N} = (1,1)$ harmonic superspace [14].

The hidden supersymmetry tells us nothing about the precise coefficients before the $\mathcal{N} = (1,1)$ invariants constructed in one or another way. To determine them, one should reproduce them from the superfield perturbation theory. The first steps towards this goal were recently undertaken in [15] - [17].

In the talk I will briefly address all these issues.

2. $6D$ superspaces and superfields

2.1. Basic superspaces

- Standard $\mathcal{N} = (1,0), 6D$ superspace [18] is defined as the following set of coordinates:

$$z = (x^M, \theta^a_i)\,, \quad M = 0,\ldots,5\,,\ a = 1,\ldots,4\,,\ i = 1,2\,, \tag{1}$$

  with Grassmann pseudoreal θ^a_i.

- Harmonic $\mathcal{N} = (1,0), 6D$ superspace [9, 10] is obtained by adding $SU(2)$ harmonics to (1):

$$Z := (z, u) = (x^M, \theta^a_i, u^{\pm i})\,, \quad u^-_i = (u^+_i)^*, u^{+i}u^-_i = 1\,,\ u^{\pm i} \in SU(2)_R/U(1)\,. \tag{2}$$

- *Analytic* $\mathcal{N} = (1,0), 6D$ superspace has half the number of Grassmann coordinates as compared to (2):

$$\zeta := (x^M_{(\text{an})}, \theta^{+a}, u^{\pm i}) \subset Z\,, \quad x^M_{(\text{an})} = x^M + \frac{i}{2}\theta^a_k \gamma^M_{ab} \theta^b_l u^{+k} u^{-l}, \quad \theta^{\pm a} = \theta^a_i u^{\pm i}\,. \tag{3}$$

Basic differential operators in the analytic basis of $6D$ HSS are defined as :

$$\begin{aligned}
D^+_a &= \partial_{-a}\,, \quad D^-_a = -\partial_{+a} - 2i\theta^{-b}\partial_{ab}\,, \\
D^0 &= u^{+i}\frac{\partial}{\partial u^{+i}} - u^{-i}\frac{\partial}{\partial u^{-i}} + \theta^{+a}\partial_{+a} - \theta^{-a}\partial_{-a} \\
D^{++} &= \partial^{++} + i\theta^{+a}\theta^{+b}\partial_{ab} + \theta^{+a}\partial_{-a}\,,\ D^{--} = \partial^{--} + i\theta^{-a}\theta^{-b}\partial_{ab} + \theta^{-a}\partial_{+a}\,,
\end{aligned} \tag{4}$$

where $\partial_{\pm a}\theta^{\pm b} = \delta^b_a$ and $\partial^{++} = u^{+i}\frac{\partial}{\partial u^{-i}}, \quad \partial^{--} = u^{-i}\frac{\partial}{\partial u^{+i}}$.

2.2. Basic superfields

- The basic geometric object of $\mathcal{N} = (1,0)$ SYM theory is the analytic gauge connection V^{++}:

$$\nabla^{++} = D^{++} + V^{++}\,, \quad \delta V^{++} = -\nabla^{++}\Lambda\,, \quad \Lambda = \Lambda(\zeta)\,.$$

[2] To avoid a possible confusion, by this we merely mean that the on-shell vanishing counterterms can be absorbed into the microscopic action by a field redefinition. No equations of motion are actually assumed for the involved (super)fields.

- The second harmonic (non-analytic) connection V^{--} covariantizes the second harmonic derivative:
$$\nabla^{--} = D^{--} + V^{--}, \quad \delta V^{--} = -\nabla^{--}\Lambda.$$
It s related to V^{++} by the harmonic flatness condition
$$[\nabla^{++}, \nabla^{--}] = D^0 \Rightarrow D^{++}V^{--} - D^{--}V^{++} + [V^{++}, V^{--}] = 0$$
$$\Rightarrow V^{--} = V^{--}(V^{++}, u^{\pm}).$$
One can make use of the analytic gauge freedom to choose the Wess-Zumino gauge:
$$V^{++} = \theta^{+a}\theta^{+b}A_{ab} + 2(\theta^+)^3_a\lambda^{-a} - 3(\theta^+)^4\mathcal{D}^{--}.$$
Here A_{ab} is the gauge field, $\lambda^{-a} = \lambda^{ai}u^-_i$ is the gaugino and $\mathcal{D}^{--} = \mathcal{D}^{ik}u^-_i u^-_k$, where $\mathcal{D}^{ik} = \mathcal{D}^{ki}$, are the auxiliary fields. This is just the irreducible contents of the $\mathcal{N} = (1,0)$ vector (gauge) multiplet.
- Having V^{--}, it is straightforward to define the covariant spinor and vector derivatives
$$\nabla^-_a = [\nabla^{--}, D^+_a] = D^-_a + \mathcal{A}^-_a, \qquad \nabla_{ab} = \frac{1}{2i}[D^+_a, \nabla^-_b] = \partial_{ab} + \mathcal{A}_{ab},$$
$$\mathcal{A}^-_a(V) = -D^+_a V^{--}, \quad \mathcal{A}_{ab}(V) = \frac{i}{2}D^+_a D^+_b V^{--},$$
$$[\nabla^{++}, \nabla^-_a] = D^+_a, \quad [\nabla^{++}, D^+_a] = [\nabla^{--}, \nabla^-_a] = [\nabla^{\pm\pm}, \nabla_{ab}] = 0,$$
and, next, the covariant superfield strengths
$$[D^+_a, \nabla_{bc}] = \frac{i}{2}\varepsilon_{abcd}W^{+d}, \qquad [\nabla^-_a, \nabla_{bc}] = \frac{i}{2}\varepsilon_{abcd}W^{-d},$$
$$W^{+a} = -\frac{1}{6}\varepsilon^{abcd}D^+_b D^+_c D^+_d V^{--}, \quad W^{-a} := \nabla^{--}W^{+a},$$
$$\nabla^{++}W^{+a} = \nabla^{--}W^{-a} = 0, \quad \nabla^{++}W^{-a} = W^{+a},$$
$$D^+_b W^{+a} = \delta^a_b F^{++}, \quad F^{++} = \frac{1}{4}D^+_a W^{+a} = (D^+)^4 V^{--},$$
$$\nabla^{++}F^{++} = 0, \quad D^+_a F^{++} = 0.$$
- The hypermultiplet superfield has off shell an infinite number of auxiliary fields coming from its expansion over harmonic variables
$$q^{+A}(\zeta) = q^{iA}(x)u^+_i - \theta^{+a}\psi^A_a(x) + \text{An infinite tail of auxiliary fields}, \ \mathrm{A} = 1, 2.$$

2.3. $\mathcal{N} = (1,0)$ superfield actions

The $\mathcal{N} = (1,0)$ SYM action in $6D$ HSS was invented by Zupnik [10]:
$$S^{SYM} = \frac{1}{f^2}\sum_{n=2}^{\infty}\frac{(-1)^{n+1}}{n}\mathrm{Tr}\int d^6x\, d^8\theta\, du_1 \dots du_n \frac{V^{++}(z,u_1)\dots V^{++}(z,u_n)}{(u^+_1 u^+_2)\dots(u^+_n u^+_1)},$$
$$\delta S^{SYM} = 0 \Rightarrow F^{++} = 0,$$
where $1/(u^+_1 u^+_2), \dots$ are the harmonic distributions defined in [11, 12].

The hypermultiplet action (with q^{+A} in the adjoint representation of gauge group) reads
$$S^q = -\frac{1}{2f^2}\mathrm{Tr}\int d\zeta^{(-4)} q^{+A}\nabla^{++}q^+_A, \quad \nabla^{++}q^+_A = D^{++}q^+_A + [V^{++}, q^+_A],$$
$$\delta S^q = 0 \Rightarrow \nabla^{++}q^{+A} = 0.$$

The $\mathcal{N}=(1,0)$ superfield form of the $\mathcal{N}=(1,1)$ SYM action is just a sum of the two actions defined above:

$$S^{(V+q)} = S^{SYM} + S^{q} = \frac{1}{f^2}\left(\int dZ\mathcal{L}^{\rm SYM} - \frac{1}{2}{\rm Tr}\int d\zeta^{(-4)} q^{+A}\nabla^{++}q^{+}_{A}\right),$$

$$\delta S^{(V+q)} = 0 \Rightarrow F^{++} + \frac{1}{2}[q^{+A}, q^{+}_{A}] = 0\,, \quad \nabla^{++}q^{+A} = 0\,.$$

It respects invariance under the second hidden $\mathcal{N}=(0,1)$ supersymmetry:

$$\delta V^{++} = \epsilon^{+A}q^{+}_{A}\,, \quad \delta q^{+A} = -(D^{+})^4(\epsilon^{-}_{A}V^{--})\,, \quad \epsilon^{\pm}_{A} = \epsilon_{aA}\theta^{\pm a}\,.$$

3. $\mathcal{N}=(1,0)$ and $\mathcal{N}=(1,1)$ invariants of higher dimension

The problem was how to construct higher-dimension $\mathcal{N}=(1,1)$ invariants from the $\mathcal{N}=(1,0)$ gauge superfield strength W^{+a} and the hypermultiplet superfield q^{+A}. Firstly, it was solved by direct calculations.

- $\underline{d=6}$: In the pure SYM case the invariant of this dimension is uniquely constructed as [19]

$$S^{(6)}_{SYM} = \frac{1}{2g^2}{\rm Tr}\int d\zeta^{(-4)}du\,\left(F^{++}\right)^2 \sim {\rm Tr}\int d^6x[(\nabla^M F_{ML})^2 + \ldots]$$

  Does its off-shell completion to an off-shell $\mathcal{N}=(1,1)$ invariant exist? The answer is NO, only an expression whose $\mathcal{N}=(0,1)$ variation vanishes *on-shell* can be found. It is unique up to two real parameters

$$\mathcal{L}^{d=6} = \frac{c_0}{2g^2}{\rm Tr}\int du d\zeta^{(-4)}\left(F^{++} + \frac{1}{2}[q^{+A}, q^{+}_{A}]\right)\left(F^{++} + 2\beta[q^{+A}, q^{+}_{A}]\right)$$

  But it vanishes on-shell by itself! Thus the non-vanishing on-shell counterterms of canonical dimension 6 are absent, and this proves the *one-loop finiteness* of $\mathcal{N}=(1,1)$ SYM.

  Recently, $\underline{d=6}$ counterterms were found by the explicit quantum calculations in $\mathcal{N}=(1,0)$ superspace [15] - [17]. It was shown that they vanish *off-shell*, without any use of the equations of motion, just due to vanishing of the corresponding numerical coefficients!
- $\underline{d=8}$: All $\mathcal{N}=(1,0)$ superfield terms of such dimension in the pure $\mathcal{N}=(1,0)$ SYM theory prove to vanish on the gauge fields mass shell, in accord with the old statement of ref. [21]. Could adding the hypermultiplet terms somehow change this?

  Our analysis showed that there exist NO $\mathcal{N}=(1,0)$ supersymmetric off-shell invariants of the dimension 8 which would respect the on-shell $\mathcal{N}=(1,1)$ invariance.

  This means that $\mathcal{N}=(1,1)$ SYM theory is at least *on-shell* finite at two loops. It is still an open question whether it is *off- shell* finite, i.e. whether the coefficients of the candidate counterterms are vanishing, like at one loop (now under examination).

Surprisingly, the $\underline{d=8}$ superfield expression which is non-vanishing on shell and respects an on-shell $\mathcal{N}=(1,1)$ supersymmetry can be constructed by *giving up* the requirement of *off-shell* $\mathcal{N}=(1,0)$ supersymmetry.

An example of such an invariant in $\mathcal{N}=(1,0)$ SYM is very simple

$$\tilde{S}^{(8)}_{1} \sim {\rm Tr}\int d\zeta^{(-4)}\,\varepsilon_{abcd}W^{+a}W^{+b}W^{+c}W^{+d}\,.$$

IOP Conf. Series: Journal of Physics: Conf. Series **965** (2018) 012021 doi:10.1088/1742-6596/965/1/012021

Since $D^+_a W^{+b} = \delta^b_a F^{++}$, this derivative vanishes on shell, *i.e.* when $F^{++} = 0$. Thus, W^{+a} is an analytic superfield, when disregarding the terms proportional to the equations of motion, and the above action respects $\mathcal{N} = (1,0)$ supersymmetry on shell. Also, a double-trace on-shell invariant exists. Both such on-shell invariants admit $\mathcal{N} = (1,1)$ completions.

Though the nontrivial on-shell $\underline{d = 8}$ invariants exist, they cannot appear as counterterms for the $\mathcal{N} = (1,1)$ SYM theory. The reason is that they do *not* possess the off-shell $\mathcal{N} = (1,0)$ supersymmetry which the physically relevant counterterms should obey within the manifestly $\mathcal{N} = (1,0)$ invariant supergraph technique. The non-existence of such counterterms agrees with the old component consideration of ref. [22].

Apart from the fact that such $\underline{d = 8}$ terms cannot appear as *counterterms* in $\mathcal{N} = (1,1)$ SYM theory, they can appear, e.g., as *finite quantum corrections* to the effective Wilsonian action. For the pure $\mathcal{N} = (1,0)$ SYM theory this was observed in [23, 24].

It was desirable to work out some simple and systematic way of constructing higher-order on-shell $\mathcal{N} = (1,1)$ supersymmetric invariants. This becomes possible within the on-shell harmonic $\mathcal{N} = (1,1)$ superspace.

4. $\mathcal{N} = (1,1)$ on-shell harmonic superspace

The first step in constructing such a superspace is to extend $\mathcal{N} = (1,0)$ superspace to $\mathcal{N} = (1,1)$ one,

$$z = (x^{ab}, \theta^a_i) \ \Rightarrow\ \hat{z} = (x^{ab}, \theta^a_i, \hat{\theta}^A_a).$$

The double set of covariant spinor derivatives appears,

$$\nabla^i_a = \frac{\partial}{\partial\theta^a_i} - i\theta^{bi}\partial_{ab} + \mathcal{A}^i_a\,,\ \hat{\nabla}^{aA} = \frac{\partial}{\partial\hat{\theta}_{Aa}} - i\hat{\theta}^A_b\partial^{ab} + \hat{\mathcal{A}}^{aA}\,.$$

The defining constraints of $\mathcal{N} = (1,1)$ SYM in this extended superspace read [18, 21]:

$$\{\nabla^{(i}_a, \nabla^{j)}_b\} = \{\hat{\nabla}^{a(A}, \hat{\nabla}^{bB)}\} = 0\,, \quad \{\nabla^i_a, \hat{\nabla}^{bA}\} = \delta^b_a \phi^{iA}$$
$$\Rightarrow \quad \nabla^{(i}_a \phi^{j)A} = \hat{\nabla}^{a(A}\phi^{B)i} = 0 \quad \text{(By Bianchis)}\,.$$

Next, we are led to define $\mathcal{N} = (1,1)$ HSS with the double set of $SU(2)$ harmonics [14]:

$$Z = (x^{ab}, \theta^a_i, u^\pm_k) \ \Rightarrow\ \hat{Z} = (x^{ab}, \theta^a_i, \hat{\theta}^A_b, u^\pm_k, u^{\hat{\pm}}_A)$$

Then we pass to the analytic basis and choose the "hatted" spinor derivatives short, $\nabla^{\hat{+}a} = D^{\hat{+}a} = \frac{\partial}{\partial\theta^{\hat{-}}_a}$. The $\mathcal{N} = (1,1)$ SYM constraints are rewritten in $\mathcal{N} = (1,1)$ HSS as

$$\{\nabla^+_a, \nabla^+_b\} = 0\,, \quad \{D^{\hat{+}a}, D^{\hat{+}b}\} = 0\,, \quad \{\nabla^+_a, D^{\hat{+}b}\} = \delta^b_a \phi^{+\hat{+}}\,,$$
$$[\nabla^{\hat{+}\hat{+}}, \nabla^+_a] = 0\,,\ [\tilde{\nabla}^{++}, \nabla^+_a] = 0\,,\ [\nabla^{\hat{+}\hat{+}}, D^{a\hat{+}}] = 0\,,\ [\tilde{\nabla}^{++}, D^{a\hat{+}}] = 0\,,$$
$$[\tilde{\nabla}^{++}, \nabla^{\hat{+}\hat{+}}] = 0\,.$$

Here

$$\nabla^+_a = D^+_a + \mathcal{A}^+_a(\hat{Z})\,,\ \tilde{\nabla}^{++} = D^{++} + \tilde{V}^{++}(\hat{\zeta})\,, \quad \nabla^{\hat{+}\hat{+}} = D^{\hat{+}\hat{+}} + V^{\hat{+}\hat{+}}(\hat{\zeta})\,,$$
$$\hat{\zeta} = (x^{ab}_{\text{an}}, \theta^{\pm a}, \theta^{\hat{+}}_c, u^\pm_i, u^{\hat{\pm}}_A)\,.$$

The starting point of solving these constraints is to fix, using the $\Lambda(\hat{\zeta})$ gauge freedom, the WZ gauge for the second harmonic connection $V^{\hat{+}\hat{+}}(\hat{\zeta})$

$$V^{\hat{+}\hat{+}} = i\theta^{\hat{+}}_a\theta^{\hat{+}}_b\hat{\mathcal{A}}^{ab} + \varepsilon^{abcd}\theta^{\hat{+}}_a\theta^{\hat{+}}_b\theta^{\hat{+}}_c\varphi^A_d u^{\hat{-}}_A + \varepsilon^{abcd}\theta^{\hat{+}}_a\theta^{\hat{+}}_b\theta^{\hat{+}}_c\theta^{\hat{+}}_d\mathcal{D}^{AB}u^{\hat{-}}_A u^{\hat{-}}_B\,,$$

with $\hat{\mathcal{A}}^{ab}, \varphi^A_d$ and $\mathcal{D}^{(AB)}$ being some $\mathcal{N} = (1,0)$ harmonic superfields.

Then the above constraints are reduced to some harmonic equations which can be explicitly solved. The crucial point is the requirement that the vector $6D$ connections in the sectors of hatted and unhatted variables **are identical** to each other.

As the final result, we have obtained that the first harmonic connection V^{++} coincides precisely with the standard $\mathcal{N} = (1,0)$ one, $V^{++} = V^{++}(\zeta)$, while the dependence of all other geometric $\mathcal{N} = (1,1)$ objects on the variables with "hat" proves to be strictly fixed

$$V^{\hat{+}\hat{+}} = i\theta^{\hat{+}}_a\theta^{\hat{+}}_b\mathcal{A}^{ab} - \frac{1}{3}\epsilon^{abcd}\theta^{\hat{+}}_a\theta^{\hat{+}}_b\theta^{\hat{+}}_c D^{+}_d q^{-\hat{-}} + \frac{1}{8}\epsilon^{abcd}\theta^{\hat{+}}_a\theta^{\hat{+}}_b\theta^{\hat{+}}_c\theta^{\hat{+}}_d[q^{+\hat{-}}, q^{-\hat{-}}]$$
$$\phi^{+\hat{+}} = q^{+\hat{+}} - \theta^{\hat{+}}_a W^{+a} - i\theta^{\hat{+}}_a\theta^{\hat{+}}_b\nabla^{ab}q^{+\hat{-}} + \frac{1}{6}\varepsilon^{abcd}\theta^{\hat{+}}_a\theta^{\hat{+}}_b\theta^{\hat{+}}_c[D^{+}_d q^{-\hat{-}}, q^{+\hat{-}}]$$
$$+\frac{1}{24}\varepsilon^{abcd}\theta^{\hat{+}}_a\theta^{\hat{+}}_b\theta^{\hat{+}}_c\theta^{\hat{+}}_d[q^{+\hat{-}}, [q^{+\hat{-}}, q^{-\hat{-}}]]\,.$$

Here, $q^{+\hat{\pm}} = q^{+A}(\zeta)u^{\hat{\pm}}_A$, $q^{-\hat{\pm}} = q^{-A}(\zeta)u^{\hat{\pm}}_A$ and $W^{+a}, q^{\pm A}$ are just the $\mathcal{N} = (1,0)$ superfields used previously. In the process of solving the constraints, there appeared the analyticity conditions for q^{+A}, as well as the full set of the superfield equations of motion

$$\nabla^{++}q^{+A} = 0\,, \quad F^{++} = \frac{1}{4}D^{+}_a W^{+a} = -\frac{1}{2}[q^{+A}, q^{+}_A]\,.$$

The basic advantage of using the constrained $\mathcal{N} = (1,1)$ strengths $\phi^{\pm\hat{+}}$ for constructing various invariants is their extremely simple transformation rules under the hidden $\mathcal{N} = (0,1)$ supersymmetry

$$\delta\phi^{\pm\hat{+}} = -\epsilon^{\hat{+}}_a\frac{\partial}{\partial\theta^{\hat{+}}_a}\phi^{\pm\hat{+}} - 2i\epsilon^{\hat{-}}_a\theta^{\hat{+}}_b\partial^{ab}\phi^{\pm\hat{+}} - [\Lambda^{(comp)}, \phi^{\pm\hat{+}}]\,,$$

where $\Lambda^{(comp)}$ is some common composite gauge parameter which does not contribute under Tr.

5. Invariants in $\mathcal{N} = (1,1)$ superspace

The previous single-trace on-shell $\underline{d=8}$ invariant admits a simple rewriting in $\mathcal{N} = (1,1)$ superspace

$$S_{(1,1)} = \int du d\zeta^{(-4)}\mathcal{L}^{+4}_{(1,1)}\,, \quad \mathcal{L}^{+4}_{(1,1)} = -\mathrm{Tr}\,\frac{1}{4}\int d\hat{\zeta}^{(-4)}d\hat{u}\,(\phi^{+\hat{+}})^4, \; d\hat{\zeta}^{(-4)} \sim (D^{\hat{-}})^4$$

$$\delta\mathcal{L}^{+4}_{(1,1)} = -2i\partial^{ab}\mathrm{Tr}\int d\hat{\zeta}^{(-4)}d\hat{u}\left[\epsilon^{\hat{-}}_a\theta^{\hat{+}}_b\frac{1}{4}(\phi^{+\hat{+}})^4\right].$$

The double-trace $\underline{d=8}$ invariant can also be easily constructed.

Now it is easy to construct the single- and double-trace $\underline{d=10}$ invariants possibly responsible for the 3-loop counterterms

$$S^{(10)}_1 = \mathrm{Tr}\int dZ d\hat{\zeta}^{(-4)}d\hat{u}\,(\phi^{+\hat{+}})^2(\phi^{-\hat{+}})^2, \quad \phi^{-\hat{+}} = \nabla^{--}\phi^{+\hat{+}}\,,$$
$$S^{(10)}_2 = -\int dZ d\hat{\zeta}^{(-4)}d\hat{u}\,\mathrm{Tr}\left(\phi^{+\hat{+}}\phi^{-\hat{+}}\right)\mathrm{Tr}\left(\phi^{+\hat{+}}\phi^{-\hat{+}}\right).$$

IOP Conf. Series: Journal of Physics: Conf. Series **965** (2018) 012021 doi:10.1088/1742-6596/965/1/012021

These are $\mathcal{N} = (1,1)$ extensions of the $\mathcal{N} = (1,0)$ SYM invariants $\sim \varepsilon_{abcd}\mathrm{Tr}\,(W^{+a}W^{-b}W^{+c}W^{-d})$, $\sim \varepsilon_{abcd}\,\mathrm{Tr}\,(W^{+a}W^{-b})\,\mathrm{Tr}\,(W^{+c}W^{-d})$.

It is notable that the single-trace $\underline{d = 10}$ invariant admits a representation as an integral over the *full* $\mathcal{N} = (1,1)$ superspace

$$S_1^{(10)} \sim \mathrm{Tr} \int dZ d\hat{Z}\; \phi^{++\hat{+}}\phi^{--\hat{-}}\,, \quad \phi^{--\hat{-}} = \nabla^{\hat{-}\hat{-}}\phi^{-\hat{+}}\,.$$

On the other hand, the double-trace $\underline{d = 10}$ invariant *cannot* be written as the full integral and so it looks as being UV protected.

This could partly explain why in the perturbative calculations of the amplitudes in the $\mathcal{N} = (1,1)$ SYM single-trace 3-loop divergence is seen, while no double-trace structures at the same order were observed [6, 7]

However, this does not seem to be like the standard non-renormalization theorems because the quantum calculation of $\mathcal{N} = (1,0)$ supergraphs should generate invariants in the off-shell $\mathcal{N} = (1,0)$ superspace, not in the on-shell $\mathcal{N} = (1,1)$ superspace. So some additional piece of reasoning is needed to explain the absence of the double-trace divergences.

6. Quantum $\mathcal{N} = (1,0)$ and $\mathcal{N} = (1,1)$ SYM

For calculating various $\mathcal{N} = (1,0)$ and $\mathcal{N} = (1,1)$ invariants, including counterterms, there was an urgent need to formulate self-consistent $\mathcal{N} = (1,0)$ superfield perturbation techniques: superpropagators, background field method, etc. All that was recently given in a few papers by Buchbinder, Ivanov, Merzlikin and Stepanyantz, [15] - [17]. These methods were used to prove the one-loop **off-shell** finiteness of $\mathcal{N} = (1,1)$ SYM theory formulated in terms of $\mathcal{N} = (1,0)$ superfields.

The basic idea of the background field method is to split the relevant superfields into the sum of the "background" superfields V^{++}, Q^+ and the "quantum" ones v^{++}, q^+,

$$V^{++} \to V^{++} + f v^{++}, \qquad q^+ \to Q^+ + q^+\,, \tag{5}$$

and then to expand the action in a power series in quantum fields.

By skipping details, the $\mathcal{N} = (1,0)$, $6D$ SYM theory in the background field approach is described by the three quantum superfield ghosts: two fermionic Faddeev-Popov ghosts $\mathbf{b}$ and $\mathbf{c}$ together with the single bosonic Nielsen-Kallosh ghost φ, in addition to the quantum v^{++} and q^+ superfields. We started with the model in which hypermultiplet belongs to an arbitrary representation R of gauge group, not just to adjoint.

After integrating, in the functional integral, over quantum superfields, the following representation for the one-loop quantum correction to the classical action is obtained

$$\Gamma^{(1)}[V^{++}, Q] = \frac{i}{2}\mathrm{Tr}\ln\Big\{ \widehat{\Box}^{AB} - 2f^2 \widetilde{Q}^{+m}(T^A G_{(1,1)} T^B)_m{}^n Q^+_n \Big\} - \frac{i}{2}\mathrm{Tr}\ln\widehat{\Box}$$
$$-i\mathrm{Tr}\ln(\nabla^{++})^2_{\mathrm{Adj}} + \frac{i}{2}\mathrm{Tr}\ln(\nabla^{++})^2_{\mathrm{Adj}} + i\mathrm{Tr}\ln\nabla^{++}_{\mathrm{R}}\,,$$

where subscripts Adj and R mean that the corresponding operators are taken in the adjoint representation and that of the hypermultiplet and

$$\widehat{\Box} = \frac{1}{2}(D^+)^4(\nabla^{--})^2$$

is the covariant Box operator.

The complete one-loop divergent part of the effective action reads

$$\Gamma^{(1)}_{div}[V^{++},Q^{+}] = \frac{C_2 - \frac{d_R}{d_G}C_2(R)}{3(4\pi)^3\varepsilon}\,\mathrm{tr}\int d\zeta^{(-4)}du\,(F^{++})^2 - \frac{2if^2}{(4\pi)^3\varepsilon}\int d\zeta^{(-4)}du\,\tilde{Q}^{+}(C_2 - C_2(R))F^{++}Q^{+}.$$

The coefficients of the $(F^{++})^2$ and $\tilde{Q}^{+}F^{++}Q^{+}$ terms in the divergent part of one-loop effective action are proportional to the differences between the second order Casimir operator C_2 for adjoint representation of gauge group and the operators $T(R) = \frac{d_R}{d_G}C_2(R)$ and $C_2(R)$ for the hypermultiplet representation R, respectively.

Since $\mathcal{N} = (1,1)$, $6D$ supersymmetric Yang-Mills theory involves the hypermultiplet in **adjoint** representation of gauge group, with $d_R = d_G$, $C_2(R) = C_2$, the divergent part vanishes for this case. Hence, the $\mathcal{N} = (1,1)$ SYM theory is one-loop finite, and there is no need to use the equations of motion to prove this property.

For any other choice of the hypermultiplet irrep, it does not vanish even on shell, so in general the theory is divergent already at one loop. The pure $\mathcal{N} = (1,0)$ SYM corresponds to $C_2(R) = 0$ and the one-loop divergency is vanishing on the equation of motion $F^{++} = 0$, in accord with the old result by Howe and Stelle [21].

7. Summary and outlook

- Off-shell $\mathcal{N} = (1,0)$ and on-shell harmonic $\mathcal{N} = (1,1)$, $6D$ superspaces can be efficiently used to construct higher-dimensional invariants in the $\mathcal{N} = (1,0)$ and $\mathcal{N} = (1,1)$ SYM theories.
- $\mathcal{N} = (1,1)$ SYM constraints were solved in terms of harmonic $\mathcal{N} = (1,0)$ superfields. This allowed to explicitly construct the full set of the superfield dimension $d = 8$ and $d = 10$ invariants with $\mathcal{N} = (1,1)$ on-shell supersymmetry.
- All $\underline{d = 6}$ $\mathcal{N} = (1,1)$ invariants are at least on-shell vanishing, proving the UV finiteness of $\mathcal{N} = (1,1)$ SYM at one loop.
- The off-shell $\underline{d = 8}$ $\mathcal{N} = (1,1)$ invariants are absent. Assuming that the $\mathcal{N} = (1,0)$ supergraphs yield integrals over the full $\mathcal{N} = (1,0)$ harmonic superspace, this means the absence of two-loop counterterms.
- Two $\underline{d = 10}$ invariants were explicitly constructed as integrals over the whole $\mathcal{N} = (1,0)$ harmonic superspace. The single-trace invariant can be rewritten as an integral over $\mathcal{N} = (1,1)$ superspace, while the double-trace one cannot. This property combined with an additional reasoning could explain why the double-trace invariant is UV protected.
- The quantum techniques for $\mathcal{N} = (1,0)$ SYM theory was worked out and used to show that $\mathcal{N} = (1,1)$ SYM theory is one-loop finite off shell, without need in eqs. of motion.

7.1. Further lines of study

In conclusion, we outline some further possible lines of study:

(**a**) To construct the next $d \geq 12$ invariants in the $\mathcal{N} = (1,1)$ SYM theory with the help of the on-shell $\mathcal{N} = (1,1)$ harmonic superspace techniques (Buyukli & Ivanov, in preparation);

(**b**) To reproduce higher-dimensional invariants from the quantum superfield perturbation theory, to examine whether $\mathcal{N} = (1,1)$ SYM theory is two-loop finite off shell (Buchbinder *et al*, in preparation);

(**c**) To work out the quantum superfield perturbation theory directly in $\mathcal{N} = (1,1)$ double-harmonized superspace;

(**d**) To apply the same methods for constructing the Born-Infeld action with manifest off-shell $\mathcal{N} = (1,0)$ and hidden on-shell $\mathcal{N} = (0,1)$ supersymmetries. To check the hypothesis that such an action could be identified with the full quantum effective action of the $\mathcal{N} = (1,1)$ SYM;

(**e**) Applications in supergravity?

Acknowledgments
The author thanks the organizers of the Conference ISQS25 for the kind hospitality in Prague.

References

[1] Beisert N 2010 *On Yangian Symmetry in Planar $\mathcal{N} = 4$ SYM* (*Preprint* arXiv:1004.5423 [hep-th])
[2] Dennen T and Huang Y 2011 Dual Conformal Properties of Six-Dimensional Maximal Super Yang-Mills Amplitudes *JHEP* **1101** 140 (*Preprint* arXiv:1010.5874 [hep-th])
[3] Tseytlin A A 1997 On non-abelian generalization of Born-Infeld action in string theory *Nucl. Phys.* B **501** 41-52 (*Preprint* hep-th/9701125)
[4] Frampton P H and Kephart T W Explicit Evaluation of Anomalies in Higher Dimensions 1983 *Phys. Rev. Lett.* **50** 1343
[5] Bern Z, Carrasco J J M, Dixon L J, Johansson H and Roiban R 2010 The Complete Four-Loop Four-Point Amplitude in N=4 Super-Yang-Mills Theory *Phys. Rev.* D **82** 125040 (*Preprint* arXiv:1008.3327 [hep-th])
[6] Berkovits N, Green M B, Russo J G and Vanhove P 2009 Non-renormalization conditions for four-gluon scattering in supersymmetric string and field theory *JHEP* **0911** 063 (*Preprint* arXiv:0908.1923 [hep-th])
[7] Bjornsson J 2011 Multi-loop amplitudes in maximally supersymmetric pure spinor field theory *JHEP* **1101** 002 (*Preprint* arXiv:1009.5906 [hep-th])
[8] Cederwall M 2013 *Pure spinor superfields - an overview* (*Preprint* arXiv:1307.1762 [hep-th])
[9] Howe P S, Stelle K S and West P C 1985 $\mathcal{N} = 1, d = 6$ harmonic superspace *Class. Quant. Grav.* **2** 815
[10] Zupnik B M 1986 Six-dimensional Supergauge Theories in the Harmonic Superspace *Sov. J. Nucl. Phys.* **44** 512
[11] Galperin A S, Ivanov E A, Kalitzin S, Ogievetsky V I and Sokatchev E S 1984 Unconstrained $\mathcal{N} = 2$ matter, Yang-Mills and supergravity theories in harmonic superspace *Class. Quant. Grav.* **1** 469-498
[12] Galperin A S, Ivanov E A, Ogievetsky V I and Sokatchev E S 2001 *Harmonic Superspace* (Cambridge University Press 306 p.)
[13] Bossard G, Ivanov E and Smilga A 2015 Ultraviolet behavior of 6D supersymmetric Yang-Mills theories and harmonic superspace *JHEP* **1512** 085 (*Preprint* arXiv:1509.08027 [hep-th])
[14] Bossard G, Howe P S and Stelle K S 2009 The Ultra-violet question in maximally supersymmetric field theories 2009 *Gen. Rel. Grav.* **41** 919 (*Preprint* arXiv:0901.4661 [hep-th])
[15] Buchbinder I L, Ivanov E A, Merzlikin B S and Stepanyantz K V 2016 One-loop divergences in the $6D, \mathcal{N} = (1,0)$ abelian gauge theory *Phys. Lett.* B **763** 375-381 (*Preprint* arXiv:1609.00975 [hep-th])
[16] Buchbinder I L, Ivanov E A, Merzlikin B S and Stepanyantz K V 2017 One-loop divergences in $6D, \mathcal{N} = (1,0)$ SYM theory *JHEP* **1701** 128 (*Preprint* arXiv:1612.03190 [hep-th])
[17] Buchbinder I L, Ivanov E A, Merzlikin B S and Stepanyantz K V 2017 Supergraph analysis of the one-loop divergences in $6D, \mathcal{N} = (1,0)$ and $\mathcal{N} = (1,1)$ gauge theories *Nucl. Phys.* B **921** 127-158 (*Preprint* arXiv:1704.02530 [hep-th])
[18] Howe P S, Sierra G and Townsend P K 1983 Supersymmetry in six dimensions *Nucl. Phys.* B **221** 331-348
[19] Ivanov E A, Smilga A V and Zupnik B M 2005 Renormalizable supersymmetric gauge theory in six dimensions *Nucl. Phys.* B **726** 131-148 (*Preprint* hep-th/0505082)
[20] Ivanov E A and Smilga A V 2006 Conformal properties of hypermultiplet actions in six dimensions *Phys. Lett.* B **637** 374-381 (*Preprint* hep-th/0510273)
[21] Howe P S and Stelle K S 1984 Ultraviolet Divergences in Higher Dimensional Supersymmetric Yang-Mills Theories *Phys. Lett.* B **137** 175-180
[22] Markus N and Sagnotti A 1984 A test of finiteness predictions for supersymmetric theories *Phys. Lett.* B **135** 85-90
[23] Buchbinder I L and Pletnev N G 2015 Construction of $6D$ supersymmetric field models in $\mathcal{N} = (1,0)$ harmonic superspace *Nucl.Phys.* B **892** 21-48 (*Preprint* arXiv:1411.1848 [hep-th])

IOP Conf. Series: Journal of Physics: Conf. Series **965** (2018) 012021 doi:10.1088/1742-6596/965/1/012021

[24] Buchbinder I L and Pletnev N G 2015 Leading low-energy effective action in the 6D hypermultiplet theory on a vector/tensor background *Phys. Lett.* B **744** 125-130 (*Preprint* arXiv:1502.03257 [hep-th])

Operator algebra as an application of logarithmic representation of infinitesimal generators

Yoritaka Iwata
Institute of Innovative Research, Tokyo Institute of Technology, Tokyo, Japan
Department of Mathematics, Shibaura Institute of Technology, Tokyo, Japan
E-mail: iwata_phys@08.alumni.u-tokyo.ac.jp

Abstract. The operator algebra is introduced based on the framework of logarithmic representation of infinitesimal generators. In conclusion a set of generally-unbounded infinitesimal generators is characterized as a module over the Banach algebra.

1. Introduction

The logarithmic representation of infinitesimal generator is introduced in Ref. [1]. Its role is further discussed in Ref. [2]. Let X be a Banach space, and Y be a dense Banach subspace of X. Let $\{U(t,s)\}_{-T\le t,s\le T}$ be an evolution family of operators satisfying the semigroup property (for example, see [3]):

$$U(t,r)U(r,s) = U(t,s),$$
$$U(s,s) = U(s,t)U(t,s) = I$$

on X. Let the infinitesimal generator of $U(t,s)$ be denoted by $A(t)$ from Y to X. The logarithmic representation for the infinitesimal generator is

$$A(t)\ u_s = (I + \kappa U(s,t))\ \partial_t \mathrm{Log}\ (U(t,s) + \kappa I)\ u_s, \tag{1}$$

where u_s is an element in Y, $\kappa \neq 0$ is a certain complex number, and Log denotes a principal branch of the logarithm. ∂_t stands for a kind of weak differential defined in Ref. [1]. For this definition, operators $U(t,s)$ and $A(t)$ are assumed to commute [1]. In particular the representation plays a role of extracting the bounded part of the infinitesimal generator. In this article a formulation of operator algebra is established for generally-unbounded infinitesimal generators.

2. Basic relation

The relation between exponential and logarithm functions of operators is shown under the validity of logarithmic representation (1). According to Eq. (1) the logarithm function in this case is given by Log $(U(t,s)+\kappa I)$, and is formally or expectantly equal to the indefinite integral

$$\mathrm{Log}\ (U(t,s) + \kappa I) = \int (I + \kappa U(s,t))^{-1} A(t)\ dt.$$

Its exponential function can be calculated as

$$\exp[\text{Log } (U(t,s)+\kappa I)] = U(t,s)+\kappa I \tag{2}$$

$\text{Log} \exp[\text{Log } (U(t,s)+\kappa I)] = \text{Log}(U(t,s)+\kappa I)$ is always satisfied, but the above formal equality $\int (I+\kappa U(s,t))^{-1} A(t)\, dt = \text{Log } (U(t,s)+\kappa I)$ is not necessarily satisfied because the limited range of imaginary spectral distribution is true only for the right hand side. In this sense $\text{Log } (U(t,s)+\kappa I)$ corresponds to the extracted bounded part of the infinitesimal generator $A(t)$.

A set of these operators is a candidate for bounded-operator algebra. Indeed, if κ is taken from the resolvent set of operator $U(t,s)$, $U(t,s)+\kappa I$ and $(U(t,s)+\kappa I)^{-1}$ are necessarily bounded on X, and then the boundedness of $\text{Log } (U(t,s)+\kappa I)$ follows. That is, $\text{Log } (U(t,s)+\kappa I)$ and $U(t,s)+\kappa I$ are related by the exponential function of operators which can be defined by the convergent power series. Meanwhile the logarithm function $\text{Log } (U(t,s)+\kappa I)$ cannot necessarily be defined by the convergent power series which is too restrictive to represent the function of $U(t,s)$ (i.e., $|U(t,s)-1|<1$ is required).

3. Main results

Let $\text{Log}(U(t,s)+\kappa I)$ be denoted by

$$a(t,s) = \text{Log } (U(t,s)+\kappa I), \tag{3}$$

where some algebraic properties of $a(t,s)$ can be found in Ref. [2]. Since $a(t,s)$ is bounded on X, $e^{a(t,s)}$ is well-defined by the convergent power series. Consequently the exponentiability is reduced to $\text{Log } (U(t,s)+\kappa I)$. Using Eq. (2),

$$U(t,s) = e^{a(t,s)} - \kappa I,$$

follows for a certain complex number κ. It is understood by this equation that, with respect to the mathematical property of infinitesimal generators represented by Eq. (1), it is sufficient to consider $a(t,s)$ instead of $A(t)$.

Theorem 1. Let $U_i(t,s)$ be evolution operators satisfying Eq. (1), and $\text{Log } U_i(t,s)$ be well-defined for any $t,s \in [-T,T]$ and $i=1,2,\cdots,n$. $\text{Log}U_i(t,s)$ are assumed to commute with each other.

$$V_{Lg}(X) := \{k \text{Log } U_i(t,s);\ k \in \boldsymbol{C},\ t,s \in [-T,T]\} \ \subset B(X)$$

is a normed vector space over the complex number field, where $B(X)$ denotes a set of all the bounded operators on X, and the operator norm is equipped with $B(X)$.

Theorem 2. Let $U_i(t,s)$ be evolution operators satisfying Eq. (1) for any $t,s \in [-T,T]$ and $i=1,2,\cdots,n$. For a certain $K \in B(X)$, let a subset of $B(X)$ in which each element is assumed to commute with $\text{Log } (U_i(t,s)+K)$ be $B_{ab}(X)$. $\text{Log}(U_i(t,s)+K)$ are assumed to commute with each other.

$$B_{Lg}(X) := \{\mathcal{K}\text{Log } (U_i(t,s)+K);\ \mathcal{K} \in B_{ab}(X),\ K \in B(X),\ t,s \in [-T,T]\} \ \subset B(X)$$

is a module over the Banach algebra $B(X)$.

IOP Conf. Series: Journal of Physics: Conf. Series **965** (2018) 012022 doi:10.1088/1742-6596/965/1/012022

4. Outlines of the proofs

4.1. A normed vector space

The proof of Theorem 1 is presented. In case of $\kappa = 0$ the operator $a(t, s)$ is reduced to

$$\text{Log } U(t,s) \in B(X).$$

The operator sum is calculated using Dunford-Riesz integral

$$\begin{aligned}
&\text{Log } U(t,r) + \text{Log } U(r,s) \\
&= \frac{1}{2\pi i}\int_\Gamma \text{Log}\lambda \ (\lambda - U(t,r))^{-1} d\lambda \ + \frac{1}{2\pi i}\int_{\Gamma'} \text{Log}\lambda' \ (\lambda' - U(r,s))^{-1} d\lambda' \\
&= \frac{1}{(2\pi i)^2}\int_\Gamma\int_{\Gamma'} (\text{Log}\lambda + \text{Log}\lambda') \ (\lambda - U(t,r))^{-1} \ (\lambda' - U(r,s))^{-1} \ d\lambda' d\lambda \\
&= \frac{1}{(2\pi i)^2}\int_\Gamma\int_{\Gamma'} (\text{Log}\lambda\lambda') \ (\lambda - U(t,r))^{-1} \ (\lambda' - U(r,s))^{-1} \ d\lambda' d\lambda \\
&= \text{Log } [U(t,r)U(r,s)] = \text{Log } U(t,s),
\end{aligned} \tag{4}$$

then the sum closedness is clear. Here Γ' is assumed to be included in Γ, and this condition is not so restrictive in the present setting. In a different situation, when $U(t,r)$ and $V(t,r)$ commute for the same t and r, another kind of sum is calculated as

$$\begin{aligned}
&\text{Log } U_1(t,r) + \text{Log } U_2(t,r) \\
&= \frac{1}{2\pi i}\int_\Gamma \text{Log}\lambda \ (\lambda - U_1(t,r))^{-1} d\lambda \ + \frac{1}{2\pi i}\int_{\Gamma'} \text{Log}\lambda' \ (\lambda' - U_2(t,r))^{-1} d\lambda' \\
&= \frac{1}{(2\pi i)^2}\int_\Gamma\int_{\Gamma'} (\text{Log}\lambda + \text{Log}\lambda') \ (\lambda - U_1(t,r))^{-1} \ (\lambda' - U_2(t,r))^{-1} \ d\lambda' d\lambda \\
&= \frac{1}{(2\pi i)^2}\int_\Gamma\int_{\Gamma'} (\text{Log}\lambda\lambda') \ (\lambda - U_1(t,r))^{-1} \ (\lambda' - U_2(t,r))^{-1} \ d\lambda' d\lambda \\
&= \text{Log } [U_1(t,r)U_2(t,r)],
\end{aligned} \tag{5}$$

where, for $W(t,r) = U_1(t,r)U_2(t,r)$, the semigroup property is satisfied as

$$\begin{aligned}
&W(t,r)W(r,s) = U_1(t,r)U_2(t,r)U_1(r,s)U_2(r,s) = U_1(t,r)U_1(r,s)U_2(t,r)U_2(r,s) = W(t,s), \\
&W(s,s) = W(s,t)W(t,s) = I,
\end{aligned}$$

and then the sum closedness is clear. Although the logarithm function is inherently a multi-valued function, the uniqueness of sum operation is ensured by the single-valued property of the principal branch "Log". Consequently, since the closedness for scaler product is obvious,

$$V_{Lg}(X) = \{k\text{Log } U(t,s); \ k \in \mathbf{C}, \ t,s \in [-T,T]\} \subset B(X)$$

is a normed vector space over the complex number field. In particular the zero operator $\text{Log}I$ is included in $V_{Lg}(X)$. Theorem 1 has been proved.

4.2. $B(X)$-module

The proof of Theorem 2 is presented. It is worth generalizing the above normed vector space. In this sense, utilizing a common operator $K \in B(X)$, components are changed to $\text{Log } (U_i(t,r)+K)$.

The operator sum is calculated as

$$\begin{aligned}
&\text{Log } (U(t,r)+K) + \text{Log } (U(r,s)+K) \\
&= \tfrac{1}{2\pi i} \int_\Gamma \text{Log}\lambda \ (\lambda - U(t,r) - K)^{-1} d\lambda \ + \tfrac{1}{2\pi i} \int_{\Gamma'} \text{Log}\lambda' \ (\lambda' - U(r,s) - K)^{-1} d\lambda' \\
&= \tfrac{1}{(2\pi i)^2} \int_\Gamma \int_{\Gamma'} (\text{Log}\lambda + \text{Log}\lambda') \ (\lambda - U(t,r) - K)^{-1} \ (\lambda' - U(r,s) - K)^{-1} \ d\lambda' d\lambda \\
&= \tfrac{1}{(2\pi i)^2} \int_\Gamma \int_{\Gamma'} (\text{Log}\lambda\lambda') \ (\lambda - U(t,r) - K)^{-1} \ (\lambda' - U(r,s) - K)^{-1} \ d\lambda' d\lambda \\
&= \text{Log } [(U(t,r)+K)(U(r,s)+K)] \\
&= \text{Log } [U(t,s) + KU(t,r) + KU(r,s) + K^2].
\end{aligned} \tag{6}$$

After introducing a certain K with sufficient large $|K|$, it is always possible to take integral path Γ' to be included in Γ. Since the part "$KU(t,r) + KU(r,s) + K^2 I$" is included in $B(X)$, the sum-closedness is clear. In a different situation, when $U_1(t,r)$ and $U_2(t,r)$ commute for the same t and r, another kind of sum is calculated as

$$\begin{aligned}
&\text{Log } (U_1(t,r)+K) + \text{Log } (U_2(t,r)+K) \\
&= \tfrac{1}{2\pi i} \int_\Gamma \text{Log}\lambda \ (\lambda - U_1(t,r) - K)^{-1} d\lambda \ + \tfrac{1}{2\pi i} \int_{\Gamma'} \text{Log}\lambda' \ (\lambda' - U_2(t,r) - K)^{-1} d\lambda' \\
&= \tfrac{1}{(2\pi i)^2} \int_\Gamma \int_{\Gamma'} (\text{Log}\lambda + \text{Log}\lambda') \ (\lambda - U_1(t,r) - K)^{-1} \ (\lambda' - U_2(t,r) - K)^{-1} \ d\lambda' d\lambda \\
&= \tfrac{1}{(2\pi i)^2} \int_\Gamma \int_{\Gamma'} (\text{Log}\lambda\lambda') \ (\lambda - U_1(t,r) - K)^{-1} \ (\lambda' - U_2(t,r) - K)^{-1} \ d\lambda' d\lambda \\
&= \text{Log } [U_1(t,r)+K)(U_2(t,r)+K)] \\
&= \text{Log } [W(t,r) + KU_1(t,r) + KU_2(t,r) + K^2 I].
\end{aligned} \tag{7}$$

Since the part "$KU_1(t,r) + KU_2(t,r) + K^2$" is included in $B(X)$, the sum-closedness is clear.

The product $\mathcal{K}$Log $(U_i(t,s) + K_2) \in B(X)$ is justified by the operator product equipped with $B(X)$. Consequently, since the closedness for operator product within $B(X)$ is obvious,

$$B_{Lg}(X) = \{\mathcal{K}\text{Log } (U(t,s)+K); \ \mathcal{K} \in B_{ab}(X), \ K \in B(X), \ t,s \in [-T,T]\}$$

is a module over the Banach algebra ($B(X)$-module). In particular relation $V_{Lg}(X) \subset B_{Lg}(X)$ is satisfied. Theorem 2 has been proved.

For the structure of $B_{Lg}(X)$, a certain originally unbounded part can be classified to Log $(U(t,s)+K) \in B(X)$, and the rest part to $\mathcal{K} \in B_{ab}(X)$. Here the terminology "originally unbounded" is used, because some unbounded operators are reduced to bounded operators under the validity of the logarithmic representation.

Acknowledgement
The author is grateful to Prof. Emeritus Hiroki Tanabe for fruitful comments. This work was supported by grant-in-aid research (C) 17K05440.

References
[1] Iwata Y 2017 Methods Funct. Anal. Topology **23** 1 26
[2] Iwata Y 2017 Journal of Applied Mathematics and Physics **5** 822
[3] Tanabe H 1979 Equations of Evolution. Pitman

IOP Conf. Series: Journal of Physics: Conf. Series **965** (2018) 012023 doi:10.1088/1742-6596/965/1/012023

Quantum models based on finite groups

Vladimir V Kornyak
Laboratory of Information Technologies, Joint Institute for Nuclear Research
141980 Dubna, Russia

E-mail: vkornyak@gmail.com

Abstract. We consider a constructive modification of quantum-mechanical formalism. Replacement of a general unitary group by unitary representations of finite groups makes it possible to reproduce quantum formalism without loss of its empirical content. Since any linear representation of a finite group can be implemented as a subrepresentation of a permutation representation, quantum-mechanical problems can be formulated in terms of permutation groups. Reproducing quantum behavior in the framework of permutation representations of finite groups makes it possible to clarify the meaning of a number of physical concepts.

1. Introduction

Physics, being an empirical science, is insensitive to the replacement of infinite concepts in physical theories by finite ones. The reason for such a replacement is that effective modeling is possible only if the problem is formulated in finite terms. Moreover, it is preferable to formulate physical problems in finite terms from the very beginning, since infinities can lead to both descriptive losses and artifacts. In this context, Paul Dirac made the often quoted statement that the most important challenge in physics is "*to get rid of infinity*". In this paper, we describe a constructive version of quantum formalism that does not involve any concepts associated with actual infinities.

The main part of the paper begins with a summary of the basic concepts of the standard quantum mechanics with emphasis on the aspects important for our purposes.

Further we describe a constructive modification of the quantum formalism. We start with replacing a unitary group — a continuous group of symmetries of quantum states — by a finite group. The natural consequence of this replacement is unitarity: any linear representation of a finite group is unitary. Any finite group is naturally associated with some cyclotomic field — an extension of the rational numbers by roots of unity. Generally, a cyclotomic field is a dense subfield of the field of complex numbers. This can be regarded as an explanation of the presence of complex numbers in the quantum formalism. Any linear representation of a finite group over a cyclotomic field associated with the group can be obtained from a permutation action of the group on vectors with natural components by projection into suitable invariant subspace. This allows us to reproduce all the elements of quantum formalism in invariant subspaces of permutation representations.

We consider a model of quantum evolution inspired by the quantum Zeno effect[1] — the most convincing manifestation of the role of observation in the dynamics of quantum systems. The

[1] The *quantum Zeno effect* (also known as the *Turing paradox*) is an experimentally verifiable property of a quantum system to follow a prescribed evolution when a properly prepared sequence of observations is applied.

model represents the quantum evolution as a sequence of observations with unitary transitions between them. Standard quantum mechanics assumes a single deterministic unitary transition between observations. We generalize this assumption. We treat a unitary transition as a kind of gauge connection — a way of identifying indistinguishable entities at different times. *A priori*, any unitary transformation can be used as a data identification rule. So, we assume that all unitary transformations participate in transitions between observations with appropriate weights. We call a unitary evolution *dominant* if it provides the maximum transition probability.[2] The Monte Carlo simulation shows a sharp dominance of such evolutions over other evolutions. We also present the Lagrangian of the continuum approximation of the model in order to illustrate the typical assumptions made in the transition to a continuous description.

2. Basic concepts of quantum mechanics

Here is a brief outline of the basic concepts of quantum mechanics. We divide these concepts into three categories: *states*, *observations and measurements*, and *time evolution*.

2.1. Quantum states

Pure states: A *pure quantum state* is a ray in a Hilbert space $\mathcal{H}$ over the field of complex numbers $\mathbb{C}$. A ray is an *equivalence class* of vectors $\left|\psi\right\rangle \in \mathcal{H}$ with respect to the *equivalence relation* $\left|\psi\right\rangle \sim a\left|\psi\right\rangle,\ a \in \mathbb{C}\setminus\{0\}$.

We can reduce the equivalence classes by the *normalization*: $\left|\psi\right\rangle \sim \mathrm{e}^{\mathrm{i}\alpha}\left|\psi\right\rangle, \left\|\psi\right\| = 1, \alpha \in \mathbb{R}$.

The phase "degree of freedom" α can be eliminated by transition to the rank one *projector* $\Pi_\psi = \left|\psi\right\rangle\left\langle\psi\right|$, which is a special case of a *density matrix*.

Mixed states: A *mixed quantum state* is described by a *general density matrix* ρ characterized by the properties: (a) $\rho = \rho^\dagger$, (b) $\left\langle\psi\right|\rho\left|\psi\right\rangle \geq 0$ for any $\left|\psi\right\rangle \in \mathcal{H}$, (c) $\operatorname{tr}\rho = 1$. The set of all density matrices will be denoted by $\mathcal{D}(\mathcal{H})$.

Any mixed state is a weighted mixture of pure states, that is, any density matrix can be represented in the form $\rho = \sum_k p_k \Pi_{e_k}$, where $\{p_k\}$ is a probability distribution in the ensemble of pure states $\Pi_{e_k} = \left|e_k\right\rangle\left\langle e_k\right|$, $\{e_k\}$ is an orthonormal basis in $\mathcal{H}$.

On the other hand, any mixed state can be obtained from a pure state in a larger Hilbert space by taking a partial trace. This is called *'purification'* and can be done as follows. Let $\rho \cong \operatorname{diag}(p_1, p_2, \ldots, p_n)$ be a density matrix in a Hilbert space $\mathcal{H}_X$, $\dim\mathcal{H}_X = n$. If $\mathcal{H}_Y$ is another n-dimensional Hilbert space, then $\rho = \operatorname{tr}_Y\left(\left|\psi\right\rangle\left\langle\psi\right|\right)$, where $\left|\psi\right\rangle = \sum_k \sqrt{p_k}\left|e_k\right\rangle \otimes \left|f_k\right\rangle$ is a pure state in the Hilbert space $\mathcal{H}_X \otimes \mathcal{H}_Y$, tr_Y denotes the trace over $\mathcal{H}_Y$, $\{e_k\}$ and $\{f_k\}$ are orthonormal bases in $\mathcal{H}_X$ and $\mathcal{H}_Y$, respectively.

States of composite system: The Hilbert space of a *composite system* is the tensor product of Hilbert spaces of the components: $\mathcal{H} = \mathcal{H}_1 \otimes \mathcal{H}_2 \otimes \cdots$. The *states of composite system* are classified into two types: *separable* and *entangled* states.

The set of ***separable*** states, $\mathcal{D}_\mathrm{S}(\mathcal{H})$, consists of the states $\rho \in \mathcal{D}(\mathcal{H})$ that can be represented as weighted sums of the tensor products of states of the constituents:

$$\rho = \sum_k w_k\left(\rho_1^k \otimes \rho_2^k \otimes \cdots\right),\ w_k \geq 0,\ \sum_k w_k = 1,\ \rho_1^k \in \mathcal{D}(\mathcal{H}_1),\ \rho_2^k \in \mathcal{D}(\mathcal{H}_2),\ \ldots .$$

The set of ***entangled*** states, $\mathcal{D}_\mathrm{E}(\mathcal{H})$, is by definition the complement of $\mathcal{D}_\mathrm{S}(\mathcal{H})$ in the set of all states: $\mathcal{D}_\mathrm{E}(\mathcal{H}) = \mathcal{D}(\mathcal{H}) \setminus \mathcal{D}_\mathrm{S}(\mathcal{H})$.

[2] The principle of least action in physical theories actually implies the selection of dominant evolutions among all possible ("virtual") evolutions. The apparent determinism of these selected evolutions can be explained by the sharpness of their dominance.

2.2. Observations and measurements

The terms 'observation' and 'measurement' are often regarded as being synonymous. However, it makes sense to separate these concepts. We treat observation as a more general concept which does not imply, in contrast to measurement, obtaining numerical information.

Observation is the detection ("click of detector") of a system, that is in the state ρ, in the subspace $\mathcal{S} \leq \mathcal{H}$. The mathematical abstraction of the "detector in the subspace" $\mathcal{S}$ of a Hilbert space is the operator of projection, $\Pi_{\mathcal{S}}$, into this subspace. The result of quantum observation is random and its statistics is described by a probability measure defined on subspaces of the Hilbert space. Any such legitimate measure $\mu(\cdot)$ must be additive on any set of mutually orthogonal subspaces of a Hilbert space: if, e.g., $\mathcal{A}$ and $\mathcal{B}$ are mutually orthogonal subspaces, then $\mu(\mathrm{span}(\mathcal{A},\mathcal{B})) = \mu(\mathcal{A}) + \mu(\mathcal{B})$. Gleason proved [1] that, excepting the case $\dim\mathcal{H} = 2$, the only such measures have the form $\mu_\rho(\mathcal{S}) = \mathrm{tr}(\rho\Pi_{\mathcal{S}})$, where ρ is an arbitrary density matrix.[3] If, in particular, ρ describes a pure state, $\rho = |\psi\rangle\langle\psi|$, and $\mathcal{S}$ is one-dimensional, $\mathcal{S} = \mathrm{span}(|\varphi\rangle)$, we come to the familiar Born rule: $\mathrm{tr}(\rho\Pi_{\mathcal{S}}) = \mathbf{P}_{\mathrm{Born}} = |\langle\varphi \mid \psi\rangle|^2$.

Measurement is a special case of observation, when the partition of a Hilbert space into mutually orthogonal subspaces is specified by a Hermitian operator A, called "observable". Let $\{e_k\}$ be an *orthonormal basis of eigenvectors* of A and let $\{a_k\}$ be the *spectrum* of A. In the basis $\{e_k\}$ the operator is written as $A = \sum_k a_k\Pi_{e_k}$, where $\Pi_{e_k} = |e_k\rangle\langle e_k|$. "Click of the detector" Π_{e_k} is interpreted as that the *eigenvalue* $a_k \in \mathbb{R}$ is the result of the measurement. The mean for multiple measurements tends to the *expectation value* of A in the state ρ: $\langle A\rangle_\rho = \mathrm{tr}(\rho A)$.

Note that *observation* can formally be treated as a *measurement* if we regard the projector $\Pi_{\mathcal{S}}$ as an observable with eigenvalues belonging to the Boolean domain $\{0,1\}$ (or $\{\mathrm{no},\mathrm{yes}\}$).

2.3. Time evolution

Observation and measurement are timeless concepts in the standard quantum formalism. Time appears as a parameter in describing the unitary transformation of data between observations. The unitary evolution of the density matrix has the form

$$\rho_{t'} = U_{t't}\rho_t U_{t't}^\dagger, \tag{1}$$

where ρ_t is the state *after* observation (sometimes called in this context "*preparation*") at the time t, $\rho_{t'}$ is the state *before* observation at the time t', and $U_{t't}$ is the unitary transition between the observations. In standard quantum formalism, time is considered as a continuous parameter, and relation (1) becomes the *von Neumann equation* in the infinitesimal limit. The evolution of a pure state can be written as

$$|\psi_{t'}\rangle = U_{t't}\,|\psi_t\rangle\,,$$

and the corresponding infinitesimal limit is the *Schrödinger equation.* The operator of unitary evolution can be represented as

$$U_{t't} = \mathrm{e}^{2\pi\mathrm{i}H\Delta t}, \tag{2}$$

where H is a Hermitian operator called the *Hamiltonian*, $\Delta t = t' - t$. The concept of a Hamiltonian is especially useful in the case of continuous time. In the discrete time Δt is a natural number, $\Delta t \in \mathbb{N}$, and instead of (2) we can write

$$U_{t't} = U_1^{\Delta t},$$

[3] In fact, Gleason's theorem establishes a one-to-one correspondence between measures and quantum states in the case $\dim\mathcal{H} \neq 2$. In the two-dimensional case (the case of a qubit) the set of admissible measures is "greater" than the set of states, but the measures that are not associated with quantum states are considered non-physical.

IOP Conf. Series: Journal of Physics: Conf. Series **965** (2018) 012023 doi:10.1088/1742-6596/965/1/012023

where U_1 is the operator of evolution for a unit time step.

To emphasize the role of observation in quantum physics, we accentuate that unitary evolution is simply a change of coordinates in Hilbert space, and it alone is insufficient to describe observable physical phenomena.

2.4. Emergence of geometry in large Hilbert space due to entanglement

Quantum-mechanical theory does not need a geometric space as a fundamental concept — everything can be formulated using only the Hilbert space formalism. In this view, the observed geometry must emerge as an approximation. The currently popular idea [2, 3, 4] of the emergence of geometry within a Hilbert space is based on the notion of entanglement. Briefly, the scheme of extracting geometric manifold from the entanglement structure of a quantum state ρ in a Hilbert space $\mathcal{H}$ is as follows:

- The Hilbert space decomposes into a large number of tensor factors: $\mathcal{H} = \otimes_x \mathcal{H}_x,\ x \in X$. Each factor is treated as a point (or bulk) of geometric space to be built. A graph G — called *tensor network* — with vertices $x \in X$ and edges $(x, y) \in X \times X$ is introduced.
- The edges of G are assigned *weights* based on a *measure of entanglement*, a function that vanishes on separable states and is positive on entangled states. A typical such measure is the *mutual information*: $I(\rho_{xy}) = S(\rho_x) + S(\rho_y) - S(\rho_{xy})$, where ρ_x denotes the result of taking traces of ρ over all tensor factors excepting the x-th (and similarly for ρ_y and ρ_{xy}); $S(a) = -\operatorname{tr}(a \log a)$ is the *von Neumann entropy*. The graph G is supplied with a metric derived from the weights of the edges.
- Finally, the graph G is approximately isometrically embedded in a smooth metric manifold of as small as possible dimension using algorithms like *multidimensional scaling* (MDS).

3. Constructive modification of quantum formalism

David Hilbert, a prominent advocate of the free use of the concept of infinity in mathematics, wrote the following about the relation of the infinite to the reality: "*Our principal result is that the infinite is nowhere to be found in reality. It neither exists in nature nor provides a legitimate basis for rational thought — a remarkable harmony between being and thought.*" Adopting this view, we can not but conclude that all the concepts of quantum mechanics can be formulated in constructive finite terms without distorting the empirical content [5, 6, 7].

3.1. Losses due to continuum and differential calculus

Differential calculus — including differential equations, differential geometry, etc. — underlies mathematical methods in physics. The applicability of differential calculus is based on the assumption that any relevant function can be approximated by linear relations on a small scale. This assumption simplifies many problems in physics and mathematics, but at the cost of loss of completeness.

As an example, consider the problem of classifying simple groups. The concept of a group is an abstraction of the properties of *permutations* (also called *one-to-one mappings* or *bijections*) of a set. Namely, an abstract group is a set with an *associative* operation, an *identity* element, and an *invertibility* for each element. Additional assumptions allow to make the concept of a group more meaningful. The most natural of these assumptions is: (a) *the group is finite.* It is clear that empirical physics is insensitive to this assumption: ultimately, any empirical description is reduced to a finite set of data. A simple complement to assumption (a) — *the group is infinite* — is of little use for a number of reasons, e.g., there are one-to-one mappings between infinite sets and their proper subsets. Thus, more restrictive assumptions are used, the most important of which is: (b) *the group is a differentiable manifold* — such a group is called *Lie group.* Assumption (b) imposes severe constraints on possible physical models.

IOP Conf. Series: Journal of Physics: Conf. Series **965** (2018) 012023 doi:10.1088/1742-6596/965/1/012023

The problem of classification of simple groups[4] under assumption (b) turned out to be rather easy and was solved by two people (Killing and Cartan) in a few years. The result is four infinite series: A_n, B_n, C_n, D_n; and five exceptional groups: E_6, E_7, E_8, F_4, G_2.

The solution of the classification problem under assumption (a) required the efforts of about a hundred people for over a hundred years [8]. But the result — *"the enormous theorem"* — is much richer. The list of finite simple groups contains $16+1+1$ infinite series:

- *groups of Lie type*:
  $A_n(q)$, $B_n(q)$, $C_n(q)$, $D_n(q)$, $E_6(q)$, $E_7(q)$, $E_8(q)$, $F_4(q)$, $G_2(q)$,
  ${}^2A_n\left(q^2\right)$, ${}^2B_n\left(2^{2n+1}\right)$, ${}^2D_n\left(q^2\right)$, ${}^3D_4\left(q^3\right)$, ${}^2E_6\left(q^2\right)$, ${}^2F_4\left(2^{2n+1}\right)$, ${}^2G_2\left(3^{2n+1}\right)$;
- *cyclic groups of prime order*, $\mathbb{Z}_p$;
- *alternating groups*, A_n, $n \geq 5$;

and 26 *sporadic groups*: M_{11}, M_{12}, M_{22}, M_{23}, M_{24}, J_1, J_2, J_3, J_4, Co_1, Co_2, Co_3, Fi_{22}, Fi_{23}, Fi_{24}, HS, McL, He, Ru, Suz, $O'N$, HN, Ly, Th, B, M.
Note that finite groups have an advantage over Lie groups in the sense that in empirical applications any Lie group can be modeled by some finite group, but not vice versa.

3.2. Replacing unitary group by finite group
The main non-constructive element of the standard quantum formalism is the unitary group $\mathsf{U}(n)$, a set of cardinality of the continuum.

Formally, the group $\mathsf{U}(n)$ can be replaced by some finite group which is empirically equivalent to $\mathsf{U}(n)$ as follows. From the theory of quantum computing it is known that $\mathsf{U}(n)$ contains a dense finitely generated — and, hence, countable — matrix subgroup $\mathsf{U}_*(n)$. The group $\mathsf{U}_*(n)$ is *residually finite* [9], i.e. it has a reach set of non-trivial homomorphisms to finite groups [10].

In essence, instead of deriving a finite group from an infinite group, it is more natural to assume that the fundamental symmetries are presented by finite groups, and $\mathsf{U}(n)$'s are just continuum approximations of their unitary representations.

The following properties of finite groups are important for our purposes:

- any finite group is a subgroup of a *symmetric group*,
- any linear representation of a finite group is *unitary*,
- any linear representation is subrepresentation of some *permutation representation*.

3.3. "Physical" numbers and finite groups
The basic number system in quantum formalism is the complex field $\mathbb{C}$. This non-constructive field can be obtained as a metric completion of many algebraic extensions of rational numbers. We consider here constructive numbers that are closely related to finite groups and are based on two primitives with a clear intuitive meaning:

(i) *natural numbers* ("counters"): $\mathbb{N} = \{0, 1, \ldots\}$;
(ii) kth *roots of unity*[5] ("algebraic form of the idea of k-periodicity"): $\mathsf{r}_k \mid \mathsf{r}_k^k = 1$.

These basic concepts are sufficient to represent all physically meaningful numbers.

We start by introducing $\mathbb{N}[\mathsf{r}_k]$, the extension of the *semiring* $\mathbb{N}$ by primitive kth root of unity. The extension $\mathbb{N}[\mathsf{r}_k]$ is a *ring* if $k \geq 2$. This construction allows, in particular, to add *negative numbers* to the naturals — the ring of integer numbers is the extension of $\mathbb{N}$ by the primitive square root of unity: $\mathbb{Z} = \mathbb{N}[\mathsf{r}_2]$. Further, by a standard mathematical procedure (applicable to any *integral domain* — a commutative ring without zero divisors), we construct

[4] *Simple groups*, that is, not containing non-trivial *normal subgroups*, are "building blocks" for all other groups.
[5] There are k different kth roots of unity. A kth root of unity is called *primitive* if $\mathsf{r}_k^m = 1$ only for $m \equiv 0 \mod k$.

the kth *cyclotomic field* as the *fraction field* of the ring $\mathbb{N}[\mathsf{r}_k]$: $\mathbb{Q}(\mathsf{r}_k) = \mathrm{Frac}(\mathbb{N}[\mathsf{r}_k])$. For $k = 2$ we have the field of rational numbers: $\mathbb{Q} = \mathrm{Frac}(\mathbb{N}[\mathsf{r}_2])$. If $k \geq 3$, then the field $\mathbb{Q}(\mathsf{r}_k)$ is a *dense subfield* of $\mathbb{C}$, i.e. (constructive) cyclotomic fields are empirically indistinguishable from the (non-constructive) complex field.

The importance of cyclotomic numbers for constructive quantum mechanics is explained by the following. Let us recall some terms. The *exponent* of a group G is the least common multiple of the orders of its elements. A *splitting field* for a group G is a field that allows to split completely any linear representation of G into irreducible components. A *minimal splitting field* is a splitting field that does not contain proper splitting subfields. Although minimal splitting field for a given group G may be non-unique, any minimal splitting field is a subfield of some cyclotomic field $\mathbb{Q}(\mathsf{r}_k)$, where k is a divisor of the exponent of G. If we want to minimize computations, we can select the field $\mathbb{Q}(\mathsf{r}_k)$ with the smallest k among cyclotomic splitting fields. Thus, to work with any representation of G it is sufficient to use the kth cyclotomic field, where k is determined by the structure of G.

An important field associated with a group is a so-called 'sufficiently large field'. A *sufficiently large field* for G is defined as a field that contains all the mth roots of unity, where m is the exponent of G. In particular, the mth cyclotomic field $\mathbb{Q}(\mathsf{r}_m)$ is a sufficiently large field. A sufficiently large field for G is a splitting field for *every subgroup* of G. This fact is useful in considering a symmetry breaking, which is actually a transition to a subgroup.

3.4. Constructive representations of a finite group

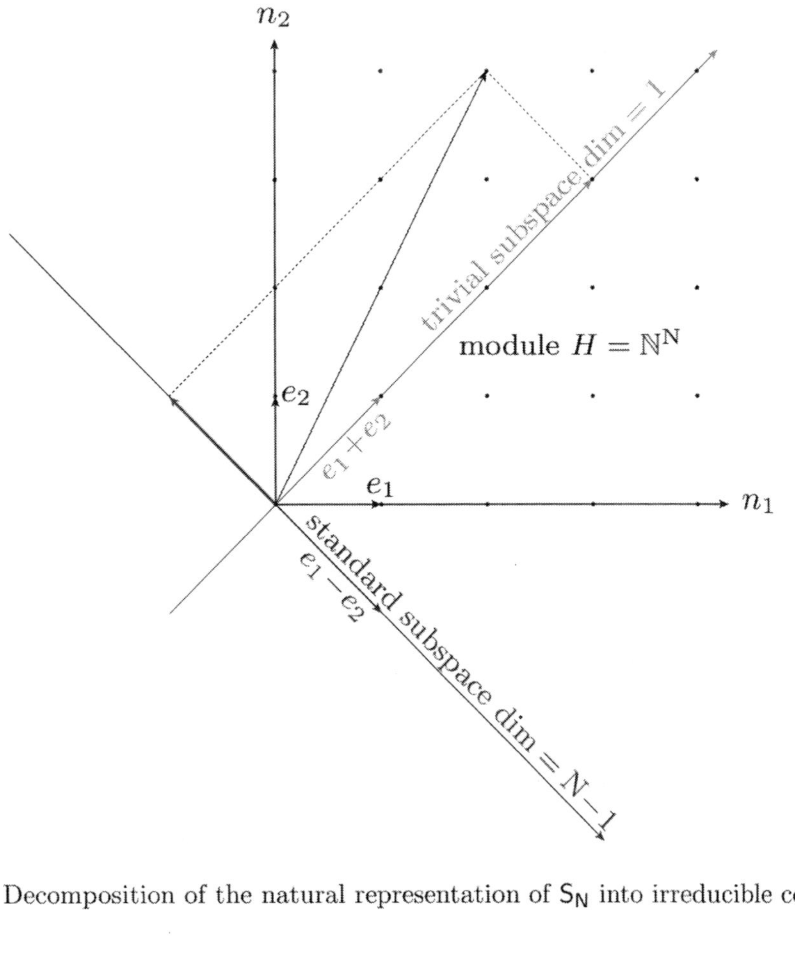

Figure 1. Decomposition of the natural representation of S_N into irreducible components

IOP Conf. Series: Journal of Physics: Conf. Series **965** (2018) 012023 doi:10.1088/1742-6596/965/1/012023

Let a group G act by permutations on a set Ω, $|\Omega| = \mathsf{N}$. If we assume that the elements of Ω are "types" of some discrete entities ("ontological entities", "elements of reality"), then collections of these entities can be described as elements of the module $H = \mathbb{N}^{\mathsf{N}}$ with the basis Ω over the semiring of natural numbers $\mathbb{N}$. The decomposition of the action of G in the module H into irreducible components reflects the structure of the invariants of the action. In order for the decomposition to be complete, it is necessary to extend the semiring $\mathbb{N}$ to a splitting field, e.g., to a cyclotomic field $\mathbb{Q}(\mathsf{r}_k)$, where k is a suitable divisor of the exponent of G. With such an extension of the scalars, the module H is transformed into a Hilbert space $\mathcal{H}$ over $\mathbb{Q}(\mathsf{r}_k)$. This construction, with a suitable choice of the permutation domain Ω, allows us to obtain *any representation* of the group G in some invariant subspace of the Hilbert space $\mathcal{H}$. We obtain "quantum mechanics" within an invariant subspace if, in addition to unitary evolutions, projective measurements are also restricted by this subspace.

The above is illustrated in figure 1 by the example of the natural action of the symmetric group S_N on the set $\Omega = \{e_1, \ldots, e_\mathsf{N}\}$. The *natural* representation of S_N decomposes into two irreducible components: one-dimensional *trivial* and $(\mathsf{N}-1)$-dimensional *standard* representations. The spaces of these representations have, respectively, the following canonical bases: $e_1+e_2+\cdots+e_\mathsf{N}$ and $e_1 - e_2, e_2 - e_3, \ldots, e_{\mathsf{N}-1} - e_\mathsf{N}$. Note, that any symmetric group is a so-called *rational-representation* group, i.e. the field of rational numbers $\mathbb{Q}$ is a splitting field for S_N.

4. Modeling quantum evolution

4.1. Time

For description of evolution, we use a discrete model of time. We define the fundamental ("Planck") time as the ordered sequence of natural numbers: $\mathcal{T} = \mathbb{N}$. We introduce the "empirical time" as a finite sequence of "instants of observations":

$$\mathrm{T} = [t_0, t_1, \ldots, t_{k-1}, t_k, \ldots, t_\mathrm{n}]. \tag{3}$$

The simplest assumption is that T is a subsequence of $\mathcal{T}$. A more realistic model (which we will not develop here) could be a sequence of distributions[6] around points $t_k \in \mathcal{T}$, like, e.g., the *binomial distribution*

$$K_\sigma(\tau - t_k) = \frac{(2\sigma)!}{4^\sigma\,(\sigma - t_k + \tau)!\,(\sigma + t_k - \tau)!}, \quad t_k - \sigma \le \tau \le t_k + \sigma.$$

4.2. Model of quantum evolution

The data of the model include the sequence of the length $\mathrm{n}+1$ for states

$$[\rho_0, \rho_1, \ldots, \rho_{k-1}, \rho_k, \ldots, \rho_\mathrm{n}] \tag{4}$$

and the sequence of the length n for unitary transitions between observations

$$[U_1, \ldots, U_k, \ldots, U_\mathrm{n}]. \tag{5}$$

Standard quantum mechanics presupposes a single unitary evolution, U_k, between observations at times t_{k-1} and t_k. The *single-step transition probability* takes the form

$$\mathbf{P}_k = \mathrm{tr}\left(U_k \rho_{k-1} U_k^\dagger \rho_k\right).$$

[6] The smallest uncertainty in fixing time instants available in modern physics is about 10^{26} Planck units.

The evolution can be expressed in terms of the Hamiltonian H_k: $U_k = \mathrm{e}^{-2\pi\mathrm{i}H_k(t_k-t_{k-1})}$. In physical theories, Hamiltonians are usually derived from *the principle of least action*. Like any extremal principle, it implies the selection of a small subset of the dominant elements in a large set of candidates. Thus it is natural to assume that any unitary evolution takes part in the transition between observations with appropriate weight, but only the dominant evolutions are manifested in observations. Therefore, in our model, we use the following modification of the single-step transition probability

$$\mathbf{P}_k = \sum_{g\in\mathsf{G}} w_{kg}\,\mathrm{tr}\big(\mathsf{U}(g)\,\rho_{k-1}\mathsf{U}(g^{-1})\,\rho_k\big)\,,$$

where G is a finite group, U is a unitary representation of G, w_{kg} is the weight of $g\in\mathsf{G}$ at kth transition: $w_{kg}\geq 0$, $\sum_{g\in\mathsf{G}} w_{kg}=1$.

The *single-step entropy* is defined as

$$\Delta\mathbf{S}_k = -\log\mathbf{P}_k. \tag{6}$$

Continuum approximation of (6) leads to the *Lagrangian* $\mathcal{L}$.

The probability of the whole trajectory is

$$\mathbf{P}_{0\to\mathrm{n}} = \prod_{k=1}^{\mathrm{n}}\mathbf{P}_k. \tag{7}$$

Taking the logarithm of (7), we arrive at the *entropy of trajectory*

$$\mathbf{S}_{0\to\mathrm{n}} = \sum_{k=1}^{\mathrm{n}}\Delta\mathbf{S}_k,$$

the continuum approximation of which is the *action* $\mathcal{S}=\int\mathcal{L}dt$.

4.3. Continuum approximation of discrete model

To build a continuum approximation, we immerse sequence of observation times (3) in the continuous time interval $[t_0,t_\mathrm{n}]\subseteq\mathbb{R}$, replace sequences (4) and (5) with the continuous differentiable functions $\rho(t)$ and $U(t)$, and go to the limit $\mathrm{n}\to\infty$, $\Delta t_k = t_k - t_{k-1}\to 0$. Using the Taylor expansion of $\rho(t)$ and $U(t)$ at the point t_{k-1}, we can compute the approximation for the single-step transition probability:

$$\mathbf{P}_k \approx P_k^{(0)} - P_k^{(1)}\Delta t_k - P_k^{(2)}\Delta t_k^2 - \cdots .$$

We see that product (7) can have a nonzero limit only if $P_k^{(0)}=1$ for all but a finite number of factors. This is an artificial restriction imposed by the introduction of infinity. The term $P_k^{(0)}$ has the meaning of the "zero-step transition probability". Its value is $P_k^{(0)}=\mathrm{tr}\big(\rho_{k-1}^2\big)$. Since $\mathrm{tr}\big(\rho^2\big)=1$ only for pure states, $\rho=|\psi\rangle\langle\psi|$, we can not construct a reasonable continuum approximation in the case of general mixed states for which $\mathrm{tr}\big(\rho^2\big)<1$. Calculation for the case of a pure state shows that

$$\mathbf{P}_k \approx 1 - P_k^{(2)}\Delta t_k^2 - \cdots ,$$

and the single-step entropy has the following approximation

$$\Delta\mathbf{S}_k = -\log\mathbf{P}_k \approx P_k^{(2)}\Delta t_k^2 + \cdots . \tag{8}$$

The leading-order (i.e. the second-order) term in expansion (8) defines a Lagrangian. For further calculations, we use the following simplifying assumptions:

- Assuming that unitary matrix U belongs to a unitary representation of a Lie group, we use the Lie algebra approximation, $U \approx \mathbb{1} + \mathrm{i}A$, where A is a Hermitian matrix. Thus, we replace the function $U(t)$ by a continuous function $A(t)$, which can be regarded as a gauge field (gauge connection) in the $(0+1)$-dimensional "space-time".
- We introduce derivatives $\dot{A}(t)$ and $\dot{\psi}(t)$ as the coefficients in the linear approximations $\Delta A_k \approx \dot{A}\Delta t_k$ and $\Delta\psi_k \approx \dot{\psi}\Delta t_k$, where $\Delta A_k = A_k - A_{k-1}$ and $\Delta\psi_k = \psi_k - \psi_{k-1}$. The time derivative $\dot{A}(t)$ is an analog of the gauge curvature in $(0+1)$ dimensions.

Applying the above assumptions and approximations to single-step entropy (8) and taking the infinitesimal limit, we obtain the Lagrangian:

$$\mathcal{L} = \underbrace{\left\langle\psi\middle|\dot{A}^2\middle|\psi\right\rangle - \left\langle\psi\middle|\dot{A}\middle|\psi\right\rangle^2}_{\text{dispersion of } \dot{A} \text{ in state } \psi} - \mathrm{i}\Big(\left\langle\dot{\psi}\middle|\dot{A}\middle|\psi\right\rangle - \left\langle\psi\middle|\dot{A}\middle|\dot{\psi}\right\rangle + 2\left\langle\psi\middle|\dot{A}\middle|\psi\right\rangle\left\langle\psi\mid\dot{\psi}\right\rangle\Big) - \left\langle\psi\mid\dot{\psi}\right\rangle^2. \tag{9}$$

Note that the first two terms of (9) form the *dispersion* of the "gauge curvature" in the state ψ: $\left(\Delta_\psi\dot{A}\right)^2 = \left\langle\psi\middle|\dot{A}^2\middle|\psi\right\rangle - \left\langle\psi\middle|\dot{A}\middle|\psi\right\rangle^2$, where the value $\Delta_\psi\dot{A}$ is called the *standard deviation.*

4.4. Dominant evolutions

Let U be a unitary representation of a group G in a Hilbert space $\mathcal{H}$. For an arbitrary pair of quantum states, $\rho_1, \rho_2 \in \mathcal{D}(\mathcal{H})$, we define an element $g_* \in \mathsf{G}$ by

$$g_* = \underset{g\in\mathsf{G}}{\arg\max}\ \mathrm{tr}\big(\mathsf{U}(g)\,\rho_1\mathsf{U}(g^{-1})\,\rho_2\big)\,.$$

The operator $\mathsf{U}(g_*)$ maximizes the probability associated with unitary transition between the states ρ_1 and ρ_2. We call such operators *dominant evolutions.*

In the case of natural and standard representations of symmetric groups, the dominant evolutions can be found by a simple algorithm. Namely, the dominant evolutions from the group S_N between the pure states represented by the vectors from the module $H = \mathbb{N}^\mathsf{N}$ can be computed as follows. Let $|n\rangle = \begin{pmatrix} n_1 \\ \vdots \\ n_\mathsf{N}\end{pmatrix}$, $|m\rangle = \begin{pmatrix} m_1 \\ \vdots \\ m_\mathsf{N}\end{pmatrix}$, $|1\rangle = \begin{pmatrix} 1 \\ \vdots \\ 1\end{pmatrix}$ be N-dimensional vectors with natural components. The Born probabilities for the pair $|n\rangle$ and $|m\rangle$ are

$$\mathbf{P}_{\mathrm{nat}}\left(|n\rangle, |m\rangle\right) = \frac{\langle n\mid m\rangle^2}{\langle n\mid n\rangle\langle m\mid m\rangle}$$

and

$$\mathbf{P}_{\mathrm{std}}\left(|n\rangle, |m\rangle\right) = \frac{\left(\langle n\mid m\rangle - \frac{1}{\mathsf{N}}\langle n\,|1\rangle\langle 1|\,m\rangle\right)^2}{\left(\langle n\mid n\rangle - \frac{1}{\mathsf{N}}\langle n\,|1\rangle\langle 1|\,n\rangle\right)\left(\langle m\mid m\rangle - \frac{1}{\mathsf{N}}\langle m\,|1\rangle\langle 1|\,m\rangle\right)} \tag{10}$$

for the natural and standard representations, respectively.

Let R_a denote the permutation (as well as its unitary representation), that sorts the components of the vector $|a\rangle$ in some (ascending or descending) order. It is not hard to show that the unitary operator $U = R_m^{-1}R_n$ maximizes the probability $\mathbf{P}_{\mathrm{nat}}(U|n\rangle, |m\rangle)$, if the permutations R_n and R_m sort the vectors $|n\rangle$ and $|m\rangle$ *identically* — i.e., either both in ascending or both in descending order. The probability $\mathbf{P}_{\mathrm{std}}(U|n\rangle, |m\rangle)$ is maximized by the operator $U = R_m^{-1}R_n$, if R_n and R_m sort the vectors either *identically* or *oppositely* — to make selection one should compare the values of the numerator in (10) for both possibilities (the denominator is invariant under any permutations).

4.5. Energy of permutation

Planck's formula, $E = h\nu$, relates energy to frequency. This relation is reproduced by the quantum-mechanical definition of energy as an eigenvalue of a Hamiltonian. The Hamiltonian associated with a unitary transformation U can be written as $H = \mathrm{i}\hbar \ln U$.

Consider the energy spectrum of a unitary operator defined by a permutation. Let p be a permutation of the cycle type $\{\ell_1^{m_1}, \ldots, \ell_k^{m_k}, \ldots, \ell_K^{m_K}\}$, where ℓ_k and m_k represent lengths and multiplicities of cycles in the decomposition of p into disjoint cycles. Decomposition into disjoint cycles implies the representation of a permutation by a block diagonal matrix, where the block corresponding to the cycle $(1, 2, \ldots, \ell)$ is the $\ell \times \ell$ matrix

$$C_\ell = \begin{pmatrix} 0 & 1 & 0 & \cdots & 0 \\ 0 & 0 & 1 & \cdots & 0 \\ \vdots & \vdots & \vdots & \vdots & \vdots \\ 1 & 0 & 0 & \cdots & 0 \end{pmatrix}.$$

This matrix can be diagonalized, in particular, by the unitary matrix of the discrete Fourier transform

$$F_\ell = \frac{1}{\sqrt{\ell}} \begin{pmatrix} 1 & 1 & 1 & \cdots & 1 \\ 1 & \mathsf{r} & \mathsf{r}^2 & \cdots & \mathsf{r}^{\ell-1} \\ 1 & \mathsf{r}^2 & \mathsf{r}^4 & \cdots & \mathsf{r}^{2(\ell-1)} \\ \vdots & \vdots & \vdots & \cdots & \vdots \\ 1 & \mathsf{r}^{\ell-1} & \mathsf{r}^{2(\ell-1)} & \cdots & \mathsf{r}^{(\ell-1)(\ell-1)} \end{pmatrix},$$

where r is a primitive ℓth root of unity. The diagonalized matrix has the form

$$F_\ell^{-1} C_\ell F_\ell = \begin{pmatrix} 1 & 0 & 0 & \cdots & 0 \\ 0 & \mathsf{r} & 0 & \cdots & 0 \\ 0 & 0 & \mathsf{r}^2 & \cdots & 0 \\ \vdots & \vdots & \vdots & \vdots & \vdots \\ 0 & 0 & 0 & \cdots & \mathsf{r}^{\ell-1} \end{pmatrix}.$$

After taking the logarithm, we see that the Hamiltonian[7] of the permutation p has the following diagonal form

$$H_p = \begin{pmatrix} \mathbb{1}_{m_1} \otimes H_{\ell_1} & & \\ & \ddots & \\ & & \mathbb{1}_{m_K} \otimes H_{\ell_K} \end{pmatrix}, \quad \text{where} \quad H_{\ell_k} = \frac{1}{\ell_k} \begin{pmatrix} 0 & & & \\ & 1 & & \\ & & \ddots & \\ & & & \ell_k - 1 \end{pmatrix}.$$

We shall call the least nonzero energy of a permutation the *base energy*:

$$\varepsilon = \frac{1}{\max(\ell_1, \ldots, \ell_K)}.$$

Simulation shows that the base *("ground state", "zero-point", "vacuum")* energy is statistically more significant than other energy levels.

4.6. Monte Carlo simulation of dominant evolutions

We can compare the contributions of different evolutions using a Monte Carlo simulation. The simulation shows that the contributions of non-dominant evolutions are negligible for sufficiently large groups.

[7] We define the Hamiltonian via the relation $U = \mathrm{e}^{2\pi \mathrm{i} H}$ to ensure rationality of eigenvalues of H.

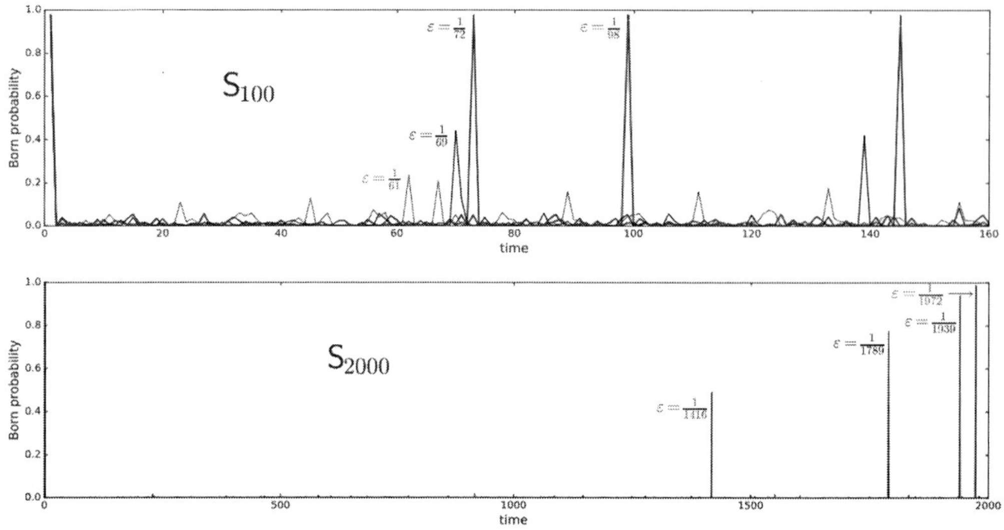

Figure 2. Dominant evolutions between randomly generated states. Born probability vs time

Figure 2 shows several dominant evolutions for the standard representation of the symmetric groups S_{100} and S_{2000}. Each graph represents the time dependencies of Born's probabilities for the dominant evolutions between four randomly generated pairs of natural vectors. The dominant evolutions are marked by labeling their peaks with their base energies, ε, which are listed in table 1. The table contains also the orders[8] of the dominant evolutions.

Table 1. Base energies and orders of dominant evolutions

	S_{100}, $\dim\mathcal{H}=99$				S_{2000}, $\dim\mathcal{H}=1999$			
ε	$\frac{1}{61}$	$\frac{1}{69}$	$\frac{1}{72}$	$\frac{1}{98}$	$\frac{1}{1416}$	$\frac{1}{1789}$	$\frac{1}{1939}$	$\frac{1}{1972}$
ord	4026	4830	72	98	3161864216280	1061485260	93072	53244

We compared randomly generated evolutions with the dominant ones. Such non-dominant evolutions are not shown in the figure, since they are almost invisible against the sharp peaks of dominant evolutions. The superiority of the dominant evolutions increases with the group size.

Summary

(i) The general scheme of quantum mechanics can be completely reproduced in terms of projections of permutation representations of finite groups into invariant subspaces.

(ii) Quantum randomness is a consequence of the fundamental impossibility of tracing the individuality of indistinguishable entities in their evolution: the only available objective information about such entities is reduced to the invariants of their symmetry group, so information about individual events can only be probabilistic.

[8] The *order* (or *period*) of an element g of a group is the smallest integer $m > 0$ such that $g^m = e$, where e denotes the identity of the group.

(iii) The native number systems for quantum formalism are cyclotomic fields, and the field of complex numbers is just a non-constructive metric completion of cyclotomic fields.
(iv) Observable behavior of quantum system is determined by the dominants among all possible evolutions.
(v) The principle of least action is a continuum approximation of the principle of selection of the most probable trajectories.

Acknowledgments
I am grateful to Yu.A. Blinkov, V.P. Gerdt and S.I. Vinitsky for fruitful discussions.

References
[1] Gleason A M 1957 Measures on the closed subspaces of a Hilbert space *Indiana Univ. Math. J.* **6** 885–893
[2] Van Raamsdonk M 2010 Building up spacetime with quantum entanglement, *Gen. Rel. Grav.* **42** 2323–2329
[3] Maldacena J and Susskind L 2013 Cool horizons for entangled black holes *Fortschr. Phys.* **61** 781–811
[4] Cao C, Carroll S M and Michalakis S 2017 Space from Hilbert space: recovering geometry from bulk entanglement *Phys. Rev. D* **95** 024031
[5] Kornyak V V 2013 Classical and quantum discrete dynamical systems *Phys. Part. Nucl.* **44** No 1 47–91
[6] Kornyak V V 2015 Discrete dynamical models: combinatorics, statistics and continuum approximations *Math. Model. Geom.* **3** No 1 1–24
[7] Kornyak V V 2016 Combinatorial approach to modeling quantum systems *EPJ Web Conf.* **108** 01007
[8] Solomon R 2001 A brief history of the classification of the finite simple groups *Bull. Amer. Math. Soc. (N.S.)* **38** No 3 315–352
[9] Malcev A 1940 On isomorphic matrix representations of infinite groups *Rec. Math. [Mat. Sbornik] N.S.* **8(50)** No 3 405–422
[10] Magnus W 1969 Residually finite groups *Bull. Amer. Math. Soc.* **75** No 2 305–316

IOP Conf. Series: Journal of Physics: Conf. Series **965** (2018) 012024 doi:10.1088/1742-6596/965/1/012024

Magnetic monopoles and symmetries in noncommutative space

Samuel Kováčik

Dublin Institute for Advanced Studies, 10 Burlington Road, Dublin 4, Ireland

E-mail: skovacik@stp.dias.ie

Peter Prešnajder

Faculty of Mathematics, Physics and Informatics, Comenius University Bratislava, Mlynská dolina, Bratislava, 842 48, Slovakia

E-mail: presnajder@fmph.uniba.sk

Abstract. In this paper, we review the progress in the analysis of magnetic monopoles as generalized states in quantum mechanics. We show that the considered model contains rich algebraic structure that generates symmetries which have been utilized in different physical contexts. Even though are we focused on quantum mechanics in noncommutative space $\mathbf{R}^3_\lambda$, the results can be reconstructed in ordinary quantum mechanics in $\mathbf{R}^3$ as well.
Keywords: Magnetic monopoles, quantum mechanics, noncommutative space

1. Introduction

It is generally believed that the Planck units describe important scales where aspects of both the quantum world and the curved space-time being relevant. It is perfectly safe to assume that the laws of nature looked different when the Universe had the Planck temperature. Similarly, we can expect the space to have a nontrivial structure on the scale of the Planck length. Quantum corrections to gravity cannot be neglected and vice versa.

The existence of a fundamental length scale is predicted by most of the candidate theories of quantum gravity. That it shall be no surprise can be argued from already well-known facts. For example, the original argument of Snyder [1] states that if you want to distinguish two space points separated by the Planck length you need to use a photon of a similar wavelength. However, such a photon would be hidden under its event horizon – a black hole would form and no information could be obtained. It is natural to abandon the notion of exact space(time) points in a similar manner as we disregarded the classical notion of position and momentum of a particle when formulating quantum mechanics.

Correspondingly to the phase space in quantum mechanics, a space whose coordinates cannot be distinguished is described by a general noncommutativity (NC) relation

$$[x_i, x_j] = i\lambda\Theta_{ij}(x), \tag{1}$$

where Θ_{ij} is antisymmetric in indices i, j and λ describes the scale of noncommutavity, which is often assumed to be of the order of the Planck length. There are many possible choices for the range (and structure) of indices i, j and for Θ. A popular choice is to take $i, j = 1, 2$ and constant Θ, the so called Groenewold-Moyal plane [2, 3]. However, we will be interested in the case of $\Theta_{ij}(x) = 2\varepsilon_{ijk}x_k$, where the indices i, j, k take values of $1, 2, 3$ and ε is the Levi-Civita symbol. Therefore, our starting point is the following NC relation

$$[x_i, x_j] = 2i\lambda\varepsilon_{ijk}x_k. \tag{2}$$

This choice describes a 3D rotationally invariant space $\mathbf{R}^3_\lambda$, which can be viewed as a sequence of concentric fuzzy spheres $\mathbf{S}^2_\lambda$ of different radii. There are many ways of constructing NC coordinates, we will utilize the construction using (two sets of) creation and annihilation (c/a) operators satisfying the usual relations

$$[a_\alpha, a^+_\beta] = \delta_{\alpha\beta}, \quad [a_\alpha, a_\beta] = [a^+_\alpha, a^+_\beta] = 0 \tag{3}$$

and acting in an auxiliary Fock space $\mathcal{F}$ spanned on normalized states

$$|n_1, n_2\rangle = \frac{(a^+_1)^{n_1}\,(a^+_2)^{n_2}}{\sqrt{n_1!\,n_2!}}\;|0\rangle. \tag{4}$$

The NC coordinates are defined as

$$x^i = \lambda\sigma^i_{\alpha\beta}a^+_\alpha a_\beta. \tag{5}$$

Each part of this expressions serves its own purpose: the c/a operators bring in the NC property, the Pauli matrices σ^i generate the $su(2)$ structure with coefficients $f_{ijk} = \varepsilon_{ijk}$ and λ carries the length scale. We also define the radial distance coordinate as

$$r = \lambda\left(a^+_\alpha a_\alpha + 1\right), \tag{6}$$

where the $+1$ was added to ensure that $r^2 - x_i^2 = O(\lambda^2)$ (so there is no term linear in λ).

This construction of 3D rotational invariant space $\mathbf{R}^3_\lambda$ was developed in [4] and explored by the present authors and their collaborator in [5, 6, 7, 8, 9, 10, 11].

2. Hilbert space $\mathcal{H}$ and subspace $\mathcal{H}_0$

We can now formulate quantum mechanics (QM) in $\mathbf{R}^3_\lambda$. This theory, which is being referred to as NC QM, will differ from the ordinary QM unless we send $\lambda \to 0$, which is often being referred to as taking the commutative limit.

A natural choice for the Hilbert space $\mathcal{H}$ is the set of functions of the c/a operators (3). Most of the previous work was focused on states containing an equal number of creation as annihilation operators. Restricting only on this subspace, which is denoted $\mathcal{H}_0$, makes a perfect sense. Note that in the definition of NC coordinates (5) each of the coordinates contains exactly one creation and one annihilation operator. Therefore any state of the form $\Psi(x)$ should be in $\mathcal{H}_0$ (and actually nothing else is).

The Hilbert space $\mathcal{H}$ contains states of the form $\Psi = \Psi(a^+, a)$ and is equipped with a norm

$$\|\Psi\|^2 = 4\pi\,\lambda^2\,\mathrm{Tr}[\Psi^\dagger\hat{r}\Psi], \tag{7}$$

where $\hat{r}\Psi = \frac{1}{2}\,(r\Psi + \Psi r)$. Let us now recall the definition of some important operators. Because of the $su(2)$ structure appearing in (2) it is not hard to define the generators of rotations $\hat{L}_i$ satisfying the $su(2)$ relations, they follow as

$$\hat{L}_i\Psi = \frac{1}{2\lambda}[x_i, \Psi], \tag{8}$$

$$\hat{X}_i\Psi = \frac{1}{2}\{x_i, \Psi\}, \tag{9}$$

where in the second line we have defined the symmetrical action of the coordinate operator. In [7] it has been shown that these two can be combined into a single object $\hat{L}_{ij}$ generating $SO(4)$ rotations.

The free Hamiltonian has been defined in [5] as

$$\hat{H}_0\Psi = \frac{1}{2m\lambda\hat{r}}\,[a^+_\alpha, [a_\alpha, \Psi]], \tag{10}$$

where it was used to define the hydrogen atom problem, which was solved in the same paper by directly solving the corresponding Schrödinger equation and in [5, 6] algebraically by the means of Laplace-Runge-Lenz vector.

This operator was also used in [7] to define the velocity operator (by the Heisenberg relation $\hat{V}_i = i[\hat{H}_0, \hat{X}_i]$) and $\hat{V}_4$ as

$$\hat{V}_i\Psi = \frac{i}{2\hat{r}}\sigma^i_{\alpha\beta}\left(a^+_\alpha\Psi a_\beta - a_\beta\Psi a^+_\alpha\right), \hat{V}_4 = \frac{1}{2\hat{r}}\left(a^+_\alpha\Psi a_\alpha + a_\alpha\Psi a^+_\alpha\right) = \lambda^{-1}\left(1 - \lambda^2\hat{H}_0\right)\Psi. \tag{11}$$

It was shown in the same paper that the four-vector $\hat{V}_a$ transforms under the $SO(4)$ rotations generated by (8) and squares to a constant $\hat{V}_a^2 = \lambda^2$, which demonstrates the UV regularization of the theory. This result holds only for states in $\mathcal{H}_0$, but it has been generalized to $\mathcal{H}_\kappa$ (which is defined in the next section) in [11].

Another important result holding only for $\mathcal{H}_0$ is that the components of $\hat{V}_i$ do commute with each other, even though the coordinate operators do not.

The eigenspectrum of this operator has been found in [9], the eigenfunctions have the same form as in the ordinary QM (that is $e^{i\vec{k}\cdot\vec{r}}$), but the eigenvalues have a(n upper) cut-off $|k| \to \lambda^{-1}\sin\lambda|k|$.

It should be noted that all of the presented results were exact (and not only perturbative), which signalizes symmetries are at work (and perhaps not all of them have been revealed yet). A downside of the model is that some of the famous problems we would expect to solve easily have not been solved as of now – the linear harmonic oscillator being a prominent example. Perhaps, the relevant symmetry is yet to be found.

3. Generalized Hilbert space $\mathcal{H}_\kappa$

Let us now move our attention to a generalized class of states which was investigated in [11]. It contains functions containing an unequal number of creation and annihilation operators, with their difference held fixed and equal to κ

$$\Psi_\kappa(e^{-i\gamma}a^+, e^{i\gamma}a) = e^{-i\gamma\kappa}\Psi_\kappa(a^+, a),\ \gamma \in \mathbf{R},\ \kappa \in \mathbf{Z}, \tag{12}$$

the corresponding Hilbert space is denoted $\mathcal{H}_\kappa$. We can use the same physical operators as for $\mathcal{H}_0$, yet there is one crucial difference now. From the fact that $[r, x_i] = 0$ it follows immediately that $[\Psi(x_i), r] = 0$, so when working in $\mathcal{H}_0$ multiplying the states with r from the left and from the right is the same and the calculations can often be simplified. Now the left and the right multiplication differ

$$\hat{r}_L - \hat{r}_R = \lambda\kappa \neq 0$$

and one has to go through all of the calculations and correct them correspondingly.

Let us now write down directly the results. For the sake of the point we will be making shortly we gather the relations describing a QM system containing a magnetic monopole of a charge μ as derived in [12] on the left-hand side and the relations holding for $\mathcal{H}_\kappa$ on the right-hand side

$$\begin{aligned}
[\hat{x}_i, \hat{x}_j] = 0 \quad &\leftrightarrow \quad [\hat{X}_i, \hat{X}_j] = \lambda^2 \varepsilon_{ijk} \hat{L}_k, \\
[\hat{x}_i, \hat{p}_j] = i\delta_{ij} \quad &\leftrightarrow \quad [\hat{X}_i, \hat{V}_j] = i\delta_{ij}\left(1 - \lambda^2 \hat{H}_0\right), \\
[\hat{p}_i, \hat{p}_j] = i\mu\varepsilon_{ijk}\frac{\hat{x}_k}{r^3} \quad &\leftrightarrow \quad \left[\hat{V}_i, \hat{V}_j\right] = i\frac{-\kappa}{2}\varepsilon_{ijk}\frac{\hat{X}_k}{\hat{r}(\hat{r}^2 - \lambda^2)}, \\
\hat{C}_1 = -\mu q \quad &\leftrightarrow \quad \hat{C}_1 = \frac{\kappa}{2} q, \\
\hat{C}_2 = q^2 + (\mu)^2(-2E) \quad &\leftrightarrow \quad \hat{C}_2 = q^2 + \left(\frac{\kappa}{2}\right)^2 (-2E + \lambda^2 E^2).
\end{aligned} \tag{13}$$

$\hat{C}_1, \hat{C}_2$ are the Casimir operators for the hydrogen atom problem, which can be used to directly derive the energy spectrum without needing to solve the corresponding Schrödinger equation.

If we set $\lambda = 0$ it is obvious that the relations coincide if we identify $\mu = -\frac{\kappa}{2}$. We have to be careful since μ has to satisfy the Dirac quantization condition [13]. However, the identification is perfect since κ describes the difference between the number of creation and annihilation operators it follows that $\frac{\kappa}{2} \in \mathbf{Z}/2$, which is the same condition as has to be satisfied by μ. It was therefore concluded in [11] that $\mathcal{H}_\kappa$ is the Hilbert space of monopoles states of strength $-\frac{\kappa}{2}$.

Monopole states can in some scenarios cause the non-associativity of the algebra [14, 15]. This is however not the case here since it can be shown that

$$\varepsilon_{ijk}[\hat{V}_i, [\hat{V}_j, \hat{V}_k]] = 0. \tag{14}$$

This results is obvious as the operator product in a Hilbert space is associative, but is not very straightforward to prove as the zero comes from a cancellation of two rather nontrivial terms.

4. Underlying algebraic structure

The fact that the generalization of the results could be carried out in such a simple way suggests that the underlying algebraic structures remain (nearly) unaffected by the presence the monopole states. Let us now investigate this issue.

The NC space was defined using the c/a operators (3) and so was the Hilbert space $\mathcal{H}$ and operators on it. In general, the c/a operators can act on the states in $\mathcal{H}$ either from the left or from the right. Let us denote these actions as

$$\begin{aligned}
\hat{a}_\alpha \Psi = a_\alpha \Psi, \quad &\hat{a}^+_\alpha \Psi = a^+_\alpha \Psi, \\
\hat{b}_\alpha \Psi = \Psi a_\alpha, \quad &\hat{b}^+_\alpha \Psi = \Psi a^+_\alpha.
\end{aligned} \tag{15}$$

We now have 4 annihilation (and 4 creation) operators, two acting from the left and two from the right. Let us combine them into a single object

$$\hat{A} = (\hat{a}_1, \hat{a}_2, \hat{b}_1, \hat{b}_2)^T, \ \hat{A}^+ = (\hat{a}^+_1, \hat{a}^+_2, \hat{b}^+_1, \hat{b}^+_2) \tag{16}$$

Since multiplying from the right reverse the order the sign of the commutators is different

$$[\hat{a}_\alpha, \hat{a}^+_\beta] = -[\hat{b}_\alpha, \hat{b}^+_\beta] = \delta_{\alpha\beta} \tag{17}$$

and therefore $[\hat{A}_a, \hat{A}^+_b] \neq \delta_{ab}$ where $a, b = 1, ..., 4$. To achieve this we need to add a matrix $\Gamma = \sigma_3 \otimes \mathbf{1}_2$ flipping the corresponding sings. Then it holds that

$$[\hat{A}_a, \Gamma_{bc}\hat{A}^+_c] = \delta_{ab}. \tag{18}$$

We can now introduce a set of $su(2,2)$ matrices $S_{AB} = -S_{BA}$, where $A, B = 0, ..., 5$, as

$$\begin{aligned} S_{ij} &= \frac{1}{2}\varepsilon_{ijk}\begin{pmatrix} \sigma_k & 0 \\ 0 & \sigma_k \end{pmatrix}, & S_{k4} &= \frac{1}{2}\begin{pmatrix} \sigma_k & 0 \\ 0 & \sigma_k \end{pmatrix}, \\ S_{k5} &= \frac{1}{2}\begin{pmatrix} 0 & \sigma_k \\ -\sigma_k & 0 \end{pmatrix}, & S_{45} &= \frac{1}{2}\begin{pmatrix} 0 & i \\ i & 0 \end{pmatrix}, \\ S_{0k} &= \frac{1}{2}\begin{pmatrix} 0 & i\sigma_k \\ i\sigma_k & 0 \end{pmatrix}, & S_{04} &= \frac{1}{2}\begin{pmatrix} 0 & 1 \\ -1 & 0 \end{pmatrix}, \\ S_{05} &= \frac{1}{2}\begin{pmatrix} 1 & 0 \\ 0 & -1 \end{pmatrix} = \frac{1}{2}\Gamma \end{aligned} \tag{19}$$

to define operators

$$\hat{S}_{AB} = \hat{A}^{+}\Gamma S_{AB}\hat{A} \tag{20}$$

forming a representation of the same algebra.

All of the essential objects in the considered model of NC QM can be (in hindsight) expressed using $\hat{S}_{AB}$, as is expressed in the following table:

$\hat{S}_{AB}$	$\hat{S}_{ij}$	$\hat{S}_{k4}$	$\hat{S}_{05}$	$\hat{S}_{k5}$	$\hat{S}_{45}$	$\hat{S}_{0k}$	$\hat{S}_{04}$
$\sim$	$\varepsilon_{ijk}\hat{L}_k$	$\hat{X}_k$	$\hat{r}$	$\hat{r}\bar{V}_k$	$\hat{r}\hat{V}$	$\hat{r}\hat{V}_k$	$\hat{r}\hat{H}_0$

where $\hat{V}$ is the dilatation operator and $\bar{V}_k\Psi = \frac{1}{2\hat{r}}\sigma^i_{\alpha\beta}\left(a^+_\alpha\Psi a_\beta + a_\beta\Psi a^+_\alpha\right)$. Up to a numerical factor, $\bar{V}_k$ is proportional to $\hat{\zeta}_k$, which was used to write down the Laplace-Runge-Lenz vector in [8].

The operators $\hat{L}_k$ and $\hat{X}_k$ commute with the weight factor $\hat{r}$ in (7) and consequently they are Hermitian with respect to this norm. On the other hand, the operators $\hat{V}_k$, $\bar{V}_k$ and $\hat{H}_0$ do not commute with it and so the factor $\hat{r}^{-1}$ has to be added to ensure this.

The crucial realization is that this structure is completely indifferent of κ. Its only presence is in the quadratic operator $\hat{\mathcal{L}} = \hat{a}^+_\alpha\hat{a}_\alpha - \hat{b}_\alpha\hat{b}^+_\alpha$ that counts the difference between the creation and annihilation operators and is therefore proportional to κ.

It shall be also noted that the $su(2,2)$ is the least you can utilize within the model, not the most. For example adding a factor $f(\hat{r})$ to $\hat{S}_{AB}$ differs the considered commutation relations a lot. Let us demonstrate it on the following example. We take two of the elements of the $su(2,2)$ and commute them

$$[\hat{S}_{0i}, \hat{S}_{0j}] \propto \hat{S}_{ij}. \tag{21}$$

The result is, obviously, in the same algebra. However, if we add the factor $\hat{r}^{-1} \propto \hat{S}_{05}$ it holds that

$$\lambda^{-1}[\hat{r}^{-1}\hat{S}_{0i}, \hat{r}^{-1}\hat{S}_{0j}] \propto \frac{\kappa}{\hat{r}\,(\hat{r}^2 - \lambda^2)}\hat{S}_{k4}, \tag{22}$$

or in other words, the third equation in (13). Thus, the monopole magnetic field appears as a correction due to the factors $\hat{r}^{-1}$ in the commutator (22). An analogous effects is present in all other commutators $[\hat{r}^{-1}\hat{S}_{i5}, \hat{r}^{-1}, \hat{S}_{j5}], [\hat{r}^{-1}\hat{S}_{0i}, \hat{r}^{-1}, \hat{S}_{j5}], \ldots$ associated with all noncompact $so(4,1)$ generators $\hat{S}_{0i}, \hat{S}_{04}, \hat{S}_{i5}$ and $\hat{S}_{45}$. Various components has been utilized this way, for example in [7] to explore the Euclidean kinematic $E(4)$ symmetry or in [8] to describe the $SO(4)$ and $SO(1,3)$ symmetries of the hydrogen atom.

5. Conclusions

We have reviewed the progress in the study of NC QM, focusing mostly on the very natural appearance of magnetic monopole states. The novel results presented here is firstly the $su(2,2)$ symmetry, which appears as a building block of the theory and is indifferent of considering the monopole states and secondly the fact that neither the nonassociativity of the model is violated by it (14).

The results of this paper can be reconstructed in QM in ordinary (commutative) space. One first needs to formulate the theory in $\mathbf{C}^2$ instead of $\mathbf{R}^3$ and replace the c/a operators with (complex) coordinates $\sqrt{\lambda}a_\alpha \to z_\alpha$, $\sqrt{\lambda}a^+_\alpha \to \bar{z}_\alpha$ and the commutators with the Poisson brackets $[\ ,\ ] \to -i\{\ ,\ \}$. By considering states of the form $\Psi(x)$, where $x_i = \bar{z}_\alpha \sigma^i_{\alpha\beta} z_\beta$, that contain an equal number of z and $\bar{z}$, one reproduces the ordinary QM in $\mathbf{R}^3$. Allowing states with a fixed difference in the number of z and $\bar{z}$ (for example of the form $\Psi(x) \cdot z_1^\kappa$) results into introducing monopole states of an arbitrary field strength $\mu = -\frac{\kappa}{2}$, more details can be found in [11].

There are different lines of research currently being investigated. For example, the operator $\hat{S}_{05} = \lambda^{-1}\hat{r}$ plays a crucial role in the presented model. It can be used to define a dual velocity operator $\bar{V}_a = i\lambda^{-1}[\hat{r}, \hat{V}_a], a = 1, ..., 4$, that can be used to enclose a larger algebraic structure with particular $\hat{r}$-dependent coefficients. Another, physically very appealing, option is to try to define a relativistic version of the model – a daunting task finally starting to seem possible using the presented underlying symmetries.

In [10], a Schwarzschild black hole with NC smeared singularity was analyzed, leading to multiple interesting results: infinite Hawking temperatures are avoided, a microscopic black hole does not evaporate completely, but leaves a Planck size remnant, possibly contributing to the observed dark matter density and also being able to generate ultra-high-energy rays. It might be possible to follow up on this research by analyzing a black hole with a nonzero magnetic charge.

An important property of the presented model of NC QM is that the obtained results were always exact, there was no need for a perturbative approach. This signalizes a rich structure of underlying symmetries at work – some of which have been revealed now. From this, it seems obvious that achieving any further development of the model should be done by utilizing and generalizing the presented algebraic frameworks.

Acknowledgment
This research was partially supported by COST Action MP1405 and project VEGA 1/0985/16.

References

[1] Snyder S 1947 Quantized space-time *Phys. Rev.* **71** 38
[2] Groenewold H J 1946 On the Principles of elementary quantum mechanics *Physica* **12** 405
[3] Bartlett M S, Moyal J E 1949 Quantum mechanics as a statistical theory *Math. Proc. of the Camb. Phil. Soc.* **45** 99
[4] Hammou A B, Lagraa M and Sheikh-Jabbari M M 2002 Coherent state induced star product on R^λ_3 and the fuzzy sphere *Phys. Rev.* **D66** 025025
[5] Gáliková V and Prešnajder P 2012 Hydrogen atom in fuzzy spaces - Exact solution *J. Phys.: Conf. Ser.* **343** 012096
[6] Gáliková V and Prešnajder P 2013 Coulomb problem in non-commutative quantum mechanics *Jour. of Math. Phys.* **54** 052102
[7] Kováčik S and Prešnajder P 2013 The velocity operator in quantum mechanics in noncommutative space *Jour. of Math. Phys.* **54** 102103
[8] Gáliková V, Kováčik S and Prešnajder P 2013 Laplace-Runge-Lenz vector in quantum mechanics in noncommutative space *Journal of Mathematical Physics* **54** 122106
[9] Kováčik S and Prešnajder P 2016 NC plane waves, Casimir effect and flux tube potential with Lüscher terms (*Preprint quant-ph/*1603.00292)
[10] Kováčik S 2017 R^λ_3-inspired black holes *Mod.Phys.Lett.* **A32** 1750130

IOP Conf. Series: Journal of Physics: Conf. Series **965** (2018) 012024 doi:10.1088/1742-6596/965/1/012024

[11] Kováčik S and Prešnajder P 2017 Magnetic monopoles in noncommutative quantum mechanics in *J.Math.Phys.* **58** 012101
[12] Zwanziger D 1968 Exactly Soluable Nonrelativistic Model of Particles with Both Electric and Magnetic Charges *Phys. Rev.* **176** 1480
[13] Dirac P A M 1931 Quantised Singularities in the Electromagnetic Field *Proc. Roy. Soc.* **A133** 60
[14] Lassig C C and Jos G C 1997 Magnetic charge in a nonassociative field theory *Phys.Lett.* **B400**
[15] Lust D, Malek E, Szabo R J 2017 Non-geometric Kaluza-Klein monopoles and magnetic duals of M-theory R-flux backgrounds (*Preprint hep-ph/*1705.09639)

IOP Conf. Series: Journal of Physics: Conf. Series **965** (2018) 012025 doi:10.1088/1742-6596/965/1/012025

Oscillators from nonlinear realizations

N Kozyrev and S Krivonos

Bogoliubov Laboratory for Theoretical Physics, JINR, 141980 Dubna, Russia

E-mail: nkozyrev,krivonos@theor.jinr.ru

Abstract. We construct the systems of the harmonic and Pais-Uhlenbeck oscillators, which are invariant with respect to arbitrary noncompact Lie algebras. The equations of motion of these systems can be obtained with the help of the formalism of nonlinear realizations. We prove that it is always possible to choose time and the fields within this formalism in such a way that the equations of motion become linear and, therefore, reduce to ones of ordinary harmonic and Pais-Uhlenbeck oscillators. The first-order actions, that produce these equations, can also be provided. As particular examples of this construction, we discuss the $so(2,3)$ and $G_{2(2)}$ algebras.

1. Introduction

In the papers [1] and [2], the dynamical systems, which provide nonlinear realizations [3-6] of the $su(1,2)$, $so(2,3)$ and $G_{2(2)}$ algebras, were considered. The equations of motion of these systems were obtained within the coset approach as the conditions on some of the Cartan forms of the specifically chosen coset spaces. Other forms of these cosets were used to construct the invariant actions. The equations of motion, obtained in [1], [2], can be viewed as deformations of the equations of motion of usual harmonic oscillators. The ability to obtain the oscillator equations of motion is strongly related to the fact that these algebras admit the 5-graded decomposition. And, due to the fact that at least one real form of any simple Lie algebra, aside of $sl(2)$, admits this decomposition [7], it is possible to realize all these algebras in terms of time and fields, that satisfy the deformed oscillator equations of motion.

The idea of the papers [1], [2] can be traced back to the paper [8], where the deformations of the ℓ-conformal Galilei algebras were considered and it was noted that the deformed commutation relations resemble ones of a very general class of Lie algebras. In this sense, the appearance of the oscillators is not surprising, as for the ℓ-conformal Galilei algebras the nonlinear realizations that produce equations of motion of the harmonic and Pais-Uhlenbeck oscillators, are known [9], [10].

An interesting feature of the construction of the invariant oscillators in [2] was that the second-order equations of motion of the $so(2,3)$ invariant system were still linear, though both $so(2,3)$ and $G_{2(2)}$-invariant actions were deformed, and required an introduction of some additional semi-dynamical variables. Moreover, in [1] the redefinition of the coordinates and the momenta was found, which brought the equations of motion and the Hamiltonian of the $su(1,2)$ deformed oscillators to the standard form. As, obviously, the exact form of the equations of motion heavily depends on the parametrization of the coset space, one may ask the question whether these equations can be made linear by the change of variables.

The purpose of this paper is to prove that for any simple Lie algebra it is possible to construct parametrization, which ensures linearity of equations of motion. This parametrization can be called homogeneous, as it is just required to place all the dynamical fields together with their generators into a single exponent. We also consider a general recipe for construction of the action and analyze how it produces the expected oscillator equations of motion. Then we consider, for illustrative purposes, the example of $so(2,3)$ in homogeneous parametrization and find another ("matryoshka") parametrization of the $G_{2(2)}$ coset, which also produces linear equations of motion.

2. General construction

2.1. Linear equations of motion

As was already noted, we consider algebras with the structure [8], [2]

$$\begin{aligned} &\big[L_m, L_n\big] = (m-n)L_{m+n},\ m,n=-1,0,1;\ \ [M_a, M_b] \sim M_c, \\ &\big[L_m, G_{r,\mu}\big] = \big(\ell m - r\big)G_{r+m,\mu}, r=-\ell,\ldots,\ell;\ \ [M,G]\sim G, \\ &\big[G_{r,\mu}, G_{s,\nu}\big] = \delta_{r+s,0} f^a(r,\mu,\nu)M_a + g^m(r,s,\mu,\nu)L_m, \end{aligned} \tag{1}$$

where ℓ is a half-integer to have the finite number of $G_{r,\mu}$ and L_m form the conformal algebra in one dimension. This structure of the algebra is called graded, as all generators can be classified with respect to the action of L_0. The grade of the algebra (1) depends on ℓ. It was proven in [7] that a 5-graded structure, which corresponds to $\ell = \frac{1}{2}$, exists for at least one real form of any simple Lie algebra, though different graded structures are known for various algebras. In this paper, we concentrate on the case $\ell = \frac{1}{2}$.

The equations that are deformed equations of motion of harmonic oscillators can be obtained as conditions $\Omega^{r,\mu} = 0$ on the $G_{r,\mu}$ Cartan forms of the coset spaces

$$g = e^{t\left(L_{-1}+\omega^2 L_1\right)} \times \text{exponents of } u^{r,\mu}G_{r,\mu}. \tag{2}$$

There are a lot of coset spaces with this structure, which differ by exact arrangement of $G_{r,\mu}$ generators. However, it is always possible, by redefining t and $u^{r,\mu}$, to bring all the $G_{r,\mu}$ inside a single exponent. Let us show that in this case the resulting equations of motion are necessarily linear and for $\ell = \frac{1}{2}$ coincide with those of harmonic oscillators of identical frequency. Indeed, using standard techniques, one may calculate the Cartan forms of the coset element

$$\begin{aligned} g &= e^{t\left(L_{-1}+\omega^2 L_1\right)} e^{u\cdot G},\ u\cdot G = \sum_{r,\mu} u^{r,\mu}G_{r,\mu} \Rightarrow \\ g^{-1}dg &= e^{-u\cdot G} dt\big(L_{-1}+\omega^2 L_1\big) e^{u\cdot G} + e^{-u\cdot G} d e^{u\cdot G} = \\ &= e^{-u\cdot G} \wedge dt\big(L_{-1}+\omega^2 L_1\big) + \frac{1-e^{-u\cdot G}}{u\cdot G} \wedge \big(du\cdot G\big). \end{aligned} \tag{3}$$

Here $\wedge$ is the Zumino notation for the nested commutators: $X\wedge Y = [X,Y]$, $X^2\wedge Y = [X,[X,Y]]$, e.t.c.

Rewriting each of the exponents as a sum of even and odd part $e^X = \cosh X + \sinh X$,

$$\begin{aligned} &e^{-X}\wedge Y = \cosh X\wedge Y - \sinh X\wedge Y = \cosh X\wedge Y - \frac{\sinh X}{X}\wedge [X,Y], \\ &\frac{1-e^{-X}}{X}\wedge dX = \frac{1-\cosh X}{X^2}\wedge [X,dX] + \frac{\sinh X}{X}\wedge dX, \end{aligned} \tag{4}$$

the Cartan form can be represented as

$$\Omega = dt\big(L_{-1}+\omega^2 L_1\big) - \frac{\cosh(u\cdot G)-1}{(u\cdot G)^2}\wedge\big[u\cdot G, du\cdot G - dt\big[u\cdot G, L_{-1}+\omega^2 L_1\big]\big] + \\ +\frac{\sinh(u\cdot G)}{(u\cdot G)}\wedge\big(du\cdot G - dt\big[u\cdot G, L_{-1}+\omega^2 L_1\big]\big). \quad (5)$$

As the structure of the algebra (1) suggests, the expression $du\cdot G - dt\big[u\cdot G, L_{-1}+\omega^2 L_1\big]$ is composed only of G generators. Moreover, calculating N nested commutators of this expression with $u\cdot G$, one may find that the result is proportional to G again if N is even but contains only L and M generators if N is odd. Therefore, one may conclude that $G_{r,\mu}$ appear only in the last line of (5) and

$$\sum_{r,\mu}\Omega^{r,\mu}G_{r,\mu} = \frac{\sinh(u\cdot G)}{(u\cdot G)}\wedge\big(du\cdot G - dt\big[u\cdot G, L_{-1}+\omega^2 L_1\big]\big). \quad (6)$$

As the $\sinh(x)/x$ begins with 1, its adjoint action is always invertible and the equations, which follow from $\Omega^{r,\mu}=0$, are equivalent to

$$\dot u\cdot G - \big[u\cdot G, L_{-1}+\omega^2 L_1\big] = 0. \quad (7)$$

For $\ell=\frac{1}{2}$, r takes only values $-\frac{1}{2},\frac{1}{2}$, and it is not difficult to evaluate the expression (7) explicitly. It would result in the equations of motion of a few (depending on the representation of M) harmonic oscillators

$$\dot u^{-1/2,\mu}+u^{1/2,\mu}=0,\;\; \dot u^{1/2,\mu}-\omega^2 u^{-1/2,\mu}=0 \;\Rightarrow\; \ddot u^{-1/2,\mu}+\omega^2 u^{-1/2,\mu}=0. \quad (8)$$

If there exists a basis for the algebra with different ℓ, one may obtain the Pais-Uhlenbeck oscillators.

Equations (7) explicitly linear and, therefore, similar to ones obtainable from the Galilei algebras. Let us also note that M_a generators appear only in the second term in the first line of (5), and their forms

$$\sum_a \Omega^a M_a = -\frac{\cosh(u\cdot G)-1}{(u\cdot G)^2}\wedge\big[u\cdot G, du\cdot G - dt\big[u\cdot G, L_{-1}+\omega^2 L_1\big]\big]|_M, \quad (9)$$

where $|_M$ implies that we consider only M_a generators, also vanish on equations of motion.

2.2. The action

The Cartan forms of $G_{r,\mu}$ generators were already used in the derivation of the equations of motion, and the only ingredients to construct the action appear to be the forms of M_a. Let us note, however, that due to the general transformation law of the Cartan forms

$$g_0 g = g'h \;\Rightarrow\; \Omega' = (g')^{-1}dg' = h\Omega h^{-1} - dhh^{-1}, \quad (10)$$

the forms of the small subgroup transform non-homogeneously. This is not a problem if the small subgroup is a product of $U(1)$'s, as the dhh^{-1} term in this case will be full differential and its integral would vanish, making integral of the whole form invariant. While it is not possible to use this directly in the case of a non-Abelian small subgroup, one may still split its generators $\{M_a\}\to\{M_0=u(1), M_A\}$, move the M_A to the coset space with parameters Λ^A and use the

IOP Conf. Series: Journal of Physics: Conf. Series **965** (2018) 012025 doi:10.1088/1742-6596/965/1/012025

integral of the form of M_0 as the action. This is possible, because following modification of the coset would not spoil the equations of motion:

$$\begin{aligned} \tilde{g} &= e^{t\left(L_{-1}+\omega^2 L_1\right)} e^{u\cdot G} e^{\Lambda\cdot M} = g e^{\Lambda\cdot M} \\ \tilde{g}^{-1}d\tilde{g} &= e^{-\Lambda\cdot M} \wedge g^{-1}dg + e^{-\Lambda\cdot M} de^{\Lambda\cdot M}. \end{aligned} \tag{11}$$

As the second term in $\tilde{g}^{-1}d\tilde{g}$ does not contain $G_{r,\mu}$ generators, and the adjoint action of the exponent is invertible, the conditions $\widetilde{\Omega}^{r,\mu} = 0$ are equivalent to $\Omega^{r,\mu} = 0$. Moreover, as M forms in $g^{-1}dg$ vanish on these equations, the equations of motion for Λ^A come only from the second term in $\tilde{g}^{-1}d\tilde{g}$, and, therefore, read just as $\dot{\Lambda}^A = 0$.[1] Let us note that this only happens if the $\Omega^{r,\mu} = 0$ forms vanish on the $u^{r,\mu}$ equations of motion, which was not the case of the paper [2].

Let us mention some properties of the action constructed this way.

- The usual inverse Higgs effect [11] is typically absent in such systems, as its presence in the case of spontaneously broken spacetime symmetries [6] actually imposes strong conditions on the structure constants of the algebra [12]. Though it is possible to put covariantly to zero not all $\Omega^{r,\mu}$ forms, they in general contain differentials of all variables. This means that resulting equations contain both fields and their derivatives, and equations cannot be solved algebraically for fields representing some of them in terms of others. (This does not preclude solving the whole set of equations for *derivatives* of fields).
- The action, constructed in the described way, contains derivatives of all variables, and is necessarily linear in them. It should be considered, therefore, in the Hamiltonian sense, with $u^{r,\mu}$ variables (parameters of $G_{r,\mu}$) representing both coordinates and momenta and Λ^A variables being auxiliary to ensure invariance.
- Proposed symmetry of systems of oscillators is dynamical, it cannot be realized on coordinates only.
- The main variables $u^{r,\mu}$ parameterize the coset space

$$\frac{\mathcal{G}}{SO(1,2)\times\mathcal{M}}.$$

The action is made of connections on this space, which are related to the algebra $\mathcal{M}$.

2.3. The equations of motion from the action

It would be natural to expect that due to invariance of the action and the equations of motion, the necessary equations follow from the proposed action. Indeed, let us consider variation of this action

$$\begin{aligned} S = \int \widetilde{\Omega}_{M_0} = \int \tilde{g}^{-1}d\tilde{g}|_{M_0}, \quad \delta S = \int \left(\tilde{g}^{-1}d\delta\tilde{g} - \tilde{g}^{-1}\delta\tilde{g}\tilde{g}^{-1}d\tilde{g}\right)|_{M_0} = \\ = \int \left(\tilde{g}^{-1}d\tilde{g}\,\tilde{g}^{-1}\delta\tilde{g} - \tilde{g}^{-1}\delta\tilde{g}\,\tilde{g}^{-1}d\tilde{g}\right)|_{M_0}. \end{aligned} \tag{12}$$

One may notice that this variation is an integral of the commutator of the Cartan form of the extended coset and the forms, where differentials $du^{r,\mu}$, $d\lambda^A$ are replaced by the variations $\delta u^{r,\mu}$, $\delta\lambda^A$ (and $\delta t = 0$, as t is not varied). Therefore, the coefficients of $\delta u^{r,\mu}$, $\delta\lambda^A$ can only be combinations of expected equations of motion for $u^{r,\mu}$ and Λ^A. Some additional analysis would

[1] The equations of motion for $u^{r,\mu}$ do not change if the exponents of M generators are placed after the exponents of G, regardless of how exactly the G's are arranged. In this case, the equations of motion for Λ can be different, however.

be required to find whether all of them are reproduced, and the $u(1)$ generator should be chosen carefully. In particular, if the $\mathcal{M}$ algebra is $so(1,2)$, $[M_a, M_b] = (a-b)M_{a+b}$, $a = -1, 0, 1$, it is required to put $M_{\pm 1}$ generators in the coset space, while the action will be obtained as the integral of the form of M_0. This will be illustrated by the following examples.

3. Two harmonic oscillators and $so(2,3)$ algebra

To make the statements above more clear, let us consider the example that can be treated in the homogenous parametrization completely. This is the system of two harmonic oscillators with identical frequency, which provide a realization of the $so(2,3)$ algebra (already considered in [2], but in different parametrization). The commutation relations of this algebra read

$$\begin{aligned}
&\big[L_m, L_n\big] = (m-n)L_{m+n},\ m,n = -1,0,1;\ \big[M_a, M_b\big] = (a-b)M_{a+b},\ a,b = -1,0,1,\\
&\big[L_m, G_{r,\mu}\big] = \big(\tfrac{1}{2}m - r\big)G_{m+r,\mu},\ r = -\tfrac{1}{2},\tfrac{1}{2},\quad \big[M_a, G_{r,\mu}\big] = \big(\tfrac{1}{2}a - \mu\big)G_{m,a+\mu},\ \mu = -\tfrac{1}{2},\tfrac{1}{2},\\
&\big[G_{r,\mu}, G_{s,\nu}\big] = 2\big(r\delta_{r+s,0}M_{\mu+\nu} + \mu\delta_{\mu+\nu,0}L_{r+s}\big).
\end{aligned} \tag{13}$$

It is evident that it is 5-graded in this basis (a basis with 3-grading also exists, but we do not consider it here). For our purposes, it would be convenient to rewrite them in the $SO(1,2)$ spinor notation, with $i,j,k,l = 1,2$ and $\alpha,\beta,\mu,\nu = 1,2$:

$$\begin{aligned}
&[L_{ij}, L_{kl}] = \epsilon_{ik}L_{jl} + \epsilon_{jk}L_{il} + \epsilon_{il}L_{jk} + \epsilon_{jl}L_{ik},\\
&[M_{\alpha\beta}, M_{\mu\nu}] = \epsilon_{\alpha\mu}M_{\beta\nu} + \epsilon_{\beta\mu}M_{\alpha\nu} + \epsilon_{\alpha\nu}M_{\beta\mu} + \epsilon_{\beta\nu}M_{\alpha\mu},\\
&[L_{ij}, G_{k\alpha}] = \epsilon_{ik}G_{j\alpha} + \epsilon_{jk}G_{i\alpha},\quad [M_{\alpha\beta}, G_{i\gamma}] = \epsilon_{\alpha\gamma}G_{i\beta} + \epsilon_{\beta\gamma}G_{i\alpha},\\
&[G_{i\alpha}, G_{j\beta}] = -\epsilon_{ij}M_{\alpha\beta} - \epsilon_{\alpha\beta}L_{ij}.
\end{aligned} \tag{14}$$

Here, $\epsilon_{ij} = -\epsilon_{ji}$, $\epsilon_{\alpha\beta} = -\epsilon_{\beta\alpha}$, $\epsilon_{12} = 1$, $\epsilon^{ij}\epsilon_{jk} = \delta^i_k$, $\epsilon^{\alpha\beta}\epsilon_{\beta\gamma} = \delta^\alpha_\gamma$.

The coset space can be parameterized as

$$g = e^{-\frac{1}{2}t\left(L_{11} + \omega^2 L_{22}\right)} e^{u^{k\alpha}G_{k\alpha}}. \tag{15}$$

One may introduce the standard stereographical coordinates on $S^{2,2}$

$$z^{i\alpha} = \frac{\tanh\sqrt{\frac{u\cdot u}{2}}}{\sqrt{\frac{u\cdot u}{2}}}u^{i\alpha},\quad u\cdot u = u^{i\alpha}u_{i\alpha},\ z\cdot z = z^{i\alpha}z_{i\alpha}. \tag{16}$$

and write down the Cartan forms of interest as

$$\begin{aligned}
\big(\Omega_G\big)^{1\alpha} &= \tfrac{dz^{1\alpha} - dt z^{2\alpha}}{1-\frac{1}{2}z\cdot z},\ \big(\Omega_G\big)^{2\alpha} = \tfrac{dz^{2\alpha} + dt\omega^2 z^{1\alpha}}{1-\frac{1}{2}z\cdot z},\\
\big(\Omega_M\big)^{\alpha\beta} &= \frac{1}{2}\frac{z^{i(\alpha}dz_i^{\beta)} + dt\big(z^{2\alpha}z^{2\beta} + \omega^2 z^{1\alpha}z^{1\beta}\big)}{1-\frac{1}{2}z\cdot z}.
\end{aligned} \tag{17}$$

From the conditions $\big(\Omega_G\big)^{i\alpha} = 0$ one may obtain the equations

$$\dot z^{1\alpha} = z^{2\alpha},\ \dot z^{2\alpha} = -\omega^2 z^{1\alpha}\ \Rightarrow \ddot z^{1\alpha} + \omega^2 z^{1\alpha} = 0, \tag{18}$$

which are the equations of motion of two harmonic oscillators. Let us note that $u^{i\alpha}$ satisfy similar equations, as $u\cdot u$ and $z\cdot z$ are the constants of motion.

To construct the action, one should extend the coset space by two more generators

$$\tilde g = g\, e^{\Lambda^{11}M_{11} + \Lambda^{22}M_{22}},\quad \lambda^{\alpha\beta} = \Lambda^{\alpha\beta}\frac{\tan 2\sqrt{\Lambda^{11}\Lambda^{22}}}{2\sqrt{\Lambda^{11}\Lambda^{22}}}. \tag{19}$$

The action is an integral of the $(\widetilde{\Omega}_M)^{12}$ Cartan form of this extended coset

$$S = \int \Big\{ \frac{\lambda^{22} d\lambda^{11} - \lambda^{11} d\lambda^{22}}{1+4\lambda^{11}\lambda^{22}} + (\Omega_M)^{12} \frac{1-4\lambda^{11}\lambda^{22}}{1+4\lambda^{11}\lambda^{22}} + 2\frac{\lambda^{11}(\Omega_M)^{22} - \lambda^{22}(\Omega_M)^{11}}{1+4\lambda^{11}\lambda^{22}} \Big\} \tag{20}$$

It can be directly checked that the equations of motion (18) as well as $\dot{\lambda}^{\alpha\beta}=0$ follow from this action. The equations $\dot{\lambda}^{\alpha\beta}=0$ also follow from the conditions $(\widetilde{\Omega}_M)^{11}=0$, $(\widetilde{\Omega}_M)^{22}=0$ on the new Cartan forms

$$\begin{aligned}
(\widetilde{\Omega}_M)^{11} &= \frac{d\lambda^{11}}{1+4\lambda^{11}\lambda^{22}} + \frac{(\Omega_M)^{11} - 4\lambda^{11}(\Omega_M)^{12} + 4(\lambda^{11})^2(\Omega_M)^{22}}{1+4\lambda^{11}\lambda^{22}}, \\
(\widetilde{\Omega}_M)^{22} &= \frac{d\lambda^{22}}{1+4\lambda^{11}\lambda^{22}} + \frac{(\Omega_M)^{22} + 4\lambda^{22}(\Omega_M)^{12} + 4(\lambda^{22})^2(\Omega_M)^{11}}{1+4\lambda^{11}\lambda^{22}},
\end{aligned} \tag{21}$$

as expected.

The action (20) is of the first order and should be treated as the Hamiltonian one with the factor of dt in (20) being the Hamiltonian itself and the rest being the source of the Poisson brackets. In describing the Hamiltonian and brackets, it is very useful to define set of functions

$$\begin{aligned}
h^{11} &= \frac{4\lambda^{11}}{(1+4\lambda^{11}\lambda^{22})(1-\frac{z\cdot z}{2})}, \quad h^{22} = \frac{-4\lambda^{22}}{(1+4\lambda^{11}\lambda^{22})(1-\frac{z\cdot z}{2})}, \\
h^{12} = h^{21} &= \frac{-(1-4\lambda^{11}\lambda^{22})}{(1+4\lambda^{11}\lambda^{22})(1-\frac{z\cdot z}{2})},
\end{aligned} \tag{22}$$

which the action (20) is composed of. Then the action can be rewritten as

$$S = \int \Big\{ \frac{\lambda^{22} d\lambda^{11} - \lambda^{11} d\lambda^{22}}{1+4\lambda^{11}\lambda^{22}} + \frac{1}{4} h_{\alpha\beta}\, z^{i\alpha} dz_i^{\beta} - dtH \Big\}, \quad H = -\frac{1}{2} z^{2\alpha} z^{2\beta} h_{\alpha\beta} - \frac{\omega^2}{2} z^{1\alpha} z^{1\beta} h_{\alpha\beta}. \tag{23}$$

Differential of the first part of this action provides the brackets between variables. The brackets between λ^{11} and λ^{22}, as well as between $z^{i\alpha}$ are reasonably simple

$$\{\lambda^{11}, \lambda^{22}\} = -\frac{1}{4}\Big(1-\frac{z\cdot z}{2}\Big)(1+4\lambda^{11}\lambda^{22})^2, \quad \{z^{i\alpha}, z^{j\beta}\} = \Big(1-\frac{z\cdot z}{2}\Big)^3 \epsilon^{ij}\, h^{\alpha\beta}. \tag{24}$$

The brackets between $\lambda^{\alpha\beta}$ and $z^{i\gamma}$ are not that systematic. They, however, imply relatively simple brackets of $z^{i\alpha}$ with the functions $h^{\mu\nu}$, as well as between $h^{\alpha\beta}$ and $h^{\mu\nu}$:

$$\{h^{\alpha\beta}, h^{\mu\nu}\} = \epsilon^{\alpha\mu} h^{\beta\nu} + \epsilon^{\beta\mu} h^{\alpha\nu} + \epsilon^{\alpha\nu} h^{\beta\mu} + \epsilon^{\beta\nu} h^{\alpha\mu}, \quad \{h^{\alpha\beta}, z^{i\gamma}\} = \epsilon^{\alpha\gamma} z^{i\beta} + \epsilon^{\beta\gamma} z^{i\alpha}. \tag{25}$$

As expected, these brackets and Hamiltonian (23) reproduce the expected equations of motion (18), along with $\dot{\lambda}^{11}=0$, $\dot{\lambda}^{22}=0$. One may finally note that the following quantities reproduce the $so(2,3)$ algebra with respect to the Poisson brackets:

$$\widetilde{L}_{ij} = z_{i\alpha} z_{j\beta} h^{\alpha\beta}, \quad \widetilde{M}_{\alpha\beta} = -h_{\alpha\beta}, \quad \widetilde{G}_{i\alpha} = z_i^{\beta} h_{\alpha\beta}. \tag{26}$$

4. $G_{2(2)}$ oscillators

Let us note that the considered homogeneous parametrization is not only one that allows to obtain linear equations of motion. For example, let us consider the exceptional $G_{2(2)}$ algebra. Commutation relations of its 14 generators read

$$\begin{aligned}
&[L_m, L_n] = (m-n)L_{m+n}, \; m,n=-1,0,1, \; [M_a, M_b] = (a-b)M_{a+b}, \; a,b=-1,0,1, \\
&[L_m, G_{r,\mu}] = \Big(\frac{m}{2}-r\Big)G_{r+m,\mu}, \; [M_a, G_{r,\mu}] = \Big(\frac{3a}{2}-r\Big)G_{r,\mu+a}, \; r=-\frac{1}{2},\frac{1}{2}, \; \mu=-\frac{3}{2},-\frac{1}{2},\frac{1}{2},\frac{3}{2}, \\
&[G_{r,\mu}, G_{s,\nu}] = r\,\delta_{r+s,0}(6\mu^2+6\nu^2-8\mu\nu-9)\,M_{\mu+\nu} + 3\,\mu(4\mu^2-5)\delta_{\mu+\nu,0} L_{r+s}.
\end{aligned} \tag{27}$$

Looking at the commutators (27), one may note that this algebra contains two different $sl(2)$ algebras and can be viewed as 5- or 7-graded, depending on whether L or M algebra is considered as the primary one. We may, therefore, consider two coset spaces

$$\hat{g}_1 = e^{t(L_{-1}+\omega^2 L_1)} \prod_{\mu=-3/2}^{3/2} e^{u_\mu G_{-1/2,\mu}+v_\mu G_{1/2,\mu}}, \tag{28}$$

$$\hat{g}_2 = e^{t(M_{-1}+\omega^2 M_1)} e^{\sum_{\mu=-3/2}^{3/2} x_\mu G_{-1/2,\mu}} e^{\sum_{\mu=-3/2}^{3/2} y_\mu G_{1/2,\mu}}. \tag{29}$$

The generators G in these parameterizations are arranged in such a way that 1) generators inside each exponent form closed subsets under the action of the conformal generators, 2) exponents themselves are placed so that second subset of generators can be produced from the first by the action of M_1 (or L_1), and so on. For this reason, this structure of the coset can be called "matryoshka". It can be checked that in both cases the equations of motion are again linear.

4.1. Four harmonic oscillators

Let us briefly consider the coset space $\hat{g}_1$ (28). Evaluating its Cartan forms, one may note that all forms of $G_{r,\mu}$ are composed of subforms

$$U_\mu = du_\mu - v_\mu\, dt,\; V_\mu = dv_\mu + \omega^2 u_\mu\, dt. \tag{30}$$

Therefore, the equations $\Omega_{r,\mu} = 0$ are the linear system of equations w.r.t. U_μ, V_μ with zero right-hand side. As this system appears to be non-degenerate, it implies

$$U_\mu = V_\mu = 0 \;\Rightarrow\; \ddot{u}_\mu + \omega^2 u_\mu = 0, \tag{31}$$

which are just the equations of motion of four harmonic oscillators with the same frequency. The system of only four equations $\Omega_{-1/2,\mu} = 0$ cannot be solved for v_μ, and the inverse Higgs effect is also absent in this basis.

The action for this system can also be constructed by extending the coset space $\hat{g}_1 \to \tilde{\hat{g}}_1 = \hat{g}_1 \exp(\Lambda_{-1}L_{-1} + \Lambda_1 L_1)$. It reads

$$S = \int \widetilde{\Omega}_{M_0} = \int \Big[\frac{\lambda_{-1}d\lambda_1 - \lambda_1 d\lambda_{-1}}{1+\lambda_1\lambda_{-1}} + \frac{(1-\lambda_1\lambda_{-1})\Omega_{M_0} + 2\lambda_{-1}\Omega_{M_1} - 2\lambda_1\Omega_{M_{-1}}}{1+\lambda_1\lambda_{-1}}\Big], \tag{32}$$

where we defined

$$\lambda_{\pm 1} = \Lambda_{\pm 1}\frac{\tan\sqrt{\Lambda_{-1}\Lambda_1}}{\sqrt{\Lambda_{-1}\Lambda_1}} \tag{33}$$

and Ω_{M_a} are the M_a forms of the $\hat{g}_1$ coset. It can be checked explicitly that the action (32) produces the expected equations of motion $\dot{u}_\mu = v_\mu$, $\dot{v}_\mu = -\omega^2 u_\mu$, $\lambda_{\pm 1} = 0$.

4.2. Two Pais-Uhlenbeck oscillators

The coset $\hat{g}_2$ (28) also leads to linear equations. The Cartan forms in this case are composed of subforms

$$\begin{gathered} X_{-3/2} = dx_{-3/2} - x_{1/2}dt,\;\; X_{-1/2} = dx_{-1/2} - 2x_{1/2}dt + 3\omega^2 x_{3/2}dt, \\ X_{1/2} = dx_{1/2} - 3x_{3/2}dt + 2\omega^2 x_{-1/2}dt,\;\; X_{3/2} = dx_{3/2} + \omega^2 x_{1/2}dt, \\ Y_{-3/2} = dy_{-3/2} - y_{1/2}dt,\;\; Y_{-1/2} = dy_{-1/2} - 2y_{1/2}d\tau + 3\omega^2 y_{3/2}dt, \\ Y_{1/2} = dy_{1/2} - 3y_{3/2}dt + 2\omega^2 y_{-1/2}dt,\;\; Y_{3/2} = dy_{3/2} + \omega^2 y_{1/2}dt. \end{gathered} \tag{34}$$

Here, it makes sense to introduce completely symmetric 3-spinor notation by the relation

$$x_{-3/2} = \frac{1}{3}x^{1,1,1}, \; x_{-1/2} = x^{1,1,2}, \; x_{1/2} = x^{1,2,2}, \; x_{3/2} = \frac{1}{3}x^{2,2,2}. \tag{35}$$

As a result, the $G_{r,\mu}$ and L_m Cartan forms can be written as

$$\begin{aligned} &\Omega_{L_{-1}} = xX, \; \Omega_{L_0} = 2yX, \; \Omega_{L_1} = yY + 2(y^3X) - \tfrac{1}{2}y^4 \cdot xX, \\ &\Omega^{\alpha\beta\gamma}_{-1/2} = X^{\alpha\beta\gamma} + y^{\alpha\beta\gamma} \cdot xX, \; \Omega^{\alpha\beta\gamma}_{1/2} = Y^{\alpha\beta\gamma} - 3(yXy)^{\alpha\beta\gamma} + y^{\alpha\beta\gamma}(yX) + (y^3)^{\alpha\beta\gamma} \cdot xX. \end{aligned} \tag{36}$$

Here, we used the notation

$$\begin{aligned} &xy = x^{\alpha\beta\gamma}y_{\alpha\beta\gamma}, \; y^4 = y^{\alpha\mu\nu}y_{\mu\nu\sigma}y^{\sigma\beta\gamma}y_{\alpha\beta\gamma}, \\ &(xyz)^{\alpha\beta\gamma} = \frac{1}{3}\left(x^{\alpha\mu\nu}y_{\mu\nu\sigma}z^{\sigma\beta\gamma} + x^{\beta\mu\nu}y_{\mu\nu\sigma}z^{\sigma\alpha\gamma} + x^{\gamma\mu\nu}y_{\mu\nu\sigma}z^{\sigma\alpha\beta}\right). \end{aligned} \tag{37}$$

It is obvious to show that the equations $\Omega^{\alpha\beta\gamma}_r = 0$ imply $X^{\alpha\beta\gamma} = 0$ and $Y^{\alpha\beta\gamma} = 0$. Excluding from them all variables in terms of derivatives of $x_{-3/2}$ and $y_{-3/2}$, one may find

$$\left(\frac{d}{dt^2} + (3\omega)^2\right)\left(\frac{d}{dt^2} + \omega^2\right)x_{-3/2} = 0, \; \left(\frac{d}{dt^2} + (3\omega)^2\right)\left(\frac{d}{dt^2} + \omega^2\right)y_{-3/2} = 0 \tag{38}$$

-the equations of motion of two Pais-Uhlenbeck oscillators with frequencies ω and 3ω.

One may check explicitly that the action

$$S = \int \tilde{\Omega}_{L_0} = \int \left[\frac{\lambda_{-1}d\lambda_1 - \lambda_1 d\lambda_{-1}}{1 + \lambda_{-1}\lambda_1} + \frac{(1 - \lambda_{-1}\lambda_1)\Omega_{L_0} + 2\lambda_{-1}\Omega_{L_1} - 2\lambda_1\Omega_{L_{-1}}}{1 + \lambda_{-1}\lambda_1}\right] \tag{39}$$

reproduces the equations of motion (38). As before, it can be obtained by extending the coset as $\hat{g}_2 \to \tilde{\hat{g}}_2 = \hat{g}_2 \exp(\Lambda_{-1}L_{-1} + \Lambda_1 L_1)$, evaluating the modified L_0 Cartan form and performing substitution (33).

5. Conclusion

In this paper, we demonstrated that for any noncompact simple Lie algebras one may construct a dynamical realization, then the conditions on the Cartan forms of the specifically chosen coset space reproduce the equations of motion of the harmonic and/or Pais-Uhlenbeck oscillators. Also, we showed that for each appearing system of oscillators, one may provide the first-order action, which is invariant with respect to the whole group and produces the expected equations of motion.

The studied examples show that one may parameterize the coset space in a few different ways and still obtain the linear equations of motion. One way was used in general proof and can be called "homogeneous parametrization", as it involves placing of all dynamical variables and their generators into just a single exponent. Another parametrization, which can be called "matryoshka", was found for the $G_{2(2)}$ algebra and involves placing generators into the exponents according to their properties with respect to the conformal and automorphism $sl(2)$ algebras. Let us note that such a parametrization is also known for the $so(2,3)$ algebra, and it is possible that it is another universal parametrization which ensures the linearity of the equations of motion.

Each of these two parameterizations has its own advantages and disadvantages. The homogeneous parametrization is suitable for general proofs, can be constructed for any group, the fields in it are in linear representation of conformal and automorphism groups, which provides a more natural connection to the geometry. However, the calculations in this parametrization involve infinite matrix power series, which sometimes are very difficult to evaluate explicitly. Contrary to this, in the "matryoshka" parametrization the Cartan forms involve the finite number of terms, which are always possible to find explicitly. The variables involved, however, do not form a linear representation of the automorphism group, which can make the Cartan forms longer, complicated and difficult to deal with.

A typical feature of the considered parameterizations is the absence of the usual inverse Higgs effect, which necessitates the related actions to be considered as Hamiltonian ones, with $u^{r,\mu}$ variables representing both coordinates and momenta. The symmetry then will also be dynamical, which is impossible to realize off-shell in terms of coordinates only.

Acknowledgments
The work of N.K. and S.K. was supported by the RSCF, grant 14-11-00598.

References

[1] Krivonos S and Nersessian A 2017 SU(1,2) invariance in two-dimensional oscillator *JHEP* **1702** 006 (*Preprint* arXiv:1610.02499)
[2] Krivonos S, Lechtenfeld O and Sorin A 2017 Hidden symmetries of deformed oscillators *Nucl. Phys.* **B924** 33-46 (*Preprint* arXiv:1612.07832)
[3] Coleman S, Wess J and Zumino B 1969 Structure of phenomenological lagrangians. 1 *Phys. Rev.* **177** 2239,
[4] Callan C, Coleman S, Wess J and Zumino B 1969 Structure of phenomenological lagrangians. 2 *Phys. Rev.* **177** 2247
[5] Volkov D 1973 Phenomenological lagrangians *Sov. J. Part. Nucl.* **4** 3
[6] Ogievetsky V 1974 Nonlinear realizations of internal and spacetime symmetries *Acta Universitatis Wratislaviensis* **207** 117 (Proc. of the X'th Winter school of theoretical physics in Karpacz)
[7] Bina B and Günadin M 1997 Real forms of nonlinear superconformal and quasi-superconformal algebras and their unified realization *Nucl. Phys.* **B502** 713 (*Preprint* hep-th/9703188)
[8] Krivonos S, Lechtenfeld O and Sorin A 2016 Minimal realization of ℓ-conformal Galilei algebra, Pais-Uhlenbeck oscillators and their deformation *JHEP* **1610** 078 (*Preprint* arXiv:1607.03756)
[9] Galajinsky A 2010 Conformal mechanics in Newton-Hooke spacetime *Nucl. Phys.* **B832** 586 (*Preprint* arXiv:1002.2290)
[10] Andrzejewski K, Galajinsky A, Gonera J and Masterov I 2014 Conformal Newton-Hooke symmetry of Pais-Uhlenbeck oscillator *Nucl. Phys.* **B885** 150-162 (*Preprint* arXiv:1402.1297)
[11] Ivanov E and Ogieevetsky A 1975 The Inverse Higgs Phenomenon in Nonlinear Realizations *Teor.Mat.Phys* **25** 164-177
[12] Klein R, Roest D and Stefanyszyn D 2017 Spontaneously Broken Spacetime Symmetries and the Role of Inessential Goldstones (*Preprint* arXiv:1709.03525)

Curved WDVV equation and supersymmetric mechanics

N Kozyrev[1], S Krivonos[1], O Lechtenfeld[2], A Nersessian[3,4] and A Sutulin[1]
[1] Bogoliubov Laboratory of Theoretical Physics, JINR, 141980 Dubna, Russia
[2] Institut für Theoretische Physik and Riemann Center for Geometry and Physics, Leibniz Universität Hannover, Appelstrasse 2, D-30167 Hannover, Germany
[3] Yerevan State University, 1 Alex Manoogian St., Yerevan, 0025, Armenia
[4] Tomsk Polytechnic University, Lenin Ave. 30, 634050 Tomsk, Russia

E-mail: nkozyrev@theor.jinr.ru, krivonos@theor.jinr.ru, lechtenf@itp.uni-hannover.de, arnerses@ysu.am, sutulin@theor.jinr.ru

Abstract. We extend the relation between the Witten-Dijkgraaf-Verlinde-Verlinde equation and $N=4$ supersymmetric mechanics to arbitrary curved spaces. The resulting curved WDVV equation is written in terms of the third rank Codazzi tensor. We provide the solutions of the curved WDVV equation for the $so(n)$ symmetric conformally flat metrics. We also explicitly demonstrate how each solution of the flat WDVV equation can be lifted up to the curved WDVV solution on the conformally flat spaces.

1. Introduction

One of the areas where the famous Witten-Dijkgraaf-Verlinde-Verlinde (WDVV) equation [1] and [2] plays a significant role is the n-particle $N=4$ supersymmetric mechanics. Indeed, it was firstly demonstrated in [3] that if we insist that the $N=4$ supercharges $Q^a, \overline{Q}_b$, $(a,b=1,2)$

$$Q^a = p_i\,\psi^{ai} + \mathrm{i}\widetilde{F}_{ijk}\,\psi^{bi}\,\psi^j_b\,\bar\psi^{ak} \quad \overline{Q}_a = p_i\,\bar\psi^i_a + \mathrm{i}\widetilde{F}_{ijk}\,\bar\psi^i_b\,\bar\psi^{bj}\,\psi^k_a \tag{1.1}$$

form the $N=4$ super Poincaré algebra

$$\left\{Q^a, Q^b\right\} = 0,\ \{\overline{Q}_a, \overline{Q}_b\} = 0, \qquad \{Q^a, \overline{Q}_b\} = \frac{\mathrm{i}}{2}\,\delta^a_b\,H, \tag{1.2}$$

then the totally symmetric over the indices structure functions $\widetilde{F}_{ijk}(x)$ entering into supercharges (1.1) have to obey the equations

$$\partial_i\widetilde{F}_{jkl} - \partial_j\widetilde{F}_{ikl} = 0, \quad \widetilde{F}_{ikp}\delta^{pq}\widetilde{F}_{jmq} - \widetilde{F}_{jkp}\delta^{pq}\widetilde{F}_{imq} = 0. \tag{1.3}$$

The second equation in (1.3) in which the evident solution of the first equation

$$\widetilde{F}_{ijk} = \frac{\partial^3}{\partial x^i \partial x^j \partial x^k}\widetilde{F} \tag{1.4}$$

is substituted, is just the WDVV equation.

While evaluating the brackets between the supercharges in (1.2), it was supposed that the basic variables obey the standard Dirac brackets

$$\{x^i, p_j\} = \delta^i_j, \ \{\psi^{ai}, \bar\psi^j_b\} = \frac{\mathrm{i}}{2}\delta^a_b\delta^{ij}. \tag{1.5}$$

Therefore, the bosonic part of the Hamiltonian H in (1.2)

$$H = \delta^{ij} p_i p_j \tag{1.6}$$

describes the motion over a flat space. Clearly, this is not the most general case and the interesting question is what will be the analog of the WDVV equations (1.3) in the case of $N = 4$ supersymmetric mechanics defined on an arbitrary space with the metric g_{ij}? The main goal of this Letter is to answer this question by providing the corresponding curved WDVV equation and discussing its solutions.

2. Supercharges and generalized WDVV equation

In the case of the spaces with the metric g_{ij} the Dirac brackets between the basic variables have to be written in a covariant way as

$$\begin{aligned} &\{x^i, p_j\} = \delta^i_j, \;\; \{\psi^{ai}, \bar\psi^j_b\} = \frac{\mathrm{i}}{2}\delta^a_b\, g^{ij}, \;\; \{p_i, \psi^{aj}\} = \Gamma^j_{ik}\psi^{ak}, \;\; \{p_i, \bar\psi^j_a\} = \Gamma^j_{ik}\bar\psi^k_a, \\ &\{p_i, p_j\} = -2\mathrm{i} R_{ijkm}\psi^{ak}\bar\psi^m_a. \end{aligned} \tag{2.1}$$

Here, g_{ij} is a metric on the bosonic space, while the Levi-Civita connection Γ^k_{ij} and the curvature tensor $R^i{}_{jkl}$ are defined as

$$\begin{aligned} &\Gamma^k_{ij} = \frac{1}{2}\, g^{km}\left(\partial_i g_{jm} + \partial_j g_{im} - \partial_m g_{ij}\right), \\ &R^i{}_{jkl} = \partial_k \Gamma^i_{jl} - \partial_l \Gamma^i_{jk} + \Gamma^m_{jl}\,\Gamma^i_{mk} - \Gamma^m_{jk}\,\Gamma^i_{ml}. \end{aligned} \tag{2.2}$$

It should be noted that the structure of the brackets (2.1) is completely fixed by the Jacobi identities as soon as we defined the first two brackets in (2.1).

To construct the supercharges $Q^a, \overline{Q}_b$, we will use the same Ansatz (1.1) :

$$Q^a = p_i\,\psi^{ai} + \mathrm{i} F_{ijk}\,\psi^{bi}\,\psi^j_b\,\bar\psi^{ak}, \quad \overline{Q}_a = p_i\,\bar\psi^i_a + \mathrm{i} F_{ijk}\,\bar\psi^i_b\bar\psi^{bj}\,\psi^k_a\,, \tag{2.3}$$

where F_{ijk} is an arbitrary, for the time being, function depending on the n-coordinates x^i and is, by construction, symmetric over the first two indices. Now it is a matter of straightforward calculations to check that the conditions that these supercharges span the $N = 4$ super Poincaré algebra (1.2), with the new Dirac brackets (2.1) taken into account, result in the following equations on the function F_{ijk}:

$$F_{ijk} - F_{ikj} = 0, \;\Rightarrow\; F_{ijk} \text{ is totally symmetric over the indices} \tag{2.4}$$

$$\nabla_i F_{jkm} - \nabla_j F_{ikm} = 0, \tag{2.5}$$

$$F_{ikp} g^{pq} F_{jmq} - F_{jkp} g^{pq} F_{imq} + R_{ijkm} = 0, \tag{2.6}$$

where, as usual,

$$\nabla_i F_{jkl} = \partial_i F_{jkl} - \Gamma^m_{ij} F_{klm} - \Gamma^m_{ik} F_{jlm} - \Gamma^m_{il} F_{jkm}. \tag{2.7}$$

Once the equations (2.4)-(2.6) are satisfied, the Hamiltonian H acquires the form

$$H = g^{ij} p_i p_j - 2\left[\nabla_i F_{kmj} + \nabla_m F_{ijk} + 2R_{imjk}\right]\psi^{ci}\bar\psi_c{}^m\,\psi^{dj}\bar\psi_d{}^k. \tag{2.8}$$

Equations (2.4)-(2.6) are just the curved WDVV equations we were looking for.

3. Particular solutions of the curved WDVV equation

The serious problem with construction of the solutions of the curved WDVV equations (2.4)-(2.6) is that even for the first equation (2.5), which defines the so called third rank Codazzi tensor, the general solution, to the best of our knowledge, is known only for the constant curvature spaces [4]

$$\begin{aligned} F_{ijk} \;=\;& \frac{1}{3}\left(\nabla_i\nabla_j\nabla_k F + \nabla_j\nabla_i\nabla_k F + \nabla_k\nabla_i\nabla_j F\right) + \frac{4R}{3n(n-1)}\left(g_{ij}\nabla_k F + g_{ik}\nabla_j F + g_{jk}\nabla_i F\right) = \\ & \nabla_i\nabla_j\nabla_k F + \frac{R}{n(n-1)}\left(2g_{jk}\partial_i F + g_{ij}\partial_k F + g_{ik}\partial_j F\right), \quad R = g^{jk} R^i_{jik} = const. \end{aligned} \tag{3.1}$$

The pre-potential F has to be determined by the second equation (2.6). Clearly, this is a very complicated task. In what follows, we will present particular solutions of the curved WDVV equations for some special metrics g_{ij}.

3.1. Real Kähler metrics

In a series of papers [5, 6] the many particle $N = 4$ supersymmetric mechanics was considered within the superfield formalism. One of the main results obtained in these papers is the conclusion that the resulting bosonic metrics is the real Kähler one, i.e.

$$g_{ij} = \frac{\partial^2}{\partial x^i \partial x^j} \mathcal{A}. \tag{3.2}$$

Let us analyze how this case fits in our scheme.

With the metrics (3.2), the Levi-Civita connection with all lower indices is fully symmetric, and the Riemann tensor does not contain the fourth and higher derivatives of $\mathcal{A}(x)$:

$$\Gamma_{ijk} = \frac{1}{2} \frac{\partial^3 A(x)}{\partial x^i \partial x^j \partial x^k}, \quad R_{ijkm} = g^{pq}\left(\Gamma_{imp}\Gamma_{jkq} - \Gamma_{ikp}\Gamma_{jmq}\right). \tag{3.3}$$

With such structure of the Levi-Civita connection and Riemann tensor, it is rather easy to check that the choice

$$F_{ijk} = \pm \Gamma_{ijk} \tag{3.4}$$

solves both curved WDVV equations (2.5), (2.6).

3.2. $so(n)$-invariant metrics

Careful analysis of equation (2.5) shows that it admits a series of solutions in case of the conformally flat metrics depending on x^2 only, i.e. in the case when

$$g_{ij} = \frac{1}{f^2[x^2]} \delta_{ij}. \tag{3.5}$$

Indeed, starting from the following Ansatz on the F_{ijk}:

$$(F_1)_{ijk} = A[x^2]\delta_{ii'}\delta_{jj'}\delta_{kk'}x^{i'}x^{j'}x^{k'} + B[x^2]\left(\delta_{ij}\delta_{kk'}x^{k'} + \delta_{jk}\delta_{ii'}x^{i'} + \delta_{ik}\delta_{jj'}x^{j'}\right), \tag{3.6}$$

one may check that equation (2.5) is satisfied if the functions A and B obey the equation

$$A\left(2f'x^2 - f\right) + 8Bf' + 2fB' = 0. \tag{3.7}$$

Here, prime means differentiation over x^2.

Besides the solution (3.7), there are also solutions which generalize any solution $\widetilde{F}_{ijk}$ of the flat WDVV equations (1.3) which additionally obey the condition (see, e.g., (2.15) from [13])

$$x^i \widetilde{F}_{ijk} = \delta_{jk}. \tag{3.8}$$

This new solution of equation (2.5) reads

$$(F_2)_{ijk} = \frac{1}{f^2} \widetilde{F}_{ijk}. \tag{3.9}$$

Combining now these two solutions, one may check that the second curved WDVV equation (2.6) is satisfied by the choice

$$F_{ijk} = A\delta_{ii'}\delta_{jj'}\delta_{kk'}x^{i'}x^{j'}x^{k'} + B\left(\delta_{ij}\delta_{kk'}x^{k'} + \delta_{jk}\delta_{ii'}x^{i'} + \delta_{ik}\delta_{jj'}x^{j'}\right) + \frac{1}{f^2}\widetilde{F}_{ijk} \tag{3.10}$$

if the functions A and B are further constrained to obey the equations

$$A + f^2 B\left(x^2 A + B\right) = -4\frac{f''}{f^3}, \quad B\left(2 + x^2 f^2 B\right) = \frac{4f'}{f^4}\left(x^2 f' - f\right). \tag{3.11}$$

One may check that equation (3.7) is satisfied automatically if the functions A and B obey the equations (3.11).

Equations (3.11) may be easily solved as

$$A = -\frac{f^5 B^2 + 4f''}{f^3\left(1 + x^2 f^2 B\right)}, \quad B = -\frac{f \pm \left(f - 2x^2 f'\right)}{x^2 f^3}. \tag{3.12}$$

Thus, for the $so(n)$-invariant conformally flat metrics (3.5) the curved WDVV equations (2.4), (2.5), (2.6) admit a quite general solution (3.10).

4. Conclusion

In the present Letter we have used the relation between the WDVV equation and n-particle $N = 4$ supersymmetric mechanics to find the curved WDVV equations (2.4)-(2.6) defined on arbitrary spaces with the metric g_{ij}. In these generalized equations, the third derivative of the pre-potential $\widetilde{F}_{ijk} = \partial^3_{ijk}\widetilde{F}$ is replaced by the third rank Codazzi tensor F_{ijk}, while the WDVV equation itself acquires the non-trivial r.h.s. equal to the curvature tensor. Due to existing problems with the explicit structure of the third rank Codazzi tensors on arbitrary spaces, we have found the solutions of the curved WDVV equations on the $so(n)$-invariant conformally flat spaces. A nice feature of these solutions is that they include an arbitrary solution of the flat WDVV equation. Thus, any solution found for the flat WDVV equations in [7]-[11] (with the constraint (3.8) being satisfied) can be lifted up to the solution of the curved WDVV equations.

As concerns the applications of the obtained results to the $N = 4$ supersymmetric mechanics, the urgent task is to extend the supercharges by the potential terms (as it was done in [3], [12], [13] for the flat space case) and to find admissible potentials. This task will be considered elsewhere.

Acknowledgments

We are grateful to M. Feigin and A. Veselov for stimulating discussions. The work of N.K. and S.K. was partially supported by RSCF grant 14-11-00598 and the Heisenberg-Landau program. The work of A.S. was partially supported by RFBR grant 15-02-06670 and the Heisenberg-Landau program. The work of A.N. was partially supported by Tomsk Polytechnic University competitiveness enhancement program, by the grant of Armenian State Committee of Science 15T-1C367 and by the ICTP Network project NT-04. This article is based upon work from COST Action MP1405 QSPACE, supported by COST (European Cooperation in Science and Technology).

[1] Witten E 1990 *Nucl. Phys.* B **340** 281
[2] Dijkgraaf R, Verlinde H and Verlinde E (1991) *Nucl. Phys.* B **352** 59
[3] Wyllard N 2000 *J.Math.Phys.* **41** 2826; `arXiv:hep-th/9910160`
[4] Liu H L, Simon U and Wang C P 1998 *Contributions to Algebra and Geometry* **39** 329
[5] Donets E, Pashnev A, Rosales J Juan and Tsulaia M 2000 *Phys.Rev.* **D61** 043512; `arXiv:hep-th/9907224`
[6] Donets E E, Pashnev A, Rivelles V O, Sorokin D and Tsulaia M 2000 *Phys.Lett.* **B484** 337; `arXiv:hep-th/0004019`
[7] Martini R and Gragert P K H 1999 *J. Nonlin. Math. Phys.* **6** 1; `arXiv:hep-th/9901166`
[8] Veselov A P 1999 *Phys. Lett.* A **261** 297; `arXiv:hep-th/9902142`
[9] Chalykh O A and Veselov A P 2001 *Phys. Lett.* A **285** 339; `arXiv:math-ph/0105003`
[10] Feigin M V and Veselov A P 2007 *Adv. Math.* **212** 143; `arXiv:math-ph/0512095`
[11] Feigin M V and Veselov A P 2007; `arXiv:0710.5729 [math-ph]`
[12] Galajinsky A, Lechtenfeld O and Polovnikov K 2007 *JHEP* **0711** 008; `arXiv:0708.1075[hep-th]`
[13] Krivonos S anf Lechtenfeld O 2012 *JHEP* **1102** 042; `arXiv:1012.4639[hep-th]`

Spectral Structure of Elastic Neumann–Poincaré Operators

Yoshihisa Miyanishi[1], **Kazunori Ando**[2] **and Hyeonbae Kang**[3]

[1] Center for Mathematical Modeling and Data Science, Osaka University, Osaka 560-8631, Japan

[2] Department of Electrical and Electronic Engineering and Computer Science, Ehime University, Ehime 790-8577, Japan

[3] Department of Mathematics and Institute of Applied Mathematics, Inha University, Incheon 22212, S. Korea

E-mail: miyanishi@sigmath.es.osaka-u.ac.jp

Abstract. In [1, 2], it is proved that the elastic Neumann–Poincaré operator defined on the smooth boundary of a bounded domain, which is known to be non-compact, is in fact polynomially compact. As a consequence, it is shown that the spectrum of the elastic Neumann-Poincaré operator consists of non-empty sets of eigenvalues accumulating to certain numbers determined by Lamé parameters. The purpose of this paper is to review these results and their proofs, and to discuss about some questions related to these results.

1. Introduction: Polynomial compactness and spectral structure

The elastic Neumann–Poincaré (abbreviated by NP) operator is a boundary integral operator which appears naturally when solving classical boundary value problems for the Lamé system [4, 6]. The Lamé system is defined in terms of a pair of Lamé constants (λ, μ) satisfying the strong convexity condition: $\mu > 0$ and $d\lambda + 2\mu > 0$, where d denotes the space dimension, i.e., $d = 2, 3$. The isotropic elasticity tensor $\mathbb{C} = (C_{ijkl})_{i,j,k,l=1}^{d}$ and the corresponding Lamé system $\mathcal{L}_{\lambda,\mu}$ are defined by

$$C_{ijkl} := \lambda\, \delta_{ij}\delta_{kl} + \mu\, (\delta_{ik}\delta_{jl} + \delta_{il}\delta_{jk}) \tag{1}$$

and

$$\mathcal{L}_{\lambda,\mu} u := \nabla \cdot \mathbb{C}\widehat{\nabla} u = \mu\Delta u + (\lambda + \mu)\nabla\nabla \cdot u, \tag{2}$$

where $\widehat{\nabla}$ denotes the symmetric gradient, namely,

$$\widehat{\nabla} u := \frac{1}{2}\left(\nabla u + \nabla u^T\right) \quad (T \text{ for transpose}).$$

Let $\mathbf{\Gamma}(x) = (\Gamma_{ij}(x))_{i,j=1}^{d}$ be the Kelvin matrix of the fundamental solution to the Lamé operator $\mathcal{L}_{\lambda,\mu}$, namely,

$$\Gamma_{ij}(x) = \begin{cases} -\dfrac{\alpha_1}{4\pi}\dfrac{\delta_{ij}}{|x|} - \dfrac{\alpha_2}{4\pi}\dfrac{x_i x_j}{|x|^3}, & \text{if } d = 3, \\ \dfrac{\alpha_1}{2\pi}\delta_{ij}\log|x| - \dfrac{\alpha_2}{2\pi}\dfrac{x_i x_j}{|x|^2}, & \text{if } d = 2 \end{cases} \tag{3}$$

where

$$\alpha_1 = \frac{1}{2}\left(\frac{1}{\mu} + \frac{1}{2\mu+\lambda}\right) \quad \text{and} \quad \alpha_2 = \frac{1}{2}\left(\frac{1}{\mu} - \frac{1}{2\mu+\lambda}\right). \tag{4}$$

The NP operator for the Lamé system is defined by

$$\mathbf{K}[f](x) := \text{p.v.} \int_{\partial\Omega} \partial_{\nu_x}\mathbf{\Gamma}(x-y)f(y)d\sigma(y) \quad \text{a.e. } x \in \partial\Omega. \tag{5}$$

Here, p.v. stands for the Cauchy principal value, and the conormal derivative $\partial_{\nu_x}\mathbf{\Gamma}(x-y)$ of the Kelvin matrix with respect to x-variables is defined by

$$\partial_{\nu_x}\mathbf{\Gamma}(x-y)b = \partial_{\nu_x}(\mathbf{\Gamma}(x-y)b) \tag{6}$$

for any constant vector b (We note that our definitiion of $\mathbf{K}$ is the L^2-adjoint of that in [1]).

The NP operator $\mathbf{K}$ is related to the boundary value problem in the following way. Let Ω be a bounded domain in $\mathbf{R}^d$ with the smooth boundary, and consider the Neumann boundary value problem with the given traction g on $\partial\Omega$: $\mathcal{L}_{\lambda,\mu}u = 0$ in Ω and $\partial_\nu u = g$ on $\partial\Omega$. Here and throughout this paper, ∂_ν denotes the conormal derivative on $\partial\Omega$, namely,

$$\partial_\nu u := (\mathbb{C}\widehat{\nabla}u)n = \lambda(\nabla\cdot u)n + 2\mu(\widehat{\nabla}u)n \quad \text{on } \partial\Omega, \tag{7}$$

where n is the outward unit normal to $\partial\Omega$. One may seek the solution in the form $u = \mathbf{S}[f]$ for some vector-valued function f on $\partial\Omega$, where $\mathbf{S}$ is the single layer potential, namely,

$$\mathbf{S}[f](x) := \int_{\partial\Omega} \mathbf{\Gamma}(x-y)f(y)d\sigma(y) \quad x \in \Omega. \tag{8}$$

Then u satisfies the equation: $\mathcal{L}_{\lambda,\mu}u = 0$ in Ω. The conormal derivative of the single layer potential is related to the NP operator by the jump relation:

$$\partial_\nu\mathbf{S}[f](x) = \left(-\frac{1}{2}\mathbf{I} + \mathbf{K}\right)[f](x) \quad \text{a.e. } x \in \partial\Omega, \tag{9}$$

where $\mathbf{I}$ is the identity operator. It means that to solve the Neumann boundary value problem, it suffices to solve the integral equation on $\partial\Omega$:

$$\left(-\frac{1}{2}\mathbf{I} + \mathbf{K}\right)[f] = g. \tag{10}$$

Here arises the crux of this paper's investigation. Unlike the electro-static NP operator, the elastic NP operator is a genuine singular integral operator and is not a compact operator even if $\partial\Omega$ is smooth (see section 2), which implies that the Fredholm index theory cannot be applied to solve (10). It is proved in [4] that $-1/2\mathbf{I} + \mathbf{K}$ is invertible on $H_0^{-1/2}(\partial\Omega)^d$. Here $H^{-1/2}$ denotes the Sobolev space of order $-1/2$, and the subscript 0 means that the members are orthogonal to rigid motions.

Since the elastic NP operator $\mathbf{K}$ is *not* compact even on smooth domains, a naturally arising question is how its spectrum looks like. We emphasize that even though $\mathbf{K}$ is *not* self-adjoint on $L^2(\partial\Omega)^d$, it can be realized as a self-adjoint operator on $H^{-1/2}(\partial\Omega)^d$ (see [1, 5]). Thus its spectrum on $H^{-1/2}(\partial\Omega)^d$ consists of discrete and continuous ones. In relation to the spectrum of $\mathbf{K}$, it is recently proved in [1, 2] that $\mathbf{K}$ is in fact polynomially compact. An operator $\mathbf{A}$ is said to be polynomially compact if there is a polynomial p such that $p(\mathbf{A})$ is compact. In fact, the following theorem is obtained. Throughout this paper we let $\mathcal{H} := H^{-1/2}(\partial\Omega)^d$ for ease of notation.

IOP Conf. Series: Journal of Physics: Conf. Series **965** (2018) 012027 doi:10.1088/1742-6596/965/1/012027

Theorem 1.1 ([1, 2]). *Let*

$$k_0 = \frac{\mu}{2(2\mu + \lambda)}. \tag{11}$$

Then we have

(i) In two dimensions $\mathbf{K}^2 - k_0^2\mathbf{I}$ *is compact on* $\mathcal{H}$ *if* $\partial\Omega$ *is* $C^{1,\alpha}$ *for some* $\alpha > 0$,

(ii) In three dimensions $\mathbf{K}(\mathbf{K}^2 - k_0^2\mathbf{I})$ *is compact on* $\mathcal{H}$ *if* $\partial\Omega$ *is* C^∞*-smooth. Moreover,* $\mathbf{K}(\mathbf{K} - k_0\mathbf{I})$, $\mathbf{K}(\mathbf{K} + k_0\mathbf{I})$ *and* $\mathbf{K}^2 - k_0^2\mathbf{I}$ *are* not *compact.*

As a consequence, the following theorem regarding the spectral structure of the elastic NP operator.

Theorem 1.2 ([1, 2]). *Let* $\mathbf{K}$ *be the elastic NP operator on* $\partial\Omega$.

(i)′ In two dimensions the spectrum of $\mathbf{K}$ *on* $\mathcal{H}$ *consists of two non-empty sets of eigenvalues accumulating at* k_0 *and* $-k_0$*, respectively (provided that* $\partial\Omega$ *is* $C^{1,\alpha}$ *for some* $\alpha > 0$*),*

(ii)′ In three dimensions the spectrum of $\mathbf{K}$ *on* $\mathcal{H}$ *consists of three non-empty sets of eigenvalues accumulating at* k_0*, 0 and* $-k_0$*, respectively (provided that* $\partial\Omega$ *is* C^∞*).*

Some questions arise from above mentioned results. An apparent question is why the smoothness of $\partial\Omega$, in stead of $C^{1,\alpha}$ for two dimensions, is required in three dimensions. In fact, this assumption is forced because of the method of proofs in [2]. A less apparent but quite interesting question is whether there is a physical meaning on those eigenvalues converging to different accumulation points. Another question is about the invertibility of $-1/2\mathbf{I} + \mathbf{K}$. As mentioned before, the Fredholm theory cannot be applied since $\mathbf{K}$ is not compact. But, what if $\mathbf{K}$ is polynomially compact? We discuss about these questions in this paper. For that purpose we first review the main ideas of proofs of the above mentioned results.

We end the introduction with a remark. Recently there is rapidly growing interest in the spectral properties of the NP operator (electro-static and elastic) in relation to plasmon resonance and stress concentration. For these relations we refer to [1].

2. Review of proofs

Straightforward computations using the definition (6) yield that

$$\partial_{\nu_x}\mathbf{\Gamma}(x-y) = -k_0\mathbf{K}_1(x,y) + \mathbf{K}_2(x,y), \tag{12}$$

where

$$\mathbf{K}_1(x,y) = \frac{2[n_x(x-y)^T - (x-y)n_x^T]}{\omega_d|x-y|^d}, \tag{13}$$

$$\mathbf{K}_2(x,y) = \frac{\mu}{2\mu+\lambda}\frac{(x-y)\cdot n_x}{\omega_d|x-y|^d}I + \frac{2(\mu+\lambda)}{2\mu+\lambda}\frac{(x-y)\cdot n_x}{\omega_d|x-y|^{d+2}}(x-y)(x-y)^T, \tag{14}$$

where ω_d is 2π if $d = 2$ and 4π if $d = 3$, I is the $d \times d$ identity matrix. The superscript T denotes transpose.

Note that if $\partial\Omega$ is $C^{1,\alpha}$ for some $\alpha > 0$, then

$$|(x-y)\cdot n_x| \le C|x-y|^{1+\alpha}$$

for some constant C, and hence

$$\left|\frac{(x-y)\cdot n_x}{|x-y|^d}\right| \le \frac{C}{|x-y|^{d-1-\alpha}}$$

for all $x, y \in \Omega$, in other words, it is weakly singular. It means that the integral operator on $\partial\Omega$ defined by $\mathbf{K}_2$ is compact, and the integral operator defined by $\mathbf{K}_1(x, y)$ is responsible for non-compactness of $\mathbf{K}$.

Let

$$\mathbf{T}[f](x) := \text{p.v.} \int_{\partial\Omega} \mathbf{K}_1(x, y) f(y)\, d\sigma(y), \quad x \in \partial\Omega. \tag{15}$$

Then we have

$$\mathbf{K} \equiv -k_0 \mathbf{T}. \tag{16}$$

Here and throughout the expression $\mathbf{A} \equiv \mathbf{B}$ for operators $\mathbf{A}$ and $\mathbf{B}$ on $\mathcal{H}$ indicates that $\mathbf{A} - \mathbf{B}$ is compact on $\mathcal{H}$. We emphasize that $\mathbf{T}$ is a singular integral operator and bounded on $\mathcal{H}$ as well as on $L^2(\partial\Omega)^d$ (see [3]).

In two dimensions $\mathbf{K}_1$ takes the form

$$\mathbf{K}_1(\mathbf{x}, \mathbf{y}) = \frac{1}{\pi|x-y|^2} \begin{bmatrix} 0 & K(x,y) \\ -K(x,y) & 0 \end{bmatrix}, \tag{17}$$

where

$$K(x, y) := -n_2(x)(x_1 - y_1) + n_1(x)(x_2 - y_2).$$

Let $\mathcal{R}$ is the Hilbert transformation on $\partial\Omega$, namely,

$$\mathcal{R}[f] = g \quad \text{on } \partial\Omega,$$

where g is the boundary value of $u^\perp$ which is a harmonic conjugate of u such that $\Delta u = 0$ in Ω and $u = f$ on $\partial\Omega$. It is proved in [1] that

$$\mathbf{T} \equiv \begin{bmatrix} 0 & -\mathcal{R} \\ \mathcal{R} & 0 \end{bmatrix}. \tag{18}$$

Since $\mathcal{R}^2 = -\mathbf{I}$, we have $\mathbf{T}^2 - \mathbf{I} \equiv 0$ and (i) of Theorem 1.1 follows.

Since existence of harmonic conjugates is used for proofs in two dimensions, the same argument may not be applied to the three dimensional case. In three dimensions $\mathbf{K}_1$ takes the form

$$\mathbf{K}_1(x, y) = \frac{1}{2\pi|x-y|^3} \begin{bmatrix} 0 & K_{12}(x,y) & K_{13}(x,y) \\ -K_{12}(x,y) & 0 & K_{23}(x,y) \\ -K_{13}(x,y) & -K_{23}(x,y) & 0 \end{bmatrix}, \tag{19}$$

where

$$K_{12}(x, y) = n_1(x)(x_2 - y_2) - n_2(x)(x_1 - y_1),$$
$$K_{13}(x, y) = n_1(x)(x_3 - y_3) - n_3(x)(x_1 - y_1),$$
$$K_{23}(x, y) = n_2(x)(x_3 - y_3) - n_3(x)(x_2 - y_2).$$

Let

$$T_{ij}[f](x) := \int_{\partial\Omega} \frac{K_{ij}(x,y)}{2\pi|x-y|^3} f(y) d\sigma(y), \tag{20}$$

so that

$$\mathbf{T} = \begin{bmatrix} 0 & T_{12} & T_{13} \\ -T_{12} & 0 & T_{23} \\ -T_{13} & -T_{23} & 0 \end{bmatrix}. \tag{21}$$

It is helpful to consider the case when Ω is the upper half-space even if it is not a bounded domain. In this case, $n_x = (0, 0, -1)^T$, and hence

$$\mathbf{T} = \begin{bmatrix} 0 & 0 & R_1 \\ 0 & 0 & R_2 \\ -R_1 & -R_2 & 0 \end{bmatrix}, \tag{22}$$

where R_j is the Riesz transform, *i.e.*,

$$R_j[f](x_1, x_2) = \frac{1}{2\pi} \int_{\mathbf{R}^2} \frac{x_j - y_j}{[(x_1 - y_1)^2 + (x_2 - y_2)^2]^{3/2}} f(y_1, y_2)\, dy_1 dy_2, \quad j = 1, 2.$$

Since $R_1^2 + R_2^2 = -\mathbf{I}$ (see [7]), one can see easily from (22) that $\mathbf{T}^3 - \mathbf{T} = 0$.

It is shown in [2] that the operator $\mathbf{T}$ on a bounded surface $\partial\Omega$ can be well approximated by the surface Riesz transforms which is defined by local coordinates charts. Let U be a coordinate chart in $\partial\Omega$ so that there is an open set D in $\mathbf{R}^2$ and a parametrization $\Phi : D \to U$, namely,

$$\mathbf{x} = (x(s,t), y(s,t), z(s,t)), \quad \mathbf{x} \in U, \quad (s,t) \in D. \tag{23}$$

Then the metric tensor of the surface, denoted by $(g_{ij}(s,t))_{i,j=1}^2$, is given by

$$\begin{aligned} dx^2 + dy^2 + dz^2 &= (x_s^2 + y_s^2 + z_s^2)ds^2 + 2(x_s x_t + y_s y_t + z_s z_t) ds \otimes dt + (x_t^2 + y_t^2 + z_t^2)dt^2 \\ &= g_{11} ds^2 + g_{12} ds \otimes dt + g_{21} dt \otimes ds + g_{22} dt^2. \end{aligned}$$

The unit normal vector is given by

$$\mathbf{n}(s,t) = (x_s(s,t), y_s(s,t), z_s(s,t)) \times (x_t(s,t), y_t(s,t), z_t(s,t)) / |(x_s, y_s, z_s) \times (x_t, y_t, z_t)|$$

and the surface element is given by $d\sigma(s,t) = |(x_s, y_s, z_s) \times (x_t, y_t, z_t)| ds \wedge dt$.

Let $x' = (x_1, x_2) = (s_1, t_1)$ and $y' = (y_1, y_2) = (s_2, t_2)$, and let

$$L(x', x' - y') = \frac{1}{[g_{11}(x')(x_1 - y_1)^2 + 2g_{12}(x')(x_1 - y_1)(x_2 - y_2) + g_{22}(x')(x_2 - y_2)^2]^{3/2}}.$$

Then the surface Riesz transform R_j^g, $j = 1, 2$, is defined by

$$R_j^g[f](x') = \frac{1}{2\pi} \int_{\mathbf{R}^2} (x_j - y_j) L(x', x' - y') f(y') dy'. \tag{24}$$

It is then shown that

$$\begin{aligned} T_{12} &\equiv (z_t g_{11} - z_s g_{12})\, R_1^g - (z_s g_{22} - z_t g_{21})\, R_2^g, \\ T_{13} &\equiv (y_t g_{11} - y_s g_{12})\, R_1^g - (y_s g_{22} - y_t g_{21})\, R_2^g, \\ T_{23} &\equiv (x_t g_{11} - x_s g_{12})\, R_1^g - (x_s g_{22} - x_t g_{21})\, R_2^g. \end{aligned}$$

We emphasize that above identities hold locally on coordinate charts.

To show that

$$\mathbf{T}^3 - \mathbf{T} \equiv 0 \tag{25}$$

in three dimensions, calculus of pseudo-differential operator is adapted. For this reason the smoothness assumption on $\partial\Omega$ is required. In fact, it is shown that the symbol of R_j^g is given by

$$\frac{-i}{\sqrt{\det(g_{jk}(x))}} \frac{\sum_k g^{jk}(x')\xi_k}{\sqrt{\sum_{i,j} g^{jk}(x')\xi_j \xi_k}}. \tag{26}$$

Using this it is shown that the following identity holds locally: Here $Op(Z)$ denotes the ψDO defined by the symbol Z.

$$\mathbf{K} \equiv \begin{bmatrix} 0 & -Op(Z) & -Op(Y) \\ Op(Z) & 0 & -Op(X) \\ Op(Y) & Op(X) & 0 \end{bmatrix},$$

where

$$X = \frac{i\sqrt{\det(g^{jk}(x))}(x_t\xi_1 - x_s\xi_2)}{\sqrt{\sum_{j,k} g^{jk}(x)\xi_j\xi_k}},$$

$$Y = \frac{i\sqrt{\det(g^{jk}(x))}(y_t\xi_1 - y_s\xi_2)}{\sqrt{\sum_{j,k} g^{jk}(x)\xi_j\xi_k}},$$

$$Z = \frac{i\sqrt{\det(g^{jk}(x))}(z_t\xi_1 - z_s\xi_2)}{\sqrt{\sum_{j,k} g^{jk}(x)\xi_j\xi_k}}.$$

Since $X^2 + Y^2 + Z^2 = -1$, (25) follows, and so does (ii) of Theorem 1.1.

These are the main ideas of proofs in [1, 2], even though there are several technical difficulties which are mentioned in this review.

3. Discussions

It is now apparent why the smoothness assumption on $\partial\Omega$ is required in three dimensions: it is because calculus of ψDO is adapted. It is likely that the result for the three dimensional case is valid for domains with $C^{1,\alpha}$ boundaries like the two dimensional case. To prove it, it is necessary to compute the compositions of surface Riesz potentials, which are singular integral operators. We will pursue this in future.

Theorem 1.2 shows that there are two groups of eigenvalues converging to k_0 and $-k_0$, respectively, in two dimensions, and three groups of eigenvalues converging to k_0, 0, and $-k_0$, respectively, in three dimensions. It is quite interesting to find out relevant physical meanings of the corresponding eigenspaces. In this respect the following remark is instructive. It is proved in [1] that k_0 and $-k_0$ are actually eigenvalues (not accumulation points of eigenvalues) of $\mathbf{K}^*$ when Ω is a disk (there are two more eigenvalues of multiplicities one). The eigenfunctions corresponding to k_0 and $-k_0$ are

- k_0:

$$\varphi_m^{(1)} = \begin{bmatrix} \cos m\theta \\ \sin m\theta \end{bmatrix}, \quad \varphi_m^{(2)} = \begin{bmatrix} -\sin m\theta \\ \cos m\theta \end{bmatrix}, \quad m = 2, 3, \ldots, \tag{27}$$

- $-k_0$:

$$\psi_m^{(1)} = \begin{bmatrix} \cos m\theta \\ -\sin m\theta \end{bmatrix}, \quad \psi_m^{(2)} = \begin{bmatrix} \sin m\theta \\ \cos m\theta \end{bmatrix}, \quad m = 1, 2, \ldots. \tag{28}$$

If we apply the single layer potential operator $\mathbf{S}$ to these eigenfunctions, we obtain

$$u_m^{(1)} := \mathbf{S}[\varphi_m^{(1)}] = (-1/2 + k_0) \begin{bmatrix} r^m \cos m\theta \\ r^m \sin m\theta \end{bmatrix}, \quad u_m^{(2)} := \mathbf{S}[\varphi_m^{(2)}] = (-1/2 + k_0) \begin{bmatrix} -r^m \sin m\theta \\ r^m \cos m\theta \end{bmatrix},$$

$$v_m^{(1)} := \mathbf{S}[\psi_m^{(1)}] = (-1/2 - k_0) \begin{bmatrix} r^m \cos m\theta \\ -r^m \sin m\theta \end{bmatrix}, \quad v_m^{(2)} := \mathbf{S}[\psi_m^{(2)}] = (-1/2 - k_0) \begin{bmatrix} r^m \sin m\theta \\ r^m \cos m\theta \end{bmatrix}.$$

If we identify the two-dimensional vector $(a_1, a_2)^T$ with the complex number, then $u_m^{(j)}$ is analytic and $v_m^{(j)}$ is anti-analytic for $j = 1, 2$. In particular,

$$\operatorname{div} v_m^{(j)} = 0, \tag{29}$$

in other words, $v_m^{(j)}$ is a shear displacement. We also mention that the convergence rates of eigenvalues at k_0 and $-k_0$ are different on ellipses (see [1]).

The third question we discuss is regarding the Fredholm alternative, which says that $\mathbf{I} + \mathbf{A}$ is invertible if and only if it is injective provided that $\mathbf{A}$ is a compact operator. What if $\mathbf{A}$ is polynomially compact, not compact?

Acknowledgements

This work is supported by grant A3 Foresight Program among China (NSF), Japan (JSPS), and Korea (NRF 2014K2A2A6000567). Work of H.K. is supported by NRF 2016R1A2B4011304. Work of Y.M. is partially supported by a MEXT Grant-in-Aid for Scientific Research on Innovative Areas (16H06576).

References

[1] Ando K, Ji Y, Kang H, Kim K and Yu S 2017 Spectral properties of the Neumann-Poincaré operator and cloaking by anomalous localized resonance for the elasto-static system, *Euro. J. Appl. Math.* S095672517000080 (*Preprint* arXiv: 1510.00989)

[2] Ando K, Kang H and Miyanishi Y 2017 Elastic Neumann–Poincaré operators on three dimensional smooth domains: Polynomial compactness and spectral structure: Elastic Neumann–Poincaré operators on three dimensional smooth domains, *Int. Math. Res. Not.* rnx258 (*Preprint* arXiv: 1702.03415)

[3] Coifman R R, McIntosh A and Meyer Y 1982 *Ann. Math.* **116** 361

[4] Dahlberg B E J, Kenig C E and Verchota G C 1988 *Duke Math. J.* **57**(3) 795

[5] Khavinson D, Putinar M and Shapiro H S 2007 *Arch. Rational Mech. Anal.* **185** 143

[6] Kupradze V D 1965 *Potential methods in the theory of elasticity* (New York: Daniel Davey & Co.)

[7] Stein E M and Weiss G 1971 *Introduction to Fourier analysis on Euclidean spaces* (Princeton: Princeton University Press)

IOP Conf. Series: Journal of Physics: Conf. Series **965** (2018) 012028 doi:10.1088/1742-6596/965/1/012028

Comparison of realizations of Lie algebras

Maryna Nesterenko[1] **and Severin Pošta**[2]

[1] Institute of Mathematics of NAS of Ukraine, 3 Tereshchenkivska Str., Kyiv-4, 01601 Ukraine

[2] Department of Mathematics, Faculty of Nuclear Sciences and Physical Engineering, Czech Technical University in Prague, Trojanova 13, CZ-120 00 Prague, Czech Republic

E-mail: maryna@imath.kiev.ua, severin.posta@fjfi.cvut.cz

Abstract. The notion of equivalence of Lie algebra realizations is revisited and the quantities stable under the equivalence transformations are proposed. As a result we formulate a practical algorithm that allows to establish the existence of equivalence between any two realizations of a Lie algebra. Several illustrative examples are considered.

1. Introduction

Notion of "realization" is widely explored in mathematical and physical literature and denotes a number of different objects, in particular, there are several definitions and types of Lie algebra realizations, e.g. in [3], realizations are considered in form of derivations of formal power series over a field of zero characteristic, and in [1] and [2] two different "physical" types of realizations are considered. The "Lie algebra realization" term also appears in different contexts, for example, it can be connected to a realization of the corresponding group action, see e.g. [5]. In this paper we consider realizations of Lie algebras by vector fields, i.e., representations by first-order differential operators.

This type of realizations was first considered by S. Lie and has a major place in modern group analysis of differential equations (general theory, group classification and integration of differential equations). Representation of Lie algebras by vector fields is also of great interest and widely applicable in classification of gravity fields of a general form with respect to motion groups, in geometric control theory and in the theory of systems with superposition principles and in difference schemes for numerical solutions of differential equations.

One of the key problems that usually emerge with realizations is to detect if two given realizations are the same (equivalent) or not and this forms the main objective of the paper. To start with we give all necessary notions and definitions in the first section, then we introduce a number of quantities and objects that are invariant and helpful in the establishment of equivalence, and, finally, we propose the practical algorithm that allows to detect the equivalence existence.

2. Definitions and conventions

Let V be an n-dimensional vector space over the field of real numbers. Consider a Lie algebra $\mathfrak{g}$ on V spanned by a basis $\{e_1, e_2, \dots, e_n\}$ with the structure constants $C_{ij}^k \in \mathbb{R}$, here and below $i, j, k = 1, 2, \dots, n$. We denote an open domain of $\mathbb{R}^m$ as M and $\mathrm{Vect}(M)$ is the Lie algebra of smooth vector fields on M with the Lie product defined as commutator (i.e. the Lie algebra of first-order linear differential operators with analytical function coefficients).

Definition 1. A *realization of a Lie algebra* $\mathfrak{g}$ *in vector fields on* M is a homomorphism $R(\mathfrak{g}) = R\colon \mathfrak{g} \to \mathrm{Vect}(M)$. The realization is called *faithful* if $\ker R = \{0\}$ and *unfaithful* otherwise.

In Lie theory realizations are considered locally at some neighborhood $U_x \subset M \subset \mathbb{R}^m$ of a point $x \in M$ and in most of the cases without loss of generality the realization can be considered in a neighborhood of a zero point $x = 0$.

Definition 2. A *local realization* $R|_x$ of a Lie algebra $\mathfrak{g}$ at $x \in M$ is the realization R of $\mathfrak{g}$ in vector fields actions of which are restricted to U_x.

Denote local coordinates of a point $x \in M$ as $(x_1, \ldots, x_m)$, then in coordinate form a realization $R(\mathfrak{g})$ is performed by the images $\Xi_i(x)$ of the basis elements e_i of a general form

$$\Xi_i(x) = R(e_i) = \sum_{l=1}^{m} \xi_{il}(x_1, x_2, \ldots, x_m)\partial_l, \tag{1}$$

hereafter $\partial_l = \frac{\partial}{\partial x_l}$ and the coefficients $\xi_{il}(x_1, x_2, \ldots, x_m)$ are smooth (analytic) functions.

Let us fix a point $x \in M$ and let R_x be a realization of $\mathfrak{g}$ at this point. Consider the linear map $R_x : \mathfrak{g} \to Vect(M)(x)$ that transforms a vector $v \in \mathfrak{g}$ to it's image $R(v(x))$ at x. The matrix that corresponds to this linear map is the n by m matrix ξ formed by the coefficients of the realization (1)

$$\xi(x) = \begin{pmatrix} \xi_{11}(x) & \xi_{12}(x) & \ldots & \xi_{1m}(x) \\ \xi_{21}(x) & \xi_{22}(x) & \ldots & \xi_{2m}(x) \\ \vdots & \vdots & \ddots & \vdots \\ \xi_{n1}(x) & \xi_{n2}(x) & \ldots & \xi_{nm}(x) \end{pmatrix}. \tag{2}$$

Definition 3. The rank of the linear map R_x, or, equivalently, the rank of the matrix $\xi(x)$ at a point x is called a *rank of realization* R at point x and is denoted $\mathrm{rank}\, R_x$.

Remark 1. The realization rank value possess the obvious inequality $0 \leq \mathrm{rank}\, R_x \leq n$, where n is the dimension of a Lie algebra $\mathfrak{g}$. The second inequality is dictated by the number of rows in matrix ξ, which is equal to the number of basis vector fields of $\mathfrak{g}$.

Definition 4. A realization R of a Lie algebra $\mathfrak{g}$ is called *transitive* if the action of the local Lie group corresponding to R is transitive. Or, equivalently (see [4]), a realization R of a Lie algebra $\mathfrak{g}$ is called *transitive* if $\mathrm{rank}\, R_p = m$ for all $p \in M$.

Definition 5. A point $x \in M$ is called a *regular point* of a realization R if there exists a neighborhood U_x of x, such that $\mathrm{rank}\, R_y$ is constant for all $y \in U_x$. A point that is not regular is called a *singular point*.

3. Equivalence of realizations

For many practical applications it is necessary to define if two given sets of first order differential operators (with the isomorphic commutation relations) can be transformed to each other or not. In particular this problem arises as subproblem in methodization and matching of differential equations and in classification of realizations of a given Lie algebra. This task is rather complicated even in the case of small number of operators and variables.

Example 1. Consider two realizations of the smallest Poincaré algebra $p(1,1)$:
$R_1\colon \{\partial_1,\ \ \cosh x_1\partial_2,\ \ -\sinh x_1\partial_2\}$ and
$R_2\colon \{\partial_1,\ \ x_2\partial_1,\ \ x_1x_2\partial_1 - (1 - x_2^2)\partial_2\}$.
This realizations are equivalent and connected by the change of variables $y_2 = -\tanh x_2$.

So the main target of this work is the constructive procedure matching the two given realizations. For further study let us introduce an equivalence relation on the set of all possible realizations of a fixed Lie algebra.

Denote $\mathrm{Aut}(\mathfrak{g}) \subset GL_n$ the whole automorphism group of $\mathfrak{g}$ and let $\mathrm{Int}(\mathfrak{g}) \subset \mathrm{Aut}(\mathfrak{g})$ denote the inner automorphism group of $\mathfrak{g}$.

Consider notions of "weak" and "strong" equivalences introduced in [11].

Let A be a subgroup of $\mathrm{Aut}(\mathfrak{g})$. The realizations $R_1\colon A \to \mathrm{Vect}(M_1)$ and $R_2\colon A \to \mathrm{Vect}(M_2)$ are called *A-equivalent* if there exist $\alpha \in A$ and a diffeomorphism f from M_1 to M_2 such that $R_2(v) = f_* R_1(\alpha(v))$ for all $v \in A$. Here f_* is the isomorphism from $\mathrm{Vect}(M_1)$ to $\mathrm{Vect}(M_2)$ induced by f. If A contains only the identical transformation, the realizations are called *strongly equivalent*. The realizations are *weakly equivalent* if $A = \mathrm{Aut}(\mathfrak{g})$. All the diffeomorphisms are local.

The above definition of equivalence does not work well in singular points, to demonstrate this let us consider the one-dimensional Lie algebra span $\{e_1\}$ and its realization $R(e_1(x)) = x_1\partial_1$ on some open domain M of $\mathbb{R}^m$, such that $0 \in M$. Since the application of Lie algebras to differential equations [6, 10] is based on the *local* correspondence between Lie groups and Lie algebras, in Lie theory it is supposed that any one-dimensional vector field is equivalent to the vector field corresponding to shifts (locally the rectification theorem is supposed to hold). In our case this means that $x_1\partial_1$ has to be equivalent to the operator ∂_1. But there is no necessary diffeomorphism at zero point and these two realizations are inequivalent in frames of the above definition. To avoid the equivalence collisions at the singular points we introduce a relation of local equivalence.

Consider a regular point $p \in M$ and two systems $(x_1, \ldots, x_m)$ and $(y_1, \ldots, y_m)$ of local coordinates of the mentioned point p on M. Then there exists a local diffeomorphism $\varphi : U(p_x) \to U(p_y)$, such that $y_i = \varphi_i(x_1, \ldots, x_m)$. Here $U(p_x) \in \mathbb{R}^m$ and $U(p_y) \in \mathbb{R}^m$ are the neighborhoods of the images (under the coordinate transformation) p_x and p_y of the point p in "x" and "y" coordinate spaces $\mathbb{R}^m$ respectively. Since the manifold M is an open subspace in $\mathbb{R}^m$ we do not need to consider additional coordinate spaces and can introduce the diffeomorphism φ providing the change of coordinates directly on the manifold M between the neighborhoods of points $x \in M$ and $y \in M$, that is $\varphi : U_x \to U_y$.

Definition 6. The realization $R_1\colon \mathfrak{g} \to \mathrm{Vect}(M)$ at a point $x \in M$ and the realization $R_2\colon \mathfrak{g} \to \mathrm{Vect}(M)$ at a point $y \in M$ are *locally equivalent* if there exist two neighborhoods U_x and U_y and a diffeomorphism $\varphi : U_x \to U_y$, such that $y = \varphi(x)$ and $R_2(v(\varphi(x))) = R_1(v(x))$ for all $v \in \mathfrak{g}$.

The introduced notion of local equivalence is indeed an equivalence relation, as far as it is symmetric, reflexive and transitive.

Definition 7. The realization $R_1\colon \mathfrak{g} \to \mathrm{Vect}(M)$ at a point $x \in M$ and the realization $R_2\colon \mathfrak{g} \to \mathrm{Vect}(M)$ at a point $y \in M$ are called Aut-*equivalent* if there exist $\alpha \in Aut(\mathfrak{g})$, two neighborhoods U_x and U_y and a diffeomorphism $\varphi : U_x \to U_y$, such that $y = \varphi(x)$ and $R_2(v(\varphi(x))) = R_1(\alpha v(x))$ for all $v \in \mathfrak{g}$.

Definition 8. The realization $R_1\colon \mathfrak{g} \to \mathrm{Vect}(M)$ at a point $x \in M$ and the realization $R_2\colon \mathfrak{g} \to \mathrm{Vect}(M)$ at a point $y \in M$ are called Int-*equivalent* if there exist $\alpha \in Int(\mathfrak{g})$, two neighborhoods U_x and U_y and a diffeomorphism $\varphi : U_x \to U_y$, such that $y = \varphi(x)$ and $R_2(v(\varphi(x))) = R_1(\alpha v(x))$ for all $v \in \mathfrak{g}$.

If it is necessary to consider realizations not only on a neighborhood of a point, but on M, then the equivalence has to be established for all regular points (set of regular points of a vector field is an open domain in $\mathbb{R}^m$). Another possibility is to take an analytic continuation of the local realization. Hereafter the term "equivalence" means the Aut-equivalence if the type of equivalence is not indicated explicitly.

3.1. *Induced faithful realizations and essential variables*

Roughly speaking we can conclude that two realizations are equivalent, if they can be transformed to the identical form by means of non-singular automorphic basis changes ($e_i \mapsto \tilde{e}_i$) and 1 to 1 changes of variables ($x_l \mapsto y_l = \varphi_l(x)$) with non-zero Jacobi determinant.

Let us have a diffeomorphism of M such that for the corresponding $x, y \in M$ we have $y_1 = \varphi_1(x_1, \ldots, x_m)$, $y_2 = \varphi_2(x_1, \ldots, x_m)$, ..., $y_m = \varphi_m(x_1, \ldots, x_m)$. Then the realization of the form (1) transforms to the following:

$$\tilde{R}(e_i) = \sum_{l=1}^{m} \tilde{\xi}_{il}(y)\partial_{y_l} = \sum_{l=1}^{m} \left(\sum_{l'=1}^{m} \tilde{\xi}_{il'}(x) \frac{\partial \varphi_l(x)}{\partial x_{l'}} \right) \partial_{y_l}. \tag{3}$$

Note, that the coefficients $\tilde{\xi}_{il}(y)$ are written in terms of y using the inverse transformation φ^{-1}.

From the definition of equivalence immediately follows that *none of faithful realizations of* $\mathfrak{g}$ *is equivalent to an unfaithful realization of* $\mathfrak{g}$.

Therefore we will not consider unfaithful realizations at all, they have to be replaced by the faithful ones using the following induction rule. Let $R(\mathfrak{g})$ be an unfaithful realization of the algebra $\mathfrak{g}$ with the basis elements $\{R(e_1) \neq 0, \ldots, R(e_k) \neq 0, R(e_{k+1}) \equiv 0, \ldots R(e_n) \equiv 0\}$, then it can be replaced by the faithful realization $R(\mathfrak{g}_0) = \text{span}\,\{R(e_1), \ldots, R(e_k)\}$ of the smaller Lie algebra $\mathfrak{g}_0 = \text{span}\,\{e_1, \ldots, e_k\}$, which is a subalgebra of $\mathfrak{g}$. The realization $R(\mathfrak{g}_0)$ of $\mathfrak{g}_0$ is called *a realization induced by* $R(\mathfrak{g})$.

It is obvious that application of transformations from $\text{Aut}(\mathfrak{g})$ to the realization R does not change the rank of R, and none of diffeomorphisms of M can change the realization rank either. Therefore the equivalent realizations have the same ranks.

Lemma 1. *Let a realization* $R(x)\colon \mathfrak{g} \to \text{Vect}(M)$ *has a rank* $r = \text{rank}\, R < m$ *at a regular point* $x \in M$, *where* $m = \dim M$. *then there exists a locally equivalent realization* $\tilde{R}(y)\colon \mathfrak{g} \to \text{Vect}(M)$ *at a regular point* $y \in M$ *such that the coefficients of basic vector fields* $\tilde{\xi}_{il}(y) = 0$ *for all* $i = 1, \ldots, n$, $l = r+1, \ldots, m$.

Proof. To prove the lemma let us construct the desired diffeomorphism. Since the realization rank is equal to r it is known from the theory of invariants [10] that there are $m - r$ functionally independent invariants $J_1(x_1, \ldots, x_m), \ldots, J_{m-r}(x_1, \ldots, x_m)$ of the realization R. The diffeomorphism of the form $y_a = x_a$, $a = 1, \ldots, r$; $y_{r+b} = J_b$, $b = 1, \ldots, m-r$ gives the following zero coefficients of the realization $\tilde{R}$: $\tilde{\xi}_{i(r+b)}(y) = R(e_i)(J_b) = 0$ for all $i = 1 \ldots, n$, $b = 1, \ldots, m - r$. □

Definition 9. The above variables $y_1, \ldots, y_r$ are called *essential* and the rest of non-zero variables from $y_{r+1}, \ldots, y_m$ are called *additional*. The equivalence transformation described in Lemma 1 is called *reduction of realization to the essential variables.*

Example 2. Consider two-dimensional abelian Lie algebra $2A_1$. It is well-known that the basis elements of this algebra can be realized by two operators of translations

$$R_1(e_1) = \partial_1, \quad R_1(e_2) = \partial_2.$$

It was shown in [11] that there are exactly two inequivalent realizations of $2A_1$, and the second one is

$$R_2(e_1) = \partial_1, \quad R_2(e_2) = x_2\partial_1.$$

In these cases $\text{rank}\, R_1 = 2$ and $\text{rank}\, R_2 = 1$.

Consider the formal sum of these realizations $R_3 = R_1 + R_2$ (R_1 for the variables (x_1, x_2) and R_2 for the variables (x_3, x_4)), namely

$$R_3(e_1) = \partial_1 + \partial_3, \quad R_3(e_2) = \partial_2 + x_4\partial_3.$$

As far as $[\partial_1 + \partial_3, \partial_2 + x_4\partial_3] = 0$, then R_3 do realize the Lie algebra $2A_1$ in the space of four variables $(x_1,\ x_2,\ x_3,\ x_4)$ and rank $R_3 = 3$, what means that the number of essential variables is equal to 3 and Lemma 1 can be applied.

Indeed, the diffeomorphism φ given by the non-singular functions

$$\varphi_1(x_1,\dots,x_4) = x_1,\ \varphi_2(x_1,\dots,x_4) = x_2,\ \varphi_3(x_1,\dots,x_4) = x_1 - x_3 + x_2x_4,\ \varphi_4(x_1,\dots,x_4) = x_4$$

transforms the realization R_3 to the equivalent realization R_1 in essential variables.

Corollary 1. *In transitive realizations all variables are essential and, since* rank $R \leq n$, *any transitive realization of a Lie algebra is realized in not more then n variables.*

3.2. Correspondence between realization and subalgebra

Two recent papers on the construction of vector fields [7] and classification of realizations [4] establish one-to-one correspondence between inequivalent transitive realizations of a Lie algebra $\mathfrak{g}$ and Int-inequivalent subalgebras of $\mathfrak{g}$. Moreover, this relation was extended to the non-transitive case as well. To use this correspondence as a criteria of equivalence of two considered realizations we have to define explicitly the subalgebra that corresponds to the given realization. In practice the desired subalgebra can be found using the following lemma.

Lemma 2. *The subalgebra that corresponds to the given realization is the kernel of its linear map at the origin of coordinates.*

In other words at the point $x = 0 \in \mathbb{R}^m$ the realization vectors that form a basis of corresponding subalgebra are identically equal to zero.

The proof of the lemma follows from the algorithm of construction of vector fields proposed in [7].

Example 3. Consider the realizations
$R_1\colon e_1 = \partial_1,\quad e_2 = x_2\partial_1,\quad e_3 = x_1\partial_1 + 2x_2\partial_2$ and
$R_2\colon e_1 = \partial_1,\quad e_2 = x_1\partial_1 - x_2\partial_2,\quad e_3 = \partial_2$.
At the origin of coordinates $x = 0$ their basis vectors have the form
$R_1(x = 0)\colon e_1 = \partial_1,\quad e_2 = 0,\quad e_3 = 0$ and
$R_2(x = 0)\colon e_1 = \partial_1,\quad e_2 = 0,\quad e_3 = \partial_2$.
Therefore the realization R_1 corresponds to the subalgebra $\langle e_2, e_3\rangle$ and R_2 corresponds to $\langle e_2\rangle$.

4. Match algorithm

Let us sum up all the above statements and formulate the procedure allowing us to compare and match two finite n-sets of linearly independent vector fields, namely

$$S_1 = \{\sum_{j=1}^{m} \xi_{ij}(x_1, x_2, \dots, x_m)\partial_{x_j}\} \text{ and } S_2 = \{\sum_{j=1}^{m} \eta_{ij}(x_1, x_2, \dots, x_m)\partial_{x_j}\},\ i = 1, 2, \dots, n.$$

4.1. Preliminary testing

The first thing that should be checked is if these two sets *do realize* a Lie algebra (differential operators have to be closed with respect to the Lie product).

The next question is if the considered realizations do realize *the same* Lie algebra. At this stage the structure constants for both sets have to be calculated and then, if they do not coincide, the isomorphism transformation from GL_n has to be found. The *construction of isomorphism* is rather complicated problem, but it can be simplified in some cases: for the low-dimensional $(n \leq 5)$ algebras the known classification of Lie algebras can be used as well as for semi-simple Lie algebras; for Lie algebras with the abelian ideal of codimension one the Jordan canonical form of the adjoint action of the "non-commuting" basis element can be used.

The last preparatory procedure is the *elimination of unfaithful* realizations from consideration or replacement of them by the faithful induced realization.

4.2. Reduction to essential variables
The second step is to calculate ranks of realizations, to compare them and, if the ranks coincide, to reduce realizations to the essential variables. Then one should test the transitivity of realizations (all variables have to be essential) and note that a transitive realization is not equivalent to a non-transitive one.

4.3. Subalgebra detection
Now one can define the corresponding subalgebras and establish their equivalence transformation using Int. Once the subalgebras coincide (note that in the non-transitive case all parameters of subalgebras have to coincide completely) then the task moves to the next step.

4.4. Construction of diffeomorphism and automorphism
At the last stage we are able to construct diffeomorphism by solving the system of differential equations arising from (3) explicitly. It is important that any partial solution of this system fits our requirement and in the most cases the construction of the solution is not necessary at all, since it is enough to know that realizations are equivalent. Sometimes the diffeomorphism transformation has to be followed by simple automorphism (e.g., sign change or sum of two basis elements).

Acknowledgments
MN is grateful for the hospitality extended to her at the Department of mathematics, FNSPE, Czech Technical University in Prague.

References
[1] Burdet G and Perrin M 1975 Realizations of the central extension of the inhomogeneous symplectic algebra as time dependent invariance algebras of nonrelativistic quantum systems *J. Math. Phys.* **16** (8) 1692–1703
[2] Burdík Č 1988 A new class of realisations of the Lie algebra so$(q, 2n-q)$ *J. Phys. A: Math. Gen.* **21** 284–295
[3] Draisma J 2012 Transitive Lie Algebras of Vector Fields: An Overview. *Qual. Theory Dyn. Syst.* **11** 39–60
[4] Gromada D and Pošta S 2017 On classification of Lie algebra realizations math-ph/1703.00808v1 20 pages
[5] Krivonos S, Lechtenfeld O and Sorin A 2016 Minimal realization of l-conformal Galilei algebra, Pais-Uhlenbeck oscillators and their deformation *J. High Energ. Phys.* **78**
[6] Lie S 1888, 1890, 1893 Theorie der Transformationsgruppen **1–3**, Leipzig
[7] Magazev A, Mikheyev V and Shirokov I 2015 Computation of composition functions and invariant vector fields in terms of structure constants of associated Lie algebras *SIGMA* **11**:066 17 pages
[8] Nesterenko M 2015 The Poincaré algebras $p(1,1)$ and $p(1,2)$: realizations and deformations *J. Phys. Conf. Ser.* **621**:012009
[9] Nesterenko M, Pošta S and Vaneeva O 2016 Realizations of Galilei algebras *J. Phys. A: Math. Theor.*, **49**(11) 115203 26 pages
[10] Olver P 1994 Differential invariants and invariant differential equations, *Lie Groups and Appl.* **1** 177–192
[11] Popovych R, Boyko V, Nesterenko M and Lutfullin M 2003 Realizations of real low-dimensional Lie algebras *J. Phys. A*, 36(26):7337–7360. See math-ph/0301029 for the revised and extended version.
[12] Shirokov I V 1997 Construction of Lie algebras of first-order differential operators. *Izv. Vyssh. Uchebn. Zaved. Fiz.* **40**(6) 25–32

On curvature squared invariants in 6D supergravity[1]

J Novak[1] **and G Tartaglino-Mazzucchelli**[2]
[1]Max-Planck-Institut für Gravitationsphysik, Albert-Einstein-Institut, Am Mühlenberg 1, D-14476 Golm, Germany
[2]Instituut voor Theoretische Fysica, KU Leuven, Celestijnenlaan 200D, B-3001 Leuven, Belgium
E-mail: joseph.novak@aei.mpg.de, gabriele.tartaglino-mazzucchelli@kuleuven.be

Abstract.
We review recent developments in the construction of curvature squared invariants in off-shell $\mathcal{N} = (1, 0)$ supergravity in six dimensions.

1. Introduction and Motivations

It is well known that the low-energy effective action of string theory is described by supergravity modified by an infinite series of higher-derivative quantum corrections, which can schematically be represented by the Lagrangian

$$\mathcal{L}^{\text{low}}_{\text{string}} = \mathcal{L}_{\text{SG}} + \sum \mathcal{D}^p\mathcal{R}^q + \text{forms contributions} + \text{susy completion} . \tag{1.1}$$

Besides the purely gravitational higher-derivative terms, there are contributions involving the extra fields of the low-energy spectrum of string theory, including the tensor hierarchy of forms, and the fermions needed for local supersymmetry invariance of the action. The supersymmetric higher-derivative terms are in general poorly understood but are of crucial importance in various contexts. For instance, the contributions depending on the p-forms play an important role in the dynamics of the moduli in compactified string theory and the low-energy description of string dualities; see, e. g., [1, 2, 3]. Moreover, higher-derivative terms control the stringy corrections to the entropy of black-holes (see for example [4, 5, 6, 7]) and they are needed for precision tests within the context of the AdS/CFT correspondence. Locally supersymmetric higher-derivative terms also describe counterterms for UV divergencies in supergravity and their classification is intimately connected to the understanding of surprising quantum properties of extended supergravities such as in the case of 4D $\mathcal{N} = 8$ supergravity (see [8] for a short review and a list of references).

Even the supersymmetric extension of curvature squared invariants, which would be the next to leading order in the expansion (1.1), is in general not fully understood. However, non-supersymmetric curvature squared gravity has attracted much attention for over 50 years, irrespective of string theory motivations, for various reasons. For instance, the renormalization of quantum field theories in curved spacetime requires counterterms containing curvature squared terms [9]. In four dimensions, these govern the structure of conformal anomalies that are relevant

[1] Based on the talk presented by GT-M at ISQS25 (Prague, Czech Republic, 6 – 10 June, 2017).

in studying renormalization group flows, see for example the a-theorem of Komargodski and Schwimmer [10]. Forty years ago it was also realized that Lagrangians defined by a linear combination of the Riemann tensor squared, the Ricci tensor squared and the scalar curvature squared, $\alpha(\mathcal{R}_{abcd})^2 + \beta(\mathcal{R}_{ab})^2 + \gamma\mathcal{R}^2$, lead to renormalizable but not unitary theories of gravity [11]. Recently, it was also realized that the Starobinsky model of inflation [12], based on the addition of a $\mathcal{R}^2$ term to the Einstein-Hilbert gravity, was proposed to be a promising inflationary candidate consistent with the Cosmic-Microwave-Background (CMB) data.

An interesting curvature squared invariant is described by the Gauss-Bonnet (GB) combination

$$\mathcal{L}_{\rm GB} = \mathcal{R}_{abcd}\mathcal{R}^{abcd} - 4\mathcal{R}_{ab}\mathcal{R}^{ab} + \mathcal{R}^2 = 6\mathcal{R}_{[ab}{}^{ab}\mathcal{R}_{cd]}{}^{cd} . \tag{1.2}$$

In 4D it is a topological term, the Euler characteristic, associated with the type A conformal anomaly [13]. In D-dimensions the GB combination, which is ghost free, is actually expected to govern the first order α' corrections in string theory [14, 15]. In general its structure for any spacetime dimensions and number of supersymmetry charges is not known. In particular, the dependence upon the extra supergravity fields, i.e. the dilaton and the various p-forms, is still not in general understood. In 4D and 5D the Gauss-Bonnet was constructed off-shell and completely classified, see [16, 17, 18, 19, 20, 21, 22, 23] and [24, 25, 26], respectively. In the 6D case only partial results were obtained 30 years ago [27, 28, 29, 30] and a classification of curvature squared invariants has been absent until new insights this year. In particular, the GB invariant in six dimensions has been constructed in our recent paper [31]. One of the main aims of this note is to review this result.

An obvious question that one may ask is: how should one efficiently construct the higher-derivative supergravity invariants? A useful approach would be a formalism that guarantees manifest supersymmetry in a model independent way. Off-shell approaches to supergravity, when available, can indeed be used to construct general supergravity-matter couplings possessing model independent supersymmetry. Two possibilities have been largely developed in the literature: (i) a component field approach (superconformal tensor calculus), see for example the recent "Supergravity" book by Freedman and Van Proeyen [32] for an exhaustive review and references; and (ii) superspace approaches, see the classic books [33, 34, 35, 36].

It turns out that the two approaches can be linked and powerfully used together through the so called "conformal superspace." In this formalism the entire superconformal algebra is gauged in superspace combining the main advantages of both approaches. Conformal superspace was first introduced by D. Butter for 4D $\mathcal{N} = 1$ [37] and $\mathcal{N} = 2$ [38] supergravity (see also the seminal work by Kugo and Uehara [39]) and it was developed and extended to 3D $\mathcal{N}$−extended supergravity [40], 5D $\mathcal{N} = 1$ supergravity [26], and recently to 6D $\mathcal{N} = (1,0)$ supergravity [41], see also [42].

In this paper, we review the description of curvature squared invariants for 6D $\mathcal{N} = (1,0)$ off-shell supergravity and show how using conformal superspace it is possible to efficiently: describe off-shell supermultiplets through simple differential constraints; provide manifestly supersymmetric off-shell action principles by using powerful cohomological "superform" techniques; and reduce the results to component fields and the superconformal tensor calculus. With these techniques at hand, one has a systematic approach to construct higher-derivative off-shell invariants.

This paper is organized as follows. In section 2 we review how Poincaré gravity can be obtained by gauge fixing conformal gravity. By using the same logic, we review basic aspects of 6D conformal superspace and show how to construct 6D $\mathcal{N} = (1,0)$ Poincaré supergravity from superspace in section 3. Section 4 is devoted to the construction of curvature squared invariants, while in section 5 we analyze Einstein-Gauss-Bonnet supergravity, which is related to the low-energy description of 6D compactified string theory. Conclusions are presented in section 6.

2. 6D conformal and Poincaré gravity

Before turning to supergravity, it is useful to remind the reader about conformal gravity and how Poincaré gravity can be viewed as a gauge fixed limit of it. As we will see in the next section, the same principle generalizes to the supersymmetric case.

Conformal gravity in six dimensions may be constructed by gauging the entire conformal group SO(6,2), which possesses the generators $X_{\tilde{a}} = \{P_a, M_{ab}, \mathbb{D}, K_a\}$. To each generator we then associate a connection. The vielbein $e_m{}^a$ and its inverse $e_a{}^m$ are associated with the momentum P_a generating local-translations/diffeomorphism. Together with the other connections for the Lorentz ($\omega_m{}^{bc}$), dilatation (b_m) and special conformal ($f_m{}^b$) generators, one can introduce the covariant derivatives

$$\nabla_a = e_a{}^m \Big(\partial_m - \frac{1}{2}\omega_m{}^{bc}M_{bc} - b_m\mathbb{D} - f_m{}^b K_b\Big) . \tag{2.1}$$

The covariant derivative algebra is constrained to be

$$[\nabla_a, \nabla_b] = -\frac{1}{2}C_{ab}{}^{cd}M_{cd} - \frac{1}{6}\nabla^d C_{abcd}K^c . \tag{2.2}$$

The algebra is expressed entirely in terms of a single primary dimension two tensor $C_{ab}{}^{cd}$,

$$K_f C_{ab}{}^{cd} = 0 , \quad \mathbb{D}C_{ab}{}^{cd} = 2C_{ab}{}^{cd} , \tag{2.3}$$

satisfying the following algebraic constraints

$$C_{abcd} = C_{[ab][cd]} , \quad \eta^{ac}C_{abcd} = 0 , \quad C_{[abc]d} = 0 . \tag{2.4}$$

The solution of the previous constraints determine $\omega_m{}^{bc} = \omega_m{}^{bc}(e)$ and $f_m{}^b = f_m{}^b(e)$ as composite functions of the vielbein $e_m{}^a$ while b_m proves to be a pure gauge degree of freedom. The tensor $C_{ab}{}^{cd}$ coincides with the 6D Weyl tensor: $C_{ab}{}^{cd} = \mathcal{R}_{ab}{}^{cd} - \delta_{[a}{}^{[c}\mathcal{R}_{b]}{}^{d]} + \frac{1}{10}\delta_{[a}{}^{[c}\delta_{b]}{}^{d]}\mathcal{R}$ with $\mathcal{R}_{ab}{}^{cd} = \mathcal{R}_{ab}{}^{cd}(e)$ being the standard Riemann tensor, $\mathcal{R}_a{}^b := \mathcal{R}_{ad}{}^{bd}$ the Ricci tensor and $\mathcal{R} := \mathcal{R}_a{}^a$ the scalar curvature.

So far we have considered the gauging of the full conformal algebra. Standard Poincaré gravity can be described by coupling conformal gravity to a primary dimension two scalar field ϕ. If one chooses a gauge for dilatation and special conformal transformations in which $b_m = 0$ and $\phi = 1$ then what it is left is standard gravity, gauging only diffeomorphisms and Lorentz transformations. Along this line, the Einstein-Hilbert action of general relativity is described by the action of the dilaton field ϕ in a conformal gravity background

$$S_{\rm EH} = -\frac{5}{2}\int \mathrm{d}^6x\, e\, \phi\nabla^a\nabla_a\phi . \tag{2.5}$$

In a gauge where $b_m = 0$ and $\phi = 1$ the previous action reduces to the standard Einstein-Hilbert term

$$S_{\rm EH} = -\frac{1}{2}\int \mathrm{d}^6x\, e\, \mathcal{R} . \tag{2.6}$$

Although the previous approach might seem cumbersome in the case of gravity, its supersymmetric analogue proves to be a natural way to describe in a uniform framework different multiplets of off-shell Poincaré supergravity, and general supergravity-matter couplings. The strategy is to first describe off-shell conformal supergravity and then break unnecessary local symmetries, as dilatations and local superconformal transformations, by coupling to compensating multiplets. Let us show how this works in the case of 6D $\mathcal{N} = (1,0)$ supergravity.

3. 6D $\mathcal{N} = (1,0)$ conformal and Poincaré supergravity

The standard Weyl multiplet of $\mathcal{N} = (1,0)$ conformal supergravity is associated with the local off-shell gauging of $\mathrm{OSp}(6,2|1)$, the $\mathcal{N} = (1,0)$ superconformal group in 6D [43]. The multiplet contains 40 + 40 physical components described by the following independent gauge fields: the vielbein $e_m{}^a$ and a dilatation connection b_m (as in conformal gravity); the gravitino $\psi_m{}_i^\alpha$, associated to the gauging of Q-supersymmetry; and $\mathrm{SU}(2)_R$ gauge fields $\mathcal{V}_m{}^{ij}$. The other gauge fields associated with the remaining generators of $\mathrm{OSp}(6,2|1)$ are composite fields. In addition to the independent gauge connections, the standard Weyl multiplet comprises a set of covariant matter fields given by an anti-self-dual tensor T^-_{abc}, a real scalar field D and a chiral fermion χ^i. These extra independent degrees of freedom are necessary to close the local $\mathrm{OSp}(6,2|1)$ algebra of transformations without imposing any equation of motions.

So far we have described the Weyl multiplet in components. Let us review the basic aspects of its description in the 6D $\mathcal{N} = (1,0)$ conformal superspace developed in [41]. In this paper, we adopt the notation and conventions of [41, 42]. Take a $\mathcal{N} = (1,0)$ curved superspace $\mathcal{M}^{6|8}$ parametrised by local coordinates

$$z^M = (x^m, \theta^\mu_{\mathrm{i}}) , \qquad m = 0,1,2,3,4,5 , \quad \mu = 1,2,3,4 , \quad \mathrm{i} = 1,2 . \tag{3.1}$$

The structure group of conformal superspace, denoted by X, contains the generators for $\mathrm{SO}(5,1)$ Lorentz, $\mathrm{SU}(2)_R$ R-symmetry, dilatation, S-supersymmetry and conformal boosts transformations. The superspace covariant derivatives are

$$\nabla_A = E_A{}^M \partial_M - \frac{1}{2}\Omega_A{}^{ab} M_{ab} - \Phi_A{}^{ij} J_{ij} - B_A \mathbb{D} - \mathfrak{F}_{AB} K^B , \tag{3.2}$$

where $E_A{}^M(z)$ is the supervielbein associated with translations in superspace, $P_A = (P_a, Q^i_\alpha)$, and $\partial_M = \partial/\partial z^M$; $\Omega_A{}^{cd}(z)$ is the Lorentz connection; $\Phi_A{}^{ij}(z)$ is the $\mathrm{SU}(2)_R$ connection; B_A is the dilatation connection; and $\mathfrak{F}_{AB}$ is the special superconformal connection, with $K^A = (K^a, S^\alpha_i)$. The supergravity gauge group, which is associated with the gauging of the $\mathrm{OSp}(6,2|1)$ transformations, is generated by the following transformations

$$\delta_{\mathcal{K}} \nabla_A = [\mathcal{K}, \nabla_A] , \quad \mathcal{K} := \xi^A \nabla_A + \frac{1}{2}\Lambda^{bc} M_{bc} + \Lambda^{ij} J_{ij} + \tau \mathbb{D} + \Lambda_A K^A . \tag{3.3}$$

Similarly to conformal gravity, to describe the 6D (1,0) Weyl multiplet, one constrains the algebra of covariant derivatives

$$\begin{aligned}[\nabla_A, \nabla_B\} = &-T_{AB}{}^C \nabla_C - \frac{1}{2} R(M)_{AB}{}^{cd} M_{cd} - R(J)_{AB}{}^{kl} J_{kl} \\ &- R(\mathbb{D})_{AB} \mathbb{D} - R(S)_{AB}{}^k_\gamma S^\gamma_k - R(K)_{ABc} K^c \end{aligned} \tag{3.4}$$

to be completely determined in terms of the super-Weyl tensor $W^{\alpha\beta} = (\tilde{\gamma}^{abc})^{\alpha\beta} W_{abc}$ (see, e.g. [44, 41] for its definition) describing in superspace the 6D $\mathcal{N} = (1,0)$ Weyl multiplet of conformal supergravity. The details of the algebra are more intricate than in the non-supersymmetric case and we refer the reader to [41] for details. It is only relevant to mention that $W^{\alpha\beta}$ is a dimension-1 primary superfield

$$K^A W^{\beta\gamma} = 0 , \quad \mathbb{D} W^{\alpha\beta} = W^{\alpha\beta} , \tag{3.5}$$

and satisfies the Bianchi identities

$$\nabla^{(i}_\alpha \nabla^{j)}_\beta W^{\gamma\delta} = -\delta^{(\gamma}_{[\alpha} \nabla^{(i}_{\beta]} \nabla^{j)}_\rho W^{\delta)\rho} , \qquad \nabla^k_\alpha \nabla_{\gamma k} W^{\beta\gamma} - \frac{1}{4}\delta^\beta_\alpha \nabla^k_\gamma \nabla_{\delta k} W^{\gamma\delta} = 8\mathrm{i} \nabla_{\alpha\gamma} W^{\gamma\beta} . \tag{3.6}$$

IOP Conf. Series: Journal of Physics: Conf. Series **965** (2018) 012029 doi:10.1088/1742-6596/965/1/012029

It can be shown that the component fields of the standard Weyl multiplet can be identified as $\theta = 0$ projections of the superspace one-forms and descendants of $W^{\alpha\beta}$, see [42].

So far we have considered only the standard Weyl multiplet possessing the covariant component fields: T^-_{abc}, $\chi^{\alpha i}$ and D. However, there exists a variant multiplet for conformal supergravity called the dilaton-Weyl multiplet [43], which plays an important role in the formulation of 6D Poincaré supergravity. The dilaton-Weyl multiplet is obtained by coupling the standard Weyl multiplet to a (on-shell) tensor multiplet. In superspace, the tensor multiplet can be described by a scalar superfield Φ embedded in a gauge super two-form B_2. Its field strength $H_3 = \mathrm{d}B_2$ is constrained to have the following component superfields

$$H^{i\,j\,k}_{\alpha\beta\gamma} = 0 \ , \quad H_{a\alpha}{}^{i}{}_{\beta}^{j} = 2\mathrm{i}\varepsilon^{ij}(\gamma_a)_{\alpha\beta}\Phi \ , \quad H_{ab}{}^{i}_{\alpha} = (\gamma_{ab})_\alpha{}^\beta \nabla^i_\beta \Phi \ , \tag{3.7a}$$

$$H_{abc} = -\tfrac{\mathrm{i}}{8}(\tilde\gamma_{abc})^{\gamma\delta}\nabla^k_\gamma \nabla_{\delta k}\Phi - 4W_{abc}\Phi \ . \tag{3.7b}$$

Here Φ is a primary superfield of dimension 2 satisfying the differential constraint $\nabla^{(i}_\alpha \nabla^{j)}_\beta \Phi = 0$. Equation (3.7b) shows that one can redefine the super-Weyl tensor as a composite of the tensor multiplet

$$W_{abc} = -\frac{1}{4}H_{abc} - \frac{\mathrm{i}}{32}(\tilde\gamma_{abc})^{\gamma\delta}\nabla^k_\gamma \nabla_{\delta k}\Phi \ , \tag{3.8}$$

which ends up being the dynamical multiplet for conformal supergravity in the dilaton-Weyl formulation. At the component level, the previous result is equivalent to showing that the covariant fields of the standard Weyl multiplet, T^-_{abc}, $\chi^{\alpha i}$ and D, are exchanged with the component fields of the tensor multiplet, σ, ψ^i_α and b_{mn} in the following way[2]

$$T^-_{abc} = \frac{1}{2\sigma}H^-_{abc} \ , \quad D = \frac{15}{4\sigma}\Big(\nabla^a\nabla_a\sigma + \frac{1}{3}T^{-abc}H_{abc}\Big) + \text{fermions} \ . \tag{3.9}$$

So far we have discussed kinematical properties of 6D supergravity. Let us turn to constructing dynamical systems. To do so, we will use the superform approach to engineer invariant actions [45, 46, 47, 48]. This approach has been rediscovered a number of times in the literature and it has been developed and used in, e. g., the study of the properties of UV counterterms in supersymmetric Yang-Mills and supergravity theories, see [49, 50, 51, 52, 53]. Since 2009 it has been employed and developed also by the two of us together with our collaborators to construct off-shell higher-derivative invariants in various dimensions, see e. g. [54, 55, 56, 57].

In six dimensions, the construction of a supersymmetric invariant from a superform goes as follows. Consider a closed super 6-form $J = \frac{1}{6!}\mathrm{d}z^{M_6}\wedge\cdots\wedge\mathrm{d}z^{M_1}J_{M_1\cdots M_6}$,

$$\mathrm{d}J = \frac{1}{6!}\mathrm{d}z^{M_6}\wedge\cdots\wedge\mathrm{d}z^{M_0}\partial_{M_0}J_{M_1\cdots M_6} = 0 \ . \tag{3.10}$$

By using J one can define the following action principle

$$S = \int \mathrm{d}^6x\, {}^*J|_{\theta=0} \ , \qquad {}^*J = \frac{1}{6!}\varepsilon^{mnpqrs}J_{mnpqrs} \ . \tag{3.11}$$

Under superdiffeomorphisms, with $\xi = \xi^A E_A = \xi^M \partial_M$, a closed 6-form satisfies

$$\delta_\xi J = \mathcal{L}_\xi J \equiv i_\xi \mathrm{d}J + \mathrm{d}i_\xi J = \mathrm{d}i_\xi J \ , \tag{3.12}$$

[2] Here we focus on just the bosonic fields.

which implies that (up to boundary terms) the action (3.11) is invariant under local supersymmetry transformations. Expressing the action in terms of the tangent frame gives

$$\begin{aligned} S &= \int \mathrm{d}^6x\,\frac{1}{6!}\varepsilon^{m_1\cdots m_6}E_{m_6}{}^{A_6}\cdots E_{m_1}{}^{A_1}J_{A_1\cdots A_6}|_{\theta=0}\;, \\ &\propto \int \mathrm{d}^6x\,\mathrm{e}\,\varepsilon^{a_1\cdots a_6}\Big[J_{a_1\cdots a_6}+3\psi_{a_1}{}^{\alpha}_{i}J^{i}_{\alpha a_2\cdots a_6}+\frac{15}{4}\psi_{a_2}{}^{\beta}_{j}\psi_{a_1}{}^{\alpha}_{i}J^{ij}_{\alpha\beta a_3\cdots a_6} \\ &\quad+\frac{5}{2}\psi_{a_3}{}^{\gamma}_{k}\psi_{a_2}{}^{\beta}_{j}\psi_{a_1}{}^{\alpha}_{i}J^{ijk}_{\alpha\beta\gamma a_4a_5a_6}+\mathcal{O}(\psi^4)\Big]|_{\theta=0}\;, \end{aligned} \tag{3.13}$$

which makes evident how to obtain from (3.11) the component expression of supergravity actions as a power series expansion in the gravitini. To impose invariance under the entire local supergravity gauge group, it is necessary to also require the action to be invariant under the structure group X and any possible additional gauge transformations. This means that J should transform by (at most) an exact 6-form under these transformations

$$\delta_X J = \mathrm{d}\Xi\;, \qquad \text{for some} \quad 5-\text{form}\;\; \Xi\;. \tag{3.14}$$

The classification of X-exact closed super 6-forms J is equivalent to the classification of supersymmetric invariants.

Now that we have identified a powerful way to construct supersymmetric invariants, it is natural to ask: how can we describe the $\mathcal{N}=(1,0)$ extension of the Einstein-Hilbert term of [43] in superspace? To describe this invariant one has to make use of the dilaton-Weyl multiplet coupled to a linear-multiplet conformal compensator. This is based on a scalar isotriplet superfield $G^{ij}(z)$, constrained by $\nabla_\alpha^{(i}G^{jk)}=0$. Remarkably, this multiplet can be engineered as a closed super 5-form such that $H_5=\mathrm{d}B_4$ for some gauge super 4-form and whose lowest dimension nonvanishing component is

$$H_{abc\alpha\beta}^{\;\;\;\;\;\,i\,j} = -2\mathrm{i}(\gamma_{abc})_{\alpha\beta}G^{ij}\;. \tag{3.15}$$

By using the linear multiplet 5-form, one can construct an invariant action in the following way. First, in a conformal supergravity background, consider a (1,0) vector multiplet described by a spinor superfield $W^{\alpha i}$ obeying the Bianchi identities $\nabla_\gamma^k W_k^\gamma=0$, $\nabla_\alpha^{(i}W^{\beta j)}=\frac{1}{4}\delta_\alpha^\beta\nabla_\gamma^{(i}W^{\gamma j)}$ and appropriately embedded in a closed super two-form F_2. Next, consider the following closed 6-form $J_{B_4\wedge F_2}$

$$J_{B_4\wedge F_2}=B_4\wedge F_2-\Sigma_{B_4\wedge F_2}\;,\qquad \mathrm{d}\Sigma_{B_4\wedge F_2}=-F_2\wedge H_5\;, \tag{3.16}$$

with $\Sigma_{B_4\wedge F_2}$ a covariant 6-form constructed entirely in terms of the superfields $W^{\alpha i}$ and G^{ij} in F_2 and H_5.[3] By using the closed 6-form $J_{B_4\wedge F_2}$, together with the superform action principle (3.11), one obtains the following $B_4\wedge F_2$ action principle[4]

$$S_{B_4\wedge F_2}=\frac{1}{2}\int \mathrm{d}^6x\,e\left(\frac{1}{4!}\varepsilon^{abcdef}f_{ab}b_{cdef}+X^{ij}G_{ij}+\text{fermions}\right)\;, \tag{3.17}$$

with f_{ab} the component field strength and X^{ij} the scalar component of the vector multiplet, while b_{abcd} is the component gauge 4-form of the linear multiplet. Note that f_{ab} and b_{abcd} are directly related to the $\theta=0$ component of the field strength $F_{mn}(z)$ and the gauge superfield $B_{mnpq}(z)$ respectively. The previous action reproduces the one first constructed in [43].

[3] Its existence is guaranteed by the fact that $F_2\wedge H_5$ is Weil trivial.

[4] The procedure here very much follows the one adopted in [58] in 4D.

We are now ready to introduce the (1,0) Poincaré supergravity EH invariant. Consider the following composite vector multiplet built by using the linear multiplet

$$\begin{aligned}\mathbb{W}^{\alpha i} &= \frac{1}{G}\nabla^{\alpha\beta}\chi^i_\beta + \frac{4}{G}\big(W^{\alpha\beta}\chi^i_\beta + 10\mathrm{i}X^\alpha_j G^{ji}\big) - \frac{1}{2G^3}G_{jk}(\nabla^{\alpha\beta}G^{ij})\chi^k_\beta \\ &\quad + \frac{1}{2G^3}G^{ij}E^{\alpha\beta}\chi_{\beta j} + \frac{\mathrm{i}}{16G^5}\varepsilon^{\alpha\beta\gamma\delta}\chi_{\beta j}\chi_{\gamma k}\chi_{\delta l}G^{ij}G^{kl}\,, \end{aligned} \tag{3.18}$$

where

$$\nabla^{(i}_\alpha G^{jk)} = 0\,, \quad \chi^i_\alpha := \frac{2}{3}\nabla_{\alpha j}G^{ij}\,, \quad E^{\alpha\beta} := \frac{\mathrm{i}}{8}\varepsilon^{\alpha\beta\gamma\delta}\nabla^k_{[\gamma}\chi_{\delta]k}\,. \tag{3.19}$$

By plugging back this composite vector multiplet in the $B_4 \wedge F_2$ action, and gauge fixing dilatation, conformal-boosts and $\mathrm{SU}(2)_R$ symmetry, one ends up with the (1,0) Poincaré supergravity action

$$e^{-1}\mathcal{L}_{\rm EH} = -\frac{1}{2}e^{-2\varphi}[\mathcal{R} - 4\partial_m\varphi\partial^m\varphi + \tfrac{1}{12}H_{abc}H^{abc}] + \text{fermions}\,, \tag{3.20}$$

where φ is the only component of G^{ij} surviving the $\mathrm{SU}(2)_R \to \mathrm{U}(1)_R$ gauge fixing condition. The results for the EH terms obtained from superspace coincide with the original results of [43] but the path we have followed will prove more useful for higher-derivative actions.

4. Curvature squared invariants

Let us show how it is possible to construct supersymmetric extensions of all curvature squared Lagrangians in superspace. We are searching for extensions of a Lagrangian that is schematically the following linear combination of Weyl squared, Riemann squared and the scalar curvature squared

$$\mathcal{L}_{\rm curvature-squared} \propto a\, C^{abcd}C_{abcd} + b\,\mathcal{R}^{abcd}\mathcal{R}_{abcd} + c\,\mathcal{R}^2 + \cdots\,. \tag{4.1}$$

In constructing the supersymmetric extension of the EH term a crucial role was played by the $B_4 \wedge F_2$ action principle. It turns out that we can construct the two known invariants plus one new invariant by using a new action principle corresponding to the supersymmetrization of a $B_2 \wedge H_4$ terms where B_2 is the gauge 2-form of a tensor multiplet with field strength $H_3 = \mathrm{d}B_2$ and H_4 is a closed 4-form, $\mathrm{d}H_4 = 0$, possessing the following components

$$H^{\,i\,j\,kl}_{\alpha\beta\gamma\delta} = H_a{}^{\,j\,kl}_{\beta\gamma\delta} = 0\,, \quad H_{ab}{}^{kl}_{\gamma\delta} = \mathrm{i}(\gamma_{abc})_{\gamma\delta}B^{ckl}\,, \tag{4.2a}$$

$$H_{abc}{}^l_\delta = -\tfrac{1}{12}\varepsilon_{abcdef}(\gamma^{de})_\delta{}^\rho\nabla_{\rho p}B^{flp}\,, \quad H_{abcd} = \tfrac{\mathrm{i}}{48}\varepsilon_{abcdef}(\tilde\gamma^e)^{\alpha\beta}\nabla_{\alpha k}\nabla_{\beta l}B^{fkl}\,. \tag{4.2b}$$

The superfield $B^{\alpha\beta ij} = (\tilde\gamma^a)^{\alpha\beta}B^{ij}_a$, is a dimension three primary superfield satisfying the following differential constraints induced by the constraint $\mathrm{d}H_4 = 0$

$$\nabla^{(i}_\alpha B^{\beta\gamma jk)} = -\frac{2}{3}\delta^{[\beta}_\alpha\nabla^{(i}_\delta B^{\gamma]jk)}\,, \quad [\nabla^{(i}_\alpha, \nabla_{\beta k}]B^{\alpha\beta j)k} = -8\mathrm{i}\nabla_{\alpha\beta}B^{\alpha\beta ij}\,. \tag{4.3}$$

We can now engineer the new action principle by using the superform approach. We start with the following closed 6-form $J_{B_2\wedge H_4}$

$$J_{B_2\wedge H_4} = B_2 \wedge H_4 - \Sigma_{B_2\wedge H_4}\,, \qquad \mathrm{d}\Sigma_{B_2\wedge H_4} = H_3 \wedge H_4 \tag{4.4}$$

with $\Sigma_{B_2 \wedge H_4}$ a covariant super 6-form constructed only in terms of the covariant superfields Φ and B_a^{ij}, respectively associated with the field strengths H_3 and H_4. Once one explicitly constructs $J_{B_2 \wedge H_4}$ and plugs the result into (3.11) we obtain the new $B_2 \wedge H_4$ action principle

$$S_{B_2 \wedge H_4} = \int \mathrm{d}^6 x \, e \left\{ \frac{1}{4} (b_{ab} - \eta_{ab} \sigma) C^{ab} + \text{fermions} \right\} , \tag{4.5}$$

which proves to be locally superconformal invariant. Here $\sigma := \Phi|_{\theta=0}$ and b_{ab} is related to the $\theta = 0$ component of the gauge superfield $B_{mn}(z)$, while C_{ab} is

$$C_{ab} := \frac{\mathrm{i}}{12} (\tilde{\gamma}_a)^{\alpha\beta} \nabla_{\alpha k} \nabla_{\beta l} B_b^{kl} |_{\theta=0} . \tag{4.6}$$

Given a B_a^{ij} superfield satisfying (4.3), which plays the role of a Lagrangian for the $B_2 \wedge H_4$ action principle, we will automatically have a $(1,0)$ locally superconformal invariant. Let us describe relevant examples.

A locally supersymmetric extension of the Riemann squared term was first constructed in 1987 in [30]. We can reproduce it by using the $B_2 \wedge H_4$ action principle and the composite superfield

$$B^{\alpha\beta ij} = -\frac{\mathrm{i}}{2} \Lambda^{\alpha(i}{}_{\gamma}{}^{\delta} \Lambda^{\beta j)}{}_{\delta}{}^{\gamma} \tag{4.7}$$

with $\Lambda^{\alpha i}{}_{\gamma}{}^{\delta}$ given by the following dimension-3/2 primary superfield

$$\begin{aligned} \Lambda^{\alpha i}{}_{\beta}{}^{\gamma} = {} & X_{\beta}^{i\,\alpha\gamma} - \frac{1}{3} \delta_{\beta}^{\alpha} X^{\gamma i} + \frac{1}{12} \delta_{\beta}^{\gamma} X^{\alpha i} + \frac{\mathrm{i}}{4} \Phi^{-1} \psi_{\beta}^{i} W^{\alpha\gamma} + \frac{\mathrm{i}}{12} \Phi^{-1} \delta_{\beta}^{\alpha} W^{\gamma\delta} \psi_{\delta}^{i} \\ & - \frac{\mathrm{i}}{12} \Phi^{-1} \delta_{\beta}^{\gamma} W^{\alpha\delta} \psi_{\delta}^{i} + \frac{\mathrm{i}}{12} \varepsilon^{\alpha\gamma\delta\rho} \Phi^{-1} \nabla_{\delta(\rho} \psi_{\beta)}^{i} - \frac{\mathrm{i}}{8} \varepsilon^{\alpha\gamma\delta\rho} \Phi^{-2} (\nabla_{\delta(\rho} \Phi) \psi_{\beta)}^{i} \\ & + \frac{\mathrm{i}}{32} \varepsilon^{\alpha\gamma\delta\rho} \Phi^{-2} H_{\rho\beta} \psi_{\delta}^{i} - \frac{1}{16} \varepsilon^{\alpha\gamma\delta\rho} \Phi^{-3} \psi_{\delta}^{i} \psi_{(\rho}^{k} \psi_{\beta)k} , \end{aligned} \tag{4.8}$$

where

$$\psi_{\alpha}^{i} = \nabla_{\alpha}^{i} \Phi , \quad \nabla_{\alpha}^{i} \psi_{\beta}^{j} = -\frac{\mathrm{i}}{2} \varepsilon^{ij} (\gamma^{abc})_{\alpha\beta} H_{abc}^{+} - \mathrm{i} \varepsilon^{ij} (\gamma^{a})_{\alpha\beta} \nabla_a \Phi . \tag{4.9}$$

In the gauge $\Phi = 1$ and $B_A = 0$, the composite (4.7) together with the action principle (4.5) reproduces the Riemann squared invariant of [30] whose leading term is

$$e^{-1} \mathcal{L}_{\text{Riem}^2} = \mathcal{R}^{abcd} \mathcal{R}_{abcd} + \cdots . \tag{4.10}$$

A scalar curvature squared invariant, with leading term $\mathcal{R}^2$, was constructed in components in [59] by using results of the superconformal tensor calculus of [43]. We can reproduce it from superspace by using the $B_2 \wedge H_4$ action principle where

$$B^{\alpha\beta ij} = -\frac{\mathrm{i}}{2} \mathbb{W}^{\alpha(i} \mathbb{W}^{\beta j)} , \tag{4.11}$$

and $\mathbb{W}^{\alpha i}$ is the composite vector multiplet (3.18) used to construct the EH term.

Remarkably, by using the $B_2 \wedge H_4$ action principle we can construct a new curvature squared invariant. A composite built entirely out of the super-Weyl tensor can be constructed [41]

$$B^{\alpha\beta\, ij} = -4 W^{\gamma[\alpha} Y_{\gamma}{}^{\beta]ij} - 32\mathrm{i}\, X_{\gamma}{}^{\alpha\delta(i} X_{\delta}{}^{\beta\gamma j)} + 10\mathrm{i}\, X^{\alpha(i} X^{\beta j)} , \tag{4.12}$$

where $X^{\alpha i}$, $X_\alpha{}^{\beta\gamma i}$ and $Y_\alpha{}^{\beta ij}$ are descendant superfields constructed in terms of spinor derivatives of the super-Weyl tensor $W^{\alpha\beta}$. The previous composite superfield was originally constructed to describe one of the two $(1,0)$ conformal supergravity actions [41, 42]. On the other hand, by plugging B_a^{ij} of (4.12) in the $B_2 \wedge H_4$ action we obtain the following new curvature squared invariant

$$\begin{aligned} e^{-1}\mathcal{L}_{\rm new} \;=\; & \frac{1}{32}\Big\{\sigma C_{ab}{}^{cd}C_{cd}{}^{ab} - 3\sigma\mathcal{R}_{ab}{}^{ij}\mathcal{R}^{ab}{}_{ij} + \frac{4}{15}\sigma D^2 - 8\sigma T^{-dab}\nabla_d\nabla^c T^-_{abc} \\ & + 4\sigma(\nabla_c T^{-abc})\nabla^d T^-_{abd} + 4\sigma T^{-abc}T^-_{ab}{}^d T^{-ef}{}_c T^-_{efd} - \frac{8}{45}H_{abc}T^{-abc}D \\ & - 2H_{abc}C^{ab}{}_{de}T^{-cde} + 4H_{abc}T^-_d{}^{ab}\nabla_e T^{-cde} - \frac{4}{3}H_{abc}T^{-dea}T^{-bcf}T^-_{def} \\ & - \frac{1}{4}\varepsilon^{abcdef}b_{ab}\big(C_{cd}{}^{gh}C_{efgh} - \mathcal{R}_{cd}{}^{ij}\mathcal{R}_{ef\,ij}\big)\Big\} + \text{fermions}\;, \end{aligned} \tag{4.13}$$

where $\mathcal{R}_{ab}{}^{ij}$ is the $SU(2)_R$ field strength. In the gauge $\sigma = 1$ and $b_a = 0$, equation (4.13) reduces to

$$e^{-1}\mathcal{L}_{\rm new} = \frac{1}{32}\mathcal{R}_{ab}{}^{cd}\mathcal{R}_{cd}{}^{ab} - \frac{1}{32}\mathcal{R}_b{}^d\mathcal{R}_d{}^b + \frac{1}{128}\mathcal{R}^2 + \cdots\;. \tag{4.14}$$

5. Applications: Einstein-Gauss-Bonnet supergravity

It turns out that the supersymmetric GB invariant can be constructed by taking a linear combination of the new curvature squared invariant (4.13) and the Riemann squared invariant. By comparing (1.2) with (4.10) and (4.14) it is simple to see that the following combination

$$S_{\rm GB} = -3S_{\rm Riem^2} + 128S_{\rm new} \tag{5.1}$$

describes a 6D $(1,0)$ off-shell supersymmetric extension of the Gauss-Bonnet term. In the gauge $\sigma = 1$ and $b_a = 0$, the resulting Lagrangian becomes

$$\begin{aligned} e^{-1}\mathcal{L}_{\rm GB} \;=\; & \mathcal{R}_{abcd}\mathcal{R}^{abcd} - 4\mathcal{R}_{ab}\mathcal{R}^{ab} + \mathcal{R}^2 \\ & -\tfrac{1}{2}\mathcal{R}_{abcd}H^{abe}H^{cd}{}_e + \mathcal{R}^{ab}H^2_{ab} - \tfrac{1}{6}\mathcal{R}H^2 + \tfrac{1}{144}(H^2)^2 - \tfrac{1}{8}(H^2_{ab})^2 + \tfrac{5}{24}H^4 \\ & +\epsilon^{abcdef}b_{ab}\Big(-\tfrac{1}{4}\mathcal{R}_{cd}{}^{gh}(\tilde\omega)\mathcal{R}_{efgh}(\tilde\omega) + \mathcal{R}_{cd}{}^{ij}\mathcal{R}_{ef\,ij}\Big) + \text{fermions}\;, \end{aligned}$$

where we have introduced the torsionfull connection $\tilde\omega$ and various contractions as

$$\tilde\omega_m{}^{cd} := \omega_m{}^{cd} - \tfrac{1}{2}e_m{}^a H_a{}^{cd}\;, \tag{5.2a}$$

$$H^2 := H_{abc}H^{abc}\;, \quad H^2_{ab} := H_a{}^{cd}H_{bcd}\;, \quad H^4 := H_{abe}H_{cd}{}^e H^{acf}H^{bd}{}_f\;. \tag{5.2b}$$

There are various advantages in having an off-shell version of the $(1,0)$ Gauss-Bonnet invariant. First of all, it is possible to couple it to general supergravity-matter couplings without having to modify the supersymmetry transformations. Secondly, we have a complete off-shell descriptions of the dependance on the NSNS 2-form gauge potential. Let us show how these advantages can be used in a relevant case.

Consider the combination

$$\mathcal{L} = \mathcal{L}_{\rm EH} + \frac{1}{16}\alpha'\mathcal{L}_{\rm GB}\,. \tag{5.3}$$

This turns out to be an off-shell extension of the α' corrected low-energy effective action of string theory compactified to six dimensions. More precisely, once one properly integrated out the $\mathrm{SU}(2)_R$ gauge connection and an auxiliary field of the linear multiplet compensator, it can be shown that the previous Lagrangian reduces to

$$\begin{aligned} 2\kappa^2 e^{-1}\mathcal{L} &= e^{-2\varphi}[-\mathcal{R} + 4\partial_m\varphi\partial^m\varphi - \tfrac{1}{12}H_{abc}H^{abc}] \\ &\quad + \tfrac{1}{16}\alpha'\Big[6\mathcal{R}_{[ab}{}^{ab}\mathcal{R}_{cd]}{}^{cd} - \tfrac{1}{6}\mathcal{R}H^2 + \mathcal{R}^{ab}H^2_{ab} + \tfrac{5}{24}H^4 + \tfrac{1}{144}(H^2)^2 - \tfrac{1}{8}(H^2_{ab})^2 \\ &\quad - \tfrac{1}{2}\mathcal{R}_{abcd}H^{abe}H^{cd}{}_e + \tfrac{1}{4}\epsilon^{abcdef}b_{ab}R_{cd}{}^g{}_h(\tilde\omega)R_{ef}{}^h{}_g(\tilde\omega)\Big] + \text{fermions} \end{aligned} \quad (5.4)$$

which precisely matches the on-shell Einstein-Gauss-Bonnet supergravity action that was first described in [3]. In [3], Liu and Minasian constructed the Lagrangian (5.4) by fixing the one-loop four-derivative corrections in six dimensions by means of a K3 reduction of the R^4 corrected type IIA strings and requiring that the dyonic string remains a solution, as well as the duality of this model to heterotic strings compactified on T^4. Our result relies only on supersymmetry and can be considered as an important consistency check of the results and assumptions of [3].

6. Conclusion

The examples reviewed in this paper demonstrate that there exist powerful "top-down" approaches based on superspace to construct off-shell higher-derivative supergravity invariants. By using these techniques we can potentially improve the classification of higher-derivative invariants in various dimensions. For instance, the new 6D curvature squared invariant completes a piece that has been missing for three decades. The supersymmetric extension of the Gauss-Bonnet invariant is of importance in the study of the low-energy limit of string theory and for α'-corrected AdS/CFT tests, and might find various interesting applications. For instance, the Einstein-Gauss-Bonnet action (5.1) possesses a supersymmetric $\mathrm{AdS}_3 \times \mathrm{S}^3$ solution analogue of the famous $\mathrm{AdS}_5 \times \mathrm{S}^5$ solution in IIB string theory. Since we know the full off-shell description of (5.1), in [31] we were able to compute the α'-corrected Kaluza-Klein spectrum of fluctuations around $\mathrm{AdS}_3 \times \mathrm{S}^3$ and show that it organizes in short and long multiplets of its super-isometry group, $\mathrm{SU}(1,1|2) \times \mathrm{SL}(2,\mathbb{R}) \times \mathrm{SU}(2)$. Our results could be used to better understand the dynamics of strings in $\mathrm{AdS}_3 \times \mathrm{S}^3 \times \mathrm{K3}(\mathrm{T}^4)$ backgrounds and for applications to AdS/CFT.

Acknowledgments

We are grateful to D. Butter, M. Ozkan and Y. Pang for collaboration. JN is supported by a Humboldt research fellowship of the Alexander von Humboldt Foundation. GT-M was supported in part by the Interuniversity Attraction Poles Programme initiated by the Belgian Science Policy (P7/37) and in part by the KU Leuven C1 grant ZKD1118 C16/16/005.

References

[1] Antoniadis I, Ferrara S, Minasian R and Narain K S 1997 *Nucl. Phys.* B **507** 571
[2] Antoniadis I, Minasian R, Theisen S and Vanhove P 2003 *Class. Quant. Grav.* **20** 5079
[3] Liu J T and Minasian R 2013 *Nucl. Phys.* B **874** 413
[4] Maldacena J M, Strominger A and Witten E 1997 *JHEP* **9712** 002
[5] Vafa C 1998 *Adv. Theor. Math. Phys.* **2** 207
[6] Lopes Cardoso G,de Wit B and Mohaupt T 1999 *Phys. Lett.* B **451** 309
[7] Lopes Cardoso G, de Wit B, Kappeli J and Mohaupt T 2000 *JHEP* **0012**, 019
[8] Kallosh R arXiv:1412.7117 [hep-th]
[9] Utiyama R and DeWitt B S 1962 *J. Math. Phys.* **3** 608
[10] Komargodski Z and Schwimmer A 2011 *JHEP* **1112** 099
[11] Stelle K S 1977 *Phys. Rev.* D **16**, 953
[12] Starobinsky A A 1980 *Phys. Lett.* B **91** 99
[13] Deser S and Schwimmer A 1993 *Phys. Lett.* B **309** 279
[14] B. Zwiebach 1985 *Phys. Lett.* B **156** 315
[15] S. Deser and A. N. Redlich 1986 *Phys. Lett.* B **176** 350 Erratum: [1987 *Phys. Lett.* B **186** 461]

[16] Cecotti S, Ferrara S, Girardello L and Porrati M 1985 *Phys. Lett.* B **164** 46
[17] Theisen S 1986 *Nucl. Phys.* B **263**, 687
[18] Buchbinder I L and Kuzenko S M, 1988 *Nucl. Phys.* B **308**, 162
[19] Cecotti S, Ferrara S, Girardello L, Porrati M and Pasquinucci A 1986 *Phys. Rev.* D **33** 2504
[20] Ferrara S, Sabharwal S and Villasante M 1988 *Phys. Lett.* B **205**, 302
[21] Ferrara S and Villasante M 1989 *J. Math. Phys.* **30**, 104
[22] Le Du R 1998 *Eur. Phys. J.* C **5**, 181
[23] Butter D, de Wit B, Kuzenko S M and Lodato I 2013 *JHEP* **1312** 062
[24] Ozkan M and Pang Y 2013 *JHEP* **1303** 158 Erratum: [2013 *JHEP* **1307** 152]
[25] Ozkan M and Pang Y 2013 *JHEP* **1308**, 042
[26] Butter D, Kuzenko S M, Novak J and Tartaglino-Mazzucchelli G 2015 *JHEP* **1502** 111
[27] Bergshoeff E, Salam A and Sezgin E 1986 *Phys. Lett.* B **173**, 73
[28] Bergshoeff E, Salam A and Sezgin E 1987 *Nucl. Phys.* B **279**, 659
[29] Nishino H and Gates S J Jr. 1986 *Phys. Lett.* B **173** 417
[30] Bergshoeff E and Rakowski M 1987 *Phys. Lett.* B **191**, 399
[31] Novak J, Ozkan M, Pang Y and Tartaglino-Mazzucchelli G 2017 *Phys. Rev. Lett.* **119** no.11, 111602
[32] Freedman D Z and Van Proeyen A 2012 Cambridge Univ. Press
[33] Gates S J Jr., Grisaru M T, Rocek M and Siegel W 1983 *Front. Phys.* **58** 1
[34] Wess J and Bagger J 1992 *Princeton, USA: Univ. Pr.* 259 p
[35] Buchbinder I L and Kuzenko S M 1998 *Bristol, UK: IOP* 656 p
[36] Galperin A S, Ivanov E A, Ogievetsky V I and Sokatchev E S 2001 *Cambridge, UK: Univ. Pr.* 306 p
[37] Butter D 2010 *Annals Phys.* **325**, 1026
[38] Butter D 2011 *JHEP* **1110**, 030
[39] Kugo T and Uehara S 1985 *Prog. Theor. Phys.* **73** 235
[40] Butter D, Kuzenko S M, Novak J and Tartaglino-Mazzucchelli G 2013 *JHEP* **1309** 072
[41] Butter D, Kuzenko S M, Novak J and Theisen S 2016 *JHEP* **1612**, 072
[42] Butter D, Novak J and Tartaglino-Mazzucchelli G 2017 *JHEP* **1705**, 133
[43] Bergshoeff E, Sezgin E and Van Proeyen A 1986 *Nucl. Phys.* B **264**, 653, Erratum: [2001 *Nucl. Phys.* B **598**, 667]
[44] Linch W D III and Tartaglino-Mazzucchelli G 2012 *JHEP* **1208**, 075
[45] Castellani L, D'Auria R and Fre P 1991 *World Scientific, Singapore* pp. 680–684
[46] Hasler M F 1998 *Eur. Phys. J.* C **1**, 729
[47] Gates S J Jr. 1999 *Nucl. Phys.* B **541**, 615
[48] Gates S J Jr., Grisaru M T, Knutt-Wehlau M E and Siegel W 1998 *Phys. Lett.* B **421**, 203
[49] Bossard G, Howe P S and Stelle K S 2009 *Gen. Rel. Grav.* **41**, 919
[50] Bossard G, Howe P S and Stelle K S 2009 *Phys. Lett.* B **682**, 137
[51] Bossard G, Howe P S and Stelle K S 2011 *JHEP* **1101**, 020
[52] Bossard G, Howe P S, Lindstrom U, Stelle K S and Wulff L, 2011 *JHEP* **1105** 021
[53] Bossard G, Howe P S and Stelle K S 2013 *JHEP* **1307**, 117
[54] Butter D, Kuzenko S M, Novak J and Tartaglino-Mazzucchelli G 2015 *JHEP* **1502**, 111
[55] Butter D, Kuzenko S M, Novak J and Tartaglino-Mazzucchelli G 2013 *JHEP* **1310**, 073
[56] Kuzenko S M, Novak J and Tartaglino-Mazzucchelli G 2014 *JHEP* **1401**, 121
[57] Kuzenko S M, Novak J and Tartaglino-Mazzucchelli G 2015 *JHEP* **1509**, 081
[58] Butter D, Kuzenko S M, Novak J 2012 *JHEP* **1209**, 131
[59] Ozkan M 2013 *PhD Thesis, Texas A&M University*

IOP Conf. Series: Journal of Physics: Conf. Series **965** (2018) 012030 doi:10.1088/1742-6596/965/1/012030

Transfer matrix spectrum for cyclic representations of the 6-vertex reflection algebra by quantum separation of variables

Baptiste Pezelier

Univ Lyon, Ens de Lyon, Univ Claude Bernard Lyon 1,
CNRS, UMR 5672 Laboratoire de Physique, F-69342 Lyon, France

E-mail: baptiste.pezelier@ens-lyon.fr

Abstract. In this proceeding, we recall the notion of quantum integrable systems on a lattice and then introduce the Sklyanin's Separation of Variables method. We sum up the main results for the transfer matrix spectral problem for the cyclic representations of the trigonometric 6-vertex reflection algebra associated to the Bazanov-Stroganov Lax operator. These results apply as well to the spectral analysis of the lattice sine-Gordon model with open boundary conditions. The transfer matrix spectrum (both eigenvalues and eigenstates) is completely characterized in terms of the set of solutions to a discrete system of polynomial equations. We state an equivalent characterization as the set of solutions to a Baxter's like T-Q functional equation, allowing us to rewrite the transfer matrix eigenstates in an algebraic Bethe ansatz form.

1. Introduction

One of the main tasks of statistical mechanics is to understand macroscopic behaviours, like transport properties, specific heat or susceptibility, from microscopic models. The main theoretical tool turns out to be the correlation function, from which the physical quantities can be derived.

Starting with a quantum system described by an Hamiltonian H, the research program is clearly established : the first step is to compute the spectrum, *ie* both the eigenvalues and the eigenstates of the Hamiltonian. Then, the computation of form factors (the matrix elements of an operator on a left and a right eigenstate) should be achieved. At zero temperature, the correlation function can be expressed as the mean value of a string of operators on the ground state, and can be decomposed in terms of form factors. In this proceeding, we focus on the first step of such a research program, for models associated to cyclic representations of the 6-vertex reflection algebra [1, 2].

The study of quantum models with integrable open boundary conditions has attracted a large research interest, as they are of physical interest to describe both equilibrium and out of equilibrium physics. See [1, 2] and the references therein. In this proceeding, we analyze the class of open integrable quantum models associated to cyclic representations of the 6-vertex reflection algebra. Let us recall that in [3], Sklyanin has shown how to construct classes of quantum integrable models with integrable boundaries in the framework of the so-called Quantum Inverse Scattering Method [8,9], constructing families of commuting transfer matrices (conserved charges of the model). In fact, Sklyanin's construction allows to use solutions of

IOP Conf. Series: Journal of Physics: Conf. Series **965** (2018) 012030 doi:10.1088/1742-6596/965/1/012030

the Yang-Baxter equation to generate new solutions of the reflection equation [4] once a scalar solution of this last equation is known. Then as a consequence of the reflection equation these new solutions generate commuting transfer matrices [3]. More generally, one can associate to any closed integrable quantum model (characterized by a solution of the Yang-Baxter equation) new open integrable quantum models (characterized by the associated Sklyanin's solutions of the reflection equation).

The literature of integrable quantum models associated to cyclic representations of the 6-vertex reflection algebra is so far rather sparse, with the exception of some special representations and boundary conditions, like the open XXZ chains at the roots of unity, that can be traced back to these representations under some special constraints, and for which some results are known in the framework of algebraic Bethe ansatz (ABA) [8, 9]. In order to study the class of models in the Sklyanin's construction associated to general scalar solutions of the 6-vertex reflection equation [5, 6] and the general Bazhanov-Stroganov cyclic solution of the 6-vertex Yang-Baxter algebra [12], we have to go beyond traditional methods [8,9] which do not apply for these general settings. This is done by developing the Sklyanin Separation of Variables (SoV) method [10] for this class of models, a method that has the advantage to lead (mainly by construction) to the complete characterization of the spectrum and has proven to be applicable for a large variety of integrable quantum models where traditional methods fail.

The aim of this proceeding is to give a short introduction to the works [1, 2], where we have generalized this type of results to the 6-vertex cyclic representations of the reflection algebra. In section 2 we recall generalities about the notion of quantum integrability and focus on integrable systems on a lattice. We also introduce the cyclic 6-vertex Yang-Baxter algebra. In section 3, we deal with closed and open chains, introducing representations of the cyclic 6-vertex reflection algebra. The section 4 introduces the main tool, *ie* Sklyanin's Separation of Variables, while section 5 is dedicated to the results, giving the spectrum characterization of the considered representations.

2. Generalities on quantum integrability

2.1. Quantum integrability

Let a quantum system be described by an Hamiltonien H in a Hilbert space $\mathcal{H}$.

Definition 2.1 (Quantum integrability). *This system is quantum integrable if there exists an operator family $T(\lambda)$ acting on $\mathcal{H}$, parametrized thanks to $\lambda \in \mathbb{C}$, such that :*

- $\forall(\lambda, \mu) \in \mathbb{C}^2 \quad [T(\lambda), T(\mu)] = 0$
- $\forall\lambda \in \mathbb{C} \quad [T(\lambda), H] = 0$
- $\forall\lambda \in \mathbb{C}$, $T(\lambda)$ *is diagonalizable and with non degenerated spectrum.*

With this definition, the set of the eigenstates of $T(\lambda)$ gives an eigenbasis of the space $\mathcal{H}$ and thus we can characterize the spectrum of the Hamiltonian of a quantum integrable system by characterizing the spectrum of the family $T(\lambda)$.

Starting with a given quantum system, it is difficult to directly exhibit such a commuting family $T(\lambda)$. The standard computation consists in rather constructing a family of commuting operators, thanks to an algebraic structure, and then to give Hamiltonians which commute with this family. These Hamiltonians are constructed from the elements of the family, and depending on the previous algebraic structure, one can reproduce physical models, such as spin chain Hamiltonians for example.

The heart of integrability is thus the above-mentioned algebraic structure, the so-called Yang-Baxter algebra :

Let $\mathcal{R}(x_1, x_2) \in \mathcal{A} \otimes \mathcal{A}$, a matrix solution of the Yang-Baxter equation

$$\mathcal{R}_{12}(\lambda, \mu)\mathcal{R}_{13}(\lambda, \nu)\mathcal{R}_{23}(\mu, \nu) = \mathcal{R}_{23}(\mu, \nu)\mathcal{R}_{13}(\lambda, \nu)\mathcal{R}_{12}(\lambda, \mu) \quad \in \mathcal{A} \otimes \mathcal{A} \otimes \mathcal{A} \tag{1}$$

The Yang-Baxter algebra associated to $\mathcal{R}$ is then

$$\mathcal{R}_{12}(\lambda,\mu)M_{1Q}(\lambda,\xi)M_{2Q}(\mu,\xi) = M_{2Q}(\mu,\xi)M_{1Q}(\lambda,\xi)\mathcal{R}_{12}(\lambda,\mu) \quad \in \mathcal{A}\otimes\mathcal{A}\otimes\mathcal{H} \tag{2}$$

The elements of the matrix M are operators on the total Hilbert space $\mathcal{H}$: $M_{0Q}(\lambda) \in End(\mathcal{A}\otimes\mathcal{H})$.

By taking the trace on the spaces 1 and 2 of the equation (2), it is very easy to show the following proposition :

Proposition 2.1. *If the R-matrix $\mathcal{R}_{12}$ is invertible, then a commuting family of operators $T(\lambda)$ on $\mathcal{H}$ is given by :*

$$T(\lambda) = tr_0(M_{0Q}(\lambda)) \tag{3}$$

The question is now how to find a matrix M satisfying the Yang-Baxter algebra. It might be involved, as the matrix elements are acting on the total Hilbert space $\mathcal{H}$.

2.2. Integrable lattice systems

Let us study systems on a lattice, which are decomposed in N sites, with degrees of freedom on each site. The Hilbert space takes the form of a tensor product of N local Hilbert spaces

$$\mathcal{H} = \otimes_{n=1}^{N}\mathcal{H}_n \tag{4}$$

For example, on each site, the local operators can be taken as canonical operators $\hat{x}$ and $\hat{p}$ (satisfying $[\hat{x},\hat{p}] = i$), or Weyl operators u_n and v_n (satisfying $u_n v_m = q^{\delta_{mn}} v_m u_n$, $q \in \mathbb{C}$) or also spin operators S_n^α , $\alpha \in \{1,2,3\}$ (satisfying $[S_m^\alpha, S_n^\beta] = i\varepsilon_{\alpha\beta\gamma}S_n^\gamma\delta_{nm}$), acting in $\mathcal{H}_n$.

With this formalism, one can describe the Heinsenberg spin chains or the discrete sine-Gordon 1D model.

Thanks to the work of Fadeev,Takhtajan and Sklyanin [8,9], we know that given an R-matrix satisfying (1) we can obtain a solution to the Yang-Baxter algebra (2) by considering the (local) Yang-Baxter algebra :

$$\mathcal{R}_{12}(\lambda,\mu)L_{1n}(\lambda,p_n)L_{2n}(\mu,p_n) = L_{2n}(\mu,p_n)L_{1n}(\lambda,p_n)\mathcal{R}_{12}(\lambda,\mu) \quad in\ \mathcal{A}\otimes\mathcal{A}\otimes\mathcal{H}_n \tag{5}$$

$L_{0n}(\lambda)$ are matrices called the Lax operators, whith matrix elements acting on the local Hilbert space $\mathcal{H}_n$: $L_{0n}(\lambda) \in End(\mathcal{A}\otimes\mathcal{H}_n)$.

The next proposition gives the expression of a solution to the Yang-Baxter algebra using solutions to the local algebra :

Proposition 2.2 (The monodromy matrix). *For a system on a lattice, a solution to the Yang-Baxter algebra associated to $\mathcal{R}$ can be computed by taking the ordered product of solutions of the local Yang-Baxter algebras on each site :*

$$M_{0Q}(\lambda) = L_{0,q_N}(\lambda)L_{0,q_{N-1}}(\lambda)...L_{0,q_1}(\lambda) \tag{6}$$

This matrix is then called the monodromy matrix.

2.3. The cyclic 6-vertex Yang-Baxter algebra

As described before, an $\mathcal{R}$ matrix and a Lax operator are enought to describe an integrable system on a lattice. In this proceeding, we consider

- the standar trigonometric 6-vertex $\mathcal{R}$ matrix : $(\mathcal{A} = \mathbb{C}^2)$

$$\mathcal{R}(\lambda,\mu) = \begin{pmatrix} q\lambda/\mu - (q\lambda/\mu)^{-1} & 0 & 0 & 0 \\ 0 & \lambda/\mu - (\lambda/\mu)^{-1} & q - q^{-1} & 0 \\ 0 & q - q^{-1} & \lambda/\mu - (\lambda/\mu)^{-1} & \\ 0 & 0 & 0 & q\lambda/\mu - (q\lambda/\mu)^{-1} \end{pmatrix} \tag{7}$$

- the Bazhanov-Stroganov Lax operator [12] :

$$L_{0n}(\lambda) = \begin{pmatrix} \lambda\alpha_n \mathfrak{v}_n - \frac{\beta_n}{\lambda}\mathfrak{v}_n^{-1} & \mathfrak{u}_n \left(q^{-1/2}a_n\mathfrak{v}_n + q^{1/2}b_n\mathfrak{v}_n^{-1}\right) \\ \mathfrak{u}_n^{-1}\left(q^{1/2}c_n\mathfrak{v}_n + q^{-1/2}d_n\mathfrak{v}_n^{-1}\right) & \frac{\gamma_n}{\lambda}\mathfrak{v}_n - \lambda\delta_n\mathfrak{v}_n^{-1} \end{pmatrix}_{[0]} \tag{8}$$

The free parameters $(a_n, b_,c_n, d_,\alpha_n, \beta_n, \gamma_n, \delta_n)$ have to satisfy the constraints

$$a_n c_n = \alpha_n\gamma_n \text{ and } b_n d_n = \beta_n\delta_n \tag{9}$$

to satisfy the Yang Baxter equation.

The local operators $\mathfrak{u}_n$ and $\mathfrak{v}_n$ are generators of a Weyl algebra of dimension p :

$$\mathfrak{u}_n\mathfrak{v}_m = q^{\delta_{nm}}\mathfrak{v}_m\mathfrak{u}_n \;\; ; \;\; \mathfrak{u}_n^p = \mathfrak{v}_n^p = 1 \; ; \; q^p = 1 \tag{10}$$

We emphasize the fact that the free parameter q is taken to be a pth root of unit.

The considered Lax operator is thus the cyclic local generator of the trigonometric 6-vertex Yang-Baxter algebra.

For different choices of the free parameters, one can describe the discretized sine-Gordon model, the chiral-Potts model, or the XXZ spin s chain at p^{th}-root of unity, with $2s + 1 = p$.

3. On the integrable boundary conditions

3.1. Closed chains

Thanks to propositions 2.1 and 2.2, the transfer matrix

$$T(\lambda) = tr_0\left(M_{0Q}(\lambda)\right) \tag{11}$$

defines a one parameter family of commuting operators. However, when constructing the local Hamiltonians from this transfer matrix, the chain is closed. (cf figure 1.)

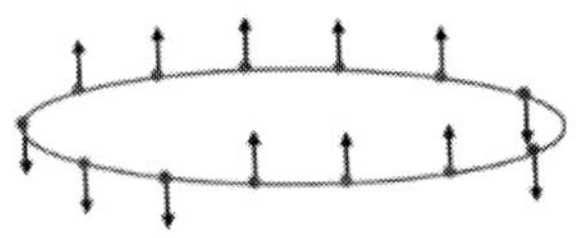

Figure 1. A closed spin chain. The last site N and the first one are neighbours.

Considering a scalar solution P of the Yang-Baxter algebra (2), one can describe twisted integrable boundary conditions, by using the commuting family :

$$T(\lambda) = tr_0\left(P_0.M_{0Q}(\lambda)\right) \tag{12}$$

Different matrices P lead to the anti-periodic chain or more generally to twisted chains.

3.2. More general integrable boundary conditions

In order to describe more general integrable boundary conditions [3, 4], we follow the general procedure introduced by Sklyanin, associating to any solution $M_{0Q}(\lambda) \in End(\mathbb{C}^2 \otimes \mathcal{H})$ of the 6-vertex Yang-Baxter equation, a solution $\mathcal{U}_{-0Q}(\lambda) \in End(\mathbb{C}^2 \otimes \mathcal{H})$ of the 6-vertex reflection equation :

$$\mathcal{R}_{12}(\frac{\lambda}{\mu})\mathcal{U}_{-\ 1Q}(\lambda)\mathcal{R}_{12}(\frac{\lambda\mu}{q})\mathcal{U}_{-\ 2Q}(\mu) = \mathcal{U}_{-\ 2Q}(\mu)\mathcal{R}_{12}(\frac{\lambda\mu}{q})\mathcal{U}_{-\ 1Q}(\lambda)\mathcal{R}_{12}(\frac{\lambda}{\mu}) \tag{13}$$

The solution, called open monodromy matrix, takes the form :

$$\mathcal{U}_{-\ 0Q}(\lambda) = M_{0Q}(\lambda) K_{-\ 0}(\lambda) M_{0Q}^{-1}(\frac{1}{\lambda}) \tag{14}$$

where $K_-(\lambda)$ is a scalar solution of the reflection algebra.
Moreover, we can consider the following dual 6-vertex reflection equation :

$$\mathcal{R}_{12}(\frac{\mu}{\lambda}) \mathcal{U}_{+\ 1Q}^{t_1}(\lambda) \mathcal{R}_{12}(\frac{1}{\lambda\mu q^2}) \mathcal{U}_{+\ 2Q}^{t_2}(\mu) = \mathcal{U}_{+\ 2Q}^{t_2}(\mu) \mathcal{R}_{12}(\frac{1}{\lambda\mu q^2}) \mathcal{U}_{+\ 1Q}^{t_1}(\lambda) \mathcal{R}_{12}(\frac{\mu}{\lambda}) \tag{15}$$

Proposition 3.1 (Boundary transfer matrix)**.** *If $K_{+\ 0}(\lambda)$ is a scalar solution of the dual 6-vertex reflection equation, a one parameter family of commuting transfer matrices is given by*

$$T(\lambda) = tr_0\,(K_{+\ 0}(\lambda) \mathcal{U}_{-\ 0Q}(\lambda)) \tag{16}$$

$T(\lambda)$ is then called the boundary transfer matrix.

Thanks to this boundary transfer matrix, one can describe integrable quantum Hamiltonians with general integrable boundary conditions, linked with the $K_\pm$ matrices. For example, in the case of a XXZ spin chain, diagonal boundary matrices describe z-oriented boundary magnetic fields.

Figure 2. An open chain : the boundary sites are exposed to external actions. The study of such systems is more complicated than in closed systems (the algebra is quadratic in the R matrix), but is worth as it can lead to the description of non equilibrium physics.

Here, we are interested in the most general scalar (boundary matrices) solutions of the trigonometric 6-vertex reflection equation and its dual [5, 6]

$$K_{-\ 0}(\lambda) = \begin{pmatrix} a_-(\lambda) & b_-(\lambda) \\ c_-(\lambda) & d_-(\lambda) \end{pmatrix}_{[0]} = \frac{1}{\xi_- - 1/\xi_-} \begin{pmatrix} \frac{\lambda\xi_-}{q^{1/2}} - \frac{q^{1/2}}{\lambda\xi_-} & \kappa_- e^{\tau_-}\left(\frac{\lambda^2}{q} - \frac{q}{\lambda^2}\right) \\ \kappa_- e^{-\tau_-}\left(\frac{\lambda^2}{q} - \frac{q}{\lambda^2}\right) & \frac{q^{1/2}\xi_-}{\lambda} - \frac{\lambda}{q^{1/2}\xi_-} \end{pmatrix}_{[0]} \tag{17}$$

$$K_{+\ 0}(\lambda) = \begin{pmatrix} a_+(\lambda) & b_+(\lambda) \\ c_+(\lambda) & d_+(\lambda) \end{pmatrix}_{[0]} = \frac{1}{\xi_+ - 1/\xi_+} \begin{pmatrix} \lambda q^{1/2}\xi_+ - \frac{1}{\lambda q^{1/2}\xi_+} & \kappa_+ e^{\tau_+}\left(q\lambda^2 - \frac{1}{q\lambda^2}\right) \\ \kappa_+ e^{-\tau_+}\left(q\lambda^2 - \frac{1}{q\lambda^2}\right) & \frac{\xi_+}{\lambda q^{1/2}} - \frac{\lambda q^{1/2}}{\xi_+} \end{pmatrix}_{[0]} \tag{18}$$

Let us use the following stansars notation for the monodromy matrices :

$$M_{0Q}(\lambda) = \begin{pmatrix} A(\lambda) & B(\lambda) \\ C(\lambda) & D(\lambda) \end{pmatrix}_{[0]} \quad \text{and} \quad \mathcal{U}_{-\ 0Q}(\lambda) = \begin{pmatrix} \mathcal{A}(\lambda) & \mathcal{B}(\lambda) \\ \mathcal{C}(\lambda) & \mathcal{D}(\lambda) \end{pmatrix}_{[0]} \tag{19}$$

Using Sklyanin procedure, equation (14) and the quantum determinant, the generators of the open monodromy matrix are quadratic in the generators of the closed monodromy matrix. For example,

$$\mathcal{B}(\lambda) = -a_-(\lambda)A(\lambda)B(\frac{1}{\lambda}) + b_-(\lambda)A(\lambda)A(\frac{1}{\lambda}) - c_-(\lambda)B(\lambda)B(\frac{1}{\lambda}) + d_-(\lambda)B(\lambda)A(\frac{1}{\lambda})$$

Then, the boundary transfer matrix has the expression

$$T(\lambda) = a_+(\lambda)\mathcal{A}(\lambda) + d_+(\lambda)\mathcal{D}(\lambda) + b_+(\lambda)\mathcal{C}(\lambda) + c_+(\lambda)\mathcal{B}(\lambda) \tag{20}$$

The characterization of the spectrum (eigenvalues and eigenstates) of this class of transfer matrices is the main subject of our works [1, 2] and of this proceeding. In order to give this characterization, *ie* to diagonalize this $T(\lambda)$ operator, let us introduce the main tool : the quantum separation of variables.

4. The quantum separation of variables

This method, introduced by Sklyanin, is an alternative to the Bethe ansatz.

Definition 4.1 (The separate basis). *Let $\mathcal{O}(\lambda)$ be a co-diagonalizable one parameter family of operators on $\mathcal{H}$. If*

- *the spectrum of the $\mathcal{O}(\lambda)$ operators is non degenerated*
- *and the spectral problem for the transfer matrix $T(\lambda)$, in the eigenbasis of $\mathcal{O}(\lambda)$, separates into several one variable independant problems*

then the eigenbasis of $\mathcal{O}(\lambda)$ constitutes a separate basis.

In the case of the Yang-Baxter algebra, Sklyanin's idea is to use the operator $B(\lambda)$ of the monodromy matrix to construct a separate basis [10]. This idea is in fact generalizable to the reflection algebra case for a certain class of models [7, 11]. For example, $\mathcal{B}(\lambda)$ can be used for models with a lower triangular boundary matrix K_+.

Let indeed $\langle \xi_1, ..., \xi_n, ..., \xi_M|$ be the eigenbasis of the $\mathcal{B}(\lambda)$, parametrized such that :

$$\langle \xi_1, ..., \xi_n, ..., \xi_M|\,\mathcal{B}(\lambda) = b_0 \prod_{k=1}^{M} \left(\frac{\lambda}{\xi_k} - \frac{\xi_k}{\lambda}\right)\left(\lambda\xi_k - \frac{1}{\lambda\xi_k}\right) \langle \xi_1, ..., \xi_n, ..., \xi_M| \tag{21}$$

For a lower triangular K_+, the boundary transfer matrix reads

$$T(\lambda) = a_+(\lambda)\mathcal{A}(\lambda) + d_+(\lambda)\mathcal{D}(\lambda) + c_+(\lambda)\mathcal{B}(\lambda) \tag{22}$$

and using now (21) and the reflection algebra commutation relations, one gets

$$\langle \xi_1, ..., \xi_n, ..., \xi_M|\,T(\xi_n) = a(\xi_n)\,\langle \xi_1, ..., q\xi_n, ..., \xi_M| + d(\xi_n)\,\langle \xi_1, ..., \xi_n/q, ..., \xi_M| \tag{23}$$

where $a(\lambda)$ and $d(\lambda)$ are known functions. The essential point to derive (23) is that in the zeros of $\mathcal{B}(\lambda)$, the actions of $\mathcal{A}(\lambda)$ and $\mathcal{D}(\lambda)$ are just shifts on the spectral parameter.

Now, let $|t\rangle$ be an eigenvector of the boundary transfer matrix with eigenvalue $t(\lambda)$

$$T(\lambda)\,|t\rangle = t(\lambda)\,|t\rangle \tag{24}$$

and let us denote by Ψ_t the wave functions associated to $t(\lambda)$

$$\Psi_t(\xi_1, ..., \xi_M) = \langle \xi_1, ..., \xi_M|t\rangle \tag{25}$$

Projecting the equation (23) on an eigenvector $|t\rangle$, one gets the Baxter equation for the wave functions :

$$t(\xi_n)\Psi_t(\xi_1, ..., \xi_M) = a(\xi_n)\Psi_t(\xi_1, ..., q\xi_n, ..., \xi_M) + d(\xi_n)\Psi_t(\xi_1, ..., \xi_n/q, ..., \xi_M) \tag{26}$$

The Baxter equation clearly shows that the spectral problem for the transfer matrix $T(\lambda)$, in the eigenbasis of $\mathcal{B}(\lambda)$, separates in 1 variable independant problems. The variables $(\xi_n)_{1\leq n\leq M}$ are called separate variables.

Moreover, let $Q_t(\xi_n)$ be the solution of the one variable problem (26). We have the following representation in separate variables

$$\Psi_t(\xi_1, ..., \xi_M) = \prod_{k=1}^{M} Q_t(\xi_k) \tag{27}$$

In the next section, I explain how to implement this method to solve the spectral problem of the 6-vertex cyclic representations of the reflection algebra. For brievity, in this proceeding we only state the main propositions of [1] and [2].

5. Characterization of the transfer matrix spectrum

Here we present the complete characterization of the transfer matrix spectrum using the separation of variables. We first exhibit the separate basis, and then give two different characterizations

- in terms of the solution to a system of polynomial equations
- in terms of the solution to a functional Baxter type equation

5.1. The quantum separation of variables for the considered systems

5.1.1. For a lower triangular K_+ Following the idea exposed in section 4, we first focus on a triangular K_+ boundary matrix [1]. The proposition 5.1 allows us to indeed use the quantum separation of variables technique :

Proposition 5.1 (Existence of separate basis). *For general bulk parameters, the $\mathcal{B}(\lambda)$ operators are co-diagonalizable and with simple spectrum.*

Moreover, the following proposition gives the form of the SoV basis :

Proposition 5.2 ((Left) separate basis). *For almost every parameters, the states*

$$\left\langle \overrightarrow{h} \right| = \langle h_1, ..., h_N| = \mathcal{N} \langle \Omega| \prod_{n=1}^{N} \prod_{k_n=1}^{h_n} \mathcal{A}(\xi_n^{(k_n)}) \quad ; \quad h_i \in [[0, p-1]] \tag{28}$$

define a $\mathcal{B}(\lambda)$-eigenbasis of $\mathcal{H}^*$: $\left\langle \overrightarrow{h} \right| \mathcal{B}(\lambda) = \mathrm{B}_{\overrightarrow{h}}(\lambda) \left\langle \overrightarrow{h} \right|$.

In this general case, the reference state $\langle \Omega|$ is given by a recursion formula. The factor $\mathcal{N}$ is a normalization.

In order to make completely explicit the SoV basis, we impose here the following constraint on the parameters of the representation at any quantum site

$$\forall n \in [[1, N]] \ , \quad a_n^p + b_n^p = 0 \tag{29}$$

With the extra condition (29), we have an explicit expression for the reference state :

$$\langle \Omega| = \bigotimes_{n=1}^{\mathrm{N}} \langle j_n - 1, n| \text{ where } j_n \text{ is given by } b_n = -q^{2j_n - 1} a_n. \tag{30}$$

The vectors $\langle k, n|$ are a basis for the operators $\mathfrak{u}_n$ and $\mathfrak{v}_n$:

$$\mathfrak{v}_n |k, n\rangle = q^k |k, n\rangle \text{ and } \mathfrak{u}_n |k, n\rangle = |k+1, n\rangle \quad \forall (n, k) \in \{1, ..., \mathrm{N}\} \times [[0, p-1]]. \tag{31}$$

The eigenvalues have the following expression :

$$\mathrm{B}_{\overrightarrow{h}}(\lambda) = \frac{\kappa_- e^{\tau_-}}{\xi_- - \frac{1}{\xi_-}} \left(\frac{\lambda^2}{q} - \frac{q}{\lambda^2} \right) \mathrm{A}_{\overrightarrow{h}}(\lambda) \mathrm{A}_{\overrightarrow{h}}(\frac{1}{\lambda}) \text{ with } \mathrm{A}_{\overrightarrow{h}}(\lambda) = (-1)^N \prod_{n=1}^{N} (\alpha_n \beta_n)^{1/2} \left(\frac{\lambda}{\xi_n^{(h_n)}} - \frac{\xi_n^{(h_n)}}{\lambda} \right) \tag{32}$$

IOP Conf. Series: Journal of Physics: Conf. Series **965** (2018) 012030 doi:10.1088/1742-6596/965/1/012030

5.1.2. For the most general K_+ Now we consider the question of the existence of a separate basis for the most general case, with K_+ the most general scalar solution of the trigonometric 6-vertex reflection equation. In this proceeding, we just highlight the main steps of [2].

- First, we adapted a gauge transformation originally introduced for spin 1/2 generators by Baxter [13]. This introduces three free gauge parameters, refered later as β, and we thus constructed gauged generators for a gauged Yang-Baxter algebra and then for a gauged reflection algebra.
- The key point with this transformation is that the gauged generators of the gauged reflection algebra, $\mathcal{A}(\lambda|\beta)$, $\mathcal{B}(\lambda|\beta)$, $\mathcal{C}(\lambda|\beta)$ and $\mathcal{D}(\lambda|\beta)$, satisfy similar commutation relations to the ungauged case.
- Then, the gauged boundary matrix $K_+(\lambda|\beta)$ is put lower triangular, specifying an appropriate choice of β. This way, refering to the ungauged case, the operator $\mathcal{B}(\lambda|\beta)$ is a natural candidate to give a separate basis.
- And indeed, we have the following :

  Proposition 5.3. *A separate basis for the general case is the pseudo-eigenbasis of the gauged operator $\mathcal{B}(\lambda|\beta)$:*

$$\langle \beta, h_1, ..., h_{\mathsf{N}}| \quad = \quad \frac{1}{\mathrm{N}_\beta}\langle \Omega_\beta| \prod_{n=1}^{\mathsf{N}} \prod_{k_n=1}^{h_n} \mathcal{A}_-(\xi_n^{(k_n)})|\beta q^2) \tag{33}$$

$$\langle \beta, h_1, ..., h_{\mathsf{N}}|\mathcal{B}_-(\lambda|\beta) \quad = \quad \mathrm{B}_{\boldsymbol{h}}(\lambda|\beta)\langle \beta/q^2, h_1, ..., h_{\mathsf{N}}| \tag{34}$$

  The SoV basis has the same form as in the ungauged case, and as before, $\langle \Omega_\beta|$ is in general known by recursion, but can be explicited under some constraints.

Using this separate basis, we can give the complete characterization of the transfer matrix spectrum, without any constraints neither on the bulk parameters, nor on the boundary matrices.

5.2. A first characterization of the spectrum

Proposition 5.4 (Eigenvalues characterization I). *Let $t(\lambda)$ be an eigenvalue of the transfer matrix. Then $t(\lambda)$ is an even polynomial of degree 2N+2 in λ, invariant under $\lambda \to \frac{1}{\lambda}$, given by an interpolation formula with N coefficients $t(\xi_n)$ which have to satisfy the system :*

$$det\,(D_t(\xi_n)) = 0 \qquad \forall n \in [[1, N]] \tag{35}$$

$D_t(\lambda)$ is the $p \times p$ matrix, quasi tri-diagonal, which encodes the Baxter equation (26). *The equation* (35) *can be understood as the compatibility condition, ie the existence of a non trivial solution to the Baxter equation.*

Proposition 5.5 (Eigenvectors characterization I). *In the previously introduced separate basis, the eigenvectors $|t\rangle$ have the following form :*

$$\langle \beta, h_1, ..., h_{\mathsf{N}}\,|t\rangle = \prod_{n=1}^{N} q_{t,n,h_n} \tag{36}$$

The coefficients q_{t,n,h_i} are the unique solution, up to normalisation, of the system :

$$D_t(\xi_n)\begin{pmatrix} q_{t,n,0} \\ \vdots \\ q_{t,n,p-1} \end{pmatrix} = \begin{pmatrix} 0 \\ \vdots \\ 0 \end{pmatrix} \tag{37}$$

5.3. *A second characterization of the spectrum*

The previous discrete characterization of the spectrum can be reformulated in terms of a Baxter type TQ-functional equation.

Proposition 5.6 (Eigenvalues characterization II : inhomogeneous Baxter equation). *Let $t(\lambda)$ be an entire function. Then $t(\lambda)$ is an eigenvalue if and only if there exists an even polynomial $Q(\lambda)$, of degree $N_Q = 2(p-1)N$ in λ, invariant under $\lambda \to \frac{1}{\lambda}$, which satisfies the functional equation :*

$$t(\lambda)Q(\lambda) = \bar{a}(\lambda)Q\left(\frac{\lambda}{q}\right) + \bar{a}(\frac{1}{\lambda})Q(q\lambda) + G(\lambda) \tag{38}$$

The functions $\bar{a}(\lambda)$ and $G(\lambda)$ are known functions of the parameters. The function $G(\lambda)$ is here to ensure the same asymptotic behaviour for the both sides of the equation. It determines the inhomogeneous term in the Baxter equation.

Remark 1. *The polynomial Q interpolates the coefficients q_{t,n,h_n} previously introduced in* (37) *: $Q(\xi_n^{(h_n)}) = q_{t,n,h_n}$.*

This type of formulation for the spectrum allows us to give an algebraic Bethe ansatz like formulation for the eigenstates, as stated in the next proposition.

Proposition 5.7 (Eigenvectors characterization II). *We can construct a vector $\langle\omega|$, whose components are 1 in the separate basis, such that :*

$$\langle t| = \langle\omega| \prod_{b=1}^{N_Q} \mathcal{B}(\lambda_b) \tag{39}$$

where the λ_b are linked to the roots of $Q(\lambda)$.

This last characterization makes a link between the Bethe ansatz methods and the quantum separation of variables method.

Moreover, we give here a characterization of the eigenvalues thanks to an inhomogeneous Baxter functional equation. There is quite a sparse literature on this topic, compared to the rich one about the homogeneous Baxter equations, and it is noticeable that with only 1 constraint on the parameters the equation becomes homogeneous.

6. Conclusions

In this proceeding I have reported about the works [1, 2] on the transfer matrix spectrum of the class of cyclic representations of the 6-vertex reflection algebra in the case of completely general boundary matrices and for general bulk parameters. Our result gives the complete characterization of the spectrum (eigenvalues and eigenstates) of this class of models both in terms of a discrete system of Baxter's like second order difference equations and through a single inhomogeneous TQ-functional equation within a class of polynomial Q-functions.

Acknowledgments

The author, supported by ENS de Lyon and ENS de Cachan, would like to thank Jean-Michel Maillet and Giuliano Niccoli for their supervision and for fruitful discussions.

References

[1] Maillet J M, Niccoli G and Pezelier B 2017 *SciPost Phys.* **2** 009
[2] Maillet J M, Niccoli G and Pezelier B *to appear*
[3] Sklyanin E K 1988 *J. Phys.* A **21** 2375.
[4] Cherednik I V 1984 *Theor. Math. Phys.* **61** 977.

[5] de Vega H J and González-Ruiz A 1993 *J. Phys.* A **26** L519.
[6] Ghoshal S and Zamolodchikov A B 1994 *Int. J. Mod. Phys.* A **9** 3841 ; **9** 4353
[7] Frahm H, Seel A and Wirth T 2008 *Nucl. Phys.* B **802** 351.
[8] Sklyanin E K and Faddeev L D 1978 *Sov. Phys. Dokl.* **23** 902.
[9] Sklyanin E K, Takhtajan L A and Faddeev L D 1980 *Theor. Math. Phys.* **40** 688.
[10] Sklyanin E K 1992, *Quantum inverse scattering method. Selected topics.* in : Quantum groups and quantum integrable systems, World Scientific **63**.
[11] Niccoli G 2012, *J. Stat. Mech.* P10025
[12] Bazhanov V V and Stroganov Y G 1990 Chiral Potts model as a descendant of the six vertex model *J. Stat. Phys.* **59** 799.
[13] Baxter R J 1972 *Ann. Phys.* **70** 193-228 ;

Solvable spectral problems from 2d CFT and $\mathcal{N} = 2$ gauge theories

M R Piątek[1,3] and A R Pietrykowski[2,3]

[1] Institute of Physics, University of Szczecin, Wielkopolska 15, 70–451 Szczecin, Poland
[2] Institute of Theoretical Physics, University of Wrocław, M. Borna 9, 50–204 Wrocław, Poland
[3] Bogoliubov Laboratory of Theoretical Physics, JINR, 141980 Dubna, Russia

E-mail: piatek@fermi.fiz.univ.szczecin.pl, radekpiatek@gmail.com, pietrie@theor.jinr.ru

Abstract. The so-called 2d/4d correspondences connect two-dimensional conformal field theory (2d CFT), $\mathcal{N} = 2$ supersymmetric gauge theories and quantum integrable systems. The latter in the simplest case of the SU(2) gauge group are nothing but the quantum-mechanical systems. In the present article we summarize our recent results and list open problems concerning an application of the aforementioned dualities in the studies of spectral problems for some Schrödinger operators with Mathieu–type periodic, periodic PT–symmetric and (Heun's) elliptic potentials.

1. Introduction

They have long been known connections between $\mathcal{N} = 2$ supersymmetric Yang–Mills theories and integrable models. Let us consider perhaps the simplest example of such a relationship, i.e., a link between the SU(2) pure gauge Seiberg–Witten theory and the one-dimensional sine-Gordon model (cf. [1]). Concretely, the statement here is that the Bohr–Sommerfeld periods

$$\Pi(\Gamma) \equiv \oint_{\Gamma} P_0(\varphi)\,\mathrm{d}\varphi = \oint_{\Gamma} \sqrt{2(u - \hat{\Lambda}^2\cos\varphi)}\,\mathrm{d}\varphi$$

for the classical sine-Gordon model defined by $\mathcal{L}_{\rm sG} = \frac{1}{2}\dot{\varphi}^2 - \hat{\Lambda}^2\cos\varphi$, and for two complementary contours $\Gamma = A, B$ encircling two turning points $\pm\arccos(u/\hat{\Lambda}^2)$ define the Seiberg–Witten system [2]: $a = \Pi(A)$, $\partial\mathcal{F}(a)/\partial a = \Pi(B)$. Here, a is a modulus and $\mathcal{F}(a)$ denotes the Seiberg–Witten prepotential determining the low energy effective dynamics of the four-dimensional $\mathcal{N} = 2$ supersymmetric SU(2) pure gauge theory.

As has been observed in [3] (see also [4, 5]) the above statement has its 'quantum analogue' or 'quantum extension' which can be formulated as follows. Namely, the monodromies (*exact* BS periods)

$$\tilde{\Pi}(\Gamma) \equiv \oint_{\Gamma} P(\varphi,\hbar)\,\mathrm{d}\varphi, \qquad P(\varphi,\hbar) = P_0(\varphi) + P_1(\varphi)\hbar + P_2(\varphi)\hbar^2 + \ldots$$

of the exact WKB solution

$$\psi(\varphi) = \exp\left\{\frac{i}{\hbar}\int^{\varphi} P(\rho,\hbar)\,\mathrm{d}\rho\right\} \quad \text{to the eq.} \quad \boxed{\left[-\frac{\hbar^2}{2}\frac{\partial^2}{\partial\varphi^2} + \hat{\Lambda}^2\cos\varphi\right]\psi(\varphi) \;=\; \mathrm{E}\,\psi(\varphi)} \qquad (1)$$

define the Nekrasov–Shatashvili system [6]: $a = \widetilde{\Pi}(A)$, $\partial\mathcal{W}(\hat{\Lambda}, a, \hbar)/\partial a = \widetilde{\Pi}(B)$.[1] Here, $\mathcal{W}(\hat{\Lambda}, a, \hbar) = \mathcal{W}_{\rm pert}(\hat{\Lambda}, a, \hbar) + \mathcal{W}_{\rm inst}(\hat{\Lambda}, a, \hbar)$ is the effective twisted superpotential of two-dimensional SU(2) pure gauge (Ω-deformed) SYM theory defined in [6] as the following Nekrasov–Shatashvili (NS) limit

$$\mathcal{W}(\hat{\Lambda}, a, \hbar) \equiv \lim_{\epsilon_2\to 0} \epsilon_2 \log \mathcal{Z}(\hat{\Lambda}, a, \epsilon_1{=}\,\hbar, \epsilon_2)$$

of the Nekrasov partition function $\mathcal{Z}(\hat{\Lambda}, a, \epsilon_1, \epsilon_2) = \mathcal{Z}_{\rm pert}(\hat{\Lambda}, a, \epsilon_1, \epsilon_2)\mathcal{Z}_{\rm inst}(\hat{\Lambda}, a, \epsilon_1, \epsilon_2)$ [8, 9].

Interestingly, on the other hand the Mathieu equation written in (1) is nothing but the Schrödinger equation for the sine-Gordon model. One can show that the energy eigenvalue is determined by the SU(2) pure gauge twisted superpotential ($\hbar = \epsilon_1$) [3, 10]:

$$\begin{aligned} 2{\rm E}_{\rm sG} = 2{\rm E}(a) &= a^2 + \hat{\Lambda}^4\left(\frac{1}{2\,a^2} + \frac{\epsilon_1^2}{8\,a^4} + \frac{\epsilon_1^4}{32\,a^6} + \frac{\epsilon_1^6}{128\,a^8}\right) \\ &+ \hat{\Lambda}^8\left(\frac{5}{32\,a^6} + \frac{21\,\epsilon_1^2}{64\,a^8} + \frac{219\,\epsilon_1^4}{512\,a^{10}} + \frac{121\,\epsilon_1^6}{256\,a^{12}}\right) \\ &+ \hat{\Lambda}^{12}\left(\frac{9}{64\,a^{10}} + \frac{55\,\epsilon_1^2}{64\,a^{12}} + \frac{1495\,\epsilon_1^4}{512\,a^{14}} + \frac{4035\,\epsilon_1^6}{512\,a^{16}}\right) + \mathcal{O}(\hat{\Lambda}^{16}) \\ &= \frac{1}{2}\epsilon_1\hat{\Lambda}\partial_{\hat{\Lambda}}\mathcal{W}^{N_f=0,{\rm SU}(2)}. \end{aligned}$$

The fact that the eigenvalue of the Mathieu operator is given by[2] $\hat{\Lambda}\partial_{\hat{\Lambda}}\mathcal{W}^{N_f=0,{\rm SU}(2)}$ is an example of the Bethe/gauge correspondence discovered by Nekrasov and Shatashvili [6]. The Bethe/gauge correspondence maps supersymmetric vacua of the $\mathcal{N} = 2$ theories to Bethe states of some quantum integrable systems (QIS). A result of that duality is that the twisted superpotentials for SU(N) theories are identified with the Yang–Yang (YY) functions which describe spectra of the corresponding N–particle quantum integrable systems.[3]

As has been already mentioned the twisted superpotentials are defined as the NS limit of the Nekrasov partition functions. The latter play also a key role in another relation — the celebrated AGT correspondence [11]. Indeed, let $C_{g,n}$ denotes the Riemann surface with genus g and n punctures. A significant part of the AGT conjecture is an exact correspondence between the Virasoro conformal blocks on $C_{g,n}$ and the instanton sectors of the Nekrasov partition functions of certain SU(2) quiver gauge theories belonging to some class $S_{g,n}$. It turns out that the NS limit of the Nekrasov functions corresponds to the so-called classical limit of conformal blocks [12]. Hence, combining the classical/NS limit of the AGT duality and the Bethe/gauge correspondence one thus gets the triple correspondence which links the classical Virasoro blocks to the SU(2) twisted superpotentials and then to spectra of some Schrödinger operators (see Fig.1). Indeed, let us note that 2-particle QIS are nothing but the quantum–mechanical systems.

Motivated by the aforementioned correspondences, we have studied in [10, 14, 15, 16] examples of a direct link between 'classical' 2d CFT (i.e., the classical, large central charge limit of the BPZ equations [17] obeyed by certain *degenerate* conformal blocks) and the corresponding quantum–mechanical models (= certain stationary Schrödinger equations), cf. Fig.1. In sections 2 and 3 of

[1] For SU(N) generalization of this result, see [7].

[2] Precisely, the canonical form of the Mathieu equation is $\boxed{\psi''(x) + \left(\lambda - 2h^2\cos 2x\right)\psi(x) = 0}$. Hence, comparing it to eq. (1) one gets

$$2x = \varphi, \quad h^2 = \frac{4\hat{\Lambda}^2}{\hbar^2} = \frac{4\hat{\Lambda}^2}{\epsilon_1^2}, \quad \lambda = \frac{8{\rm E}}{\hbar^2} = \frac{8{\rm E}}{\epsilon_1^2}.$$

[3] The Yang–Yang functions are potentials for Bethe equations.

IOP Conf. Series: Journal of Physics: Conf. Series **965** (2018) 012031 doi:10.1088/1742-6596/965/1/012031

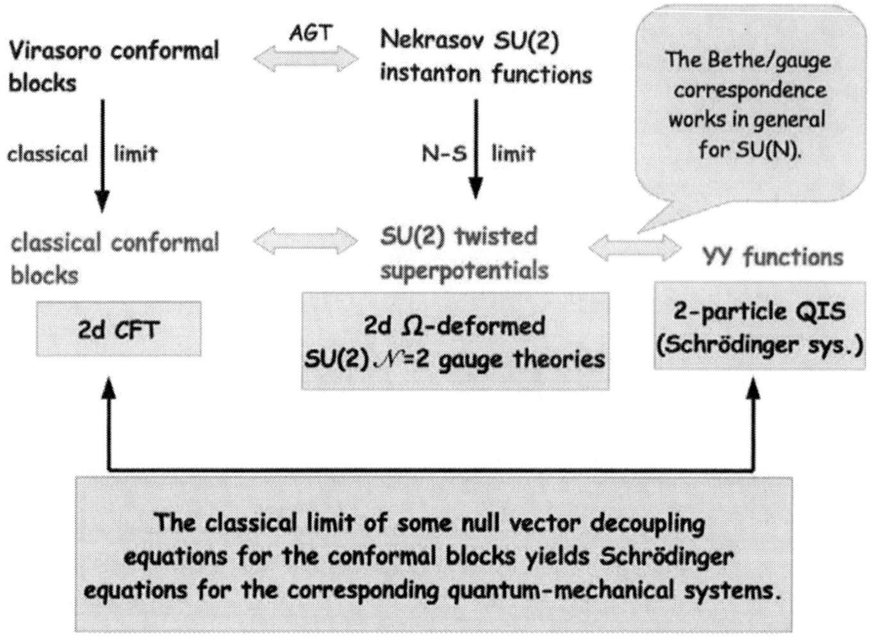

Figure 1. The triple correspondence in the case of the Virasoro classical conformal blocks links the latter to SU(2) instanton twisted superpotentials which describe the spectra of some quantum–mechanical systems. The Bethe/gauge correspondence on the r.h.s. connects the SU(N) $\mathcal{N}=2$ SYM theories with the N–particle quantum integrable systems. An extension of the above triple relation to the case N > 2 needs to consider on the l.h.s. the classical limit of the $W_{\rm N}$ symmetry conformal blocks according to the known extension [13] of the AGT conjecture.

the present article we review main results of these investigations. In conclusions we explain our motivations and present some open problems for further studies.

In the rest of the introduction, in order to spell out our results, we remind a definition and some properties of quantum and classical conformal blocks. To begin with, let us recall that basic ingredients of any CFT model defined on Riemann surface $C_{g,n}$ are the correlation functions of the primary fields [17, 18]. Thanks to conformal symmetry any correlation function of the primary and descendants fields can be derived once the conformal blocks and structure constants are known. The conformal blocks are model independent CFT 'special functions' defined entirely within a representation theoretic framework.

Indeed, let $\mathcal{V}^n_{c,\Delta}$ denotes the vector space generated by all vectors of the form:

$$|\,\Delta^n_I\,\rangle = L_{-I}|\,\Delta\,\rangle \equiv L_{-k_1}\ldots L_{-k_{\ell(I)}}|\,\Delta\,\rangle, \qquad n = k_1 + \ldots + k_{\ell(I)} =: |I|, \tag{2}$$

where $I = (k_1 \geqslant \ldots \geqslant k_{\ell(I)} \geqslant 1)$ is a partition of n,[4] L_n's are the Virasoro generators obeying

$$[L_n, L_m] = (n-m)L_{n+m} + \frac{c}{12}(n^3 - n)\delta_{n+m,0}, \tag{3}$$

and $|\,\Delta\,\rangle$ is the highest weight state with the following property:

$$L_0|\,\Delta\,\rangle = \Delta|\,\Delta\,\rangle, \qquad L_n|\,\Delta\,\rangle = 0, \quad \forall\, n > 0. \tag{4}$$

[4] We will use the notation $I \vdash n$.

The representation of the Virasoro algebra on the space:

$$\mathcal{V}_{c,\Delta} = \bigoplus_{n=0}^{\infty} \mathcal{V}^{n}_{c,\Delta}, \qquad \mathcal{V}^{0}_{c,\Delta} = \mathbb{R}|\,\Delta\,\rangle$$

defined by the relations (3), (4) is called the Verma module with the central charge c and the highest weight Δ. It is clear that $\dim \mathcal{V}^{(n)}_{c,\Delta} = \mathsf{p}(n)$, where $\mathsf{p}(n)$ is the number of partitions of n (with the convention $\mathsf{p}(0) = 1$). On $\mathcal{V}^{(n)}_{c,\Delta}$ exists symmetric bilinear form $\langle\,\cdot\,|\,\cdot\,\rangle$ uniquely defined by the relations $\langle\,\Delta\,|\,\Delta\,\rangle = 1$ and $(L_n)^{\dagger} = L_{-n}$.

Let $|\,0\,\rangle$ denotes the vacuum state, i.e., the highest weight state in the vacuum module with the highest weight $\Delta = 0$. The conformal blocks on the Riemann sphere are defined as the matrix elements

$$\langle\,0\,|V_{\Delta_n}(z_n)\ldots V_{\Delta_1}(z_1)|\,0\,\rangle_{\text{sphere}}$$

of compositions of the primary chiral vertex operators (CVO's):

$$\begin{aligned} V_{\Delta_j}(z) &\equiv V^{\alpha_j}_{\alpha_k,\alpha_i}(z) : \mathcal{V}_{\Delta_i} \longrightarrow \mathcal{V}_{\Delta_k}, \qquad \Delta_l = \Delta_{\alpha_l} = \alpha_l(Q-\alpha_l), \\ [L_n, V_\Delta(z)] &= z^n\left(z\frac{\mathrm{d}}{\mathrm{d}z} + (n+1)\Delta\right)V_\Delta(z), \qquad n \in \mathbb{Z}. \end{aligned}$$

acting between the Verma modules. The conformal blocks on the torus are traced cylinder matrix elements of 2d Euclidean 'space-time' translation operators and CVO's insertions, i.e., 'chiral partition functions'. The conformal blocks on the higher genus Riemann surfaces can be constructed by making use of the so-called *sewing* or *gluing* procedure.

By inserting projection operators[5] $\mathbb{P}_{\tilde{\Delta}_p}$ on the intermediate conformal weights $\tilde{\Delta}_p$, $p = 1,\ldots,3g-3+n$ into the internal channels of the conformal blocks one gets the latter in terms of the formal power series. In the simplest cases, namely, for the 4-point block on the sphere and the 1-point block on the torus the coefficients of the power series are implicitly defined via recursive relations. Up to now these coefficient are not been computed in closed form in the general case.

Taking into account a number of recent applications one of central issues concerning the conformal blocks is the existence of their classical limit.[6] This is the limit in which all parameters of the conformal blocks tend to infinity in such a way that their ratios are fixed:

$$\Delta_i,\ \tilde{\Delta}_p,\ c \longrightarrow \infty, \qquad \frac{\Delta_i}{c} = \frac{\tilde{\Delta}_p}{c} = \text{const.}\ ,$$

$i = 1,\ldots,n$, $p = 1,\ldots,3g-3+n$. For the standard parametrization of the central charge $c = 1+6Q^2$, where $Q = b+\frac{1}{b}$ and for 'heavy' weights $(\tilde{\Delta}_p, \Delta_i) = \frac{1}{b^2}(\tilde{\delta}_p, \delta_i)$ with $\tilde{\delta}_p, \delta_i = \mathcal{O}(b^0)$ the classical limit corresponds to $b \to 0$. There exist many convincing arguments, but there is no proof, that in the classical limit the conformal blocks exponentiate to the functions $f_{\tilde{\delta}_p}(\delta_i, Z)$ known as the classical conformal blocks [12, 19], e.g.:

$$\langle\,0\,|V_{\Delta_n}(z_n)\mathbb{P}_{\tilde{\Delta}_{n-1}}\cdots\mathbb{P}_{\tilde{\Delta}_2}V_{\Delta_1}(z_1)|\,0\,\rangle_{\text{sphere}} \overset{b\to 0}{\sim} \mathrm{e}^{\frac{1}{b^2}f_{\tilde{\delta}_p}(\delta_i,Z)}, \qquad Z := (z_n,\ldots,z_1).$$

The conformal blocks may also include the 'light' conformal weights Δ_{light} which are defined by the property $\lim_{b\to 0} b^2\Delta_{\mathsf{light}} = 0$. It is known, but not proven in general, that light insertions

[5] $\mathbb{P}_{\tilde{\Delta}_p}$ are identity operators in $\mathcal{V}_{c,\tilde{\Delta}_p}$ built out of the basis vectors (2) and their duals.
[6] For a list of some applications, see introduction in [16].

have no influence to the classical limit, i.e., do not contribute to the classical blocks:

$$\left\langle V_{\Delta_{\rm light}}(w)\prod_{i=1}^{n}V_{\Delta_{\rm heavy}}(z_i)\right\rangle_{\rm sphere} \stackrel{b\to 0}{\sim} \Psi(w)\, {\rm e}^{\frac{1}{b^2}f_{\tilde{\delta}_p}(\delta_i,Z)}. \tag{5}$$

Due to the discovery of the AGT correspondence a considerable progress in the theory of conformal blocks has been recently achieved. In particular, Gaiotto analyzing an extension of the AGT conjecture to the class of the so-called 'non-conformal' $\mathcal{N}=2$, SU(2) super Yang–Mills theories has postulated the existence of the *irregular* conformal blocks [20]. These new types of the conformal blocks were introduced in Gaiotto's work [20] as products of some new states belonging to the Hilbert space of 2d CFT. The novel irregular Gaiotto states are kind of coherent vectors for some Virasoro generators. It is also known that irregular blocks can be obtained from standard (regular) conformal blocks in properly defined *decoupling limits* of the external conformal weights, cf. [21, 22]. Furthermore, the Gaiotto vectors can be understood as a result of suitable defined *collision limit* of locations of vertex operators in their operator product expansion, cf. [23]. Interestingly, also the classical limit makes sense in the case of the irregular blocks. This claim first time has clearly appeared in [10] as a result of the non-conformal AGT relations and observations made in [6].

2. Classical limit of irregular blocks and Hill's–type equations

2.1. Hermitian spectral problems

As mentioned above some 1–dimensional stationary Schrödinger equations can be obtained entirely within the framework of 2d CFT as the classical limit of the null vector decoupling (NVD) equations obeyed by certain *degenerate* conformal blocks.[7] This procedure yields equations some of which are well known in mathematics and physics. For instance, the celebrated Mathieu and Whittaker–Hill equations, and some of their solutions have realizations in 2d CFT.

Let $|\Delta,\Lambda^2\rangle$ denotes a 'pure gauge' or rank $\frac{1}{2}$ irregular vector defined by the eqs.:

$$L_0|\Delta,\Lambda^2\rangle = \left(\Delta+\frac{\Lambda}{2}\partial_\Lambda\right)|\Delta,\Lambda^2\rangle, \qquad L_1|\Delta,\Lambda^2\rangle = \Lambda^2|\Delta,\Lambda^2\rangle,$$

$$L_n|\Delta,\Lambda^2\rangle = 0, \qquad n\geqslant 2.$$

One can find that the representation of $|\Delta,\Lambda^2\rangle$ in terms of the basis vectors (2) in $\mathcal{V}_{c,\Delta}$ reads as follows:

$$|\Delta,\Lambda^2\rangle = \sum_{n\geqslant 0}\Lambda^{2n}\sum_{I\vdash n}\left(G_{c,\Delta}\right)^{(1^n)I}L_{-I}|\Delta\rangle,$$

where $\left(G_{c,\Delta}\right)^{IJ}$ is the inverse of the *Gram matrix* (or Shapovalov matrix)

$$\left(G_{c,\Delta}\right)_{IJ} = \langle\Delta|L_I L_{-J}|\Delta\rangle.$$

The product $\langle\,\Delta,\Lambda^2\,|\,\Delta,\Lambda^2\,\rangle$ of pure gauge irregular vectors yields the $N_f=0$ irregular block:

$$\langle\,\Delta,\Lambda^2\,|\,\Delta,\Lambda^2\,\rangle = \sum_{n\geqslant 0}\Lambda^{4n}\left(G_{c,\Delta}\right)^{(1^n)(1^n)}, \quad \Lambda=\frac{\hat{\Lambda}}{\epsilon_1\epsilon_2}. \tag{6}$$

[7] It should be emphasized that this claim is consistent with the correspondences described above, cf. Fig.1. Moreover, let us notice that due to the identification between the classical blocks, the twisted superpotentials and Yang-Yang functions, eigenvalues of mentioned Schrödinger operators can be found by solving appropriate Bethe-like equations. Interestingly, very similar in a spirit to what we observe is the so-called ODE/IM correspondence [24, 25, 26]. It should be noted here also recently observed link between quantum mechanics and topological string theory [27].

The $N_f = 0$ irregular block in the case of the heavy conformal weight $\Delta \sim b^{-2}\delta$ and for $\epsilon_2/\epsilon_1 = b^2$ exponentiates in the classical limit $b \to 0$ to the classical $N_f = 0$ irregular block:

$$f_\delta\left(\hat{\Lambda}/\epsilon_1\right) = \lim_{b\to 0} b^2 \log\langle\, \Delta, \Lambda^2 \,|\, \Delta, \Lambda^2 \,\rangle = \sum_{n\geqslant 1}\left(\hat{\Lambda}/\epsilon_1\right)^{4n} f_\delta^{(n)}.$$

The coefficients $f_\delta^{(n)}$ above can be computed order by order from the semi-classical asymptotic and the expansion (6), e.g.:

$$f_\delta^{(1)} = \frac{1}{2\delta}, \qquad f_\delta^{(2)} = \frac{5\delta-3}{16\delta^3(4\delta+3)}, \qquad f_\delta^{(3)} = \frac{9\delta^2-19\delta+6}{48\delta^5(4\delta+3)(\delta+2)},$$

$$f_\delta^{(4)} = \frac{5876\delta^5-16489\delta^4-22272\delta^3+17955\delta^2+9045\delta-4050}{512\delta^7(\delta+2)(4\delta+3)^3(4\delta+15)}, \quad \ldots .$$

Let

(i) ν denotes the Floquet characteristic exponent defined by the property $\psi(x+\pi) = \exp(i\pi\nu)\psi(x)$ of the Floquet solution to the Mathieu equation:

$$\psi''(x) + \left[\lambda - 2h^2\cos 2x\right]\psi(x) = 0\;; \tag{7}$$

(ii) $\tilde{\Delta}$ and Δ' denote the heavy conformal weights related by the fusion rule:

$$\text{(I)}\quad \tilde{\Delta} = \Delta\left(\sigma - \frac{b}{4}\right), \qquad \Delta' = \Delta\left(\sigma + \frac{b}{4}\right),$$

where $\Delta(\sigma) = \frac{Q^2}{4} - \sigma^2$;

(iii) $V_+(z)$ denotes the degenerate primary chiral vertex operator with the light degenerate conformal weight:

$$\Delta_+ := \Delta_{21} = -\frac{3}{4}b^2 - \frac{1}{2}.$$

The non-integer order ($\nu \notin \mathbb{Z}$) Floquet solution of the Mathieu equation (7) is [10, 14]:

$$\begin{aligned}\psi_{\rm I}(x) &= {\rm e}^{i\nu x}\,\Psi_{\rm I}(h/2, {\rm e}^{2ix}) = {\rm me}_\nu(x,h)\\ &= {\rm e}^{i\nu x} + \frac{h^2}{4}\left(\frac{{\rm e}^{(\nu-2)ix}}{\nu-1} - \frac{{\rm e}^{(\nu+2)ix}}{\nu+1}\right) + \frac{h^4}{32}\left(\frac{{\rm e}^{(\nu+4)ix}}{(\nu+1)(\nu+2)} + \frac{{\rm e}^{(\nu-4)ix}}{(\nu-2)(\nu-1)}\right) + \ldots,\end{aligned}$$

where

$$\Psi_{\rm I}(\hat{\Lambda}/\epsilon_1, z) = \lim_{b\to 0}\left(1 + \frac{\Phi_{\rm I}^{(m\neq n)}(\Lambda, z)}{\Phi_{\rm I}^{(m=n)}(\Lambda)}\right), \quad z = {\rm e}^{2ix}, \quad \frac{\epsilon_2}{\epsilon_1} = b^2, \quad \Lambda = \frac{\hat{\Lambda}}{\epsilon_1\epsilon_2}, \quad h = \pm\frac{2\hat{\Lambda}}{\epsilon_1}$$

and

$$\Phi_{\rm I}^{(m=n)}(\Lambda) = \sum_{n\geqslant 0}\Lambda^{4n}\sum_{|I|=n}\left(G^n_{c,\Delta'}\right)^{(1^n)I}\langle\,\Delta'|L_I V_+(1) L_{-I}|\,\tilde{\Delta}\rangle\left(G^n_{c,\tilde{\Delta}}\right)^{I\,(1^n)},$$

$$\Phi_{\rm I}^{(m\neq n)}(\Lambda, z) = \sum_{\substack{m\neq n\\ m,n\geqslant 0}}\Lambda^{2(m+n)}z^{m-n}\sum_{\substack{|I|=m\\ |J|=n}}\left(G^m_{c,\Delta'}\right)^{(1^m)\,I}\langle\,\Delta'|L_I V_+(1) L_{-J}|\,\tilde{\Delta}\rangle\left(G^n_{c,\tilde{\Delta}}\right)^{J\,(1^n)}.$$

IOP Conf. Series: Journal of Physics: Conf. Series **965** (2018) 012031 doi:10.1088/1742-6596/965/1/012031

The index I labels first independent solution and means that the fusion rule (I) is assumed. The corresponding Mathieu eigenvalue λ is determined by the $N_f = 0$ classical irregular block:

$$\begin{aligned}\lambda =& -\hat{\Lambda}\,\partial_{\hat{\Lambda}}\, f_\delta\left(\hat{\Lambda}/\epsilon_1\right) + 4\xi^2\\ =& -\frac{4h^4}{16}\, f^{(1)}_{\frac{1}{4}-\frac{\nu^2}{4}} - \frac{8h^8}{256}\, f^{(2)}_{\frac{1}{4}-\frac{\nu^2}{4}} - \frac{12h^{12}}{4096}\, f^{(3)}_{\frac{1}{4}-\frac{\nu^2}{4}} - \ldots + 4\left(\frac{\nu^2}{4}\right)\\ =& \nu^2 + \frac{h^4}{2\left(\nu^2-1\right)} + \frac{\left(5\nu^2+7\right)h^8}{32\left(\nu^2-4\right)\left(\nu^2-1\right)^3} + \frac{\left(9\nu^4+58\nu^2+29\right)h^{12}}{64\left(\nu^2-9\right)\left(\nu^2-4\right)\left(\nu^2-1\right)^5} + \ldots\,,\end{aligned}$$

where $\delta = \frac{1}{4} - \xi^2$ and $\xi = \nu/2$.

The above result one gets by considering the classical limit of the NVD equation obeyed by the $N_f = 0$ degenerate 3-point irregular block $\langle\,\Delta', \Lambda^2\,|V_+(z)|\tilde{\Delta}, \Lambda^2\,\rangle$. A key point in a derivation is the semi-classical asymptotical behavior of the type (5), which in the case under consideration takes the form:

$$z^{-\kappa}\,\langle\,\Delta', \Lambda^2\,|\,V_+(z)\,|\,\tilde{\Delta}, \Lambda^2\,\rangle \overset{b\to 0}{\sim} \Psi(\hat{\Lambda}/\epsilon_1, z)\,\exp\left\{\frac{1}{b^2} f_\delta(\hat{\Lambda}/\epsilon_1)\right\},$$

$\kappa = \Delta' - \Delta_+ - \tilde{\Delta}$. Let us stress that the matrix element $\langle\,\Delta', \Lambda^2\,|V_+(z)|\tilde{\Delta}, \Lambda^2\,\rangle$ obeys the NVD equation provided that the fusion rules: (I) or

$$\text{(II)}\quad \tilde{\Delta} = \Delta\left(\sigma + \frac{b}{4}\right), \quad \Delta' = \Delta\left(\sigma - \frac{b}{4}\right)$$

are assumed. The second possibility determines the second linearly independent solution $\psi_{\rm II}(x)$, which can be obtained from $\psi_{\rm I}(x)$ by the substitution $\nu \to -\nu$.

Analogous result holds for the Whittaker–Hill equation [15]:

$$\left[-\frac{{\rm d}^2}{{\rm d}x^2} + \frac{1}{2}h^2\cos 4x + 4h\mu\cos 2x\right]\psi^{\mathbf{2}}_\xi = \lambda^{\mathbf{2}}_\xi\,\psi^{\mathbf{2}}_\xi. \tag{8}$$

Here new objects enter the game, i.e.:

- the rank 1 irregular state $|\,\Delta, \Lambda, m\,\rangle$ defined by

$$\begin{aligned} L_0|\,\Delta,\Lambda,m\,\rangle &= \left(\Delta + \Lambda\frac{\partial}{\partial\Lambda}\right)|\,\Delta,\Lambda,m\,\rangle, & L_1|\,\Delta,\Lambda,m\,\rangle &= m\Lambda|\,\Delta,\Lambda,m\,\rangle,\\ L_2|\,\Delta,\Lambda,m\,\rangle &= \Lambda^2|\,\Delta,\Lambda,m\,\rangle, & L_n|\,\Delta,\Lambda,m\,\rangle &= 0\quad \forall\, n\geqslant 3\end{aligned}$$

$$\Longleftrightarrow\quad |\,\Delta,\Lambda,m\,\rangle = \sum_{n\geqslant 0}\Lambda^n\sum_{p=0}^{[\frac{n}{2}]} m^{n-2p}\sum_{I\vdash n}\left(G^n_{c,\Delta}\right)^{(1^{n-2p}\,2^p)I} L_{-I}|\,\Delta\,\rangle;$$

- the $N_f = 2$ irregular block,

$$\langle\Delta, \tfrac{1}{2}\Lambda, 2m_1|\Delta, \tfrac{1}{2}\Lambda, 2m_2\rangle = \sum_{n\geqslant 0}\left(\frac{\Lambda}{2}\right)^{2n}\sum_{p,p'=0}^{[\frac{n}{2}]}(2m_1)^{n-2p}\left(G^n_{c,\Delta}\right)^{(1^{n-2p}\,2^p)\,(1^{n-2p}\,2^p)}(2m_2)^{n-2p'}.$$

Indeed, if we assume the fusion rules (I) or (II) then the equation (8) is derived in the classical limit $b \to 0$ from the NVD equation obeyed by the 3-point degenerate irregular block,

$$\langle \Delta', \tfrac{1}{2}\Lambda, 2m_1 | V_+(z) | \tilde{\Delta}, \tfrac{1}{2}\Lambda, 2m_2 \rangle$$

with

$$z = \mathrm{e}^{-2ix}, \qquad \frac{h}{2b} = \frac{\hat{\Lambda}}{\epsilon_1 b} = \Lambda, \qquad m_i = \frac{\hat{m}_i}{\epsilon_1 b}, \qquad \mu = \frac{\hat{m}_1}{\epsilon_1} = \frac{\hat{m}_2}{\epsilon_1}, \qquad \xi = \nu/2.$$

The spectrum $\lambda_\xi^{\mathbf{2}}$ in eq. (8) is given in terms of the $N_f = 2$ classical irregular block $f_\delta^{\mathbf{2}}(\cdot,\cdot,\cdot)$, i.e.:

$$\begin{aligned}\lambda_\xi^{\mathbf{2}} \equiv \lambda_\nu^{\mathbf{2}}(h,\mu) &= \nu^2 - 2h\frac{\partial}{\partial h} f^{\mathbf{2}}_{\frac{1}{4}(1-\nu^2)}\left(\tfrac{1}{2}h,\mu,\mu\right) \\ &= \nu^2 + \frac{2}{\nu^2-1}\mu^2 h^2 + \ldots, \qquad\qquad \nu \notin \mathbb{Z},\end{aligned}$$

where in the equation above

$$\begin{aligned} f_\delta^{\mathbf{2}}\left(\hat{\Lambda}/\epsilon_1, \hat{m}_1/\epsilon_1, \hat{m}_2/\epsilon_1\right) &= \lim_{b\to 0} b^2 \log \langle \Delta, \tfrac{1}{2}\Lambda, 2m_1 | \Delta, \tfrac{1}{2}\Lambda, 2m_2\rangle \\ &= \sum_{n\geqslant 1} \left(\hat{\Lambda}/\epsilon_1\right)^{2n} f_\delta^{\mathbf{2},n}\left(\frac{\hat{m}_1}{\epsilon_1}, \frac{\hat{m}_2}{\epsilon_1}\right), \\ f_\delta^{\mathbf{2},1}\left(\frac{\hat{m}_1}{\epsilon_1}, \frac{\hat{m}_2}{\epsilon_1}\right) &= \frac{1}{2\delta}\frac{\hat{m}_1}{\epsilon_1}\frac{\hat{m}_2}{\epsilon_1}, \\ f_\delta^{\mathbf{2},2}\left(\frac{\hat{m}_1}{\epsilon_1}, \frac{\hat{m}_2}{\epsilon_1}\right) &= \frac{\delta^2\left(\delta - 3\left(\frac{\hat{m}_2}{\epsilon_1}\right)^2\right) + \left(\frac{\hat{m}_1}{\epsilon_1}\right)^2\left((5\delta-3)\left(\frac{\hat{m}_2}{\epsilon_1}\right)^2 - 3\delta^2\right)}{16\delta^3(4\delta+3)}, \; \ldots .\end{aligned}$$

The corresponding non-integer order linearly independent solutions $(\psi_{\mathrm{I}}^{\mathbf{2}}, \psi_{\mathrm{II}}^{\mathbf{2}})$ are determined by the fusion rules (I), (II), and are computable in the same way as in the case of the Mathieu equation.

2.2. Non-Hermitian PT-symmetric spectral problems with real eigenvalues

It turns out that analyzing the classical limit of the NVD equations obeyed by the 3-point degenerate irregular blocks one can get certain solutions to the eigenvalue problems

$$\left[-\frac{\mathrm{d}^2}{\mathrm{d}x^2} + Q(x)\right]\psi = \lambda\psi$$

with *new* complex periodic PT–symmetric[8] potentials, $\overline{Q(x)} = Q(-x)$, which yield *real* spectra λ. The simplest example of such novel class of non-Hermitian and solvable potentials reads as follows:

$$Q(x;h,\mu) = \frac{1}{4}h^2\mathrm{e}^{-4ix} + h\mu\,\mathrm{e}^{-2ix} + h^2\mathrm{e}^{2ix}. \tag{9}$$

The eigenvalue problem with the potential (9) is the classical limit of the NVD equation fulfilled by the $N_f = 1$ degenerate irregular block,

$$\langle \Delta', \frac{1}{2}\Lambda, 2m\,| V_+(z) |\, \tilde{\Delta}, \Lambda^2 \rangle,$$

[8] I.e. invariant under a parity (P) reflection and a time (T) reversal.

IOP Conf. Series: Journal of Physics: Conf. Series **965** (2018) 012031 doi:10.1088/1742-6596/965/1/012031

where

$$z = \mathrm{e}^{-2ix}, \qquad \Lambda = \frac{\hat{\Lambda}}{\epsilon_1 b}, \qquad m = \frac{\hat{m}}{\epsilon_1 b}, \qquad \frac{h}{2} = \frac{\hat{\Lambda}}{\epsilon_1}, \qquad \frac{\mu}{2} = \frac{\hat{m}_1}{\epsilon_1}.$$

The potential (9) is PT-symmetric, $\overline{Q(x;h,\mu)} = Q(-x;h,\mu)$, for $x \in \mathbb{R}$ and for real couplings $h,\mu \in \mathbb{R}$. In particular, from (9) one gets

- for $\mu = h$ — the PT–symmetric (for $h^2, x \in \mathbb{R}$) periodic potential discussed in [15]:

$$Q(x;h,h) \;=\; \frac{1}{4}h^2\mathrm{e}^{-4ix} + 2h^2\cos 2x = \frac{1}{4}h^2\cos 4x + 2h^2\cos 2x - i\frac{1}{4}h^2\sin 4x, \qquad (10)$$

- for $\mu = 0$ — the PT–symmetric (for $h^2, x \in \mathbb{R}$) periodic potential:

$$Q\,(x;h,0) \;=\; \frac{1}{4}h^2\mathrm{e}^{-4ix} + h^2\mathrm{e}^{2ix} = h^2\left(\frac{1}{4}\cos 4x + \cos 2x\right) + ih^2\left(\sin 2x - \frac{1}{4}\sin 4x\right).$$

In [15] we have computed the non-integer order fundamental solutions $(\psi_{\mathrm{I}}^{\mathbf{1}}, \psi_{\mathrm{II}}^{\mathbf{1}})$ of the Schrödinger equation with the potential (10). In particular, we have found that the associated eigenvalue $\lambda := \lambda_\nu^{\mathbf{1}}(h)$ is expressed in terms of the $N_f = 1$ classical irregular block $f_\delta^{\mathbf{1}}(\cdot,\cdot)$, i.e.:

$$\begin{aligned}\lambda_\nu^{\mathbf{1}}(h) &= \nu^2 - \frac{4h}{3}\frac{\partial}{\partial h} f^{\mathbf{1}}_{\frac{1}{4}(1-\nu^2)}\left(\tfrac{1}{2}h, \tfrac{1}{2}h\right) \\ &= \nu^2 + \frac{2}{3(\nu^2-1)}h^4 + \frac{3}{32(\nu^2-4)(\nu^2-1)}h^6 + \frac{5\nu^2+7}{8(\nu^2-4)(\nu^2-1)^3}h^8 + \ldots, \qquad (11)\end{aligned}$$

where $\nu \notin \mathbb{Z}$ and

$$\begin{aligned} f_\delta^{\mathbf{1}}\left(\hat{\Lambda}/\epsilon_1, \hat{m}/\epsilon_1\right) &= \lim_{b\to 0} b^2 \log\langle\, \Delta, \frac{1}{2}\Lambda, 2m\,|\Delta, \Lambda^2\,\rangle = \sum_{n\geqslant 1}\left(\hat{\Lambda}/\epsilon_1\right)^{3n} f_\delta^{\mathbf{1},n}\left(\frac{\hat{m}}{\epsilon_1}\right), \\ f_\delta^{\mathbf{1},1}\left(\frac{\hat{m}}{\epsilon_1}\right) &= \frac{1}{2\delta}\frac{\hat{m}}{\epsilon_1}, \\ f_\delta^{\mathbf{1},2}\left(\frac{\hat{m}}{\epsilon_1}\right) &= \frac{5\delta-3}{16\delta^3(4\delta+3)}\left(\frac{\hat{m}}{\epsilon_1}\right)^2 - \frac{3}{16\delta(4\delta+3)}, \\ f_\delta^{\mathbf{1},3}\left(\frac{\hat{m}}{\epsilon_1}\right) &= \frac{\delta(9\delta-19)+6}{48\delta^5(\delta+2)(4\delta+3)}\left(\frac{\hat{m}}{\epsilon_1}\right)^3 + \frac{6-7\delta}{48\delta^3(\delta+2)(4\delta+3)}\frac{\hat{m}}{\epsilon_1}, \ \ldots\,. \end{aligned}$$

The first few terms in the expansion (11) suggest that the spectrum $\lambda_\nu^{\mathbf{1}}(h)$ is real for $h^2 \in \mathbb{R}$ and $\nu \in \mathbb{R}\backslash\mathbb{Z}$. Indeed, this observation is true and in elementary way follows from the definition of the $N_f = 1$ irregular block. It should be stressed that an alternative 'reality proof' is available here as well. Really, one may use the 'classical' AGT relation and methods of the dual $\mathcal{N} = 2$ gauge theory employed in the calculation of the twisted superpotentials.

3. Classical limit of regular spherical blocks and the Huen eqution

In [16] we have continued the line of research described in the previous section, this time examining the 2d CFT realization of the Heun equation. Here, one can show that the classical limit of the second order BPZ NVD equation for the simplest two 5-point degenerate spherical blocks:

$$\mathscr{F}_{\pm}(z,x) \;:=\; \langle\alpha_4|V^{\alpha_3}_{\alpha_4,\beta_\pm}(1)V^{-b/2}_{\beta_\pm,\beta}(z)V^{\alpha_2}_{\beta,\alpha_1}(x)|\alpha_1\rangle, \qquad \beta_\pm = \beta \pm \frac{b}{2}, \qquad \Delta_j = \alpha_j\,(Q-\alpha_j)$$

with $V_+(z) := V^{-b/2}_{\beta_\pm,\beta}(z)$ yields:

(i) the normal form of the Heun equation, i.e.,

$$\frac{\mathrm{d}^2\Psi}{\mathrm{d}z^2}+\left[\frac{\delta_1}{z^2}+\frac{\delta_2}{(z-x)^2}+\frac{\delta_3}{(1-z)^2}+\frac{\delta_1+\delta_2+\delta_3-\delta_4}{z(1-z)}+\frac{x(1-x)c_2(x)}{z(z-x)(1-z)}\right]\Psi=0 \quad (12)$$

with the holomorphic accessory parameter $c_2(x)$ determined by the classical 4-point block on the sphere,

$$c_2(x)=\frac{\partial}{\partial x}\,f_\delta\left[\begin{smallmatrix}\delta_3 & \delta_2\\ \delta_4 & \delta_1\end{smallmatrix}\right](x),$$

where $(\Delta,\Delta_i)\sim\frac{1}{b^2}(\delta,\delta_i)$ and

$$\mathcal{F}_{c,\Delta_\beta}\left[\begin{smallmatrix}\Delta_3 & \Delta_2\\ \Delta_4 & \Delta_1\end{smallmatrix}\right](x):=\langle\alpha_4|V^{\alpha_3}_{\alpha_4,\beta}(1)V^{\alpha_2}_{\beta,\alpha_1}(x)|\alpha_1\rangle \overset{b\to 0}{\sim} \exp\left\{\frac{1}{b^2}f_\delta\left[\begin{smallmatrix}\delta_3 & \delta_2\\ \delta_4 & \delta_1\end{smallmatrix}\right](x)\right\};$$

(ii) the pair of the Floquet–type (path-multiplicative) linearly independent solutions:

$$\Psi_\pm(z,x)=\lim_{b\to 0}\frac{\mathscr{F}_\pm(z,x)}{\mathcal{F}_{c,\Delta_\beta}\left[\begin{smallmatrix}\Delta_3 & \Delta_2\\ \Delta_4 & \Delta_1\end{smallmatrix}\right](x)}\,. \quad (13)$$

The first point in the claim above is a well known fact, cf. e.g. [28, 29]. The second point is an original result of [16]. More concretely, in [16] we have derived the formula (13) and explicitly computed the limit $b\to 0$ by looking into depths of the heavy–light factorization of $\mathscr{F}_\pm(z,x)$, i.e.:

$$\mathscr{F}_\pm(z,x)\overset{b\to 0}{\sim}\Psi_\pm(\infty,1,z,x,0)\;\mathrm{e}^{\frac{1}{b^2}f_\delta\left[\begin{smallmatrix}\delta_3 & \delta_2\\ \delta_4 & \delta_1\end{smallmatrix}\right](x)}. \quad (14)$$

In addition we have computed in [16] the limit $x\to 0$ of the Heun's solutions (13) within 2d CFT framework. Work is in progress to answer the question whether this limit solves the trigonometric Pöschl–Teller potential.

4. Discussion

The results quoted in section 2 are consequences of taking the classical limit of the NVD equations obeyed by certain 3-point degenerate irregular blocks defined as matrix elements of $V_+(z)$ between some 'simple' (lower-rank) Gaiotto states. One can extend the above analysis to the cases of (i) the irregular blocks being matrix elements of $V_+(z)$ between the higher-rank irregular vectors, and (ii) the irregular blocks built out of multi-point insertions of $V_+(z)\equiv V_{\Delta_{21}}(z)$ or $V_{\Delta_{12}}(z)$ and generic primary CVO's. It seems that in the first case we should get the Schrödinger equation with (a) the Hermitian potential built out of higher cosine terms if $N_f=\mathsf{even}$; (b) the non-Hermitian PT–symmetric potential being generalization of (9) if $N_f=\mathsf{odd}$.[9]

In section 2 we have reported about the weak coupling non-integer order solutions, i.e., the solutions which make sense for small couplings h, μ and $\nu\notin\mathbb{Z}$. Therefore, two interesting questions emerge at this point: (i) how to derive the solutions in the other regions of the spectra, for instance, for large couplings;[10] (ii) how to extract from the irregular blocks the solutions with integer values of the Floquet parameter ν? We suspect that the answer to the second question hides in the study of the degenerate intermediate conformal weight limit of the solutions we

[9] Let us notice that the complex periodic PT–symmetric Hamiltonians with real–valued spectra have fascinating applications in the context known as 'PT–symmetric complex crystals', cf. [30, 31]. Moreover, the latter idea has amazing experimental realizations in optics, cf. e.g. [31].

[10] Note that analogous question has been studied for example in [32] by making use of the $\mathcal{N}=2$ gauge theory tools.

have obtained. Our idea of how to tackle the first problem is based on the bootstrap technics. We believe, that the relations connecting weak and strong coupling eigenvalue expansions for Mathieu, Whittaker–Hill and periodic PT–symmetric operators are encoded in the classical and decoupling limits of *braiding relation* [33] for the 4-point spherical block. Interestingly, this idea is *not* hopelessly technically difficult to verify in the case when one of the external weights is heavy and degenerate.[11]

Finally, using the results quoted in section 3 we plan to find 2d CFT realizations of the equations closely related to the Heun equation, i.e., the Schrödinger equations with the so-called Treibich–Verdier, Pöschl–Teller and Lamé potentials. The latter have also an interesting link with the KdV equation, according to the inverse scattering method. These investigations should shed some new light on (i) the long-known duality between the torus and sphere correlation functions of Liouville theory, and on (ii) the connection problem for the Heun equation and its application in black hole physics.

References

[1] Gorsky A, Krichever I, Marshakov A, Mironov A and Morozov A 1995 *Phys. Lett.* B **355** 466-474
[2] Seiberg N, Witten E 1994 *Nucl. Phys.* B **426** 19-52
[3] Mironov A, Morozov A 2010 *J. High Energy Phys.* JHEP04(2010)040
[4] He W 2010 *Phys. Rev.* D **81** 105017
[5] Maruyoshi K, Taki M 2010 *Nucl. Phys.* B **841** 388-425
[6] Nekrasov N A, Shatashvili S L 2009 Quantization of Integrable Systems and Four Dimensional Gauge Theories (*Preprint* 0908.4052v1)
[7] Mironov A, Morozov A 2010 *J. Phys.* A **43** 195401
[8] Nekrasov N A, Okounkov A 2003 Seiberg-Witten Theory and Random Partitions (*Preprint* hep-th/0306238v2)
[9] Nekrasov N A 2004 *Adv. Theor. Math. Phys.* **7** 831-864
[10] Piatek M, Pietrykowski A R 2014 *J. High Energy Phys.* JHEP12(2014)032
[11] Alday L F, Gaiotto D and Tachikawa Y 2010 *Lett. Math. Phys.* **91** 167-197
[12] Zamolodchikov A B, Zamolodchikov A B 1996 *Nucl. Phys.* B **477** 577-605.
[13] Wyllard N 2009 *J. High Energy Phys.* JHEP11(2009)002
[14] Piatek M, Pietrykowski A R 2016 *J. High Energy Phys.* JHEP01(2016)115
[15] Piatek M, Pietrykowski A R 2016 *J. High Energy Phys.* JHEP07(2016)131
[16] Piatek M, Pietrykowski A R 2017 Solving Heun's equation using conformal blocks (*Preprint* 1708.06135)
[17] Belavin A, Polyakov A M and Zamolodchikov A 1984 *Nucl. Phys.* B **241** 333-380
[18] Eguchi T, Ooguri H 1987 *Nucl. Phys.* B **282** 308-328
[19] Hadasz L, Jaskolski Z and Piatek M 2005 *Nucl. Phys.* B **724** 529-554
[20] Gaiotto D 2009 Asymptotically free $N = 2$ theories and irregular conformal blocks (*Preprint* 0908.0307)
[21] Marshakov A, Mironov A and Morozov A 2009 *Phys. Lett.* B **682** (2009) 125-129
[22] Alba V, Morozov A 2009 *JETP Lett.* **90** 708-712
[23] Gaiotto D, Teschner J 2012 *J. High Energy Phys.* JHEP12(2012)050
[24] Dorey P, Tateo R 1999 *J. Phys.* A **32** L419-L425
[25] Bazhanov V V, Lukyanov S L and Zamolodchikov A B 2001 *J. Statist. Phys.* **102** 567-576
[26] Dorey P, Dunning C and Tateo R 2007 *J. Phys.* A **40** R205
[27] Codesido S, Marino M 2016 Holomorphic Anomaly and Quantum Mechanics (*Preprint* 1612.07687)
[28] Litvinov A, Lukyanov S, Nekrasov N and Zamolodchikov A 2014 *J. High Energy Phys.* JHEP07(2014)144
[29] Ferrari F, Piatek M 2012 *J. High Energy Phys.* JHEP05(2012)025
[30] Bender C M 2007 *Rept. Prog. Phys.* **70** 947
[31] Longhi S 2011 *J. Phys. A: Math. Theor.* **44** 485302
[32] Basar G, Dunne G V 2015 *J. High Energy Phys.* JHEP02(2015)160
[33] Hadasz L, Jaskolski Z, Piatek M 2005 *Acta Phys. Polon.* B **36** 845-864
[34] Jackson S, McGough L, Verlinde H 2015 *Nucl. Phys.* B **901** 382-429

[11] Let us note that even in the general case of non-rational 2d CFT the classical limit of *braiding matrix* is already known, see [34].

IOP Conf. Series: Journal of Physics: Conf. Series **965** (2018) 012032 doi:10.1088/1742-6596/965/1/012032

Quantum spaces, central extensions of Lie groups and related quantum field theories[1]

Timothé Poulain, Jean-Christophe Wallet
Laboratoire de Physique Théorique, Bât. 210, CNRS and Université Paris-Sud 11, 91405 Orsay Cedex, France

E-mail: timothe.poulain@th.u-psud.fr, jean-christophe.wallet@th.u-psud.fr

Abstract. Quantum spaces with $\mathfrak{su}(2)$ noncommutativity can be modelled by using a family of $SO(3)$-equivariant differential *-representations. The quantization maps are determined from the combination of the Wigner theorem for $SU(2)$ with the polar decomposition of the quantized plane waves. A tracial star-product, equivalent to the Kontsevich product for the Poisson manifold dual to $\mathfrak{su}(2)$ is obtained from a subfamily of differential *-representations. Noncommutative (scalar) field theories free from UV/IR mixing and whose commutative limit coincides with the usual ϕ^4 theory on $\mathbb{R}^3$ are presented. A generalization of the construction to semi-simple possibly non simply connected Lie groups based on their central extensions by suitable abelian Lie groups is discussed.

1. Introduction

In the recent years, deformations of $\mathbb{R}^3$ for which the algebra of coordinates forms a $\mathfrak{su}(2)$ Lie algebra have received some interest. This was in particular related to developments in 3-d gravity, viewing $\mathbb{R}^3$ as the dual algebra of the relativity group (see e.g [1–3] and references therein), and to constructions of analytic formulas for the star-products (for a recent review on star-products and their applications, see e.g [4]) defining these deformations as well as investigations of their structural properties [5–8] (see also [9–11]). Field theories built on these noncommutative (quantum) spaces have been shown to have a perturbative quantum behaviour different from the one of field theories built on Moyal spaces, at least regarding renormalizability as well as UV/IR mixing [12–16]. For earlier works on noncommutative field theories (NCFT) on Moyal spaces, see e.g [17–24] and references therein.

The purpose of this paper is to select salient features developed in our recent works [25, 26] on quantum spaces with $\mathfrak{su}(2)$ noncommutativity, hereafter denoted generically by $\mathbb{R}^3_\theta$ (see below). It appears that these latter can be modelled conveniently by exploiting a family of $SO(3)$-equivariant differential *-representations as we will show in a while. It turns out that the use of differential representations [7, 27, 28] may prove useful in the construction of star-products whenever the noncommutativity is of a Lie algebra type as it is the case here. Consistency of the construction definitely requires that one works with *-representations. Note that a similar

[1] *Based on a talk presented by Poulain T at the XXVth International Conference on Integrable Systems and Quantum symmetries (ISQS-25), Prague, June 6-10 2017.*

IOP Conf. Series: Journal of Physics: Conf. Series **965** (2018) 012032 doi:10.1088/1742-6596/965/1/012032

construction can also be applied to the kappa-Minkowski spaces which are related to (the universal enveloping algebra of) a solvable Lie algebra. For constructions of star-products within kappa-Minkowski spaces, see e.g [29, 30].

The characterization of the related quantization maps defining the quantum spaces $\mathbb{R}^3_\theta$ can be achieved from a natural combination of the polar decomposition of the quantized plane waves with the Wigner theorem for $SU(2)$. Recall that the quantized plane waves are defined by the action of the quantization map on the usual plane waves. From this follows the characterization of the star-products. The use of star-product formulation of NCFT is convenient for fast construction of functional actions. However, it may lead to difficulties whenever the star-product is represented by a complicated formula and/or is not closed for a trace functional. In this respect, we then construct a tracial star-product with respect to the usual Lebesgue measure on $\mathbb{R}^3$, equivalent to the Kontsevich product [31] for the Poisson manifold dual to $\mathfrak{su}(2)$, thanks to a suitable use of the Harish-Chandra map [32–34]. We then present noncommutative (scalar) field theories which are free from UV/IR mixing and whose commutative limit is the usual ϕ^4 theory on $\mathbb{R}^3$. We discuss the generalization of the construction to semi-simple, possibly non simply connected, Lie groups based on their central extensions by suitable abelian Lie groups. Considering central extensions of Lie groups is the natural framework to deal with this problem, while the classification of the extensions needs to consider different types of cohomology including suitable group cohomologies. For some details on various relevant cohomologies in physics, see e.g [35–40].

Notations: We will denote generically the involutions by the symbol †. The actual nature of each involution should be clear from the context. $\mathcal{S}(\mathbb{R}^3)$ and $\mathcal{M}(\mathbb{R}^3)$ are respectively the algebra of Schwartz functions on $\mathbb{R}^3$ and its multiplier algebra. $\langle \cdot, \cdot \rangle$ is the hermitian product, i.e, $\langle f, g\rangle := \int d^3x\ \bar{f}(x)g(x)$ for any $f, g \in \mathcal{M}(\mathbb{R}^3)$ ($\bar{f}(x)$ is the complex conjugate of $f(x)$). $\tilde{f}(p) = \int d^3x f(x)e^{-ipx}$ is the Fourier transform of f. $\mathcal{L}(\mathcal{M}(\mathbb{R}^3))$ denotes the set of linear operators acting on $\mathcal{M}(\mathbb{R}^3)$.

We will deal with a family of deformations of $\mathbb{R}^3$, indexed by 3 functional $f,\ g,\ \ell$ depending on the Laplacian of $\mathbb{R}^3$, Δ, and a positive real parameter θ (the so-called deformation parameter), i.e $\mathbb{R}^3_{\theta,f,g,\ell} := (\mathcal{F}(\mathbb{R}^3), \star_{\theta,f,g,\ell})$ (with $\mathcal{F}(\mathbb{R}^3)$ a suitable linear space of functions to be characterized below) where $\star_{\theta,f,g,\ell}$ is the deformed product. To simplify the notations, any element of this family will be denoted by $\mathbb{R}^3_\theta = (\mathcal{F}(\mathbb{R}^3), \star)$, the actual nature of the objects indexing the family should be clear from the context.

2. $\mathfrak{su}(2)$-noncommutativity and differential *-representations

It is convenient to represent the abstract *-algebra $\mathbb{A}[\hat{X}_\mu]$ generated by the self-adjoint coordinate operators $\hat{X}_\mu$ fulfilling $[\hat{X}_\mu, \hat{X}_\nu] = i2\theta\varepsilon_{\mu\nu}{}^\rho\hat{X}_\rho$ $(\mu,\nu,\rho = 1,2,3)$ by making use of the poly-differential *-representation $\pi : \mathbb{A}[\hat{X}_\mu] \to \mathcal{L}(\mathcal{M}(\mathbb{R}^3))$,

$$\pi : \hat{X}_\mu \mapsto \pi(\hat{X}_\mu) =: \hat{x}_\mu(x,\partial) = x_\nu\varphi^\nu{}_\mu(\partial) + \chi_\mu(\partial), \tag{2.1}$$

where the functionals $\varphi^\nu{}_\mu(\partial)$ and $\chi_\mu(\partial)$ are viewed as formal expansions in the usual derivatives of $\mathbb{R}^3$, ∂_μ, $\mu = 1,2,3$.

Since by assumption π is a morphism of *-algebra, one has $[\hat{x}_\mu, \hat{x}_\nu] = i2\theta\varepsilon_{\mu\nu}{}^\rho\hat{x}_\rho$ together with $\langle f, \hat{x}_\mu g\rangle = \langle \hat{x}_\mu f, g\rangle$ for any $f, g \in \mathcal{M}(\mathbb{R}^3)$, stemming from the self-adjointness of $\hat{x}_\mu$ so that $\hat{x}^\dagger_\mu = \hat{x}_\mu$. By combining these two latter conditions with (2.1) and using $[x^\lambda, h(x,\partial)] = -\frac{\partial h}{\partial(\partial_\lambda)}$, which holds true for any functional h of x^μ and ∂_μ, together with $\partial^\dagger_\mu = -\partial_\mu$, $h^\dagger(\partial) = \bar{h}(-\partial)$, a standard computation gives rise to the following functional differential equations constraining

${\varphi^\nu}_\mu$ and χ_μ:

$$i2\theta\varphi_{\alpha\rho} = {\varepsilon_\rho}^{\mu\nu}\frac{\partial\varphi_{\alpha\mu}}{\partial(\partial_\beta)}\varphi_{\beta\nu}, \tag{2.2}$$

$$\varphi^\dagger_{\alpha\rho} = \varphi_{\alpha\rho} \tag{2.3}$$

$$i2\theta\chi_\rho = {\varepsilon_\rho}^{\mu\nu}\frac{\partial\chi_\mu}{\partial(\partial_\alpha)}\varphi_{\alpha\nu}, \tag{2.4}$$

$$\frac{\partial\varphi^\dagger_{\alpha\rho}}{\partial(\partial_\alpha)} = \chi_\rho - \chi^\dagger_\rho, \tag{2.5}$$

where use has been made of the algebraic relation $\delta_{\mu\gamma}{\delta_\nu}^\sigma - {\delta_\mu}^\sigma\delta_{\nu\gamma} = {\varepsilon_{\mu\nu}}^\rho{\varepsilon_{\rho\gamma}}^\sigma$.

In view of $\mathbb{R}^3_\theta \subsetneq U(\mathfrak{su}(2)) \cong \mathbb{A}[\hat{X}_\mu]/[\hat{X}_\mu, \hat{X}_\nu]$, see e.g [16,25], where $U(\mathfrak{su}(2))$ is the universal enveloping algebra of $\mathfrak{su}(2)$, there is a natural action of $SU(2)/\mathbb{Z}_2 \simeq SO(3)$ on any $\mathbb{R}^3_\theta$. This selects $SO(3)$-equivariant *-representations among those defined by (2.2)-(2.5). A mere application of the Schur-Weyl decomposition theorem shows that the $SO(3)$-equivariance of the representation can be achieved whenever the functional ${\varphi^\nu}_\mu$ and χ_μ have the following form:

$$\varphi_{\alpha\mu}(\partial) = f(\Delta)\delta_{\alpha\mu} + g(\Delta)\partial_\alpha\partial_\mu + ih(\Delta){\varepsilon_{\alpha\mu}}^\rho\partial_\rho, \tag{2.6}$$

$$\chi_\mu(\partial) = \ell(\Delta)\partial_\mu\ , \tag{2.7}$$

where the *real* $f(\Delta)$, $g(\Delta)$, $h(\Delta)$ and *complex* $\ell(\Delta)$ $SO(3)$-invariant functionals are constrained by (2.2)-(2.5).
By solving the constraints, one easily finds that the admissible solutions (i.e those $\hat{x}_\mu$ admitting an expansion of the form $\hat{x}_\mu = x_\mu + \mathcal{O}(\theta)$) are such that $h = \theta$ and form a family indexed by 3 functional f, g and ℓ defined by

$$\hat{x}_\mu = x^\alpha\left[f(\Delta)\delta_{\alpha\mu} + g(\Delta)\partial_\alpha\partial_\mu + i\theta{\varepsilon_{\alpha\mu}}^\rho\partial_\rho\right] + \ell(\Delta)\partial_\mu, \tag{2.8}$$

$$2\left[(f+g\Delta)' + g\right] = \ell + \ell^\dagger\ , \tag{2.9}$$

$$2(f+g\Delta)f' = gf + \theta^2. \tag{2.10}$$

An interesting subfamily of poly-differential *-representations arises whenever $f + g\Delta = 1$, so that, setting $g(\Delta) := \frac{\theta^2}{3}G(2\theta^2\Delta)$, (2.9), (2.10) reduce to

$$l + l^\dagger \ = \ 2g(\Delta)\ , \tag{2.11}$$

$$0 \ = \ 2t\frac{dG}{dt} + 3\left(G(t)+1\right) - \frac{t}{6}G^2(t),\ , \tag{2.12}$$

for which the Ricatti equation (2.12) is solved by $G(t) = -6\sum_{n=1}^\infty \frac{2^n B_{2n}}{(2n)!}t^{n-1}$ where B_n are Bernoulli numbers. The resulting subfamily[2] takes the form

$$\hat{x}_\mu = x^\alpha\left[(1 - g(\Delta)\Delta)\delta_{\alpha\mu} + g(\Delta)\partial_\alpha\partial_\mu + i\theta{\varepsilon_{\alpha\mu}}^\rho\partial_\rho\right] + g(\Delta)\partial_\mu. \tag{2.13}$$

We will derive, Section 5, a tracial star-product from this subfamily equivalent to the Kontsevich product for the Poisson manifold dual to the Lie algebra $\mathfrak{su}(2)$.

[2] Notice that consistency with (3.8) below will require ℓ to be real.

3. Quantization maps, deformations of $\mathbb{R}^3$ and extensions

Let Q denotes the quantization map, i.e an invertible, linear, *-algebra morphism

$$Q : (\mathcal{M}(\mathbb{R}^3), \star) \to (\mathcal{L}(\mathcal{M}(\mathbb{R}^3)), \cdot), \tag{3.1}$$

where "·" is the product between differential operators omitted from now on, such that for any $f,\ g \in \mathcal{M}(\mathbb{R}^3)$,

$$f \star g := Q^{-1}\left(Q(f)Q(g)\right),\ Q(1) = \mathbb{I},\ Q(\bar{f}) = (Q(f))^\dagger\,, \tag{3.2}$$

$$Q(f) \triangleright 1 = f(x)\,, \tag{3.3}$$

so that $Q^{-1}\left(Q(f)\right) = Q(f) \triangleright 1$, where "$\triangleright$" is the left action of operators. Therefore the star-product can be expressed as

$$(f \star g)(x) = \int \frac{d^3p}{(2\pi)^3}\frac{d^3q}{(2\pi)^3}\tilde{f}(p)\tilde{g}(q)Q^{-1}\left(E_p(\hat{x})E_q(\hat{x})\right)\,, \tag{3.4}$$

for any $f, g \in \mathcal{M}(\mathbb{R}^3)$, where the quantized plane waves $E_p(\hat{x})$ are defined by

$$E_p(\hat{x}) := Q(e^{ipx}). \tag{3.5}$$

The quantized plane waves (3.5) define a map $E : SU(2) \to \mathcal{L}(\mathcal{M}(\mathbb{R}^3))$,

$$E : g \mapsto E(g) := E_p(\hat{x}), \tag{3.6}$$

$$E(g^\dagger) = E^\dagger(g)\,, \tag{3.7}$$

for any $g \in SU(2)$. Using polar decomposition, one can write

$$E(g) = U(g)|E(g)|\,, \tag{3.8}$$

where $|E(g)| := \sqrt{E^\dagger(g)E(g)} \neq 0$ and the unitary operator $U : SU(2) \to \mathcal{L}(\mathcal{M}(\mathbb{R}^3))$ can be expressed as

$$U(g) = e^{i\xi_g^\mu \hat{x}_\mu}, \tag{3.9}$$

in view of the Stone's theorem, where $\xi_g^\mu \in \mathbb{R}$. Then, the Baker-Campbell-Hausdorff formula for $\mathfrak{su}(2)$ gives

$$e^{i\xi_{g_1}\hat{x}}e^{i\xi_{g_2}\hat{x}} = e^{iB(\xi_{g_1},\xi_{g_2})\hat{x}}\,, \tag{3.10}$$

where the infinite expansion $B(\xi_{g_1}, \xi_{g_2})$ fulfils

$$B(\xi_{g_1}, \xi_{g_2}) = -B(-\xi_{g_2}, -\xi_{g_1}),\ B(\xi_g, 0) = \xi_g\,. \tag{3.11}$$

Observe that $U(g)$ and $E(g)$ define representations of $SU(2)$. Then, one has for any $g_1, g_2 \in SU(2)$

$$U(g_1)U(g_2) = U(g_1g_2)\,, \tag{3.12}$$

which holds true (up to unitary equivalence) as a mere application of the Wigner theorem for $SU(2)$, while we demand

$$E(g_1)E(g_2) = \Omega(g_1, g_2)E(g_1g_2) \tag{3.13}$$

where $\Omega(g_1, g_2)$ is to be determined. Combining $E(g^\dagger g) = E(\mathbb{I}) = \mathbb{I}$ with Eqn. (3.13), one obtains $E(g^\dagger)E(g) = \Omega(g^\dagger, g)\mathbb{I}$, for any $g \in SU(2)$. Therefore $|E(g)| = \sqrt{\Omega(g^\dagger, g)}\mathbb{I}$, so that

$$\omega_g := \sqrt{\Omega(g^\dagger, g)} \in \mathbb{R}, \ \omega_g > 0, \tag{3.14}$$

together with

$$[|E(g)|, U(g)] = 0. \tag{3.15}$$

Using (3.15) and (3.8), one get

$$E(g_1)E(g_2) = |E(g_1)||E(g_2)|U(g_1g_2) = |E(g_1)||E(g_2)||E(g_1g_2)|^{-1}E(g_1g_2)\ , \tag{3.16}$$

where the 2nd equality stems from (3.12), which combined with the expression for $|E(g)|$ yields

$$E(g_1)E(g_2) = (\omega_{g_1}\omega_{g_2}\omega_{g_1g_2}^{-1})E(g_1g_2)\ , \tag{3.17}$$

where

$$E(g_1g_2) = \omega_{g_1g_2}e^{iB(\xi_{g_1},\xi_{g_2})\hat{x}}. \tag{3.18}$$

Note that (3.17) insures the associativity of the star-product (3.4), since the 2-cocycle $\Omega(g_1, g_2) := \omega_{g_1}\omega_{g_2}\omega_{g_1g_2}^{-1}$ obeys $\Omega(g_1, g_2)\Omega(g_1g_2, g_3) = \Omega(g_1, g_2g_3)\Omega(g_2, g_3)$, for any $g_1, g_2, g_3 \in SU(2)$. From eqn. (3.12), one infers that any unitary equivalent representations, U and U' correspond to unitary equivalent products. Indeed, one has $U'(g) = e^{i\gamma(g)}U(g) = e^{i\gamma(g)}e^{i\xi_g\hat{x}}$, where γ is a real function, which implies the following equivalence relation $T(f \star' g) = Tf \star Tg$ where T is defined by $E'_k(\hat{x}) \equiv Q'(e^{ikx}) := Q \circ T(e^{ikx}) = e^{i\gamma(k)}Q(e^{ikx}) = e^{i\gamma(k)}E_k(\hat{x})$. Informally speaking, the star-product which will be determined in a while is essentially unique up to unitary equivalence.

This result can be understood [26] within the more general framework of central extensions of Lie groups which in addition offers a convenient way for generalizations of the present work. In the following $\mathcal{A}$ is a 1-d abelian Lie group which we will assumed to be $\mathcal{A} = \mathbb{R}/\mathcal{D}$ where $\mathcal{D}$ is a discrete group of $\mathbb{R}$, i.e $\mathcal{D} = p\mathbb{Z}$. Let G be a connected Lie group with Lie algebra $\mathfrak{g}$ whose action on $\mathcal{A}$, $\rho : G \times \mathcal{A} \to \mathcal{A}$, is assumed to be trivial ($\rho(g, a) = a$, for any $g \in G$, $a \in \mathcal{A}$). $\mathcal{E}$ is a central extension of G if $\mathcal{A}$ is isomorphic to a subgroup of the center of G, $\mathcal{Z}(G)$ and $G \simeq \mathcal{E}/\mathcal{A}$ as group isomorphism. Recall that inequivalent central extensions of groups encoded by the short exact sequence $\mathbb{I} \to \mathcal{A} \to \mathcal{E} \stackrel{\pi}{\to} G \to \mathbb{I}$ (where π is the canonical projection, with in addition $Im(\mathcal{A}) \subset \mathcal{Z}(\mathcal{E})$), are classified by $H^2(G, \mathcal{A})$, the 2nd cohomology group of G with values in $\mathcal{A}$. Actually, up to additional technical requirements, $\mathcal{E}$ defines a principal fiber bundle over G with 1-form connection and structure group $\mathcal{A}$. For *simply connected* Lie groups G, one has

$$H^2(G, \mathbb{R}/\mathcal{D}) \simeq H^2_{alg}(\mathfrak{g}, \mathbb{R}), \tag{3.19}$$

see e.g [39, 40], where $H^2_{alg}(\mathfrak{g}, \mathbb{R})$ is the 2nd group of (real) cohomology of the Lie algebra $\mathfrak{g}$.
When G is in addition semi simple, which is for instance the case for $SU(2)$, one has $H^2_{alg}(\mathfrak{g}, \mathbb{R}) = \{0\}$ so that $H^2(G, \mathbb{R}/\mathcal{D})$ is trivial. When $\mathcal{D} = \mathbb{Z}$, one has $\mathbb{R}/\mathcal{D} = U(1)$, hence the triviality of $H^2(SU(2), U(1))$ which explains the uniqueness of the above star-products up to unitary equivalence. The above conclusion extends to the central extension of any semi-simple and simply connected Lie group by $\mathbb{R}/\mathcal{D}$ so that the extension of the present construction to spaces with noncommutativity based on the corresponding Lie algebra should produce uniqueness of the star-product (up to equivalence).
When G is semi-simple but not necessarily simply connected, its inequivalent central extensions

by $\mathbb{R}/\mathcal{D}$ are classified (up to additional technical requirements) by $H^1_{\check{C}}(G,\mathbb{R}/\mathcal{D})$, where $H^\bullet_{\check{C}}$ denotes the Čech cohomology. The following isomorphism holds true

$$H^1_{\check{C}}(G,\mathbb{R}/\mathcal{D}) \simeq Hom(\pi_1(G) \to \mathbb{R}/\mathcal{D}). \tag{3.20}$$

Now assume $G = SL(2,\mathbb{R})$ and $\mathcal{D} = \mathbb{Z}$ so that $\mathbb{R}/\mathcal{D} \simeq U(1)$. The use of Iwazawa decomposition yields $SL(2,\mathbb{R}) \simeq \mathbb{R}^2 \times \mathbb{S}^1$, which combined with $\pi_1(X \times Y) = \pi_1(X) \times \pi_1(Y)$ for any topological spaces X and Y implies $\pi_1(SL(2,\mathbb{R})) \simeq \mathbb{Z}$. Hence, $Hom(\mathbb{Z} \to U(1)) \simeq U(1)$ which classifies the inequivalent extensions of $SL(2,\mathbb{R})$ by $U(1)$. Accordingly, one expect that the extension of the present construction of $SL(2,\mathbb{R})$ gives rise to inequivalent classes of star-products.

4. Quantized plane waves for $\mathbb{R}^3_\theta$

The determination of (3.6) which takes the generic form

$$E_p(\hat{x}) = \omega(p)e^{i\xi(p)\hat{x}}, \tag{4.1}$$

can be achieved through a standard albeit tedious computation whose main steps are summarized now.

First, using $e^{i\xi(p)\hat{x}} \triangleright 1 = \frac{e^{ipx}}{\omega(p)}$, one easily infers that

$$e^{-i\xi(\lambda p)\hat{x}}\partial_\mu e^{i\xi(\lambda p)\hat{x}} = (i\lambda p_\mu)\mathbb{I}, \tag{4.2}$$

which holds for any $\lambda \in \mathbb{R}$. Then, the combination of $[\partial_\mu, \hat{x}_\nu] = [\partial_\mu, x^a\varphi_{a\nu}] = \varphi_{\mu\nu}$ with the functional derivative of (4.2) with respect to λ yields

$$\varphi_{\mu\nu}(i\lambda p)\frac{d}{d\lambda}\left[\xi^\nu(\lambda p)\right] = p_\mu. \tag{4.3}$$

From $SO(3)$-covariance requirement, it can be shown [26] that this latter functional differential equation is solved by

$$\xi^\mu(p) = \int_0^1 d\lambda(\varphi^{-1})^{\mu\nu}_{|i\lambda p}p_\nu, \tag{4.4}$$

$$(\varphi^{-1})^{\mu\nu}(ip) = \frac{1}{f^2+\theta^2p^2}\left(f\delta^{\mu\nu} + 2f'p^\mu p^\nu + \theta\varepsilon^{\mu\nu\rho}p_\rho\right), \tag{4.5}$$

where f and f', (2.8)-(2.10), depend on $(-p^2)$. ξ_μ can be verified to be an injective antisymmetric real-valued function. Finally, Eqn. (4.4) can be recast as a Volterra integral given by

$$\xi^\mu(p) = \int_{-p^2}^0 \frac{dt}{2|p|\sqrt{-t}}\frac{f(t)-2tf'(t)}{f^2(t)-\theta^2t}p^\mu. \tag{4.6}$$

It can be verified that (4.6) simplifies to $\xi^\mu = p^\mu$ when use is made of the subfamily of *-representations (2.13).

To determine $\omega(p)$, one observes that

$$\frac{d}{d\lambda}\left[e^{i\xi(\lambda p)\hat{x}}\right] = i(\varphi^{-1})^{\mu\nu}_{|i\lambda p}p_\nu\hat{x}_\mu e^{i\xi(\lambda p)\hat{x}}, \tag{4.7}$$

where (4.4) has been used, implying

$$\frac{d}{d\lambda}\left[e^{i\xi(\lambda p)\hat{x}}\right] \triangleright 1 = i\left(x^\nu + \chi_\mu(\varphi^{-1})^{\mu\nu}_{|i\lambda p}\right)p_\nu\frac{e^{i\lambda px}}{\omega(\lambda p)}. \tag{4.8}$$

Besides, the following relation holds true

$$\frac{d}{d\lambda}\left[\frac{e^{i\lambda px}}{\omega(\lambda p)}\right] = \left(ix^\nu p_\nu - \frac{1}{\omega(\lambda p)}\frac{d}{d\lambda}\left[\omega(\lambda p)\right]\right)\frac{e^{i\lambda px}}{\omega(\lambda p)}. \tag{4.9}$$

Now, one can verify that $\frac{d}{d\lambda}\left[\hat{A}f(x)\right] = \frac{d\hat{A}}{d\lambda}f(x)$ where $\hat{A}$ and f are any operator and function suitably chosen for the latter expression to be well defined. Hence

$$\frac{d}{d\lambda}\left[e^{i\xi(\lambda p)\hat{x}} \triangleright 1\right] = \frac{d}{d\lambda}\left[e^{i\xi(\lambda p)\hat{x}}\right] \triangleright 1, \tag{4.10}$$

$$i\left(x^\nu + \chi_\mu(\varphi^{-1})^{\mu\nu}_{|_{i\lambda p}}\right)p_\nu = ix^\nu p_\nu - \frac{1}{\omega(\lambda p)}\frac{d}{d\lambda}\left[\omega(\lambda p)\right], \tag{4.11}$$

implying

$$\frac{1}{\omega(\lambda p)}\frac{d}{d\lambda}\left[\omega(\lambda p)\right] = -i\chi_\mu(\varphi^{-1})^{\mu\nu}_{|_{i\lambda p}}p_\nu, \tag{4.12}$$

which is solved by [26] $\omega(p) = e^{-i\int_0^1 d\lambda\ \chi_\mu(i\lambda p)(\varphi^{-1})^{\mu\nu}_{|_{i\lambda p}} p_\nu}$. This latter expression can be recast as a Volterra integral given by

$$\omega(p) = e^{\int_{-p^2}^0 dt\ \frac{f(t)-2tf'(t)}{f^2(t)-\theta^2 t}\ell(t)}. \tag{4.13}$$

Consistency with (3.14) requires $\omega(p)$ to be a positive real quantity therefore constraining ℓ to be a real functional, $\ell^\dagger = \ell$. It follows that (2.9) reduces to $\ell = (f + g\Delta)' + g$, which thus constraints ℓ once f and g fulfilling (2.10) are obtained.

To summarize the above analysis, given a *-representation belonging to the family (2.8)-(2.10), eqns. (4.6) and (4.13) fully characterize the corresponding quantization map Q (3.1), (3.3) together with its star-product (3.4). Two remarks are in order:
i) One can show [26] that for the family of poly-differential *-representations considered in this note, Q *cannot* be the Weyl quantization map W related to the symmetric ordering for operators. In fact, the only poly-differential representation compatible with the Weyl quantization is defined by $\hat{x}_\mu = x^\alpha\left((1-g\Delta)\delta_{\alpha\mu} + g\partial_\alpha\partial_\mu + i\theta\varepsilon_{\alpha\mu\rho}\partial_\rho\right)$ (namely with $\chi = 0$), where g has been defined above, which however is not a *-representation.
ii) One can verify that the star-product used in [5] and some ensuing works does not belong to the general family of star-products related to (2.8). The former star-product, defining a particular deformation of $\mathbb{R}^3$ called $\mathbb{R}^3_\lambda$, can be related to the Wick-Voros product [9–11, 41] stemming from a twist. So far, whether or not the present family of star-products also admits a representation in terms of a twist is not known.

5. Tracial star-product and related quantum field theories

In this section, we restrict ourselves to the particular subfamily of *-representations (2.13). By combining this latter with eqns. (4.6) and (4.13), one easily finds that the quantized plane waves are defined by $Q(e^{ipx}) \equiv E_p(\hat{x}) = \left(\frac{\sin(\theta|p|)}{\theta|p|}\right)^2 e^{ip\hat{x}}$, from which follows

$$(f \star_Q g)(x) = \int \frac{d^3p}{(2\pi)^3}\frac{d^3q}{(2\pi)^3}\tilde{f}(p)\tilde{g}(q)\mathcal{W}^2(p,q)e^{iB(p,q)x}, \tag{5.1}$$

for any $f, g \in \mathcal{M}(\mathbb{R}^3)$, with $\mathcal{W}(p,q) := \frac{|B(p,q)|}{\theta|p||q|}\frac{\sin(\theta|p|)\sin(\theta|q|)}{\sin(\theta|B(p,q)|)}$ where $B(p,q)$ is given by (3.10). Now, introduce a new quantization map $\mathcal{K} : \mathcal{M}(\mathbb{R}^3) \to \mathcal{L}(\mathcal{M}(\mathbb{R}^3))$ through

$$\mathcal{K} := Q \circ H, \tag{5.2}$$

IOP Conf. Series: Journal of Physics: Conf. Series **965** (2018) 012032 doi:10.1088/1742-6596/965/1/012032

where the operator H acting on $\mathcal{M}(\mathbb{R}^3)$ is given by

$$H := \frac{\theta\sqrt{\Delta}}{\sinh(\theta\sqrt{\Delta})}, \tag{5.3}$$

and such that

$$H(f \star_{\mathcal{K}} g) = H(f) \star_Q H(g), \tag{5.4}$$

for any $f, g \in \mathcal{M}(\mathbb{R}^3)$. One obtains from a standard computation

$$\mathcal{K}(e^{ipx}) = \frac{\sin(\theta|p|)}{\theta|p|} e^{ip\hat{x}}, \tag{5.5}$$

from which it is easy to obtain finally

$$(f \star_{\mathcal{K}} g)(x) = \int \frac{d^3p}{(2\pi)^3} \frac{d^3q}{(2\pi)^3} \tilde{f}(p)\tilde{g}(q)\mathcal{W}(p,q)e^{iB(p,q)x}\ , \tag{5.6}$$

for any $f, g \in \mathcal{M}(\mathbb{R}^3)$, where $\mathcal{W}(p,q)$ has been given above.

The star-product $\star_{\mathcal{K}}$ (5.6) is nothing but the Kontsevich product [31] related to the Poisson manifold dual to the finite dimensional Lie algebra $\mathfrak{su}(2)$. This can be realized [26] by noticing that (5.5) can be recast into the form $\mathcal{K}(e^{ipx}) = W(j^{\frac{1}{2}}(\Delta)(e^{ipx}))$ where W is the Weyl quantization map, $W(e^{ipx}) = e^{ip\hat{x}}$ and

$$j^{\frac{1}{2}}(\Delta) = \frac{\sinh(\theta\sqrt{\Delta})}{\theta\sqrt{\Delta}}\ , \tag{5.7}$$

is the Harish-Chandra map [32–34]. Hence,

$$\mathcal{K} = W \circ j^{\frac{1}{2}}(\Delta)\ , \tag{5.8}$$

which coincides with the Kontsevich product in the present case, see e.g [8]. Note that (5.3) and (5.7) imply

$$j^{\frac{1}{2}}(\Delta) = H^{-1}, \tag{5.9}$$

so that H (5.3) is the inverse of the Harish-Chandra map. The star-product $\star_{\mathcal{K}}$ (5.6) is tracial with respect to the trace functional which in the present case is simply defined by the Lebesgue integral on $\mathbb{R}^3$. Namely, one has

$$\int d^3x (f \star_{\mathcal{K}} g)(x) = \int d^3x f(x)g(x). \tag{5.10}$$

This last interesting property can be exploited to built NCFT admitting standard (i.e commutative) massive real or complex scalar field theories with quartic interaction as formal commutative limits [7, 25], namely

$$S_1 = \int d^3x \big[\frac{1}{2}\partial_\mu\phi \star_{\mathcal{K}} \partial_\mu\phi + \frac{1}{2}m^2\phi \star_{\mathcal{K}} \phi + \frac{\lambda}{4!}\phi \star_{\mathcal{K}} \phi \star_{\mathcal{K}} \phi \star_{\mathcal{K}} \phi\big], \tag{5.11}$$

$$S_2 = \int d^3x \big[\partial_\mu\Phi^\dagger \star_{\mathcal{K}} \partial_\mu\Phi + m^2\Phi^\dagger \star_{\mathcal{K}} \Phi + \lambda\Phi^\dagger \star_{\mathcal{K}} \Phi \star_{\mathcal{K}} \Phi^\dagger \star_{\mathcal{K}} \Phi\big]. \tag{5.12}$$

It turns out that (5.11) and (5.12) do not have (perturbative) UV/IR mixing, as shown in [25]. Indeed, the relevant contributions to the 2-point function can be written as $\Gamma_2^{(I)} = \int d^3x\ \phi(x)\phi(x)\omega_I$ and $\Gamma_2^{(II)} = \int \frac{d^3k_1}{(2\pi)^3}\frac{d^3k_1}{(2\pi)^3}\ \tilde{\phi}(k_1)\tilde{\phi}(k_2)\omega_{II}(k_1,k_2)$, with

$$\omega_I \sim \frac{4}{\theta^2}\int \frac{d^3p}{(2\pi)^3}\ \frac{\sin^2(\frac{\theta}{2}|p|)}{p^2(p^2+m^2)} = \frac{1-e^{-\theta m}}{2m\pi\theta^2} \tag{5.13}$$

$$\omega_{II} \sim \int d^3x \frac{d^3p}{(2\pi)^3}\ \frac{1}{p^2+m^2}(e^{ipx} \star_{\mathcal{K}} e^{ik_1x} \star_{\mathcal{K}} e^{-ipx} \star_{\mathcal{K}} e^{ik_2x}). \tag{5.14}$$

When $\theta \neq 0$, ω_I is obviously finite even for $m=0$ while $\omega_{II}(0,k_2) \sim \delta(k_2)\omega_I$. Similar expression holds for $\omega_{II}(k_1,0)$ so that no IR singularity occurs within (5.11). This signals the absence of UV/IR mixing. One can check [25] the UV one-loop finiteness of ω_{II}. Note that similar conclusions hold true for the complex scalar field case (5.12).

The origin of the absence of UV/IR mixing as well as the present mild (finite) UV behaviour (which should extend to all orders) is likely due to the Peter-Weyl decomposition[3] of the algebra modeling the noncommutative space, i.e $\mathbb{R}^3_\theta = \oplus_{j\in\frac{\mathbb{N}}{2}}\mathbb{M}_{2j+1}(\mathbb{C})$, which reflects the relationship between the algebra and the convolution algebra of $SU(2)$, see [16]. This is particularly apparent in a class of NCFT [12–14], with however kinetic operators different from the usual Laplacian used here, for which the theory splits into an infinite tower of (matrix) field theories, each on a finite geometry with the radius $\sim j$ of $\mathbb{M}_{2j+1}(\mathbb{C})$ serving as a natural cut-off. The use of the standard Laplacian in (5.11), (5.12) complicates the analysis of the UV behaviour to arbitrary orders but one can reasonably conjecture that both (5.11) and (5.12) NCFT are (UV) finite to all orders.

An interesting issue would be to extend the above analysis to the case of noncommutative gauge theories whose commutative limit would reproduce the usual gauge (Yang-Mills) theory on $\mathbb{R}^3$. Such an extension would presumably exclude the choice of a derivation-based differential calculus [42–44] since this latter would produce natural Laplacians without (analog of) radial dependence. Note that one proposal aiming to include radial dependence presented in [11] amounts to enlarge the initial algebra by incorporating the deformation parameter itself.

Acknowledgments

Wallet J-C is grateful to F. Besnard, N. Franco and F. Latrémolière for discussions on various topics related to the present work. Poulain T thanks COST Action MP1405 QSPACE for partial financial support. This work is partially supported by H2020 Twinning project No. 692194, RBI-T-WINNING.

References

[1] Guedes C, Oriti D and Raasakka M 2013 *J. Math. Phys.* **54** 083508
[2] Freidel L and Livine E 2006 *Phys. Rev. Lett.* **96** 221301
[3] Baratin A and Oriti D 2010 *Phys. Rev. Lett.* **105** 221302
[4] Lizzi F and Vitale P 2014 *SIGMA* **10** 086
[5] Hammou A B, Lagraa M and Sheikh-Jabbari M M 2002 *Phys. Rev.* D **66** 025025
[6] Gracia-Bondía J M, Lizzi F, Marmo G and Vitale P 2002 *J. High Energy Phys.* JHEP**04**(2002)026
[7] Kupriyanov V G and Vitale P 2015 *J. High Energy Phys.* JHEP**08**(2015)024

[3] This should hold provided the kinetic operator has a reasonable behaviour, i.e has compact resolvant insuring a sufficient decay of the propagator.

[8] Freidel L and Majid S 2008 *Class. Quant. Grav.* **25** 045006
[9] Galluccio S, Lizzi F and Vitale P 2008 *Phys.Rev.* D **78** 085007
[10] Aschieri P, Lizzi F and Vitale P 2008 *Phys.Rev.* D **77** 025037
[11] Gracia-Bondía J M, Lizzi F, Ruiz Ruiz F and Vitale P 2006 *Phys.Rev.* D **74** 025014
[12] Vitale P and Wallet J-C 2013 *J. High Energy Phys.* JHEP**04**(2013)115
[13] Géré A, Vitale P and Wallet J-C 2014 *Phys. Rev.* D **90** 045019
[14] Géré A, Jurić T and Wallet J-C 2015 *J. High Energy Phys.* JHEP**12**(2015)045
[15] Vitale P 2014 *Fortschr. Phys.* **62** 825-834
[16] Wallet J-C 2016 *Nucl. Phys.* B **912** 354
[17] Grosse H and Wulkenhaar R 2003 *J. High Energy Phys.* JHEP**0312**(2003)019
[18] Grosse H and Wulkenhaar R 2005 *Commun. Math. Phys.* **256** 305
[19] de Goursac A, Tanasa A and Wallet J-C 2008 *Eur. Phys. J.* C **53** 459
[20] de Goursac A, Wallet J-C and Wulkenhaar R 2008 *Eur. Phys. J.* C **56** 293
[21] Martinetti P, Vitale P and Wallet J-C 2013 *J. High Energy Phys.* JHEP**09**(2013)051
[22] de Goursac A and Wallet J-C 2011 *J. Phys.* A: Math. Theor. **44** 055401
[23] Wallet J-C 2008 *J. Phys.: Conf. Ser.* **103** 012007
[24] de Goursac A, Wallet J-C and Wulkenhaar R 2007 *Eur. Phys. J.* C **51** 977
[25] Jurić T, Poulain T and Wallet J-C 2016 *J. High Energy Phys.* JHEP**05**(2016)146
[26] Jurić T, Poulain T and Wallet J-C 2017 *J. High Energy Phys.* JHEP**07**(2017)116
[27] Durov N, Meljanac S, Samsarov A and Skoda Z 2007 *J. Algebra* **309** 318
[28] Kupriyanov V G and Vassilevich D V 2008 *Eur. Phys. J.* C **58** 627
[29] Meljanac S, Samsarov A, Stojic M and Gupta K S 2008 *Eur. Phys. J.* C **53** 295
[30] Borowiec A and Pachol A 2009 *Phys.Rev.* D **79** 045012
[31] Kontsevich M 2003 *Lett. Math. Phys.* **66** 157
[32] Harish-Chandra 1951 *Trans. Amer. Math. Soc.* **70** 28-96
[33] Duflo M 1977 *Ann. Sc. Ec. Norm. Sup.* **10** 265-288
[34] Duflo M 1970 *Ann. Sc. Ec. Norm. Sup.* **3** 23-74
[35] Brown K S 1982 *Cohomology of Groups* (Berlin: Springer-Verlag)
[36] Stora R, Thuillier F and Wallet J-C 1995 *Infinite dimensional geometry, non commutative geometry, operator algebras, fundamental interactions* (Cambridge Press) pp 266-297
[37] Wallet J-C 1990 *Phys. Lett.* B **235** 71
[38] Baulieu L, Bellon M, Ouvry S and Wallet J-C 1990 *Phys.Lett.* B **252** 387
[39] Hochschild G 1951 *Ann. of Math.* **54** 96
[40] Shapiro A 1949 *Ann. of Math.* **50** 581
[41] Wick-Voros A 1989 *Phys. Rev.* A **40** 6814
[42] Dubois-Violette M 2001 *Noncommutative Differential Geometry and Its Applications to Physics* Mathematical Physics Studies vol 23, ed Maeda Y et al. (Dordrecht: Springer) pp 245-306
[43] Wallet J-C 2009 *SIGMA* **5** 013
[44] Cagnache E, Masson T and Wallet J-C 2011 *J. Noncommut. Geom.* **5** 39-67

IOP Conf. Series: Journal of Physics: Conf. Series **965** (2018) 012033 doi:10.1088/1742-6596/965/1/012033

Integrals of motion and trajectories for the Perlick system I: an algebraic approach

O Ragnisco[1], S Kuru[2], J Negro[3]

[1] Istituto Nazionale di Fisica Nucleare sezione di Roma Tre, Via Vasca Navale 84, I-00146 Roma, Italy

[2] Department of Physics, Faculty of Science, Ankara University, 06100 Ankara, Turkey

[3] Departamento de Física Teórica, Atómica y Óptica, Universidad de Valladolid, 47011 Valladolid, Spain

E-mail: ragnisco@fis.uniroma3.it, kuru@science.ankara.edu.tr, jnegro@fta.uva.es

Abstract. In this paper we tersely recall the main algebraic and geometric properties of the maximally superintegrable system known as "Perlick System Tipe I", considering all possible values of the relevant parameters. We will follow a classical variant of the so called factorization method, emphasizing the role played the Poisson Algebra of the constants of motion in sheding light on the geometric features of the trajectories.

1. Introduction

In his famous paper dating back to 1992 ([1]), V. Perlick identified two multi-parametric families of Lagrangean systems on curved manifolds complying with the requirements of Bertrand's Theorem ([2]), namely (i) the existence of stable circular orbits, and (ii) the fact that all bounded trajectories are closed.

The two families are deformation either of the Kepler-Coulomb system (Family I) or of the isotropic harmonic oscillator (Family II), the corresponding classical Bertrand systems being recovered in the Euclidean limit. Moreover both families can be extended to an arbitrary number of dimensions by (hyper) spherical symmetry, and are amenable to 'an exact' quantum-mechanical treatment.

It turns out that Perlick Systems are the most general example of (hyper-)spherically symmetric "Maximally Superintegrable" hamiltonian systems ([3]). Indeed, in the case of N degrees of freedom, they are equipped with the maximum number $(2N-1)$ of independent integrals of motion. Such extra integrals are of course related to *additional* dynamical symmetries. They are associated to the *Laplace-Runge-Lenz* vector in the case of Family (I) and to the *Demkov-Fradkin* tensor [4],[5], [6] in the case of Family (II). Moreover, maximal superintegrability entails a further crucial property, namely that *the trajectories can be determined without resorting to analytic calculations, thanks to the underlying symmetry algebra*

Here we restrict our analysis to Family (I). Most of the results will be presented in a sketchy way; for further details we refer the interested reader to the letter [8].

IOP Conf. Series: Journal of Physics: Conf. Series **965** (2018) 012033 doi:10.1088/1742-6596/965/1/012033

2. Factorization Method in a Classical Context

2.1. The Perlick Hamiltonian: radial and angular parts

We recall Perlick I Hamiltonian in spherical coordinates (r, θ, φ) :

$$\widetilde{H}^{\pm} = \beta^2(1 + K\, r^2)\,\frac{p_r^2}{2} + \frac{\mathbf{L}^2}{2\, r^2} + G \pm \frac{1}{r}\,\sqrt{1 + K\, r^2}\,, \tag{2.1}$$

where $\mathbf{L}$ is the angular momentum and $\mathbf{L}^2$ can be considered as an angular Hamiltonian $H_{\theta\varphi}$ in the variables (θ, φ), having the form

$$\mathbf{L}^2 = H_{\theta\varphi} = p_\theta^2 + \frac{p_\varphi^2}{\sin^2\theta} \tag{2.2}$$

being K, G real constants and $\beta = m/n$ a rational number.

We notice that p_φ^2 can be seen as a free Hamiltonian H_φ in the variables (φ, p_φ), defined on the unit circle. The radial variable has different ranges depending on the sign of K. Indeed, If $K \geq 0$, then $0 < r < \infty$, while if $K < 0$, r must be restricted to the finite interval $0 < r < 1/\sqrt{-K}$. The Perlick choice was $\widetilde{H}^-$, to include bounded motion, but in the case $K < 0$ we need both Hamiltonians $\widetilde{H}^{\pm}$ to have a well defined Hamiltonian on the full three dimensional sphere. The associated metric reads:

$$ds^2 = \frac{dr^2}{\beta^2\,(1 + K\, r^2)} + r^2\,(d\theta^2 + \sin^2\theta\, d\varphi^2)\,. \tag{2.3}$$

The "radial" hamiltonian $\widetilde{H}_r^{\pm}$ is simply obtained by setting in 2.1 $G = 0$, and letting $\mathbf{L}^2 \to \ell^2$:

$$\widetilde{H}_r^{\pm} = \beta^2(1 + K\, r^2)\,\frac{p_r^2}{2} + \frac{\ell^2}{2\, r^2} \pm \frac{1}{r}\,\sqrt{1 + K\, r^2}\,. \tag{2.4}$$

Since $\{p_\varphi, \widetilde{H}^{\pm}\} = 0$, p_φ is another constant of motion: $p_\varphi = \ell_z = const.$ Then from the angular Hamiltonian $H_{\theta\varphi}$ we get the effective polar Hamiltonian H_θ depending on the single variable θ:

$$H_\theta = p_\theta^2 + \frac{\ell_z^2}{\sin^2\theta}\,.$$

Notice that H_θ is singular at the angles $\theta = 0$ and $\theta = \pi$, so the trajectories lie between these two values (i.e. the North and South poles cannot be reached unless $\ell_z = 0$). Once fixed the values of ℓ, ℓ_z, the turning points for θ are given by the solutions of $\ell^2 = \ell_z^2/\sin^2\theta$.

2.2. New Coordinates

To describe both signs of K in a unified way we introduce κ-dependent trigonometric functions:

$$\mathrm{C}_\kappa(u) \equiv \begin{cases} \cos\sqrt{\kappa}\, u & \kappa > 0 \\ 1 & \kappa = 0 \\ \cosh\sqrt{-\kappa}\, u & \kappa < 0 \end{cases}, \qquad \mathrm{S}_\kappa(u) \equiv \begin{cases} \frac{1}{\sqrt{\kappa}}\sin\sqrt{\kappa}\, u & \kappa > 0 \\ u & \kappa = 0 \\ \frac{1}{\sqrt{-\kappa}}\sinh\sqrt{-\kappa}\, u & \kappa < 0 \end{cases}.$$

The κ-tangent is defined by

$$\mathrm{T}_\kappa(u) \equiv \frac{\mathrm{S}_\kappa(u)}{\mathrm{C}_\kappa(u)}\,.$$

Useful relations among these κ-functions are:

$$\mathrm{C}_\kappa^2(u) + \kappa\, \mathrm{S}_\kappa^2(u) = 1, \qquad \mathrm{C}_\kappa'(u) = -\kappa\, \mathrm{S}_\kappa(u), \qquad \mathrm{S}_\kappa'(u) = \mathrm{C}_\kappa(u)\,.$$

Let us define:

$$K = -\kappa\,, \qquad r = \mathrm{S}_\kappa(\xi), \qquad p_r = \frac{p_\xi}{\mathrm{C}_\kappa(\xi)}\,,$$

where the range of ξ is as follows:

- for $\kappa \leq 0$, $0 < \xi < \infty$;
- for $\kappa > 0$, we have two natural choices: $0 < \xi < \pi/(2\sqrt{\kappa})$ or $\pi/(2\sqrt{\kappa}) < \xi < \pi/\sqrt{\kappa})$.

The variables (ξ, θ, φ) are angular coordinates of the points of a (pseudo) sphere in $\mathbb{R}^4$.

(i) when $\kappa < 0$ they parametrize the points of one sheet of a three dimensional (3D) hyperboloid;

(ii) when $\kappa = 0$ the coordinates (ξ, θ, φ) represent the spherical coordinates of the points of $\mathbb{R}^3$;

(iii) finally, for $\kappa > 0$ these variables with $0 < \xi < \pi/(2\sqrt{\kappa})$ parametrize the points of one half (the north hemisphere) of a deformed 3D sphere, and the values $\pi/(2\sqrt{\kappa}) < \xi < \pi/\sqrt{\kappa}$ give the points of the south hemisphere.

In the Perlick's original case, $\kappa < 0$, this choice of coordinates is not important since the north and south hemispheres of a hyperboloid are not connected.

We end this subsection by making a few remarks about terminology.

In the sequel, we will speak about *shift* and *ladder* functions, with some abuse of language. It is a terminology coming from the quantum context that we import by analogy to the classical setting. A good description can be found for instance in [9].

In a not too rigorous way, in the language of Supersymmetrc Quantum Mechanics, *shift* operators are those associated with a factorization of a second order differential operator (typically the Schroedinger operator) such that, by interchanging the factors, we pass from a given potential to a novel one having the same functional form (*shape*) but a different (i.e. *shifted*) coupling constant. In principle the process can be indefinitely iterated, yielding a sequence of *almost isospectral* operators (at each stage we "miss" the ground state). Conversely, *ladder* operators (as the usual " raising" and " lowering" operators) do not change the original Schroedinger operator, but act on each eigenstate transforming it in a new one associated to a higher (or lower) eigenvalue.

3. Constants of Motion

In this model, we already have three independent constants of motion: $H, H_{\theta\varphi}, H_\varphi$ with constant values given by E, ℓ^2, ℓ_z^2. Our aim is to find additional constants of motion by means of the factorization method in order to show the maximal superintegrability of the system. We recall that the application of the factorisation method to the classical realm was first proposed in [12] and then developed in [13], [14],[15], [16], [17], [18]. A first pair of constants $X^\pm$ will be obtained from the effective Hamiltonians H_ξ and H_θ, and a second pair $Y^\pm$ from the next two effective Hamiltonians, H_θ and H_φ.

(i) The constants of motion $X^\pm$

The first set of constants $X^\pm$ are constructed in terms of the shift functions of H_ξ and the ladder functions of H_θ which are obtained as follows.

- *Shift functions of H_ξ*

  The Hamiltonian H_ξ can be factorized as

$$H_\xi = B^+ B^- + \lambda_\xi ,$$

where

$$B^\pm = \frac{1}{\sqrt{2}} \left(\mp i\,\beta\, p_\xi + \frac{\ell}{\mathrm{T}_\kappa(\xi)} - \frac{1}{\ell} \right), \qquad \lambda_\xi = -\frac{1}{2}\left(\frac{1}{\ell^2} - \kappa\,\ell^2 \right).$$

The shift functions $B^\pm$ are complex conjugate of each other and they satisfy the following PBs together with the Hamiltonian H_ξ

$$\{B^-, B^+\} = i\,\beta\,\frac{\ell}{\mathrm{S}_\kappa{}^2(\xi)}, \qquad \{H_\xi, B^\pm\} = \pm i\,\beta\,\frac{\ell}{\mathrm{S}_\kappa{}^2(\xi)}\,B^\pm .$$

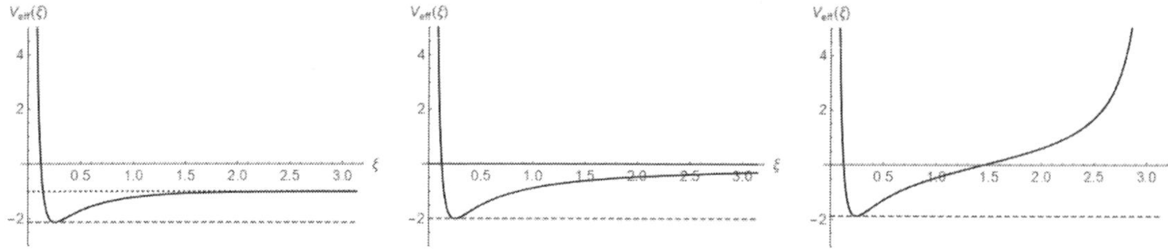

Figure 1. Plot of the effective potential $V_{\rm eff}(\xi)$ for $\kappa = -1$ (left), $\kappa = 0$ (center) and $\kappa = 1$ (right) together with the limiting values of the energy for $\ell = 0.5$. The range of κ for $\kappa \leq 0$ is $0 < \xi < \infty$ and for $\kappa > 0$ it is $0 < \xi < \pi/\sqrt{\kappa}$.

The second PB implies that

$$\{H, B^{\pm}\} = \pm i\,\beta\,\frac{\sqrt{H_{\theta\varphi}}}{\mathrm{S}_\kappa{}^2(\xi)}\,B^{\pm}\,.$$

From the above factorization or the effective potential, we conclude that the energy E of the total Hamiltonian H_ξ for bounded motions must satisfy the following inequalities depending on κ:

$$\begin{aligned} &\kappa < 0\,, \qquad && -\sqrt{|\kappa|} > E \geq -\frac{1}{2}\left(\frac{1}{\ell^2} + |\kappa|\ell^2\right),\\ &\kappa = 0\,, && 0 > E \geq -\frac{1}{2\,\ell^2}\,,\\ &\kappa > 0\,, && \infty > E \geq -\frac{1}{2}\left(\frac{1}{\ell^2} - \kappa\ell^2\right). \end{aligned}$$

- *Ladder functions of H_θ*
  In order to find the ladder functions for the angular Hamiltonian H_θ, first we multiply it by $\sin^2\theta$ and after rearranging, we get

$$-\ell_z^2 = p_\theta^2 \sin^2\theta - H_\theta \sin^2\theta\,.$$

Now, we can factorize the right hand side of this equality in terms of complex conjugate ladder functions

$$-\ell_z^2 = A^+ A^- + \lambda_\theta\,,$$

where

$$A^{\pm} = \mp i\,\sin\theta\,p_\theta + \sqrt{H_\theta}\cos\theta\,, \qquad \lambda_\theta = -H_\theta\,.$$

These ladder functions $A^{\pm}$ together with the Hamiltonian H_θ satisfy the following PBs

$$\{A^-, A^+\} = 2\,i\,\sqrt{H_\theta}\,, \qquad \{H_\theta, A^{\pm}\} = \pm 2\,i\,\sqrt{H_\theta}\,A^{\pm}\,.$$

Therefore, the PB with H is

$$\{H, A^{\pm}\} = \mp i\,\frac{\sqrt{H_{\theta\varphi}}}{\mathrm{S}_\kappa{}^2(\xi)}\,A^{\pm}\,.$$

Notice that the factorization implies the following inequality:

$$\ell^2 \geq \ell_z^2$$

which is reasonable in view of the physical interpretation of ℓ (the total angular momentum) and ℓ_z (the component of the angular momentum in the z-direction).

- *Construction of $X^\pm$*
  From the PBs between B^+ and B^-, taking into account that $\beta = m/n$, we obtain
  $$\{H, X^\pm\} = 0\,, \qquad X^\pm = (A^\pm)^m (B^\mp)^n\,.$$
  Hence
  $$X^\pm = q_{\rm x}\, e^{\pm i\,\alpha_{\rm x}}, \qquad 0 \leq q_{\rm x} < \infty, \quad -\pi \leq \alpha_{\rm x} < \pi.$$
  The absolute value $q_{\rm x}$ is directly obtained from the factorization properties of $A^\pm$ and $B^\pm$:
  $$q_{\rm x} = |X^\pm| = \left(\ell^2 - \ell_z^2\right)^{m/2} \left(E + \frac{1}{2}\left(\frac{1}{\ell^2} - \kappa\,\ell^2\right)\right)^{n/2}.$$

By means of these constants of motion, one can easily find the relation between the variables ξ (or r) and θ along the trajectories, as well as the associate frequencies. However, when $\ell = \ell_z$, this relation breaks, because $p_\theta = 0$ and $\theta = \pi/2$. In other words, if $\ell = \ell_z$, the motion takes place in the horizontal plane $z = 0$. This particular case will be considered in a separate subsection.
Notice that $X^\pm$ are not polynomial in momenta, since $A^\pm$ and $B^\pm$ depend on ℓ which is a square root . However, expanding $m-th$ and $n-th$ powers, and taking real and imaginary parts,one gets constants that are polynomial in momenta.

(ii) *The constants of motion $Y^\pm$*
This new set of constants of motion is derived from the shift functions of H_θ and the ladder functions of H_φ. They are obtained in a similar way as the previous set.

- *Shift functions of H_θ*
  The angular Hamiltonian is factorized as
  $$H_\theta = p_\theta^2 + \frac{\ell_z^2}{\sin^2\theta} = C^+C^- + \lambda_\ell\,,$$
  where
  $$C^\pm = \mp i\, p_\theta + \ell_z \cot\theta\,, \qquad \lambda_\ell = \ell_z^2\,.$$
  These factor functions $C^\pm$ together with the Hamiltonian H_θ satisfy the following PBs
  $$\{C^-, C^+\} = 2\,i\,\frac{\ell_z}{\sin^2\theta}\,, \qquad \{H_\theta, C^\pm\} = \pm 2\,i\,\frac{\ell_z}{\sin^2\theta}\,C^\pm\,.$$
  The second PB implies that
  $$\{H_{\theta\varphi}, C^\pm\} = \pm 2\,i\,\frac{\sqrt{H_\varphi}}{\sin^2\theta}\,C^\pm\,.$$

- *Ladder functions of H_φ*
  The Hamiltonian $H_\varphi = p_\varphi^2$ is factorized as
  $$H_\varphi = D^+D^-\,, \qquad D^\pm = \sqrt{H_\varphi}\,e^{\mp i\varphi}\,.$$
  These ladder functions $D^\pm$ together with the Hamiltonian H_φ satisfy the following PBs
  $$\{D^-, D^+\} = 2\,i\,\sqrt{H_\varphi}, \qquad \{H_\varphi, D^\pm\} = \pm 2\,i\,\sqrt{H_\varphi}\,D^\pm\,.$$
  Then, the PB with the Hamiltonian $H_{\theta\varphi}$ is
  $$\{H_{\theta\varphi}, D^\pm\} = \pm 2\,i\,\frac{\sqrt{H_\varphi}}{\sin^2\theta}\,D^\pm\,.$$

IOP Conf. Series: Journal of Physics: Conf. Series **965** (2018) 012033 doi:10.1088/1742-6596/965/1/012033

- *Final expressions for* $Y^{\pm}$
The Hamiltonian $H_{\theta\varphi}$ has the constants of motion $Y^{\pm}$.

$$\{H_{\theta\varphi}, Y^{\pm}\} = 0\,, \qquad Y^{\pm} = (C^{\pm})(D^{\mp})\,.$$

The geometric meaning of these constants of motion can be appreciated by rewriting them in the form

$$Y^{\pm} = -L_z(L_x \pm iL_y)\,,$$

where L_x, L_y and L_z are the components of the angular momentum in the x, y and z direction, respectively:

$$L_x = -\sin\varphi\, p_\theta - \cot\theta\,\cos\varphi\, p_\varphi\,, \quad L_y = \cos\varphi\, p_\theta - \cot\theta\,\sin\varphi\, p_\varphi\,, \quad L_z = p_\varphi\,.$$

On the other hand they can be cast in the form

$$Y^{\pm} = q_{\rm y}\, e^{\pm i\alpha_{\rm y}}, \qquad 0 \le q_{\rm y} < \infty, \quad \pi \le \alpha_{\rm y} < \pi\,,$$

with

$$q_{\rm y} = |Y^{\pm}| = \ell_z(\ell^2 - \ell_z^2)^{1/2}\,,$$

while the angle $\alpha_{\rm y}$ coincides with the azimuthal angle of the angular momentum vector $\boldsymbol{L} = (L_x, L_y, L_z)$.
The value of this constant of motion fixes the relation of the angles θ and φ along a trajectory. For example, for $\ell_y = 0$, $\alpha_{\rm y} = \pi$, one gets the following expression for $\theta(\varphi)$ (or $\varphi(\theta)$)

$$\cot\theta = -\frac{\ell_x}{\ell_z}\cos\varphi\,.$$

However, as mentioned before, this relation also breaks when $\ell = \ell_z$, since in this case $\ell_x = 0$ and therefore $\theta = \pi/2$.
It turns out that $Y^{\pm}$ are polynomial in the momentum variables.
Thus, we have arrived at an obvious result: the Hamiltonian $\mathbf{L}^2$ has the symmetries L_z and $Y^{\pm}$ or equivalently L_x, L_y and L_z. The basis L_z, $Y^{\pm}$ will be the most adequate to write the symmetry algebra.

4. The Trajectories

From the results derived in the previous sections it follows that, once the values of the constants of motion E, ℓ, ℓ_z have ben fixed, the new constants of motion $X^{\pm}$ and $Y^{\pm}$ determine the relation between the coordinates r-θ and θ-φ, respectively. In this way, one can obtain all the trajectories of the system. In order to characterize those trajectories we will use the properties of the ladder and shift functions.
The ladder functions $B^{\pm}$, and the shift functions, $A^{\pm}$, can be expressed as

$$B^{\pm}(\xi, p_\xi) = \left(E + \tfrac{1}{2}\left(\tfrac{1}{\ell^2} - \kappa\,\ell^2\right)\right)^{1/2} e^{\pm i\, b(\xi, p_\xi)}\,,$$
$$A^{\pm}(\theta, p_\theta) = \left(\ell^2 - \ell_z^2\right)^{1/2} e^{\pm i\, a(\theta, p_\theta)}\,,$$

where $b(\xi, p_\xi)$ and $a(\theta, p_\theta)$ are real phase functions that depend also on the constants of motion E, ℓ and ℓ_z.

The effective Hamiltonian H_ξ depends of the variables (ξ, p_ξ). When the energy E satisfies the proper restrictions the motion is periodic between two turning points. The variables (θ, p_θ)

are described by the effective Hamiltonian H_θ. For $\ell_z \neq 0$, the motion of these variables will be periodic, and the range of θ is determined by its corresponding turning points.
As a consequence, the functions $b(\xi, p_\xi)$ and $a(\theta, p_\theta)$ will also be periodic. Now, taking into account the constants of motion $X^\pm$, the phases of $A^\pm$, $B^\pm$ and $X^\pm$ are related as follows

$$m\, a(\theta, p_\theta) - n\, b(\xi, p_\xi) = \alpha_{\rm x}, \qquad \xi_1 \le \xi \le \xi_2, \quad \theta_1 \le \theta \le \theta_2 .$$

This equation fixes the relation of the variables (ξ, p_ξ) and (θ, p_θ) along the motion. Therefore, if we time-differentiate the above relation between the phases, we get

$$m\, \dot{a}(\theta, p_\theta) - n\, \dot{b}(\xi, p_\xi) = 0.$$

This implies that the frequencies ω_ξ and ω_θ are related by

$$m\, \omega_\theta - n\, \omega_\xi = 0.$$

In the same way, we can write the functions $C^\pm$ and $D^\pm$ in the form

$$C^\pm(\theta, p_\theta) = (\ell^2 - \ell_z^2)^{1/2} e^{\pm i\, c(\theta, p_\theta)} ,$$

$$D^\pm(\varphi, p_\varphi) = \ell_z\, e^{\pm i\, d(\varphi, p_\varphi)} ,$$

where $c(\theta, p_\theta)$ and $d(\varphi, p_\varphi)$ are real phase functions. Due to the constants of motion $Y^\pm$ these variables are related by

$$c(\theta, p_\theta) - d(\varphi, p_\varphi) = \alpha_{\rm y}, \qquad \theta_1 \le \theta \le \theta_2 , \quad -\pi \le \varphi < \pi .$$

As the motion of the (θ, p_θ) variables and the (φ, p_φ) is periodic, differentiating the above equality with respect to time, we obtain

$$\omega_\theta - \omega_\varphi = 0.$$

Hence, the frequencies of the angular variables θ and φ are equal (except for the case $\ell = \ell_z$).

In conclusion, when the energy E satisfies the above restrictions, the motion is bounded, and the frequencies of the three variables are related as we have shown. This implies that the bounded motion is periodic and the trajectories are closed. In conclusion, we have checked in this case that the Bertrand's theorem is satisfied.

In Fig. 2 and Fig. 3 some examples of the trajectories for different values of the energies E and of the parameter β corresponding to bounded and unbounded motions are shown for the case $\kappa = -1$. The trajectories of the initial Hamiltonian are plotted in a three dimensional space $(r \sin\theta \cos\varphi, r \sin\theta \sin\varphi, r\cos\theta)$, where (r, φ, θ) are the spherical coordinates in $\mathbb{R}^3$.

5. Algebra of the constants of motion for the Perlick system type I

As we have seen in the previous sections, we have obtained seven constants of motion for the three dimensional Perlick system: $H, H_{\theta\varphi}, H_\varphi, X^\pm, Y^\pm$. Of course, only five of these constants are independent. For example we can choose: $H, H_{\theta\varphi}, H_\varphi$, X^+, Y^+. Therefore, the Perlick system type I for a rational $\beta = m/n$ is a maximally superintegrable system. It turns out that, all the aforementioned seven constants are useful to express in a simple form the algebraic

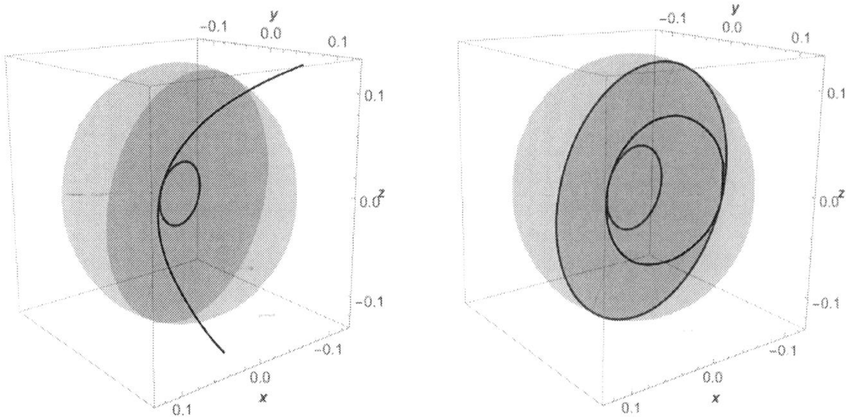

Figure 2. Plot of the trajectories of $\beta = 1$ for the values $E = -1$ (left), $E = -6$ (right), $\kappa = -1$, $\ell = 0.25$ and $\ell_z = 0.1$.

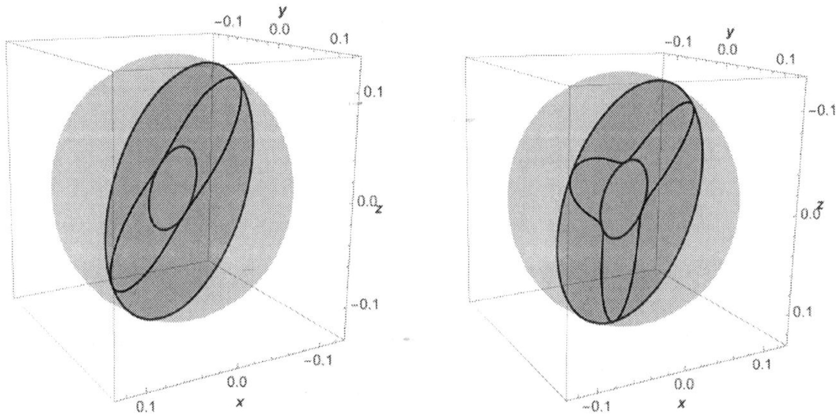

Figure 3. Plot of the trajectories of $\beta = 2$ (left), $\beta = 3$ (right) for the values $E = -6$, $\kappa = -1$, $\ell = 0.25$ and $\ell_z = 0.1$.

structure defined in terms of Poisson brackets:

$$\{H, \cdot\} = 0, \qquad \{H_{\theta\varphi}, Y^{\pm}\} = 0, \qquad \{H_{\theta\varphi}, X^{\pm}\} = \pm 2\, i\, m \sqrt{H_{\theta\varphi}}\, X^{\pm},$$

$$\{H_{\varphi}, Y^{\pm}\} = \mp 2\, i \sqrt{H_{\varphi}} Y^{\pm}, \qquad \{H_{\varphi}, X^{\pm}\} = 0, \qquad \{Y^{+}, Y^{-}\} = 2\, i \sqrt{H_{\varphi}}\, (H_{\theta\varphi} - 2H_{\varphi}),$$

$$\{X^{+}, X^{-}\} = i\, m\, n\, (H_{\theta\varphi} - H_{\varphi})^{m}\, (H_{\xi} + \frac{1}{2}(\frac{1}{H_{\theta\varphi}} - \kappa\, H_{\theta\varphi}))^{n-1} (\kappa \sqrt{H_{\theta\varphi}} + \frac{1}{H_{\theta\varphi}^{3/2}})$$
$$-2\, i\, m^{2} (H_{\theta\varphi} - H_{\varphi})^{m-1}\, (H_{\xi} + \frac{1}{2}(\frac{1}{H_{\theta\varphi}} - \kappa\, H_{\theta\varphi}))^{n} \sqrt{H_{\theta\varphi}},$$

$$\{X^{\pm}, Y^{\pm}\} = \mp \frac{i\, m}{\sqrt{H_{\theta\varphi}} + \sqrt{H_{\varphi}}} X^{\pm} Y^{\pm}, \qquad \{X^{\pm}, Y^{\mp}\} = \mp \frac{i\, m}{\sqrt{H_{\theta\varphi}} - \sqrt{H_{\varphi}}} X^{\pm} Y^{\mp}.$$

In this way, we have found the algebra of the constants of motion for any two coprime integers m and n and any value of the constant κ. The corresponding polynomial algebras can also be found. .
The complex constants of motion $X^{\pm}$ and $Y^{\pm}$ allow to get real constants: $Y_R = \mathrm{Re}(Y^{\pm}), Y_I =$

$\mathrm{Im}(Y^{\pm})$, $X_R = \mathrm{Re}(X^{\pm})$, $X_I = \mathrm{Im}(X^{\pm})$, which will close a real algebra. Besides, for $\beta = 1, \kappa = 0$, the complex constants of motion $X^{\pm}$ can be expressed in terms of the angular momentum $\mathbf{L} = (L_x, L_y, L_z)$ and Runge-Lenz $\boldsymbol{\mathcal{A}} = (\mathcal{A}_x, \mathcal{A}_y, \mathcal{A}_z)$ vectors,

$$X^{\pm} = \mathrm{Re}(X^{\pm}) \pm i\,\mathrm{Im}(X^{\pm}) = \frac{\mathcal{A}_z}{\sqrt{2}} \pm i\frac{(\mathbf{L}\times\boldsymbol{\mathcal{A}})_z}{\sqrt{2}\,\ell}\,,$$

where

$$\boldsymbol{\mathcal{A}} = \mathbf{p}\times\mathbf{L} - \hat{\mathbf{r}}\,.$$

Here, we have used the expressions of $\mathbf{r}, \mathbf{p}, \mathbf{L}$ in spherical coordinates. Notice that in the context of the analysis of the Perlick system, a generalization of the Runge-Lenz vector has been proposed in [7]. In the next section we will consider the motion in the equatorial plane $\theta = \pi/2$.

6. Perlick I on the Equatorial Plane

By choosing the plane described by the initial position and velocity of the "planet" to be the Horizontal plane $z = 0$ (or the "equatorial" plane $\theta = \pi/2$) the expression of the relevant constants of motion and of the corresponding trajectories simplify dramatically.
Indeed, by taking $\theta = \pi/2$ so that $\mathbf{L} = (0, 0, L_z)$ ($\ell = \ell_z$). the corresponding Hamiltonian in the remaining r, φ variables is

$$\widetilde{H}^{\pm} = \beta^2(1 + K\,r^2)\,\frac{p_r^2}{2} + \frac{{p_\varphi}^2}{2\,r^2} \pm \frac{1}{r}\sqrt{1 + K\,r^2}\,.$$

Now, we have two trivial constants of motion: the total energy E and ℓ_z. Following the procedure outlined in the previous sections, letting $(r, p_r) \to (\xi, p_\xi)$, we get the Hamiltonian

$$H(\xi, \theta, \varphi) = \beta^2\,\frac{p_\xi^2}{2} + \frac{{p_\varphi}^2}{2}\frac{1}{\mathrm{S}_\kappa^2(\xi)} - \frac{1}{\kappa(\xi)}\,,$$

with the same range of the variable ξ. In this case the effective Hamiltonian H_ξ is given by

$$H_\xi = \beta^2\,\frac{p_\xi^2}{2} + \frac{{\ell_z}^2}{2}\frac{1}{\mathrm{S}_\kappa^2(\xi)} - \frac{1}{\mathrm{T}_\kappa(\xi)} = \beta^2\,\frac{p_\xi^2}{2} + V_{\rm eff}(\xi)\,.$$

We have two additional constants of motion $Z^{\pm}$ for that Hamiltonian:

$$\{H, Z^{\pm}\} = 0\,, \qquad Z^{\pm} = (D^{\pm})^m (B^{\mp})^n\,,$$

where

$$B^{\pm} = \frac{1}{\sqrt{2}}\left(\mp i\,\beta\,p_\xi + \frac{\ell_z}{\mathrm{T}_\kappa(\xi)} - \frac{1}{\ell_z}\right), \qquad D^{\pm} = \sqrt{H_\varphi}\,e^{\mp i\varphi}\,,$$

and

$$H_\xi = B^+B^- + \lambda_\xi\,, \qquad \lambda_\xi = -\frac{1}{2}\left(\frac{1}{\ell_z^2} - \kappa\,\ell_z^2\right),$$

$$H_\varphi = p_\varphi^2 = D^+D^-\,.$$

These constants of motion $Z^{\pm}$ have complex values denoted by

$$Z^{\pm} = q_z\,e^{\pm i\,\varphi_z}\,,$$

where $-\pi \leq \varphi_z < \pi$ and $0 \leq q_z < \infty$. The modulus q_z has a value determined by the other constants of motion E and ℓ_z,

$$q_z = |Z^{\pm}| = \ell_z^m \left(E + \frac{1}{2}\left(\frac{1}{\ell_z^2} - \kappa, \ell_z^2\right)\right)^{n/2}.$$

The functions $B^{\pm}$ and $D^{\pm}$ can be written as

$$B^{\pm}(\xi, p_\xi) = \left(E + \frac{1}{2}\left(\frac{1}{\ell_z^2} - \kappa\,\ell_z^2\right)\right)^{1/2} e^{\pm i\, b(\xi, p_\xi)},$$
$$D^{\pm}(\varphi, p_\varphi) = \ell_z\, e^{\pm i\, d(\varphi, p_\varphi)},$$

where $b(\xi, p_\xi)$ and $d(\varphi, p_\varphi)$ are the corresponding real phase functions. When the energy E satisfies the bound-state restriction, the bounded motion of the variables (ξ, p_ξ) is periodic. Besides, the motion of the variables (φ, p_φ) is also periodic, due to the angular character of φ. Then we get

$$m\, d(\varphi, p_\varphi) - n\, b(\xi, p_\xi) = \varphi_z, \qquad \xi_1 \leq \xi \leq \xi_2, \quad -\pi \leq \varphi < \pi.$$

Whence, by a t-differentiation:

$$m\,\omega_\varphi - n\,\omega_\xi = 0.$$

The constant of motion $Z^{\pm}$ reads:

$$(\ell_z\, e^{\mp i\,\varphi})^m\, (\mp i\,\frac{\beta}{\sqrt{2}}\, p_\xi + \frac{\ell_z}{\sqrt{2}}\frac{1}{\mathrm{T}_\kappa(\xi)} - \frac{1}{\ell_z\sqrt{2}})^n = q_z\, e^{\pm i\,\varphi_z}.$$

Taking the real and imaginary parts of the equation above we obtain the relation between ξ and φ along the trajectories:

$$\cos\left(\frac{1}{n}\varphi_z + \frac{m}{n}\varphi\right) = \frac{\dfrac{\ell_z^2}{\mathrm{T}_\kappa(\xi)} - 1}{\sqrt{2E\,\ell_z^2 + 1 - \kappa\,\ell_z^4}}.$$

which becomes the well known conic section equation for the values $\kappa = 0$, $\beta = m/n = 1$ and $\varphi_z = 0$:

$$\frac{\alpha}{\xi} = 1 + \varepsilon\,\cos\varphi,$$

where $\varepsilon^2 = 2E\,\ell_z^2 + 1$ (eccentricity) and $\alpha = \ell_z^2$ (semi-latus rectum) [10]. In this case, the problem is reduced to the Kepler-Coulomb system in the Euclidean plane. If $\kappa = -1$ ($\kappa = 1$) and $\beta = 1$, then we get the conic section equation for the hyperboloid (sphere) [11]

$$\kappa = -1, \quad \text{(Hyperbolic)} \qquad \frac{\alpha}{\tanh\xi} = 1 + \sqrt{\varepsilon^2 + \alpha^2}\,\cos\varphi,$$

$$\kappa = 1, \quad \text{(Spherical)} \qquad \frac{\alpha}{\tan\xi} = 1 + \sqrt{\varepsilon^2 - \alpha^2}\,\cos\varphi.$$

So, we may say that our trajectory equation can be considered as a generalized conic section equation.

In the special case where $\kappa = 0$ and $\beta = m/n = 1$, the constants of motion $Z^{\pm}$ can also be expressed in terms of the Runge-Lenz vector $\boldsymbol{\mathcal{A}} = (\mathcal{A}_x, \mathcal{A}_y, 0)$,

$$Z^{\pm} = \mathrm{Re}(Z^{\pm}) \pm i\,\mathrm{Im}(Z^{\pm}) = \frac{\mathcal{A}_x}{\sqrt{2}} \mp i\frac{\mathcal{A}_y}{\sqrt{2}}.$$

For the other cases ($\kappa = \pm 1$), the constants of motion $Z^{\pm}$ can be written in terms of a type of generalized Runge-Lenz vector [7] Note that the constants H, H_φ, $Z^{\pm}$ close a similar algebra as in the general case.

7. Concluding Remarks and Future Perspectives

In the above pages we have tersely recalled the algebraic and geometrical properties of the classical Perlick system of type I, which have been described in term of the factorization approach. The further challenging task is performing a similar construction for the full Perlick family II, on which some preliminary results (not yet published) have been already derived.

References

[1] Perlick V 1992 *Class. Quantum Grav.* **9** 1009.
[2] Bertrand J 1873 *C. R. Acad. Sci. Paris* **77** 849
[3] Evans NW 1990 *Phys. Rev. A* **41** 5666
[4] Ballesteros A, Enciso A, Herranz FJ, Ragnisco O and Riglioni D 2011 *Sigma* **7** 048
[5] Fradkin DM 1965 *Amer. J. Phys.* **33** 207
[6] Ballesteros A and Herranz FJ 2009 *J. Phys. A: Math.Theor.* **42** 245203
[7] Ballesteros A, Enciso A, Herranz FJ and Ragnisco O 2009 *Comm. Math. Phys.* **290** 1033
[8] Kuru Ş, Negro J and Ragnisco O 2017 *Phys. Lett. A* **381** 3355
[9] Cooper JL 1993 *J. Phys. A: Math. Gen.* **26** 1601
[10] Goldstein H, Poole C.P. and Safko JL 2001 *Classical Mechanics (3rd Edition)* (New York: Addison-Wesley)
[11] Cariñena JF, Rañada MF and Santander M 2005 *J. Math. Phys.* **46** 052702
[12] Kuru Ş and Negro J 2008 *Ann. Phys.* **323** 413
[13] Kuru Ş and Negro J 2012 *J. Phys.: Conf. Ser.* **343** 012063
[14] Celeghini E, Kuru Ş, Negro J and del Olmo MA 2013 *Ann. Phys.* **332** 27
[15] Calzada J.A., Kuru Ş and Negro J 2014 *Eur. Phys. J. Plus* **129** 164.
[16] Ballesteros A, Herranz FJ, Kuru Ş and Negro J 2016 *Ann. Phys.* **373** 399
[17] Ballesteros A, Herranz FJ, Kuru Ş and Negro J 2017 *Phys. At. Nucl.* **80** 389
[18] Latini D and Ragnisco O 2015 *J. Phys. A: Math. Theor.* **48** 175201

IOP Conf. Series: Journal of Physics: Conf. Series **965** (2018) 012034 doi:10.1088/1742-6596/965/1/012034

Generalized Pareto optimum and semi-classical spinors

M Rouleux

Aix-Marseille University, Université de Toulon, CNRS, CPT, Marseille, France

E-mail: rouleux@univ-tln.fr

Abstract. In 1971, S.Smale presented a generalization of Pareto optimum he called the critical Pareto set. The underlying motivation was to extend Morse theory to several functions, i.e. to find a Morse theory for m differentiable functions defined on a manifold M of dimension ℓ. We use this framework to take a 2×2 Hamiltonian $\mathcal{H} = \mathcal{H}(p) \in C^\infty(T^*\mathbf{R}^2)$ to its normal form near a singular point of the Fresnel surface. Namely we say that $\mathcal{H}$ has the Pareto property if it decomposes, locally, up to a conjugation with regular matrices, as $\mathcal{H}(p) = u'(p)C(p)(u'(p))^*$, where $u : \mathbf{R}^2 \to \mathbf{R}^2$ has singularities of codimension 1 or 2, and $C(p)$ is a regular Hermitian matrix ("integrating factor"). In particular this applies in certain cases to the matrix Hamiltonian of Elasticity theory and its (relative) perturbations of order 3 in momentum at the origin.

1. Matrix valued Hamiltonian systems and crossing of modes

Here we recall some well known facts about generic normal forms for a matrix Hamiltonian near a crossing of modes. Consider a real valued symmetric Hamiltonian $\mathcal{H} \in C^\infty(T^*\mathbf{R}^d) \otimes \mathbf{R}^n$. We may also replace $\mathcal{H}(x,\xi)$ by a more general symbol (in a suitable class) with asymptotic sum $\mathcal{H}(x,\xi;h) = \mathcal{H}_0(x,\xi) + h\mathcal{H}_1(x,\xi) + \cdots$, and consider its semi-classical Weyl quantization

$$\mathcal{H}^w(x,hD_x;h)U(x;h) = (2\pi h)^{-d}\int\int e^{i(x-y)\eta/h}\mathcal{H}(\frac{x+y}{2},\eta;h)U(y)\,dy\,d\eta$$

This defines an operator essentially self-adjoint on the space of square-integrable "spinors" $L^2(\mathbf{R}^d) \otimes \mathbf{C}^n$. For simplicity we keep denoting by $\mathcal{H}$ the principal symbol $\mathcal{H}_0$ which actually plays the main role in the present analysis.

Let $\Delta(x,\xi) = \det \mathcal{H}(x,\xi)$, and $N = \{(x,\xi) \in T^*\mathbf{R}^d, \Delta(x,\xi) = 0\}$ be the characteristic variety.

At points of N, the polarisation space $\ker H(x,\xi)$ has positive dimension $k \geq 1$.

For $k = 1$, $\{(x,\xi) \in N : d\Delta \neq 0\}$ is a smooth hypersurface, and modulo an elliptic factor, we can reduce $\mathcal{H}$ to a scalar symbol with simple characteristics; the corresponding operator, of real principal type, can again be locally conjugated to hD_{x_1} by an elliptic Fourier integral operator (FIO); this induces well-known polarization properties for solutions of the original system.

Effects which are truly specific for systems occur therefore when $k \geq 2$; let $\Sigma = \{(x,\xi) \in N : d\Delta = 0\}$ be the singular part of N.

For generic Hamiltonian $\mathcal{H}$, Σ is a stratified set. Splitting off elliptic summands, H can be (generically) reduced to a 2×2 system. For a 2×2 symmetric system $\begin{pmatrix} a+f & b \\ b & -a+f \end{pmatrix}$, one should generically move 2 parameters to bring it to a matrix with an eigenvalue of multiplicity 2, and one more to bring this eigenvalue to 0; thus the top (larger) stratum of Σ is of codim 3. On directions transverse to Σ, N looks like a quadratic cone. In the next item we review some known results, making these observations more precise.

a) Classical and semi-classical normal forms for generic H

We first mention Arnold's [1] normal forms for (generic) $\Delta = \det \mathcal{H}$ near the stratum of Σ of codim 3: at a conical point (i.e. where $a = b = c = 0$), there are symplectic coordinates (p, q) such that modulo an elliptic factor, we can bring Δ to one of the following expressions :

$$\Delta(x,\xi) \equiv p_1 q_1 - p_2^2 \quad \text{hyperbolic conical point}$$
$$\Delta(x,\xi) \equiv p_1^2 + q_1^2 - p_2^2 \quad \text{elliptic conical point}$$

When $d \geq 3$ Braam and Duistermaat [6] [7] have further brought $\mathcal{H}$ to its normal form near the stratum of Σ of codim 3 at a conical point : Modulo elliptic summands, we can reduce to the case where $\mathcal{H}$ is a 2×2 real symmetric matrix, and in the sense of quadratic forms, i.e. by replacing $\mathcal{H}$ by $\mathcal{A}^*\mathcal{H}\mathcal{A}$, where $\mathcal{A}$ is elliptic, there are symplectic coordinates (p, q) such that

$$\mathcal{A}^*\mathcal{H}\mathcal{A} = \begin{pmatrix} -p_1 + p_2 & q_1 p_3 \\ q_1 p_3 & p_1 + p_2 \end{pmatrix} \quad \text{hyperbolic conical point}$$
$$\mathcal{A}^*\mathcal{H}\mathcal{A} = \begin{pmatrix} p_1 - p_2 & q_2 p_3 \\ q_2 p_3 & p_1 + p_2 \end{pmatrix} \quad \text{elliptic conical point}$$

Moreover these normal forms extend naturally to quantization of Weyl symbols, composing $\mathcal{A}$ by a suitable FIO $\mathcal{U}$ microlocally defined near the conical point, and associated with the canonical change of coordinates $(x, \xi) \mapsto (q, p)$. Namely, in the hyperbolic case

$$\mathcal{U}^*(\mathcal{A}^*\mathcal{H}\mathcal{A})^w\mathcal{U} = \begin{pmatrix} -hD_{q_1} + hD_{q_2} & q_1 hD_{q_3} \\ q_1 hD_{q_3} & hD_{q_1} + hD_{q_2} \end{pmatrix}$$

and similarly in the elliptic case.

Normal forms are useful for studying the Hamiltonian flow of $\mathcal{H}(x, \xi)$ or finding quasi-modes for $\mathcal{H}(x, hD_x)$ in terms of classical functions. These normal forms are structurally stable, i.e. they are not affected by small perturbations of a generic $\mathcal{H}$ in the $C^\infty(T^*\mathbf{R}^d)\otimes\mathbf{R}^n$ topology. Even more precise results are available in 1-D, see e.g. [15] in the framework of Born-Oppenheimer approximation, or [8], [3] for Bogoliubov-de Gennes Hamiltonian.

b) Particular cases: $k = 2$, codim 3 singularities

In many physical situations however, the genericity assumption is not fulfilled, and special reductions should be carried out :

• *Conical refraction in 3-D* Recall from [11] in dimension $d = 3$ that the normal form is given by the symmetric matrix

$$\mathcal{H} = \begin{pmatrix} p_1 + p_2 & p_3 \\ p_3 & p_1 - p_2 \end{pmatrix} \tag{1}$$

which is independent of space variables. This is not structurally stable, because the singular part Σ of N is involutive.

- *Graphene Hamiltonian in 2-D.* The Hamiltonian is a complex Hermitian matrix [9]:

$$\mathcal{H}(p) = \begin{pmatrix} 0 & f(p_1, p_2) \\ \overline{f}(p_1, p_2) & 0 \end{pmatrix} = |\lambda(p)| \begin{pmatrix} 0 & e^{-i\theta(p)} \\ e^{-i\theta(p)} & \end{pmatrix} \tag{2}$$

Here $p = (p_1, p_2) \in \mathbf{R}^2$ are quasi-momenta, and the eigenvalues are explicitely given by

$$\lambda(p) = \pm t\sqrt{3 + 2\cos(\sqrt{3}p_1 a) + 4\cos(\sqrt{3}p_2 a/2)\cos(3p_2 a/2)}$$

Energy vanishes at Dirac points (2 Dirac points per hexagonal cell). The linearization at the Dirac point is of the form

$$\mathcal{H}(p) = v \begin{pmatrix} 0 & k_1 - ik_2 \\ k_1 + ik_2 & 0 \end{pmatrix}, \quad v = 3t/(2a) \tag{3}$$

c) Particular case: higher order singularities

Here we are concerned in the case where Δ vanishes of order 4 at the conical point (leaving open the case $k = 3$). The physical example is the Hamiltonian quadratic in the momenta from Elasticity Theory [5] in (2+1)-D, that is obtained in the following way.

On the set of maps $\Gamma : \mathbf{R}_t \times \mathbf{R}^2_x \to \mathbf{R}^2, (t, x, y) \mapsto (\phi(x, y, t), \psi(x, y, t))$ with Sobolev regularity $H^1(\mathbf{R}^3)$, we consider the Lagrangian density:

$$\mathcal{L}(\begin{pmatrix} x \\ y \end{pmatrix}, \begin{pmatrix} \phi_t \\ \psi_t \end{pmatrix}, \begin{pmatrix} \phi_x \\ \psi_x \end{pmatrix}, \begin{pmatrix} \phi_y \\ \psi_y \end{pmatrix}) = T - V = \frac{1}{2}(\phi_t^2 + \psi_t^2) - \frac{1}{2}(\phi_x^2 + \psi_y^2) - \frac{1}{2}A(x, y)(\phi_y + \psi_x)^2$$

Euler-Lagrange equations from the variational principle $\delta \int \mathcal{L}\, dt\, dx\, dy = 0$ lead to extremal curves (rays) in the (t, x, y)-space, and the set of points (t, x, y) connected to the origin in $\mathbf{R}^3$ by such a ray is called the "world front". A matter of interest are the singularities of the world front.

Applying Fourier transformation with respect to (x, y, t)

$$\phi \mapsto \widehat{\phi}, \ \phi_x \mapsto \xi\widehat{\phi}, \ \phi_y \mapsto \eta\widehat{\phi}, \ \phi_t \mapsto \tau\widehat{\phi}, \ \text{etc.}\ldots$$

switches from Lagrangian formulation to Hamiltonian formulation and leads to a Pseudo-differential Hamiltonian system (as a quadratic form in $(\widehat{\phi}, \widehat{\psi})$) with principal symbol

$$\mathcal{H} = \begin{pmatrix} \tau^2 & 0 \\ 0 & \tau^2 \end{pmatrix} - \begin{pmatrix} \xi^2 + A(x, y)\eta^2 & A(x, y)\xi\eta \\ A(x, y)\xi\eta & A(x, y)\xi^2 + \eta^2 \end{pmatrix} \tag{4}$$

Genericity properties for this Hamiltonian imply as above that the singular set Σ is of codimension 3, so that it can be brought to one of Arnold-Braam-Duistermaat normal forms (in 2-D) near Σ. However, genericity breaks down in the case of constant coefficient $A(x, y) =$ Const., which justify a direct approach.

In particular, for $A = \frac{1}{2}$, the spatial part has determinant

$$\Delta(\xi, \eta) = \frac{1}{2}(\xi^2 + \eta^2)^2$$

which vanishes of order 4 at $(\xi, \eta) = 0$.

Our purpose, precisely in this case, is to provide a normal form for $\mathcal{H}$ near $\tau = \xi = \eta = 0$. This could be achieved by a straightforward diagonalisation of $\mathcal{H}$, but our method carries naturally to (relative) perturbations of this Hamiltonian, depending on (ξ, η) alone; more naively this example is intended as an illustration of the role played by the generalized Pareto set.

2. Generalized Pareto optimum

A central problem in Differential Calculus consists in maximizing a function: Morse theory on a smooth manifold provides a globalization of this problem.

Economists are rather concerned in maximizing simultenaously several "utility functions", obtaining this way the notion of Pareto optimum in a free exchange economy.

In 1971, S.Smale presented a generalization of Pareto optimum he called the *critical Pareto set* [16]. The main mathematical motivation is to find a Morse theory for m differentiable functions defined on a manifold M of dimension ℓ. Note that this problem is distinct from this of relative extrema of a single function, which is relevant to Lagrange multipliers.

a) "Classical" Pareto optimum in a free exchange economy

Consider a free exchange economy consisting in m consumers $i = 1, \cdots, m$, and for each i, let $x_i = (x_i^{(1)}, \cdots, x_i^{(J)})$ represent the (positive) amount of goods $j = 1, \cdots, J$, with $x_i^{(j)} \geq 0$. We define the "commodity space" as $P = \{x = (x^{(1)}, \cdots, x^{(J)}) \in \mathbf{R}^J; x^{(j)} \geq 0\}$. An unrestricted state of the economy is a point $\vec{x} = (x_1, \cdots, x_m) \in P^m$, but we may restrict to the affine space $M = \{\vec{x} \in P^m : \sum_{i=1}^m x_i^{(j)} = C^{(j)}, j = 1, \cdots, J\}$ with total ressource $C^{(j)}$ of good j.

Each consumer is supposed to have an utility function $u_i : P \to \mathbf{R}$ (generally an homogenous function of $x^{(1)}, \cdots, x^{(J)}$). Thus consumer i prefers x_i to x_i' iff $u_i(x_i) > u_i(x_i')$. The level sets $u_i^{-1}(c)$, $c > 0$ are called "indifference surfaces".

One considers exchanges in M which will increase each u_i on M. A state $\vec{x} \in M$ is called "Pareto optimal" iff there is no $\vec{x}' \in M$ such that $u_i(x_i) \geq u_i(x_i')$ for all i, and $u_j(x_j) > u_j(x_j')$ for some j. If $\vec{x} \in M$ is not Pareto, $\vec{x}$ is not economically stable. For $m = 1$ Pareto optimum equals the usual notion of a (constrained) maximum of $u : M \to \mathbf{R}, x \mapsto u(x)$.

Physicists would replace everywhere the words "maximizing" by "minimizing", and "total resource" by "total energy".

b) Generalized Pareto optimum in the sense of Smale.

Here we do not only consider (joined) maxima, but also (joined) critical points.

Let M be a smooth manifold, dim $M = \ell$, and $u : M \to \mathbf{R}^m$ be m smooth functions defined by $u = (u_1, \cdots, u_m)$ (vector of "utilities").

Let $H(x) = (u'(x))^{-1}(\mathbf{R}_+^m)$, where $u'(x) : T_xM \to \mathbf{R}_+^m$ denotes the derivative (Jacobian) of u at $x \in M$, and $\mathbf{R}_+^m$ the set of $v = (v_1, \cdots, v_m) \in \mathbf{R}^m$ with $v_j > 0$ all j.

Definition 1 [16]: We call $\Theta = \{x \in M : H(x) = \emptyset\}$ the Pareto critical set.

Thus the relation $x \in \Theta$ means that there is no smooth curve $\gamma :]-\epsilon, \epsilon[\to M$ starting at x, and such that $t \mapsto u_i \circ \gamma(t)$ increases for all i's (gradient flow dynamics). For $m = 1$, Pareto critical set is just the set of critical points of u.

If u is a (single) Morse function on M, Θ is a discrete set, and the Hessian $u''(x)$ is defined intrinsically on Θ. For $m > 1$, Θ need not be discrete; but when it consists, as is usual, of a submanifold of dim $m - 1$, then $u''(x)$ is still intrinsically defined on Θ and valued in the 1-D space $\mathbf{R}^m/\mathrm{Ran}\,u'(x)$.

The open subset $\Theta_S \subset \Theta$ of stable points (classical Pareto set), which reduces for $m = 1$ to the set of local maxima of u, plays a special rôle.

- *Paradigm of Pareto critical in 2-D*: the immersed Klein bottle in $\mathbf{R}^3$.

The paradigm of a Morse function on a compact 2-manifold is the "height function" on the embedded torus, and its (Pareto) critical points are the minimum, 2 saddles, and maximum. Similarly Pareto critical set for 2 functions of 2 variables will be obtained from Klein bottle [Wan], by projecting the immersed bottle in $\mathbf{R}^3$ onto a suitable plane, and looking at the "joined extrema" of the coordinates functions on the projection plane. Another, more convenient immersion of Klein bottle in $\mathbf{R}^3$ (though not so easy to visualize) is the so called "figure eight" or "bagel" immersion, given by

$$x_1 = (r + \cos\tfrac{\theta}{2}\sin v - \sin\tfrac{\theta}{2}\sin 2v)\cos\theta$$
$$x_2 = (r + \cos\tfrac{\theta}{2}\sin v - \sin\tfrac{\theta}{2}\sin 2v)\sin\theta$$
$$x_3 = \sin\tfrac{\theta}{2}\sin v + \cos\tfrac{\theta}{2}\sin 2v$$

with $r > 2$ a parameter, $-\pi \leq \theta < \pi, 0 \leq v < 2\pi$ are the variables. It is obtained by gluing two Möbius bands along their edges. Then the map with typical critical Pareto set is given by $u = (\sqrt{x_1^2 + x_2^2}, x_3)$.

Θ is a stratified set, consisting in a finite number of segments (codimension 1 strata) where $\mathrm{rank}(u') = 1$, terminating at codimension 2 strata (isolated points) where $\mathrm{rank}(u') = 0$.

We will only consider $\ell = m$ which leads to the simplest topology.

- *Elementary Pareto sets*: the case of quadratic polynomials in 2-D.

As in the one-dimensional case, quadratic polynomials in $\mathbf{R}^\ell$ provide useful examples of maps with a typical critical Pareto set. For $m = \ell = 2$ we may take

$$u_1(x) = \tfrac{1}{2}(x_1^2 + x_2^2),\ u_2(x) = \tfrac{1}{2}(x_1^2 + x_1x_2 + x_2^2), \quad \Theta = \{0\}$$
$$u_1(x) = x_1x_2,\ u_2(x) = \tfrac{1}{2}(x_1^2 - x_2^2), \quad \Theta = \{0\}$$
$$u_1(x) = \tfrac{1}{2}(x_1^2 + x_2^2),\ u_2(x) = \tfrac{1}{2}(x_1^2 - x_2^2), \quad \Theta = \{x_1 = 0\}$$

More generally, the critical Pareto set of two elliptic or two hyperbolic quadratic linearly independent polynomials reduces to the origin, and to a line for a "mixed" pair.

3. Matrix valued Hamiltonian systems with the Pareto property

Since we are interested in 2-spinors, we work with $m = \ell = 2$ and $\mathcal{H}(x,p) = \mathcal{H}(p)$.

Definition 2: Let $\mathcal{H} \in C^\infty(\mathbf{R}^2) \otimes \mathbf{R}^2 \otimes \mathbf{R}^2$ be a (real) Hermitian matrix. We say that $\mathcal{H}$ has the *Pareto property* iff there exists a smooth map (but with possibly degenerate derivative) $u : \mathbf{R}^2 \to \mathbf{R}^2$, and a (regular) Hermitian matrix $C(p)$ such that, locally, and up to conjugation with an elliptic factor of the form $\mathcal{H}(p) \mapsto \mathcal{A}(p)^*\mathcal{H}(p)\mathcal{A}(p)$, we have:

$$\mathcal{H}(p) = u'(p)C(p)(u'(p))^*$$

Remark: A hint on this definition is the following: let $p \in \mathbf{R}^2$ such that $u_1'(p) = 0$, then the "pure classical state" $\binom{1}{0} \in N(p)$, and similarly for $\binom{0}{1}$ when $u_2'(p) = 0$. So any "classical state" $\binom{x}{y}$ is a superposition of "classical pure states".

These decompositions are local and sometimes can be checked only in the sense of germs. If exact, we say that $\mathcal{H}$ is "integrable in Pareto sense". Only in 1-D problems, one can consider general Hamiltonians of the form $H(x,p)$.

4. Pareto property and the quadratic Hamiltonian of Elasticity Theory

Because Hamiltonians depending on p alone are not structurally stable, few Hamiltonian systems verify Pareto property. We have :

Theorem 1: For $A = 1/2$, the quadratic Hamiltonian of Elasticity Theory (4) is integrable in Pareto sense: with $u_1(\xi,\eta) = \frac{1}{2\sqrt{2}}(\xi^2-\eta^2)$, $u_2(\xi,\eta) = \frac{1}{\sqrt{2}}\xi\eta$, we have $\mathcal{H}(\xi,\eta) = u'(\xi,\eta)\mathrm{diag}(2,1)(u'(\xi,\eta))^*$ with $u'(u')^* = \frac{1}{2}(\xi^2+\eta^2)\mathrm{Id}$. Let $\sigma_2 = \begin{pmatrix} 0 & i \\ -i & 0 \end{pmatrix}$, we have the "skew-diagonalization"

$$\frac{1}{2}\sigma_2(\mathcal{H}(\xi,\eta)-\tau^2)\sigma_2^* = u'(\xi,\eta)\big(\mathrm{diag}(\frac{1}{2},0)-(\xi^2+\eta^2)^{-1}\tau^2\mathrm{Id}\big)(u'(\xi,\eta))^* \tag{5}$$

From this we can construct by inverse Fourier transformation semi-classical spinors ${}^t(\phi,\psi)(x,y;\tau)$ that verify near "Helmholtz equation" corresponding to (4), τ^2 standing for the energy parameter. We write $\widetilde{\phi}$ for the Fourier transform w.r.t the space variables. There are two linearly independent solutions ${}^t(\phi_1,\psi_1)(x,y;\tau)$ of the equation

$$\big(\mathrm{diag}(\frac{1}{2},0)-(\xi^2+\eta^2)^{-1}\tau^2\big){}^t(\widetilde{\phi}_1,\widetilde{\psi}_1) = 0$$

given by

$$\phi_1(x,y,\tau) = J_0(\sqrt{x^2+y^2}\frac{\sqrt{2}\tau}{h}), \quad \psi_1 = J_0(\sqrt{x^2+y^2}\frac{\tau}{h}) \tag{6}$$

and ϕ,ψ are derived from (5) by convolution integrals. From this it is standard to deduce the spectral properties of $\mathcal{H}$.

- *Relative stability of the Pareto property*

Theorem 2: Let

$$\mathcal{H}(\xi,\eta) = \begin{pmatrix} \xi^2+\frac{1}{2}\eta^2 & \frac{1}{2}\xi\eta \\ \frac{1}{2}\xi\eta & \frac{1}{2}\xi^2+\eta^2 \end{pmatrix} + \mathcal{O}((\xi,\eta)^3)$$

Then $\mathcal{H}(\xi,\eta)$ has the Pareto property (at least in the sense of germs at 0), with $C(\xi,\eta) = C_0+\mathcal{O}(\xi,\eta)$, $u(\xi,\eta) = u_0(\xi,\eta)+\mathcal{O}((\xi,\eta)^2)$. Moreover there is a skew-diagonalization of type (5), and we can find a set of solutions to the "Helmholtz equation" as in (6).

The proof goes as in Poincaré-Dulac theorem [2]: Namely we seek for a "new" u of the form $v = u+f$, with $f = \mathcal{O}(|\xi,\eta|^3)$, and a "new" C of the form $\mathrm{diag}(2,1)+\begin{pmatrix} a & b \\ b & d \end{pmatrix}$. The upper-left matrix element of $v'\mathcal{H}(v')^*$ is given by $\sqrt{2}L_1f_1+(\xi^2+\frac{\eta^2}{2})+\frac{1}{2}(a\xi^2-2b\xi\eta+d\eta^2)$, with $L_1 = 2x\frac{\partial}{\partial x}-y\frac{\partial}{\partial y}$, which has a resonance 2:1. Its lower-right matrix element is given by $\sqrt{2}L_2f_2+(\eta^2+\frac{\xi^2}{2})+\frac{1}{2}(a\eta^2+2b\xi\eta+d\xi^2)$ with $L_2 = 2y\frac{\partial}{\partial x}+x\frac{\partial}{\partial y}$. The off-diagonal terms involve the term $L_1f_2+L_2f_1$. It turns out that we can solve (at least perturbatively) this system, the coefficients $a = a_1\xi+a_2\eta+\cdots$, $b = b_1\xi+b_2\eta+\cdots$, $d = d_1\xi+d_2\eta+\cdots$ being determined to fulfill the compatibility conditions. The skew-diagonalization of type (5) follows from the fact that $v'(\xi,\eta)(v'(\xi,\eta))^* = \frac{1}{2}(\xi^2+\eta^2)^2(\mathrm{Id}+\mathcal{O}(|\xi,\eta|)$ is close to a multiple of Id. We can still construct semi-classical solutions, and their asymptotics in h for small E, obtained by varying the argument of the Bessel functions.

5. Conclusion

We have made an attempt to extend the notion of "non-degenerate critical point" for a scalar Hamiltonian to the notion of "Pareto critical set" Θ for a 2×2 Hamiltonian system. We focussed to the case where Θ reduces to a point. Our analysis applies to the Hamiltonian $\mathcal{H}(p)$ quadratic in momenta arising in Elasticity theory, for a particular value of the coefficients. It allows to account for the spectral properties of $\mathcal{H}$ together with the semi-classical spectral asymptotics of (relative, i.e. depending again only on p) perturbations of $\mathcal{H}$ near $E = 0$. These Hamiltonians present a codimension 3 singularity of order 4 at $\Theta = \{0\}$. It seems difficult to extend this purely algebraic approach for general Hamiltonians of the form $\mathcal{H}(x, p)$: namely, when trying a map $u : \mathbf{R}^4 \to \mathbf{R}^2$, eigenvectors of $\mathcal{H}$ are given from those of $C(x, p)$ by applying a pseudo-inverse of $(u')^*$, which is not uniquely defined.

Acknowledgements: I thank Ilya Bogaevsky for interesting discussions. This work has been partially supported by the grant PRC CNRS/RFBR 2017-2019 No.1556 "Multi-dimensional semi-classical problems of Condensed Matter Physics and Quantum Dynamics".

[1] Arnold V 1990 *Singularities of Caustics and Wave Fronts* (Kluwer, Dordrecht)
[2] Arnold V 1983 *Geometrical methods in the theory of ordinary differential equations* (Springer, Berlin)
[3] Bensouissi A, M'hadbi N and Rouleux M 2011 *Proc. "Days of Diffraction 2011" (Saint-Petersburg)* (IEEE 101109/DD.2011.6094362) 39
[4] Bogaevsky I 2016 *Int. Conf. Diff. Eq. Dyn. Systems (Suzdal, Russia)*
[5] Bogaevsky I 2003 *Moscow Math. J.* **3** 807
[6] Braam P J and Duistermaat J J 1993 *Indag. Math. N.S.* **4** 407
[7] Braam P J and Duistermaat J J 1995 *Panorama of Math. Banach Center Publ.* **34** 29
[8] Duncan K P and Györffy B L 2002 *Annals of. Phys.* **298** 273.
[9] Fefferman C and Weinstein M 2012 *J. American Math. Soc.* **25(4)** 1169
[10] Fermanian Kammerer Cl and Mehats F "A kinetic model for the transport of electrons in a graphene layer", arXiv: 1605.02451v1
[11] Ivrii V Ya 1979 *Siberian Math. J.* **20** 390
[12] Kaidi N and Rouleux M 1993 *Comptes Rendus Acad. Sci. Paris* **317(I)** 359
[13] Mnëv N 1985 *Russ. Math. Surv.* **40** 137
[14] Ralston J 1992 *Astérisque* **210** (Soc. Math. France, Paris)
[15] Rouleux M 1999 *Tunneling effects for h-Pseudodifferential Operators, Feshbach resonances and the Born-Oppenheimer approximation* (Adv. Part. Diff. Eq. **16**) ed M Demuth, E Schrohe *et al.* (Wiley VCH, Berlin)
[16] Smale S 1973 Global Analysis and Economics I: Pareto Optimum and a Generalization of Morse Theory *Proc. Salvador Symposium on Dynamical Systems 1971 (Salvador, Bahia)* (Dynamical Systems) ed M Peixoto (Acad. Press, New-York)
[17] Wan H Y 1975 *Topology* **14** 217

IOP Conf. Series: Journal of Physics: Conf. Series **965** (2018) 012035 doi:10.1088/1742-6596/965/1/012035

Dark and bright solitons for the two-dimensional complex modified Korteweg-de Vries and Maxwell-Bloch system with time-dependent coefficient

G. Shaikhova, N. Ozat, K. Yesmakhanova, G. Bekova

Eurasian International Center for Theoretical Physics and Department of General Theoretical Physics, Eurasian National University, Astana, 010008, Kazakhstan.

E-mail: `g.shaikhova@gmail.com`

Abstract. In this work, we present Lax pair for two-dimensional complex modified Korteweg-de Vries and Maxwell-Bloch (cmKdV-MB) system with the time-dependent coefficient. Dark and bright soliton solutions for the cmKdV-MB system with variable coefficient are received by Darboux transformation. Moreover, the determinant representation of the one-fold and two-fold Darboux transformation for the cmKdV-MB system with time-dependent coefficient is presented.

1. Introduction

The nonlinear phenomena play a important rule in a variety of scientific fields such as optical fibers, fluid mechanics, chemical kinetics, solid state physics. It is modeled by nonlinear partial differential equations (NPDEs) [1-4]. Finding exact solutions of NPDEs is crucial, because, exact solutions help one to well understand the mechanism of the complicated physical phenomena and dynamical processes modeled by these NPDEs. Moreover, the real physical space-time being in higher-dimensional and more complex, research generalized integrable equations in higher dimensions is still open and attractive [5-15]. One of the well-known NPDE is the complex modified Korteweg-de Vries equation. This equation and its generalization have been derived and studied for the dynamical evolution of nonlinear lattices, fluid dynamics, ultra-short pulses in nonlinear optics, plasma physics, nonlinear transmission lines [16-17] .

In this paper, we present the two-dimensional cmKdV-MB system with time-dependent coefficient as the following form

$$q_t + q_{xxy} + wq_x + w_x q + iqv + \gamma(t)p = 0, \quad (1)$$

$$v_x - 2iqr_{xy} + 2irq_{xy} = 0, \quad (2)$$

$$w_x - 2rq_y - 2qr_y = 0, \quad (3)$$

$$\eta_x + qk + rp = 0, \quad (4)$$

$$p_x - 2ip\omega - 2q\eta = 0, \quad (5)$$

where q, v, w, η, p are unknown potentials and $r = q*, k = p*$. The symbol $*$ denotes the complex conjugate. The time-dependent coefficient $\gamma(t)$ describes the averaging with respect to inhomogeneous broadening of the resonant frequency. The real parameter ω is a constant corresponding to the frequency. In the equations (1)-(5), x and y represent the normalized distance and t represented normalized time respectively.

The equations (1)-(5) admit the following integrable reduction, if $p = \eta = 0$:

$$q_t + q_{xxy} + wq_x + w_x q + iqv = 0, \tag{6}$$
$$v_x - 2iqr_{xy} + 2irq_{xy} = 0, \tag{7}$$
$$w_x - 2rq_y - 2qr_y = 0, \tag{8}$$

it is the usual two-dimensional cmKdV equations [18].

If we let $\gamma = -2$ the equations (1)-(5) will be reduced to two-dimensional cmKdV-MB equations as the following form [19]

$$q_t + q_{xxy} + wq_x + w_x q + iqv - 2p = 0, \tag{9}$$
$$v_x - 2iqr_{xy} + 2irq_{xy} = 0, \tag{10}$$
$$w_x - 2rq_y - 2qr_y = 0, \tag{11}$$
$$\eta_x + qk + rp = 0, \tag{12}$$
$$p_x - 2ip\omega - 2q\eta = 0. \tag{13}$$

The paper is organized as follows. In section 2, Lax pair for the two-dimensional cmKdV-MB system with time-dependent coefficient is presented. We derive the one-fold Darboux transformation of the cmKdV-MB system with time-dependent coefficient in section 3. The determinant-formed generalization of one-fold and two-fold Darboux transformation for the two-dimensional cmKdV-MB system with time-dependent coefficient will be given in section 4. In section 5, we receive one-soliton, two-soliton solutions by assuming trivial seed solutions. Section 6 is devoted to conclusion.

2. Lax representation

The corresponding Lax representation of the two-dimensional cmKdV-MB system with time-dependent coefficient can be expressed as follows:

$$\Psi_x = A\Psi, \tag{14}$$
$$\Psi_t = 4\lambda^2 \Psi_y + B\Psi, \tag{15}$$

where

$$\Psi = \begin{pmatrix} \psi_1 \\ \psi_2 \end{pmatrix}, \tag{16}$$

and

$$A = -i\lambda \begin{pmatrix} 1 & 0 \\ 0 & -1 \end{pmatrix} + \begin{pmatrix} 0 & q \\ -r & 0 \end{pmatrix} = -i\lambda\, \sigma_3 + A_0, \tag{17}$$

$$B = \lambda \begin{pmatrix} iw & 2iq_y \\ 2ir_y & -iw \end{pmatrix} + \begin{pmatrix} -\frac{i}{2}v & -q_{xy} - wq \\ r_{xy} + wr & \frac{i}{2}v \end{pmatrix} \tag{18}$$

$$+ V_1 \begin{pmatrix} \eta & -p \\ -k & -\eta \end{pmatrix} \tag{19}$$

$$= \lambda B_1 + B_0 + V_1 B_{-1}, \tag{20}$$

IOP Conf. Series: Journal of Physics: Conf. Series **965** (2018) 012035 doi:10.1088/1742-6596/965/1/012035

with

$$B_{-1} = \begin{pmatrix} \eta & -p \\ -k & -\eta \end{pmatrix}, V_1 = -\frac{i\gamma(t)}{2(\lambda + \omega)}. \tag{21}$$

The compatible condition of equations (1)-(5) is

$$A_t - B_x - 4\lambda^2 A_y + [A, B] = 0, \tag{22}$$

where $[A, B] = AB - BA$. By direct calculation of above equation, we can yield the two-dimensional cmKdV-MB system with time-dependent coefficient.

3. Darboux transformation

In this section, we consider the following transformation of equations (14)-(15) based on the Darboux transformation for Ablowitz-Kaup-Newell-Segur system

$$\Psi^{'} = T\Psi = (\lambda I - M)\Psi, \tag{23}$$

where

$$M = \begin{pmatrix} m_{11} & m_{12} \\ m_{21} & m_{22} \end{pmatrix}, \quad I = \begin{pmatrix} 1 & 0 \\ 0 & 1 \end{pmatrix}. \tag{24}$$

New function $\Psi^{'}$ satisfies

$$\Psi^{'}_x = A^{'}\Psi^{'}, \tag{25}$$
$$\Psi^{'}_t = 4\lambda^2\Psi^{'}_y + B^{'}\Psi^{'}, \tag{26}$$

where A' and B' depend on q', v', w', η', p' and λ. In order to hold equation (25)-(26), the T must satisfies the system

$$T_x + TA = A^{'}T, \tag{27}$$
$$T_t + TB = 4\lambda^2 T_y + B^{'}T. \tag{28}$$

The relation between q, p, v, w, η and $q^{[1]}, p^{[1]}, v^{[1]}, w^{[1]}, \eta^{[1]}$ can be obtained from (27)-(28), which is the Darboux transformation of cmKdV-MB system with time-dependent coefficient. Combining coefficients of the different powers of λ^i of the equation (27) and doing some calculations we obtain

$$q^{'} = q - 2im_{12}, \tag{29}$$
$$r^{'} = r - 2im_{21}, \tag{30}$$

and $m_{12}^* = -m_{21}$. At the same time, from the equation (28) we can get

$$v^{'} = v - 4(m_{12}r_y - m_{21}q_y + 2im_{11}m_{11y} + m_{21}m_{12y}), \tag{31}$$
$$w^{'} = w - 4im_{11y} = w + 4im_{22y}, \tag{32}$$
$$B^{'}_{-1} = (M + \omega)B_{-1}(M + \omega)^{-1} \tag{33}$$

and we additionally have $m_{11}^* = m_{22}$. The equations (29)-(33) gives one-fold Darboux transformation of the two-dimensional cmKdV-MB system with time-dependent coefficient.

We now assume that

$$M = H\Lambda H^{-1}, \tag{34}$$

where $\Lambda = \begin{pmatrix} \lambda_1 & 0 \\ 0 & \lambda_2 \end{pmatrix}, H = \begin{pmatrix} \psi_1(\lambda_1;t,x,y) & \psi_1(\lambda_2;t,x,y) \\ \psi_2(\lambda_2;t,x,y) & \psi_2(\lambda_2;t,x,y) \end{pmatrix} := \begin{pmatrix} \psi_{1,1} & \psi_{1,2} \\ \psi_{2,1} & \psi_{2,2} \end{pmatrix}$. In order to satisfy the constraints of M and B'_{-1} as mentioned above, we first notes that

$$\begin{aligned} \lambda_2 &= \lambda_1^*, m_{11}^* = m_{22}, &(35)\\ H &= \begin{pmatrix} \psi_1(\lambda_1,t,x,y) & -\psi_2^*(\lambda_1,t,x,y) \\ \psi_2(\lambda_1,t,x,y) & \psi_1^*(\lambda_1,t,x,y) \end{pmatrix}. &(36) \end{aligned}$$

In next part, the determinant form of one-fold Darboux transformation of cmKdV-MB system with time-dependent coefficient will be given.

4. Determinant representation of Darboux transformation

In this section, we will consider the first two Darboux transformation of the cmKdV-MB system with time-dependent coefficient. Firstly, we introduce n eigenfunctions $\begin{pmatrix} \psi_{1,i} \\ \psi_{2,i} \end{pmatrix} = \psi(\lambda = \lambda_i), i = 1,2$ with the constraints on eigenvalues as $\lambda_{2i-1} = \lambda_{2i}^*$ and reduction conditions as $\psi_{2,2i} = \psi_{1,2i-1}^*, \psi_{2,2i-1} = -\psi_{1,2i}^*$ [20-23].

4.1. One-fold Darboux transformation

As the simplest Darboux transformation, the determinant of one-fold Darboux transformation of cmKdV-MB system with time-dependent coefficient given in the following form [20-23]

$$T_1(\lambda;\lambda_1,\lambda_2) = \frac{1}{\Delta_1}\begin{pmatrix} (T_1)_{11} & (T_1)_{12} \\ (T_1)_{21} & (T_1)_{22} \end{pmatrix}, \tag{37}$$

where

$$(T_1)_{11} = \begin{vmatrix} 1 & 0 & \lambda \\ \psi_{1,1} & \psi_{2,1} & \lambda_1\psi_{1,1} \\ \psi_{1,2} & \psi_{2,2} & \lambda_2\psi_{1,2} \end{vmatrix}, (T_1)_{12} = \begin{vmatrix} 0 & 1 & 0 \\ \psi_{1,1} & \psi_{2,1} & \lambda_1\psi_{1,1} \\ \psi_{1,2} & \psi_{2,2} & \lambda_2\psi_{1,2} \end{vmatrix}, \tag{38}$$

$$(T_1)_{21} = \begin{vmatrix} 1 & 0 & 0 \\ \psi_{1,1} & \psi_{2,1} & \lambda_1\psi_{2,1} \\ \psi_{1,2} & \psi_{2,2} & \lambda_2\psi_{2,2} \end{vmatrix}, (T_1)_{22} = \begin{vmatrix} 0 & 1 & \lambda \\ \psi_{1,1} & \psi_{2,1} & \lambda_1\psi_{2,1} \\ \psi_{1,2} & \psi_{2,2} & \lambda_2\psi_{2,2} \end{vmatrix}, \tag{39}$$

and

$$A^{[1]} = A - [T_1, \sigma_3], \quad B_{-1}^{[1]} = T_1|_{\lambda=-\omega} B_{-1} T_1^{-1}|_{\lambda=-\omega}, \tag{40}$$

Then one-fold DT can be presented by next form

$$\begin{aligned} q^{[1]} &= q - 2i\frac{(T_1)_{12}}{\Delta_1}, &(41)\\ v^{[1]} &= v + 4\left(q_y^*\frac{(T_1)_{12}}{\Delta_1} - q_y\frac{(T_1)_{21}}{\Delta_1} + 2i\frac{(T_1)_{11}}{\Delta_1}\left(\frac{(T_1)_{11}}{\Delta_1}\right)_y + 2i\frac{(T_1)_{21}}{\Delta_1}\left(\frac{(T_1)_{21}}{\Delta_1}\right)_y\right), &(42)\\ w^{[1]} &= w - 4i\left(\frac{(T_1)_{11}}{\Delta_1}\right)_y, &(43)\\ p^{[1]} &= \frac{2\eta(T_1)_{11}(T_1)_{12} - p^*(T_1)_{12}(T_1)_{12} + p(T_1)_{11}(T_1)_{11}}{(T_1)_{11}(T_1)_{22} - (T_1)_{12}(T_1)_{21}}|_{\lambda=-\omega}, &(44)\\ \eta^{[1]} &= \frac{\eta\left((T_1)_{11}(T_1)_{12} + (T_1)_{21}(T_1)_{12}\right) + p(T_1)_{11}(T_1)_{21} - p^*(T_1)_{12}(T_1)_{22}}{(T_1)_{11}(T_1)_{22} - (T_1)_{12}(T_1)_{21}}|_{\lambda=-\omega}. &(45) \end{aligned}$$

We can find the transformation T_1 has following property

$$T_1(\lambda;\lambda_1,\lambda_2)|_{\lambda=\lambda_i}\begin{pmatrix}\psi_{1,i}\\ \psi_{2,i}\end{pmatrix}=0, \tag{46}$$

where $i=1,2$.

4.2. Two-fold Darboux transformation

The two-fold Darboux transformation of cmKdV-MB system with time-dependent coefficient is as following [20-23]

$$T_2(\lambda;\lambda_1,\lambda_2,\lambda_3,\lambda_4)=\frac{1}{\Delta_2}\begin{pmatrix}(T_2)_{11} & (T_2)_{12}\\ (T_2)_{21} & (T_2)_{22}\end{pmatrix}, \tag{47}$$

where

$$\Delta_2=\begin{vmatrix}\psi_{1,1} & \psi_{2,1} & \lambda_1\psi_{1,1} & \lambda_1\psi_{2,1}\\ \psi_{1,2} & \psi_{2,2} & \lambda_2\psi_{1,2} & \lambda_2\psi_{2,2}\\ \psi_{1,3} & \psi_{2,3} & \lambda_3\psi_{1,3} & \lambda_3\psi_{2,3}\\ \psi_{1,4} & \psi_{2,4} & \lambda_4\psi_{1,4} & \lambda_4\psi_{2,4}\end{vmatrix}, \tag{48}$$

$$(T_2)_{11}=\begin{vmatrix}1 & 0 & \lambda & 0 & \lambda^2\\ \psi_{1,1} & \psi_{2,1} & \lambda_1\psi_{1,1} & \lambda_1\psi_{2,1} & \lambda_1^2\psi_{1,1}\\ \psi_{1,2} & \psi_{2,2} & \lambda_2\psi_{1,2} & \lambda_2\psi_{2,2} & \lambda_2^2\psi_{1,2}\\ \psi_{1,3} & \psi_{2,3} & \lambda_3\psi_{1,3} & \lambda_3\psi_{2,3} & \lambda_3^2\psi_{1,3}\\ \psi_{1,4} & \psi_{2,4} & \lambda_4\psi_{1,4} & \lambda_4\psi_{2,4} & \lambda_4^2\psi_{1,4}\end{vmatrix},\ (T_2)_{12}=\begin{vmatrix}0 & 1 & 0 & \lambda & 0\\ \psi_{1,1} & \psi_{2,1} & \lambda_1\psi_{1,1} & \lambda_1\psi_{2,1} & \lambda_1^2\psi_{1,1}\\ \psi_{1,2} & \psi_{2,2} & \lambda_2\psi_{1,2} & \lambda_2\psi_{2,2} & \lambda_2^2\psi_{1,2}\\ \psi_{1,3} & \psi_{2,3} & \lambda_3\psi_{1,3} & \lambda_3\psi_{2,3} & \lambda_3^2\psi_{1,3}\\ \psi_{1,4} & \psi_{2,4} & \lambda_4\psi_{1,4} & \lambda_4\psi_{2,4} & \lambda_4^2\psi_{1,4}\end{vmatrix},$$

$$(T_2)_{21}=\begin{vmatrix}1 & 0 & \lambda & 0 & 0\\ \psi_{1,1} & \psi_{2,1} & \lambda_1\psi_{1,1} & \lambda_1\psi_{2,1} & \lambda_1^2\psi_{2,1}\\ \psi_{1,2} & \psi_{2,2} & \lambda_2\psi_{1,2} & \lambda_2\psi_{2,2} & \lambda_2^2\psi_{2,2}\\ \psi_{1,3} & \psi_{2,3} & \lambda_3\psi_{1,3} & \lambda_3\psi_{2,3} & \lambda_3^2\psi_{2,3}\\ \psi_{1,4} & \psi_{2,4} & \lambda_4\psi_{1,4} & \lambda_4\psi_{2,4} & \lambda_4^2\psi_{2,4}\end{vmatrix},\ (T_2)_{22}=\begin{vmatrix}0 & 1 & 0 & \lambda & \lambda^2\\ \psi_{1,1} & \psi_{2,1} & \lambda_1\psi_{1,1} & \lambda_1\psi_{2,1} & \lambda_1^2\psi_{2,1}\\ \psi_{1,2} & \psi_{2,2} & \lambda_2\psi_{1,2} & \lambda_2\psi_{2,2} & \lambda_2^2\psi_{2,2}\\ \psi_{1,3} & \psi_{2,3} & \lambda_3\psi_{1,3} & \lambda_3\psi_{2,3} & \lambda_3^2\psi_{2,3}\\ \psi_{1,4} & \psi_{2,4} & \lambda_4\psi_{1,4} & \lambda_4\psi_{2,4} & \lambda_4^2\psi_{2,4}\end{vmatrix}.$$

We can find

$$T_2(\lambda;\lambda_1,\lambda_2,\lambda_3,\lambda_4)|_{\lambda=\lambda_i}\begin{pmatrix}\psi_{1,i}\\ \psi_{2,i}\end{pmatrix}=0,$$

where $i=1,2,3,4$. Similarly, for transformation T_2, following transformation formula holds

$$T_{2x}+T_2A = A^{[2]}T_2, \tag{49}$$

$$T_{2t}+T_2B = 4\lambda^2T_{2y}+B^{[2]}T_2. \tag{50}$$

Then the relation between q,p,v,w,η and $q^{[2]},p^{[2]},v^{[2]},w^{[2]},\eta^{[2]}$ will be got in the following equations

$$A^{[2]} = A-[T_2,\sigma_3], \tag{51}$$

$$B^{[2]}_{-1} = T_2|_{\lambda=-\omega}B_{-1}T_2^{-1}|_{\lambda=-\omega}. \tag{52}$$

Finally, we can get from (49)-(50) the following two-fold Darboux transformation in detail,

$$q^{[2]} = q - 2i\frac{(T_2)_{12}}{\Delta_2}, \tag{53}$$

$$v^{[2]} = v + 4\left(q_y^*\frac{(T_2)_{12}}{\Delta_2} - q_y\frac{(T_2)_{21}}{\Delta_2} + 2i\frac{(T_2)_{11}}{\Delta_2}\left(\frac{(T_2)_{11}}{\Delta_2}\right)_y + 2i\frac{(T_2)_{21}}{\Delta_2}\left(\frac{(T_2)_{21}}{\Delta_2}\right)_y\right), \tag{54}$$

$$w^{[2]} = w - 4i\left(\frac{(T_2)_{11}}{\Delta_2}\right)_y, \tag{55}$$

$$p^{[2]} = \frac{2\eta(T_2)_{11}(T_2)_{12} - p^*(T_2)_{12}(T_2)_{12} + p(T_2)_{11}(T_2)_{11}}{(T_2)_{11}(T_2)_{22} - (T_2)_{12}(T_2)_{21}}|_{\lambda=-\omega}, \tag{56}$$

$$\eta^{[2]} = \frac{\eta\left((T_2)_{11}(T_2)_{12} + (T_2)_{21}(T_2)_{12}\right) + p(T_2)_{11}(T_2)_{21} - p^*(T_2)_{12}(T_2)_{22}}{(T_2)_{11}(T_2)_{22} - (T_2)_{12}(T_2)_{21}}|_{\lambda=-\omega}. \tag{57}$$

The equations (53)-(57) can be used to generate two-soliton solutions of the two-dimensional cmKdV-MB system with time-dependent coefficient later.

5. Dark and bright soliton solutions

Having the explicit form of the DT, we are ready to construct exact solutions of the two-dimensional cmKdV-MB system with time-dependent coefficient (1)-(5). We assume trivial seed solutions as $q = p = w = v = 0, \eta = 1$.

Then the corresponding associated linear system takes the form

$$\Psi_x = A\Psi, \tag{58}$$

$$\Psi_t = 4\lambda^2\Psi_y + B\Psi, \tag{59}$$

where

$$A = \begin{pmatrix} -i\lambda & 0 \\ 0 & i\lambda \end{pmatrix}, \tag{60}$$

$$B = \begin{pmatrix} -\frac{i\gamma(t)}{2(\lambda+\omega)} & 0 \\ 0 & \frac{i\gamma(t)}{2(\lambda+\omega)} \end{pmatrix}. \tag{61}$$

The system (58)-(59) admits the following exact solutions

$$\psi_1 = \exp(-i\lambda x + \mu y + \int(4\lambda^2\mu - \frac{i\gamma}{2(\lambda+\omega)})dt + \delta_0), \tag{62}$$

$$\psi_2 = \exp(i\lambda x - \mu y - \int(4\lambda^2\mu - \frac{i\gamma}{2(\lambda+\omega)})dt + \delta_0 + i\delta_1), \tag{63}$$

where δ_0 and δ_1 are all arbitrary fixed real constants. Choosing $\lambda = a_1 + b_1 i, \omega = 1.5, \mu = c_1 + d_1 i, \delta_0 = 0, \delta_1 = 0$ and substituting (62)-(63) into the one-fold Darboux transformation (41)-(45) and we can get the one-soliton solutions in the following form:

$$q^{[1]} = 8i\cosh(\frac{\theta_2}{A_1})\exp(\frac{-i\theta_1}{A_1}), \tag{64}$$

$$v^{[1]} = \frac{A_2}{A_1}\cosh^{-2}\left(\frac{\theta_2}{A_1}\right), \tag{65}$$

$$w^{[1]} = \frac{A_3}{A_1}\cosh^{-2}\left(\frac{\theta_2}{A_1}\right), \tag{66}$$

$$\eta^{[1]} = \frac{(\omega^2+2\,\omega\,a_1+{a_1}^2+{b_1}^2)\cosh(\frac{\theta_2}{A_1})+2\,\omega^2+4\,\omega\,a_1+2\,{a_1}^2-6\,{b_1}^2}{\left(\omega^2+2\,\omega\,a_1+{a_1}^2+{b_1}^2\right)\cosh(\frac{\theta_2}{A_1})+2\,\omega^2+4\,\omega\,a_1+2\,{a_1}^2+2\,{b_1}^2}, \tag{67}$$

$$p^{[1]} = \frac{2\sinh(\frac{\theta_3}{A_1})+(2\omega+2a_1)\cosh(\frac{\theta_3}{A_1})}{(2\omega^2+2\omega a_1+2a_1^2+2b_1^2)\cosh(\frac{\theta_2}{A_1})}, \tag{68}$$

where $\theta_1 = 8\,\omega^2 t{a_1}^2 d_1+16\,\omega^2 ta_1b_1c_1-8\,\omega^2 t{b_1}^2 d_1+16\,\omega\,t{a_1}^3 d_1+32\,\omega\,t{a_1}^2 b_1c_1-16\,\omega\,ta_1{b_1}^2 d_1+8\,t{a_1}^4 d_1+16\,t{a_1}^3 b_1c_1+16\,ta_1{b_1}^3 c_1-8\,t{b_1}^4 d_1-2\,\omega^2 xa_1+2\,\omega^2 yd_1-4\,\omega\,x{a_1}^2+4\,\omega\,ya_1d_1-2\,x{a_1}^3-2\,xa_1{b_1}^2+2\,y{a_1}^2 d_1+2\,y{b_1}^2 d_1-\omega\,t\gamma-ta_1\gamma$,

$\theta_2 = 8\,\omega^2 t{a_1}^2 c_1-16\,\omega^2 ta_1b_1d_1-8\,\omega^2 t{b_1}^2 c_1+16\,\omega\,t{a_1}^3 c_1-32\,\omega\,t{a_1}^2 b_1d_1-16\,\omega\,ta_1{b_1}^2 c_1+8\,t{a_1}^4 c_1-16\,t{a_1}^3 b_1d_1-16\,ta_1{b_1}^3 d_1-8\,t{b_1}^4 c_1+2\,\omega^2 xb_1+2\,\omega^2 yc_1+4\,\omega\,xa_1b_1+4\,\omega\,ya_1c_1+2\,x{a_1}^2 b_1+2\,x{b_1}^3+2\,y{a_1}^2 c_1+2\,y{b_1}^2 c_1-t\gamma b_1$,

$\theta_3 = 16\,i\omega^2 ta_1b_1c_1+32\,i\omega\,t{a_1}^2 b_1c_1+4\,\omega\,xa_1b_1+4\,\omega\,ya_1c_1+8\,\omega^2 t{a_1}^2 c_1-8\,\omega^2 t{b_1}^2 c_1+16\,\omega\,t{a_1}^3 c_1-16\,t{a_1}^3 b_1d_1-16\,ta_1{b_1}^3 d_1+2\,iy{b_1}^2 d_1+8\,it{a_1}^4 d_1+2\,i\omega^2 yd_1+2\,iy{a_1}^2 d_1+4\,i\omega\,ya_1d_1+16\,ita_1{b_1}^3 c_1+16\,it{a_1}^3 b_1c_1+16\,i\omega\,t{a_1}^3 d_1+8\,i\omega^2 t{a_1}^2 d_1-16\,i\omega\,ta_1{b_1}^2 d_1-2\,ix{a_1}^3-8\,it{b_1}^4 d_1-2\,i\omega^2 xa_1-4\,i\omega\,x{a_1}^2-2\,ixa_1{b_1}^2-8\,i\omega^2 t{b_1}^2 d_1-32\,\omega\,t{a_1}^2 b_1d_1-16\,\omega\,ta_1{b_1}^2 c_1-i\omega\,t\gamma-16\,\omega^2 ta_1b_1d_1-it\gamma a_1+2\,x{b_1}^3+2\,y{a_1}^2 c_1+2\,x{a_1}^2 b_1-8\,t{b_1}^4 c_1+2\,\omega^2 yc_1+2\,y{b_1}^2 c_1-t\gamma b_1+2\,\omega^2 xb_1+8\,t{a_1}^4 c_1$, $A_1 = (ib_1+a_1+\omega)\,(ib_1-\omega-a_1)$, $A_2 = 64\left({b_1}^2+(\omega+a_1)^2\right)b_1\left((\omega+2\,a_1-d_1)\,b_1+c_1\,(\omega+a_1)\right)$, $A_3 = 32\,c_1\left(\omega^2+2\,\omega\,a_1+{a_1}^2+{b_1}^2\right)b_1$.

By substituting eigenfunctions (62) and (63) into Darboux transformation (41)-(45), one-soliton solutions of the system (1)-(5) are obtained as Fig.1,Fig.2 and Fig.3. These present that $q^{[1]}, p^{[1]}, w^{[1]}$ and $v^{[1]}$ are bright solitons, $\eta^{[1]}$ is dark soliton.

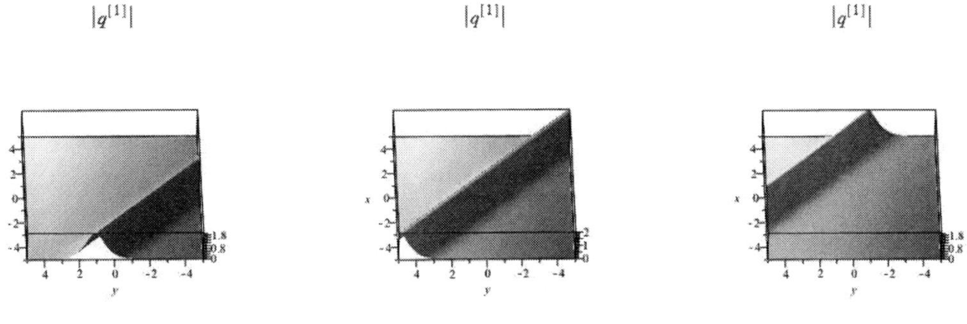

a) t=-4 b) t=0 c) t=4

Figure 1. Evolution of bright one-soliton solution $q^{[1]}$ when $a_1 = 1, b_1 = 0.5, c_1 = 1, d_1 = -0.5, \omega = 1.5, \gamma(t) = 8t$.

In order to construct the two-soliton solutions of the two-dimensional cmKdV-MB system with time-dependent coefficient we use four spectral parameters $\lambda_1 = a_1+b_1 i, \lambda_2 = a_2+b_2 i, \mu_1 = c_1+d_1 i, \mu_2 = c_2+d_2 i$. The two-soliton solutions of the two-dimensional cmKdV-MB system with time-dependent coefficient can be obtained by taking $\omega = 1.5, a_1 = 0.5, b_1 = 2, c_1 = 0.5, d_1 = -0.5, a_2 = 1, b_2 = 1.5, c_2 = 0.5, d_2 = 1, \gamma(t) = 8t$ and using two-fold Darboux transformation (53)-(57). The pictorial representation for interaction of two-solitons of the two-dimensional cmKdV-MB system with time-dependent coefficient is shown in Fig.4, Fig.5 and Fig.6.

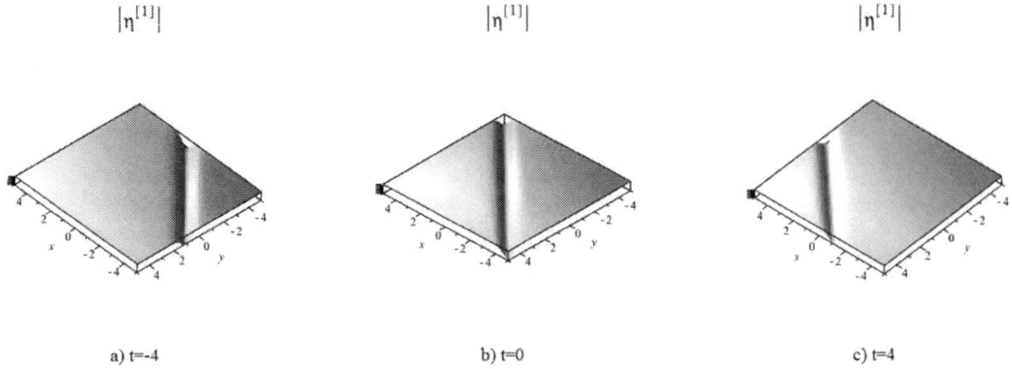

Figure 2. Evolution of dark one-soliton solution $\eta^{[1]}$ when $a_1 = 1, b_1 = 0.5, c_1 = 1$, $d_1 = -0.5, \omega = 1.5, \gamma(t) = 8t$.

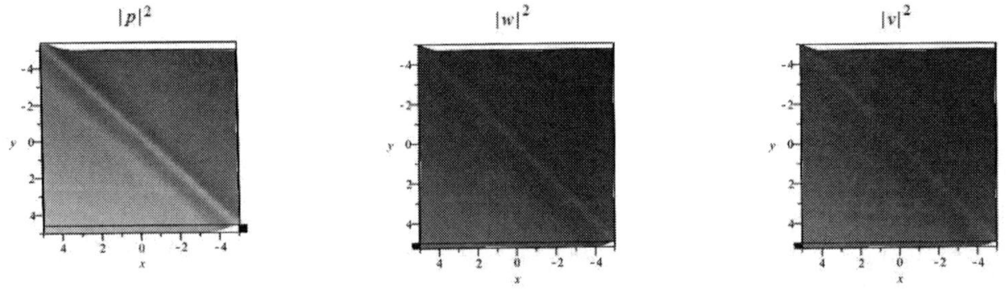

Figure 3. Bright one-soliton solutions $p^{[1]}, w^{[1]}, v^{[1]}$ when $a_1 = 1, b_1 = 0.5, c_1 = 1$, $d_1 = -0.5, \omega = 1.5, \gamma(t) = 8t$.

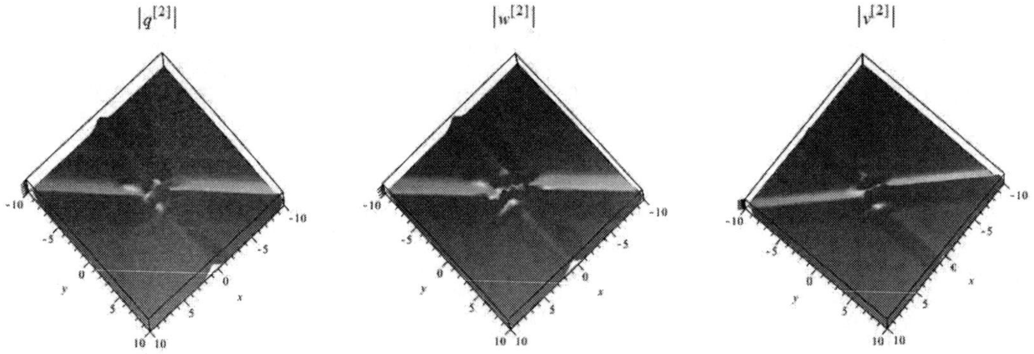

Figure 4. Bright two-soliton solutions $q^{[2]}, w^{[2]}, v^{[2]}$ when $a_1 = 1, b_1 = 0.5, \omega = 1.5, c_1 = 1$, $d_1 = -0.5, a_2 = 1, b_2 = 1.5, c_2 = 0.5, d_2 = 1, \gamma(t) = cos(t)$.

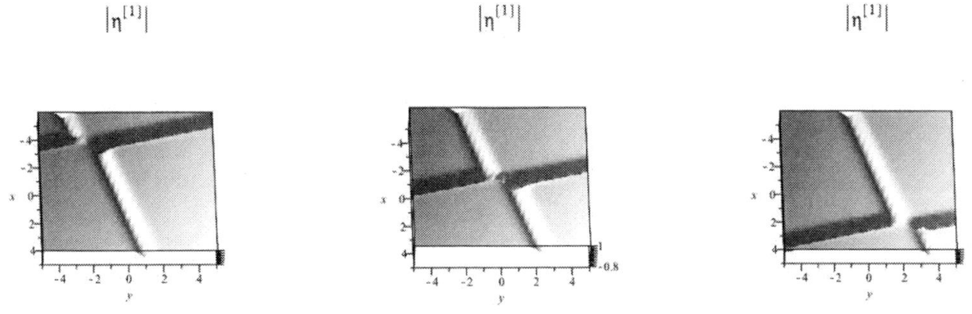

a) t=-2 b) t=0 c) t=4

Figure 5. Evolution of dark two-soliton solutions $\eta^{[2]}$ when $a_1 = 1, b_1 = 0.5, \omega = 1.5$, $c_1 = 1, d_1 = -0.5, a_2 = 1, b_2 = 1.5, c_2 = 0.5, d_2 = 1, \gamma(t) = cos(t)$.

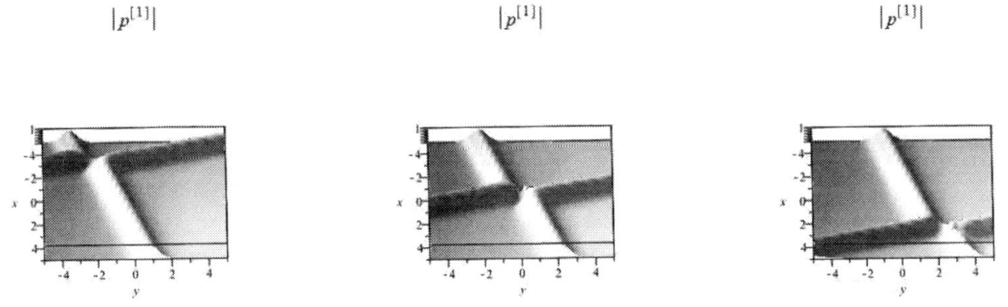

a) t=-2 b) t=0 c) t=4

Figure 6. Evolution of bright two-soliton solutions $p^{[2]}$ when $a_1 = 1, b_1 = 0.5, \omega = 1.5$, $c_1 = 1, d_1 = -0.5, a_2 = 1, b_2 = 1.5, c_2 = 0.5, d_2 = 1, \gamma(t) = cos(t)$.

6. Conclusion

In this work, we presented the Lax representation and Darboux transformaion of the two-dimensional complex modified Korteweg-de Vries and Maxwell-Bloch system with time-dependent coefficient. Also, dark and bright soliton solutions of the cmKdV-MB system with time-dependent coefficient have been constructed explicity by assuming trivial seed solutions. Further, the figures of soliton solutions for the two-dimensional cmKdV-MB system with time-dependent coefficient are given. We note that using the above presented Darboux transformation, one can also construct other type exact solutions of the two-dimensional cmKdV-MB system with time-dependent coefficient.

The work of G.Sh., O.N., K.Y.and G.B. is supported by the Ministry of Education and Science of the Republic of Kazakhstan (the base part of the state task No.0888/GF4)

References

[1] Debnath L 1997 *Nonlinear partial differential equations for scientist and engineers* (Boston: Birkhauser)
[2] Wadati M 2001 *Pramana: Journal of Physics* **57** 841-847
[3] Wadati M 1972 *J. Phys Soc. Jpn* **32** 1681-1687
[4] Wazwaz A M 2009 *Partial Differential Equations and Solitary Waves Theory* (Berlin: Springer)
[5] Peng Y Z 2010 *Commun. Theor. Phys.* **54** 863-865
[6] Lu X, Ma W X, Khalique C M 2015 *Appl. Math. Lett.* **50** 37-42

[7] Wazwaz A M 2011 *Phys. Scr.* **84** 035010
[8] Grses M, Pekcan A 2011 *J. Math. Phys.* **52** 083516
[9] Hereman W, Nuseir A 1997 *Math. Comput. Simul.* **43** 1327
[10] Lou S 1997 *Commun. Theor. Phys* **28** 4150
[11] Biswas A 2009 *Nonlinear Dyn.* **58** 345-348
[12] Ma W X, Zhu Z 2012 *Appl. Math. and Comput.* **218** 11871-11879
[13] Ma W X, Abdeljabbar A and Assad M G 2011 *Appl. Math. Comput.* **217** 10016-10023
[14] El-Tantawy S A, Moslem W M 2014 *Phys. Plasma* **21** 052112
[15] Yesmakanova K R, Shaikhova G N, Bekova G T 2016 *AIP Conf. Proceedings* **1759** 020147
[16] He J, Wang L,Li L, Porseizian K and Erdely R *Preprint* 1405.7845
[17] Bekova G T, Shaikhova G N, Yesmakanova K R, Myrzakulov R 2017 *J. Phys.: Conf. Series* **804** 012004
[18] Myrzakulov R, Mamyrbekova G K, Nugmanova G N, Lakshmanan M 2015 *Symmetry* **7** 1352-1375
[19] Yesmakanova K R, Shaikhova G N, Bekova G T, Myrzakulov R 2016 *J. Phys.: Conf. Series* **738** 012018
[20] Yesmakanova K R, Shaikhova G N , Bekova G T, Myrzakulova Zh R 2016 *Advances in Intelligent Systems and Computing* **441** 183-198
[21] Li C, He J *Preprint* 1210.2501v1
[22] Yang J, Li Ch,Li T, Cheng Zh 2013 *Chinese Physics Letters* **30** 104201
[23] Li Ch, He J, Porsezian K 2013 *Physical Review E* **87** 012913

IOP Conf. Series: Journal of Physics: Conf. Series **965** (2018) 012036 doi:10.1088/1742-6596/965/1/012036

Conformal amplitude hierarchy and the Poincaré disk

Hirohiko Shimada

Mathematical and Theoretical Physics Unit, Okinawa Institute of Science and Technology Graduate University, Onna, Okinawa, 904-0495, Japan

E-mail: hirohiko.shimada@oist.jp

Abstract. The amplitude for the singlet channels in the 4-point function of the fundamental field in the conformal field theory of the 2d $O(n)$ model is studied as a function of n. For a generic value of n, the 4-point function has infinitely many amplitudes, whose landscape can be very spiky as the higher amplitude changes its sign many times at the simple poles, which generalize the unique pole of the energy operator amplitude at $n = 0$. In the stadard parameterization of n by angle in unit of π, we find that the zeros and poles happen at the rational angles, forming a hierarchical tree structure inherent in the Poincaré disk. Some relation between the amplitude and the Farey path, a piecewise geodesic that visits these zeros and poles, is suggested. In this hierarchy, the symmetry of the congruence subgroup $\Gamma(2)$ of $SL(2, \mathbb{Z})$ naturally arises from the two clearly distinct even/odd classes of the rational angles, in which one respectively gets the truncated operator algebras and the logarithmic 4-point functions.

1. Introduction

The amplitudes in CFT$_d$ In general, a CFT has the 3-point functions $\langle\phi_i\phi_j\phi_k\rangle$ of the fields with the scaling dimensions $(\Delta_i, \Delta_j, \Delta_k)$ taking the power law form $\langle\phi_i(x_1)\phi_j(x_2)\phi_k(x_3)\rangle = C_{ijk}$ $|x_{12}^2|^{\frac{\Delta_i+\Delta_j-\Delta_k}{2}}$ $|x_{23}^2|^{\frac{\Delta_j+\Delta_k-\Delta_i}{2}}$ $|x_{31}^2|^{\frac{\Delta_k+\Delta_i-\Delta_j}{2}}$ with a set of unknown numbers C_{ijk} and the 4-point function $\langle\phi\phi\phi\phi\rangle$ of the same field with the scaling dimension Δ_ϕ taking the form $\langle\phi(x_1)\phi(x_2)\phi(x_3)\phi(x_4)\rangle = g(u,v)(x_{12}^2 x_{34}^2)^{-\Delta_\phi} = g(v,u)(x_{14}^2 x_{23}^2)^{-\Delta_\phi}$ with an unknown function $g(u,v)$ of the cross ratios $u = \frac{x_{12}^2 x_{34}^2}{x_{13}^2 x_{24}^2}$ and $v = \frac{x_{14}^2 x_{23}^2}{x_{13}^2 x_{24}^2}$ [1]. Although the function $g(u,v) = g_\phi(u,v)$ depends on the choice of the field ϕ via the possible selection rule, hereafter the subscript ϕ is omitted in the simplest setting, where we are mostly interested in the 4-point function of the fundamental field only.

On top of these kinematical constraints, the dynamical constraints come from the associativity of the operator algebra, which can be formulated as the crossing symmetry of the 4-point function

$$v^{\Delta_\phi} g(u,v) = u^{\Delta_\phi} g(v,u) \tag{1}$$

The function $g(u,v)$ starts from 1 in the factorization limit $\langle\phi\phi\phi\phi\rangle \to \langle\phi\phi\rangle\langle\phi\phi\rangle$, and the subleading terms are a sum over the contribution from the intermediate channel labeled by k:

$$g(u,v) = 1 + \sum_{k=1}^{\infty} \left[C_{\phi\phi}^{k}\right]^2 G_k(u,v) \tag{2}$$

Each term is a conformal block G_k which comes with some amplitude in the form of C^2 as a result of applying twice the operator product expansions (OPE) [1]. Then the crossing symmetry (1) gives highly non-trivial constraints on the amplitudes $\left[C^k_{\phi\phi}\right]^2$. Only later in Section 4.2, we will need the conformal block G_k. For the time being, we focus on the amplitude.

The positivity of the amplitudes Another interesting extra assumption is the unitarity, which requires all the amplitudes be positive: $\forall k \quad \left[C^k_{\phi\phi}\right]^2 > 0$. Indeed, the nice developments comes from this unitarity restriction: By using the linear functionals [2, 3], it is possible to judge whether any *unitary* solutions to (1) can exist for a given set of the lowest scaling dimensions, namely, that of the spin operator and the energy operator. With some reflection on the low-lying spectrum, one certainly expect some boundary curve will appear. In retrospect, this is not surprising. What is surprising is that this bound is not monotone, but it has a pronounced kink, which turns out to be located at the scaling dimensions where the 3D Ising model has to be [2, 4]. We have yet to understand the real background of this phenomenon.

If we try to avoid being tied down to the kink, and instead, try to generalize the problem as far as possible, one would recognize that the perfect unitarity is very special, isolated phenomenon. Unitarity is always violated except the case where both d and n are integers. In $d = 4 - \epsilon$ dimensions [5], it is suggested that the unitarity is violated very weakly in the sense that the negative amplitude occurs only at the operators with high scaling dimensions, and if it occurs, with very small amplitudes. Such scenario is consistent with the observation that one could locate the 3d $O(n)$ CFTs rather successfully until it gets really close to $n = 0$ [6]. Here we will discuss (1) the unitarity violation occurs at $n = 0$ at the main singlet amplitude (i.e. the energy operator), and (2) it is the root of the other relatively minor violations.

This note is intended for giving crude ideas and tangible examples. In Section 2, after some warm up, the computation of certain amplitudes in the 2d 4-point function $\langle\sigma\sigma\sigma\sigma\rangle$ is outlined omitting the details. In Section 3, all the location of the zeros and poles are identified: the Stern-Brocot tree structure and the $\Gamma(2)$ symmetry [7] of a hyperbolic space arise naturally. In Section 4, we see there are clearly distinct even/odd classes of the 4-point functions. The even one is rather trivial, and is related to the finite truncation of the OPE, while the odd one has more interesting structure leaving the logarithmic contribution to the 4-point functions.

2. Amplitudes as functions on $(d, n) \in \mathbb{R}^2$

2.1. a global view of the universality class

The main body of this note deals with an interesting analytic structure of the conformal amplitude in $d = 2$. Yet, in order not to overlook the important physics underlying the conformal amplitude, it is perhaps good to start with the Ising model, with which most readers are familiar with. It has the Hamiltonian,

$$H = J \sum_{\langle ij \rangle} s_i \cdot s_j \tag{3}$$

where s_i is the spin variable on site i taking the value $+1$ or -1, and $\langle ij \rangle$ stands for the nearest-neighbor sites on a d-dimensional lattice. At the critical point this model shows the scale invariant pattern, and the continuum Hamiltonian with a double-well potential (φ^4 model) well describes the vicinity of the critical point. A slight generalization of the internal symmetry:

[1] The OPE coefficient $C^k_{\phi\phi}$ is the same thing as C_{ijk} in the 3-point function of $\phi_i = \phi, \phi_j = \phi$, and ϕ_k under the natural normalization of the 2-point functions $\langle\phi(x_1)\phi(x_2)\rangle = (x_{12}^2)^{-\Delta_\phi}$

$\mathbb{Z}^2 \to O(n)$ yields a Hamiltonian,

$$H = \int d^d x \; (\partial\vec{\varphi})^2 - m^2\vec{\varphi}^2 + \lambda(\vec{\varphi}^2)^2, \tag{4}$$

where the fundamental field is the n-component vector $\vec{\varphi} = \phi_a$ $(a = 1, \cdots n)$. The model at $n = 1$ corresponds to the Ising model, and the interacting CFT is realized as the IR fixed point of the renormalization group.

Now a particularly important idea is to regard the space dimension d as a continuous variable [8]. This led us to the idea of non-trivial fixed points below $d = 4$, which passes through $d = 2$ as the well-understood $c = 1/2$ CFT (Majorana fermion). On the other hand, a detailed analysis of 3d CFTs is still challenging; in the long run, it would be nice if one could somehow extend the analytic techniques in 2d to 3d.

Consider the global view of the CFTs on (d, n)-plane. Here it is very important to regard also n as a continuous parameter, For instance, the $n \to 0$ limit describes the physics of the polymer in a good solvent [9]. This problem is revisited [6] by using the modern technique of the conformal bootstrap to obtain the estimate for the fractal dimensions $d_F = 1.7016(36)$.

In 2d, beyond $n = 2$ the target space $S^{n-1} = O(n)/O(n-1)$ becomes essentially non-abelian, resulting in a massive theory. The one parameter family of CFT [11, 12] is studied for $|n| \leqslant 2$. Whole family can be studied in the Schramm Loewner Evolution [13], which is an attempt at describing various critical fractals by using the conformal map driven by the stochastic Brownian motion of strength κ. There is also nice formula for the fractal dimension proved by Beffara [14]. Along with the central charge, they are given by

$$d_F = 1 + \frac{\kappa}{8}, \quad n = -2\cos 4\pi/\kappa, \quad \kappa/4 = \rho = 1/g, \quad 2 \leqslant \kappa \leqslant 4, \quad c = 1 - 6(g-1)^2/g, \tag{5}$$

where we have also indicated the relation between κ and the Coulomb gas couplings $g = \alpha_+^2$, $\rho = \alpha_-^2$. In Section 3.3, the coupling g measures the angle along the boundary of the Poincaré disk.

2.2. *The 4-point function of the fundamental fields $\langle\sigma\sigma\sigma\sigma\rangle$ is non-degenerate*

Let us then turn to our current focus, the 4-point function in $d = 2$. BPZ shows that a 4-point function satisfies a differential equation of order $r \cdot s$ [15] if it contains at least one *degenerate* field $\phi_{r,s}$ labeled by a pair of positive integers (r, s) (the Kac labels). But unfortunately, we do not know how to compute the 4-point function of the fundamental fields, that lies at the heart of the bootstrap program. This is because, according to the spectrum obtained from the torus partition function [11], the spin field is identified with $(r, s) = (1/2, 0)$: $\sigma \to \phi_{\frac{1}{2},0}$, where the label is not a pair of positive integers. So they are non-degenerate.

On the other hand the energy field is identified with $(r, s) = (1, 3)$: $\varepsilon \to \phi_{1,3}$. They are degenerate, and satisfies 3-rd order differential equation. Thus, we also know $\langle\sigma\sigma\varepsilon\varepsilon\rangle$ has exactly three channels [16]. In contrast, $\langle\sigma\sigma\sigma\sigma\rangle$ may have infinitely many intermediate states in general. Such cases is more difficult, but they deserve more attention as the interacting CFT in $d > 2$ has typically infinitely many channels.

2.3. *Double factorization of the integral*

The main issue here is that there seems to be no simple way [2] of working out the fusion rule between the non-degenerate fields $\phi_{r,s}$ with $(r, s) = (1/2, 0)$. Still one should try to infer the

[2] As a technical remark, if $(2r, 2s)$ were a pair of *positive* integers such as the case with $(r, s) = (1/2, 3/2)$, one could still devise an integral formula by introducing the integer numbers of the screening charges $\alpha_\pm$. Such cases had been known to be important in the geometrical lattice models (the polymer CFT [17] for example) and has been revisited for the percolation CFT [18].

fusion rule indirectly. When the energy operator $\varepsilon = \phi_{1,3}$ is fused by itself, it creates a new higher operator $\phi_{1,5}$ as well as the identity operator $\phi_{1,1}$ and the energy operator itself since $\phi_{1,3} \times \phi_{1,3} = \phi_{1,1} + \phi_{1,3} + \phi_{1,5}$. By repeated fusions, it is natural to think that the algebra in the $O(n)$ CFT for generic $n \in \mathbb{R}$ may contain $\varepsilon^{(p)} = \phi_{1,2p+1}$ $(p = 0, 1, \cdots)$. In terms of the fields in the Hamiltonian (4), the correspondence should look like $\varepsilon^{(p)} =: (\vec{\varphi}^2)^p :$. By symmetry, it is natural to have

$$\sigma \times \sigma = \varepsilon^{(0)} + \varepsilon^{(1)} + \varepsilon^{(2)} + \varepsilon^{(3)} + \cdots \tag{6}$$

Note that this expression focus on the $O(n)$ *singlet* channels, and the contribution of the other types, if exists, may be omiteed.

We take the following strategy used in Dotsenko-Fateev: instead of computing $\langle\sigma\sigma\sigma\sigma\rangle$, we first compute the amplitudes B_k in the mixed correlator $\langle\sigma\sigma\varepsilon^{(p)}\varepsilon^{(p)}\rangle$, which contains the both the *non-degenerate* field σ and the degenerate field $\varepsilon^{(p)}$, and the other amplitudes A_k in the 4-point function $\langle\sigma\sigma\varepsilon^{(p)}\varepsilon^{(p)}\rangle$ of the degenerate field $\varepsilon^{(p)}$ only. Concretely, these two correlation function reads,

$$\langle\sigma(0,0)\sigma(z,\bar{z})\varepsilon^{(p)}(1,1)\varepsilon^{(p)}(\infty)\rangle = |z|^{2\Delta_\sigma}|z-1|^{2\Delta_{\varepsilon^{(p)}}} \sum_{k=0}^{N} B_k \, |z|^{2\lambda_k^{(B)}} f_k^{(B)}(z,\bar{z}), \tag{7}$$

$$\langle\varepsilon^{(p)}(0,0)\varepsilon^{(p)}(z,\bar{z})\varepsilon^{(p)}(1,1)\varepsilon^{(p)}(\infty)\rangle = |z(z-1)|^{2\Delta_{\varepsilon^{(p)}}} \sum_{k=0}^{N} A_k \, |z|^{2\lambda_k^{(A)}} f_k^{(A)}(z,\bar{z}), \tag{8}$$

where $f_k(z,\bar{z}) = |\mathcal{F}(c, \lambda_k, \{\Delta_i\}|z)|^2$ is a product of two chiral conformal blocks. The function $f_k(z,\bar{z}) = 1 + \mathcal{D}(z,\bar{z})$ is a universal object, with which the form of the descendant contribution $\mathcal{D}(z,\bar{z})$ is entirely determined from the conformal invariance as an infinite series in z and $\bar{z}$, and carries no dynamical information. The exponent λ_k is the scaling dimension $\Delta_{\varepsilon^{(k)}}$, which plays no role until Section 4.2. The main objects here are A_k and B_k, which are products of two OPE coefficients: $A_k = \left[C_{\sigma\sigma}^{\varepsilon^{(p)}}\right]^2$ and $B_k = \left[C_{\sigma\sigma}^{\varepsilon^{(p)}} C_{\varepsilon^{(p)}\varepsilon^{(p)}}^{\varepsilon^{(p)}}\right]$. Then one may take the ratio and eliminates the unwanted OPE coefficient $C_{\varepsilon^{(p)}\varepsilon^{(p)}}^{\varepsilon^{(p)}}$, yielding the amplitude of our interest,

$$\left[C_{\sigma\sigma}^{\varepsilon^{(p)}}\right]^2 = B_k^2/A_k. \tag{9}$$

The normalization to $\left[C_{\sigma\sigma}^{\varepsilon^{(0)}}\right]^2 = 1$ can be done by further divide this by B_0^2/A_0. Both A_k and B_k can be computed as the same coefficient $Q_k^{(N)}$ in the expansion of the following symmetric complex integral:

$$I_N(a,b,c,\rho;z,\bar{z}) = \int \prod_{j=0}^{N-1} \left[d^2t_j |t_j|^{2a} |1-t_j|^{2b} |t_j - z|^{2c} \prod_{i(\neq j)}^{N-1} |t_i - t_j|^{2\rho} \right] \tag{10}$$

$$= \sum_{k=0}^{N} Q_k^{(N)} |z|^{2\lambda_k} \left(f_k(z,\bar{z})\right), \tag{11}$$

where, though the detail of the vertex operator representation is omitted, the exponents satisfy[3] $a + c = 2 \cdot 2\alpha_0\alpha_- = 2\rho - 2$ and $b = 2\alpha_{1,N+1}\alpha_- = -N\rho$ with $N = 2p$. The Coulomb gas screening charges $\alpha_+ > 0$ and $\alpha_- < 0$ are given in (5).

[3] If one adopts another known identification [16] $\sigma \to \phi_{\frac{m-1}{2},\frac{m+1}{2}}$ with $\rho = \frac{m}{m+1}$, the exponents a and c are interchanged, giving the same amplitudes. Indeed, these two fields are related by the symmetry (24).

We find a particularly simple way of taming this integral, where we choose k out of N variables and scale it by z. The result is the factorization of $Q_k^{(N)}$ into the two integrals,

$$Q_k^{(N)} = \frac{N!}{k!(N-k)!}\mathcal{C}_k(1+a, 1+c, \rho)\mathcal{C}_{N-k}(1+a+c+2k\rho, 1+b, \rho). \tag{12}$$

The complex Selberg integral $\mathcal{C}_K$ is defined as

$$\mathcal{C}_K(\alpha,\beta,\gamma) = \int d^2t_1 \cdots d^2t_K \prod_{j=0}^{K-1} |t_j|^{2(\alpha-1)}|1-t_j|^{2(\beta-1)} \cdot \prod_{0 \leqslant i<j<K} |t_i - t_j|^{4\gamma}, \tag{13}$$

which, in turn, factorizes [19] into the square of the (usual) chiral Selberg integral [20] times an interesting trigonometric factor,

$$\mathcal{C}_K(\alpha,\beta,\gamma) = S_K^2(\alpha,\beta,\gamma)\frac{1}{K!}\prod_{j=0}^{K-1}\frac{\sin\pi(\alpha+j\gamma)\sin\pi(\beta+j\gamma)\sin\pi(j+1)\gamma}{\sin\pi(\alpha+\beta+(K+j-1)\gamma)\sin\pi\gamma}, \tag{14}$$

$$\begin{aligned} S_K(\alpha,\beta,\gamma) &= \int_0^1 \cdots \int_0^1 \prod_{j=1}^{K-1} t_i^{\alpha-1}(1-t_i)^{\beta-1} \prod_{0\leqslant i<j<K} |t_i - t_j|^{2\gamma}\, dt_0 \cdots dt_{K-1} \\ &= \prod_{j=0}^{K-1} \frac{\Gamma(\alpha+j\gamma)\Gamma(\beta+j\gamma)\Gamma(1+(j+1)\gamma)}{\Gamma(\alpha+\beta+(K+j-1)\gamma)\Gamma(1+\gamma)}. \end{aligned} \tag{15}$$

This factor guarantees the monodromy invariance of the 4-point function solved for the simplest case by Dotsenko-Fateev. The monodromy constraints was also solved in a useful form [21]. In contrast, the integral formula (10) is based on the full vertex operator, and then the monodromy invariance follows automatically. The amplitude in $\langle\sigma\sigma\sigma\sigma\rangle$ yields,

$$\left[C_{\sigma\sigma}^{\varepsilon(p)}\right]^2 = \frac{\mathcal{Z}(\rho)}{\gamma((2p+1)\rho)}\prod_{j=1}^{p-1}\frac{\gamma^2((1+j)\rho)}{\gamma^2((p+j)\rho)}\,\underline{\gamma^2\left(j\rho-\frac{1}{2}\right)\gamma^2\left(j\rho+\frac{1}{2}\right)}, \tag{16}$$

with the less important factor $\mathcal{Z}(\rho)$, which has no singularities in $\frac{1}{2} < \rho < 1$ ($|n| < 2$). The result is expressed by the useful combination $\gamma(x) = \Gamma(x)/\Gamma(1-x)$, which has zeros at positive integers. The factor indicated by the underline is unique to the 4-point function of the spin $\phi_{\frac{1}{2},0}$.

For the $O(n)$ CFT to be unitary, this amplitude should be positive for all $p \in \mathbb{N}$. In $d = 2$, it had been known [11] that the perfect positivity is realized only at $n = 1$ (Ising model) and at $n = 2$ (XY model). We are more interested in how severely such positivity is violated for generic n. The formula (16) should capture the primary part of the positivity condition used in the conformal bootstrap in the simplest setting using $\langle\sigma\sigma\sigma\sigma\rangle$. Understanding the sign changes of (16) is therefore important. Identifying the location of all the singularities in the next section is an essential step to achieve this.

3. An anatomy of the conformal amplitude

3.1. General structure: the spiky landscape with fast decay as the depth p increases

In Figure 2-(b), the amplitudes (16) is plotted in log-scale, in which one sees many spikes. These spikes are either zeros or poles, and the dashed curve represents negative amplitudes. The horizontal axis is the number of components n in the parameter ρ as in (5). This amplitude decays very fast as p increases. For instance, $p = 4$ for generic $n > 0$ is already the scale of the inverse Avogadro number $\sim 10^{-23}$, thus the amplitudes with $p > 4$ are literally microscopic for statistical physics application. Although the decay of the absolute value is also of interest to quantify the degree of the unitarity violation [5, 6], in the following, we focus our attention to the sign behavior of the amplitude (16).

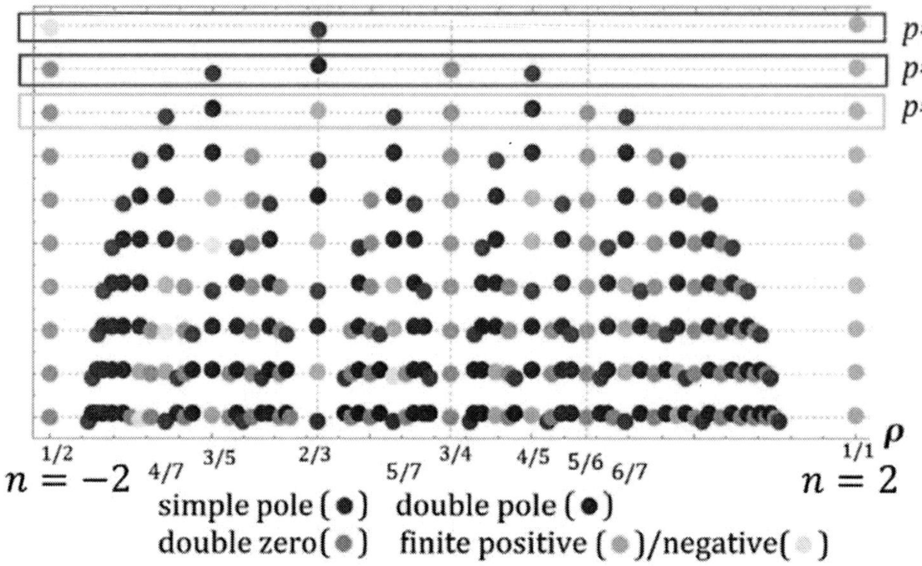

Figure 1. Locations of all the zeros and poles for the amplitude with the level $p \leqslant 10$. At given level p, they may be found when the parameter ρ belongs to the Farey sequence F_{2p+1}. There are no simple zeros. The zeros of even order (green) lie at the ρ with even denominators while simple (red) and double (blue) poles lie at ρ with odd denominators. In F_{2p+1}, only limited points are finite whose sign can be either positive (light-blue) or negative (yellow). The vertical shifts of the markers within the marker size are just for avoiding the overlaps.

3.2. Stern-Brocot tree and the even/odd denominator classes

With more detailed inspection, one would notice that the location of the singularities in the ρ-coordinates may be organized in a tree structure, namely, the Stern-Brocot tree. In this tree, the next level is generated by the mediant rule[4],

$$\frac{b}{a} \dot{+} \frac{d}{c} = \frac{b+d}{a+c}, \tag{17}$$

where one sums respectively the numerators and denominators. Then each row consists of irreducible fractions with denominator less than a certain integer, namely, the Farey sequences. The locations of singularities are summarized in the diagram in Figure 1. The amplitude at level p in (16) has the zeros and poles on the Farey sequence F_{2p+1}: the set of irreducible fractions $\rho = m/m'$ with $m' \leqslant 2p+1$. The point of this note is considering both the horizontal (deforming the CFT continuously) and the vertical (going higher levels of operator in a given CFT) directions of this diagram.

Horizontally, one finds no zeros of odd order including simple zeros Thus the unitarity violation is triggered not by the zeros, but by the poles of the amplitude. At the $n = 2$ (XY model), since the CFT is unitary, all the amplitudes for $p \in \mathbb{N}$ are positive, which may have relation with the multiple-vortex spectrum. The amplitude for the energy operator ($p = 1$) changes its sign at the unique simple pole at $n = 0$ indicated by the red dot at $\rho = 2/3$. The positivity violation starts from here, and the others for higher p may be considered as the generalization of it. The amplitude at some limited members of the Farey fraction is finite, whose sign is determined by the number of the simple poles in the right of it.

[4] Or the Farey addition rule, see a nice exposition [23] for the mathematical background beyond puzzle games.

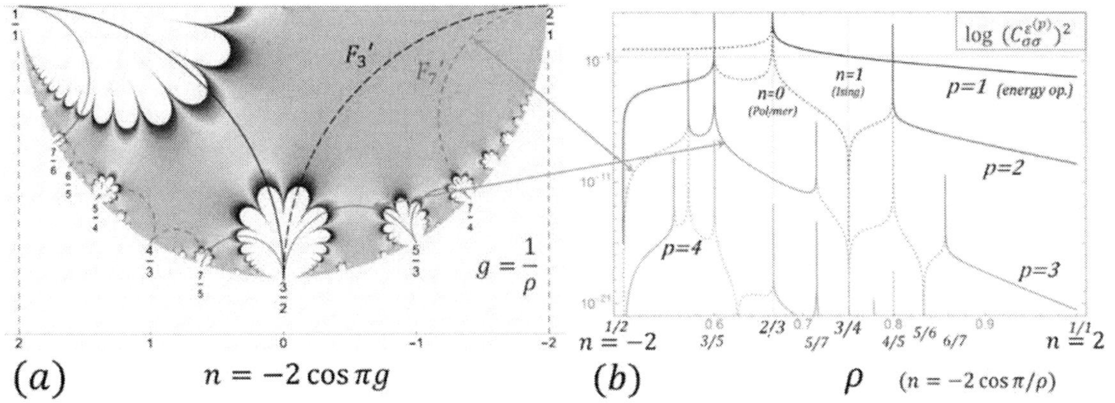

Figure 2. (a) The Poincaré disk and the Farey paths, F_3' (blue) and F_7' (gray). Each rational angle g indicated by a fraction corresponds to a special CFT. A portion of the disk is colored red (blue) if $\mathrm{Re}\,\xi(q)$ is finitely positive (negative), or colored green (light-gray) if the absolute value is almost infinitesimal (infinite). (b) The amplitudes in log-scale for $p = 1$ (blue), $p = 1$ (red), $p = 3$ (gray), and $p = 4$ (orange) as a function of the parameter ρ. The positive (negative) amplitude is represented by the solid (dashed) curve. The up-(down-)ward spikes are the poles (zeros). The two-sided gray arrows indicate the possible map between the F_7' and the $p = 3$ amplitude. A crudest, yet intuitive, way to transform the raw plot (b) to the geometrical picture (a) would be rotating the former by 180° and flipping the numerators/denominators of ρ to obtain g.

Vertically, for an even denominator ρ, we see a series of the infinite zeros of even order (hereafter double poles), corresponding to a finite truncation of the operator algebra as detailed in Section 4.1. On the other hand, for an odd denominator ρ, we see both the double and simple poles. The simple poles are responsible for the unitarity violation at each level, and the double ones lead to the logarithmic 4-pt functions. This is outlined in Section 4.2.

3.3. Hierarchical geodesics on the Poincare disk: the Farey paths and the finite-p amplitudes
Let us now describe some possible analogy between our problem and certain geodesic paths in the hyperbolic geometry. One could motivate this by following physicists' habit of taking some extremal case to identify the symmetry of the problem. Let us imagine $p \to \infty$ limit, where we anticipate the spikes everywhere, at every rational points of ρ. Apparently, such a function might look peculiar, or even pathological. In retrospect, however, such an accumulation of the dense singularities would be rather commonly found as a section of the automorphic function near the infinity of the hyperbolic space. An automorphic function is a function which is invariant under the action of some discrete groups such as $SL(2,\mathbb{Z})$. It is essentially defined on the single fundamental tile, and this tile is tessellated by iterating certain reflections. As in the Escher's famous "Circle Limit", these tiles are accumulated around the infinity.

Here the real-axis interval $|n| \leqslant 2$ is projected back to the lower boundary circle of the Poincaré disk as in Figure 2-(a), and we measure its angle in the unit of π by g. This angle is nothing but the Coulomb gas coupling [11], which takes the value $1 < g < 2$ and $g = \rho^{-1}$. In order to regularize the infinite-p amplitude to get a finite-p amplitude, it is necessary to escape from the singularities by entering inside the disk. To this end, the nome q is introduced by

$$q = e^{i\pi g} = \frac{1}{2}\left(n - i\sqrt{4-n^2}\right), \tag{18}$$

which is the same as the parameter in the quantum group $U_q(sl_2)$ [22]. Although the g-n

projection relation in (18) is equivalent to the renowned one in (5), to our knowledge, the direct embedding construction such as the one in Figure 2-(a) has not been discussed.

By virtue of the clear distinction between the spectrum at the even/odd denominator ρ observed in Section 3.2, among the many possible automorphic functions, the elliptic modular function [7] $\lambda(q)$ may be useful. Further, the $\xi(q)$ function is introduced by

$$\xi(q) = 2 - (\lambda(q) + \lambda(q)^{-1}), \quad \lambda(q) = \left(\frac{\theta_2(q)}{\theta_3(q)}\right)^4 = 16q - 128q^2 + 704q^3 - 3072q^4 + \cdots. \tag{19}$$

Both function has the following invariance under the transformations of the angle g,

$$\lambda(g) = \lambda(g+2), \qquad \lambda(g) = \lambda\left(\frac{g}{2g+1}\right) \tag{20}$$

corresponding to the two generators $\begin{pmatrix}1 & 2\\ 0 & 1\end{pmatrix}$ and $\begin{pmatrix}1 & 0\\ 2 & 1\end{pmatrix}$ defining a congruence subgroup of the $SL(2,\mathbb{Z})$ denoted by $\Gamma(2)$. Along the edges of the tiles generated by $\Gamma(2)$, the imarginary part of $\xi(q)$ vanishes as Im $\xi(q) =$ Im $\lambda(q) = 0$. At the corners of the fundamental triangle tile, $\lambda(q)$ and $\xi(q)$ respectively take the set of values $\{0, 1, \infty\}$ and $\{\infty, 0, \infty\}$, where the sign of ∞ depends on the incoming direction to the corner. The choice of $\xi(q)$ is such that it has a double zero (up to reparameterization) at ρ with an even denominator instead of any singularities, but is still ad-hoc. Accordingly, the following correspondence between the path and the amplitude may have minor exceptions in an idealized rule.

To connect the zeros and the poles of the amplitude (16) for finite-p, we introduce the Farey path in the following. In the idealized rule, it is simply a path that connects the neighboring fractions of g by the unique geodesic semicircle. Each broccoli-like structure in Figure 2-(a) happens at the angle where infinitely many tiles share the corners, and represents an essential singularity of $\xi(q)$. Such an existence of the essential singularity at an odd denominator ρ makes the sign along the path diverse and interesting. For instance, if a Farey path enters the broccoli from a red region and escapes to a blue region, we interpret it that the corresponding amplitude has a chance to cross a simple pole at that particular angle since $\xi(q)$ changes its sign from $+\infty$ to $-\infty$. Similarly, the path enters it and escapes from it passing the same color, the amplitude has a chance to hit a double pole, or just remains finite and the same sign. For the latter case, one need to regularize the (idealized geodesic) path within the broccoli region to avoid the singularity.

The first non-trivial examples of the Farey paths visit the angles in the following order:

$$p = 1 \longrightarrow F_3: \quad \frac{1}{1}, \frac{\hat{3}}{2}, \frac{2}{1} \tag{21}$$

$$p = 2 \longrightarrow F_5: \quad \frac{1}{1}, \frac{\hat{5}}{4}, \frac{4}{3}, \frac{3}{2}, \frac{\hat{5}}{3}, \frac{2}{1} \tag{22}$$

$$p = 3 \longrightarrow F_7: \quad \frac{1}{1}, \frac{\hat{7}}{6}, \frac{6}{5}, \frac{5}{4}, \frac{4}{3}, \frac{\hat{7}}{5}, \frac{3}{2}, \frac{5}{3}, \frac{\hat{7}}{4}, \frac{2}{1}, \tag{23}$$

where the hats are on the highest odd numerators for a given p, which corresponds to the simple poles of the amplitude.

The simplest case F_3 ($p = 1$) corresponds to the amplitude of the energy operator. The amplitude starts from $g = 1/1$ ($n = 2$, $c = 1$) with a positive value, changes its sign at the unique simple pole at $g = 3/2$ ($n = 0$), and terminates at $g = 2/1$ ($n = -2$, $c = -2$) with a negative value. Similarly, along F_3, $\xi(q) \in \mathbb{R}$ changes its sign from $+\infty$ to $-\infty$ at $g = 3/2$. As

$g = 3/2$ is an essential singularity, the behavior of $\xi(q)$ along F_3 is the same as the amplitude for $p = 1$ up to a monotonic reparameterization along the path to make it like a simple pole.

The Farey path F_7 is also shown (gray) as an example at the higher level, which amounts to the depth $p = 3$ (gray box) in Figure 2-(a). The amplitude for $p = 3$ has three simple poles and three zeros. In this example, the amplitude and the $\xi(q)$ function along the Farey path F_7 share exactly the same sign and the same location of the double zeros.

4. Even and odd classes of the special points in the $O(n)$ CFT

4.1. Even denominator ρ: an algebra embedding to the minimal model

Let us now start our case study of even integer denominators. This is very simple, but important. It is well-known that the Kac scaling dimension $\Delta_{r,s}$ is invariant under the reflection of the (r, s)-plane with respect to the line $s = (m/m')r$ [15]. This also leads to a secret symmetry

$$\Delta_{r,s} \Longleftrightarrow \Delta_{\frac{m}{2}-r,\frac{m'}{2}-s} \tag{24}$$

in addition to the standard identification $\Delta_{r,s} \Leftrightarrow \Delta_{m-r,m'-s}$ in the minimal model [15]. Now let $\rho = m/m'$ be an irreducible fraction with a denominator m' even integer. This assumption yields a pair of integers $(\tilde{r}, \tilde{s}) = ((m-1)/2, m'/2) \in \mathbb{N}^2$. The spin field at $(r, s) = (1/2, 0)$ then becomes a degenerate field $(\tilde{r}, \tilde{s})$ and the operator algebra closes. In other words, one can embed the operator algebra into that of a certain minimal model $M_{m,m'}$ at these special cases of ρ if m' is an even integer. According to (5), for instance, the $O(n)$ CFT for $n = 1$ is at $\rho = 3/4$, and has the spin field at $(r, s) = (1, 2)$:

$$\sigma = \phi_{\frac{1}{2},0} \Longleftrightarrow \phi_{1,2} \qquad \text{at} \ \ \rho = 3/4 \ \ (n = 1). \tag{25}$$

In this case, the standard fusion rule $\phi_{1,2} \times \phi_{1,2} \sim \phi_{1,1} + \phi_{1,3}$ corresponds to,

$$\sigma \times \sigma = \varepsilon^{(0)} + \varepsilon^{(1)}. \tag{26}$$

The infinite operator algebra (6) for the $O(n)$ CFT truncates at $p = 1$. Here one has only one OPE coefficient to be determined. In passing, the formula (16) naturally reproduces the famous result $[C^{\varepsilon}_{\sigma\sigma}]^2 = \frac{1}{4}$ in the 2d Ising model as it should. It is easy, but important, to show that each of the zeros for $p \geqslant 2$,

$$0 = \left[C^{\varepsilon^{(2)}}_{\sigma\sigma}\right]^2 = \cdots = \left[C^{\varepsilon^{(\infty)}}_{\sigma\sigma}\right]^2 \tag{27}$$

is not a simple zero, but the degree is even. Thus, all the amplitude in (16) do not change their sign at $n = 1$. In summary, these infinitely many zeros realize a finite algebra at ρ with an even denominator while they do not trigger the unitarity violation as n crosses the corresponding value $n = -2\cos\pi m/m'$.

4.2. Odd denominator ρ: the logarithmic 4-point function $\langle\sigma\sigma\sigma\sigma\rangle_{phys}$

The odd denominator case is much more interesting. Consider an example the CFT at $\rho = 4/5 + \delta$, for which one has $n = \sqrt{2}$ as $\delta \to 0$. One may observe the resonance between different channels caused by the situation that there exists a pair of the scaling dimensions that differ by some integers. This can be easily seen by organizing the right hand side of (6) by using the field identification $\phi_{r,s} = \phi_{m-r,m'-s}$ which holds exactly at $\delta = 0$:

$$\sigma \times \sigma = \phi_{1,1} + \phi_{1,3} + \phi_{1,5} + \phi_{1,\,7} + \phi_{1,\,9} + \cdots \tag{28}$$

$$\overset{\delta=0}{=} \phi_{3,4} + \phi_{3,2} + \underset{\sim\sim}{\phi_{3,0}} + \phi_{3,-2} + \phi_{3,-4} + \cdots \tag{29}$$

IOP Conf. Series: Journal of Physics: Conf. Series **965** (2018) 012036 doi:10.1088/1742-6596/965/1/012036

By the well-known relation $\Delta_{r,-s} = \Delta_{r,s} + rs$, one has $\Delta_{1,9} = \Delta_{1,1} + 12$, $\Delta_{1,7} = \Delta_{1,3} + 6$, thereby the claimed resonances happen. For instance, a primary contribution of the 5-th channel ($\varepsilon^{(4)} = \phi_{1,9}$) starts to mix with a level-12 descendant contribution of the 1-st channel ($\varepsilon^{(0)} = \phi_{1,1}$; the identity operator) as shown in (30). A "primary" contribution here means that it comes from the leading contibution in the block $|F_{\Delta_{1,9}}(z)|^2 \sim |z|^{-2\Delta_{1,9}}$. Similarly, the 2-nd ($\varepsilon^{(1)} = \phi_{1,3}$; the energy operator) and 4-th channel ($\varepsilon^{(3)} = \phi_{1,7}$) mix with each other at level 6 as in (31). These resonances are summarized as follows.

$$\begin{aligned}
\langle\sigma\sigma\sigma\sigma\rangle = & (C^{1,1}_{\sigma\sigma})^2|F_{\Delta_{1,1}}(z)|^2 && \Longleftarrow \mathcal{O}(1) = \left(C^{1,1}_{\sigma\sigma}\right)^2,\ \Delta_{1,9} = \Delta_{1,1} + 12 + \mathcal{O}(\delta), && p = 0\ (\spadesuit) \quad (30)\\
& + (C^{1,3}_{\sigma\sigma})^2|F_{\Delta_{1,3}}(z)|^2 && \Longleftarrow \mathcal{O}(1) = \left(C^{1,3}_{\sigma\sigma}\right)^2,\ \Delta_{1,7} = \Delta_{1,3} + 6 + \mathcal{O}(\delta), && p = 1\ (\clubsuit) \quad (31)\\
& + (C^{1,5}_{\sigma\sigma})^2|F_{\Delta_{1,5}}(z)|^2 && \Longleftarrow \mathcal{O}(\delta^{-1}) = (C^{1,5}_{\sigma\sigma})^2, && p = 2\ (\diamondsuit) \quad (32)\\
& + (C^{1,7}_{\sigma\sigma})^2|F_{\Delta_{1,7}}(z)|^2 && \Longleftarrow -(A_1/\delta)^2 + \mathcal{O}(\delta^{-1}) = \left(C^{1,7}_{\sigma\sigma}/C^{1,3}_{\sigma\sigma}\right)^2, && p = 3\ (\clubsuit) \quad (33)\\
& + (C^{1,9}_{\sigma\sigma})^2|F_{\Delta_{1,9}}(z)|^2 && \Longleftarrow -(A_2/\delta)^2 + \mathcal{O}(\delta^{-1}) = \left(C^{1,9}_{\sigma\sigma}/C^{1,1}_{\sigma\sigma}\right)^2, && p = 4\ (\spadesuit) \quad (34)\\
& + \cdots ,
\end{aligned}$$

where the simplified notation $C^{r,s}_{\sigma\sigma} = C^{\phi_{r,s}}_{\sigma\sigma}$ is used. This shows the first resonance ($\spadesuit$) occurs between $\varepsilon^{(0)} \leftrightarrow \varepsilon^{(4)}$ and the second ($\clubsuit$) occurs between $\varepsilon^{(1)} \leftrightarrow \varepsilon^{(3)}$. Note that this phenomenon is due to the exisitence of the pivot operator $p = 2$ ($\diamondsuit$) at $\delta = 0$ as indicated by the wavy line in (29), where the duality $\phi_{1,5} \Leftrightarrow \phi_{3,0}$ holds. This operator comes as a simple pole $\mathcal{O}(\delta^{-1})$.

The amplitude ratios between these resonating channels (the ratio of the squared OPE coefficients) may be computed from (16), resulting in the leading double poles in the deviation parameter δ as in (33) and (34). The coefficients A_1 and A_2 may be obtained as rather large products of the gamma functions, which could look non-instructive at first:

$$\begin{aligned}
A_1 =& \frac{1}{60}\sqrt{\frac{-\Gamma\left(-\frac{19}{5}\right)\Gamma\left(-\frac{18}{5}\right)\Gamma\left(-\frac{14}{5}\right)\Gamma\left(-\frac{11}{5}\right)\Gamma\left(-\frac{6}{5}\right)\Gamma\left(\frac{11}{10}\right)^2\Gamma\left(\frac{8}{5}\right)^4\Gamma\left(\frac{19}{10}\right)^2\Gamma\left(\frac{21}{10}\right)^2\Gamma\left(\frac{12}{5}\right)^3\Gamma\left(\frac{29}{10}\right)^2}{\Gamma\left(-\frac{19}{10}\right)^2\Gamma\left(-\frac{7}{5}\right)^3\Gamma\left(-\frac{11}{10}\right)^2\Gamma\left(-\frac{9}{10}\right)^2\Gamma\left(-\frac{3}{5}\right)^4\Gamma\left(-\frac{1}{10}\right)^2\Gamma\left(\frac{11}{5}\right)\Gamma\left(\frac{16}{5}\right)\Gamma\left(\frac{19}{5}\right)\Gamma\left(\frac{23}{5}\right)\Gamma\left(\frac{24}{5}\right)}}\\
=& \frac{3^2 \cdot 7^1 \cdot 11^1 \cdot 19^1}{2^{23} \cdot 5^7 \cdot 13^1}, \qquad (35)\\
A_2 =& \frac{1}{60}\sqrt{\frac{\Gamma\left(-\frac{27}{5}\right)\Gamma\left(-\frac{26}{5}\right)\Gamma\left(-\frac{23}{5}\right)\Gamma\left(-\frac{22}{5}\right)\Gamma\left(-\frac{19}{5}\right)\Gamma\left(-\frac{18}{5}\right)\Gamma\left(-\frac{14}{5}\right)\Gamma\left(-\frac{6}{5}\right)\Gamma\left(-\frac{2}{5}\right)}{\Gamma\left(-\frac{27}{10}\right)^2\Gamma\left(-\frac{11}{5}\right)^3\Gamma\left(-\frac{19}{10}\right)^2\Gamma\left(-\frac{17}{10}\right)^2\Gamma\left(-\frac{7}{5}\right)^3\Gamma\left(-\frac{11}{10}\right)^2\Gamma\left(-\frac{9}{10}\right)^2\Gamma\left(-\frac{3}{5}\right)^3\Gamma\left(-\frac{3}{10}\right)^2\Gamma\left(-\frac{1}{10}\right)^2}}\\
&\cdot\sqrt{\frac{\Gamma\left(\frac{3}{10}\right)^2\Gamma\left(\frac{2}{5}\right)\Gamma\left(\frac{4}{5}\right)^3\Gamma\left(\frac{11}{10}\right)^2\Gamma\left(\frac{13}{10}\right)^2\Gamma\left(\frac{8}{5}\right)^3\Gamma\left(\frac{19}{10}\right)^2\Gamma\left(\frac{21}{10}\right)^2\Gamma\left(\frac{12}{5}\right)^3\Gamma\left(\frac{27}{10}\right)^2\Gamma\left(\frac{29}{10}\right)^2\Gamma\left(\frac{16}{5}\right)^3\Gamma\left(\frac{37}{10}\right)^2}{\Gamma\left(\frac{1}{5}\right)^3\Gamma\left(\frac{3}{5}\right)\Gamma\left(\frac{7}{10}\right)^2\Gamma\left(\frac{7}{5}\right)\Gamma\left(\frac{11}{5}\right)\Gamma\left(\frac{19}{5}\right)\Gamma\left(\frac{23}{5}\right)\Gamma\left(\frac{24}{5}\right)\Gamma\left(\frac{27}{5}\right)\Gamma\left(\frac{28}{5}\right)\Gamma\left(\frac{31}{5}\right)\Gamma\left(\frac{32}{5}\right)}}\\
=& \frac{3^3 \cdot 7^2 \cdot 11^2 \cdot 17^2 \cdot 19^1}{2^{50} \cdot 5^7 \cdot 13^3 \cdot 23^1}. \qquad (36)
\end{aligned}$$

Both these coefficients, however, suggestively reduce to nice fractions as above after some repeated uses of the elementary relation $\Gamma(x)\Gamma(1-x) = \pi/\sin\pi x$. Such reduction can not be a coincidence.

In fact, we can show that the double pole contributions at higher operators are exactly canceled out by the descendant contributions of the lower operators. The key is the Zamolodchikov recursion formula [24]. If we focus on the dependence on the scaling dimension Δ of the intermediate conformal blocks, then they have simple poles at the degenerate dimensions:

$$F_\Delta = g_\Delta + \sum_{(r,s)\in\mathbb{N}^2} \frac{R_{r,s}z^{rs}}{\Delta - \Delta_{r,s}} F_{\Delta_{r,s}+rs}, \qquad (37)$$

where the residue is proportional to the conformal block with shifted Δ with the coefficient $R_{r,s}$ conjectured by Zamolodchikov, which is the following finite product of linear combinations of the external charges $\{\alpha_i\}$ and the screening charges $\alpha_\pm$:

$$R_{r,s} = B_{r,s} \cdot P_{r,s}(c, \{\alpha_i\}), \qquad B_{r,s} = -\frac{1}{2}\prod_{j,k} \lambda_{j,k}^{-1} \quad (|j| \leqslant r,\ |k| \leqslant s,\ (j,k) \neq (0,0), (r,s)), \tag{38}$$

$$P_{r,s}(c, \{\alpha_i\}) = \prod_{\pm,p,q} \left(\alpha_1 \pm \alpha_2 + \frac{\lambda_{p,q}}{2}\right)\left(\alpha_3 \pm \alpha_4 + \frac{\lambda_{p,q}}{2}\right), \tag{39}$$

where (p,q) runs $p = -r+1, -r+3, \cdots, r-1$ and $q = -s+1, -s+3, \cdots, s-1$ and $\lambda_{p,q} = p\alpha_+ + q\alpha_-$. Importantly, our assumptions of the denominator m be odd, and of the external dimensions be all non-degenerate ($\alpha_i = \alpha_{1/2,0}$ for $i = 1,2,3,4$) guarantee non-vanishing of the residue coefficients $R_{r,s}$ for certain pairs (r,s), such as $(r,s) = (3,4)$ and $(r,s) = (3,2)$, which respectively correspond to the resonating channels in (30) and (31). And this property makes $F_{\Delta_{1,1}}$ and $F_{\Delta_{1,3}}$ the *singular* conformal blocks [25] as $\delta \to 0$:

$$F_{\Delta_{1,1}} \supset \frac{R_{3,4}\ z^{12+\mathcal{O}(\delta)}}{\Delta_{1,1} - \Delta_{3,4}} = \frac{A_2 z^{12}}{\delta}\left(1 + \mathcal{O}(\delta)\ln z\right), \ F_{\Delta_{1,3}} \supset \frac{R_{3,2}\ z^{6+\mathcal{O}(\delta)}}{\Delta_{1,3} - \Delta_{3,2}} = \frac{A_1 z^6}{\delta}\left(1 + \mathcal{O}(\delta)\ln z\right). \tag{40}$$

Then it is possible to show that the left hand side becomes simple pole whose coefficients, when squared, are exactly the same as (35) and (36) from our integral formula computation except the signs. Since the charges at a rational ρ are at most the quadratic irrational, the gamma functions can not remain in (35) and (36), which rationalizes the nice reduction we mentioned above. Due to the cancellation at the leading δ^{-2}, one should expand the resonant powers of z in (30) and (31), which results in the logarithmic 4-point function. Since all the dominant terms contribute as δ^{-1}, the physical correlation function is actually proportional to [5] $\delta\langle\sigma\sigma\sigma\sigma\rangle_{\rm CFT}$. As an important example, at the polymer CFT ($n = 0$), one has $\rho = 2/3 + \delta$ and $\delta = \mathcal{O}(n)$. Then the observable 4-point function is $\langle\sigma\sigma\sigma\sigma\rangle_{\rm phys} = n\langle\sigma\sigma\sigma\sigma\rangle_{\rm CFT}$ in $n \to 0$. Besides the application to the polymer, this gives us a clue to the way we should deal with the simple pole at $n = 0$ of the energy operator amplitude, and with all its generalization at ρ with the odd denominator.

5. Conclusion

The 4-point function $\langle\sigma\sigma\sigma\sigma\rangle$ of the $O(n)$ fundamental field plays a central role in the conformal bootstrap in general dimensions. In contrast with 4-point functions that contain degenerate fields [15, 16], this 4-point function $\langle\sigma\sigma\sigma\sigma\rangle$ remains elusive even in $d = 2$ for generic values of $n = -2\cos\pi/\rho$ since all the fields in the correlation function are non-degenerate. We have seen, however, if the standard identification [11] $\varepsilon \to \phi_{1,3}$ and $\sigma \to \phi_{1/2,0}$ is assumed, one can compute the important amplitudes (16) that correspond to the $O(n)$ singlet contribution. Seeing the unitarity from a higher perspective, it would be important to quantify to what degree such amplitudes can be negative. By brushing up the knowledge obtained here, it is possible to discuss that the positivity violation is very weak at $n = 0.99$, for instance, like it is weak at $d = 3.99$ [5]. On the practical side, such direction could be a clue for approaching the real non-unitary CFT from the (almost unique) unitary solution to the crossing symmetry obtained in non-integer d [10] and in non-integer n [6]. We have seen that the sign-change occurs only at the simple poles, whose exact locations are encoded in the behavior of the automorphic $\xi(q)$-function (19) along the Farey path. Although the correspondence between these ideal Farey paths on the Poincaré disk and the singlet comformal amplitudes beyond the simple poles is not perfect as it is, it would be interesting to make it precise or to see if such hierarchical structure is ubiquitous in the CFT with infinitely many primary operators.

[5] A similar observation has been discussed in the percolation CFT [18].

IOP Conf. Series: Journal of Physics: Conf. Series **965** (2018) 012036 doi:10.1088/1742-6596/965/1/012036

Acknowledgments
This work is supported by KAKENHI 16K05491.

References

[1] Polyakov A M 1974 Nonhamiltonian approach to conformal quantum field theory *Zh. Eksp. Teor. Fiz.* **66** 23–42
[2] El-Showk S, Paulos M, Poland D, Rychkov S, Simmons-Duffin D and Vichi A 2012 Solving the 3D Ising model with the conformal bootstrap *Phys. Rev.* D **86** 025022
[3] Kos F, Poland D and Simmons-Duffin D 2014 Bootstrapping the O(N) vector models *J. High Energy Phys.* JHEP06(2014)091
[4] El-Showk S, Paulos M, Poland D, Rychkov S, Simmons-Duffin D and Vichi A 2014 Solving the 3D Ising model with the conformal bootstrap II. c-minimization and precise critical exponents *J. Stat. Phys.* **157** 869–914
[5] Hogervorst M, Rychkov S and van Rees B C 2016 Unitarity violation at the Wilson-Fisher fixed point in $4-\epsilon$ dimensions *Phys. Rev.* D **93** 125025
[6] Shimada H and Hikami S 2016 Fractal dimensions of self-avoiding walks and Ising high-temperature graphs in 3D conformal bootstrap *J. Stat. Phys.* **165** 1006–35
[7] Simon B 2015 *Basic Complex Analysis: A Comprehensive Course in Analysis, Part 2A* (American Mathematical Society)
[8] Wilson M E and Fisher M E 1972 Critical exponents in 3.99 dimensions *Phys. Rev. Lett.* **28** 240–243
[9] de Gennes PG 1972 Exponents for the excluded volume problem as derived by the Wilson method *Phys. Lett.* A **38** 339
[10] El-Showk S, Paulos M, Poland D, Rychkov S, Simmons-Duffin D and Vichi A 2014 Conformal field theories in fractional dimensions *Phys. Rev. Lett.* **112** 141601
[11] di Francesco P, Saleur H, Zuber J B 1987 Relations between the Coulomb gas picture and conformal invariance of two-dimensional critical models *J. Stat. Phys.* **49** 57–79
[12] Jacobsen J L 2009 Conformal Field Theory Applied to Loop Models *Lect. Note. Phys.* **775** 347–424
[13] Bauer M and Bernard D 2006 2D growth processes: SLE and Loewner chains *Phys. Rep.* 432 115–221
[14] Beffara V 2008 The dimension of the SLE curves *Ann. Probab* **36** 1421–1452
[15] Belavin A A, Polyakov A M and Zamolodchikov A B 1984 Infinite conformal symmetry in two-dimensional quantum field theory *Nucl. Phys.* B **241** 333–380
[16] Dotsenko Vl S and Fateev V A 1984 Conformal algebra and multipoint correlation functions in 2D statistical models *Nucl. Phys.* B **240** 312-348
[17] Saleur H 1992 Polymers and percolation in two dimensions and twisted N=2 supersymmetry *Nucl.Phys.* B **382** 486–531
[18] Dotsenko Vl S 2016 Correlation function of four spins in the percolation model *Nucl. Phys.* B **911** 712–743
[19] Aomoto K 1987 On the complex Selberg integral *Q. J. Math.* **38** 385–399
[20] Selberg A 1944 Remarks on a multiple integral *Norsk Mat. Tidsskr.* **26** 71–78
[21] Estienne B and Ikhlef Y 2015 Correlation functions in loop models *Preprint* arXiv:1505.00585
[22] Pasquier V and Saleur H 1990 Common structures between finite systems and conformal field theories through quantum groups *Nucl. Phys.* B **330** 523–556
[23] 2002 Mumford D, Series C and Wright D *Indra's Pearls: The Vision of Felix Klein* (Cambridge University Press)
[24] Zamolodchikov Al B 1984 Conformal symmetry in two dimensions: an explicit recurrence formula for the conformal partial wave amplitude *Comm. Math. Phys.* **96** 419–422
[25] Viti J and Santachiara R 2014 Local logarithmic correlators as limits of Coulomb gas integrals *Nucl. Phys.* B **882** 229–262

IOP Conf. Series: Journal of Physics: Conf. Series **965** (2018) 012037 doi:10.1088/1742-6596/965/1/012037

A resolution of non-uniqueness puzzle of periodic orbits in the 2-dim anisotropic Kepler problem: bifurcation $U \to S + U'$

[1]T. Shimada, [1]T. Nakajima, [2]K. Sumiya, [3]K. Kubo
[1]Department of physics, School of Science and Technology,Meiji University, Higashimita 1-1-1, Tama, Kawasaki, Kanagawa 214-8571, Japan
[2]Faculty of Global Media Studies, Komazawa University, Komazawa 1-23-1, Setagaya, Tokyo 154-8525, Japan
[3]Minerva Center and Department of Physics, Bar-Ilan University, Ramat-Gan 52900, Israel

E-mail: tshimada@gravity.mind.meiji.ac.jp

Abstract. Using binary coding of orbit we introduce a finite level (N) surface over the initial value domain D of 2-dim AKP. It gives a tiling of D by base *ribbons*. The scheme of the one-time map is studied and the properness of the tiling is proved. This analysis in turn resolves the long standing puzzle in AKP—the non-uniqueness issue of a PO for a given code. We argue that the *unique* existence of a periodic orbit (PO) for a given binary code generally holds (for inverse anisotropy parameter $\gamma < 8/9$) but there is a remarkable exception in which a ribbon with a certain code escapes from shrinking at large N and embodies the Broucke-type stable PO (S). It comes along the bifurcation of an unstable PO (U): $U(R) \to S(R) + U'(NR)$ (R for retracing and NR for non-retracing). An analysis based on orbit topology clarifies the pattern of the bifurcation; we give a conjecture that it occurs among odd rank Y-symmetric POs.

1. Introduction

Anisotropic Kepler Problem(AKP), cultivated by Gutzwiller [1, 2, 3, 4], is a vital testing ground of quantum chaos; we can control strength of chaos by a single tuning a parameter—the mass anisotropy. (We use inverse of anisotropy, γ). Here we focus on 2D-AKP, where a binary coding of the orbit is possible. It is proved by Devaney [5] mathematically (and showed in physical terms by Gutzwiller [2]) that there exist at least one periodic orbit (PO) for a given rational code in the high anisotropy regin ($\gamma < 8/9$). Gutzwiller also conjectured from numerical results that the PO is unique. On the other hand, Devaney instigated Broucke to find a counterexample in the form of a stable PO [4]. Broucke reported two examples of stable PO [6], and more recently Contopoulos and Harsoula have reported rather many stable orbits, which may defy the belief that AKP is an Anosov system [7].

In this report, we focus on the coarse-grained level N Devil's Staircase Surface (DSS), which defines a tiling of the properly compactified initial value domain on the Poincaré surface of section. (Let us call it Gutzwiller's rectangle D). We give a proof that the tiling is *proper*—that is, the DSS is monotonic—by an investigation of the generating mechanism of a level $N+1$ tiling from level N one by an investigation of the action of one-time map on D. It in turn shows that there is a special case that a ribbon with certain code does not shrink at large N, and this case occurs along with a transition $U \to S + U'$ with decreasing mass anisotropy. The transition is found to occur among Y-symmetric orbits, which comes from the fact that non-shrinking overlap of ribbons is produced when future (F) and past (P) ribbons become tangent each other. A topology consideration explains the detailed pattern of the bifurcation.

2. Devil's Staircase Surface and the One-time Map

2.1. Gutzwiller's rectangle

The Hamiltonian of two-dimensional AKP is given by

$$H = \frac{u^2}{2\mu} + \frac{v^2}{2\nu} - \frac{1}{r}, \quad (r \equiv \sqrt{x^2 + y^2},\ \mu > \nu) \tag{1}$$

where $u \equiv p_x, v \equiv p_y$ and the mass anisotropy is proportional to $1 - \gamma$ with $\gamma \equiv \nu/\mu < 1$. Because $\ddot{y}/\ddot{x} = (\mu/\nu)y/x > y/x$, the orbit tends to cross the heavy x-axis more frequently. Thus, the Poincaré surface of section (PSS) is specified by the condition $y = 0$, and we can encode the orbit by the sequence of $a_i = \pm 1$, the sign of x_i at the i-th crossing of the orbit with the x-axis. The future sequence is $(a_0, a_1, \cdots)$ and the past is $(a_0, a_{-1}, \cdots)$.

The constant energy condition determines the kinematically allowed region on the PSS (the physical region for short). As the potential is a homogeneous in coordinates, the system has a scaling property. Any value for the energy is equivalent and we take $H = -1/2$ by convention.

The physical region of PSS is compactified into a rectangle

$$\mathrm{D} = \left\{ (X, U) \;\middle|\; |X| \le 2 \text{ and } |U| \le B \equiv \frac{\sqrt{\mu}\pi}{2} \right\} \tag{2}$$

by an area preserving map;

$$X = x\left(1 + u^2/\mu\right), \quad U = \sqrt{\mu}\arctan\left(u/\sqrt{\mu}\right). \tag{3}$$

The collision manifold is the I-shaped backbone of D:

$$\mathrm{I} = \{\, (X, U) \mid |X| \le 2 \text{ and } U = \pm B \,\} \cup \{\, (X, U) \mid X = 0 \text{ and } |U| \le B \,\}. \tag{4}$$

2.2. Symbolic coding of orbits and Devil's staircase surface

The future and past surfaces over D at level N are respectively described by the height functions

$$\zeta_N^F(X_0, U_0) \equiv \sum_{j=0}^{N} \frac{1}{2^{j+1}} a_j(X_0, U_0), \ \zeta_N^P(X_0, U_0) \equiv \sum_{j=0}^{N} \frac{1}{2^{j+1}} a_{-j}(X_0, U_0). \tag{5}$$

As Fig. 1 shows, the surface defined by (5) has rather simple structure. The level N surface by definition consists of 2^{N+1} distinct heights and the base of it is divided into *ribbons*. Furthermore, the height of the step is monotonously increasing, if we traverse ribbons from left to right. We express that the ribbons are tiling D *properly*.[1]

[1] The edge, separating two adjacent ribbons, can be worked out by Gutzwiller's technique of the collision parameter analysis. They are the unstable and stable manifolds of the collision for the future and past DSS respectively [2]. See related discussion for Fig. 3.

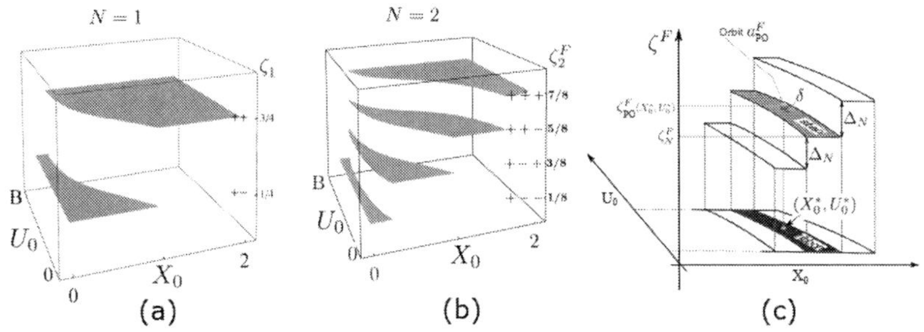

Figure 1. (a), (b):The first quadrant of future DSS. (c) PO and level N surface.

Now, the code of a PO is cyclic; $a^F_{\rm PO} = (a_0, a_1, \cdots, a_{2n-1}; a_{2n} = a_0)$; n is the rank of the PO.[2] If the initial point of the PO is $(X^*_0, U^*_0) \in$D, its height is naturally defined by

$$\zeta^F_{\rm PO}(X^*_0, U^*_0) = \frac{1}{1 - \frac{1}{2^{2n}}} \left\{ \sum_{j=0}^{2n-1} \frac{1}{2^{j+1}} a_j(X^*_0, U^*_0) \right\}. \tag{6}$$

The following holds:

A step of the level N DSS whose height is given by the first N+1 bits of the string (the primary part or its repetition) is the closest to the PO in height. See Fig. 1(c).

To see this, consider the amount of misfit between (5)and (6);

$$\delta \equiv \zeta^F_{\rm PO} - \zeta^F_N = \sum_{j=N+1}^{\infty} \frac{1}{2^{j+1}} a_j(X^*_0, U^*_0) \tag{7}$$

We find $|\delta| \leq 1/2^{N+1}$. On the other hand, the difference of height between neighboring steps is $\Delta_N = 1/2^N$. Therefore, the selected step is the closest[3]. At the large N limit, the misfit δ vanishes and the PO asymptotically sits on the selected step. ■

2.3. One-time map

We now discuss the one-time map $\mathcal{F}$:

$$\begin{array}{cccc} \mathcal{F}: & {\rm D}_0 & \longrightarrow & {\rm D}_1 \\ & \Cup & & \Cup \\ & (X_0, U_0) & \longmapsto & (X_1, U_1). \end{array} \tag{8}$$

Because the AKP flow involves both hyperbolic and elliptic singularities, separation as well as blow up inevitably occur. We have to back up the numerical calculation by the map restricted to the collision manifold I.

$$\begin{array}{cccc} & {\rm D}_0 & \longrightarrow & {\rm D}_1 \\ & \cup & & \cup \\ \mathcal{F}|_I: & I & \longrightarrow & I. \end{array} \tag{9}$$

[2] The length of PO must be even ($2n$), because it must cross the x-axis even times to complete its period.

[3] This statement holds irrespective of whether the tiling by ribbons is proper or not.

$\mathcal{F}|_I$ is worked out by Gutzwiller [2]. Polar coordinates (r,ψ) and (χ,ϑ) for both the coordinates and momenta are introduced;

$$u=\sqrt{\mu}e^{\chi}\cos\vartheta,\ v=\sqrt{\nu}e^{\chi}\sin\vartheta,\ x=r\cos\psi,\ y=r\sin\psi. \tag{10}$$

The kinetic energy is $K=e^{2\chi}/2$ and under the canonical choice of energy $H=-1/2$ it follows $r=2/(1+e^{2\chi})$. Slowing down[4] the orbit by $dt'=e^{3\chi}dt$ and taking the limit $\chi\to\infty$, the equation of motion is reduced to an autonomous form

$$\frac{d\vartheta}{dt'}=-\left(\sqrt{\mu}\cos\vartheta\sin\psi-\sqrt{\nu}\sin\vartheta\cos\psi\right),\ \frac{d\psi}{dt'}=-2\left(\sqrt{\nu}\cos\vartheta\sin\psi-\sqrt{\mu}\sin\vartheta\cos\psi\right). \tag{11}$$

This gives the map $\mathcal{M}:(\vartheta_0,\psi_0)\to(\vartheta_1,\psi_1)$, where (ϑ_0,ψ_0) and (ϑ_1,ψ_1) parameterize respectively the initial and final I in (9). This invokes a new critical angle ϑ_c, and region 2 is divided into three sub-regions $2A,2B,2C$.

	X_0	U_0		X_1	U_1	
$0\longrightarrow\vartheta_v$	1	$2\longrightarrow X_v$	B	$1'$	$2\longrightarrow X_h$	B
$\vartheta_v\longrightarrow\pi/2$	$2A$	$X_v\longrightarrow 0$	B	$2A'$	$-X_h\longrightarrow -X_c$	$-B$
$\pi/2\longrightarrow\pi-\vartheta_c$	$2B$	$0\longrightarrow X_c$	$-B$	$2B'$	$-X_c\longrightarrow 0$	$-B$
$\pi-\vartheta_c\longrightarrow\pi-\vartheta_h$	$2C$	$X_c\longrightarrow X_h$	$-B$	$2C'$	$0\longrightarrow -X_v$	B
$\pi-\vartheta_h\longrightarrow\pi$	3	$X_h\longrightarrow 2$	$-B$	$3'$	$X_v\longrightarrow 2$	$-B$

Table 1. Action of $\mathcal{F}|_I$ on five regions. $X_v=2\cos^2\vartheta_v$, $X_h=2\cos^2\vartheta_h$, $X_c=2\cos^2\vartheta_c$. $(B=\pi\sqrt{\mu}/2)$. Arrow represents the direction of increasing ϑ_0. At the separation (solid line) by H_v (H_h), the signs of both X_1 and U_1 flip, while at the sub-criticality (dashed line) by ϑ_0 (ϑ_1) passing through $\pi/2$ $(3\pi/2)$, only the U_1 sign is flipped.

The scheme of $\mathcal{F}|_I$ is shown in Table 1 and it is combined with the numerically calculated interior map in Fig. 2.

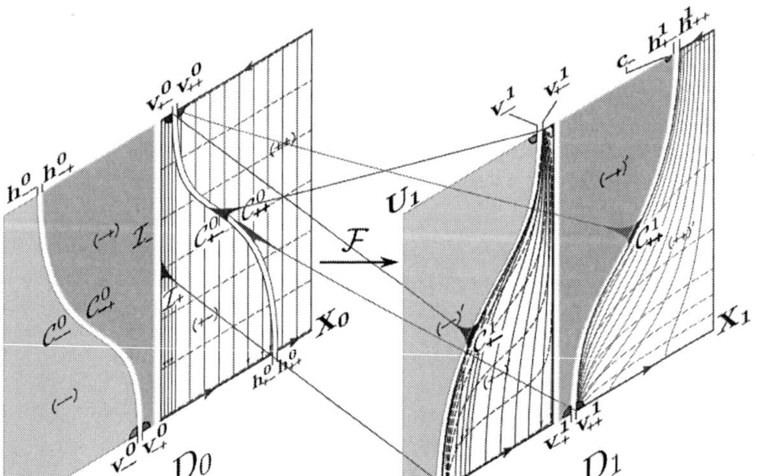

Figure 2. The scheme of one-time map. Points that blows up $(V^0_{+\pm}\to C^1_{+\pm})$ (or focused in $C^0_{+\pm}\to v^1_{+\pm}$), are depicted by quarter-circles.

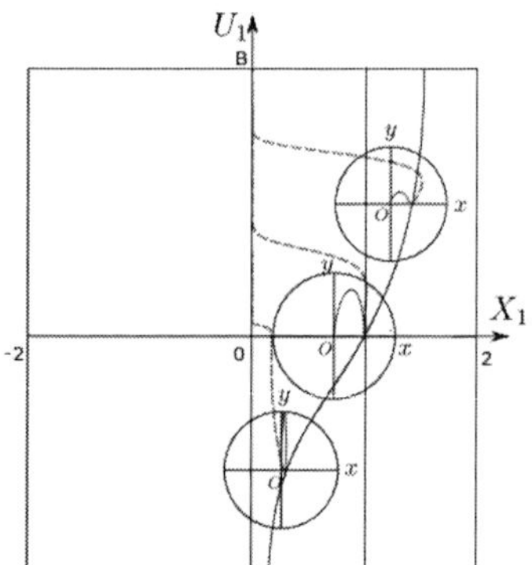

Figure 3. Inset shows orbit from collision. Dashed line is projection to (X,U) plane.

[4] The collision occurs at $\chi=\infty$. This is equivalent to the blow up technique used by Devaney [5] to remove the singularity due to the collision. See also McGehee [8].

IOP Conf. Series: Journal of Physics: Conf. Series **965** (2018) 012037 doi:10.1088/1742-6596/965/1/012037

The boundary information is faithfully reflected in the interior result. Most important is the following; the separator curve C^0 is double sided, one side is mapped to v^1_{++} and the other to v^1_{+-}. The reason why a curve is contracted into a point and a point is blown up to a curve can be understood as follows. In Fig. 3, we show the result of collision parameter analysis which follows [9]. Each orbit emanating from the collision is labeled by a parameter (corresponding to the emission angle) and the position of its first arrival on the Poincaé surface forms a curve in the (X_1, U_1)−Gutzwiller's rectangle. This is nothing but the core of the double line C^1_{++} and C^1_{-+}.

3. Generation of Ribbon-tiling and a Proof of its Properness

Let us prove the following:

For any N, the ribbons are tiled properly.

Firstly $N = 1$. It is just determined by the one-time map by itself (Fig. 2) and the heights of (future) ribbons by (5) distribute from left to right as

$$\begin{array}{lcccc} \zeta^{N=1} & -\frac{3}{4} & -\frac{1}{4} & \frac{1}{4} & \frac{3}{4} \\ & \uparrow & \uparrow & \uparrow & \uparrow \\ \text{Ribbon}:(a_0,a_1): & 1:(--) & 2:(-+) & 3:(+-) & 4:(++) \\ & L: & \longrightarrow & \longrightarrow & :R \end{array} \tag{12}$$

Clearly the ribbons are properly tiled at $N = 1$.

Now let us assume the level N tiling is proper and show that the properness of the level $N + 1$ tiling follows from it. It is a crack of nutshell as shown in Fig. 4; place D_{-1} in front of D_0, and project the level N tiling on D_0 by $\mathcal{F}^{-1}$ back to D_{-1}. Then, $\mathcal{F}^{N+1}$ induces on D_{-1} the level $N + 1$ tiling. Remarkably this way keeps the data from D_0 upto D_N intact. (By time translation symmetry, it is equivalent to the level $N + 1$ on D_0).

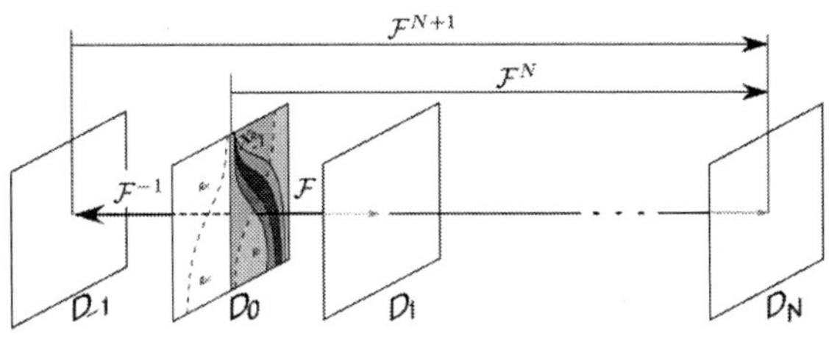

Figure 4. Construction of the level $N + 1$ tiling on D_{-1}.

In Fig. 5 the scheme[5] of $\mathcal{F}^{-1}$ is laid over the level N tiling (by 2^{N+1} ribbons) of D_0. It shows succinctly how $\mathcal{F}^{-1}$ creates the tiling of D_{-1}. Most important is the contraction of the separator lines

$$C^0_{++} \xrightarrow{\mathcal{F}^{-1}} v^{-1}_{++}, \quad C^0_{+-} \xrightarrow{\mathcal{F}^{-1}} v^{-1}_{+-}, \tag{13}$$

where the focus points v^{-1}_{++} and v^{-1}_{+-} are respectively on the right-top and left-down boundary of D_{-1}. Thus, ribbons in the right-half (left-half) of D_0 are transversely separated into two groups R_I and R_{II} (R_{III} and R_{IV}), and they create new tiling of D_{-1}. Here it should be noted that

[5] The scheme of $\mathcal{F}^{-1}$ is the reverse of $\mathcal{F}$ in Fig. 2, except that the superscript of all parts in the scheme are one step back in time.

the order of R_{II} and R_{III} is left-right switched by the action of $\mathcal{F}^{-1}$. Note that the number of ribbons are doubled and each becomes full-height extending from $U_{-1} = -B$ to B. Thus, they constitute in general finer tiling after $\mathcal{F}^{-1}$.

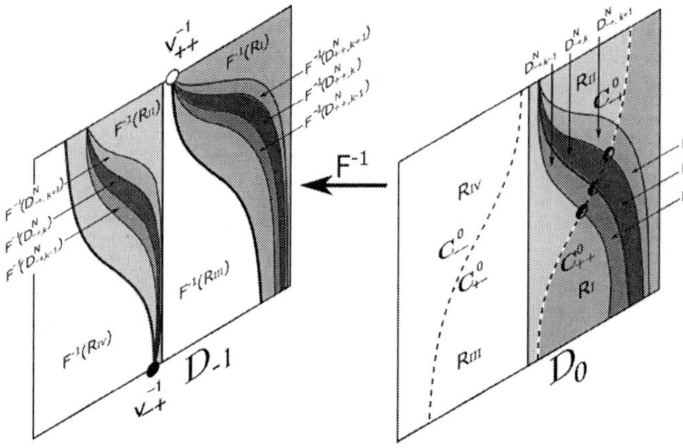

Figure 5. $\mathcal{F}^{-1} : D_0 \to D_{-1}$. Regions R_K, $K = I, \cdots, IV$ correspond to respectively $(++)'$, $(-+)'$, $(+-)'$, $(--)'$ in Fig. 2

Now, from $a_j\,(X_{-1}, U_{-1}) = a_{j-1}(X_0, U_0)$, we find

$$\zeta^{N+1}\,(X_{-1}, U_{-1}) = \frac{1}{2}\,\mathrm{sign}\,(X_{-1}) + \frac{1}{2}\zeta^{N}(X_0, U_0), \tag{14}$$

which calculates $\zeta^{N+1}(X_{-1}, U_{-1})$. It is just half of previous heights except for an overall shift by the first term. Now, because the $\mathcal{F}^{-1}$ is orientation preserving, this tells that the properness of tiling in each region is preserved via $\mathcal{F}^{-1}$. Taking into account of the first term, which is $(+1/2, -1/2, +1/2, -1/2)$ for I, II, III, IV (due to the left-right switch mentioned above), we find the height distribution of $N+1$ tiling is

$$\begin{array}{ccccc} \zeta^{N+1} \subset & (-1, -\frac{1}{2}) & (-\frac{1}{2}, 0) & (0, \frac{1}{2}) & (\frac{1}{2}, 1) \\ & \uparrow & \uparrow & \uparrow & \uparrow \\ \text{Region in } D_{-1}: & \mathcal{F}^{-1}(R_{IV}) & \mathcal{F}^{-1}(R_{II}) & \mathcal{F}^{-1}(R_{III}) & \mathcal{F}^{-1}(R_{I}). \\ & L: & \longrightarrow & \longrightarrow & :R \end{array} \tag{15}$$

Apparently, the tiling is not-only within each region proper but also globally proper. ■

4. Stable and Unstable Periodic Orbits in AKP

4.1. Use of Ribbons to Locate the Initial Point of a Periodic Orbit

Now, we are ready to resolve the long standing issue on the non-uniqueness puzzle of PO at a given code. In short, there are two cases; (A) Normally ribbons shrink roughly $\propto 1/2^{N+1}$, and the corresponding PO uniquely singled out at a cross junction of F and P asymptotic curves. See Fig. 6.

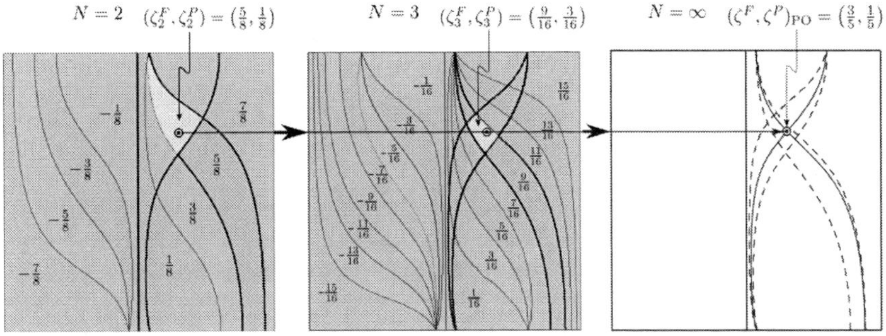

Figure 6. Longitudinal ribbon splitting process at $\gamma = 0.2$. Circle dot; PO with code $(++--)$, rank $n = 2$. (Id=3 in [3]). $(\zeta^F, \zeta^P) = (3/5, 1/5)$ by (6) and enclosed in the junction of F and P ribbons with step heights $(5/8, 1/8)$ $((9/16, 3/16))$ at $N = 2$ $(N = 3)$ as calculated by (5).

(B) Exceptional case: a certain ribbon escapes from shrinking. The best example is the so called Broucke's Island; let us start from a case study of PO3-6.

4.2. *Example:Broucke stable PO*

It is rank $n = 3$ with the code $(+--+--)$. In Fig. 7, the evolution of a ribbon which encloses the initial point of (X_0^*, U_0^*) of the Broucke's PO (call it Broucke's ribbon and give a mark 'B') along with two neighboring ribbons is tracked by successively acting $\mathcal{F}^{-1}$ on them. Remarkably, only the Broucke's ribbon survives, while others (even the adjacent ones) diminish rapidly.

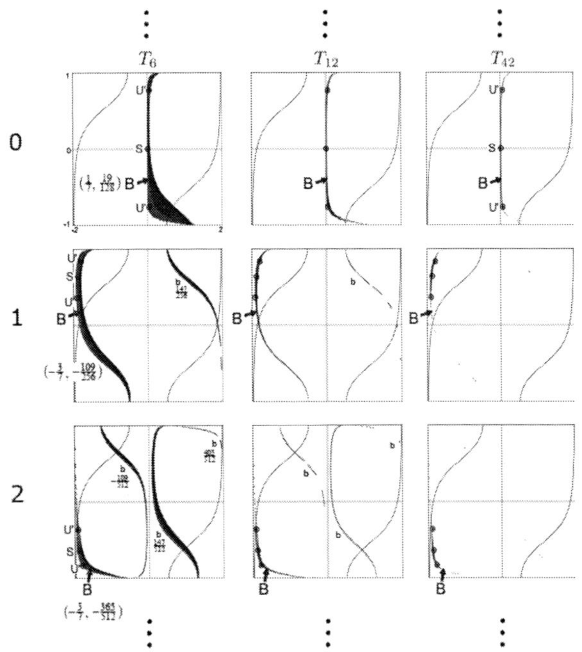

Figure 7. Evolution of future ribbons in the case of Broucke's PO3. Snap shots T_N at $N = (6, 7, 8; 12, 13, 14; 42, 43, 44)$. $\gamma = 0.611$. Circle-dots: Broucke's stable (S) and associated unstable (U') PO. 'B': the ribbon enclosing them.

Why can it survive under the chopping by the separator? It is in a very intriguing way;

Broucke's ribbon is protecting itself from shrinkage by changing at an early stage its tail to a line, so that the chopping becomes immaterial.

Now let us follow the decrease of the anisotropy by increasing γ and investigate how the Broucke stable PO3-6 comes out. As shown in Fig. 8, there is a threshold γ_{th} where the unstable PO (U) changes into the stable one (S) following the advent of non-shrinking ribbons enclosing S. At the same time, a new unstable PO (U') is born. The overlap of non-shrinking F and P ribbons is a region where the orbit from it repeats infinitely the codes, and it is natural that the stable PO is in the center of it. On the other hand, the unstable PO locates at the corner of the overlap. It must be inside the overlap, because the code must be repeated to be periodic, but it must be at the edge, because a slight shift of the initial position must lead to an exponential blow up of the shift for the PO to be unstable. Summing up, the stable PO emerges in a bifurcation process $U \to S + U'$.

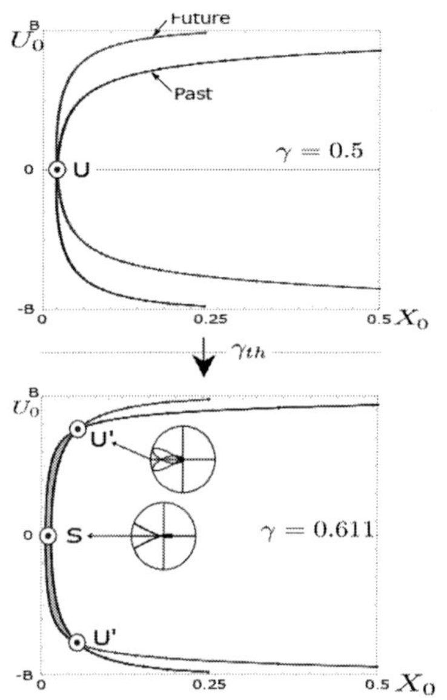

Figure 8. Broucke PO3-6($N = 48$) $U \to S + U'$. $\gamma_{\rm th} = 0.572350895$.

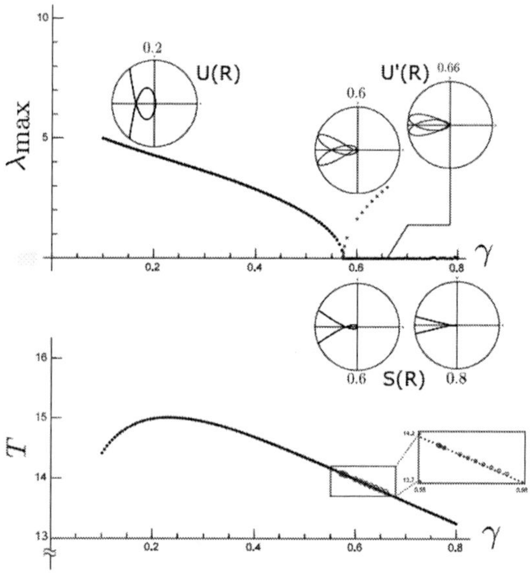

Figure 9. Broucke PO3-6: Lyapunov exponents and Period.

It is noteworthy that all of the POs (U, S, U') are symmetric under the Y transformation:

$$Y : x \to x,\ y \to -y.$$

The non-shrinking ribbons emerge just when the F and P ribbons become tangent each other at $U_0 = 0$. $U_0 = 0$ implies $p_x = 0$, therefore we consider the Y-symmetry is essential feature of the process. Also, we note that U and S are self-retracing (R) and U' is self-non-retracing (NR); this bifurcation is actually

$$U(R) \to S(R) + U'(NR), \qquad (16)$$

Finally, the bifurcation diagram is shown in Fig. 9. We observe the followings.

(1) Near the threshold, $\lambda_{max} \propto |\gamma - \gamma_{\rm th}|^{1/2}$ gives a good description.

(2) U_0^* for unstable ones also exhibits the typical threshold behaviour $U_0^* \propto (\gamma - \gamma_{\rm th})^{1/2}$.

(3) The periods of the periodic orbits are insensitive to the transition. The period of U as a function of γ below γ_{th} smoothly continues to that of S above $\gamma_{\rm th}$. This itself may be natural since both POs are of the same pattern (self-retracing), but, curiously, the period the NR PO (U') has also degenerate period in very good approximation.

4.3. Symmetry-consideration

Since all POs (U, S and U') are Y-symmetric in the transition of the Broucke type, let us now contemplate on the Y-symmetric POs. By a simple topology consideration, we can explain that the pattern $U(R) \to S(R) + U'(NR)$ is natural. We prepare two concepts. Firstly, in order to distinguish R and NR orbits, it is useful to extend the homotopy idea. We consider that a NR PO is as usual homotopic to S^1, but that R PO, where the particle is going back and forth between two turning points, is homotopic to a *squashed* S^1. See Fig. 10.

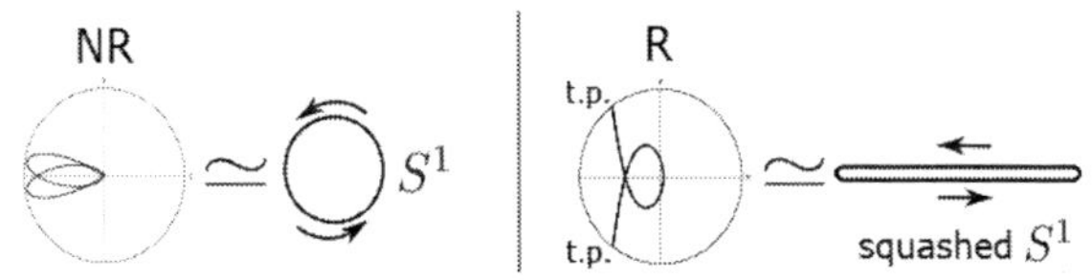

Figure 10. A NR PO (such as U' in PO3-6) is homotopic to S^1, while a R PO (U and S) to a *squashed* S^1.

Secondly, Let $n_\perp$ the number of perpendicular crossing of a Y-symmetric PO with the heavy x-axis. Then, $n_\perp$ must be either 2 or 0, because an orbit with odd $n_\perp$ cannot be closed while $n_\perp = 4, 6, \cdots$ can close but in disconnected loops.

Now we are ready to prove a remarkable fact:

Any Y-symmetric periodic orbit is subject to one of the following three classes;

(a) R with $n_\perp = 2$,

(b) NR with $n_\perp = 0$,

(c) NR with $n_\perp = 2$.

To see this, it is sufficient to consider how to realize an appropriate $n_\perp$ for the topology of R PO and NR PO respectively. See Fig. 11.

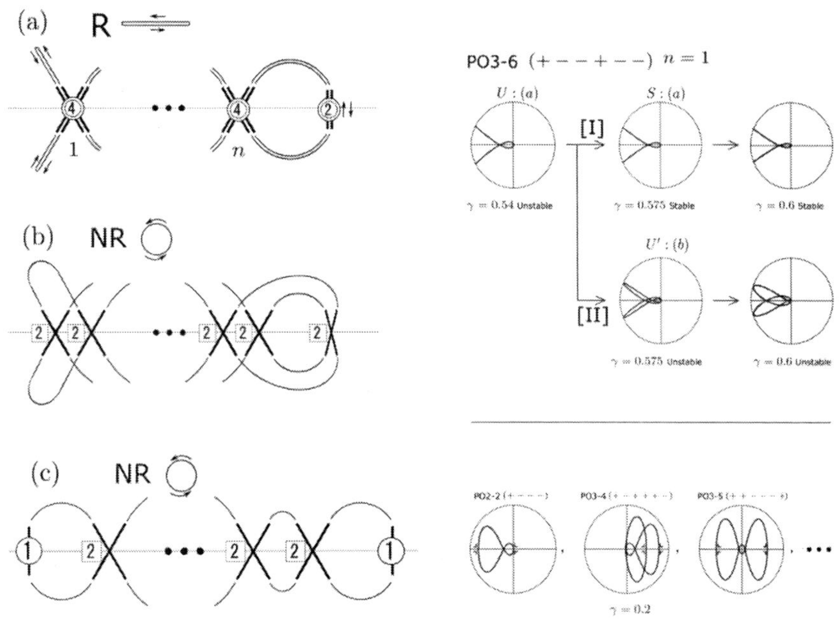

Figure 11. Left : three classes of Y-symmetric PO. Right upper diagram; Broucke PO3-6 transition $U(R) \to S(R) + U'(NR)$. Right lower; sample orbits in class (c).

(i) For R PO (a squashed S^1) to be Y-symmetric, a perpendicular crossing of the x-axis is needed. But just a single perpendicular crossing already saturates $n_\perp = 2$; the crossing is multiplicity 2 in itself. Other crossings are X-type junction each with multiplicity 4. This is class (a).

(ii) On the other hand, for a NR PO, it can be Y-symmetric without perpendicular crossing; either $n_\perp = 0$ (b) or $n_\perp = 2$ (c).
In (b), all the crossings are X-type junction with multiplicity 2.
In (c), all the crossings are X-type except for two distinct perpendicular crossings each with multiplicity one. ■

Having proved the classification, let us reconsider the Broucke transition $U \to S + U'$ in the light of it. The PO (U) at high anisotropy ($\gamma < \gamma_{th}$) is self-retracing and in class (a). Under the decrease of the anisotropy, it gradually changes its initial value $(X_0, U_0) = (X_0^*, 0)$ to remain closed. When γ exceeds γ_{th}, the PO starts transition. It can proceed in three routes:$(a) \to (a), (b), (c)$, call them type (I),(II),(III) respectively. Routes II and III involve a topology change from squashed S^1 to S^1 as shown in Fig. 12. That is, a small shift at the crossing implies global deformation of the PO.

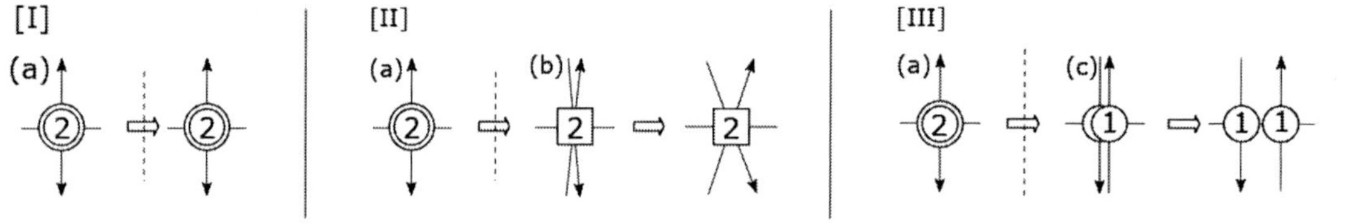

Figure 12. Class-(a)-PO can change in three routes (I),(II) and (III). Note in route (I), PO remains in class (a) and only the orbit profile changes $U(R) \to S(R)$.

Broucke-type transition occurs when the F and P ribbons become tangent each other at $U_0^* = 0$. As we saw in Fig. 8, PO (S) remains with $U_0^* = 0$ and with small ΔX_0^* enveloped by the overlap of F and P ribbons, while PO (U') is subject to rapid increase of U_0^* ($U_0^* \propto (\gamma - \gamma_{\rm th})^{1/2}$). This means $U \to S$ proceed in route (I), while $U \to U'$ in route (II). And this is exactly what is observed. See the Broucke PO3-6 transition in right diagram in Fig. 11. We can now *tell from the classification* why PO (S) is R and PO(U') is NR in (16).

For a PO in class (a), the total number of crossings of the x-axis (i.e. the length) is $4n + 2$, n is the number of cross-junction with multiplicity 4 and 2 from a single perpendicular crossing with multiplicity 2. The rank n of a PO is half of its length. Thus, the rank of class (a) PO is $2n + 1$ and we conjecture that the Broucke-type transition, associated with the non-shrinking ribbon, will occur in odd rank Y-symmetric PO. [6]

4.4. The case of PO-15

Now we apply the topology analysis to a PO at rank 15 (length 30) whose code is $(+-+-+-+--+-+-+-)^2)$. We have chosen this PO as the longest one among stable POs recently discovered by Contopoulos and Harsoula [7][7]. We note that the analysis in [7] is based on one-dimensional shooting ($p_x = 0$ assumed). Therefore, it cannot catch up the bifurcation $U \to U' + S$, where U' has non-vanishing U_0^*. The bifurcation diagram is shown in Fig. 13.

[6] Precisely, the tangency of the F and P ribbons at $U_0 = 0$ implies Y-symmetry of involved POs, and $n_\perp = 2$, but the possibility of the pre-PO (U in the Broucke case) being in class (c) is not logically excluded.

[7] We have read the $(x_0, y_0 = 0)$ by eye from their Figs. 4 and 6, and refined it by 2-dimensional shooting for higher accuracy.

Amazingly the first bifurcation $U \rightarrow U'+S$ is followed by a second one $S \rightarrow S'+S''$. (It occurs in near integrable limit $\gamma > 8/9$). The location of POs inside the relevant ribbons are depicted in Fig. 14. The orbit profiles are given in Fig. 15; remarkably, the multiplicity characteristics are just in accord with our topology analysis.

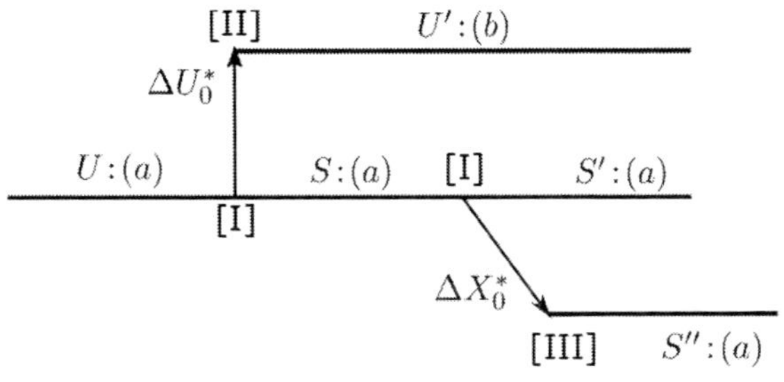

Figure 13.

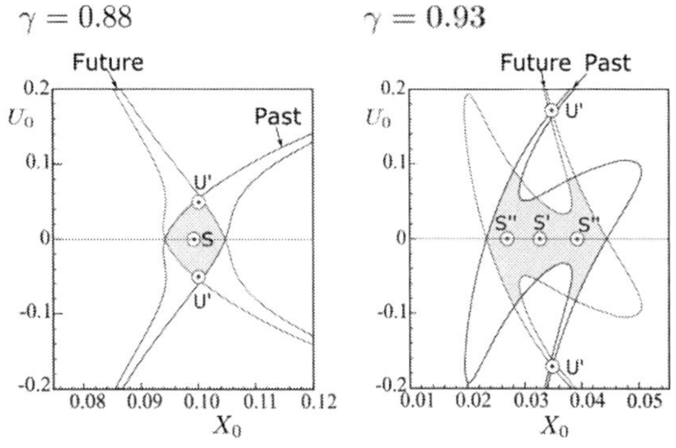

Figure 14.

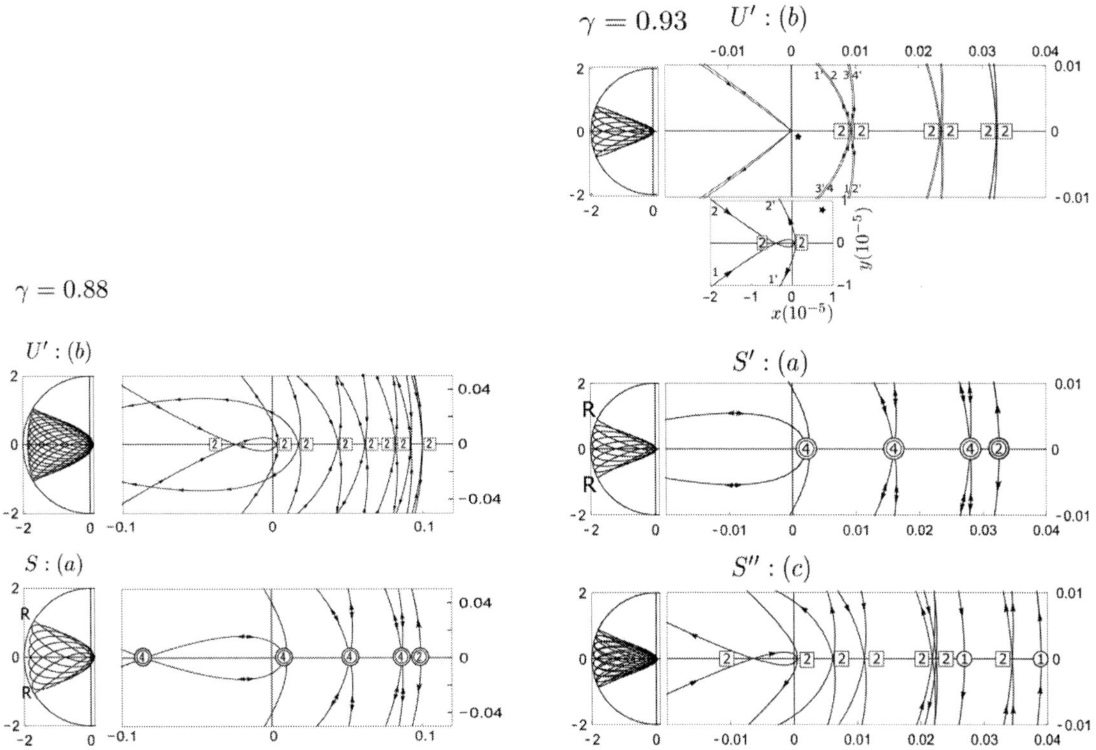

Figure 15. Fine structure of orbit profile. From the number of crossings we determine its topology. The region marked by an asterisk are resolved in the doubly magnified diagram.

5. Conclusion

The key of this work is an introduction of *tiling by ribbons* of the initial value domain D. This is a switch from a dual picture (Gutzwiller's collision parameter analysis). By the use of tiling defined at finite N, it has become tractable to track the generation of higher level DSS. This in particular flags the advent of non-shrinking ribbons. Our resolution of the non-uniqueness puzzle of POs in AKP is quite simple. When the ribbons shrink at large N, a unique PO is singled out for a given code; if the ribbons escape from shrinking, then a stable PO is associated with an unstable PO in the bifurcation process $U(R) \to S(R) + U'(NR)$. A topology argument supports the pattern of this bifurcation. A conjecture is drawn that such a code-preserving bifurcation occurs among Y-symmetric rank odd POs.

Where to proceed? First, we need a better analytic understanding of the condition for the advent of the non-shrinking ribbons. To this end, a local bifurcation analysis using the normal form will be helpful. We envisage another interesting use of it. We have successfully traced the quantum-classical correspondence in AKP in terms of inverse chaology [10]. However, if one wishes to apply the Gutzwiller trace formula directly to the bifurcation process, one instantly fails because either the Lyapunov exponent (for U) or the rotation index (S) vanishes at the bifurcation threshold. We are planning to apply the inverse chaology to this challenging bifurcation process by stepping forward to the next order in $\hbar$ and by including the satellite contribution via the normal form analysis ([11], [12] and [13]).

Acknowledgement

We thank W. Ochs at the Max-Planck Institute, München for unfailing encouragement, and P. H. Damgaard, N. E. J. Bjerrum-Bohr and P. Orland at the Niels Bohr Instute for warm hospitality. We also thank people working on dynamical systems, especially M. Saito, M. Shibayama, and K. Tanigawa for their interest in this work and various suggestions. Finally, we would like to thank, with deep respect, M. C. Gutzwiller for his kind correspondence in 2008 (just once). This work is partially supported by Research Project Grant (B) by Institute of Science and Technology, Meiji University.

References

[1] Gutzwiller M C 1971 Periodic orbits and classical quantization conditions *J. Math. Phys.* **12** 343-58
[2] Gutzwiller M C 1977 Bernoulli sequences and trajectories in the anisotropic Kepler problem *J. Math. Phys.* **18** 806-23
[3] Gutzwiller M C 1981 Periodic orbits in the anisotropic Kepler problem *Classical Mechanics and Dynamical Systems* ed R L Devaney and Z H Nitecki (New York: Marcel Dekker) pp 69-90
[4] Gutzwiller M C 1990 *Chaos in Classical and Quantum Mechanics* (New York: Springer)
[5] Devaney R L 1978 Collision orbits in the anisotropic Kepler problem *Inventiones. Math.* **45** 221-51
[6] Broucke R 1985 *Dynamical Astronomy* ed V G Szebehely and B Balazs (Austin, TX: Univiversity of Texas Press) pp 9-20
[7] Contopoulos G and Harsoula M 2005 Stability and instability in the anisotropic Kepler problem *J. Phys. A: Math. Gen.* **38** 8897-920
[8] McGehee R M 1974 Triple collision in the collinear three-body problem *Inventiones. Math.* **27** 191-227
[9] Gutzwiller M C 1988 From classical to quantum mechanics with hard chaos *J. Phys. Chem.* **92** 3154-63
[10] Kubo K and Shimada T 2014 Periodic orbit theory revisited in the anisotropic Kepler problem *Prog. Theor. Exp. Phys.* 023A06
[11] Ozorio de Almeida A M and Hannay J H 1987 Resonant periodic orbits and the semiclassical energy spectrum *J. Phys. A: Math. Gen.* **20** 5873-83
[12] Main J 1999 Use of harmonic inversion techniques in semiclassical quantization and analysis of quantum spectra *Phys. Rep.* **316** 233-338
[13] Schomerus H and Haake F 1977 Semiclassical spectra from periodic-orbit clusters in a mixed phase space *Phys. Rev. Lett.* **79** 1022-25

IOP Conf. Series: Journal of Physics: Conf. Series **965** (2018) 012038 doi:10.1088/1742-6596/965/1/012038

Spin Calogero–Moser systems for the cyclic quiver

Alexey Silantyev

BLTP, Joint Institute for Nuclear Research, Dubna, Russia

E-mail: `aleksejsilantjev@gmail.com`

Abstract. The method of construction of classical integrable systems on quiver varieties is considered in the case of cyclic quiver. For two types of framing we obtained integrable spin generalisations of the Calogero–Moser systems associated with the complex reflection groups $S_n \ltimes (\mathbb{Z}/m\mathbb{Z})^n$. This method gives Lax matrices of the systems in explicit form. In some particular cases the first Hamiltonians of the system are written explicitly. The paper is based on the work of the author with Oleg Chalykh (the University of Leeds).

1. Introduction

The first spin generalisation of the Calogero–Moser systems appeared in [7]. These were classical integrable systems defined by introducing spin interaction into the Hamiltonian of the rational Calogero–Moser system of type A_n. For the crystallographic root systems (or, equivalently, for semi-simple Lie algebras) integrable spin generalizations were found in [10]. In the rational case a grater class of integrable spin Calogero–Moser systems for classical series of root systems and, more generally, of complex reflection groups can be obtained from the representation theory of quivers (directed graphs): there are natural Poisson commuting Hamiltonians on quiver varieties. The latter were introduced by Nakajima for geometrical construction of the universal enveloping algebras for Kac–Moody algebras and their quantum versions [11].

As it is well-known, the (complexified) phase space of the Calogero–Moser system of type A_n can be completed to the so-called Calogero–Moser space introduced by Wilson in order to classify rational solutions of the Kadomtsev–Petviashvili (KP) hierarchy by using the adelic Grassmannians [14]. The Calogero–Moser spaces can be represented as quiver varieties for a one-loop quiver. More generally, the completed phase spaces of the Calogero–Moser systems for classical series (also called Calogero–Moser spaces) are isomorphic to the quiver varieties associated to the cyclic quivers [5].

This paper is based on the work [1], where the theory of quiver varieties was applied to obtain solutions of generalised KP hierarchies. The solutions of hierarchies suppose dependence of infinite number of time variables. Each time variable appears from a Hamiltonian flow on the corresponding quiver variety. Therefore, to obtain the solutions of the hierarchy we need to find infinite number of appropriate Poisson commuting Hamiltonians on the quiver variety. It is proved that there are enough number (half of dimension of the variety) of functionally independent Hamiltonians among them. The obtained integrable systems can be interpreted as spin generalisations of the Calogero–Moser systems for classical series with completed phase spaces. In the present work these systems are described without mention of hierarchies.

The quiver varieties can be formulated in terms of representations of (deformed) preprojective algebras for quivers [3]. These algebras were introduced in [2]. To define the quiver variety associated with a quiver Q one need to consider the preprojective algebra for the framed quiver Q_ζ obtained from Q by framing with respect to the framing vector $\zeta \in \mathbb{Z}^I_{\geqslant 0}$ where I is the set of the vertices of the quiver Q.

The Calogero–Moser systems for classical series are obtained by considering cyclic quivers Q and a shortest framing vector ζ, i.e. a vector with components ζ_i such that $\sum_{i\in I}\zeta_i = 1$. Replacement of this framing vector by a vector ζ such that $\sum_{i\in I}\zeta_i \geqslant 2$ gives integrable spin versions of these systems. Diversity of the obtained spin systems is mainly provided by the diversity of the forms of the vector ζ. We consider two cases in details: ζ with the identical components and ζ with only one non-zero component.

The plan of the paper is as follows. In Section 2 we recall the definition of the Calogero–Moser system associated with a root system and describe the Calogero–Moser space for A_n case in details. In Section 3 we introduce the quiver varieties in terms of quivers and their preprojective algebras. In Sections 4 and 5 we obtain the spin systems by considering the case of cyclic quiver with two types of the framing vector mentioned above.

We mainly follow the notations of the paper [13].

2. Calogero–Moser systems and Calogero–Moser spaces

First we define the rational Calogero–Moser systems. The A_{n-1} case is considered in details. We introduce the Calogero–Moser spaces and show how this spaces are related to the Calogero–Moser systems.

Let $R \subset \mathbb{R}^n \subset \mathbb{C}^n$ be a (reduced) root system in the sense of [9]. Suppose that $R \neq \emptyset$. Let W be the corresponding (real) reflection group and $c_\alpha \in \mathbb{C}$ $(\alpha \in R)$ be such that $c_{w\alpha} = c_\alpha$ for any $w \in W$, $\alpha \in R$. Denote by (x, y) the standard scalar product of the vectors $x, y \in \mathbb{C}^n$. Let p_a, x_a be the natural Darboux coordinates on the symplectic affine space $T^*\mathbb{C}^n$. Consider the function

$$H = \sum_{a=1}^{n} p_a^2 - \frac{1}{2}\sum_{\alpha\in R} c_\alpha^2 \frac{(\alpha,\alpha)}{(\alpha,x)^2}, \tag{1}$$

where $x \in \mathbb{C}^n$ is a vector with components x_a. The function H is a regular W-invariant function on the affine subvariety $T^*\mathbb{C}^n_{reg} \subset T^*\mathbb{C}^n$ where $\mathbb{C}^n_{reg} = \{x \in \mathbb{C}^n \mid (\alpha, x) \neq 0\, \forall \alpha \in R\}$. It can be regarded as a Hamiltonian of an integrable system on the phase space $T^*\mathbb{C}^n_{reg}$. Namely, there are n functionally independent regular W-invariant functions H_k^{CM} $(k = 1, \dots, n)$ on $T^*\mathbb{C}^n_{reg}$ such that $\{H_k^{\mathsf{CM}}, H_\ell^{\mathsf{CM}}\} = 0$ and H is one of these functions [12], [8]. Due to the W-invariance we can regard that the phase space of the system is the orbit space $T^*\mathbb{C}^n_{reg}/W$.

Let $\mathbb{C}[x_1, \dots, x_n]^W$ be the algebra of W-invariant regular functions on $\mathbb{C}^n$. The Chevalley's theorem says that it is freely generated by n homogeneous polynomials $P_1(x), \dots, P_n(x) \in \mathbb{C}[x_1, \dots, x_n]^W$ of degrees $d_1, \dots, d_n$ and if we suppose that $d_1 \leqslant d_2 \leqslant \dots \leqslant d_n$, then these degrees do not depend on the choice of the polynomials $P_k(x)$ (see e.g. [9]). There are unique Hamiltonians H_k^{CM} Poisson commuting with H such that each H_k^{CM} has leading term $P_k(p_1, \dots, p_n)$ with respect to momenta and is homogeneous of degree $-d_k$ with respect to the transformation $x_a \to \kappa x_a$, $p_a \to \kappa^{-1} p_a$ with $\kappa \in \mathbb{C}^\times$ (see, e.g. [6, Theorem 2.24]). Note that $d_1 = \dots = d_{n-r} = 1$ and $d_{n-r+1} = 2$, where r is the rank of R. Moreover, one can choose $P_{n-r+1}(x) = \sum_{a=1}^n x_a^2$. As consequence, $H_1^{\mathsf{CM}}, \dots, H_{n-r}^{\mathsf{CM}}$ are linear functions of p_a and $H_{n-r+1}^{\mathsf{CM}} = H$.

Consider the case of the root system of type A_{n-1}. In this case c_α does not depend on $\alpha \in R$ and can be put to be 1. The Hamiltonians are defined on the phase space $T^*\mathbb{C}^n_{reg}$ where $\mathbb{C}^n_{reg} = \{x \in \mathbb{C}^n \mid x_a \neq x_b \ (a \neq b)\}$. The leading term of H_k^{CM} with respect to momenta is

$P_k(p_1,\ldots,p_n)=\sum_{a=1}^n p_a^k$. The first two Hamiltonians have the form

$$H_1^{\mathsf{CM}}=\sum_{a=1}^n p_a, \qquad H=H_2^{\mathsf{CM}}=\sum_{a=1}^n p_a^2-2\sum_{a<b}\frac{1}{(x_a-x_b)^2}.$$

The Calogero–Moser space $\mathcal{C}_n$ defined in [14] is the quotient

$$\mathcal{C}_n:=\big\{(X,Y,v,w)\mid [X,Y]+vw=\mathbf{1}_n\big\}/\mathrm{GL}(n,\mathbb{C}), \tag{2}$$

where X,Y are $n\times n$ complex matrices, $v\in\mathbb{C}^n$, $w\in(\mathbb{C}^n)^*$ and the elements $g\in\mathrm{GL}(n,\mathbb{C})$ act as $g.(X,Y,v,w)=(gXg^{-1},gYg^{-1},gv,wg^{-1})$. This is a smooth affine variety with the symplectic structure $\omega=\mathrm{tr}(\mathrm{d}Y\wedge\mathrm{d}X)+\mathrm{d}w\wedge\mathrm{d}v$. The functions $H_k=\mathrm{tr}(Y^k)$ are regular Poisson commuting functions $\mathcal{C}_n$: $\{H_k,H_\ell\}=0$.

Let $\mathcal{C}_n'\subset\mathcal{C}_n$ be the set of points (X,Y,v,w) with diagonalisable matrix X. Substituting $X=\mathrm{diag}(x_1,\ldots,x_n)$ to the relation $[X,Y]+vw=\mathbf{1}_n$ we obtain

$$(x_a-x_b)Y_{ab}+v_aw_b=\delta_{ab}, \qquad a,b=1,\ldots,n. \tag{3}$$

For $a=b$ this gives $v_a\neq 0$, $w_a=1/v_a$ $(a=1,\ldots,n)$. Hence $x_a\neq x_b$ whenever $a\neq b$. Thus by applying $g=\mathrm{diag}(v_1^{-1},\ldots,v_n^{-1})$ we obtain

$$X=\begin{pmatrix}x_1 & & 0\\ & \ddots & \\ 0 & & x_n\end{pmatrix}, \qquad v=\begin{pmatrix}1\\ \vdots\\ 1\end{pmatrix}, \tag{4}$$

$$Y=\begin{pmatrix}p_1 & & -(x_a-x_b)^{-1}\\ & \ddots & \\ -(x_b-x_a)^{-1} & & p_n\end{pmatrix}, \qquad w=(1,\ldots,1) \tag{5}$$

for some $p_1,\ldots,p_n\in\mathbb{C}$. The variables $\{x_1,\ldots,x_n;p_1,\ldots,p_n\}$ are local Darboux coordinates on $\mathcal{C}_n$: $\omega=\sum_{a=1}^n \mathrm{d}p_a\wedge\mathrm{d}x_a$. Restricting the functions $H_k=\mathrm{tr}(Y^k)$ $(k=1,\ldots,n)$ to $\mathcal{C}_n'$ we obtain the Hamiltonians H_k^{CM} of the Calogero–Moser system of type A_{n-1}. The matrix Y defined by the formula (5) is the Lax matrix for this system.

Since the numbers $x_1,\ldots,x_n$ are pairwise different and defined up to a permutation, we obtain $\mathcal{C}_n'\simeq T^*\mathbb{C}^n_{reg}/S_n$. Thus the Calogero–Moser space $\mathcal{C}_n$ is a completed phase space of the Calogero–Moser system of type A_{n-1}.

The points of the set $\mathcal{C}_n\backslash\mathcal{C}_n'$ correspond to the non-diagonalisable X. In this case we can transform X to its Jordan normal form and each eigenvalue will correspond to only one Jordan block. These cases are interpreted as collisions of the particles on the complex plane.

The notion of the Calogero–Moser systems for root systems R (or, equivalently, for real reflection groups W) can be generalised to the case of complex reflection groups W in terms of Dunkl operators [4]. The corresponding completed phase spaces can be constructed from Cherednik algebras associated to this groups W. The classical series of the complex reflection groups are reduced to the case of the groups $W=S_n\ltimes(\mathbb{Z}/m\mathbb{Z})^n$ (see [5]). For $m=1$ and $m=2$ these are real reflection groups of the types A_{n-1} and B_n respectively. For $m\geqslant 3$ they are purely complex reflection groups. The Calogero-Moser systems for these groups have description in terms of quiver varieties for the cyclic quiver (see Section 5).

3. Quiver varieties

First we recall the notions of quiver, of its path algebra and of the preprojective algebra. Then we consider the spaces of representations of these algebras and introduce quiver varieties as in [3].

Quiver Q is a directed graph. That is $Q = (I, Q, t, h)$ where I is the set of vertices, Q is the set of edges and $t, h\colon Q \to I$ are the maps such that $t(a)$ and $h(a)$ are the tail and the head of an edge $a \in Q$. We will suppose the sets I and Q to be finite, and we will write $a\colon i \to j$ or $i \xrightarrow{a} j$, when $t(a) = i$ and $h(a) = j$.

Path algebra $\mathbb{C}Q$ is the $\mathbb{C}$-algebra generated by 1_i $(i \in I)$ and $a \in Q$ with relations $1_i 1_j = \delta_{ij} 1_i$, $a1_i = \delta_{i,t(a)}$, $1_j a = \delta_{j,h(a)} a$. Any non-zero element $a_\ell \cdots a_2 a_1 1_i \in \mathbb{C}Q$, where $a_k \in Q\,\forall k$, is called a path of the length $\ell \geqslant 0$. All the paths form a basis of $\mathbb{C}Q$. Zero length paths 1_i are called trivial paths.

To give a representation of the algebra $\mathbb{C}Q$ one need to assign a vector space V_i to each $i \in I$ and a linear map $V_a\colon V_{t(a)} \to V_{h(a)}$ to each $a \in Q$. If all V_i are finite-dimensional and $\alpha_i = \dim_{\mathbb{C}} V_i$, then the representation $V = (V_i, V_a)$ is called representation of the dimension $\alpha = (\alpha_i)_{i\in I} \in \mathbb{Z}^I_{\geqslant 0}$. For a fixed vector $\alpha = (\alpha_i)_{i\in I}$ the representations $V = (V_i, V_a)$ such that $V_i = \mathbb{C}^{\alpha_i}$ form a vector space $\mathrm{Rep}(\mathbb{C}Q, \alpha) = \prod_{a\in Q} \mathrm{Hom}(\mathbb{C}^{\alpha_{t(a)}}, \mathbb{C}^{\alpha_{h(a)}})$, where $\mathrm{Hom}(\mathbb{C}^n, \mathbb{C}^m)$ is a space of $m \times n$ complex matrices.

For a quiver $Q = (I, Q, t, h)$ consider the double quiver $\overline{Q} = (I, \overline{Q}, t, h)$, obtained by adjoining a reverse edge $a^*\colon j \to i$ for every edge $a\colon i \to j$ in Q (the extended maps $t, h\colon \overline{Q} \to I$ are denoted by the same letters). Let $\lambda \in \mathbb{C}^I$ be a vector with components λ_i $(i \in I)$. Define preprojective algebra $\Pi^\lambda(Q)$ as the quotient of the path algebra $\mathbb{C}\overline{Q}$ by the relations

$$\sum_{\substack{a\in Q\\ h(a)=i}} aa^* - \sum_{\substack{a\in Q\\ t(a)=i}} a^*a = \lambda_i 1_i \qquad (i \in I). \tag{6}$$

The set of representations of this algebra of dimension $\alpha \in \mathbb{Z}^I_{\geqslant 0}$ is an affine subvariety $\mathrm{Rep}\left(\Pi^\lambda(Q), \alpha\right) \subset \mathrm{Rep}(\mathbb{C}\overline{Q}, \alpha)$. Its points are $\mathbb{C}\overline{Q}$-modules $V = (V_i, V_a, V_{a^*})$ such that $V_i = \mathbb{C}^{\alpha_i}$ and

$$\sum_{\substack{a\in Q\\ h(a)=i}} V_a V_{a^*} - \sum_{\substack{a\in Q\\ t(a)=i}} V_{a^*} V_a = \lambda_i \mathbf{1}_{\alpha_i} \qquad (i \in I), \tag{7}$$

where $\mathbf{1}_n = \mathrm{id}_{\mathbb{C}^n}$ is the identity matrix of size n. Note that $\mathrm{Rep}\left(\Pi^\lambda(Q), \alpha\right) = \emptyset$ unless $\lambda \cdot \alpha = 0$, where $\lambda \cdot \alpha := \sum_{i\in I} \lambda_i \alpha_i$. One can prove it by taking trace in (7) and summing by $i \in I$.

Let the group $\mathrm{GL}(\alpha) := \prod_{i\in I} \mathrm{GL}(\alpha_i, \mathbb{C})$ act on $\mathrm{Rep}(\mathbb{C}\overline{Q}, \alpha)$ by the formula $V_a \to g_{h(a)} V_a g^{-1}_{t(a)}$, where $a \in \overline{Q}$ and $g_i \in \mathrm{GL}(\alpha_i, \mathbb{C})$ are components of the acting element $g \in \mathrm{GL}(\alpha)$. This action preserves the subvariety $\mathrm{Rep}\left(\Pi^\lambda(Q), \alpha\right)$. Two elements of $\mathrm{Rep}\left(\Pi^\lambda(Q), \alpha\right)$ are isomorphic as $\Pi^\lambda(Q)$-modules if and only if they belong to the same $\mathrm{GL}(\alpha)$-orbit. In general, the space of orbits $\mathrm{Rep}\left(\Pi^\lambda(Q), \alpha\right)/\mathrm{GL}(\alpha)$ do not have a structure of a variety.

For an affine variety M denote by $\mathbb{C}[M]$ the algebra of all regular functions on M. Let a group G act on M. Denote by $\mathbb{C}[M]^G$ the subalgebra of G-invariant regular function. Define categorical quotient as the affine variety $M//G := \mathrm{Spec}\left(\mathbb{C}[M]^G\right)$. In particular, the variety $N_\lambda(\alpha) = \mathrm{Rep}\left(\Pi^\lambda(Q), \alpha\right)//\mathrm{GL}(\alpha)$ can be identified with the set of $\mathrm{GL}(\alpha)$-orbits of all the semi-simple modules $V \in \mathrm{Rep}\left(\Pi^\lambda(Q), \alpha\right)$. We call the variety $N_\lambda(\alpha)$ moduli space for the quiver Q.

Let $p(\alpha) = 1 + \sum_{a\in Q} \alpha_{t(a)}\alpha_{h(a)} - \sum_{i\in I} \alpha_i^2$. The following result is known by [3, Corollary 1.4 and Lemma 6.5].

Proposition 1. *If a generic module $V \in \operatorname{Rep}\big(\Pi^\lambda(Q), \alpha\big)$ is simple, then $N_\lambda(\alpha)$ is an irreducible affine variety of dimension $2p(\alpha)$. It is smooth at the points corresponding to the simple modules. In particular, if all $V \in \operatorname{Rep}\big(\Pi^\lambda(Q), \alpha\big)$ are simple, then the variety $N_\lambda(\alpha)$ coincides with the usual orbit space $\operatorname{Rep}\big(\Pi^\lambda(Q), \alpha\big)/\mathrm{GL}(\alpha)$, it is smooth and connected in this case.*

Identifying $\operatorname{Rep}\big(\mathbb{C}\overline{Q}, \alpha\big) = T^* \operatorname{Rep}\big(\mathbb{C}Q, \alpha\big)$ we obtain a $\mathrm{GL}(\alpha)$-invariant symplectic structure on this space:

$$\omega = \sum_{a \in Q} \operatorname{tr}(\mathrm{d}V_{a^*} \wedge \mathrm{d}V_a). \tag{8}$$

The varieties $N_\lambda(\alpha)$ can be obtained by Hamiltonian reduction from the symplectic space $\operatorname{Rep}\big(\mathbb{C}\overline{Q}, \alpha\big)$ with the action of the group $\mathrm{GL}(\alpha)$ (see [3]). As consequence, when $N_\lambda(\alpha)$ is smooth it has a structure of the symplectic variety. This structure is given by the same formula (8) understood in a proper way (see [13] for details).

Define the notion of framing of a quiver $Q = (I, E, t, h)$ by a framing vector $\zeta \in \mathbb{Z}^I_{\geqslant 0}$. Let us add to the quiver Q one more vertex ∞ and ζ_i edges $b_{i,r} \colon \infty \to i$ $(r = 1, \dots, \zeta_i)$ for each $i \in I$. In this way we obtain a new quiver Q_ζ with the set of vertices $I_\infty = \{\infty\} \sqcup I$. It is called framed quiver.

Consider the case $V_\infty = \mathbb{C}^1$. Given vectors $\alpha \in \mathbb{Z}^I_{\geqslant 0}$ and $\lambda \in \mathbb{C}^I$ can be uniquely extended to the vectors $\boldsymbol{\alpha} \in \mathbb{Z}^{I_\infty}_{\geqslant 0}$ and $\boldsymbol{\lambda} \in \mathbb{C}^{I_\infty}$ such that $\boldsymbol{\alpha}_\infty = 1$ and $\boldsymbol{\lambda} \cdot \boldsymbol{\alpha} = 0$:

$$\boldsymbol{\alpha} = (1, \alpha), \qquad\qquad \boldsymbol{\lambda} = (-\lambda \cdot \alpha, \lambda). \tag{9}$$

Quiver varieties $M_\lambda(\alpha, \zeta)$ associated with the quiver Q are the moduli space $N_{\boldsymbol{\lambda}}(\boldsymbol{\alpha})$ for the quiver Q_ζ:

$$M_\lambda(\alpha, \zeta) = \operatorname{Rep}\big(\Pi^{\boldsymbol{\lambda}}(Q_\zeta), \boldsymbol{\alpha}\big)//\mathrm{GL}(\boldsymbol{\alpha}). \tag{10}$$

Since $\boldsymbol{\alpha}_\infty = 1$, we have $\mathrm{GL}(\boldsymbol{\alpha}) = \mathbb{C}^\times \times \mathrm{GL}(\alpha)$. Note that, though the subgroup $\mathrm{GL}(\boldsymbol{\alpha}_\infty, \mathbb{C}) = \mathbb{C}^\times$ acts non-trivially, the quotient over $\mathrm{GL}(\boldsymbol{\alpha})$ in (10) is reduced to the quotient over the subgroup $\mathrm{GL}(\alpha)$.

Example 1. Consider the one-loop quiver Q with one vertex 0 and one edge $a_0 \colon 0 \to 0$, and its framing for $\zeta = 1$:

By putting $\lambda = 1$, $\alpha = n$ (and $\zeta = 1$) in $M_\lambda(\alpha, \zeta)$ we obtain the Calogero–Moser space $\mathcal{C}_n$ defined in Section 2. Indeed, after the identification $X = V_{a_0}$, $Y = V_{a_0^*}$, $v = V_{b_{0,1}}$, $w = V_{b_{0,1}^*}$ the equations (7) are rewritten as $[X, Y] + vw = \mathbf{1}_n$ and $wv = n$ (the latter equation follows from the former one by taking trace). This identifies $\mathcal{C}_n$ with the quiver variety $M_\lambda(\alpha, \zeta) = M_1(n, 1)$ for the one-loop quiver Q.

4. Integrable systems for the cyclic quiver with the framing $\zeta = (d, d, \dots, d)$

Let us first consider the case of cyclic quiver (with general framing). Then we describe the quiver varieties for the particular framing vector $\zeta = (d, d, \dots, d)$ in more details.

The cyclic quiver with $m \geqslant 1$ vertices is the quiver $Q = (I, Q, t, h)$ where $I = \mathbb{Z}/m\mathbb{Z} = \{0, 1, \dots, m-1\}$, $Q = \{a_i \mid i = 0, \dots, m-1\}$, $t(a_i) = i$, $h(a_i) = i + 1$ (since I is identified with the cyclic group $\mathbb{Z}/m\mathbb{Z}$, the expressions like $i+1$ are understood modulo m):

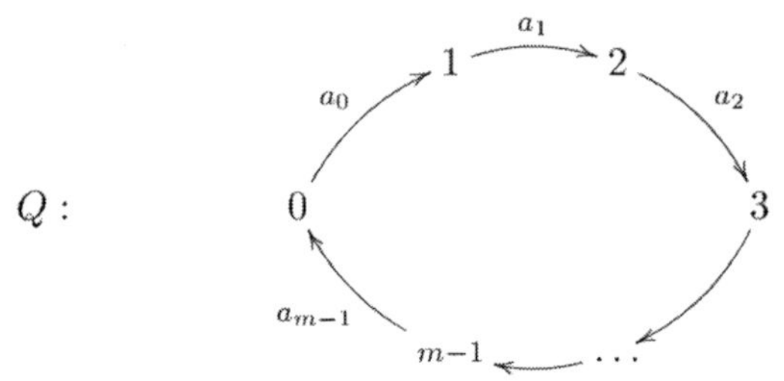

Note that for $m = 1$ the cyclic quiver is nothing but the one-loop quiver considered above.

For $\lambda = (\lambda_0, \dots, \lambda_{m-1}) \in \mathbb{C}^I$ denote $|\lambda| := \sum_{i \in I} \lambda_i$. We suppose that λ is regular in the sense that

$$n|\lambda| \neq \lambda_i + \dots + \lambda_{j-1} \qquad \text{for all } n \in \mathbb{Z} \text{ and } 1 \leqslant i \leqslant j \leqslant m \tag{11}$$

(in particular, $|\lambda| \neq 0$). Then every module $V \in \operatorname{Rep}\big(\Pi^{\boldsymbol{\lambda}}(Q_\zeta), \boldsymbol{\alpha}\big)$ is simple (see [1], [13]), so due to Proposition 1 the quiver varieties $M_\lambda(\alpha, \zeta)$ associated with the cyclic quiver Q are smooth and connected.

Consider first the framing $\zeta = (\zeta_0, \zeta_1, \dots, \zeta_{m-1})$ where $\zeta_i = d$ for all $i \in I$. That is $\zeta = d\delta$ where $\delta \in \mathbb{Z}^I_{\geqslant 0}$ has components $\delta_i = 1$. The framed quiver has the from

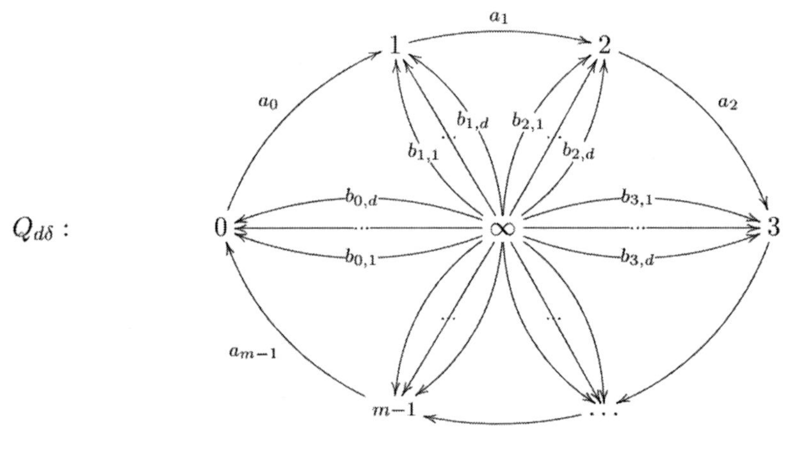

For a module $V \in \operatorname{Rep}\big(\Pi^{\boldsymbol{\lambda}}(Q_\zeta), \boldsymbol{\alpha}\big)$ consider the matrices

$$X_i = V_{a_i} \colon V_i \to V_{i+1}, \qquad Y_i = V_{a_i^*} \colon V_{i+1} \to V_i \tag{12}$$

(where $V_i = \mathbb{C}^{\alpha_i}$) and also the matrices $v_i \in \operatorname{Hom}(\mathbb{C}^d, \mathbb{C}^{\alpha_i})$, $w_i \in \operatorname{Hom}(\mathbb{C}^{\alpha_i}, \mathbb{C}^d)$ with entries $(v_i)_{l,r} = (V_{b_{ir}})_l$ and $(w_i)_{r,l} = (V_{b^*_{ir}})_l$. These are matrix-valued functions of V and each point $V \in \operatorname{Rep}\big(\Pi^{\boldsymbol{\lambda}}(Q_\zeta), \boldsymbol{\alpha}\big)$ can be given by their values. In these terms the equations (7) take the form

$$X_{i-1}Y_{i-1} - Y_iX_i + v_iw_i = \lambda_i \mathbf{1}_{\alpha_i} \qquad (i = 0, 1, \dots, m-1), \tag{13}$$

(since $\boldsymbol{\lambda} \cdot \boldsymbol{\alpha} = 0$, the equation (7) for $i = \infty$ follows from (13)). Thus a point of $M_\lambda(\alpha, d\delta)$ correspond to a collection of matrices (X_i, Y_i, v_i, w_i), satisfying (13), up to the action of $\mathrm{GL}(\alpha)$.

Consider the following regular functions on Rep $(\Pi^{\boldsymbol{\lambda}}(Q_\zeta), \boldsymbol{\alpha})$:

$$J_\ell(A) = \sum_{i \in I} \mathrm{tr}(A w_{i-\ell} Y_{i-\ell} \cdots Y_{i-1} v_i), \tag{14}$$

where $\ell \in \mathbb{Z}_{\geqslant 0}$ and A is a constant $d \times d$ matrix. Due to their $\mathrm{GL}(\alpha)$-invariance they can be considered as regular functions on $M_\lambda(\alpha, d\delta)$. Their Poisson brackets are

$$\{J_k(A), J_\ell(B)\} = -J_{k+\ell}([A, B]), \tag{15}$$

so they span a Lie subalgebra L in the Poisson algebra $\mathbb{C}[M_\lambda(\alpha, d\delta)]$. The functions (14) with diagonal matrices A span a maximal commutative subalgebra in L. Substituting $A = E_{rr}$ where E_{rr} are diagonal matrix units, we obtain the basis of this subalgebra:

$$H_{\ell,r} = J_\ell(E_{rr}) = \sum_{i \in I} w_{i-\ell,r} Y_{i-\ell} \cdots Y_{i-1} v_{i,r} \qquad (\ell \geqslant 0, \quad r = 1, \ldots, d), \tag{16}$$

where $v_{i,r} \in \mathbb{C}^{\alpha_i}$ and $w_{i,r} \in (\mathbb{C}^{\alpha_i})^*$ are vectors and covectors with components $(v_{i,r})_l = (v_i)_{lr}$, $(w_{i,r})_l = (w_i)_{rl}$. Since $\{H_{k,r}, H_{\ell,s}\} = 0$, one can consider $H_{\ell,r}$ as Hamiltonians. In this way we obtain a Hamiltonian system on the quiver variety $M_\lambda(\alpha, \zeta)$. It is convenient to regard the sums $H_{mk} = \sum_{r=1}^d H_{mk,r} = J(\mathbf{1}_r)$ $(k = 1, \ldots, n)$ as "main" Hamiltonians. By substituting (13) to $H_{mk} = \sum_{i \in I} \mathrm{tr}(Y_{i-mk} \cdots Y_{i-1} v_i w_i)$ and by taking into account $\sum_{i \in I} \mathrm{tr}(Y_{i-mk} \cdots Y_{i-1} X_{i-1} Y_{i-1}) = \sum_{i \in I} \mathrm{tr}(Y_{i-mk} \cdots Y_{i-1} Y_i X_i)$ we derive

$$H_{mk} = |\lambda| \,\mathrm{tr}\big((Y_0 Y_1 \cdots Y_{m-1})^k\big). \tag{17}$$

We also have $\sum_{r=1}^d H_{0,r} = \lambda \cdot \alpha$, so the Hamiltonians $H_{0,r}$ are algebraically dependent.

Consider the case $\alpha = n\delta$, i.e. $\alpha_i = n \;\; \forall i \in I$. Then, from Proposition 1 we obtain $\dim M_\lambda(n\delta, d\delta) = 2nmd$. In this case X_i and Y_i are square matrices of size n. For a generic point (X_i, Y_i, v_i, w_i) of $M_\lambda(n\delta, d\delta)$ the matrix $X_{m-1} \cdots X_1 X_0$ is invertible and diagonalisable. Since the collection of matrices is defined up to the action of $\mathrm{GL}(\alpha)$ we can diagonalise it: $X_{m-1} \cdots X_1 X_0 = \mathrm{diag}(x_1^m, \ldots, x_n^m)$ where x_i are non-zero complex numbers. Moreover, by action of $\mathrm{GL}(\alpha)$ one can make the matrices X_i to be

$$X_i = \begin{pmatrix} x_1 & & 0 \\ & \ddots & \\ 0 & & x_n \end{pmatrix} \qquad (i = 0, \ldots, m-1). \tag{18}$$

Let $\varphi_a \in \mathrm{Hom}(\mathbb{C}^m, \mathbb{C}^d)$ and $\psi_a \in \mathrm{Hom}(\mathbb{C}^d, \mathbb{C}^m)$ $(a = 1, \ldots, n)$ are matrices with components

$$(\varphi_a)_{ri} = (v_{i,r})_a, \qquad (\psi_a)_{ir} = (w_{i,r})_a. \tag{19}$$

Then substituting (18) and (19) to (13) we derive

$$(Y_i)_{aa} = \frac{1}{m} p_a + \frac{1}{x_a} \sum_{l=1}^{m-1} \frac{m-l}{m} \Big(\lambda_l - (\psi_a \varphi_a)_{ll}\Big) - \frac{1}{x_a} \sum_{l=1}^{i} \Big(\lambda_l - (\psi_a \varphi_a)_{ll}\Big), \tag{20}$$

$$(Y_i)_{ab} = -\sum_{j=0}^{m-1} \frac{x_a^j x_b^{m-j-1}}{x_a^m - x_b^m} (\psi_b \varphi_a)_{i-j,i-j} \qquad (a \neq b), \tag{21}$$

where $i = 0, \dots, m-1$, $a, b = 1, \dots, n$ and p_i are arbitrary complex numbers (see details in [1]). For instance, one deduces that the eigenvalues x_a^m are pairwise different. Thus we expressed the matrices X_i, Y_i, v_i, w_i through the variables

$$x_a, \quad (\varphi_a)_{ri}, \qquad p_a, \quad (\psi_a)_{ir}. \tag{22}$$

The numbers x_a and p_a are interpreted as coordinate and momentum of the a-th particle, while the matrices φ_a and ψ_a characterise its internal degrees of freedom and are interpreted as spin variables.

The total number of variables (22) is $2n + 2nmd$, which is more than the dimension of $\mathcal{M}_\lambda(n\delta, d\delta)$. In fact, we can exclude one component of each φ_a and ψ_a. Indeed, the formulae (18) and (13) also imply $\operatorname{tr}(\varphi_a\psi_a) = |\lambda|$ $\forall a = 1, \dots, n$. In a generic point, a chosen component of each φ_a, for example $(\varphi_a)_{10}$, does not vanish. The transformation $g = (g_i) \in \mathrm{GL}(\alpha)$ with $g_i = \operatorname{diag}\big((\varphi_1)_{10}^{-1}, \dots, (\varphi_n)_{10}^{-1}\big)$ does not change matrices (18). By applying g one can make these components to be unity and express the corresponding component of ψ_a:

$$(\varphi_a)_{10} = 1, \tag{23}$$

$$(\psi_a)_{01} = |\lambda| - \sum_{r=2}^{d} (\varphi_a)_{r0}(\psi_a)_{0r} - \sum_{i=1}^{m-1}\sum_{r=1}^{d} (\varphi_a)_{ri}(\psi_a)_{ir}. \tag{24}$$

In this way we obtain local coordinates (22) where $(i, r) \ne (0, 1)$. They are Darboux coordinates:

$$\omega = \sum_{i\in I} \operatorname{tr}(\mathrm{d}Y_i \wedge \mathrm{d}X_i) + \sum_{i\in I} \operatorname{tr}(\mathrm{d}w_i \wedge \mathrm{d}v_i) = \sum_{a=1}^{n} \mathrm{d}p_a \wedge \mathrm{d}x_a + \sum_{a=1}^{n} \operatorname{tr}\big(\mathrm{d}(\psi_a) \wedge \mathrm{d}(\varphi_a)\big),$$

where due to (23) the last term has the form

$$\operatorname{tr}\big(\mathrm{d}(\psi_a) \wedge \mathrm{d}(\varphi_a)\big) = \sum_{r=2}^{d} \mathrm{d}(\psi_a)_{0r} \wedge \mathrm{d}(\varphi_a)_{r0} + \sum_{i=1}^{m-1}\sum_{r=1}^{d} \mathrm{d}(\psi_a)_{ir} \wedge \mathrm{d}(\varphi_a)_{ri}.$$

By using these coordinates (or rather their dual version) one can prove the following fact [1].

Theorem 1. *The Hamiltonians $H_{\ell,r} \in \mathbb{C}\big[\mathcal{M}_\lambda(n\delta, d\delta)\big]$ ($\ell = 1, \dots, nm$, $r = 1, \dots, d$) are functionally independent, so they define an integrable system on the quiver variety $\mathcal{M}_\lambda(n\delta, d\delta)$.*

Remark 1. The Hamiltonians $H_{\ell,r}$ ($\ell = 0, \dots, nm-1$, $r = 1, \dots, d$) are algebraically dependent, since $\sum_{r=1}^{d} H_{0,r} = n|\lambda|$. One can obtain nmd independent Hamiltonians by replacing one of $H_{0,r}$ in this family, for instance, by $H_{mn} = \sum_{r=1}^{d} H_{mn,r}$.

Example 2. [1] Let $m = 2$ and $d = 1$. Then $\lambda = (\lambda_0, \lambda_1)$ and the integrable system on $\mathcal{M}_\lambda(n\delta, \delta)$ is defined by the Hamiltonians $H_{\ell,1}$ ($\ell = 1, \dots, 2n$). Let us denote $Z = \begin{pmatrix} 1 & 0 \\ 0 & -1 \end{pmatrix}$ and $F_\pm = \begin{pmatrix} 0 & \pm 1 \\ 1 & 0 \end{pmatrix}$. In this case we have $\varphi_a \in (\mathbb{C}^2)^*$ and $\psi_a \in \mathbb{C}^2$ with components $(\varphi_a)_i = (\varphi_a)_{1i}$ and $(\psi_a)_i = (\psi_a)_{i1}$ ($i = 0, 1$). The first two Hamiltonians have the form

$$H_{1,1} = w_0Y_0v_1 + w_1Y_1v_0 = \frac{1}{2}\sum_{a=1}^{n}\Big((\varphi_aF_+\psi_a)p_a + \frac{1}{x_a}(\varphi_aF_-\psi_a)\big(\lambda_1 + (\varphi_a)_1(\psi_a)_1\big)\Big) + \frac{1}{2}\sum_{a\ne b}\Big(\frac{(\varphi_aF_+\psi_b)(\varphi_b\psi_a)}{x_a - x_b} + \frac{(\varphi_aF_-\psi_b)(\varphi_bZ\psi_a)}{x_a + x_b}\Big),$$

$$H_2 = H_{2,1} = |\lambda|\,\mathrm{tr}(Y_0Y_1) = \frac{|\lambda|}{4}\sum_{a=1}^{n}\Big(p_a^2 - \frac{1}{x_a^2}\big(\lambda_1 - (\varphi_a)_1(\psi_a)_1\big)^2\Big) +$$
$$+\frac{|\lambda|}{4}\sum_{a\neq b}\Big(\frac{(\varphi_a\psi_b)(\varphi_b\psi_a)}{(x_a-x_b)^2} + \frac{(\varphi_a Z\psi_b)(\varphi_b Z\psi_a)}{(x_a+x_b)^2}\Big)+$$
$$+\frac{|\lambda|}{2}\sum_{a\neq b}\frac{(\varphi_a)_1(\psi_b)_1(\varphi_b)_0(\psi_a)_0 - (\varphi_a)_0(\psi_b)_0(\varphi_b)_1(\psi_a)_1}{x_a^2 - x_b^2}.$$

In a generic point one can take $(\varphi_a)_0 = 1$, $(\psi_a)_0 = |\lambda| - (\varphi_a)_1(\psi_a)_1$, so there are exactly two independent "spin" variables for each particle: $(\varphi_a)_1$ and $(\psi_a)_1$.

5. Framing $\zeta = (d, 0, \ldots, 0)$

Here we consider another framing for the cyclic quiver Q defined above and related integrable systems on the corresponding quiver varieties. Namely, let all the additional edges go from ∞ to only one vertex $i_0 \in I$. Without loss of generality we suppose that $i_0 = 0$. Then the corresponding framed quiver Q_ζ is

$Q_{d\varepsilon_0}$:

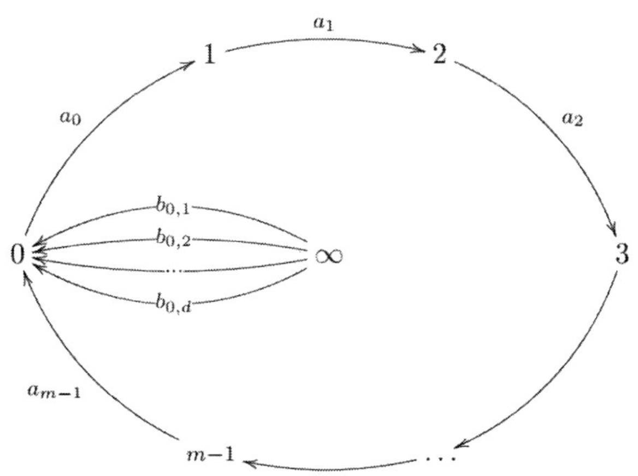

where $\varepsilon_0 \in \mathbb{Z}^I_{\geqslant 0}$ is the vector with components $(\varepsilon_0)_i = \delta_{i0}$.

As in previous case we define the matrix-valued functions $X_i \in \mathrm{Hom}(\mathbb{C}^{\alpha_i}, \mathbb{C}^{\alpha_{i+1}})$, $Y_i \in \mathrm{Hom}(\mathbb{C}^{\alpha_{i+1}}, \mathbb{C}^{\alpha_i})$, $v_0 \in \mathrm{Hom}(\mathbb{C}^d, \mathbb{C}^{\alpha_0})$, $w_0 \in \mathrm{Hom}(\mathbb{C}^{\alpha_0}, \mathbb{C}^d)$ by the formulae (12), $(v_0)_{l,r} = (V_{b_{0r}})_l$ and $(w_0)_{r,l} = (V_{b^*_{0r}})_l$. In these notations we define the following functions on $M_\lambda(\alpha, d\varepsilon_0)$:

$$J_{mk}(A) = \mathrm{tr}(Aw_0(Y_0Y_1\cdots Y_{m-1})^k v_0) \qquad (k\in\mathbb{Z}_{\geqslant 0},\quad A\in\mathrm{End}(\mathbb{C}^d)), \tag{25}$$

where $\mathrm{End}(\mathbb{C}^d)$ is the algebra of the constant complex $d\times d$ matrices.

Proposition 2. *The variety $M_\lambda(\alpha, d\varepsilon_0)$ can be identified with the subvariety in $M_\lambda(\alpha, d\delta)$ defined by the equations*

$$v_i = 0, \qquad w_i = 0 \qquad (i = 1, \ldots, m-1). \tag{26}$$

For ℓ not divisible by m the restriction of the function $J_\ell(A)$ defined by (14) vanishes on this subvariety. For $\ell = mk$ its restriction coincides with (25). The Hamiltonian flows defined by the functions (14) preserves subvariety $M_\lambda(\alpha, d\varepsilon_0)$ if $\ell = mk$.

Proof. Equations (26) define the inclusion $\mathrm{Rep}\big(\Pi^\lambda(Q_{d\varepsilon_0}),\alpha\big) \subset \mathrm{Rep}\big(\Pi^\lambda(Q_{d\delta}),\alpha\big)$. Since the action of $\mathrm{GL}(\alpha)$ preserves the equations (26), this inclusion induces the inclusion of quotients. The rest is proven by direct calculations. □

Remark 2. Though the Hamiltonians $J_\ell(A)$ vanishes on the subvariety $M_\lambda(\alpha, d\varepsilon_0)$ for ℓ not divisible by m, the corresponding flows do not preserve this subvariety (unless $A = 0$).

Proposition 2 implies that the Poisson brackets between the functions (25) can be derived by restriction of the Poisson brackets (15) to the subvariety $M_\lambda(\alpha, d\varepsilon_0) \subset M_\lambda(\alpha, d\delta)$:

$$\{J_{mk}(A), J_{m\ell}(B)\} = -J_{m(k+\ell)}([A, B]). \tag{27}$$

In particular, restriction of the Hamiltonians (16) to $M_\lambda(\alpha, d\varepsilon_0)$ gives us Poisson commuting Hamiltonians on this subvariety:

$$H_{mk,r} = J_{mk}(E_{rr}) = w_{0,r}(Y_0 Y_1 \cdots Y_{m-1})^k v_{0,r} \qquad (k \geqslant 0, \quad r = 1, \dots, d), \tag{28}$$

where $(v_{0,r})_l = (v_0)_{lr}$, $(w_{0,r})_l = (w_0)_{rl}$. Their sums $H_{mk} = \sum_{r=1}^d H_{mk,r}$ have the same form (17).

Consider again the case $\alpha = n\delta$. Proposition 1 gives $\dim M_\lambda(n\delta, d\varepsilon_0) = 2nd$. Consider generic points (X_i, Y_i, v_0, w_0) of $M_\lambda(n\delta, d\varepsilon_0)$, where $(v_{0,1})_a \neq 0$ and the matrices X_i are invertible and diagonalisable. As points of the variety $M_\lambda(n\delta, d\delta)$ they correspond to the values of the coordinates (22) such that

$$(\varphi_a)_{ri} = (\psi_a)_{ir} = 0 \qquad \text{for any } a, r \text{ and } i \neq 0. \tag{29}$$

Let $\varphi_a \in \mathbb{C}^d$ and $\psi_a \in (\mathbb{C}^d)^*$ be vectors and covectors with components

$$(\varphi_a)_r = (v_{0,r})_a, \qquad (\psi_a)_r = (w_{0,r})_a. \tag{30}$$

In accordance with the equations (29) they are obtained from those defined by (19) by deleting zero columns and rows corresponding to $i \neq 0$. This means that

$$x_a, \quad (\varphi_a)_r, \qquad p_a, \quad (\psi_a)_r \qquad (r \neq 1) \tag{31}$$

are local coordinates on $M_\lambda(n\delta, d\varepsilon_0)$. To express the matrices X_i, Y_i and rest components $(\varphi_a)_1$, $(\psi_a)_1$ through these coordinates one needs to substitute (29) to the formulae (18), (20), (21), (23), (24):

$$(X_i)_{ab} = x_a \delta_{ab}, \tag{32}$$

$$(Y_i)_{aa} = \frac{1}{m} p_a + \frac{1}{x_a}\Big(\sum_{l=1}^{m-1} \frac{m-l}{m}\lambda_l - \sum_{l=1}^{i} \lambda_l\Big), \tag{33}$$

$$(Y_i)_{ab} = -\frac{x_a^i x_b^{m-i-1}}{x_a^m - x_b^m}\psi_b \varphi_a \qquad (a \neq b), \tag{34}$$

$$(\varphi_a)_1 = 1, \tag{35}$$

$$(\psi_a)_1 = |\lambda| - \sum_{r=2}^{d} (\varphi_a)_r (\psi_a)_r, \tag{36}$$

where $i = 0, \dots, m-1$ and $a, b = 1, \dots, n$. Note that due to (35) the expression (36) can be written in the compact form: $\psi_a \varphi_a = |\lambda|$. The coordinates (31) are also Darboux coordinates:

$$\omega = \sum_{i \in I} \mathrm{tr}(\mathrm{d}Y_i \wedge \mathrm{d}X_i) + \mathrm{tr}(\mathrm{d}w_0 \wedge \mathrm{d}v_0) = \sum_{a=1}^{n} \mathrm{d}p_a \wedge \mathrm{d}x_a + \sum_{a=1}^{n}\sum_{r=2}^{d} \mathrm{d}(\psi_a)_r \wedge \mathrm{d}(\varphi_a)_r.$$

The analogue of Theorem 1 for this case is also proved in [1].

Theorem 2. *The Hamiltonians $H_{mk,r} \in \mathbb{C}[M_\lambda(n\delta, d\varepsilon_0)]$ ($k = 1,\dots,n$, $r = 1,\dots,d$) are functionally independent and, hence, they define an integrable system on $M_\lambda(n\delta, d\varepsilon_0)$.*

Consider first the case $d = 1$. In this case the quiver variety $M_\lambda(n\delta, \varepsilon_0)$ isomorphic to the Calogero–Moser space for the group $W = S_n \ltimes (\mathbb{Z}/m\mathbb{Z})^n$. If $m = 1$ then we obtain the Calogero–Moser systems on the completed phase space $M_\lambda(n,1) \simeq M_1(n,1) = \mathcal{C}_n$ ($\lambda \neq 0$) (see Example 1). If $m = 2$ then $\lambda = (\lambda_0, \lambda_1)$ and we obtain an integrable system with the Hamiltonian

$$H = \frac{4}{|\lambda|}H_2 = \sum_{a=1}^{n}\Big(p_a^2 - \frac{\lambda_1^2}{x_a^2}\Big) - 2|\lambda|^2\sum_{a<b}\Big(\frac{1}{(x_a-x_b)^2} + \frac{1}{(x_a+x_b)^2}\Big).$$

This is the Hamiltonian (1) for the root system of type B_n for some parameters c_α. The higher Hamiltonians are $H_{2k} = |\lambda|\,\mathrm{tr}\,((Y_0Y_1)^k)$ ($k = 1,\dots,n$), where

$$(Y_0)_{aa} = \frac{1}{2}\Big(p_a + \frac{\lambda_1}{x_a}\Big), \qquad (Y_1)_{aa} = \frac{1}{2}\Big(p_a - \frac{\lambda_1}{x_a}\Big), \tag{37}$$

$$(Y_0)_{ab} = -|\lambda|\frac{x_b}{x_a^2 - x_b^2}, \qquad (Y_1)_{ab} = -|\lambda|\frac{x_a}{x_a^2 - x_b^2} \qquad (a \neq b). \tag{38}$$

The role of Lax matrix is played by the $2n \times 2n$ block matrix $\mathbf{Y} = \begin{pmatrix} 0 & Y_0 \\ Y_1 & 0 \end{pmatrix}$. In particular, $H_{2k} = \frac{|\lambda|}{2}\,\mathrm{tr}(\mathbf{Y}^{2k})$. The Hamiltonians H_{2k} are proportional to the Hamiltonians H_k^{CM} defined in Section 2, so we obtain the Calogero–Moser system of type B_n with the completed phase space $M_\lambda(n\delta, \varepsilon_0)$ isomorphic to the corresponding Calogero–Moser space. For $m \geqslant 3$ the Hamiltonians (17) define the Calogero–Moser system for the complex reflection group $W = S_n \ltimes (\mathbb{Z}/m\mathbb{Z})^n$.

Now consider the case $d \geqslant 2$. In this case we have also spin variables φ_a, ψ_a with the conditions $\psi_a\varphi_a = |\lambda|$ and equivalence relations $(\varphi_a, \psi_a) \sim (\kappa_a\varphi_a, \kappa_a^{-1}\psi_a)$ where $\kappa_a \in \mathbb{C}^\times$. If $m = 1$ and $\lambda = 1$ we obtain the Gibbons–Hermsen system [7] with the "main" Hamiltonians

$$H_1 = \sum_{a=1}^{n} p_a, \qquad H_2 = \sum_{a=1}^{n} p_a^2 - 2\sum_{a<b}\frac{(\psi_a\varphi_b)(\psi_b\varphi_a)}{(x_a-x_b)^2}, \qquad H_k = \sum_{a=1}^{n} p_a^k + \dots$$

Numerator in the expression for H_2 can be interpreted as interaction of spins: $(\psi_a\varphi_b)(\psi_b\varphi_a) = \mathrm{tr}(s_a s_b)$ where $s_a = \varphi_a\psi_a$ is the spin operators. By this reason this system is also called spin Calogero–Moser system (of type A_{n-1}). The Hamiltonians of this system have the form $H_{k,r} = w_{0,r}Y^k v_{0,r}$ with the Lax matrix

$$Y = \begin{pmatrix} p_1 & & -\frac{\psi_b\varphi_a}{x_a - x_b} \\ & \ddots & \\ -\frac{\psi_a\varphi_b}{x_b - x_a} & & p_n \end{pmatrix}. \tag{39}$$

If $m = 2$ then the first Hamiltonian is

$$H = \frac{4}{|\lambda|}H_2 = \sum_{a=1}^{n}\Big(p_a^2 - \frac{\lambda_1^2}{x_a^2}\Big) - 2\sum_{a<b}\Big(\frac{1}{(x_a-x_b)^2} + \frac{1}{(x_a+x_b)^2}\Big)(\psi_a\varphi_b)(\psi_b\varphi_a).$$

The Lax matrix of this system is $\mathbf{Y} = \begin{pmatrix} 0 & Y_0 \\ Y_1 & 0 \end{pmatrix}$ where Y_0 and Y_1 are $n \times n$ matrices with the diagonal entries (37) and the following non-diagonal entries:

$$(Y_0)_{ab} = -\frac{x_b}{x_a^2 - x_b^2}\psi_b\varphi_a, \qquad (Y_1)_{ab} = -\frac{x_a}{x_a^2 - x_b^2}\psi_b\varphi_a \qquad (a \neq b). \tag{40}$$

Then the Hamiltonians are expressed as $H_{2k,r} = \mathbf{w}_r \mathbf{Y}^{2k} \mathbf{v}_r$ where $\mathbf{w}_r = (w_{0,r}, 0)$ and $\mathbf{v}_r = \begin{pmatrix} v_{0,r} \\ 0 \end{pmatrix}$. This is B_n type version of the Gibbons–Hermsen system. For $m \geqslant 3$ we obtain Gibbons–Hermsen system for the complex reflection group $S_n \ltimes (\mathbb{Z}/m\mathbb{Z})^n$.

Note that the spin generalisations of the Calogero–Moser systems obtained in this section is different from those more complicated spin generalisations obtained in Section 4.

Acknowledgements
The author is grateful to Yury Berest and Oleg Chalykh for stimulating discussions and useful comments. This paper is based on the results obtained by the author with Oleg Chalykh in the University of Leeds due to support of EPSRC under grant EP/K004999/1.

References
[1] Chalykh O A and Silantyev A V 2017 KP hierarchy for the cyclic quiver, *J. Math. Phys.* **58** no. 7 071702, 31 pp.
[2] Crawley-Boevey W and Holland M 1998 Noncommutative deformations of Kleinian singularities, *Duke Math. J.* **92** (3) 605–635
[3] Crawley-Boevey W 2001 Geometry of the moment map for representations of quivers, *Compositio Math.* **126** no. 3 257–293
[4] Dunkl C F and Opdam E M 2003 Dunkl operators for complex reflection groups, *Proc. London Math. Soc.* (3) **86** (1) 70–108
[5] Etingof P and Ginzburg V 2002 Symplectic reflection algebras, Calogero–Moser space, and deformed Harish-Chandra homomorphism, *Invent. Math.* **147** (2) 243–348
[6] Etingof P and Ma X 2010 Lecture notes on Cherednik algebras, arXiv:1001.0432
[7] Gibbons J and Hermsen T 1984 A generalisation of the Calogero–Moser system, *Physica* **11D** 337–348
[8] Heckman G J 1989 A remark on the Dunkl differential-difference operators, *Harmonic Analysis on Reductive Groups*, 181–191, (1989); *Progr. Math.* **101**, Birkhäuser Boston, (1991)
[9] Humphreys J E 1990 Reflection groups and Coxeter groups, *Cambridge University Press*
[10] Li L C and Xu P 2000 Spin Calogero-Moser systems associated with simple Lie algebras, *C. R. Acad. Sci.*, Paris, Serie I, **331** 55–60
[11] Nakajima H 1994 Instantons on ALE spaces, quiver varieties, and Kac-Moody algebras, *Duke Math. J.* **76** no. 2 365–416
[12] Olshanetsky M A and Perelomov A M 1976 Completely integrable Hamiltonian systems connected with semisimple Lie algebras, *Invent. Math.* **37** no. 2 93–108
[13] Silantyev A V: Reflection functor in the theory of representations of preprojective algebras for quivers, and integrable systems (in Russian), to appear
[14] Wilson G 1998 Collisions of Calogero–Moser particles and an adelic Grassmannian (With an appendix by I. G. Macdonald), *Invent. Math.* **133** no. 1 1–41

IOP Conf. Series: Journal of Physics: Conf. Series **965** (2018) 012039 doi:10.1088/1742-6596/965/1/012039

Quantum properties of supersymmetric theories regularized by higher covariant derivatives

Konstantin Stepanyantz
M.V.Lomonosov Moscow State University, Faculty of Physics, Department of Theoretical Physics, Leninskie Gory, Moscow, 119991, Russia

E-mail: stepan@m9com.ru

Abstract. We investigate quantum corrections in $\mathcal{N}=1$ non-Abelian supersymmetric gauge theories, regularized by higher covariant derivatives. In particular, by the help of the Slavnov–Taylor identities we prove that the vertices with two ghost legs and one leg of the quantum gauge superfield are finite in all orders. This non-renormalization theorem is confirmed by an explicit one-loop calculation. By the help of this theorem we rewrite the exact NSVZ β-function in the form of the relation between the β-function and the anomalous dimensions of the matter superfields, of the quantum gauge superfield, and of the Faddeev–Popov ghosts. Such a relation has simple qualitative interpretation and allows suggesting a prescription producing the NSVZ scheme in all loops for the theories regularized by higher derivatives. This prescription is verified by the explicit three-loop calculation for the terms quartic in the Yukawa couplings.

1. Introduction

It is well known that the ultraviolet behaviour of supersymmetric theories is better than in the non-supersymmetric case. In particular, in $\mathcal{N}=1$ supersymmetric theories the superpotential has no divergent quantum corrections [1] due to the non-renormalization theorem. The NSVZ β-function [2, 3, 4, 5] can be also considered as a non-renormalization theorem. In particular, it is closely related to the $\mathcal{N}=2$ non-renormalization theorem [6, 7, 8], see, e.g., [9, 10, 11]. Certainly, the good ultraviolet behaviour of supersymmetric theories appears due to a wide symmetry of a theory. Therefore, to reveal attractive features of supersymmetric theories, one should make manifest as many symmetries as possible [12]. From this point of view, the dimensional regularization is not convenient, because it breaks supersymmetry. Its modification called the dimensional reduction [13] is better, but due to mathematical inconsistency [14] can lead to supersymmetry breaking in higher loops [15, 16]. Unlike the dimensional technique, the higher covariant derivative regularization [17, 18, 19] is a mathematically consistent method, which can be formulated in a manifestly supersymmetric way [20, 21]. It is also possible to formulate this regularization in the $\mathcal{N}=2$ harmonic superspace [11], so that $\mathcal{N}=2$ supersymmetry remains a manifest symmetry at all stages of calculating quantum corrections.

Application of the higher derivative regularization to $\mathcal{N}=1$ supersymmetric theories revealed some interesting features of quantum corrections. For example, it was noted that the integrals giving the β-function of supersymmetric gauge theories are integrals of total derivatives [22] and even double total derivatives [23]. A lot of calculations in various supersymmetric theories confirm that this factorization seems to be a general feature of supersymmetric theories, see, e.g., [24, 25, 26, 27]. For $\mathcal{N}=1$ supersymmetric electrodynamics (SQED) it has been proven in all orders of the perturbations theory by two different methods [28, 29].[1] Such a structure of loop integrals is closely related to the NSVZ β-function. In particular, with the higher derivative regularization the renormalization group (RG) functions of $\mathcal{N}=1$ SQED defined in terms of the bare coupling constant satisfy the NSVZ relation in all loops independently of the renormalization prescription. This allows constructing the all-loop prescription giving the NSVZ scheme for the RG functions standardly defined in terms of the renormalized coupling constant [33, 34, 35], while with the dimensional technique such a scheme should be tuned in every order of the perturbation theory [36, 37, 38]. However, investigation of the non-Abelian case is much more complicated and is a subject of this paper.

2. The higher covariant derivative regularization for $\mathcal{N}=1$ supersymmetric theories

We consider the massless $\mathcal{N}=1$ SYM theory described by the action

$$S=\frac{1}{2e_0^2}\,\mathrm{Re}\,\mathrm{tr}\int d^4x\,d^2\theta\,W^aW_a+\frac{1}{4}\int d^4x\,d^4\theta\,\phi^{*i}(e^{2V})_i{}^j\phi_j+\Big(\frac{1}{6}\int d^4x\,d^2\theta\,\lambda_0^{ijk}\phi_i\phi_j\phi_k+\text{c.c.}\Big),\quad (1)$$

assuming the invariance under the gauge transformations. In the supersymmetric case the quantum-background splitting is performed by making the substitution $e^{2V}\to e^{\boldsymbol{\Omega}^+}e^{2V}e^{\boldsymbol{\Omega}}$. The background superfield $\boldsymbol{V}$ is then defined as $e^{2\boldsymbol{V}}=e^{\boldsymbol{\Omega}^+}e^{\boldsymbol{\Omega}}$. The theory is regularized by adding the gauge invariant higher derivative term

$$S_\Lambda=\frac{1}{2e_0^2}\,\mathrm{Re}\,\mathrm{tr}\int d^4x\,d^2\theta\,e^{\Omega}e^{\boldsymbol{\Omega}}W^ae^{-\boldsymbol{\Omega}}e^{-\Omega}\Big[R\Big(-\frac{\bar\nabla^2\nabla^2}{16\Lambda^2}\Big)-1\Big]_{Adj}e^{\Omega}e^{\boldsymbol{\Omega}}W_ae^{-\boldsymbol{\Omega}}e^{-\Omega}$$
$$+\frac{1}{4}\int d^4x\,d^4\theta\,\phi^+e^{\boldsymbol{\Omega}^+}e^{\Omega^+}\Big[F\Big(-\frac{\bar\nabla^2\nabla^2}{16\Lambda^2}\Big)-1\Big]e^{\Omega}e^{\boldsymbol{\Omega}}\phi,\quad (2)$$

where the functions $R(x)$ and $F(x)$ rapidly increase at infinity and are equal to 1 at $x=0$. The gauge can be fixed by adding the term invariant under the background gauge transformations,

$$S_{\text{gf}}=\frac{1}{e_0^2}\mathrm{tr}\int d^4x\,d^4\theta\Big(16\xi_0\,f^+\Big[e^{\boldsymbol{\Omega}^+}K^{-1}\Big(-\frac{\bar{\boldsymbol{\nabla}}^2\boldsymbol{\nabla}^2}{16\Lambda^2}\Big)e^{\boldsymbol{\Omega}}\Big]_{Adj}f+e^{\boldsymbol{\Omega}}fe^{-\boldsymbol{\Omega}}\boldsymbol{\nabla}^2V+e^{-\boldsymbol{\Omega}^+}f^+e^{\boldsymbol{\Omega}^+}\bar{\boldsymbol{\nabla}}^2V\Big),\quad (3)$$

where $K(0)=1$ and $K(x)$ also rapidly grows at infinity. The actions for the Faddeev–Popov and Nielsen–Kallosh ghosts have the form

$$S_{\text{FP}}=\frac{1}{e_0^2}\mathrm{tr}\int d^4x\,d^4\theta\,\Big(e^{\boldsymbol{\Omega}}\bar ce^{-\boldsymbol{\Omega}}+e^{-\boldsymbol{\Omega}^+}\bar c^+e^{\boldsymbol{\Omega}^+}\Big)$$
$$\times\Big\{\Big(\frac{V}{1-e^{2V}}\Big)_{Adj}\Big(e^{-\boldsymbol{\Omega}^+}c^+e^{\boldsymbol{\Omega}^+}\Big)+\Big(\frac{V}{1-e^{-2V}}\Big)_{Adj}\Big(e^{\boldsymbol{\Omega}}ce^{-\boldsymbol{\Omega}}\Big)\Big\};$$
$$S_{\text{NK}}=\frac{1}{2e_0^2}\mathrm{tr}\int d^4x\,d^4\theta\,b^+\Big[e^{\boldsymbol{\Omega}^+}K\Big(-\frac{\bar{\boldsymbol{\nabla}}^2\boldsymbol{\nabla}^2}{16\Lambda^2}\Big)e^{\boldsymbol{\Omega}}\Big]_{Adj}b.\quad (4)$$

The Pauli–Villars determinants should also be inserted into the generating functional to cancel the remaining one-loop divergences [19], see Ref. [27] for the discussion of details in the supersymmetric case.

[1] The same arguments allow to reveal similar features of some other theories, [30, 31, 32].

The total gauge fixed theory is invariant under the BRST transformations

$$\delta V = -\varepsilon\Big\{\Big(\frac{V}{1-e^{2V}}\Big)_{Adj}\Big(e^{-\Omega^+}c^+e^{\Omega^+}\Big) + \Big(\frac{V}{1-e^{-2V}}\Big)_{Adj}\Big(e^{\Omega}ce^{-\Omega}\Big)\Big\};\qquad \delta\phi = \varepsilon c\phi;$$
$$\delta\bar{c} = \varepsilon\bar{D}^2(e^{-2V}f^+e^{2V});\qquad \delta\bar{c}^+ = \varepsilon D^2(e^{2V}fe^{-2V});\qquad \delta c = \varepsilon c^2;\qquad \delta c^+ = \varepsilon(c^+)^2,\quad (5)$$

where ε is an anticommuting scalar parameter which is independent of the coordinates. The superfields f, b, and Ω are not transformed.

The renormalization constants are defined by the equations

$$\frac{1}{\alpha_0} = \frac{Z_\alpha}{\alpha};\qquad V = Z_V Z_\alpha^{-1/2} V_R;\qquad \bar{c}c = Z_c Z_\alpha^{-1}\bar{c}_R c_R;\qquad \phi_i = (\sqrt{Z_\phi})_i{}^j(\phi_R)_j.\qquad (6)$$

In our notation R denotes renormalized superfields, α and λ are the renormalized coupling constant and the Yukawa couplings, respectively. The renormalization constants for the Yukawa couplings and for the other parameters can be expressed in terms of the renormalization constants defined in Eq. (6).

3. Finiteness of the triple gauge-ghost vertices

First, we prove that four three-point vertices ($\bar{c}Vc$, $\bar{c}^+Vc$, $\bar{c}Vc^+$, and $\bar{c}^+Vc^+$) with two ghost legs and a single leg of the quantum gauge superfield are finite in all orders [39]. All these vertices have the same renormalization constant $Z_\alpha^{-1/2}Z_cZ_V$, so that the above statement can be rewritten as[2]

$$\frac{d}{d\ln\Lambda}(Z_\alpha^{-1/2}Z_cZ_V) = 0.\qquad (7)$$

Below we will demonstrate that in the general ξ gauge the two-point ghost Green function is divergent. This implies that all Green functions of the structure $\bar{c}V^n c$ are also divergent for all $n \neq 1$.

To prove finiteness of the $\bar{c}Vc$-vertices, we use the the Slavnov–Taylor identities for the three-point functions, which can be obtained using the standard methods [40, 41]. Taking into account that

$$\frac{\delta^2\Gamma}{\delta\bar{c}_y^{*B}\delta c_x^A} = -\frac{D_y^2\bar{D}_x^2}{16}G_c\delta^8_{xy}\delta_{AB};\qquad \frac{\delta}{\delta c_x^A}\Big\langle\delta V_y^B\Big\rangle = -\varepsilon\cdot\frac{1}{4}G_c\bar{D}^2\delta^8_{xy}\delta_{AB},\qquad (8)$$

it is possible to derive the identity

$$\varepsilon\cdot G_c(\partial_w^2/\Lambda^2)\bar{D}_w^2\frac{\delta^3\Gamma}{\delta\bar{c}_y^{*B}\delta V_w^D\delta c_z^C} - \varepsilon\cdot G_c(\partial_z^2/\Lambda^2)\bar{D}_z^2\frac{\delta^3\Gamma}{\delta\bar{c}_y^{*B}\delta V_z^C\delta c_w^D}$$
$$+\frac{1}{2}G_c\Big(\partial_y^2/\Lambda^2\Big)D_y^2\frac{\delta^2}{\delta c_z^C\delta c_w^D}\Big\langle\delta c_y^B\Big\rangle = 0.\qquad (9)$$

Explicit expression for the Green function corresponding to the triple ghost-gauge vertex in this equation can be derived using dimensional and chirality considerations,

[2] In the one-loop approximation this has first been noted in the paper [27].

$$\frac{\delta^3\Gamma}{\delta\bar{c}_x^{*A}\delta V_y^B\delta c_z^C} = -\frac{ie_0}{16}f^{ABC}\int\frac{d^4p}{(2\pi)^4}\frac{d^4q}{(2\pi)^4}\Big(f(p,q)\partial^2\Pi_{1/2}$$
$$-F_\mu(p,q)(\gamma^\mu)_{\dot{a}}{}^b\bar{D}^{\dot{a}}D_b + F(p,q)\Big)_y\Big(D_x^2\delta_{xy}^8(q+p)\,\bar{D}_z^2\delta_{yz}^8(q)\Big), \quad (10)$$

where $\partial^2\Pi_{1/2}$ is the supersymmetric projection operator and $\delta^8_{xy}(p)\equiv\delta^4(\theta_x-\theta_y)e^{ip_\alpha(x^\alpha-y^\alpha)}$.

Introducing the chiral source $\mathcal{J}$ and the source term

$$-\frac{e_0}{2}\int d^4x\,d^2\theta\,f^{ABC}\mathcal{J}^Ac^Bc^C + \text{c.c.}, \quad (11)$$

from dimensional and chirality considerations we obtain

$$\frac{\delta^2}{\delta c_z^C\delta c_w^D}\left\langle\delta c_y^B\right\rangle = -i\varepsilon\frac{\delta^3\Gamma}{\delta c_z^C\delta c_w^D\delta\mathcal{J}_y^B} = -\frac{ie_0\varepsilon}{4}f^{BCD}\int\frac{d^4p}{(2\pi)^4}\frac{d^4q}{(2\pi)^4}H(p,q)\bar{D}_z^2\delta_{zy}^8(q+p)\bar{D}_w^2\delta_{yw}^8(q), \quad (12)$$

where $H(p,q)$ is the dimensionless function and, by construction, $H(p,q)=H(p,-q-p)$.

Using explicit expressions for the Green functions, the Slavnov–Taylor identity (9) can be rewritten in the form

$$G_c(q)F(q,p)+G_c(p)F(p,q) = 2G_c(q+p)H(-q-p,q). \quad (13)$$

The function $H(p,q)$ is finite. Really, it is contributed by diagrams in which one leg corresponds to the chiral source $\mathcal{J}$ and two other legs correspond to chiral ghost superfields c. Each of these diagrams contains the expression

$$\int d^4y\,d^2\theta_y\,\mathcal{J}_y^A\cdot\frac{\bar{D}_y^2D_y^2}{4\partial^2}\delta_{y1}^8\cdot\frac{\bar{D}_y^2D_y^2}{4\partial^2}\delta_{y2}^8 = -2\int d^4y\,d^4\theta_y\,\mathcal{J}_y^A\cdot\frac{D_y^2}{4\partial^2}\delta_{y1}^8\cdot\frac{\bar{D}_y^2D_y^2}{4\partial^2}\delta_{y2}^8. \quad (14)$$

From this equation we see that all expressions for the considered superdiagrams can be presented as integrals over the total superspace, which include integration over

$$\int d^4\theta = -\frac{1}{2}\int d^2\theta\bar{D}^2 + \text{ total derivatives in the coordinate space.} \quad (15)$$

Consequently, two left spinor derivatives act to the chiral external legs, so that the non-vanishing result can be obtained only if two right spinor derivatives also act to the external legs. This implies that the result should be proportional to, at least, second degree of the external momenta and is finite in the ultraviolet region. Thus, the function $H(p,q)$ appears to be UV finite.

Multiplying the Slavnov–Taylor identity (9) to the renormalization constant Z_c and differentiating the result with respect to $\ln\Lambda$, due to finiteness of $(G_c)_R$ and H in the limit $\Lambda\to\infty$ we obtain

$$\Big((G_c)_R(q)\frac{d}{d\ln\Lambda}F(q,p)+(G_c)_R(p)\frac{d}{d\ln\Lambda}F(p,q)\Big)\Big|_{\Lambda\to\infty}=0. \quad (16)$$

For $p=-q$ this equation gives

$$\frac{d}{d\ln\Lambda}F(-q,q)\Big|_{\Lambda\to\infty}=0. \quad (17)$$

Therefore, the corresponding renormalization constant is finite,

$$\frac{d}{d\ln\Lambda}(Z_\alpha^{-1/2} Z_c Z_V) = 0, \tag{18}$$

so that the function $F(p,q)$ is also finite. This implies that all other three-point ghost-gauge vertices are finite.

4. One-loop verification of the finiteness of the $V\bar{c}c$-vertices

In the one-loop approximation the two-point ghost Green function is given by two superdiagrams presented in Fig. 1. The result for these diagrams (in the Euclidean space after the Wick rotation) can be written as

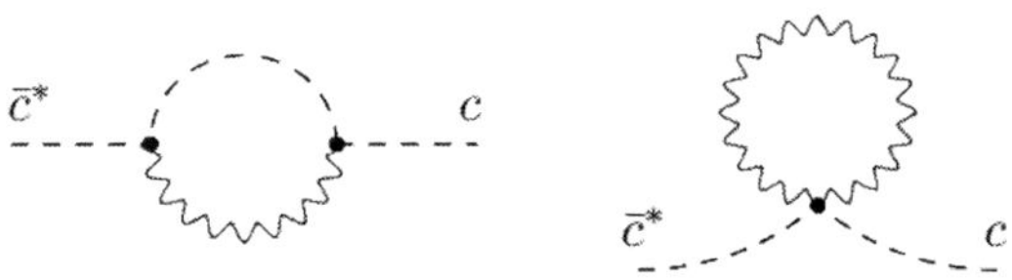

Figure 1. Two-point Green function of the Faddeev–Popov ghosts in the one-loop approximation.

$$G_c(p) = 1 + e_0^2 C_2 \int \frac{d^4k}{(2\pi)^4}\Big(\frac{\xi_0}{K_k} - \frac{1}{R_k}\Big)\Big(-\frac{1}{6k^4} + \frac{1}{2k^2(k+p)^2} - \frac{p^2}{2k^4(k+p)^2}\Big) + O(e_0^4, e_0^2\lambda_0^2), \tag{19}$$

where $R_k \equiv R(k^2/\Lambda)$ and $K_k \equiv K(k^2/\Lambda^2)$. We see that this expression is divergent in the ultraviolet region (in the limit $\Lambda \to \infty$). The anomalous dimension of the Faddeev–Popov ghosts obtained from Eq. (19) is

$$\gamma_c(\alpha_0, \lambda_0) = \frac{d\ln G_c}{d\ln\Lambda}\Big|_{p=0;\,\alpha,\lambda=\text{const}} = -\frac{\alpha_0 C_2(1-\xi_0)}{6\pi} + O(\alpha_0^2, \alpha_0\lambda_0^2). \tag{20}$$

Thus, in the general ξ-gauge the two-point Green function of the Faddeev–Popov ghosts is divergent starting from the one-loop approximation.

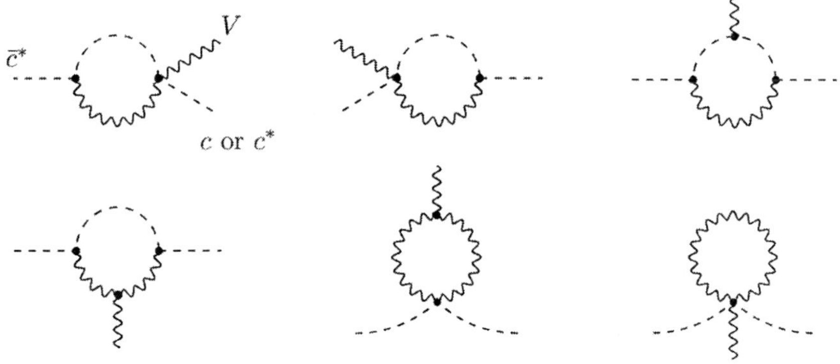

Figure 2. One-loop diagrams contributing to the three-point Green functions $\bar{c}^* V c$ and $\bar{c}^* V c^*$, where V is the quantum gauge superfield.

The three-point gauge-ghost Green functions in the one-loop approximation are contributed by the superdiagrams presented in Fig. 2. Their sum has been found in [39]. It is convenient to write the results in the form

$$\frac{ie_0}{4} f^{ABC} \int d^4\theta \frac{d^4p}{(2\pi)^4}\frac{d^4q}{(2\pi)^4}\bar{c}^{*A}(\theta,p+q)\Big(f(p,q)\partial^2\Pi_{1/2}V^B(\theta,-p) \\ +F_\mu(p,q)(\gamma^\mu)_{\dot{a}}{}^b D_b\bar{D}^{\dot{a}}V^B(\theta,-p)+F(p,q)V^B(\theta,-p)\Big)c^C(\theta,-q); \qquad (21)$$

$$\frac{ie_0}{4} f^{ABC} \int d^4\theta \frac{d^4p}{(2\pi)^4}\frac{d^4q}{(2\pi)^4}\bar{c}^{*A}(\theta,p+q)\widetilde{F}(p,q)V^B(\theta,-p)c^{*C}(\theta,-q). \qquad (22)$$

Then, in the one-loop approximation the functions F and $\widetilde{F}$ are given by the following expressions:

$$F(p,q) = 1+\frac{e_0^2C_2}{4}\int\frac{d^4k}{(2\pi)^4}\Big\{-\frac{(q+p)^2}{R_kk^2(k+p)^2(k-q)^2}-\frac{\xi_0\,p^2}{K_kk^2(k+q)^2(k+q+p)^2} \\ +\frac{\xi_0\,q^2}{K_kk^2(k+p)^2(k+q+p)^2}+\Big(\frac{\xi_0}{K_k}-\frac{1}{R_k}\Big)\Big(-\frac{2(q+p)^2}{k^4(k+q+p)^2}+\frac{2}{k^2(k+q+p)^2} \\ -\frac{1}{k^2(k+q)^2}-\frac{1}{k^2(k+p)^2}\Big)\Big\}+O(\alpha_0^2,\alpha_0\lambda_0^2). \qquad (23)$$

$$\widetilde{F}(p,q) = 1-\frac{e_0^2C_2}{4}\int\frac{d^4k}{(2\pi)^4}\Big\{\frac{p^2}{R_kk^2(k+q)^2(k+q+p)^2}+\frac{\xi_0\,(q+p)^2}{K_kk^2(k-p)^2(k+q)^2} \\ +\frac{\xi_0\,q^2}{K_kk^2(k+p)^2(k+q+p)^2}+\frac{2\xi_0}{K_kk^2(k+p)^2}-\frac{2\xi_0}{K_kk^2(k+q+p)^2}+\Big(\frac{\xi_0}{K_k}-\frac{1}{R_k}\Big) \\ \times\Big(\frac{2q^2}{k^4(k+q)^2}+\frac{1}{k^2(k+q+p)^2}-\frac{1}{k^2(k+q)^2}\Big)\Big\}+O(\alpha_0^2,\alpha_0\lambda_0^2). \qquad (24)$$

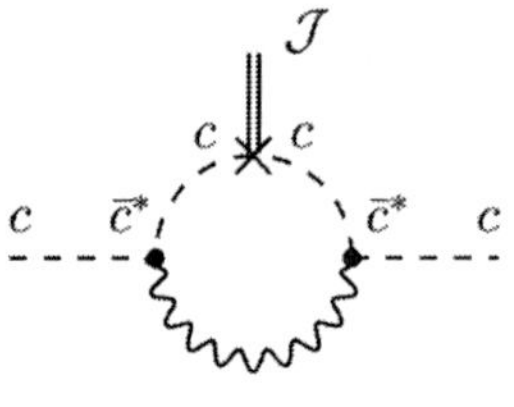

Figure 3. One-loop contribution to the function $H(p,q)$.

We see that they are finite in the ultraviolet region. Much more complicated one-loop expression for the functions $f(p,q)$ and $F_\mu(p,q)$ are presented in [39]. However, in the one-loop approximation their finiteness follows from simple dimensional considerations.

The finiteness of the function (23) and (24) confirms the general statement about the finiteness of the triple vertices with two ghost legs and one leg of the quantum gauge superfield. Note that the important ingredient of the general proof is the finiteness of the function H defined in Eq. (12). The one-loop contribution to this function is given by the diagram presented in Fig. 3 and can be written as

$$H(p,q) = 1 - \frac{e_0^2 C_2}{4}\int \frac{d^4k}{(2\pi)^4}\Bigg\{\frac{p^2}{R_k k^2(k+q)^2(k+q+p)^2} + \frac{(q+p)^2}{k^4(k+q+p)^2}\Big(\frac{\xi_0}{K_k} - \frac{1}{R_k}\Big)$$
$$+\frac{q^2}{k^4(k+q)^2}\Big(\frac{\xi_0}{K_k} - \frac{1}{R_k}\Big)\Bigg\} + O(e_0^4, e_0^2\lambda_0^2). \qquad (25)$$

From this expression we see that the function H is really finite in the ultraviolet region and is really quadratic in the external momenta p and q.

5. Finiteness of the triple ghost-gauge vertices and the NSVZ β-function

According to [2, 3, 4, 5], in $\mathcal{N} = 1$ supersymmetric theories the β-function is related to the anomalous dimension of the matter superfields by the equation

$$\beta(\alpha) = -\frac{\alpha^2\Big(3C_2 - T(R) + C(R)_i{}^j\gamma_j{}^i(\alpha)/r\Big)}{2\pi(1 - C_2\alpha/2\pi)}, \qquad (26)$$

which is called the exact NSVZ β-function. There are a lot of various arguments supporting correctness of this relation, based on instantons, anomalies, etc. For $\mathcal{N} = 1$ SQED the exact NSVZ β-function has been constructed in [42, 43]. In this simplest case the NSVZ relation was also obtained by summing supergraphs in the case of using the higher derivative regularization [28, 29]. Generalization of this result to the case of using the dimensional reduction is a complicated and so far unsolved problem [44, 45].

Qualitatively, the existence of the relation between the β-function and the anomalous dimension of the matter superfields in the Abelian case can be explained as follows [23]. Let us consider a supergraph without external lines. If two external legs of the gauge superfield $\boldsymbol{V}$ are attached to this graph by all possible ways, then one obtains diagrams contributing to the β-function. For the other hand, cutting matter lines in the original graph gives diagrams contributing to the anomalous dimension of the matter superfields. The NSVZ equation relates two these contributions if the theory is regularized by higher derivatives and the RG functions are defined in terms of the bare coupling constant.

In the non-Abelian case which is considered here the RG functions are defined in terms of the bare couplings by the equations

$$\beta(\alpha_0,\lambda_0) \equiv \frac{d\alpha_0}{d\ln\Lambda}; \qquad (\gamma_\phi)_i{}^j(\alpha_0,\lambda_0) \equiv -\frac{d\ln(Z_\phi)_i{}^j(\alpha,\lambda,\Lambda/\mu)}{d\ln\Lambda};$$
$$\gamma_V(\alpha_0,\lambda_0) \equiv -\frac{d\ln Z_V(\alpha,\lambda,\Lambda/\mu)}{d\ln\Lambda}; \qquad \gamma_c(\alpha_0,\lambda_0) \equiv -\frac{d\ln Z_c(\alpha,\lambda,\Lambda/\mu)}{d\ln\Lambda}, \qquad (27)$$

where the differentiation is made at fixed values of the couplings α and λ^{ijk}. It is well-known (see, e.g., [33, 35]) that these RG functions are scheme independent at a fixed regularization, but depend on a regularization.

Let us equivalently present the NSVZ relation for the RG functions (27) in the form

$$\frac{\beta(\alpha_0,\lambda_0)}{\alpha_0^2} = -\frac{3C_2 - T(R) + C(R)_i{}^j(\gamma_\phi)_j{}^i(\alpha_0,\lambda_0)/r}{2\pi} + \frac{C_2}{2\pi}\cdot\frac{\beta(\alpha_0,\lambda_0)}{\alpha_0}, \qquad (28)$$

and express the β-function in the right hand side in terms of the renormalization constant Z_α using the equation

$$\beta(\alpha_0,\lambda_0) = \frac{d\alpha_0(\alpha,\lambda,\Lambda/\mu)}{d\ln\Lambda}\Big|_{\alpha,\lambda=\mathrm{const}} = -\alpha_0\frac{d\ln Z_\alpha}{d\ln\Lambda}\Big|_{\alpha,\lambda=\mathrm{const}}. \qquad (29)$$

Taking into account Eq. (18), this expression can be rewritten in terms of the Faddeev–Popov ghost anomalous dimension γ_c and the quantum gauge superfield anomalous dimension γ_V defined in terms of the bare couplings,

$$\beta(\alpha_0,\lambda_0) = -2\alpha_0\frac{d\ln(Z_cZ_V)}{d\ln\Lambda}\Big|_{\alpha,\lambda=\mathrm{const}} = 2\alpha_0\Big(\gamma_c(\alpha_0,\lambda_0)+\gamma_V(\alpha_0,\lambda_0)\Big). \qquad (30)$$

Substituting this expression into the right hand side of Eq. (28) we present the NSVZ β-function in a different form,

$$\frac{\beta(\alpha_0,\lambda_0)}{\alpha_0^2} = -\frac{1}{2\pi}\Big(3C_2-T(R)-2C_2\gamma_c(\alpha_0,\lambda_0)-2C_2\gamma_V(\alpha_0,\lambda_0)+C(R)_i{}^j(\gamma_\phi)_j{}^i(\alpha_0,\lambda_0)/r\Big). \qquad (31)$$

It is interesting to note that in this form of the NSVZ equation the matter superfields and the Faddeev–Popov ghosts similarly contribute to the right hand side. Moreover, Eq. (31) admits the same simple graphical interpretation as in the Abelian case. Really, if we consider a supergraph without external lines, then attaching two external legs of the background gauge superfield produces a set of diagrams contributing to the β-function, while cutting internal lines gives superdiagrams contributing to the anomalous dimensions of the Faddeev–Popov ghosts, of the quantum gauge superfield, and of the matter superfields, see Fig. 4.

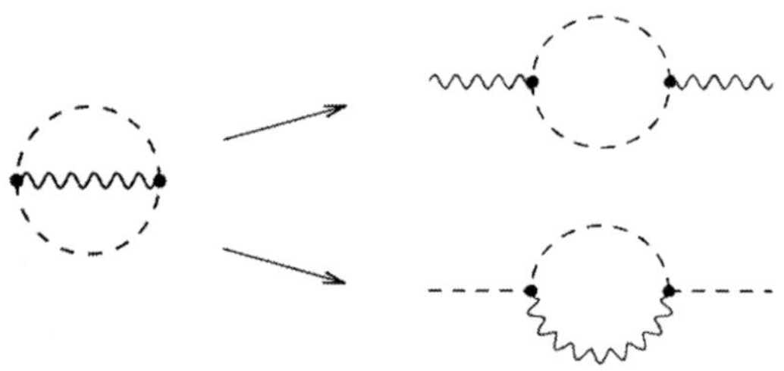

Figure 4. This picture illustrates why the NSVZ β-function appears in the non-Abelian case.

6. NSVZ scheme in the non-Abelian case

In the Abelian case the RG functions defined in terms of the bare coupling constant satisfy the NSVZ relation independently of the renormalization prescription if the theory is regularized by higher covariant derivatives. However, if the RG functions are defined in the standard way, in terms of the renormalized coupling constant, they satisfy the NSVZ relation only in a certain subtraction scheme, which is called the NSVZ scheme. In the non-Abelian case the RG functions are defined in terms of the renormalized couplings by the equations

$$\widetilde\beta(\alpha,\lambda) \equiv \frac{d\alpha}{d\ln\mu}; \qquad (\widetilde\gamma_\phi)_i{}^j(\alpha,\lambda) \equiv \frac{d\ln(Z_\phi)_i{}^j(\alpha_0,\lambda_0,\Lambda/\mu)}{d\ln\mu};$$
$$\widetilde\gamma_V(\alpha,\lambda) \equiv \frac{d\ln Z_V(\alpha_0,\lambda_0,\Lambda/\mu)}{d\ln\mu}; \qquad \widetilde\gamma_c(\alpha,\lambda) \equiv \frac{d\ln Z_c(\alpha_0,\lambda_0,\Lambda/\mu)}{d\ln\mu}, \qquad (32)$$

where the differentiation is made at fixed values of α_0 and λ_0^{ijk}. These RG functions are scheme and regularization dependent and satisfy the NSVZ relation only for a special renormalization prescription. The general equations describing the scheme dependence of the NSVZ relation have been derived in [46, 35].

Let us assume that, similarly to the Abelian case, the RG functions defined in terms of the bare coupling constant satisfy the NSVZ relation (31) with the higher covariant derivative regularization. This is reasonable, because qualitative way of its derivation looks exactly as in the Abelian case. Similarly to [33, 34, 35] it is possible to prove that in the non-Abelian case the RG functions defined in terms of the bare couplings and the ones defined in terms of the renormalized couplings coincide if the renormalization constants satisfy the conditions

$$Z_\alpha(\alpha,\lambda,x_0) = 1; \qquad (Z_\phi)_i{}^j(\alpha,\lambda,x_0) = \delta_i^j; \qquad Z_c(\alpha,\lambda,x_0) = 1, \tag{33}$$

where x_0 is a certain fixed value of $\ln\Lambda/\mu$. Therefore, under the assumption that the NSVZ relation is valid for the RG functions defined in terms of the bare couplings with the higher derivative regularization, we see that the boundary conditions (33) give the NSVZ scheme. It is easy to see that for $x_0 = 0$ in this scheme only powers of $\ln\Lambda/\mu$ are included into the renormalization constants. This looks like the minimal subtractions (which are usually used within the dimensional technique), so that the NSVZ scheme constructed according to the above prescription (for $x_0 = 0$) can be called HDMS, i.e. the higher derivative regularization supplemented by minimal subtractions. Certainly, it is also necessary to require that the renormalization constants also satisfy the condition

$$Z_V = Z_\alpha^{1/2} Z_c^{-1}, \tag{34}$$

under which the NSVZ relations (31) and (26) are equivalent.

Thus, the conditions (33) and (34) possibly give the NSVZ scheme with the higher covariant derivative regularization for non-Abelian gauge theories.

7. The NSVZ relation for the three-loop terms quartic in the Yukawa couplings

To verify the above results we consider the three-loop terms quartic in the Yukawa couplings [47]. They correspond to the graphs presented in Fig. 5. (In the three-loop approximation only these graphs give non-trivial contributions of the considered structure to the β-function.)

Figure 5. The terms in the NSVZ relation which are investigated here are obtained from these two graphs.

Attaching two external lines of the background gauge superfield we obtain a large number of two- and three-loop the diagrams contributing to the β-function, which are presented in Fig. 6. The corresponding diagrams for the anomalous dimension are obtained by cutting internal lines in the graphs presented in Fig. 5. They are shown in Fig. 7.

The calculation of the diagrams presented in Fig. 6 gives the following result for the corresponding part of the β-function defined in terms of the bare couplings:

$$\frac{\Delta\beta(\alpha_0,\lambda_0)}{\alpha_0^2} = -\frac{2\pi}{r} C(R)_i{}^j \frac{d}{d\ln\Lambda} \int \frac{d^4k}{(2\pi)^4} \frac{d^4q}{(2\pi)^4} \lambda_0^{imn} \lambda^*_{0jmn} \frac{\partial}{\partial q_\mu} \frac{\partial}{\partial q^\mu} \Big(\frac{1}{k^2 F_k\, q^2 F_q\, (q+k)^2 F_{q+k}} \Big)$$

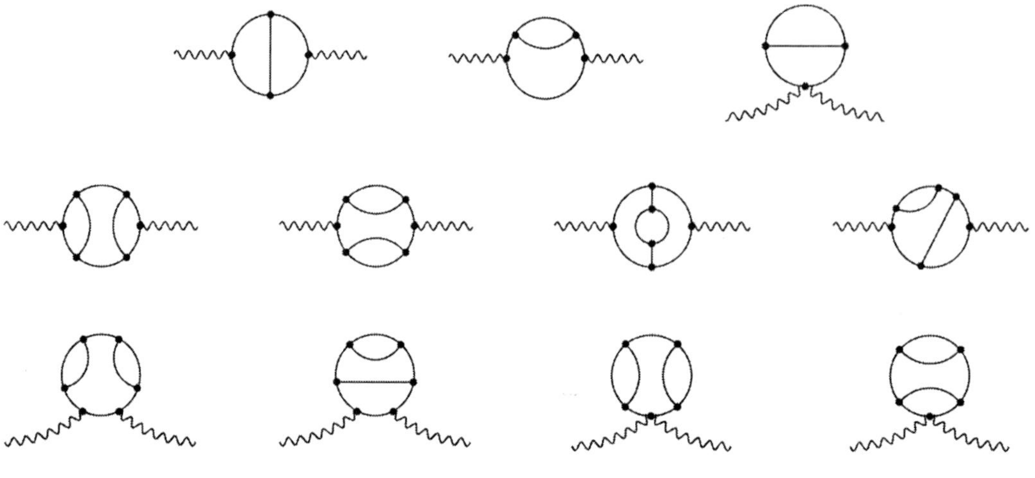

Figure 6. These superdiagrams give a part of the β-function corresponding to the supergraphs presented in Fig. 5.

Figure 7. Superdiagrams contributing to the anomalous dimension of the matter superfields which are obtained by cutting the graphs in Fig. 5.

$$+\frac{4\pi}{r}C(R)_i{}^j\frac{d}{d\ln\Lambda}\int\frac{d^4k}{(2\pi)^4}\frac{d^4l}{(2\pi)^4}\frac{d^4q}{(2\pi)^4}\Big(\lambda_0^{iab}\lambda^*_{0kab}\lambda_0^{kcd}\lambda^*_{0jcd}\Big(\frac{\partial}{\partial k_\mu}\frac{\partial}{\partial k^\mu}-\frac{\partial}{\partial q_\mu}\frac{\partial}{\partial q^\mu}\Big)+2\lambda_0^{iab}\lambda^*_{0jac}$$
$$\times\lambda_0^{cde}\lambda^*_{0bde}\frac{\partial}{\partial q_\mu}\frac{\partial}{\partial q^\mu}\Big)\frac{1}{k^2F_k^2\,q^2F_q\,(q+k)^2F_{q+k}\,l^2F_l\,(l+k)^2F_{l+k}}=-\frac{1}{2\pi r}C(R)_i{}^j\Delta\gamma_\phi(\lambda_0)_j{}^i,\quad(35)$$

where $\Delta\gamma_\phi(\lambda_0)_j{}^i$ is the part of the anomalous dimension (defined in terms of the bare couplings) corresponding to the diagrams presented in Fig. 7. From Eq. (35) we see that the considered contribution to the β-function is given by integrals of double total derivatives. This confirms the suggestion that it is a general feature of quantum corrections in supersymmetric theories. Moreover, according to Eq. (35) the NSVZ relation is satisfied for terms of the considered structure.

For the simplest regulator $F(k^2/\Lambda^2)=1+k^2/\Lambda^2$ it is possible to calculate all loop integrals. In this case

$$\Delta\gamma_\phi(\alpha_0,\lambda_0)_j{}^i=\frac{1}{4\pi^2}\lambda_0^{iab}\lambda^*_{0jab}-\frac{1}{16\pi^4}\lambda_0^{iab}\lambda^*_{0jac}\lambda_0^{cde}\lambda^*_{0bde};$$
$$\frac{\beta(\alpha_0,\lambda_0)}{\alpha_0^2}=-\frac{1}{2\pi}\Big(3C_2-T(R)\Big)-\frac{1}{2\pi r}C(R)_i{}^j\Delta\gamma_\phi(\lambda_0)_j{}^i+O(\alpha_0)+O(\lambda_0^6).\qquad(36)$$

Integrating the RG equations (27) one can find the renormalization constants,

$$(\ln Z_\phi)_j{}^i = -\frac{1}{4\pi^2}\lambda_0^{iab}\lambda^*_{0jab}\Big(\ln\frac{\Lambda}{\mu}+g_1\Big) + \frac{1}{32\pi^4}\lambda_0^{iab}\lambda^*_{0kab}\lambda_0^{kcd}\lambda^*_{0jcd}\Big(\ln^2\frac{\Lambda}{\mu}+2g_1\ln\frac{\Lambda}{\mu}+2g_1^2$$

$$-\widetilde{g}_2\Big) + \frac{1}{16\pi^4}\lambda_0^{iab}\lambda^*_{0jac}\lambda_0^{cde}\lambda^*_{0bde}\Big(\ln\frac{\Lambda}{\mu}+\ln^2\frac{\Lambda}{\mu}+2g_1\ln\frac{\Lambda}{\mu}+2g_1^2-g_2\Big)+O(\alpha_0)+O(\lambda_0^6); \qquad (37)$$

$$\frac{1}{\alpha_0} = \frac{1}{\alpha}+\frac{1}{2\pi}\Big(3C_2-T(R)\Big)\Big(\ln\frac{\Lambda}{\mu}+b_1\Big)+\frac{1}{2\pi r}C(R)_i{}^j\Big[\frac{1}{4\pi^2}\lambda^{iab}\lambda^*_{jab}\Big(\ln\frac{\Lambda}{\mu}+b_2\Big)+\frac{1}{32\pi^4}\lambda^{iab}$$

$$\times\lambda^*_{kab}\lambda^{kcd}\lambda^*_{jcd}\Big(\ln^2\frac{\Lambda}{\mu}+2g_1\ln\frac{\Lambda}{\mu}+\widetilde{b}_3\Big)+\frac{1}{16\pi^4}\lambda^{iab}\lambda^*_{jac}\lambda^{cde}\lambda^*_{bde}\Big(-\ln\frac{\Lambda}{\mu}+\ln^2\frac{\Lambda}{\mu}+2g_1\ln\frac{\Lambda}{\mu}$$

$$+b_3\Big)\Big]+O(\alpha)+O(\lambda^6). \qquad (38)$$

where g_1, b_1 etc. are arbitrary finite constants. Fixing them one fixes the subtraction scheme in the considered approximation. From these renormalization constants we construct the RG functions defined in terms of the renormalized couplings according to the prescription (32),

$$\widetilde{\gamma}_\phi(\alpha,\lambda)_j{}^i = \frac{1}{4\pi^2}\lambda^{iab}\lambda^*_{jab} - \frac{1}{16\pi^4}\lambda^{iab}\lambda^*_{jac}\lambda^{cde}\lambda^*_{bde}+O(\alpha)+O(\lambda^6); \qquad (39)$$

$$\frac{\widetilde{\beta}(\alpha,\lambda)}{\alpha^2} = -\frac{1}{2\pi}\Big(3C_2-T(R)\Big)+\frac{1}{2\pi r}C(R)_i{}^j\Big[-\frac{1}{4\pi^2}\lambda^{iab}\lambda^*_{jab}+\frac{1}{16\pi^4}\lambda^{iab}\lambda^*_{kab}\lambda^{kcd}\lambda^*_{jcd}\Big(b_2-g_1\Big)$$

$$+\frac{1}{16\pi^4}\lambda^{iab}\lambda^*_{jac}\lambda^{cde}\lambda^*_{bde}\Big(1+2b_2-2g_1\Big)\Big]+O(\alpha)+O(\lambda^6). \qquad (40)$$

Thus, the considered part of this β-function is scheme-dependent due to the presence of the finite constants b_2 and g_1. Their values can be fixed by imposing the conditions

$$Z_\phi(\alpha,\lambda,x_0)_i{}^j = \delta_i^j; \qquad Z_\alpha(\alpha,\lambda,x_0) = \alpha/\alpha_0 = 1, \qquad (41)$$

from which we obtain $g_1 = b_1 = b_2 = -x_0$, so that $b_2 - g_1 = 0$. (For $x_0 = 0$ all finite constants vanish and we obtain the HDMS scheme.) It is easy to verify that for $b_2 - g_1 = 0$ the NSVZ relation (31) is really valid for the RG functions defined in terms of the renormalized couplings,

$$\frac{\widetilde{\beta}(\alpha,\lambda)}{\alpha^2} = -\frac{1}{2\pi}\Big(3C_2-T(R)\Big)+\frac{1}{2\pi r}C(R)_i{}^j\Big[-\frac{1}{4\pi^2}\lambda^{iab}\lambda^*_{jab}+\frac{1}{16\pi^4}\lambda^{iab}\lambda^*_{jac}\lambda^{cde}\lambda^*_{bde}\Big]$$

$$+O(\alpha)+O(\lambda^6) = -\frac{1}{2\pi}\Big(3C_2-T(R)\Big)-\frac{1}{2\pi r}C(R)_i{}^j\widetilde{\gamma}_\phi(\alpha,\lambda)_i{}^j+O(\alpha)+O(\lambda^6). \qquad (42)$$

In particular, we see that the NSVZ relation appears to be valid with the higher covariant derivative regularization supplemented by minimal subtractions.

8. Conclusion

We have demonstrated that for $\mathcal{N}=1$ SYM the three-point vertices with two ghost legs and a single leg of the quantum gauge superfield are finite. Consequently, the renormalization constants can be chosen so that $Z_\alpha^{-1/2}Z_V Z_c = 1$. This equation allows to rewrite the NSVZ β-function in a new form, in terms of the anomalous dimensions of the quantum gauge superfield, of the Faddeev–Popov ghosts, and of the matter superfields. This new form of the NSVZ equation has a simple qualitative interpretation, which is similar to the Abelian case. Under the assumption that the RG functions defined in terms of the bare couplings satisfy this equation in the case of using the higher covariant derivative regularization, the NSVZ scheme in the non-Abelian

IOP Conf. Series: Journal of Physics: Conf. Series **965** (2018) 012039 doi:10.1088/1742-6596/965/1/012039

case appears to be the minimal subtraction scheme (HDMS), in which only powers of $\ln \Lambda/\mu$ are included into the renormalization constants. The non-trivial three-loop calculation for the terms quartic in the Yukawa couplings confirms the above proposal and this prescription for constructing the NSVZ scheme.

References

[1] Grisaru M T, Siegel W and Rocek M 1979 *Nucl. Phys.* B **159** 429.
[2] Novikov V A, Shifman M A, Vainshtein A I and Zakharov V I 1983 *Nucl. Phys.* B **229** 381.
[3] Jones D R T 1983 *Phys. Lett.* B **123** 45.
[4] Novikov V A, Shifman M A, Vainshtein A I and Zakharov V I 1986 *Phys. Lett.* B **166** 329; *Sov. J. Nucl. Phys.* **43** 294.
[5] Shifman M A and Vainshtein A I 1986 *Nucl. Phys.* B **277** 456; *Sov. Phys. JETP* **64** 428.
[6] Grisaru M T and Siegel W 1982 *Nucl. Phys.* B **201** 292.
[7] Howe P S, Stelle K S and Townsend P K 1984 *Nucl. Phys.* B **236** 125.
[8] Buchbinder I L, Kuzenko S M and Ovrut B A 1998 *Phys. Lett.* B **433** 335.
[9] Shifman M A and Vainshtein A I 1999 In *Shifman M A: ITEP lectures on particle physics and field theory, vol. 2* 485-647 (*Preprint* hep-th/9902018).
[10] Buchbinder I L and Stepanyantz K V 2014 *Nucl. Phys.* B **883** 20.
[11] Buchbinder I L, Pletnev N G and Stepanyantz K V 2015 *Phys. Lett.* B **751** 434.
[12] Jack I and Jones D R T 2010 *Adv. Ser. Direct. High Energy Phys.* **21** 494.
[13] Siegel W 1979 *Phys. Lett.* B **84** 193.
[14] Siegel W 1980 *Phys. Lett.* B **94** 37.
[15] Avdeev L V, Chochia G A and Vladimirov A A 1981 *Phys. Lett.* B **105** 272.
[16] Avdeev L V and Vladimirov A A 1983 *Nucl. Phys.* B **219** 262.
[17] Slavnov A A 1971 *Nucl. Phys.* B **31** 301.
[18] Slavnov A A 1972 *Theor.Math.Phys.* **13** 1064.
[19] Slavnov A A 1977 *Theor. Math. Phys.* **33** 977.
[20] Krivoshchekov V K 1978 *Theor. Math. Phys.* **36** 745.
[21] West P C 1986 *Nucl. Phys.* B **268** 113.
[22] Soloshenko A A and Stepanyantz K V 2004 *Theor. Math. Phys.* **140** 1264.
[23] Smilga A V and Vainshtein A 2005 *Nucl. Phys.* B **704** 445.
[24] Pimenov A B, Shevtsova E S and Stepanyantz K V 2010 *Phys. Lett.* B **686** 293.
[25] Stepanyantz K V 2011 "Factorization of integrals defining the two-loop β-function for the general renormalizable N=1 SYM theory, regularized by the higher covariant derivatives, into integrals of double total derivatives," (*Preprint* arXiv:1108.1491 [hep-th]).
[26] Kazantsev A E and Stepanyantz K V 2015 *J. Exp. Theor. Phys.* **120** 618.
[27] Aleshin S S, Kazantsev A E, Skoptsov M B and Stepanyantz K V 2016 *JHEP* **1605** 014.
[28] Stepanyantz K V 2011 *Nucl. Phys.* B **852** 71.
[29] Stepanyantz K V 2014 *JHEP* **1408** 096.
[30] Shifman M and Stepanyantz K 2015 *Phys. Rev. Lett.* **114** 051601.
[31] Shifman M and Stepanyantz K V 2015 *Phys. Rev.* D **91** 105008.
[32] Nartsev I V and Stepanyantz K V 2017 *JHEP* **1704** 047.
[33] Kataev A L and Stepanyantz K V 2013 *Nucl. Phys.* B **875** 459.
[34] Kataev A L and Stepanyantz K V 2014 *Phys. Lett.* B **730** 184.
[35] Kataev A L and Stepanyantz K V 2014 *Theor. Math. Phys.* **181** 1531.
[36] Jack I, Jones D R T and North C G 1996 *Phys. Lett.* B **386** 138.
[37] Jack I, Jones D R T and North C G 1997 *Nucl. Phys.* B **486** 479.
[38] Jack I, Jones D R T and Pickering A 1998 *Phys. Lett.* B **435** 61.
[39] Stepanyantz K V 2016 *Nucl. Phys.* B **909** 316.
[40] Taylor J C 1971 *Nucl. Phys.* B **33** 436.
[41] Slavnov A A 1972 *Theor. Math. Phys.* **10** 99.
[42] Vainshtein A I, Zakharov V I and Shifman M A 1985 *JETP Lett.* **42** 224.
[43] Shifman M A, Vainshtein A I and Zakharov V I 1986 *Phys. Lett.* B **166** 334.
[44] Aleshin S S, Kataev A L and Stepanyantz K V 2016 *JETP Lett.* **103** 77.
[45] Aleshin S S, Goriachuk I O, Kataev A L and Stepanyantz K V 2017 *Phys. Lett.* B **764** 222.
[46] Kutasov D and Schwimmer A 2004 *Nucl. Phys.* B **702** 369.
[47] Shakhmanov V Y and Stepanyantz K V 2017 *Nucl. Phys.* B **920** 345.

IOP Conf. Series: Journal of Physics: Conf. Series **965** (2018) 012040 doi:10.1088/1742-6596/965/1/012040

Quantum Double of Yangian of strange Lie superalgebra Q_n and multiplicative formula for universal R-matrix

Vladimir Stukopin

In memory of Ludwig Faddeev

Department of Applied Math., DSTU, Gagrin sq., 1, Rostov-on-Don, Russia; South Mathematical Institute, Vladikavkaz, Russia

E-mail: stukopin@mail.ru

Abstract. Main result is the multiplicative formula for universal R-matrix for Quantum Double of Yangian of strange Lie superalgebra Q_n type. We introduce the Quantum Double of the Yangian of the strange Lie superalgebra Q_n and define its PBW basis. We compute the Hopf pairing for the generators of the Yangian Double. From the Hopf pairing formulas we derive a factorized multiplicative formula for the universal R -matrix of the Yangian Double of the Lie superalgebra Q_n. After them we obtain coefficients in this multiplicative formula for universal R-matrix.

1. Introduction

The computation of universal R-matrix for the infinite dimensional Quantum Algebras be the one of the most impotent problems of Quantum Algebras theory for applications in mathematical physics and for the theory of exactly solvable models of statistical mechanics and quantum field theory. For example, construction of the transfer-matrix based on the determination of image of universal R-matrix of Quantum Double of Yangian by the action of tensor product of irreducible and identical representations. From the point of view of the theory of Yangians, the construction of the transfer matrix uses the image of the universal R-matrix of the Quantum Double of the Yangian under the action of the tensor product of an irreducible representation and the identity map. Computation of quantum R-matrix, spectrum of Hamiltonian and correlation functions is based also on the using universal R-matrix formula and description of irreducible representations of Yangian.

Let's note that Yangians together with quantized enveloping algebras and elliptic quantum algebras are one of the three most important examples of quantum algebras. The notion Yangian was introduced by V. Drinfel'd in honour to C.N. Yang. But algebras isomorphic to Yangian were used for investigation exactly solvable models by means of Quantum Inverse Scattering Method in the 80-th years by V. Tarasov (see, for example, [31], [32]).

Yangian theory and representations theory of Yangians of simple and reductive Lie algebras is a developed theory, which began to develop until to appearance of term "Yangian" ', as above mentioned (see [1], [10], [3], [4], [15], [16], [19]). In contrast with these theory, theory of Yangians of Lie Superalgebras is a young discipline, which appear at the beginning of 90-th years of 20 century ([17], [22]). But in the papers [33], [34], [26], [28], [30] had been done a description of finite dimensional irreducible representations in the most important cases, namely, for Yangian of general Lie superalgebra and special linear superalgebra. In the last years the number of applications of Theory of Yangians of Lie Superalgebras rise, to began research problems of Yang-Mills fields and Quantum Superstring Theory using Yangian Theory technique (see [2], [6], [20], [21]). The notion of universal R-matrix of almost cocommutative Hopf algebra was introduced by V. Drinfeld ([7], [8]). V. Drinfel'd also formulated the problem of calculating the universal R-matrix for the most important classes of quasitriangular Hopf algebras: quantized universal enveloping algebras, quantum affine algebras and Yangians.

I recall that universal R-matrix of the Yangian $Y(\mathfrak{g})$ of the simple Lie algebra $\mathfrak{g}$ was introduced by V.G. Drinfel'd (see [7], [8]) as a formal power series $R(\lambda) = 1 + \sum_{k=0}^{\infty} R_k \cdot \lambda^{-k-1}$ with coefficients $R_k \in Y(\mathfrak{g})^{\otimes 2}$, which conjugates the comultiplication Δ and opposite comultiplication $\Delta' = \tau \circ \Delta, \tau(x \otimes y) = (-1)^{p(x)p(y)}(y \otimes x)$ for Lie superalgebras. Explicit definitions will be given later in the text of this article.) More exactly, $R(\lambda)$ conjugate the images Δ and Δ' under action operator $id \otimes T_\lambda$, where T_λ is a quantum counterpart of shift operator, and id be an identical operator. The $R(\lambda)$ behave as it is image of some hypotetical R-matrix R under action $id \otimes T_\lambda$ conjugates Δ and Δ'. Drinfel'd called such formal power series $R(\lambda)$ the pseudotriangular structure and proved it existence for $Y(\mathfrak{g})$, when g be a simple Lie algebra. But explicite formula for $R(\lambda)$ it hasn't received up to now. If we shall see on classical counterparts of notions $R(\lambda)$ and R, namely, on classical r-matrices $r(\lambda)$ and r, then r be an element of a topological tensor square of a classical double and $r(\lambda) = (id \otimes T_\lambda)r$, where $T_\lambda f(u) = f(u+\lambda)$ be a shift operator. Then we can naturally to expect that and in the quantum case $R(\lambda)$ will be an image of universal R-matrix R of Quantum Double of Yangian under the action of some shift operator. When V.Drinfel'd defined pseudotriangular structure he didn't know good description of Yangian Double in terms of generators and defining relations and universal R-matrix of Yangian Double. But in the middle of 90-th S.Khoroshkin and V.Tolstoy receeived the description of Yangian Double in the terms of generators and defining relations and they computed the multiplicative formula of universal R-matrix of Yangian Double ([13]). The multiplicative formula for universal R-matrix of Quantum Double of Yangian $Y(A(m,n))$ of special linear superalgebra $A(m,n)$ was obtained in the papers [23], [24], [26]. These works was based on ideas of papers [13], [14].

The Yangian of the strange Lie superalgebra was introduced by M. Nazarov (see [18]) in the middle of 90-th years of last century. In the paper [25] the Yangian of the strange Lie superalgebra was defined in the terms of the Drinfel'd generators and in [27] Yangian of the strange Lie superalgebra was defined within a more general approach. The Quantum Double of Drinfel'd Yangian of the strange Lie superalgebra was introduced in [29]. The following natural problem is to compute the universal$R$-matrix of Quantum Double $DY(Q_n)$ is solving in this article. Let's note that in the papers [28], [30] was obtained the classification of the irreducible finite dimensional representations of Yangian $Y(A(m,n))$ of special linear superalgebra $A(m,n)$. (Let's note that $A(m,n)$ is an one more analog of general linear Lie algebra $\mathfrak{gl}_n$, along with the strange Lie superalgebra Q_n). But similar classification for finite dimensional representations of Yangian $Y(Q_n)$ of strange Lie superalgebra yet hasn't been obtain in present time. When such classification will be obtained we can be systematically describe all possible quantum R-matrices and transfer matrices connecting with Yangian symmetries of $Y(Q_n)$ type.

Some words about structure of this article. In the second section we recall definition of the Lie Superalgebra Q_n in terms of root generators and defining relations. After them we define of the Yangian of Lie Superalgebra Q_n in terms of generators and defining relations. We define it in terms of current system of generators as in the author's paper [28]. We also introduce Quantum Double $DY(Q_n)$ of Yangian of the strange Lie superalgebra. In the third section we recall construction of PBW bases and give a sketch of the proof of this theorem and describe the normal (convex) orders on the root generators of Yangian and its Quantum Double. In the forth section we formulate pairing formula for root generators of Quantum Double $DY(Q_n)$. In last section we formulate the main theorem – the multiplicative formula for the universal R-matrix of Quantum Double $DY(Q_n)$.

We'll use the following notations. We denote by $\mathbb{C}$ the field of complex numbers, by $K[u]$ the ring of polynomials with coefficients in ring K, by $\mathbb{Z}_+ = \{0, 1, 2, \ldots\}$ the set of non negative integers, which is union of zero and set of natural numbers $\mathbb{N}$, and by $\mathbb{Z}$ the set of integer numbers.

2. Definition of Yangian of Lie Superalgebra Q_n

2.1. Strange Lie Superalgebra Q_n

Let me recall basic definitions from Lie Superalgebra theory, related to the Lie Superalgebra Q_n (see also [11], [12]).

Let $C(n|n) = \mathbb{C}^n \oplus \mathbb{C}^n$ be a Z_2- graded $(n|n)$ dimensional vector space (superspace) over complex numbers field $\mathbb{C}$. Let's choose a standart basis in this superspace. We'll numerate the elements of this basis by integer numbers from segment $[-n, n]$ (without zero). So, let $(e_{-n}, e_{-n+1}, \ldots, e_{-1}, e_1, \ldots, e_n)$ be a standart basis in $C(n|n)$, and $End(C(n|n))$ be a superalgebra of linear operators acting in $C(n|n)$. Basis in $End(C(n|n))$ is formed by matrices $E_{a,b}, -n \leq a, b, \leq n, ab \neq 0$ and parity p for $E_{a,b}$ is defines by formula:

$$p(E_{a,b}) = |a| + |b|, \tag{1}$$

where $|a| = p(a) = 0$, if $a < 0$ and $|a| = p(a) = 1$, for $a > 0$, $|a|, |b| \in \mathbb{Z}_2$.

Recall, that general linear Lie superalgebra $\mathfrak{gl}(n|n)$ is defined as vector space $End(C(n|n))$ with (super)bracket $[\cdot, \cdot]$ defined by formula :

$$[x, y] = x \cdot y - (-1)^{|x||y|} y \cdot x. \tag{2}$$

Super trace on $\mathfrak{gl}(n|n)$ defined by formula $str(E_{ab}) = \delta_{ab}(-1)^{1+|a|}$ on base elements and extends linearly to all other elements of the vector $\mathfrak{gl}(n|n)$.

Recall also a definition of special linear Lie superalgebra $\mathfrak{sl}(n|n)$:

$$\mathfrak{sl}(n|n) = \{A \in \mathfrak{gl}(n|n) | str(A) = 0\}. \tag{3}$$

Let

$$\sigma : \mathfrak{gl}(n|n) \to \mathfrak{gl}(n|n),$$

be an involutive automorphism given by the formula:

$$\sigma : E_{a,b} \mapsto E_{-a,-b}.$$

Let's note that σ maps $\mathfrak{sl}(n|n)$ into $\mathfrak{sl}(n|n)$. Let's note also that on $\mathfrak{sl}(n, n)$ is defined nondegenerate bilinear invariant form $(\cdot, \cdot)$, given by formula:

$$(A, B) = str(AB). \tag{4}$$

The form (4) is different from Killing form.

Let also

$$J = \sum_{i=1}^{n} (E_{i,-i} - E_{-i,i}).$$

Definition 1 *Lie superalgebra q_n can be defined by the two following equivalent ways: as centralizer of J in $\mathfrak{gl}(n|n)$, or as the fixed points set of involutive automorphism $\sigma : \mathfrak{gl}(n|n) \to \mathfrak{gl}(n|n)$. Let also $sq_n = [q_n, q_n]$.*

Let's note, that sq_n be a Lie subsuperalgebra of Lie superalgebra q_n, consisting from $2n \times 2n$-matrices which left lower and upper right $n \times n$-blocks have a zero trace. Easy to see, that sq_n contains matrix E_{2n}.

Note, that Lie superalgebra sq_n is denoted also by $\tilde{Q}_{n-1}$.

Simple Lie superalgebra Q_{n-1} is defined (see [12]) as factor superalgebra of sq_n by one-dimensional center:

$$Q_{n-1} = sq_n/\mathbb{C}I_{2n}.$$

Let's introduce the following notations

$$\bar{h}_{i,0} := E_{i,i} + E_{-i,-i}, \quad h_{i,0} := \bar{h}_{i,0} - \bar{h}_{i+1,0}, \tag{5}$$
$$\bar{h}_{i,1} := E_{i,-i} + E_{-i,i}, \quad h_{i,1} := \bar{h}_{i,1} - \bar{h}_{i+1,1}, \tag{6}$$
$$x^+_{i,0} := E_{i+1,i} + E_{-i-1,-i}, \quad x^-_{i,0} := E_{i,i+1} + E_{-i,-i-1}, \tag{7}$$
$$x^+_{i,1} := E_{i+1,-i} + E_{-i-1,i}, \quad x^-_{i,1} := E_{i,-i-1} + E_{-i,i+1}. \tag{8}$$

Proposition 1 *Lie superalgebra sq_n isomorphic to Lie superalgebra $\mathfrak{g}$, generated by generators $x^{\pm}_{i,j}, h_{i,1}, 1 \le i \le n-1, j = 0, 1$ u $h_{i,0}, 1 \le i \le n$ (generators with second indexes $j = 1$ are odd and with second index $j = 0$ are even. These generators are satisfyed the following system of defining relations:*

$$[h_{i_1,1}, h_{i_2,1}] = 2\delta_{i_1,i_2}(h_{i_1,0} + h_{i_1+1,0}) - 2\delta_{i_1+1,i_2}h_{i_2,0}, \quad i_1 \le i_2,$$
$$[h_{i_1,j_1}, h_{i_2,j_2}] = 0, \quad (j_1, j_2) \ne (1, 1),$$
$$[h_{i_1,0}, x^{\pm}_{i_2,j_2}] = \pm(\delta_{i_1,i_2} - \delta_{i_1,i_2+1})x^{\pm}_{i_2,j_2},$$
$$[h_{i_1,1}, x^{\pm}_{i_2,j_2}] = (\pm 1)^{j_2+1}(2\delta_{i_1,i_2} \cdot \delta_{j_2,0} - (-1)^{j_2}\delta_{i_1,i_2+1} - \delta_{i_1+1,i_2})x^{\pm}_{i_2,j_2+1},$$
$$[x^+_{i_1,j_1}, x^-_{i_2,j_2}] = \delta_{i_1,i_2}h_{i_1,j_1+j_2}, \quad j_1 \ne j_2,$$
$$[x^+_{i_1,j_1}, x^-_{i_2,j_1}] = \delta_{i_1,i_2}(h_{i_1,0} - (-1)^{j_1}h_{i_1+1,0}),$$
$$[x^{\pm}_{i_1,j_1}, x^{\pm}_{i_2,j_2}] = (\pm 1)^{j_1+j_2+1}[x^{\pm}_{i_1,j_1+1}, x^{\pm}_{i_2,j_2+1}], \quad i_1 \le i_2;$$
$$ad(x^{\pm}_{i_1,j_1})^2(x^{\pm}_{i_2,j_2}) = 0, \quad 1 \le i_1, i_2 \le n-1, \quad j_1, j_2 = 0, 1, \quad |i_1 - i_2| = 1,$$
$$[x^{\pm}_{i_1,j_1}, x^{\pm}_{i_2,j_2}] = 0, \quad 1 \le i_1, i_2 \le n-1, \quad j_1, j_2 = 0, 1, \quad |i_1 - i_2| \ne 1.$$

The definitions of loop algebra $A(n,n)[u,u^{-1}]$ and twisted loop algebra $A^{(2)}(n,n)[u,u^{-1}]$ are similar (see [25], [29]).

Let be below $\mathfrak{g} = Q_n$.

2.2. Definition of Yangian

First, we recall the definition of Yangian $Y(\mathfrak{g}) = Y(Q_n)$ (see also [25], [29]).

Yangian, as other quantum algebras, is defined as a result of deformation (in the case of Yangian) of universal enveloping (super)algebra of current Lie superalgebra (wih values in Lie superalgebra) in the class of Hopf superalgebras. In the paper [25] was shown that this definition is equivalent other definition using so called current system of generators. This system of generators is convenient for representation theory of Yangians. Therefore we'll consider it further.

IOP Conf. Series: Journal of Physics: Conf. Series **965** (2018) 012040 doi:10.1088/1742-6596/965/1/012040

Definition 2 *Let $Y(\mathfrak{g})$ be an associative superalgebra (over complex number field $\mathbb{C}$), generated by generators $\bar{h}_{i,j,m}$, $\tilde{h}_{i,j,m} = \bar{h}_{i,j,m} - \bar{h}_{i+1,j,m}$, $x^{\pm}_{i,j,m}$, $i \in I = \{1, \ldots, n-1\}$, $m \in \mathbb{Z}_+$, $j \in \mathbb{Z}_2 = \{0, 1\}$, $p(x^{\pm}_{i,1,m}) = 1$, $p(x^{\pm}_{i,0,m}) = \tilde{h}_{i,j,m} = \bar{h}_{i,j,m} = 0$, which satisfied the following defining relations:*

$$[\tilde{h}_{i_1,0,m_1}, \tilde{h}_{i_2,0,m_2}] = 0, \tag{9}$$

$$[\tilde{h}_{i_1,1,2m_1}, \tilde{h}_{i_2,0,2m_2}] = 0, \tag{10}$$

$$[\tilde{h}_{i_1,1,2m_1+t}, \tilde{h}_{i_2,1,2m_2+t}] = (-1)^t 2\delta_{i_1,i_2}(\bar{h}_{i_1,0,2(m_1+m_2+t)} + \bar{h}_{i_1+1,0,2(m_1+m_2+t)}) - 2(-1)^t(\delta_{i_1+1,i_2} + \delta_{i_1-1,i_2})\bar{h}_{i_2,0,2(m_1+m_2+t)}, \tag{11}$$

$$[\tilde{h}_{i_1,1,2m_1+1}, \tilde{h}_{i_2,0,2m_2}] = [\tilde{h}_{i_1,1,2m_1}, \tilde{h}_{i_2,0,2m_2+1}] = 2\delta_{i_1,i_2}(\bar{h}_{i_1,1,2(m_1+m_2)+1} + \bar{h}_{i_1+1,1,2(m_1+m_2)+1}) - 2(\delta_{i_1+1,i_2} + \delta_{i_1-1,i_2})\bar{h}_{i_2,0,2(m_1+m_2)+1}, \tag{12}$$

$$[\tilde{h}_{i_1,1,2m_1+1}, \tilde{h}_{i_2,0,2m_2+1}] = 2\delta_{i_1,i_2}(\bar{h}_{i_1,1,2(m_1+m_2+1)} + \bar{h}_{i_1+1,1,2(m_1+m_2+1)}) - 2(\delta_{i_1+1,i_2} + \delta_{i_1-1,i_2})\bar{h}_{i_2,0,2(m_1+m_2+1)}, \tag{13}$$

$$[\bar{h}_{i_1,1,2m_1+1}, \bar{h}_{i_2,1,2m_2}] = [\bar{h}_{i_1,1,2m_1}, \bar{h}_{i_2,1,2m_2+1}] = 0, \tag{14}$$

$$[x^{+}_{i_1,j_1,m_1}, x^{-}_{i_2,j_2,m_2}] = \delta_{i_1,i_2}\tilde{h}_{i_1,j_1+j_2,m_1+m_2}, \tag{15}$$

$$[\tilde{h}_{i_1,0,m_1+1}, x^{\pm}_{i_2,j_2,m_2}] = [\tilde{h}_{i_1,0,m_1}, x^{\pm}_{i_2,j_2,m_2+1}] \pm \frac{(\alpha_{i_1}, \alpha_{i_2})}{2}(\tilde{h}_{i_1,0,m_1}x^{\pm}_{i_2,j_2,m_2} + x^{\pm}_{i_2,j_2,m_2}\tilde{h}_{i_1,0,m_1}), \tag{16}$$

$$[x^{\pm}_{i_1,0,m_1+1}, x^{\pm}_{i_2,j_2,m_2}] = [x^{\pm}_{i_1,0,m_1}, x^{\pm}_{i_2,j_2,m_2+1}] \pm \frac{(\alpha_{i_1}, \alpha_{i_2})}{2}(x^{\pm}_{i_1,0,m_1}x^{\pm}_{i_2,j_2,m_2} + x^{\pm}_{i_2,j_2,m_2}x^{\pm}_{i_1,0,m_1}), \tag{17}$$

$$[x^{\pm}_{i,1,m+1}, x^{\pm}_{j,1,r}] - [x^{\pm}_{i,1,m}, x^{\pm}_{j,1,r+1}] = \pm\frac{\alpha_j(h_i)}{2}(x^{\pm}_{i,1,m}x^{\pm}_{j,1,r} - x^{\pm}_{j,1,r}x^{\pm}_{i,1,m}) \tag{18}$$

$$[\tilde{h}_{i_1,1,m_1+1}, x^{\pm}_{i_2,j_2,m_2}] = [h_{i_1,1,m_1}, x^{\pm}_{i_2,j_2,m_2+1}] \pm \frac{(\alpha_{i_1}, \alpha_{i_2})}{2}(\tilde{h}_{i_1,1,m_1}x^{\pm}_{i_2,j_2,m_2} + x^{\pm}_{i_2,j_2,m_2}\tilde{h}_{i_1,1,m_1}), \tag{19}$$

$$[[x^{\pm}_{i_1,j_1,0}, x^{\pm}_{i_2,j_2,m_2}], x^{\pm}_{i_3,j_3,m_3}] + [x^{\pm}_{i_1,j_1,0}, [x^{\pm}_{i_2,j_2,m_2}, x^{\pm}_{i_3,j_3,m_3}]] = 0, \tag{20}$$

$$[[x^{\pm}_{i_1-1,0,t_1}, x^{\pm}_{i_1,1,0}], [x^{\pm}_{i_1,1,0}, x^{\pm}_{i_1+1,0,t_1}]] = 0, \quad t_1 \in \mathbb{Z}_+. \tag{21}$$

2.3. Definition of Yangian Double

I recall the definition of a Quantum Double of Hopf superalgebra. Let A be an arbitrary Hopf superalgebra. Denote by A^0 the dual Hopf superalgebra A^* with opposite comultiplication. Then the quantum double DA of the Hopf superalgebra A is a quasitriangular Hopf superalgebra (DA, R) such that DA contains A, A^0 as Hopf subsuperalgebras; R is the image of the canonical element in $A \otimes A^0$, corresponding to the unit operator, under the embedding of $A \otimes A^0$ in $DA \otimes DA$. Linear mapping

$$A \otimes A^0 \to DA, \quad a \otimes b \to ab$$

be an isomorphism.

We describe Quantum Double $DY(Q_n)$ of Yangian $Y(Q_n)$. First step, we describe the dual Hopf superalgebra $Y^*(Q_n)$ to Yangian $Y(Q_n)$. We'll denote the Hopf superalgebra $Y^*(Q_n)$ with the opposite comultiplication by $Y^0(Q_n)$. We acually describe the structure of the associative superalgebra on $Y^0(Q_n)$. This description will be based on the pairing formulas between the root generators of the superalgebra $Y(Q_n)$ that form the Poincare-Birkhoff-Witt basis and the dual generators of the superalgebra $Y^0(Q_n)$. But first we carried out the computation between the generators of the Hopf superalgebra $Y(Q_n)$ and the dual superalgebra $Y^0(Q_n)$.

Theorem 1 *The Quantum Double $DY(Q_n)$ of Yangian $Y(Q_n)$ be a superalgebra (over complex number field $\mathbb{C}$), generated by generators $\bar{h}_{1,j,m}$, $\tilde{h}_{i,j,m} = \bar{h}_{i,j,m} - \bar{h}_{i+1,j,m}$, $x^{\pm}_{i,j,m}$, $i \in I = \{1,\ldots,n-1\}$, $m \in \mathbb{Z}$, $j \in \mathbb{Z}_2 = \{0,1\}$, which satisfied the following above defining relations (9) - (21) (where $m \in \mathbb{Z}$).*

3. Root vectors, Poincare-Birkgoff-Witt theorem and normal orders

First of all, we describe in more detail the Poincare-Birkhoff-Witt basis (PBW-basis) for $DY(Q_n)$ (see [29]). Let, as above Δ, Δ_+ denotes the set of roots, respectively, the set of positive roots of the Lie superalgebra $A(n-1,n-1)$. We also consider the set of real roots of the corresponding affine twisted Lie superalgebra $A(n-1,n-1)^{(2)}$ (see [11]) $\hat{\Delta}^{re}$. For the generators $x^{\pm}_{i,j,k}$ of the quantum double $DY(Q_n)$ we use the following notation:

$$x_{\alpha_i+k\delta,0} = x_{\alpha_i+k\delta} := x^{+}_{i,0,k}, \quad x_{-\alpha_i+k\delta,0} = x_{-\alpha_i+k\delta} := x^{-}_{i,0,k},$$

$$x_{\alpha_i+k\delta,1} = x_{\tilde{\alpha}_i+k\delta} := x^{+}_{i,1,k},$$

$$x_{-\alpha_i+k\delta,1} = x_{-\tilde{\alpha}_i+k\delta} := x^{-}_{i,1,k}, i \in I, k \in \mathbb{Z}, \alpha_i \in \Delta_+.$$

Here (as above), we denote by $\tilde{\alpha}$ an odd root in the root system Q_n that is an odd double of the even root α.

Note that in this case $\pm\alpha_i + k\delta \in \hat{\Delta}^{re}$. Let $\Xi \subset \hat{\Delta}^{re}$. The linear order $\precneqq$ on Xi is said to be convex (normal) if for any roots $\alpha, \beta, \gamma \in \Xi$ and such that $\gamma = \alpha + \beta$ has place one of the following two relations of order:
$\alpha \precneqq \gamma \precneqq \beta$ или $\beta \precneqq \gamma \precneqq \alpha$.
Let's introduce the subsets Ξ_+, Ξ_- of set $\hat{\Delta}^{re}$:

$$\Xi_{\pm} := \{\pm\gamma + k\delta : \gamma \in \hat{\Delta}^{re}_{+}\}.$$

We introduce on Ξ_+, Ξ_- convex orders $\precneqq_+, \precneqq_-$, satisfying the following conditions:

$$\gamma + k\delta \precneqq_+ \gamma + l\delta \qquad \text{and} -\gamma + l\delta \precneqq_- -\gamma + k\delta,$$

if $k \leq l$, for $\forall \gamma \in \Delta_+$.

We now define the root vectors

$$x_{\pm\beta}, \beta \in \Xi_+ \cup \Xi_-$$

by induction as follows. Let the vectors x_{β_1}, x_{β_2} have already been constructed. If the root x_{β_3} is such that:

$$x_{\beta_1} \precneqq x_{\beta_3} \precneqq x_{\beta_2}$$

and there are no roots in the interval $(x_{\beta_1}, x_{\beta_2})$ for which the root vectors are already constructed, then we define the root vector $x_{\pm\beta_3}$ by the formulas:

$$x_{\beta_3} = [x_{\beta_1}, x_{\beta_2}], \quad x_{-\beta_3} = [x_{-\beta_2}, x_{-\beta_1}].$$

We note that a convex (normal) order is associated with the natural ordering of elements of the affine Weyl group. We need the following description of $DY(Q_n)$, which is an analog of the description of a twisted quantized affine superalgebra. First, we fix the following normal order on roots of $A(n,n)$:

$$(\epsilon_1 - \epsilon_2, \epsilon_1 - \epsilon_3, \ldots, \epsilon_1 - \epsilon_{2n+2}), (\epsilon_2 - \epsilon_3, \epsilon_2 - \epsilon_4, \ldots, \epsilon_2 - \epsilon_{2n+2}), \ldots (\epsilon_{2n+1} - \epsilon_{2n+2}).$$

Here $\epsilon_i - \epsilon_j = \alpha_i + \cdots + \alpha_{j-1}$.

We add the affine root $\alpha_0 = \delta - \theta$ to the simple roots $\alpha_1, \alpha_2, \cdots, \alpha_n, \ldots, \alpha_{2n-1}$. Here $\theta := \alpha_1 + \cdots + \alpha_{2n-1} = \epsilon_1 - \epsilon_{2n-1}$ is the highest root, δ is the minimal imaginary root. Consider the following normal order on the set $\hat{\Delta}^{re}$ of affine real roots:

$$(\alpha_1, \alpha_1+\delta, \alpha_1+2\delta, \ldots, \alpha_1+m\delta, \ldots), (\ldots \alpha_1+\alpha_2+m\delta, \ldots, \alpha_1+\alpha_2+\delta, \alpha_1+\alpha_2),$$
$$(\ldots \alpha_1+\alpha_2+\alpha_3+m\delta, \ldots, \alpha_1+\alpha_2+\alpha_3+\delta, \alpha_1+\alpha_2+\alpha_3), \ldots, (\ldots \alpha_{2n-1}+n\delta, \ldots, \alpha_{2n-1}+\delta, \alpha_{2n-1}).$$

4. Triangular decomposition and pairing formulas

We need explicit pairing formulas in the Quantum Double $DY(Q_n)$ of Yangian of the strange Lie superalgebra.

In this section we consider an analog for Yangian and Yangian Double of a Gaussian decomposition for a Lie algebra (and also a Lie superalgebra), and also for its universal enveloping algebra (superalgebra).

Let Y'_+, Y'_0, Y'_- be subsuperalgebras (without unit) of $Y(Q_n)$ are generated, correspondingly, by elements $x^+_{i,0,k} = x^+_{i,k}$, $x^+_{i,1,k} = \hat{x}^+_{i,k}$, $h_{i,0,k} = h_{i,k}$, $h_{i,1,k} = k_{i,k}$ и $x^-_{i,0,k} = x^-_{i,k}$, $x^-_{i,1,k} = \hat{x}^-_{i,k}$ $((i \in I, k \in \mathbb{Z}_+))$. Let Y_+, Y_0, Y_- are unital subsuperalgebras resulting from subsuperalgebras Y'_+, Y'_0, Y'_- adding to units.

Proposition 2 *Multiplication in $Y(Q_n)$ induces an isomorphism of vector superspaces:*

$$Y_+ \otimes Y_0 \otimes Y_- \to Y(Q_n). \tag{22}$$

We are going to describe the the Hopf superalgebra $Y^0(Q_n)$ which is dual Hopf superalgebra $Y^*(Q_n)$ (with opposite comultiplication) to the Yangian $Y(Q_n)$. We first describe the structure of the associative superalgebra on $Y^0(Q_n)$. This description will be based on the pairing formulas between the root generators of the superalgebra $Y(Q_n)$ that form the Poincare-Birkhoff-Witt basis and the dual generators of the superalgebra $Y^0(Q_n)$. But first we calculate pairing between the generators of the Hopf superalgebra $Y(Q_n)$ and the dual generators of superalgebra $Y^0(Q_n)$. We note that the pairing between these superalgebras is a Hopf pairing and the properties of the Hopf pairing are at the heart of the computations of defining relations of the dual Yangian $Y^0(Q_n)$. We calculate the pairing formulas for the root vectors. Let

$$h^*_{i,0,k}, \quad e^*_{i,0,k}, \quad f^*_{i,0,k}, h^*_{i,1,k}, \quad e^*_{i,1,k}, \quad f^*_{i,1,k}$$

are generators of subsuperalgebra $Y^* = Y_-$. Let

$$e_{i,0,k} := x^+_{i,0,k}, f_{i,0,k} := x^-_{i,0,k}, \quad e_{i,1,k} := x^+_{i,1,k}, f_{i,1,k} := x^-_{i,1,k}$$

are generators of Yangian $Y = Y(Q_n)$. Then we have the following proposition

Proposition 3 *The following two conditions are equivalent.*
1)

$$\langle e_{i,0,k}, e^*_{j,0,-l-1}\rangle = -\delta_{i,j}\delta_{k,l}; \quad \langle f_{i,0,k}, f^*_{j,0,-l-1}\rangle = -\delta_{i,j}\delta_{k,l};$$
$$\langle h_{i,0,k}, h^*_{j,0,-l-1}\rangle = -\frac{a_{ij}k!}{l!(k-l)!}; \quad \langle e_{i,1,k}, e^*_{j,1,-l-1}\rangle = -\delta_{i,j}\delta_{k,l};$$
$$\langle f_{i,1,k}, f^*_{j,1,-l-1}\rangle = -\delta_{i,j}\delta_{k,l}; \quad \langle h_{i,1,k}, h^*_{j,1,-l-1}\rangle = -\frac{a_{ij}s!}{l!(s-l)!} \qquad \textit{for} \quad s \geq l \geq 0.$$

2)

$$[h^*_{i,0,-k}, h^*_{j,0,-l}] = 0, \quad [h^*_{i,1,-k}, h^*_{j,0,-l}] = 0, \quad k-l \in 2\mathbb{Z},$$
$$[h^*_{i,0,k}, h^*_{j,1,l}] = [h^*_{i,1,k}, h^*_{j,0,l}] =$$
$$(\delta_{ij})(-1)^k(\bar{h}^*_{i,1,k+l} + \bar{h}_{i+1,1,k+l}) + \delta i+1, j\bar{h}^*_{j,1,k+l}, \quad k-l \in 2\mathbb{Z}+1,$$

$$[h^*_{i,1,k}, h^*_{j,1,l}] = \delta_{ij}(-1)^k(\bar{h}^*_{i,0,k+l} + \bar{h}_{i+1,0,k+l}) + \delta_{i+1,j}\bar{h}^*_{j,0,k+l}, \quad k-l \in 2\mathbb{Z},$$
$$[h^*_{i,1,k}, h^*_{j,1,l}] = 0, \quad k-l \in 2\mathbb{Z}+1,$$
$$\delta_{i,j}h^*_{i,0,-k-l} = [e^*_{i,0,-k}, f^*_{j,0,-l}],$$
$$[h^*_{i,0,-k-1}, e^*_{j,0,-l}] = [h^*_{i,0,-k}, e^*_{j,0,-l-1}] + (a_{ij}/2)(h^*_{i,0,-k}e_{j,0,-l} + e^*_{j,0,-l}h^*_{i,0,-k}),$$
$$[h^*_{i,0,-k-1}, f^*_{j,0,-l}] = [h^*_{i,0,-k}, f^*_{j,0,-l-1}] - (a_{ij}/2)(h^*_{i,0,-k}f_{j,0,-l} + f^*_{j,0,-l}h^*_{i,0,-k}),$$
$$[h^*_{i,1,-k-1}, e^*_{j,0,-l}] = [h^*_{i,1,-k}, e^*_{j,0,-l-1}] + (a_{ij}/2)(h^*_{i,1,-k}e_{j,0,-l} + e^*_{j,1,-l}h^*_{i,1,-k}),$$
$$[h^*_{i,1,-k-1}, f^*_{j,0,-l}] = [h^*_{i,1,-k}, f^*_{j,0,-l-1}] - (a_{ij}/2)(h^*_{i,0,-k}f_{j,0,-l} + f^*_{j,0,-l}h^*_{i,1,-k}),$$
$$[h^*_{i,0,0}, e^*_{j,r,l}] = a_{ij}e^*_{j,r,l}, \quad [h^*_{i,0,0}, f^*_{j,r,l}] = -a_{ij}f^*_{j,r,l},$$
$$[h^*_{i,1,0}, e^*_{j,0,l}] = a_{ij}e^*_{j,1,l}, \quad [h^*_{i,1,0}, f^*_{j,0,l}] = -a_{ij}f^*_{j,1,l},$$
$$[e^*_{i,0,-k+1}, e^*_{j,0,-l}] = [e^*_{i,0,-k}, e^*_{j,0,-l+1}] + (a_{ij}/2)(e^*_{i,0,-k}e^*_{j,0,-l} + e^*_{j,0,-l}e^*_{i,0,-k}),$$
$$[f^*_{i,0,-k+1}, f^*_{j,0,-l}] = [f^*_{i,0,-k}, f^*_{j,0,-l+1}] - (a_{ij}/2)(f^*_{i,0,-k}f^*_{j,0,-l} + f^*_{j,0,-l}f^*_{i,0,-k}),$$
$$[e^*_{i,1,-k+1}, e^*_{j,1,-l}] = [e^*_{i,1,-k}, e^*_{j,1,-l+1}] + (a_{ij}/2)(e^*_{i,1,-k}e^*_{j,1,-l} + e^*_{j,1,-l}e^*_{i,1,-k}),$$
$$[f^*_{i,1,-k+1}, f^*_{j,1,-l}] = [f^*_{i,1,-k}, f^*_{j,1,-l+1}] - (a_{ij}/2)(f^*_{i,1,-k}f^*_{j,1,-l} + f^*_{j,1,-l}f^*_{i,1,-k}),$$
$$[e^*_{i,1,-k+1}, e^*_{j,0,-l}] = [e^*_{i,1,-k}, e^*_{j,0,-l+1}] + (a_{ij}/2)(e^*_{i,1,-k}e^*_{j,0,-l} + e^*_{j,0,-l}e^*_{i,1,-k}),$$
$$[f^*_{i,1,-k+1}, f^*_{j,0,-l}] = [f^*_{i,1,-k}, f^*_{j,0,-l+1}] - (a_{ij}/2)(f^*_{i,1,-k}f^*_{j,0-l} + f^*_{j,0-l}f^*_{i,1,-k}),$$
$$\textstyle\sum_\sigma [e^*_{i,0,-k_{\sigma(1)}}, \cdots [e^*_{i,0,-k_{\sigma(r)}}, e^*_{j,0,-l}] \ldots] = 0, \quad i \neq j, \quad r = n_{ij} = 2,$$
$$\textstyle\sum_\sigma [f^*_{i,0,-k_{\sigma(1)}}, \cdots [f^*_{i,0,-k_{\sigma(r)}}, f^*_{j,-l}] \ldots] = 0, \quad i \neq j, \quad r = n_{ij} = 2,$$
$$\textstyle\sum_\sigma [e^*_{i,1,-k_{\sigma(1)}}, [e^*_{i,0,-k_{\sigma(1)}} \cdots [e^*_{i,0,-k_{\sigma(r)}}, e^*_{j,0,-l}] \ldots]] = 0, \quad i \neq j, \quad r = n_{ij} = 2,$$
$$\textstyle\sum_\sigma [f^*_{i,1,-k_{\sigma(1)}}, [f^*_{i,0,-k_{\sigma(1)}} \cdots [f^*_{i,0,-k_{\sigma(r)}}, f^*_{j,0,-l}] \ldots]] = 0, \quad i \neq j, \quad r = n_{ij} = 2.$$

5. Computation of Universal R-matrix of Quantum Double $DY(Q_n)$

5.1. Computation of Universal R-matrix of Quantum Double $\tilde{DY}(Q_n)$. General plan

Here we consider two Hopf superalgebra structures on $Y(Q_n)$ and study relation between these structures.

First, we recall that a universal R-matrix for a quasitriangular Hopf superalgebra A is an invertible element R lying in some completion of the tensor product $A\hat{\otimes}A$ and satisfying the following conditions:

$$\Delta^{op}(x) = R\Delta(x)R^{-l}, \qquad \forall x \in A;$$
$$(\Delta \otimes id)R = R^{13}R^{23}, (id \otimes \Delta)R = R^{13}R^{12},$$

where $\Delta^{op} = \sigma \circ \Delta, \sigma(x \otimes y) = (-1)^{p(x)p(y)} y \otimes x$, $R^{12} = R \otimes id$, $R^{23} = id \otimes R$, $R^{13} = \sum a_i \otimes id \otimes b_i$, if $R = \sum a_i \otimes b_i$.

We recall, also, the formula of universal R-matrix of Quantum Double DA of topological Hopf superalgebra A:

$$R = \sum_i e_i \otimes e^i, \tag{23}$$

here $\{e_i\}$ is a basis in topological Hopf superalgebra A, a $\{e^i\}$ is a dual basis in the Hopf superalgebra A^0. Convergence here is understood in a $\hbar$-adic topology. We will also use the triangle decomposition of the Yangian described above, and the induced decomposition of the quantum double of Yangian $DY(Q_n)$:

$$DY(Q_n) \cong Y_+ \hat{\otimes} Y_0 \hat{\otimes} Y_- \hat{\otimes} Y_-^0 \hat{\otimes} Y_0^0 \hat{\otimes} Y_+^0. \tag{24}$$

Here the isomorphism is induced simply by multiplication in the quantum double. A tensor product is a topological tensor product.

5.2. *Computation of Universal R-matrix*

First, we reformulate the above definition of the quantum double $DY(Q_n)$ in a more convenient form for us. To describe $DY(Q_n)$, it is convenient to use the following generating functions:

$$x_{i,j}^{\pm+}(u) = \sum_{k\geq 0} x_{i,j,k}^{\pm} u^{-k-1}, \quad x_{i,j}^{\pm-}(u) = -\sum_{k<0} x_{i,j,k}^{\pm} u^{-k-1},$$

$$h_{i,j}^{+}(u) = 1 + \sum_{k\geq 0} h_{i,j,k} u^{-k-1}, \quad h_{i,j}^{-}(u) = 1 - \sum_{k<0} h_{i,j,k} u^{-k-1}, \quad j \in \mathbb{Z}_2.$$

First we formulate a theorem that gives an idea of the general structure of the multiplicative formula for the universal R-matrix of the quantum double of the Yangian of the strange Lie superalgebra. We note that here we understand the tensor product of superalgebras as a topological tensor product. We will also work in the category of topological superalgebras. We understand the formula itself as a formal series.

Theorem 2 *Let $\mathfrak{g} = Q_n$. The universal R-matrix R of the Quantum Double $DY(Q_n)$ can be can be represented in the following factorized form*

$$R = R_+ R_0 R_-, \tag{25}$$

where

$$R_+ = \overrightarrow{\prod}_{\beta\in\Xi_+} \exp(-(-1)^{\theta(\beta)} a(\beta) e_\beta \otimes e_{-\beta}), \tag{26}$$

$$R_- = \overleftarrow{\prod}_{\beta\in\Xi_-} \exp(-(-1)^{\theta(\beta)} a(\beta) e_\beta \otimes e_{-\beta}), \tag{27}$$

where the product is taken in accordance with the above defined normal orders $\precneqq_+, \precneqq_-$, which is constructed from the root system of the affine Lie superalgebra $A(n,n)^{(2)}$. In addition, this ordering is convex.

The normalizing constants a(beta) can be found from the following conditions:

$$[e_\beta, e_{-\beta}] = (a(\beta))^{-1} h_\gamma \quad \text{если} \quad \beta = \gamma + n\delta \in \Xi_+, \gamma \in \Delta_+(A(n,n)^{(2)}),$$
$$[e_{-\beta}, e_\beta] = (a(\beta))^{-1} h_\gamma \quad \text{если} \quad \beta = -\gamma + n\delta \in \Xi_-, \gamma \in \Delta_+(A(n,n)^{(2)}),$$

and $\theta(\beta) = deg(e_\beta) = deg(e_{-\beta})$ denotes parity of element $e_{\pm\beta}$.

The middle term has more complicated structure.

Theorem 3 *Let* $\varphi_i^{\pm}(u) = \ln(h_{i,0}^{\pm}(u))$, $\tilde{\varphi}_i^{\pm}(u) = \ln(h_{i,1}^{\pm}(u))$, $A = (a_{ij})_{i,j=1}^n$ *– be a symmetrizable Cartan matrix for Lie algebra* A_n, $A = a_{ij}(q) = [a_{ij}]_q = \dfrac{q^{a_{ij}} - q^{-a_{ij}}}{q - q^{-1}}$, $A(q) = (a_{ij}(q))_{i,j=1}^n$. *Let also* $C(q) = (c_{ij}(q))_{i,j=1}^n$ *be a matrix with elements are proportional to corresponding elements of matrix* $A(q)^{-1}$ *and* T *be a shift operator :* $Tf(v) = f(v-1)$. *Similarly,* $\tilde{A} = (\tilde{a_{ij}})_{i,j=1}^n, \tilde{a}_{ij} = (\alpha_i, \tilde{\alpha_j}) = \delta_{i,j+1} - \delta_{i+1,j}, \tilde{a}_{ij}(q) = [\tilde{a}_{ij}]_q = \dfrac{q^{\tilde{a}_{ij}} - q^{-\tilde{a}_{ij}}}{q - q^{-1}}, \tilde{A}(q) = (\tilde{a}_{ij}(q))_{i,j=1}^n$, $\tilde{C}(q) = (\tilde{c}_{ij}(q))_{i,j=1}^n$ *is a matrix proportional to the matrix* $\tilde{A}(q)^{-1}$. *Then*

$$R_0 = R_0^e R_0^o = \prod_{m \geq 0}(\exp(\sum_{1 \leq i,j \leq n} Res_{u=v}(\varphi_i^+(u))' \otimes c_{ji}(T^{-1/2})\varphi_j^-(v + (n+1/2)))) \cdot \prod_{m \geq 0}(\exp(\sum_{1 \leq i,j \leq n} Res_{u=v}(\tilde{\varphi}_i^+(u))' \otimes \tilde{c}_{ji}(T^{-1/2})\tilde{\varphi}_j^-(v + (n+1/2)))), \quad (28)$$

where $c_{ji}(T^{-1/2})$, $\tilde{c}_{ji}(T^{-1/2})$ *substitutes* $T^{-1/2}$ *into* $c_{ji}(q)$, $\tilde{c}_{ji}(q)$, *respectively, instead of* q.

We briefly describe the plan for calculating the term R_0. We show that it is also factorized and can be represented in the form of the product $R_0 = R_0^e \cdot R_0^o$. The proof of this fact is based on the fact that the pairing between the even generators of the Cartan subsuperalgebra $h_{i,0,k}$ and the odd generators $h_{1,k}^i, i \in I, k \in \mathbb{Z}_+$ in the Cartan subsuperalgebra dual in the quantum double is zero. Therefore, we can carry out separate calculations for the terms R_0^e and R_0^o. The scheme of reasoning in both cases is very similar and essentially coincides with the calculation scheme realized in the papers [23], [24], although the formulas themselves differ somewhat.

To describe the terms R_0^e and R_0^o, we introduce the following notions.

Along with the Yangian Double $DY(\mathfrak{g})$, we consider the Hopf superalgebra $\widehat{DY}(\mathfrak{g})$, which is isomorphic as an associative superalgebra to the Quantum Double $DY(\mathfrak{g})$, but with a different structure of the cosuperalgebra determined by the following simple formulas:

$$\tilde{\Delta}(h_{i,j}^{\pm}(u)) = h_{i,j}^{\pm}(u) \otimes h_{i,j}^{\pm}(u), \quad (29)$$

$$\tilde{\Delta}(e_{i,j}(u)) = e_{i,j}(u) \otimes 1 + h_{i,j}^{-}(u) \otimes e_{i,j}(u), \quad (30)$$

$$\tilde{\Delta}(f_{i,j}(u)) = 1 \otimes f_{i,j}(u) + f_{i,j}(u) \otimes h_{i,j}^{+}(u). \quad (31)$$

We can to check that comultiplications Δ and $\tilde{\Delta}$ are conjugated by the following operator

$$\hat{t}^{\infty} := lim_{n \to \infty} \hat{t}^n,$$

$$\hat{t}(e_{i,j,k}) = e_{i,j,k+1}, \quad \hat{t}(f_{i,j,k}) = f_{i,j,k-1},$$

$$\hat{t}(h_{i,j,k}) = h_{i,j,k}, \quad j \in \{0,1\}, \quad k \in \mathbb{Z},$$

$$\tilde{\Delta}(x) = lim_{n \to \infty}(\hat{t}^n \otimes \hat{t}^n)\Delta(\hat{t}^{-n}(x)), \quad (32)$$

for $\forall x \in DY(\mathfrak{g})$.

Let also $\{\tilde{\tilde{\phi}}_{i,m}\}, \{\tilde{\phi}^{i,m}\}$ are dual bases in the spaces of primitive Cartan vectors in $\widehat{DY_0}(\mathfrak{g})$ (These subspaces we denote by $\hat{\Phi}^+, \hat{\Phi}^-$, correspondingly).

Proposition 4 *The element* R_0 *can be can be represented in the following form*

$$R_0 = R_0^e R_0^o = \exp\left(\sum_{i,m} \tilde{\phi}_{i,m} \otimes \tilde{\phi}^{i,m}\right) \exp\left(-\sum_{i,m} \tilde{\tilde{\phi}}_{i,m} \otimes \tilde{\tilde{\phi}}^{i,m}\right). \quad (33)$$

Further we deduce from the proposition 4, using the formula (32) Theorem 3. The proof of Theorem 2 is somewhat more complicated and more technical.

[1] Arakawa T 1999 Drinfeld functor and finite-dimensional representations of Yangian *Comm. Math. Phys.* **205 no 1** 1 – 18 (*Preprint: math/9807144*

[2] Arutyunov G and Frolov S 2009 Foundations of the $AdS^5 \times S^5$ Superstring. Part 1 (*Preprint: hep-th/0901.4937*)

[3] Chari V, Pressley A. A guide to quantum groups *Cambridge: Camb. Univ. Press*

[4] Chari, V., Pressley, A. 1990 *L'Enseignment Mathematique* **36** 267 – 302

[5] Chari V, Pressley A 1991 *J. Reine Angew. Math.* **417** 87 – 128

[6] L. Dolan, Ch. Nappi, E. Witten 2004 Yangian Symmetry in D=4 Superconformal Yang-Mills theory *Preprint: hep-th/0401243*

[7] Drinfeld V 1988 Quantum groups *Berkley: Proc. Int. Cong. Math.* **1** 789–820

[8] Drinfeld V 1985 *Soviet Math. Dokl.* **32** 1060 – 1064 (in Russian)

[9] Drinfeld V 1988 *Soviet Math. Dokl.* **36** 212 – 216 (in Russian)

[10] Drinfeld V 1986 *Funct. Anal. and its Applic.* **20 no 1** 58 – 60

[11] Frappat L, Sorba P 2000 Dictionary on Lie Superalgebras *London: Academic Press*

[12] Kac V 1977 A Sketch of Lie Superalgebra Theory *Commun.Math.Phys.* **53** 31-64

[13] Khoroshkin S M, Tolstoy V N 1996 Yangian Double *Lett.Math.Phys.*, **36** 373 – 402

[14] Levendorskii S, Soibelman Ya, Stukopin V 1993 Quantum Weyl group and universal R-matrix for quantum affine Lie algebra $A_1^{(1)}$ *Lett. Math. Phys.*, **27** 253 – 264

[15] Molev A 2003 Yangians and their applications, in Handbook of Algebra ed. M Hazewinkel **3** (Amsterdam: Elsevier) 907 – 959 *Preprint: math. QA/0211288*

[16] Molev A 2007 Yangians and Classical Lie algebras *Math. Survey Monogr.* **143** (Provudence, R. I.: Amer.Math. Soc.)

[17] Nazarov M 1991 Quantum Berezinian and the classical Capelly identity *Lett. Math. Phys.* **21** 123-131

[18] Nazarov M 1998 Yangian of the Queer Lie Superalgebra *Commun.Math.Phys.*, **208**, 195 – 223

[19] Nazarov M, Tarasov V 1998 On irreducibility of tensor products of Yangian modules *Internat. Math. Research Notices* 125-150 *Preprint: q-alg/9712004*

[20] Spill F Weakly coupled N = 4 Super Yang-Mills and N = 6 Chern-Simons theories from u(2|2) Yangian symmetry *Preprint: hep-th/0810.3897*

[21] Spill F, Torrielli A 2008 On Drinfeld's second realization of the *AdS/CFT* $\mathfrak{su}(2|2)$ Yangian *Preprint: hep-th/0803.3194*

[22] Stukopin V 1994 *Funct. Anal. and its Applic.* **28, no 3** 217 – 219

[23] Stukopin V 2006 *Funct. Anal. and its Applic.* **40, no 2** 155 – 158

[24] Stukopin V 2007 Quantum Double of Yangian of Lie superalgebra $A(m,n)$ and computation of Universal R-matrix *Journal of Mathematical Sciences* **142, no 2** 1989 – 2006 *Preprint: math. QA/0504302*

[25] Stukopin V 2007 *Symmetry, Integrability and Geometry: Methods and Applications*, **3**, 069

[26] Stukopin V 2002 *Asymptotic Combinatorics with Appic. to Math. Phys. Kluwer Academic Publishers* 255 – 265

[27] Stukopin V 2009 *Journal of Mathematical Sciences.*, **161**, **1** 143 – 162

[28] Stukopin V 2013 *Izvestija: Mathematics* **77 no 5** 1021 – 1043

[29] Stukopin V 2013 *Theor. and Math. Phys.* **174 no 1** 122 – 133

[30] Stukopin V 2013 On representations of Yangian of Lie Superalgebra $A(n,n)$ type *Journal of Physics. C. S.*, **411**, **1**, 012027

[31] Tarasov V 1984 *Theor. and Math. Phys.* **61 no 2** 1065 – 1072

[32] Tarasov V 1985 *Theor. and Math. Phys.* **63 no 2** 440 – 454

[33] Zhang R B 1995 Representations of super Yangian *J. Math. Phys.* **36** 3854-3865 *Preprint: hep-th/9411243*

[34] Zhang R B 1996 The $\mathfrak{gl}(M,N)$ super Yangian and its finite-dimensional representations *Lett. Math. Phys.* **37** 419 – 434 *Preprint: hep-th/9507029*

IOP Conf. Series: Journal of Physics: Conf. Series **965** (2018) 012041 doi:10.1088/1742-6596/965/1/012041

Magnetic monopoles and nonassociative deformations of quantum theory

Richard J. Szabo

Department of Mathematics, Heriot-Watt University, Colin Maclaurin Building, Riccarton, Edinburgh EH14 4AS, U.K.
Maxwell Institute for Mathematical Sciences, Edinburgh, U.K.
The Higgs Centre for Theoretical Physics, Edinburgh, U.K.

E-mail: R.J.Szabo@hw.ac.uk

Abstract. We examine certain nonassociative deformations of quantum mechanics and gravity in three dimensions related to the dynamics of electrons in uniform distributions of magnetic charge. We describe a quantitative framework for nonassociative quantum mechanics in this setting, which exhibits new effects compared to ordinary quantum mechanics with sourceless magnetic fields, and the extent to which these theoretical consequences may be experimentally testable. We relate this theory to noncommutative Jordanian quantum mechanics, and show that its underlying algebra can be obtained as a contraction of the alternative algebra of octonions. The uncontracted octonion algebra conjecturally describes a nonassociative deformation of three-dimensional quantum gravity induced by magnetic monopoles, which we propose is realised by a non-geometric Kaluza-Klein monopole background in M-theory.

1. Prologue: Three-dimensional quantum gravity

Applications of noncommutative geometry in physics are often motivated as providing a suitable mathematical framework for describing the modifications of spacetime geometry at very short distance scales which are expected in a quantum theory of gravity; the length scale at which such effects become important is usually understood to be the Planck length $\ell_{\rm P}$. However, there are relatively few precise quantitative connections between noncommutative geometry and models of quantum gravity. One notable exception is quantum gravity in three spacetime dimensions; that three-dimensional quantum gravity naturally implies a "fuzziness" to short distance spacetime structure was noted already in the early work of 't Hooft [1] and others, see e.g. [2], who observed that the spectrum of position operators in this theory is discrete. In this paper we shall be primarily concerned with the precise realisation of these structures obtained by [3], who consider a Ponzano-Regge spin foam model of three-dimensional quantum gravity coupled to spinless matter fields. After integrating out the gravitational degrees of freedom in this model, they obtain an effective scalar field theory on a noncommutative spacetime described via the deformed phase space commutation relations

$$\begin{aligned} \big[x^i\,,\,x^j\big] &= {\rm i}\,\ell_{\rm P}\,\varepsilon^{ijk}\,x^k\ , \\ \big[x_i\,,\,p_j\big] &= {\rm i}\,\sqrt{\hbar^2-\ell_{\rm P}^2\,p^2}\ \delta_{ij} - {\rm i}\,\ell_{\rm P}\,\varepsilon_{ijk}\,p_k\ , \\ \big[p_i\,,\,p_j\big] &= 0\ . \end{aligned} \tag{1.1}$$

The commutation relations among position coordinates x^i specify a Lie algebra noncommutative spacetime; the relations apply to both Euclidean signature, wherein the pertinent Lie group is $SU(2)$ and which is the main focus of this paper, and also in Minkowski signature wherein the pertinent Lie group is $SO(1,2)$.

The commutation relations (1.1) can be understood in the following way. A peculiar feature of quantum gravity in three dimensions is that the momenta p_i of a particle are bounded from above, with the bound set by the inverse of the Planck length $\ell_{\rm P}$. In particular, we can take the momentum space to be a sphere of radius $\ell_{\rm P}^{-1}$. Then the commutators in (1.1) simply reflect the fact that position coordinates act as derivations on this sphere, as implied by the usual canonical commutation relations. The deformation of the latter in (1.1) makes the noncommutative algebra invariant under a κ-deformation of the Poincaré group in three dimensions, with $\kappa = \ell_{\rm P}^{-1}$, such that the phase space commutation relations define a noncommutative but associative algebra. Setting $\ell_{\rm P} = 0$ corresponds to turning off the effects of quantum gravity and leaves an undeformed canonical phase space in (1.1) .

Many interesting physical properties of three-dimensional quantum gravity can be infered from the relations (1.1). For example, they imply a deformation of the usual addition law for Fourier wavenumbers, which in Minkowski signature leads to the modified dispersion relations

$$E^2 = \boldsymbol{p}^2\, c^2 - \Big(\frac{\sinh(\ell_{\rm P}\,\hbar^{-1}\, m\, c^2)}{\ell_{\rm P}\,\hbar^{-1}}\Big)^2 \ . \tag{1.2}$$

In the limit $\ell_{\rm P} \to 0$ wherein quantum gravity effects are neglected, this is just the usual relativistic dispersion law for a scalar particle in three spacetime dimensions. The deformation by $\ell_{\rm P} \neq 0$ is interpreted as the statement that doubly special relativity arises precisely in the low energy limit of three-dimensional quantum gravity [3].

The purpose of this paper is to describe a novel conjectural nonassociative deformation of the three-dimensional quantum gravity algebra (1.1) which is implied by certain magnetic dual analogues of the types of nonassociative spacetime geometries that are anticipated to arise in non-geometric string theory and M-theory, see e.g. [4–13]. For a while we shall therefore leave the present setting of three-dimensional quantum gravity and explain how nonassociative deformations of phase space naturally appear in a simpler elementary quantum mechanical setting of magnetic monopole distributions (Section 2). By attempting to reconcile this particular nonassocative deformation of quantum mechanics with the nonassocative algebras of observables proposed in the early days of quantum mechanics by Jordan, von Neumann, Wigner and others to study the mathematical and conceptual foundations of quantum theory, we shall arrive at a new framework that conjecturally describes magnetic monopoles in the spacetime of three-dimensional quantum gravity (Section 3). We conclude with a proposed realisation of such a system within the framework of M-theory (Section 4).

2. Magnetic monopoles and nonassociative quantum mechanics

We shall start by discussing the quantum mechanics of electrons propagating in sources of magnetic charge, focusing on the case of uniform monopole distributions which are the suitable analog systems dual to certain non-geometric string backgrounds. In this latter case we argue that the conventional framework of quantum theory nessitates a nonassociative deformation, and then demonstrate that such a model of nonassociative quantum mechanics is not only possible, but physically sensible and potentially testable.

2.1. Magnetic sources in quantum mechanics

Electrons propagating in a magnetic field $\boldsymbol{B}(\boldsymbol{x})$ under the influence of an external potential $V(\boldsymbol{x})$ in three dimensions classically obey the Lorentz force law

$$\dot{\boldsymbol{p}} = \frac{e}{2\,m\,c}\,(\boldsymbol{p}\times\boldsymbol{B}-\boldsymbol{B}\times\boldsymbol{p})+\boldsymbol{\nabla}V(\boldsymbol{x}) \tag{2.1}$$

where $\boldsymbol{p}=m\,\dot{\boldsymbol{x}}$ is the kinematical momentum (as opposed to the canonical momentum), and we have written the right-hand side in a form that generalises to the quantum theory. The corresponding Hamiltonian on phase space is the sum of kinetic and potential energies:

$$\mathsf{H}(\boldsymbol{x},\boldsymbol{p}) = \frac{1}{2m}\,\boldsymbol{p}^2+V(\boldsymbol{x}) \ . \tag{2.2}$$

Then the quantum mechanical Heisenberg equations of motion

$$-\,\mathrm{i}\,\hbar\,\dot{\boldsymbol{p}} = [\mathsf{H},\boldsymbol{p}] \quad , \quad -\,\mathrm{i}\,\hbar\,\dot{\boldsymbol{x}} = [\mathsf{H},\boldsymbol{x}] \tag{2.3}$$

are compatible with the Lorentz force law and the relation $\dot{\boldsymbol{x}}=\boldsymbol{p}/m$ only with the phase space commutation relations

$$[x^i,x^j] = 0 \quad , \quad [x^i,p_j] = \mathrm{i}\,\hbar\,\delta^i{}_j \quad , \quad [p_i,p_j] = \frac{\mathrm{i}\,\hbar\,e}{c}\,\varepsilon_{ijk}\,B^k \ , \tag{2.4}$$

which deform the canonical commutators to a noncommutative momentum space. Note that here and below everything is formulated without any reference to a vector potential, so that these considerations apply as well to the cases where the magnetic field $\boldsymbol{B}$ has sources.

Globally, the noncommutativity can be described in terms of the magnetic translation operators

$$U(\boldsymbol{a}) = \exp\big(\tfrac{\mathrm{i}}{\hbar}\,\boldsymbol{a}\cdot\boldsymbol{p}\big) \tag{2.5}$$

which generate finite translations $U(\boldsymbol{a})\,\boldsymbol{x}\,U(\boldsymbol{a})^{-1}=\boldsymbol{x}+\boldsymbol{a}$ due to the second commutator in (2.4). A simple calculation shows that they do not commute [14]:

$$U(\boldsymbol{a}_1)\,U(\boldsymbol{a}_2) = \mathrm{e}^{\frac{\mathrm{i}\,e}{\hbar\,c}\,\Phi_2(\boldsymbol{x};\boldsymbol{a}_1,\boldsymbol{a}_2)}\,U(\boldsymbol{a}_1+\boldsymbol{a}_2) = \mathrm{e}^{\frac{2\,\mathrm{i}\,e}{\hbar\,c}\,\Phi_2(\boldsymbol{x};\boldsymbol{a}_1,\boldsymbol{a}_2)}\,U(\boldsymbol{a}_2)\,U(\boldsymbol{a}_1) \ , \tag{2.6}$$

where

$$\Phi_2(\boldsymbol{x};\boldsymbol{a}_1,\boldsymbol{a}_2) = \int_{\langle\boldsymbol{a}_1,\boldsymbol{a}_2\rangle_{\boldsymbol{x}}}\boldsymbol{B}\cdot\mathrm{d}\boldsymbol{S} \tag{2.7}$$

is the magnetic flux through the plane of the triangle $\langle\boldsymbol{a}_1,\boldsymbol{a}_2\rangle_{\boldsymbol{x}}$ based at $\boldsymbol{x}\in\mathbb{R}^3$ and spanned by the vectors $\boldsymbol{a}_1$ and $\boldsymbol{a}_2$.

One can define a 3-bracket on momentum space as the combination of iterated commutators

$$[p_1,p_2,p_3] := [p_1,[p_2,p_3]]+[p_2,[p_3,p_1]]+[p_3,[p_1,p_2]] \tag{2.8}$$

that would vanish if the brackets (2.4) satisfied the Jacobi identity; it is called the 'Jacobiator'. One easily computes in this case

$$[p_1,p_2,p_3] = \frac{\hbar^2\,e}{c}\,\boldsymbol{\nabla}\cdot\boldsymbol{B} \ , \tag{2.9}$$

which expresses a further deformation of canonical phase space to a nonassociative momentum space. Globally, one easily computes using (2.6) that the magnetic translation operators thus do not associate:

$$\big(U(\boldsymbol{a}_1)\,U(\boldsymbol{a}_2)\big)\,U(\boldsymbol{a}_3) \;=\; \mathrm{e}^{\frac{\mathrm{i}\,e}{\hbar\,c}\,\Phi_3(\boldsymbol{x};\boldsymbol{a}_1,\boldsymbol{a}_2,\boldsymbol{a}_3)}\,U(\boldsymbol{a}_1)\,\big(U(\boldsymbol{a}_2)\,U(\boldsymbol{a}_3)\big)\;, \tag{2.10}$$

where $\Phi_3(\boldsymbol{x};\boldsymbol{a}_1,\boldsymbol{a}_2,\boldsymbol{a}_3)$ is the sum of the magnetic fluxes through $\langle\boldsymbol{a}_2,\boldsymbol{a}_3\rangle_{\boldsymbol{x}+\boldsymbol{a}_1}$ and $\langle\boldsymbol{a}_1,\boldsymbol{a}_2+\boldsymbol{a}_3\rangle_{\boldsymbol{x}}$, minus the fluxes through $\langle\boldsymbol{a}_1,\boldsymbol{a}_2\rangle_{\boldsymbol{x}}$ and $\langle\boldsymbol{a}_1+\boldsymbol{a}_2,\boldsymbol{a}_3\rangle_{\boldsymbol{x}}$. Together this gives the magnetic flux through the faces of the tetrahedron $\langle\boldsymbol{a}_1,\boldsymbol{a}_2,\boldsymbol{a}_3\rangle_{\boldsymbol{x}}$ based at $\boldsymbol{x}$ and spanned by the vectors $\boldsymbol{a}_1$, $\boldsymbol{a}_2$ and $\boldsymbol{a}_3$, which using the divergence theorem can be written as the magnetic charge

$$\Phi_3(\boldsymbol{x};\boldsymbol{a}_1,\boldsymbol{a}_2,\boldsymbol{a}_3) \;=\; \int_{\langle\boldsymbol{a}_1,\boldsymbol{a}_2,\boldsymbol{a}_3\rangle_{\boldsymbol{x}}} \boldsymbol{\nabla}\cdot\boldsymbol{B}\;\mathrm{d}V \tag{2.11}$$

enclosed by the tetrahedron. There are three generic situations that can now arise.

First, in Maxwell theory without magnetic sources, we have $\boldsymbol{\nabla}\cdot\boldsymbol{B}=0$. In this case all variables associate. The commutation relations (2.4) may then be represented on the Hilbert space $L^2(\mathbb{R}^3)$ of wavefunctions $\psi(\boldsymbol{x})$ in a Schrödinger polarization by the kinematical momentum operators $\boldsymbol{p}=-\mathrm{i}\,\hbar\,\boldsymbol{\nabla}-\frac{e}{c}\,\boldsymbol{A}$, where $\boldsymbol{A}$ is a globally defined vector potential for the magnetic field: $\boldsymbol{B}=\boldsymbol{\nabla}\times\boldsymbol{A}$. The vanishing of the magnetic charge, $\Phi_3=0$, implies that the phase factor Φ_2 defines a 2-cocycle of the translation group of $\mathbb{R}^3$, and the relations (2.6) reflect the fact that the magnetic translation operators only form a projective representation of this group; as usual, such projective actions on wavefunctions $\psi(\boldsymbol{x})$ are well-defined in quantum mechanics.

Next, suppose that there are magnetic sources so that $\boldsymbol{\nabla}\cdot\boldsymbol{B}\neq 0$. Then the magnetic translation operators generate a nonassociative algebra unless the magnetic charge obeys

$$\frac{e\,\Phi_3}{2\pi\,\hbar\,c} \;=\; \text{integer}\;. \tag{2.12}$$

In this case $\boldsymbol{\nabla}\cdot\boldsymbol{B}$ is necessarily a sum of delta-functions, representing a set of isolated point-like magnetic monopoles, or else the quantization condition (2.12) would be incompatible with continuous variations of the position vectors $\boldsymbol{a}_i\in\mathbb{R}^3$. This constraint is simply the celebrated Dirac quantization condition [15], as first pointed out in the present context by Jackiw [16]; with it, the magnetic translations may be represented by linear operators on a Hilbert space, which necessarily associate. At the infinitesimal level, the Jacobiator (2.9) is only non-vanishing at the loci of the magnetic sources. For isolated magnetic monopoles, one can excise their locations from position space and quantize the system in the standard way on the excised space: Classically the electrons never reach the monopole locations because of angular momentum conservation [8], while at the quantum level the wavefunctions vanish at the magnetic charge loci. Moreover, the corresponding local vector potentials $\boldsymbol{A}$, and the magnetic field $\boldsymbol{B}=\boldsymbol{\nabla}\times\boldsymbol{A}$, are singular at the monopole locations which can be interpreted in the standard way in terms of the singularities of Dirac strings which emanate from the monopoles [15]; for example, a single Dirac monopole sources the magnetic field

$$\boldsymbol{B}(\boldsymbol{x}) \;=\; \frac{\Phi_3}{4\pi}\,\frac{\boldsymbol{x}}{|\boldsymbol{x}|^3}\;. \tag{2.13}$$

Geometrically, the wavefunctions $\psi(\boldsymbol{x})$ can be treated as sections and the local monopole potential $\boldsymbol{A}$ as a connection of a non-trivial $U(1)$-bundle over the excised space [17], whose degree is precisely the magnetic charge (2.12).

In this paper we are interested in the case of a homogeneous magnetic charge density, wherein $\boldsymbol{\nabla}\cdot\boldsymbol{B}$ is constant. In this case one cannot excise the support of the magnetic charge distribution, or else one would be left with empty space, and one needs to deal directly with the nonassociativity of momentum coordinates and the magnetic translations. This is a manifestation of the type of 'local non-geometry' that arises in flux compactifications of string theory; it originates from the fact that the uniform distribution of magnetic charge can no longer be described by a connection on a $U(1)$-bundle over a non-contractible space, as now a vector potential $\boldsymbol{A}$ does not even exist locally when $\boldsymbol{\nabla}\cdot\boldsymbol{B}\neq 0$ everywhere, but rather it induces a connection $F_{ij}=\varepsilon_{ijk}\,B^k$ on a (trivial) $U(1)$-gerbe on $\mathbb{R}^3$. In particular, there is no relation between the kinematical and canonical momentum in this case. Moreover, the magnetic flux Φ_2 is a 2-cochain which no longer defines a (weak) projective representation; its coboundary is the magnetic charge Φ_3 which defines a 3-cocycle of the translation group. Nonassociativity that is induced by a 3-cocycle controlling the associativity of operators can be elegantly dealt with by working in a suitable braided monoidal category of representations [18]. The geometric meaning of such higher structures in quantization is described in [19, 20], where the sections of the trivial $U(1)$-gerbe on $\mathbb{R}^3$ with connection given by the magnetic field $\boldsymbol{B}(\boldsymbol{x})=\frac{1}{3}\,\boldsymbol{x}$ are shown to correspond to a certain 2-Hilbert space of Hermitian matrix-valued one-forms on $\mathbb{R}^3$ with bicovariantly constant matrix-valued functions on $\mathbb{R}^3$ as morphisms.

Previous algebraic characterisations of the commutation relations (2.4) have discussed their realisation as a nonassociative Malcev algebra [21, 22], and more generally as an alternative (or Jordan) algebra [23]; we shall come back to these points in Section 3, and indicate why these realisations are somewhat subtle from the perspective of this paper. Classically, the Lorentz force equations (2.1) no longer seem to be integrable, as the conservation of angular momentum of the Dirac monopole background is lost [8], while at the quantum level it is clear that some formalism of *nonassociative quantum mechanics* is needed; in the rest of this section we describe such a framework.

2.2. Phase space quantum mechanics

Operators which act on a separable Hilbert space necessarily associate, by definition. Hence the standard operator-state techniques of quantum mechanics are inadequate to handle systems with a non-trivial 3-cocycle that obstructs associativity. This can be presumably dealt with by a suitable higher operator-state formalism adapted to the quantum 2-Hilbert space of [20], but such techniques have not yet been developed. We can, however, appeal to the phase space formulation of quantum mechanics, which was succinctly developed by Grönewold [24], Moyal [25] and others, following earlier work of Weyl [26] and Wigner [27]; see e.g. [28] for a pedagogical introduction. One of the premises of this formalism is to treat position and momentum on equal footing, which are usually otherwise treated asymmetrically when choosing e.g. a Schrödinger polarization. It is through this that the formalism of star products was originally introduced.

Let us recall the basic dictionary in the case where $\boldsymbol{B}=\boldsymbol{0}$. Operators in general become functions on phase space, while observables correspond to real functions. Traces of operators are given by integrating functions over phase space. States ψ become real phase space quasi-probability distribution (Wigner) functions which are defined by Fourier transformation of their position space representations as

$$W_\psi(\boldsymbol{x},\boldsymbol{p}) \;=\; \int \frac{\mathrm{d}\boldsymbol{y}}{(2\pi)^3}\,\langle\boldsymbol{x}+\tfrac{\hbar}{2}\,\boldsymbol{y}\,|\,\psi\rangle\langle\psi\,|\,\boldsymbol{x}-\tfrac{\hbar}{2}\,\boldsymbol{y}\rangle\,\mathrm{e}^{-\mathrm{i}\,\boldsymbol{y}\cdot\boldsymbol{p}}\,, \tag{2.14}$$

such that the expectation value of an operator (function) is computed as

$$\langle A\rangle_\psi \;=\; \int \mathrm{d}\boldsymbol{x}\;\mathrm{d}\boldsymbol{p}\;W_\psi(\boldsymbol{x},\boldsymbol{p})\,A(\boldsymbol{x},\boldsymbol{p})\;. \tag{2.15}$$

The operator product, which captures the crucial noncommutativity of quantum mechanics that leads to e.g. uncertainty relations, becomes the associative noncommutative Moyal star product

$$A(\boldsymbol{x},\boldsymbol{p})\star B(\boldsymbol{x},\boldsymbol{p}) \;=\; A\big(\boldsymbol{x}-\tfrac{\mathrm{i}\,\hbar}{2}\,\boldsymbol{\nabla}_{\boldsymbol{p}}\,,\,\boldsymbol{p}+\tfrac{\mathrm{i}\,\hbar}{2}\,\boldsymbol{\nabla}_{\boldsymbol{x}}\big)B(\boldsymbol{x},\boldsymbol{p}) \tag{2.16}$$

which is defined by replacing the arguments of a function $A(\boldsymbol{x},\boldsymbol{p})$ by their Bopp shifts and letting the resulting differential operator act on the function $B(\boldsymbol{x},\boldsymbol{p})$; this can be properly defined when $A(\boldsymbol{x},\boldsymbol{p})$ is a polynomial and then continuing to power series expansions.

Given a Hamiltonian $\mathsf{H}(\boldsymbol{x},\boldsymbol{p})$ on phase space, dynamics of observables is governed by the Heisenberg-type time evolution equations

$$\frac{\mathrm{d}A}{\mathrm{d}t} \;=\; \frac{[A,\mathsf{H}]_\star}{\mathrm{i}\,\hbar} \tag{2.17}$$

where the commutator is computed using the star product (2.16), $[A,B]_\star = A\star B - B\star A$. Stationary states change simply by a time-dependent phase and are defined via the 'star-genvalue equation'

$$\mathsf{H}\star W_\psi \;=\; E\,W_\psi \ , \tag{2.18}$$

which is the time-independent Schrödinger equation determining the energy eigenvalues E. This approach to quantization can successfully capture the quantum mechanics of the standard textbook examples, but it also has various conceptual subtleties and limitations; for example, Wigner functions in general need not be positive and so do not genuinely determine probability distributions.

2.3. Nonassociative quantum mechanics

Following [29] we can apply the phase space formulation to develop a version of nonassociative quantum mechanics that rather remarkably passes all basic physical consistency tests, and moreover has great quantitative power to compute new physical consequences of nonassociativity. Developing nonassociative quantum mechanics also brings into question many foundational issues and can teach us a lot about the nature of quantum theory itself. We shall work throughout with the case where $\rho := \frac{e}{c}\,\boldsymbol{\nabla}\cdot\boldsymbol{B}$ is a constant monopole density, and choose the gauge

$$\boldsymbol{B}(\boldsymbol{x}) \;=\; \frac{\rho\,c}{3\,e}\,\boldsymbol{x} \ . \tag{2.19}$$

The first ingredient we need is a suitable modification of the Moyal star product (2.16). Such a nonassociative star product was first constructed in [7]. Here we shall use the form developed in [29] (see also [30]), and hence introduce the nonassociative phase space "monopole star product" via a simple modification of the Bopp shifts in (2.16) to 'ρ-twisted Bopp shifts' as

$$A(\boldsymbol{x},\boldsymbol{p})\star B(\boldsymbol{x},\boldsymbol{p}) \;=\; A\big(\boldsymbol{x}-\tfrac{\mathrm{i}\,\hbar}{2}\,\boldsymbol{\nabla}_{\boldsymbol{p}}\,,\,\boldsymbol{p}+\tfrac{\mathrm{i}\,\hbar}{2}\,(\boldsymbol{\nabla}_{\boldsymbol{x}}+\tfrac{1}{3}\,\rho\,\boldsymbol{x}\times\boldsymbol{\nabla}_{\boldsymbol{p}})\big)B(\boldsymbol{x},\boldsymbol{p}) \ . \tag{2.20}$$

That this provides a suitable quantization of the magnetic monopole algebra (2.4) and (2.9) can be seen by computing the corresponding star commutators of phase space coordinate functions

$$[x^i,x^j]_\star \;=\; 0 \quad , \quad [x^i,p_j]_\star \;=\; \mathrm{i}\,\hbar\,\delta^i{}_j \quad , \quad [p_i,p_j]_\star \;=\; \frac{\mathrm{i}\,\hbar}{3}\,\rho\,\varepsilon_{ijk}\,x^k \tag{2.21}$$

and the corresponding non-vanishing star Jacobiator

$$[p_1,p_2,p_3]_\star \;=\; \hbar^2\,\rho \ . \tag{2.22}$$

Globally, the monopole star product reproduces the appropriate algebra (2.6) and (2.10) of magnetic translation operators as

$$U(\boldsymbol{a}_1)\star U(\boldsymbol{a}_2) \;=\; \mathrm{e}^{\frac{\mathrm{i}}{\hbar}\,\Phi_2(\boldsymbol{x};\boldsymbol{a}_1,\boldsymbol{a}_2)}\,U(\boldsymbol{a}_1+\boldsymbol{a}_2) \quad , \quad \Phi_2(\boldsymbol{x};\boldsymbol{a}_1,\boldsymbol{a}_2) \;=\; \frac{1}{6}\,\rho\,(\boldsymbol{a}_1\times\boldsymbol{a}_2)\cdot\boldsymbol{x} \tag{2.23}$$

and

$$\begin{aligned}\big(U(\boldsymbol{a}_1)\star U(\boldsymbol{a}_2)\big)\star U(\boldsymbol{a}_3) \;&=\; \mathrm{e}^{\frac{\mathrm{i}}{\hbar}\,\Phi_3(\boldsymbol{x};\boldsymbol{a}_1,\boldsymbol{a}_2,\boldsymbol{a}_3)}\,U(\boldsymbol{a}_1)\star\big(U(\boldsymbol{a}_2)\star U(\boldsymbol{a}_3)\big) \quad , \\ \Phi_3(\boldsymbol{x};\boldsymbol{a}_1,\boldsymbol{a}_2,\boldsymbol{a}_3) \;&=\; \frac{\hbar^3}{6}\,\rho\,\boldsymbol{a}_1\cdot(\boldsymbol{a}_1\times\boldsymbol{a}_2)\ .\end{aligned} \tag{2.24}$$

The formula (2.20) is exact when ρ is constant.

The monopole star product has various noteworthy properties which are important for calculations in nonassociative quantum mechanics. First of all, since the star product $A\star B$ differs from the ordinary pointwise product of functions $A\,B$ by total derivative terms, it is '2-cyclic' in the sense that

$$\int \mathrm{d}\boldsymbol{x}\ \mathrm{d}\boldsymbol{p}\ A\star B \;=\; \int \mathrm{d}\boldsymbol{x}\ \mathrm{d}\boldsymbol{p}\ B\star A \;=\; \int \mathrm{d}\boldsymbol{x}\ \mathrm{d}\boldsymbol{p}\ A\,B\ , \tag{2.25}$$

for suitable functions A and B of Schwartz class. Hence noncommutativity at this order is washed away upon integration, i.e. "on-shell". Similarly, since the triple star product $A\star(B\star C)$ differs from $(A\star B)\star C$ by total derivative terms, it is '3-cyclic' in the sense that

$$\int \mathrm{d}\boldsymbol{x}\ \mathrm{d}\boldsymbol{p}\ A\star(B\star C) \;=\; \int \mathrm{d}\boldsymbol{x}\ \mathrm{d}\boldsymbol{p}\ (A\star B)\star C\ . \tag{2.26}$$

Hence nonassociativity at this order is also absent on-shell. However, this is not true for higher order multiple star products in general, see [29] for a general analysis of this feature. Finally, the monopole star product is Hermitian in the sense that it mimicks the usual conjugation properties of operator products,

$$(A\star B)^* \;=\; B^*\star A^*\ , \tag{2.27}$$

and it is unital in the sense that the constant function 1 serves as an identity element for the star product algebra of functions,

$$A\star 1 \;=\; A \;=\; 1\star A\ . \tag{2.28}$$

A state in nonassociative quantum mechanics is specified by a collection of L^2-normalized phase space wavefunctions $\psi_a(\boldsymbol{x},\boldsymbol{p})$,

$$\int \mathrm{d}\boldsymbol{x}\ \mathrm{d}\boldsymbol{p}\ \big|\psi_a(\boldsymbol{x},\boldsymbol{p})\big|^2 \;=\; 1\ , \tag{2.29}$$

together with statistical probabilities $\mu_a\in[0,1]$,

$$\sum_a \mu_a \;=\; 1\ . \tag{2.30}$$

Expectation values are then defined by

$$\langle A\rangle_\psi \;=\; \sum_a \mu_a\,\int \mathrm{d}\boldsymbol{x}\ \mathrm{d}\boldsymbol{p}\ \psi_a^*\star(A\star\psi_a)\ , \tag{2.31}$$

which using 2-cyclicity and 3-cyclicity of the monopole star product can be written as

$$\langle A\rangle_\psi = \int \mathrm{d}\boldsymbol{x}\ \mathrm{d}\boldsymbol{p}\ W_\psi(\boldsymbol{x},\boldsymbol{p})\,A(\boldsymbol{x},\boldsymbol{p}) \tag{2.32}$$

where

$$W_\psi = \sum_a \mu_a\,\psi_a\star\psi_a^* \quad , \quad \int \mathrm{d}\boldsymbol{x}\ \mathrm{d}\boldsymbol{p}\ W_\psi(\boldsymbol{x},\boldsymbol{p}) = 1 \tag{2.33}$$

plays the role of a Wigner distribution function and should be thought of here as a 'density operator'.

Now let us demonstrate that some of the basic physical requirements of quantum theory are satisfied by this setup. Let us first check reality of measurements. With the definition (2.31), using Hermiticity and 3-cyclicity we have

$$\langle A\rangle_\psi^* = \sum_a \mu_a \int \mathrm{d}\boldsymbol{x}\ \mathrm{d}\boldsymbol{p}\ (A\star\psi_a)^*\star\psi_a = \sum_a \mu_a \int \mathrm{d}\boldsymbol{x}\ \mathrm{d}\boldsymbol{p}\ \psi_a^*\star(A^*\star\psi_a) = \langle A^*\rangle_\psi \tag{2.34}$$

and hence observables $A = A^*$ have real expectation values; thus physical measurements are real in this quantum theory. A similar calculation establishes positivity of measurements [29].

Next we show that observables $A = A^*$ have real eigenvalues. Using Hermiticity the conjugate of the star-genvalue equation $A\star W_\psi = \lambda\,W_\psi$ is $W_\psi^*\star A = \lambda^*\,W_\psi^*$ which gives

$$W_\psi^*\star(A\star W_\psi) - (W_\psi^*\star A)\star W_\psi = (\lambda-\lambda^*)\,(W_\psi^*\star W_\psi)\ . \tag{2.35}$$

In the associative case we would be done at this stage, because the left-hand side of this equation would vanish, while the right-hand would be generically non-zero, but this is not so in the nonassociative case. However, the left-hand side vanishes on-shell by 3-cyclicity, so by integrating both sides we get

$$0 = (\lambda-\lambda^*)\int \mathrm{d}\boldsymbol{x}\ \mathrm{d}\boldsymbol{p}\ W_\psi^*\star W_\psi = (\lambda-\lambda^*)\int \mathrm{d}\boldsymbol{x}\ \mathrm{d}\boldsymbol{p}\ \big|W_\psi(\boldsymbol{x},\boldsymbol{p})\big|^2 \tag{2.36}$$

and now the right-hand side is non-zero unless $\lambda = \lambda^*$. These calculations illustrate the general features of checks in nonassociative quantum mechanics: In all cases the calculations proceed as in the associative case, but there are always a few extra steps required due to nonassociativity. But against all odds, the properties (2.25)–(2.28) of the monopole star product ensure that this nonassociative deformation of quantum theory passes all consistency checks. Various other basic features, including quantum uncertainty relations, are worked out in [29].

The key feature that makes such a nonassociative deformation tractable is the occurence of a 3-cocycle controlling associativity, as mentioned before; such an algebra should be more properly refered to as 'quasi-associative', since it would be hopeless to try to make things work in an arbitrary nonassociative algebra. In the present case it means that there is a multiplicative associator Φ which controls the rebracketing of triples of phase space functions under the monopole star product (2.20) as [10, 29]

$$A\star(B\star C) \xrightarrow{\ \Phi_{A,B,C}\ } (A\star B)\star C\ , \tag{2.37}$$

which is most easily described by passing to Fourier space and inserting the 3-cocycle phase factors

$$\Phi_{k,k',k''} = \mathrm{e}^{-\frac{\mathrm{i}\,\hbar^2}{6}\,\rho\,\boldsymbol{k_p}\cdot(\boldsymbol{k}'_{\boldsymbol{p}}\times\boldsymbol{k}''_{\boldsymbol{p}})}\ , \tag{2.38}$$

depending only on the Fourier wavenumbers $\boldsymbol{k}_{\boldsymbol{p}}$ dual to momentum coordinates $\boldsymbol{p}$, into the Fourier transformations of triple star products. The associators satisfy 'pentagon relations' which are captured by the commuting diagram

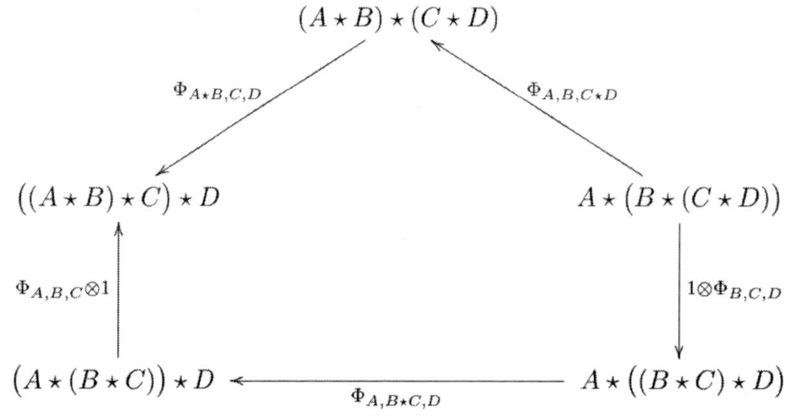

(2.39)

and Mac Lane's coherence theorem asserts that this uniquely defines the insertion of suitable associator factors into higher order iterated star products. This principle enables the extension of our considerations here to the construction of some nonassociative quantum field theories [31,32].

2.4. Momentum space quantization

Let us now examine some of the surprising physical consequences of nonassociativity. One of the standard results in ordinary quantum mechanics is the statement that a pair of observables which do not commute with each other cannot be simultaneously diagonalised. Here we find a higher version of this statement [29]: Nonassociating observables cannot have common eigenstates. In particular, this applies to the diagonalisation of the basic coordinate operators $x^I \star W_\psi = \lambda^I\, W_\psi$, with $x^I \in (\boldsymbol{x}, \boldsymbol{p})$. From (2.22), we find that the components of momentum $\boldsymbol{p}$ cannot be simultaneously measured, which implies a coarse-graining of the momentum space with a uniform monopole background.

We can quantify this quantisation more precisely by defining oriented area uncertainty operators

$$\mathcal{A}^{IJ} \;=\; \mathrm{Im}\big([\widetilde{x}^I, \widetilde{x}^J]_\star\big) \;=\; -\,\mathrm{i}\,\big(\widetilde{x}^I \star \widetilde{x}^J - \widetilde{x}^J \star \widetilde{x}^I\big) \tag{2.40}$$

which mimick the classical formula for the area of a triangle in terms of the vector product between coordinate vectors in directions I and J, where we defined the shifted coordinates $\widetilde{x}^I := x^I - \langle x^I \rangle_\psi$ appropriate to the computation of quantum uncertainties. Similarly, we define oriented volume uncertainty operators

$$\mathcal{V}^{IJK} \;=\; \frac{1}{3}\,\mathrm{Re}\big(\widetilde{x}^I \star [\widetilde{x}^J, \widetilde{x}^K]_\star + \widetilde{x}^K \star [\widetilde{x}^I, \widetilde{x}^J]_\star + \widetilde{x}^J \star [\widetilde{x}^K, \widetilde{x}^I]_\star\big) \tag{2.41}$$

which mimick the classical formula for volume of a tetrahedron in terms of the triple scalar product of coordinate vectors in directions I, J and K.

One can straightforwardly compute the expectation values of these operators to obtain the non-vanishing minimal areas [29]

$$\langle \mathcal{A}^{x^i, p_j} \rangle_\psi \;=\; \hbar\, \delta^i{}_j \quad , \quad \langle \mathcal{A}^{p_i, p_j} \rangle_\psi \;=\; \frac{\hbar}{3}\, \rho\, \varepsilon_{ijk}\, \langle x^k \rangle_\psi \;. \tag{2.42}$$

IOP Conf. Series: Journal of Physics: Conf. Series **965** (2018) 012041 doi:10.1088/1742-6596/965/1/012041

The first equation simply displays the standard Planck cells in phase space of area $\hbar$ arising from the Heisenberg uncertainty principle of quantum mechanics. The second equation is new, and is due to the uncertainty between momentum measurements proportional to the position measurement in the direction perpendicular to the plane of the momenta, which itself is subject to the standard Planck cell uncertainty. Similarly, one obtains the non-vanishing minimal volume [29]

$$\langle \mathcal{V}^{p_1,p_2,p_3} \rangle_\psi = \frac{1}{2}\,\hbar^2\,\rho \tag{2.43}$$

which clearly indicates a quantized momentum space with a quantum of minimal volume $\frac{1}{2}\,\hbar^2\,\rho$. This forbids, in particular, configurations with definite localised momentum in the monopole background.

2.5. Testable quantum effects

Let us now take a brief interlude to consider the question of whether it is possible to see experimental signatures of nonassociativity in quantum mechanics. Following [33], we add a confining force with harmonic oscillator potential

$$V(\boldsymbol{x}) = \frac{1}{2}\,m\,\omega^2\,\boldsymbol{x}^2 \ , \tag{2.44}$$

and compute the effective potential due to quantum corrections by

$$V_{\rm eff} := \langle \mathsf{H} \rangle_\psi \big|_{\langle p_i \rangle_\psi = 0} \ . \tag{2.45}$$

In the present case, the expectation values of the Hamiltonian (2.2) can be computed by using the ρ-twisted Bopp shifts in (2.20) to write its action on a phase space wavefunction $\psi(\boldsymbol{x},\boldsymbol{p})$ as

$$\mathsf{H}(\boldsymbol{x},\boldsymbol{p}) \star \psi(\boldsymbol{x},\boldsymbol{p}) = \hat{\mathsf{H}}\,\psi(\boldsymbol{x},\boldsymbol{p}) \ , \tag{2.46}$$

where $\hat{\mathsf{H}}$ is the second order differential operator

$$\begin{aligned} \hat{\mathsf{H}} = \frac{1}{2m}\Big(\boldsymbol{p}^2 &+ \mathrm{i}\,\hbar\,\boldsymbol{p}\cdot\boldsymbol{\nabla}_{\boldsymbol{x}} - \frac{\hbar^2}{4}\,\boldsymbol{\nabla}_{\boldsymbol{x}}^2 + \frac{\mathrm{i}\,\hbar}{3}\,\rho\,\boldsymbol{p}\cdot(\boldsymbol{x}\times\boldsymbol{\nabla}_{\boldsymbol{p}}) - \frac{\hbar^2}{6}\,\rho\,\boldsymbol{x}\cdot(\boldsymbol{\nabla}_{\boldsymbol{x}}\times\boldsymbol{\nabla}_{\boldsymbol{p}}) \\ &- \frac{\hbar^2}{36}\,\rho^2\,\big((\boldsymbol{x}\cdot\boldsymbol{\nabla}_{\boldsymbol{p}})^2 - \boldsymbol{x}^2\,\boldsymbol{\nabla}_{\boldsymbol{p}}^2\big)\Big) + \frac{1}{2}\,\omega^2\,\Big(\boldsymbol{x}^2 - \mathrm{i}\,\hbar\,\boldsymbol{x}\cdot\boldsymbol{\nabla}_{\boldsymbol{p}} - \frac{\hbar^2}{4}\,\boldsymbol{\nabla}_{\boldsymbol{p}}^2\Big) \ . \end{aligned} \tag{2.47}$$

One can then express the quantum corrections in terms of the fluctuations $(\Delta x^I)^2 := \big\langle \tilde{x}^I \star \tilde{x}^I \big\rangle_\psi$ to get

$$V_{\rm eff} = V\big(\langle \boldsymbol{x} \rangle_\psi\big) + \frac{1}{2m}\,(\Delta\boldsymbol{p})^2 + \frac{1}{2}\,m\,\omega^2\,(\Delta\boldsymbol{x})^2 \ . \tag{2.48}$$

Following the standard prescription known from the associative case, the uncertainty moments can be obtained by solving Ehrenfest-type equations of motion derived from the time evolution equations

$$\frac{\mathrm{d}\langle A \rangle_\psi}{\mathrm{d}t} = \frac{\langle [A,\mathsf{H}]_\star \rangle_\psi}{\mathrm{i}\,\hbar} \tag{2.49}$$

in an adiabatic approximation, and in the semi-classical limit one obtains [33]

$$V_{\rm eff} = V\big(\langle \boldsymbol{x} \rangle_\psi\big) + \frac{1}{6m}\,\hbar\,\rho\,\big|\langle \boldsymbol{x} \rangle_\psi\big| + \frac{1}{2}\,\hbar\,\omega \ . \tag{2.50}$$

IOP Conf. Series: Journal of Physics: Conf. Series **965** (2018) 012041 doi:10.1088/1742-6596/965/1/012041

Hence in addition to the usual zero point energy shift of the quantum harmonic oscillator, the effective potential demonstrates that the motion of electrons in a magnetic monopole density exhibits anharmonic deviations from the classical harmonic motion.

This effect could have potentially observable consequences in certain analogue systems of magnetic monopoles in condensed matter physics, see e.g. [34–36]. These experiments use rare earth oxide insulators $R_2M_2O_7$ where R is a magnetic ion (such as dysprosium or holmium) and M is a non-magnetic ion (such as titanium). The rare earth atoms R sit at the vertices of two intertwining pyrochlore lattices formed by corner-sharing tetrahedra, as illustrated in Figure 1. Electron spin provides a magnetic dipole at each vertex atom, which is shared by two regular tetrahedra, giving the lattice the geometry of quantum spin ice. The ground state contains two inward and two outward pointing dipoles towards the center of each tetrahedron, and hence has no magnetic charge. Flipping a magnetic dipole at one vertex produces a local topological excitation in which one of the two tetrahedra meeting there has an extra dipole pointing inwards while the other has an extra dipole pointing outwards, giving one tetrahedron three north poles and one south pole, and its neighbouring tetrahedron three south poles and one north pole.

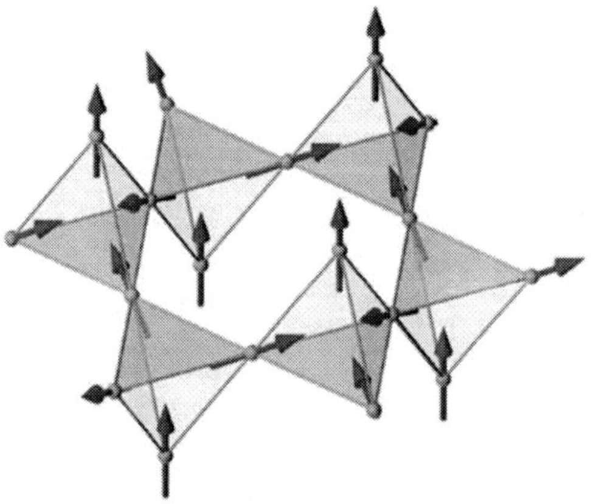

Figure 1. Spin ice pyrochlore lattice with magnetic dipoles (taken from [37]).

Iterating this process thus gives an interpretation of flipped magnetic dipoles in terms of tetrahedra containing free and unconfined magnetic monopoles, leading as before to translation group 3-cocycles which are given by the magnetic charge inside a tetrahedron. Scattering neutrons off the material reorganises the magnetic dipoles into a "spin-spaghetti" network of Dirac strings connecting pairs of monopoles, through which magnetic flux flows. These can be observed with applied magnetic fields at low temperature through interference effects by their interaction with the neutrons, which themselves carry a magnetic dipole moment. In such experimental scenarios, magnetic monopoles therefore exist as emergent states of matter. While these configurations do not involve a constant monopole density as in our theoretical framework, it is conceivable that the dynamics of electrons moving in a spin ice pyrochlore lattice could be approximated by electrons moving in a uniform magnetic charge distribution on scales larger than the lattice spacing, whereas the effects of Dirac strings may survive on averaging to a continuous distribution.

IOP Conf. Series: Journal of Physics: Conf. Series **965** (2018) 012041 doi:10.1088/1742-6596/965/1/012041

3. Jordanian quantum mechanics and the octonions

We shall now return to some general algebraic considerations and compare the magnetic monopole algebra with the nonassociative algebras that arose in the early days of quantum mechanics. This will lead us into discussing the algebra of octonions, which we shall show is related to the monopole algebra via a contraction. By suitably interpreting the former algebra in this context, we will thereby arrive at our nonassociative deformation of three-dimensional quantum gravity.

3.1. Noncommutative Jordan algebras

The idea of nonassociativity in quantum mechanics is not new, and traces back to the beginning days of quantum theory through the work of Jordan [38], who attempted to assemble quantum observables into an algebraic framework. The basic observation of Jordan is the following. If A and B are Hermitian operators, then their product $A\,B$ is not a Hermitian operator unless they commute, i.e. $A\,B$ is not observable unless A and B can be simultaneously measured; thus Hermitian operators do not close to an algebra under the usual operator product. And neither is their commutator $[A, B]$ an observable. However, the symmetrised product

$$A\circ B \ := \ \frac{1}{2}\,(A\,B+B\,A) \tag{3.1}$$

is a Hermitian operator. The product $\circ$ is evidently commutative,

$$A\circ B=B\circ A\ , \tag{3.2}$$

but it is not associative. However, it still satisfies the 'Jordan identity'

$$(A^2\circ B)\circ A \ = \ A^2\circ(B\circ A)\ . \tag{3.3}$$

A commutative nonassociative algebra over $\mathbb{R}$ whose product $\circ$ satisfies (3.3) for all elements A and B is called a (linear) Jordan algebra. If the algebra can be embedded in an associative algebra such that its nonassociative product $\circ$ is given as in (3.1) then the Jordan algebra is said to be "special". Jordan's hope was to find a system which was not derived from an associative algebra in this way, but behaved as one, in order to dispell of the insufficiencies that refer to the underlying unobservable operator algebras.

The theory of Jordan algebras was subsequently developed into a solid mathematical component of algebra, see e.g. [39, 40], wherein it was shown that it was sufficient to replace (3.3) with the weaker condition that the algebra be "alternative":

$$(A\circ B)\circ A \ = \ A\circ(B\circ A)\ , \tag{3.4}$$

and this can be used to define the notion of a noncommutative Jordan algebra. Jordan's hope of reformulating quantum mechanics in this algebraic framework was dashed by the theorem, proven originally for finite-dimensional algebras in [41] and extended to the infinite-dimensional case in [42], which essentially ruled out the existence of non-trivial Jordan algebras: The only non-special Jordan algebra (up to isomorphism) is the algebra of $3{\times}3$ Hermitian matrices over the division algebra of octonions $\mathbb{O}$, which defines the 27-dimensional "exceptional" Jordan algebra; the octonion algebra will be described in more detail below where it will play a prominent role. This somewhat exotic "octonionic quantum mechanics" satisfies the von Neumann axioms of quantum theory [43], despite the absence of a Hilbert space formulation. For reviews and further details, see e.g. [44, 45].

A natural question at this stage, in view of our previous version of nonassociative quantum mechanics, is the following: Is the magnetic monopole algebra a noncommutative Jordan algebra? This question has been answered in the negative, originally by [46], and subsequently via a more systematic treatment in [47] (see also [12]). The simplest counterexample already appears at quadratic order in the phase space coordinates [46]: Whereas one generically has the alternativity relations

$$(x^I \star x^K) \star x^I = x^I \star (x^K \star x^I) \quad , \quad \big((x^I)^{\star 2} \star x^K\big) \star x^I = (x^I)^{\star 2} \star (x^K \star x^I) \ , \tag{3.5}$$

the nonvanishing associator

$$(\boldsymbol{p}^2 \star \boldsymbol{p}^2) \star \boldsymbol{p}^2 - \boldsymbol{p}^2 \star (\boldsymbol{p}^2 \star \boldsymbol{p}^2) = -\frac{2\,\mathrm{i}}{9}\,\hbar^2\,\rho^2\,\boldsymbol{x}\cdot\boldsymbol{p} \neq 0 \tag{3.6}$$

demonstrates that the monopole star product does not define an alternative algebra.

This violation of alternativity is of course related to the volume quantization of momentum space that we observed earlier, and it brings into question if all is not lost in this theory; for example, this result would appear to rule out the existence of free stationary states. However, one only needs to ensure that $\langle \boldsymbol{x}\cdot\boldsymbol{p}\rangle_\psi = 0$ in the preparation of states for measurement. An example of such a state in the case of a free particle, $\rho = V = 0$, can be constructed from phase space wavefunctions $\psi(\boldsymbol{x},\boldsymbol{p})$ that solve the Schrödinger equation

$$\frac{1}{2m}\,\boldsymbol{p}^2 \star \psi(\boldsymbol{x},\boldsymbol{p}) = \frac{1}{2m}\,\hat{\boldsymbol{p}}^2\psi(\boldsymbol{x},\boldsymbol{p}) = E\,\psi(\boldsymbol{x},\boldsymbol{p}) \ , \tag{3.7}$$

where the first order differential operators

$$\hat{p}_i = p_i + \frac{\mathrm{i}\,\hbar}{2}\,\frac{\partial}{\partial x^i} \tag{3.8}$$

are mutually commuting Hermitian operators, so they have simultaneous eigenfunctions $\psi_{\boldsymbol{k}}(\boldsymbol{x},\boldsymbol{p})$ with

$$\hat{p}_i\psi_{\boldsymbol{k}}(\boldsymbol{x},\boldsymbol{p}) = k_i\,\psi_{\boldsymbol{k}}(\boldsymbol{x},\boldsymbol{p}) \tag{3.9}$$

and eigenvalues $k_i \in \mathbb{R}$ for $i = 1,2,3$. These three equations are solved by

$$\psi_{\boldsymbol{k}}(\boldsymbol{x},\boldsymbol{p}) = \mathrm{e}^{-\frac{2\,\mathrm{i}}{\hbar}\,(\boldsymbol{k}-\boldsymbol{p})\cdot\boldsymbol{x}}\ \varphi(\boldsymbol{p}) \tag{3.10}$$

where $\varphi(\boldsymbol{p})$ is an arbitrary function of $\boldsymbol{p}$ independent of the position coordinates $\boldsymbol{x}$. Substituting into (3.7) then gives the expected energy eigenvalues

$$E = E_{\boldsymbol{k}} = \frac{\boldsymbol{k}^2}{2m} \ , \tag{3.11}$$

and this represents the quantization of the free particle in the phase space formulation of quantum mechanics.

Equivalently, the Schrödinger equation may be cast as a star-genvalue equation

$$\frac{1}{2m}\,\boldsymbol{p}^2 \star W_\psi = E\,W_\psi \tag{3.12}$$

for the real-valued density operator $W_\psi(\boldsymbol{x},\boldsymbol{p})$ (see e.g. [28]). This collapses the equation to a pair of partial differential equations, its real and imaginary parts. The imaginary part

$$\boldsymbol{p}\cdot\boldsymbol{\nabla}_{\boldsymbol{x}}W_\psi(\boldsymbol{x},\boldsymbol{p}) = 0 \tag{3.13}$$

restricts $W_\psi(\boldsymbol{x},\boldsymbol{p}) = W_\psi(\boldsymbol{p})$ to be independent of $\boldsymbol{x}$. The real part

$$\Big(\boldsymbol{p}^2 - \frac{\hbar^2}{4}\,\boldsymbol{\nabla}_{\boldsymbol{x}}^2 - 2\,m\,E\Big)W_\psi(\boldsymbol{p}) \;=\; 0 \tag{3.14}$$

is satisfied for arbitrary real functions $W_\psi(\boldsymbol{p})$ of momentum with the energy eigenvalues

$$E \;=\; \frac{\boldsymbol{p}^2}{2m} \ . \tag{3.15}$$

When a non-vanishing background monopole charge ρ is turned on, we can consider again separately the real and imaginary parts of the corresponding star-genvalue equation (2.18). From (2.47) with $\omega = 0$ the imaginary equation can be written as

$$\frac{1}{2m}\,\boldsymbol{p}\cdot\hat{\boldsymbol{D}}\,W_\psi(\boldsymbol{x},\boldsymbol{p}) \;=\; 0 \tag{3.16}$$

where

$$\hat{\boldsymbol{D}} \;=\; \boldsymbol{\nabla}_{\boldsymbol{x}} - \frac{\rho}{3}\,\boldsymbol{x}\times\boldsymbol{\nabla}_{\boldsymbol{p}} \ . \tag{3.17}$$

The solutions $W_\psi(\boldsymbol{x},\boldsymbol{p})$ of this equation are the union of the integral curves of the vector field $(\frac{1}{m}\,\boldsymbol{p}, \frac{\rho}{3m}\,\boldsymbol{p}\times\boldsymbol{x})$ on phase space. The corresponding characteristic equations are

$$m\,\dot{\boldsymbol{x}} \;=\; \boldsymbol{p} \quad , \quad m\,\dot{\boldsymbol{p}} \;=\; \frac{\rho}{3}\,\boldsymbol{p}\times\boldsymbol{x} \ , \tag{3.18}$$

which are simply the classical equations of motion in the monopole background. One now needs to find from these flow equations a triple of classical integrals of motion $\boldsymbol{I} = (I_1, I_2, I_3)$, such that the general solution is $W_\psi(\boldsymbol{x},\boldsymbol{p}) = W_\psi(\boldsymbol{I})$, but such a complete set does not seem to exist [8] (for the free particle with $\rho = 0$ we took $\boldsymbol{I} = \boldsymbol{p}$).

3.2. Octonions and magnetic monopoles

Let us now raise the question of whether there is *any* relation of the magnetic monopole algebra to Jordanian quantum mechanics. This question was answered affirmatively by [11], but to describe that result we first need to take a slight algebraic detour. A foundational result in algebra states that there are only four normed division algebras over the field of real numbers: The real numbers $\mathbb{R}$ themselves, the complex numbers $\mathbb{C}$, the quaternions $\mathbb{H}$, and the octionions $\mathbb{O}$. They fit into a hierarchy $\mathbb{R} \subset \mathbb{C} \subset \mathbb{H} \subset \mathbb{O}$ where the first two algebras are commutative and associative, the third is noncommutative but associative, while the last one is both noncommutative and nonassociative but is alternative. An element of the octonion algebra $\mathbb{O}$ is a linear combination of generators

$$a_0\,1 + a_1\,e_1 + a_2\,e_2 + \cdots + a_7\,e_7 \tag{3.19}$$

where $a_i \in \mathbb{R}$, the element 1 is central and a unit for the algebra, and the seven imaginary unit octonions e_i, with $e_i^2 = -1$ and $e_i\,e_j = -e_j\,e_i$ for $i \neq j$, have a multiplication rule which can be represented diagrammatically through a pneumonic of the Fano plane, a finite projective plane with seven points and seven lines, as illustrated in Figure 2. Each line contains three points, and each of these triples has a cyclic ordering such that if e_i, e_j and e_k are cyclically ordered in this way then

$$e_i\,e_j \;=\; e_k \;=\; -e_j\,e_i \ . \tag{3.20}$$

IOP Conf. Series: Journal of Physics: Conf. Series **965** (2018) 012041 doi:10.1088/1742-6596/965/1/012041

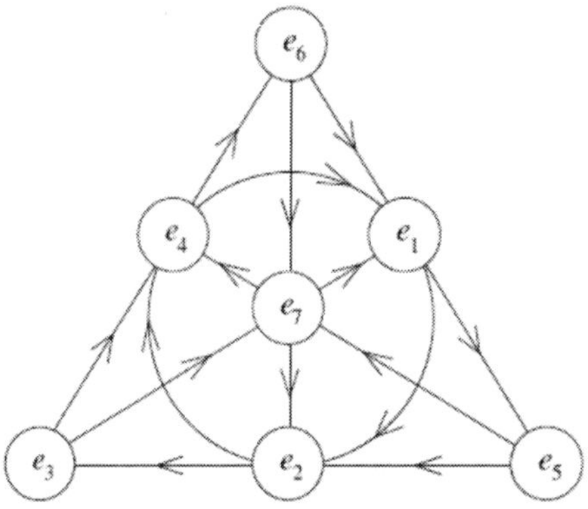

Figure 2. Fano plane pneumonic for multiplication of imaginary unit octonions $e_1, \ldots, e_7$.

This defines an alternative multiplication which makes $\mathbb{O}$ into a finite-dimensional noncommutative Jordan algebra; see e.g. [45] for a pedagogical introduction.

For our purposes it is most convenient to relabel the imaginary units as e_i, f_i, e_7 with $i = 1, 2, 3$ and $f_1, f_2, f_3 = e_4, e_5, e_6$, and to represent the multiplication rule by

$$\begin{aligned} e_i\, e_j &= -\delta_{ij} + \varepsilon_{ijk}\, e_k \quad , \quad & e_7\, e_i &= f_i \ , \\ f_i\, f_j &= -\delta_{ij} - \varepsilon_{ijk}\, e_k \quad , \quad & e_7\, f_i &= e_i \ , \\ e_i\, f_j &= \delta_{ij}\, e_7 - \varepsilon_{ijk}\, f_k \ . \end{aligned} \tag{3.21}$$

The first relation shows that e_i for $i = 1, 2, 3$ generate an associative quaternion subalgebra $\mathbb{H} \subset \mathbb{O}$. The non-vanishing Jacobiators are given by

$$\begin{aligned} [e_i, e_j, f_k] &= 4\,(\varepsilon_{ijk}\, e_7 + \delta_{kj}\, f_i - \delta_{ki}\, f_j) \ , \\ [e_i, f_j, f_k] &= -4\,(\delta_{ij}\, e_k - \delta_{ik}\, e_j) \ , \\ [f_i, f_j, f_k] &= -4\,\varepsilon_{ijk}\, e_7 \ , \\ [e_i, e_j, e_7] &= -4\,\varepsilon_{ijk}\, f_k \ , \\ [e_i, f_j, e_7] &= -4\,\varepsilon_{ijk}\, e_k \ , \\ [f_i, f_j, e_7] &= 4\,\varepsilon_{ijk}\, f_k \ . \end{aligned} \tag{3.22}$$

Following [11], let us now rescale the octonionic units and set

$$x^i = -\frac{\mathrm{i}\,\hbar}{2}\,\lambda\, e_i \quad , \quad p_i = \frac{\mathrm{i}\,\hbar}{2\sqrt{3}}\,\sqrt{\lambda\,\rho}\; f_i \quad , \quad I = \frac{\mathrm{i}\,\hbar}{2\sqrt{3}}\,\sqrt{\lambda^3\,\rho}\; e_7 \tag{3.23}$$

for a parameter $\lambda \in \mathbb{R}$, and take the contraction limit $\lambda \to 0$ of the octonionic commutation relations following from (3.21). The various commutators then contract in the following way:

$$\begin{aligned} [e_i, e_j] &= 2\,\varepsilon_{ijk}\, e_k & \overset{\lambda\to 0}{\Longrightarrow} \quad & [x^i, x^j] = 0 \ , \\ [f_i, f_j] &= -2\,\varepsilon_{ijk}\, e_k & \overset{\lambda\to 0}{\Longrightarrow} \quad & [p_i, p_j] = \frac{\mathrm{i}\,\hbar}{3}\,\rho\,\varepsilon_{ijk}\, x^k \ , \\ [e_i, f_j] &= 2\,(\delta_{ij}\, e_7 - \varepsilon_{ijk}\, f_k) & \overset{\lambda\to 0}{\Longrightarrow} \quad & [x^i, p_j] = \mathrm{i}\,\hbar\,\delta^i{}_j\, I \ , \\ [e_7, e_i] = 2\, f_i \ , \ [e_7, f_i] &= -2\, e_i & \overset{\lambda\to 0}{\Longrightarrow} \quad & [x^i, I] = 0 = [p_i, I] \ . \end{aligned} \tag{3.24}$$

Moreover, only one Jacobiator from (3.22) contracts non-trivially and is given by

$$[f_1, f_2, f_3] = -4\, e_7 \quad \overset{\lambda\to 0}{\Longrightarrow} \quad [p_1, p_2, p_3] = \hbar^2\, \rho\, I\ . \tag{3.25}$$

This demonstrates a surprising relationship between magnetic monopoles and the octonions [11]: The contracted octonion algebra is the magnetic monopole algebra given by (2.4) and (2.9), with central element I. This result was extended by [12] to a complete quantization of algebras of functions by an explicit construction of an exact (though somewhat complicated) star product which quantizes the octonions, and contracts non-trivially to the monopole star product (2.20).

3.3. Nonassociative quantum gravity

By relabelling $I = p_4$, the uncontracted octonion algebra reads as

$$\begin{aligned} [x^i, x^j] &= \mathrm{i}\,\hbar\,\lambda\,\varepsilon^{ijk}\,x^k\ , \\ [p_i, p_j] &= \frac{\mathrm{i}\,\hbar}{3}\,\rho\,\varepsilon_{ijk}\,x^k \quad , \quad [p_4, p_i] = -\frac{\mathrm{i}\,\hbar}{3}\,\lambda\,\rho\,x_i\ , \\ [x_i, p_j] &= \mathrm{i}\,\hbar\,\delta_{ij}\,p_4 - \mathrm{i}\,\hbar\,\lambda\,\varepsilon_{ijk}\,p_k \quad , \quad [x^i, p_4] = -\mathrm{i}\,\hbar\,\lambda^2\,x^i\ . \end{aligned} \tag{3.26}$$

This algebra describes a seven-dimensional nonassociative phase space with an "extra" momentum mode p_4; the non-trivial Jacobiators can be read off by substituting (3.23) into (3.22). Let us now elucidate the physical meaning of this phase space algebra, and in particular the contraction parameter λ, following [13].

Looking at the quaternionic (or $SU(2)$) subalgebra generated by the position coordinates x^i in (3.26), setting $\rho = 0$ reveals a noncommutative but associative deformation of spacetime with $[p_i, p_j] = 0$. The combination $\lambda^2\,\boldsymbol{p}^2 + p_4^2$ is straightforwardly calculated to be a central element of this algebra. Setting it to 1, and then restricting the momenta to the upper hemisphere of this sphere to eliminate p_4, the remaining non-vanishing commutation relations from (3.26) become

$$[x^i, x^j] = \mathrm{i}\,\hbar\,\lambda\,\varepsilon^{ijk}\,x^k \quad , \quad [x_i, p_j] = \mathrm{i}\,\hbar\,\sqrt{1-\lambda^2\,\boldsymbol{p}^2}\;\delta_{ij} - \mathrm{i}\,\hbar\,\lambda\,\varepsilon_{ijk}\,p_k\ . \tag{3.27}$$

We therefore recover in this way the deformed phase space commutation relations (1.1) of three-dimensional quantum gravity provided we identify

$$\lambda = \frac{\ell_{\rm P}}{\hbar}\ . \tag{3.28}$$

This observation suggests a novel nonassociative deformation of three-dimensional quantum gravity, wherein the uncontracted octonion algebra (3.26) appears to be related to monopoles in the spacetime of three-dimensional quantum gravity, with the contraction parameter identified with the Planck length through (3.28). The meaning of the contraction limit $\lambda \to 0$ is then clear: It corresponds to turning off quantum gravitational effects, leaving the nonassociative quantum mechanics of electrons in the background of a constant magnetic charge density.

4. Epilogue: Magnetic monopoles in M-theory

The interpretation of the nonassociative phase space algebra (3.26) as describing the quantum mechanics of electrons propagating simultaneously in magnetic charge and gravitational backgrounds is very tantalising in light of recent suggestions from non-geometric string theory, which anticipate a nonassociative theory of gravity to govern the low-energy effective dynamics of closed strings in certain locally non-geometric backgrounds [4, 9, 10, 32, 48, 49]. However, the appearence of the three-sphere in momentum space, which is a characteristic feature of three-dimensional quantum gravity, is put in here by hand as a restriction of a four-dimensional momentum space. This suggests that a deeper structure is inherent in the physics governed by the phase space relations (3.26), and it remains to find a concrete physical realisation.

One possible such scenario was proposed by [13] in the framework of M-theory, which by definition is a quantum theory of gravity in 11 dimensions. The starting point is to embed magnetic monopoles into IIA string theory as D6-branes, where the corresponding electron probes are D0-branes. The D6-brane lifts to M-theory as a Kaluza-Klein monopole described by the 11-dimensional metric

$$\mathrm{d}s_{11}^2 = \mathrm{d}s_7^2 + H\,\mathrm{d}\boldsymbol{x}\cdot\mathrm{d}\boldsymbol{x} + H^{-1}\left(\mathrm{d}x^4 + \boldsymbol{A}\cdot\mathrm{d}\boldsymbol{x}\right)^2 , \tag{4.1}$$

where $x^4 \in S^1$ is the M-theory direction and the harmonic function $H(\boldsymbol{x})$ is defined by

$$\boldsymbol{\nabla}\times\boldsymbol{A} = \boldsymbol{\nabla}H \quad , \quad \boldsymbol{\nabla}^2 H = \rho \, . \tag{4.2}$$

The corresponding electron probes lifting D0-branes are M-waves along S^1, which are graviton momentum modes with

$$p_4 = \frac{\hbar\, e}{R_{11}} \, . \tag{4.3}$$

Because the probes are waves, they have no local position along the M-theory circle; hence the coordinate x^4 does not appear in the phase space algebra (3.26), which we propose to describe the M-waves in the Kaluza-Klein monopole background.

The relevant parameters in IIA string theory are the string length ℓ_s and the string coupling g_s, while those of M-theory are the 11-dimensional Planck length $\ell_{\rm P}$ and the radius R_{11} of the M-theory circle. They are related through

$$\ell_s^2 = \frac{\ell_{\rm P}^3}{R_{11}} \quad , \quad g_s = \left(\frac{R_{11}}{\ell_{\rm P}}\right)^{3/2} . \tag{4.4}$$

The reduction of M-theory to IIA string theory is defined by the limit $g_s, R_{11} \to 0$ with ℓ_s finite. With the identification (3.28), this implies the contraction limit $\lambda \sim \ell_{\rm P} \sim R_{11}^{1/3} \to 0$ reducing the M-wave algebra (3.26) to the magnetic monopole algebra (2.4) of the D0-branes. In this limit the graviton momentum mode (4.3) is frozen and hence disappears from the phase space, while the Kaluza-Klein quantum number e becomes the electric charge of the D0-branes.

For a single Dirac monopole, wherein ρ is a delta-function distribution, the four-dimensional part of (4.1) is the metric of the hyper-Kähler Taub-NUT space with local coordinates $(\boldsymbol{x}, x^4) \in \mathbb{R}^3 \times S^1$. For the uniform distribution of magnetic charge that we are interested in here, ρ can be written as an integral over infinitely-many densely distributed Dirac monopoles. In this case, there is no local expression for the vector potential $\boldsymbol{A}$ and hence no local expression for the metric (4.1); the resulting solution can thereby be thought of as a 'non-geometric Kaluza-Klein monopole'. Similary to what occured before, the emergence of local non-geometry here arises because the local S^1-fibration over $\mathbb{R}^3$ described by the Taub-NUT space is replaced by an S^1-gerbe over $\mathbb{R}^3$. For further details, see [13].

Acknowledgments

The author thanks Peter Schupp and Vladislav Kupriyanov for helpful discussions, and Cestmir Burdik for the invitation to present this material at ISQS25. This work was supported by the COST Action MP1405 QSPACE, funded by the European Cooperation in Science and Technology (COST), and by the Consolidated Grant ST/L000334/1 from the UK Science and Technology Facilities Council (STFC).

References

[1] 't Hooft G 1996 *Class. Quant. Grav.* **13** 1023–1040 (*Preprint* gr-qc/9601014)
[2] Matschull H J and Welling M 1998 *Class. Quant. Grav.* **15** 2981–3030 (*Preprint* gr-qc/9708054)
[3] Freidel L and Livine E R 2006 *Phys. Rev. Lett.* **96** 221301 (*Preprint* hep-th/0512113)
[4] Blumenhagen R and Plauschinn E 2011 *J. Phys. A* **44** 015401 (*Preprint* 1010.1263)
[5] Lüst D 2010 *J. High Energy Phys.* **12** 084 (*Preprint* 1010.1361)
[6] Blumenhagen R, Deser A, Lüst D, Plauschinn E and Rennecke F 2011 *J. Phys. A* **44** 385401 (*Preprint* 1106.0316)
[7] Mylonas D, Schupp P and Szabo R J 2012 *J. High Energy Phys.* **09** 012 (*Preprint* 1207.0926)
[8] Bakas I and Lüst D 2014 *J. High Energy Phys.* **01** 171 (*Preprint* 1309.3172)
[9] Blumenhagen R, Fuchs M, Hassler F, Lüst D and Sun R 2014 *J. High Energy Phys.* **04** 141 (*Preprint* 1312.0719)
[10] Aschieri P and Szabo R J 2015 *J. Phys. Conf. Ser.* **634** 012004 (*Preprint* 1504.03915)
[11] Günaydin M, Lüst D and Malek E 2016 *J. High Energy Phys.* **11** 027 (*Preprint* 1607.06474)
[12] Kupriyanov V G and Szabo R J 2017 *J. High Energy Phys.* **02** 099 (*Preprint* 1701.02574)
[13] Lüst D, Malek E and Szabo R J 2017 Non-geometric Kaluza-Klein monopoles and magnetic duals of M-theory R-flux backgrounds (*Preprint* 1705.09639)
[14] Zak J 1964 *Phys. Rev.* **134** A1602–A1606
[15] Dirac P A M 1948 *Phys. Rev.* **74** 817–830
[16] Jackiw R 1985 *Phys. Rev. Lett.* **54** 159–162
[17] Wu T T and Yang C N 1976 *Nucl. Phys. B* **107** 365–380
[18] Barnes G E, Schenkel A and Szabo R J 2015 *J. Geom. Phys.* **89** 111–152 (*Preprint* 1409.6331)
[19] Bunk S, Saemann C and Szabo R J 2016 The 2-Hilbert space of a prequantum bundle gerbe (*Preprint* 1608.08455)
[20] Bunk S and Szabo R J 2017 *Lett. Math. Phys.* **107** 1877–1918 (*Preprint* 1612.01878)
[21] Günaydin M and Zumino B 1985 *Old and New Problems in Fundamental Physics: Meeting in Honour of G.C. Wick* pp 43–53
[22] Günaydin M and Minic D 2013 *Fortsch. Phys.* **61** 873–892 (*Preprint* 1304.0410)
[23] Bojowald M, Brahma S, Büyükçam U and Strobl T 2015 *J. High Energy Phys.* **03** 093 (*Preprint* 1411.3710)
[24] Grönewold H 1946 *Physica* **12** 405–460
[25] Moyal J E 1949 *Proc. Cambridge Phil. Soc.* **45** 99–124
[26] Weyl H 1927 *Z. Phys.* **46** 1–46
[27] Wigner E 1932 *Phys. Rev.* **40** 749–759
[28] Zachos C K 2002 *Int. J. Mod. Phys. A* **17** 297–316 (*Preprint* hep-th/0110114)
[29] Mylonas D, Schupp P and Szabo R J 2014 *J. Math. Phys.* **55** 122301 (*Preprint* 1312.1621)
[30] Kupriyanov V G and Vassilevich D V 2015 *J. High Energy Phys.* **09** 103 (*Preprint* 1506.02329)
[31] Mylonas D and Szabo R J 2014 *Fortsch. Phys.* **62** 727–732 (*Preprint* 1404.7304)
[32] Barnes G E, Schenkel A and Szabo R J 2016 *Proc. Science* **CORFU2015** 081 (*Preprint* 1601.07353)
[33] Bojowald M, Brahma S and Büyükçam U 2015 *Phys. Rev. Lett.* **115** 220402 (*Preprint* 1510.07559)
[34] Castelnovo C, Moessner R and Sondhi S L 2008 *Nature* **451** 42–45 (*Preprint* 0710.5515)
[35] Morris D J P *et al.* 2009 *Science* **326** 411–414 (*Preprint* 1011.1174)
[36] Ray M W, Ruokokoski E, Tiurev K, Möttönen and Hall D S 2015 *Science* **348** 544–547
[37] Machida Y, Nakatsuji S, Onoda S, Tayama T and Sakakibara T 2010 *Nature* **463** 210–213
[38] Jordan P 1932 *Nachr. Ges. Wiss. Göttingen* **1932** 569–575
[39] Albert A A 1946 *Trans. Amer. Math. Soc.* **59** 524–555
[40] Schafer R D 1955 *Bull. Amer. Math. Soc.* **61** 469–484
[41] Jordan P, von Neumann J and Wigner E 1934 *Ann. Math.* **35** 29–64
[42] Zelmanov E I 1983 *Sibirsk Mat. J.* **24** 89–104
[43] Günaydin M, Piron C and Ruegg H 1978 *Commun. Math. Phys.* **61** 69–85
[44] McCrimmon K 1978 *Bull. Amer. Math. Soc.* **84** 612–627
[45] Baez J C 2012 *Found. Phys.* **42** 819–855 (*Preprint* 1101.5690)
[46] Held A and Schupp P 2014 unpublished
[47] Bojowald M, Brahma S, Büyükçam U and Strobl T 2017 *J. High Energy Phys.* **04** 028 (*Preprint* 1610.08359)
[48] Barnes G E, Schenkel A and Szabo R J 2016 *J. Geom. Phys.* **106** 234–255 (*Preprint* 1507.02792)
[49] Blumenhagen R and Fuchs M 2016 *J. High Energy Phys.* **07** 019 (*Preprint* 1604.03253)

IOP Conf. Series: Journal of Physics: Conf. Series **965** (2018) 012042 doi:10.1088/1742-6596/965/1/012042

The theory of pseudo-differential operators on the noncommutative n-torus

J Tao

Department of Mathematics, Caltech MC 253-37, Pasadena, CA, 91125, USA

E-mail: jtao@caltech.edu

Abstract. The methods of spectral geometry are useful for investigating the metric aspects of noncommutative geometry and in these contexts require extensive use of pseudo-differential operators. In a foundational paper, Connes showed that, by direct analogy with the theory of pseudo-differential operators on finite-dimensional real vector spaces, one may derive a similar pseudo-differential calculus on noncommutative n-tori, and with the development of this calculus came many results concerning the local differential geometry of noncommutative tori for n=2,4, as shown in the groundbreaking paper in which the Gauss–Bonnet theorem on the noncommutative two-torus is proved and later papers. Certain details of the proofs in the original derivation of the calculus were omitted, such as the evaluation of oscillatory integrals, so we make it the objective of this paper to fill in all the details. After reproving in more detail the formula for the symbol of the adjoint of a pseudo-differential operator and the formula for the symbol of a product of two pseudo-differential operators, we extend these results to finitely generated projective right modules over the noncommutative n-torus. Then we define the corresponding analog of Sobolev spaces and prove equivalents of the Sobolev and Rellich lemmas.

1. Introduction

The methods of spectral geometry are useful for investigating the metric aspects of noncommutative geometry [1–4] and in these contexts require extensive use of pseudo-differential operators. In the foundational paper [5], Connes showed that, by direct analogy with the theory [6–8] of pseudo-differential operators on $\mathbb{R}^n$, one may derive a similar pseudo-differential calculus on noncommutative n tori $\mathbb{T}^n_\theta$. With the development of this calculus came many results concerning the local differential geometry of noncommutative tori for $n = 2, 4$, as shown in the groundbreaking paper [9] in which the Gauss–Bonnet theorem on $\mathbb{T}^2_\theta$ is proved and later papers [10–14]. In these papers, the flat geometry of $\mathbb{T}^n_\theta$ which was studied in [5] is conformally perturbed using a Weyl factor given by a positive invertible smooth element in $C^\infty(\mathbb{T}^n_\theta)$. Connes' pseudo-differential calculus is critically used to apply heat kernel techniques to geometric operators on $\mathbb{T}^n_\theta$ to derive small time heat kernel expansions that encode local geometric information such as scalar curvature. As discovered in [10, 12, 13], a purely noncommutative feature that appears in the computations and in the final formula for the curvature is the modular automorphism of the state implementing the conformal perturbation of the metric.

Certain details of the proofs in the derivation of the calculus in [5] were omitted, such as the evaluation of oscillatory integrals, so we make it the objective of this paper to fill in all the details.

After reproving in more detail the formula for the symbol of the adjoint of a pseudo-differential operator and the formula for the symbol of a product of two pseudo-differential operators, we extend these results to finitely generated projective right modules over the noncommutative n torus. Then we define the corresponding analog of Sobolev spaces for which we prove the Sobolev and Rellich lemmas.

We list these results below.

Theorem 1.1. *Suppose P is a pseudo-differential operator with symbol $\sigma(P) = \rho = \rho(\xi)$ of order M. Then the symbol of the adjoint P^* is of order M and satisfies $\sigma(P^*) \sim \sum_{\ell\in\mathbb{Z}^n_{\geq 0}} \frac{\partial^\ell\delta^\ell[(\rho(\xi))^*]}{\ell_1!\cdots\ell_n!}$.*

Theorem 1.2. *Suppose that P is a pseudo-differential operator with symbol $\sigma(P) = \rho = \rho(\xi)$ of order M_1, and Q is a pseudo-differential operator with symbol $\sigma(Q) = \phi = \phi(\xi)$ of order M_2. Then the symbol of the product QP is of order $M_1 + M_2$ and satisfies $\sigma(QP) \sim \sum_{\ell\in\mathbb{Z}^n_{\geq 0}} \frac{1}{\ell_1!\cdots\ell_n!}\partial^\ell\phi(\xi)\delta^\ell\rho(\xi)$, where $\partial^\ell := \prod_j \partial_j^{\ell_j}$ and $\delta^\ell := \prod_j \delta_j^{\ell_j}$.*

Theorem 1.3. *(a) For a pseudo-differential operator P with $r\times r$ matrix valued symbol $\sigma(P) = \rho = \rho(\xi)$, the symbol of the adjoint P^* satisfies*

$$\sigma(P^*) \sim \sum_{(\ell_1,\ldots,\ell_n)\in(\mathbb{Z}_{\geq 0})^n} \frac{\partial_1^{\ell_1}\cdots\partial_n^{\ell_n}\delta_1^{\ell_1}\cdots\delta_n^{\ell_n}(\rho(\xi))^*}{\ell_1!\cdots\ell_n!}.$$

(b) If Q is a pseudo-differential operator with $r\times r$ matrix valued symbol $\sigma(Q) = \rho' = \rho'(\xi)$, then the product PQ is also a pseudo-differential operator and has symbol

$$\sigma(PQ) \sim \sum_{(\ell_1,\ldots,\ell_n)\in(\mathbb{Z}_{\geq 0})^n} \frac{\partial_1^{\ell_1}\cdots\partial_n^{\ell_n}(\rho(\xi))\delta_1^{\ell_1}\cdots\delta_n^{\ell_n}(\rho'(\xi))}{\ell_1!\cdots\ell_n!}.$$

Theorem 1.4. *For $s > k+1$, $H^s \subseteq (A_\theta^k)^r e$.*

Theorem 1.5. *Let $\{\vec{a}_N\} \in (A_\theta^\infty)^r e$ be a sequence. Suppose that there is a constant C so that $||\vec{a}_N||_s \leq C$ for all N. Let $s > t$. Then there is a subsequence $\{\vec{a}_{N_j}\}$ that converges in H^t.*

2. Preliminaries

Fix some skew symmetric $n\times n$ matrix θ with upper triangular entries in $\mathbb{R}\backslash\mathbb{Q}$ that are linearly independent over $\mathbb{Q}$. Consider the irrational rotation C^*-algebra A_θ with n unitary generators $U_1,\ldots,U_n$ which satisfy $U_kU_j = e^{2\pi i\theta_{j,k}}U_jU_k$ and $U_j^* = U_j^{-1}$. Let $\{\alpha_s\}_{s\in\mathbb{R}^n}$ be a n-parameter group of automorphisms given by $\prod_j U_j^{m_j} \mapsto e^{is\cdot m}\prod_j U_j^{m_j}$. We define the subset A_θ^k of C^k elements of A_θ to be those $a\in A_\theta$ such that the mapping $\mathbb{R}^n \to A_\theta$ given by $s\mapsto\alpha_s(a)$ is C^k, and we define the subalgebra A_θ^∞ of smooth elements of A_θ to be those $a\in A_\theta$ such that the mapping $\mathbb{R}^n\to A_\theta$ given by $s\mapsto\alpha_s(a)$ is smooth. An alternative definition of the subalgebra A_θ^∞ of smooth elements is the elements in A_θ that can be expressed by an expansion of the form $\sum_{m\in\mathbb{Z}^n} a_m\prod_j U_j^{m_j}$, where the sequence $\{a_m\}_{m\in\mathbb{Z}^n}$ is in the Schwartz space $\mathcal{S}(\mathbb{Z}^n)$ in the sense that, for all $\alpha\in\mathbb{Z}^n$, $\sup_{m\in\mathbb{Z}^n}(\prod_j |m_j|^{\alpha_j}|a_m|) < \infty$. Define the trace $\tau : A_\theta\to\mathbb{C}$ by $\tau(\prod_j U_j^{m_j}) = 0$ for m_j not all zero and $\tau(1) = 1$ and define an inner product $\langle\cdot,\cdot\rangle : A_\theta\times A_\theta\to\mathbb{C}$ by $\langle a,b\rangle = \tau(b^*a)$ with induced norm $||\cdot|| : A_\theta\to\mathbb{R}_{\geq 0}$. Let $D_j = -i\partial_j$ and define derivations δ_j by the relations $\delta_j(U_j) = U_j$ and $\delta_j(U_k) = 0$ for $j\neq k$. For convenience, denote $\partial^\ell := \prod_j\partial_j^{\ell_j}$, $\delta^\ell := \prod_j\delta_j^{\ell_j}$, $D^\ell := \prod_j D_j^{\ell_j}$, and $\ell! = \prod_j \ell_j!$ for multi-indices $\ell = (\ell_1,\ldots,\ell_n)\in\mathbb{Z}^n$. We define a map $\psi : \rho\mapsto P_\rho$ assigning a pseudo-differential operator on A_θ^∞ to a symbol $\rho\in C^\infty(\mathbb{R}^n, A_\theta^\infty)$.

Definition 2.1. For $\rho\in C^\infty(\mathbb{R}^n, A_\theta^\infty)$, let P_ρ be the pseudo-differential operator sending arbitrary $a\in A_\theta^\infty$ to $P_\rho(a) := \frac{1}{(2\pi)^n}\int_{\mathbb{R}^n}\int_{\mathbb{R}^n} e^{-is\cdot\xi}\rho(\xi)\alpha_s(a)\,\mathrm{d}s\,\mathrm{d}\xi$.

The integral above does not converge absolutely; it is an oscillatory integral. We define oscillatory integrals below as in [7].

Definition 2.2. Let q be a nondegenerate real quadratic form on $\mathbb{R}^n$, a be a C^∞ complex-valued function defined on $\mathbb{R}^n$ such that the functions $(1+|x|^2)^{-m/2}\partial^\alpha a(x)$ are bounded on $\mathbb{R}^n$ for all $\alpha \in \mathbb{Z}^n_{\geq 0}$, and φ be a Schwartz function, i.e. the functions $x^\alpha \partial^\beta \varphi(x)$ are bounded on $\mathbb{R}^n$ for all pairs $\alpha, \beta \in \mathbb{Z}^n_{\geq 0}$. Suppose further that $\varphi(0) = 1$. Then the limit $\lim_{\epsilon\to 0}\int e^{iq(x)}a(x)\varphi(\epsilon x)\,\mathrm{d}x$ exists, is independent of φ (as long as $\varphi(0) = 1$), and is equal to $\int e^{iq(x)}a(x)\,\mathrm{d}x$ when $a \in L^1$. When $a \notin L^1$, we continue to denote this limit by $\int e^{iq(x)}a(x)\,\mathrm{d}x$, and have an estimate $\left|\int e^{iq(x)}a(x)\,\mathrm{d}x\right| \leq C_{q,m}\max_{|\alpha|\leq m+n+1}\inf\{U \in \mathbb{R} : |(1+|x|^2)^{-m/2}\partial^\alpha a| \leq U \text{ almost everywhere}\}$ where $C_{q,m}$ depends only on the quadratic form q and the order m.

As shown in [7], oscillatory integrals behave essentially like absolutely convergent integrals in that one can still make changes of variables, integrate by parts, differentiate under the integral sign, and interverse integral signs. Given certain conditions on ρ, $P_\rho(a)$ satisfies the conditions of the oscillatory integral, and we can evaluate $P_\rho(a)$.

Definition 2.3. An element $\rho = \rho(\xi) = \rho(\xi_1,\dots,\xi_n)$ of $C^\infty(\mathbb{R}^n, A^\infty_\theta)$ is a symbol of order m if and only if for all non-negative integers $i_1,\dots,i_n,j_1,\dots,j_n$, $||\delta_1^{i_1}\cdots\delta_n^{i_n}(\partial_1^{j_1}\cdots\partial_n^{j_n}\rho)(\xi)|| \leq C_\rho(1+|\xi|)^{m-|j|}$.

Example 2.6(i) of [7] gives a convenient formula for evaluating the oscillatory integrals that appear in our calculation of $P_\rho(a)$.

Proposition 2.4. *Suppose that, for some m, a is a C^∞ complex-valued function defined on $\mathbb{R}^n$ such that the functions $(1+|x|^2)^{-m/2}\partial^\alpha a(x)$ are bounded on $\mathbb{R}^n$ for all $\alpha \in \mathbb{Z}^n_{\geq 0}$. Then $\frac{1}{(2\pi)^n}\int_{\mathbb{R}^n}\int_{\mathbb{R}^n} e^{-iy\cdot\eta}a(y)\,\mathrm{d}y\,\mathrm{d}\eta = \frac{1}{(2\pi)^n}\int_{\mathbb{R}^n}\int_{\mathbb{R}^n} e^{-iy\cdot\eta}a(\eta)\,\mathrm{d}y\,\mathrm{d}\eta = a(0)$.*

We apply Proposition 2.4 to get a basic result.

Lemma 2.5. *Let $a = \sum_{m\in\mathbb{Z}^n} a_m \prod_j U_j^{m_j}$ be an arbitrary element of A^∞_θ and let $\rho \in C^\infty(\mathbb{R}^n, A^\infty_\theta)$ be a symbol of order M. Then $P_\rho(a) = \sum_{m\in\mathbb{Z}^n}\rho(m)a_m\prod_{j=1}^n U_j^{m_j}$.*

Proof. First consider the case $a = \prod_j U_j^{m_j}$. We get

$$\begin{aligned}
P_\rho\left(\prod_{j=1}^n U_j^{m_j}\right) &= \frac{1}{(2\pi)^n}\int_{\mathbb{R}^n}\int_{\mathbb{R}^n} e^{-is\cdot\xi}\rho(\xi)\alpha_s\left(\prod_{j=1}^n U_j^{m_j}\right)\mathrm{d}s\,\mathrm{d}\xi \\
&= \frac{1}{(2\pi)^n}\int_{\mathbb{R}^n}\int_{\mathbb{R}^n} e^{-is\cdot\xi}\rho(\xi)e^{is\cdot m}\prod_{j=1}^n U_j^{m_j}\,\mathrm{d}s\,\mathrm{d}\xi \\
&= \frac{1}{(2\pi)^n}\int_{\mathbb{R}^n}\int_{\mathbb{R}^n} e^{-is\cdot(\xi-m)}\rho(\xi)\,\mathrm{d}s\,\mathrm{d}\xi\prod_{j=1}^n U_j^{m_j} \\
&= \frac{1}{(2\pi)^n}\int_{\mathbb{R}^n}\int_{\mathbb{R}^n} e^{-is\cdot\eta}\rho(\eta+m)\,\mathrm{d}s\,\mathrm{d}\eta\prod_{j=1}^n U_j^{m_j} = \rho(m)\prod_{j=1}^n U_j^{m_j},
\end{aligned}$$

as desired, having substituted $\eta = \xi - m$ and applied the result of Proposition 2.4. Now consider the general case $a = \sum_{m\in\mathbb{Z}^n} a_m\prod_j U_j^{m_j}$. Since α_s is an automorphism on A_θ, we get $P_\rho(a) = \sum_{m\in\mathbb{Z}^n}\rho(m)a_m\prod_{j=1}^n U_j^{m_j}$, and we are done. □

3. Asymptotic formula for the symbol of the adjoint of a pseudo-differential operator

Here we prove the formula for the symbol of the adjoint for the noncommutative n torus, adapting the proof of Lemma 1.2.3 of [6] to the noncommutative n torus.

Theorem 3.1. *Suppose P is a pseudo-differential operator with symbol $\sigma(P) = \rho = \rho(\xi)$ of order M. Then the symbol of the adjoint P^* is of order M and satisfies $\sigma(P^*) \sim \sum_{\ell \in \mathbb{Z}^n_{\geq 0}} \frac{\partial^\ell \delta^\ell[(\rho(\xi))^*]}{\ell_1! \cdots \ell_n!}$.*

Proof. Let $a, b \in A^\infty_\theta$. We have

$$\begin{aligned}\langle P_\rho(a), b\rangle &= \tau\left(b^* \frac{1}{(2\pi)^n}\int_{\mathbb{R}^n}\int_{\mathbb{R}^n} e^{-is\cdot\xi}\rho(\xi)\alpha_s(a)\,\mathrm{d}s\,\mathrm{d}\xi\right)\\ &= \frac{1}{(2\pi)^n}\int_{\mathbb{R}^n}\int_{\mathbb{R}^n} e^{-is\cdot\xi}\tau(b^*\rho(\xi)\alpha_s(a))\,\mathrm{d}s\,\mathrm{d}\xi\\ &= \frac{1}{(2\pi)^n}\int_{\mathbb{R}^n}\int_{\mathbb{R}^n} e^{-is\cdot\xi}\tau(\alpha_{-s}(\rho(\xi)^*b)^*a)\,\mathrm{d}s\,\mathrm{d}\xi\\ &= \tau\left(\left(\frac{1}{(2\pi)^n}\int_{\mathbb{R}^n}\int_{\mathbb{R}^n} e^{+is\cdot\xi}\alpha_{-s}(\rho(\xi)^*b)\,\mathrm{d}s\,\mathrm{d}\xi\right)^* a\right) = \langle a, P^*_\rho(b)\rangle\end{aligned}$$

where

$$\begin{aligned}P^*_\rho(b) &= \frac{1}{(2\pi)^n}\int_{\mathbb{R}^n}\int_{\mathbb{R}^n} e^{+is\cdot\xi}\alpha_{-s}(\rho(\xi)^*b)\,\mathrm{d}s\,\mathrm{d}\xi = \frac{1}{(2\pi)^n}\int_{\mathbb{R}^n}\int_{\mathbb{R}^n} e^{+is\cdot\xi}\alpha_{-s}(\rho(\xi)^*)\alpha_{-s}(b)\,\mathrm{d}s\,\mathrm{d}\xi\\ &= \sum_{m,k}\frac{1}{(2\pi)^n}\int_{\mathbb{R}^n}\int_{\mathbb{R}^n} e^{+is\cdot\xi}e^{+is\cdot m}\rho_m(\xi)^*e^{-is\cdot k}b_k\,\mathrm{d}s\,\mathrm{d}\xi\, U_n^{-m_n}\cdots U_1^{-m_1}U_1^{k_1}\cdots U_n^{k_n}\\ &= \sum_{m,k}\frac{1}{(2\pi)^n}\int_{\mathbb{R}^n}\int_{\mathbb{R}^n} e^{-is\cdot\eta}\rho_m((k-m)-\eta)^*b_k\,\mathrm{d}s\,\mathrm{d}\eta\, U_n^{-m_n}\cdots U_1^{-m_1}U_1^{k_1}\cdots U_n^{k_n}\\ &= \sum_{m,k}\rho_m(k-m)^*b_kU_n^{-m_n}\cdots U_1^{-m_1}U_1^{k_1}\cdots U_n^{k_n}\\ &= \sum_{m,k}(\rho_m(k-m)U_1^{m_1}\cdots U_n^{m_n})^*(b_kU_1^{k_1}\cdots U_n^{k_n}),\end{aligned}$$

so

$$\begin{aligned}\sigma(P^*_\rho)(\xi) &= \left[\sum_m \rho_m(\xi-m)\prod_{j=1}^n U_j^{m_j}\right]^*\\ &= \left[\frac{1}{(2\pi)^n}\int_{\mathbb{R}^n}\int_{\mathbb{R}^n} e^{-ix\cdot y}\sum_m \rho_m(\xi-y)\alpha_x\left(\prod_{j=1}^n U_j^{m_j}\right)\mathrm{d}x\,\mathrm{d}y\right]^*\\ &= \left[\frac{1}{(2\pi)^n}\int_{\mathbb{R}^n}\int_{\mathbb{R}^n} e^{-ix\cdot y}\alpha_x(\rho(\xi-y))\,\mathrm{d}x\,\mathrm{d}y\right]^*.\end{aligned}$$

We have $\rho(\xi - y) = \sum_{|\ell|<N_1}\frac{(-y)^\ell}{\ell!}(\partial^\ell\rho)(\xi) + R_{N_1}(\xi, y)$ and

$$\alpha_x(\rho(\xi-y)) = \sum_{|\ell|<N_1}\sum_m \frac{(-y)^\ell}{\ell!}(\partial^\ell\rho_m)(\xi)e^{ix\cdot m}\left(\prod_{j=1}^n U_j^{m_j}\right) + \alpha_x(R_{N_1}(\xi,y))$$

where $R_{N_1}(\xi,y) = N_1 \sum_{|\ell|=N_1} \frac{(-y)^\ell}{\ell!} \int_0^1 (1-\gamma)^{N_1-1}(\partial^\ell \rho)(\xi - y\gamma)\,\mathrm{d}\gamma$ with corresponding symbol

$$\sigma(P_\rho^*)(\xi) = \left[\sum_{|\ell|<N_1}\sum_m \frac{(-m)^\ell}{\ell!}(\partial^\ell \rho_m)(\xi)\prod_{j=1}^m U_j^{m_j} + T_{N_1}(\xi,y)\right]^* = \sum_{|\ell|<N_1} \frac{\partial^\ell\delta^\ell[(\rho(\xi))^*]}{\ell!} + T_{N_1}(\xi,y)^*$$

where $T_{N_1}(\xi,y) = \frac{1}{(2\pi)^n}\int_{\mathbb{R}^n}\int_{\mathbb{R}^n} e^{-ix\cdot y}\alpha_x(R_{N_1}(\xi,y))\,\mathrm{d}x\,\mathrm{d}y$. It remains to show that this symbol is of order M. Obviously, $\sum_{N\le\ell<N_1} \frac{\partial^\ell\delta^\ell[(\rho(\xi))^*]}{\ell!} \in S^{M-N}$, so we need to show that the remainder is of order $M-N_1$. Note that $\sigma(P_\rho^*)(\xi) - \sum_{|\ell|<N_1} \frac{\partial^\ell\delta^\ell[(\rho(\xi))^*]}{\ell!} = T_{N_1}(\xi,y)^*$. Integrating by parts,

$$\begin{aligned}
&\int_{\mathbb{R}^n}\int_{\mathbb{R}^n} e^{-ix\cdot y}\frac{(-y)^\ell}{\ell!}\alpha_x\left(\int_0^1 (1-\gamma)^{N_1-1}(\partial^\ell\rho)(\xi-y\gamma)\,\mathrm{d}\gamma\right)\mathrm{d}x\,\mathrm{d}y \\
&= \frac{1}{\ell!}\int_{\mathbb{R}^n}\int_{\mathbb{R}^n} e^{-ix\cdot y}(-D_x)^\ell\alpha_x\left(\int_0^1 (1-\gamma)^{N_1-1}(\partial^\ell\rho)(\xi-y\gamma)\,\mathrm{d}\gamma\right)\mathrm{d}x\,\mathrm{d}y \\
&= \frac{1}{\ell!}\int_{\mathbb{R}^n}\int_{\mathbb{R}^n} e^{-ix\cdot y}(-\delta)^\ell\alpha_x\left(\int_0^1 (1-\gamma)^{N_1-1}(\partial^\ell\rho)(\xi-y\gamma)\,\mathrm{d}\gamma\right)\mathrm{d}x\,\mathrm{d}y \\
&= \frac{1}{\ell!}\int_{\mathbb{R}^n}\int_{\mathbb{R}^n} e^{-ix\cdot y}\int_0^1 (1-\gamma)^{N_1-1}(-\delta)^\ell\alpha_x((\partial^\ell\rho)(\xi-y\gamma))\,\mathrm{d}\gamma\,\mathrm{d}x\,\mathrm{d}y \\
&= \frac{1}{\ell!}\int_{\mathbb{R}^n}\int_{\mathbb{R}^n} e^{-ix\cdot y}(-1)^\ell\int_0^1 (1-\gamma)^{N_1-1}\alpha_x((\delta^\ell\partial^\ell\rho)(\xi-y\gamma))\,\mathrm{d}\gamma\,\mathrm{d}x\,\mathrm{d}y
\end{aligned}$$

where, for arbitrary $a = \sum_m a_m \prod_j U_j^{m_j}$,

$$\begin{aligned}
(-D_x)^\ell\alpha_x(a) &= (-D_x)^\ell\alpha_x\left(\sum_m a_m\prod_{j=1}^n U_j^{m_j}\right) = (-D_x)^\ell\sum_m a_m e^{ix\cdot m}\prod_{j=1}^n U_j^{m_j} \\
&= \sum_m a_m(-m)^\ell e^{ix\cdot m}\prod_{j=1}^n U_j^{m_j} = (-\delta)^\ell\alpha_x(a).
\end{aligned}$$

Since $\rho\in S^M$ and $|\ell| = N_1$ we have $\delta^\ell\partial^\ell\rho \in S^{M-N_1}$. We get the boundedness of $\ell^{N_1-M}(\xi)\int_{\mathbb{R}^n}\int_{\mathbb{R}^n} e^{-ix\cdot y}\alpha_x(R_{N_1}(\xi,y))\,\mathrm{d}x\,\mathrm{d}y$ because $\partial^\ell R_{N_1}(y,\xi)$ is the rest of index N_1 in the Taylor expansion of $\partial^\ell(\xi-y\gamma)$ for which one has $\partial^\ell\rho\in S^{M-N_1}$. □

4. Asymptotic formula for the symbol of a product of two pseudo-differential operators

Next we prove the formula for the product or composition of symbols for the noncommutative n torus, adapting the proof of Theorem 7.1 of [8].

Theorem 4.1. *Suppose that P is a pseudo-differential operator with symbol $\sigma(P) = \rho = \rho(\xi)$ of order M_1, and Q is a pseudo-differential operator with symbol $\sigma(Q) = \phi = \phi(\xi)$ of order M_2. Then the symbol of the product QP is of order $M_1 + M_2$ and satisfies $\sigma(QP) \sim \sum_{\ell\in\mathbb{Z}^n_{\ge 0}} \frac{1}{\ell_1!\cdots\ell_n!}\partial^\ell\phi(\xi)\delta^\ell\rho(\xi)$, where $\partial^\ell := \prod_j \partial_j^{\ell_j}$ and $\delta^\ell := \prod_j \delta_j^{\ell_j}$.*

Proof. We want to show that if $\rho:\mathbb{R}^n\to A_\theta^\infty$ is of order M_1 and $\phi:\mathbb{R}^n\to A_\theta^\infty$ is of order M_2, $P_\phi\circ P_\rho = P_\mu$ where μ is of order M_1+M_2 and has asymptotic expansion $\mu\sim\sum_\ell \frac{1}{\ell!}\partial^\ell\phi(\xi)\delta^\ell\rho(\xi)$. Let $\{\varphi_k\}$ be the partition of unity constructed in Theorem 6.1 of [8] and define $\phi_k(\xi) := \phi(\xi)\varphi_k(\xi)$. We have $P_{\phi_k}(a) = \frac{1}{(2\pi)^n}\int_{\mathbb{R}^n}\int_{\mathbb{R}^n} e^{is\cdot\xi}\phi_k(\xi)\alpha_s(a)\,\mathrm{d}s\,\mathrm{d}\xi$. Summing over k from zero to infinity and applying Fubini's Theorem, we get $\sum_{k=0}^\infty P_{\phi_k}(a) = \frac{1}{(2\pi)^n}\int_{\mathbb{R}^n}\int_{\mathbb{R}^n} e^{-is\cdot\xi}\sum_{k=0}^\infty \phi_k(\xi)\alpha_s(a)\,\mathrm{d}s\,\mathrm{d}\xi =$

$\frac{1}{(2\pi)^n}\int_{\mathbb{R}^n}\int_{\mathbb{R}^n} e^{-is\cdot\xi}\phi(\xi)\alpha_s(a)\,\mathrm{d}s\,\mathrm{d}\xi$, so $P_\phi(a) = \sum_{k=0}^{\infty} P_{\phi_k}(a)$ and the convergence of the series is absolute and uniform for all $a \in A_\theta^\infty$. We want to compute the symbol of $P_\phi \circ P_\rho$, but issues with convergence of integrals make it so we need to compute the symbol of $P_{\phi_k} \circ P_\rho$. Let $a \in A_\theta^\infty$ be arbitrary. Applying $P_{\phi_k} \circ P_\rho$ we get

$$\begin{aligned}
P_{\phi_k}(P_\rho(a)) &= \frac{1}{(2\pi)^n}\int_{\mathbb{R}^n}\int_{\mathbb{R}^n} e^{-is\cdot\xi}\phi_k(\xi)\alpha_s(P_\rho(a))\,\mathrm{d}s\,\mathrm{d}\xi \\
&= \frac{1}{(2\pi)^n}\int_{\mathbb{R}^n}\int_{\mathbb{R}^n} e^{-is\cdot\xi}\phi_k(\xi)\alpha_s\Big(\frac{1}{(2\pi)^n}\int_{\mathbb{R}^n}\int_{\mathbb{R}^n} e^{-it\cdot\eta}\rho(\eta)\alpha_t(a)\,\mathrm{d}t\,\mathrm{d}\eta\Big)\,\mathrm{d}s\,\mathrm{d}\xi \\
&= \frac{1}{(2\pi)^{2n}}\int_{\mathbb{R}^n}\int_{\mathbb{R}^n}\Big\{\int_{\mathbb{R}^n}\int_{\mathbb{R}^n} e^{-is\cdot\xi-it\cdot\eta}\phi_k(\xi)\alpha_s(\rho(\eta))\alpha_{s+t}(a)\,\mathrm{d}t\,\mathrm{d}\eta\Big\}\,\mathrm{d}s\,\mathrm{d}\xi \\
&= \frac{1}{(2\pi)^{2n}}\int_{\mathbb{R}^n}\int_{\mathbb{R}^n}\Big\{\int_{\mathbb{R}^n}\int_{\mathbb{R}^n} e^{-ix\cdot\xi-i(y-x)\cdot\eta}\phi_k(\xi)\alpha_x(\rho(\eta))\alpha_y(a)\,\mathrm{d}y\,\mathrm{d}\eta\Big\}\,\mathrm{d}x\,\mathrm{d}\xi \\
&= \frac{1}{(2\pi)^{2n}}\int_{\mathbb{R}^n}\int_{\mathbb{R}^n}\Big\{\int_{\mathbb{R}^n}\int_{\mathbb{R}^n} e^{-ix\cdot(\xi-\eta)-iy\cdot\eta}\phi_k(\xi)\alpha_x(\rho(\eta))\alpha_y(a)\,\mathrm{d}y\,\mathrm{d}\eta\Big\}\,\mathrm{d}x\,\mathrm{d}\xi \\
&= \frac{1}{(2\pi)^{2n}}\int_{\mathbb{R}^n}\int_{\mathbb{R}^n}\Big\{\int_{\mathbb{R}^n}\int_{\mathbb{R}^n} e^{-ix\cdot\sigma-iy\cdot\tau}\phi_k(\sigma+\tau)\alpha_x(\rho(\tau))\alpha_y(a)\,\mathrm{d}y\,\mathrm{d}\tau\Big\}\,\mathrm{d}x\,\mathrm{d}\sigma \\
&= \frac{1}{(2\pi)^n}\int_{\mathbb{R}^n}\int_{\mathbb{R}^n} e^{-iy\cdot\tau}\Big\{\frac{1}{(2\pi)^n}\int_{\mathbb{R}^n}\int_{\mathbb{R}^n} e^{-ix\cdot\sigma}\phi_k(\sigma+\tau)\alpha_x(\rho(\tau))\,\mathrm{d}x\,\mathrm{d}\sigma\Big\}\,\alpha_y(a)\,\mathrm{d}y\,\mathrm{d}\tau \\
&= \frac{1}{(2\pi)^n}\int_{\mathbb{R}^n}\int_{\mathbb{R}^n} e^{-iy\cdot\tau}\mu_k(\tau)\alpha_y(a)\,\mathrm{d}y\,\mathrm{d}\tau
\end{aligned}$$

where $\mu_k(\tau) = \frac{1}{(2\pi)^n}\int_{\mathbb{R}^n}\int_{\mathbb{R}^n} e^{-ix\cdot\sigma}\phi_k(\sigma+\tau)\alpha_x(\rho(\tau)))\,\mathrm{d}x\,\mathrm{d}\sigma$, having done the changes of variables $(x, y) = (s, s+t)$ and $(\sigma, \tau) = (\xi-\eta, \eta)$ and applied Proposition 2.4. This suggests that $P_\phi(P_\rho(a)) = \frac{1}{(2\pi)^n}\int_{\mathbb{R}^n}\int_{\mathbb{R}^n} e^{-iy\cdot\tau}\mu(\tau)\alpha_y(a)\,\mathrm{d}y\,\mathrm{d}\tau$ where $\mu(\tau) = \sum_{k=0}^{\infty}\mu_k(\tau)$. We need to show that μ is a symbol in $S^{M_1+M_2}$ and has our desired asymptotic expression. Define μ_k by $\mu_k(\xi) = \frac{1}{(2\pi)^2}\int_{\mathbb{R}^n}\int_{\mathbb{R}^n} e^{-ix\cdot y}\phi_k(\xi+y)\alpha_x(\rho(\xi))\,\mathrm{d}x\,\mathrm{d}y$ for all $\xi \in \mathbb{R}^n$. By Taylor's formula with integral remainder given in Theorem 6.3 of [8], we get $\phi_k(\xi+y) = \sum_{|\ell|<N_1}\frac{y^\ell}{\ell!}(\partial^\ell\phi_k)(\xi) + R_{N_1}(y,\xi)$ where $R_{N_1}(y,\xi) = N_1\sum_{|\ell|=N_1}\frac{y^\ell}{\ell!}\int_0^1(1-\gamma)^{N_1-1}(\partial^\ell\phi_k)(\xi+\gamma y)\,\mathrm{d}\gamma$ for all $y, \xi \in \mathbb{R}^2$. Substituting back into our expression for $\mu_k(\xi)$ we get $\mu_k(\xi) = \frac{1}{(2\pi)^n}\int_{\mathbb{R}^n}\int_{\mathbb{R}^n} e^{-ix\cdot y}\sum_{|\ell|<N_1}\frac{y^\ell}{\ell!}(\partial^\ell\phi_k)(\xi)\alpha_x(\rho(\xi))\,\mathrm{d}x\,\mathrm{d}y + T_{N_1}^{(k)}(\xi)$ where $T_{N_1}^{(k)}(\xi) = \frac{1}{(2\pi)^n}\int_{\mathbb{R}^n}\int_{\mathbb{R}^n} e^{-ix\cdot y}R_{N_1}(y,\xi)\alpha_x(\rho(\xi))\,\mathrm{d}x\,\mathrm{d}y$. Expressing $\rho(\xi)$ as $\rho(\xi) = \sum_m \rho_m(\xi)\prod_{j=1}^n U_j^{m_j}$, we see that $\alpha_x(\rho(\xi)) = \sum_m \rho_m(\xi)e^{ix\cdot m}\prod_{j=1}^n U_j^{m_j}$ so

$$\begin{aligned}
\mu_k(\xi) - T_{N_1}^{(k)}(\xi) &= \sum_{|\ell|<N_1}\sum_m \frac{1}{(2\pi)^n}\int_{\mathbb{R}^n}\int_{\mathbb{R}^n} e^{-ix\cdot y}\frac{y^\ell}{\ell!}(\partial^\ell\phi_k)(\xi)\rho_m(\xi)e^{ix\cdot m}\prod_{j=1}^n U_j^{m_j}\,\mathrm{d}x\,\mathrm{d}y \\
&= \sum_{|\ell|<N_1}\frac{1}{\ell!}(\partial^\ell\phi_k)(\xi)\sum_m \rho_m(\xi)\prod_{j=1}^n U_j^{m_j}\frac{1}{(2\pi)^n}\int_{\mathbb{R}^n}\int_{\mathbb{R}^n} e^{-ix\cdot(y-m)}y^\ell\,\mathrm{d}x\,\mathrm{d}y \\
&= \sum_{|\ell|<N_1}\frac{1}{\ell!}(\partial^\ell\phi_k)(\xi)\sum_m \rho_m(\xi)\prod_{j=1}^n U_j^{m_j}m^\ell = \sum_{|\ell|<N_1}\frac{1}{\ell!}(\partial^\ell\phi_k)(\xi)(\delta^\ell\rho)(\xi).
\end{aligned}$$

Let $\mu(\xi) = \sum_{k=0}^{\infty}\mu_k(\xi)$. It remains to show that μ is of order $M_1 + M_2$. Obviously, $\sum_{N\le|\ell|<N_1}\frac{y^\ell}{\ell!}(\partial^\ell\phi)(\xi)(\delta^\ell\rho)(\xi) \in S^{M_1+M_2-N}$, so we just need to show that the remainder is of order $M_1 + M_2 - N_1$. Note that $\mu(\xi) - \sum_{|\ell|<N_1}\frac{y^\ell}{\ell!}(\partial^\ell\phi)(\xi)(\delta^\ell\rho)(\xi) = \sum_{k=0}^{\infty}T_{N_1}^{(k)}(\xi)$ where $T_{N_1}^{(k)}(\xi) = \frac{1}{(2\pi)^n}\int_{\mathbb{R}^n}\int_{\mathbb{R}^n} e^{-ix\cdot y}R_{N_1}(y,\xi)\alpha_x(\rho(\xi))\,\mathrm{d}x\,\mathrm{d}y$ and $R_{N_1}(y,\xi) = N_1\sum_{|\ell|=N_1}\frac{y^\ell}{\ell!}\int_0^1(1-$

$\gamma)^{N_1-1}(\partial^\ell\phi_k)(\xi+\gamma y)\,\mathrm{d}\gamma$. Integrating by parts,

$$\begin{aligned}
&\int_{\mathbb{R}^n}\int_{\mathbb{R}^n} e^{-ix\cdot y}\frac{y^\ell}{\ell!}\int_0^1(1-\gamma)^{N_1-1}(\partial^\ell\phi_k)(\xi+\gamma y)\,\mathrm{d}\gamma\alpha_x(\rho(\xi))\,\mathrm{d}x\,\mathrm{d}y\\
&=\frac{1}{\ell!}\int_{\mathbb{R}^n}\int_{\mathbb{R}^n} e^{-ix\cdot y}y^\ell\int_0^1(1-\gamma)^{N_1-1}(\partial^\ell(\phi_k)(\xi+\gamma y)\,\mathrm{d}\gamma\alpha_x(\rho(\xi))\,\mathrm{d}x\,\mathrm{d}y\\
&=\frac{1}{\ell!}\int_{\mathbb{R}^n}\int_{\mathbb{R}^n} e^{-ix\cdot y}\int_0^1(1-\gamma)^{N_1-1}(\partial^\ell(\phi_k)(\xi+\gamma y)\,\mathrm{d}\gamma D_x^\ell\alpha_x(\rho(\xi))\,\mathrm{d}x\,\mathrm{d}y\\
&=\frac{1}{\ell!}\int_{\mathbb{R}^n}\int_{\mathbb{R}^n} e^{-ix\cdot y}\int_0^1(1-\gamma)^{N_1-1}(\partial^\ell(\phi_k)(\xi+\gamma y)\,\mathrm{d}\gamma\delta^\ell\alpha_x(\rho(\xi))\,\mathrm{d}x\,\mathrm{d}y\\
&=\frac{1}{\ell!}\int_{\mathbb{R}^n}\int_{\mathbb{R}^n} e^{-ix\cdot y}\int_0^1(1-\gamma)^{N_1-1}(\partial^\ell(\phi_k)(\xi+\gamma y)\,\mathrm{d}\gamma\alpha_x((\delta^\ell\rho)(\xi))\,\mathrm{d}x\,\mathrm{d}y
\end{aligned}$$

where for arbitrary $a=\sum_m a_m\prod_j U_j^{m_j}$ we have

$$\begin{aligned}
(D_x)^\ell\alpha_x(a)&=(D_x)^\ell\alpha_x\left(\sum_m a_m\prod_{j=1}^n U_j^{m_j}\right)=(D_x)^\ell\sum_m a_m e^{ix\cdot m}\prod_{j=1}^n U_j^{m_j}\\
&=\sum_m a_m m^\ell e^{ix\cdot m}\prod_{j=1}^n U_j^{m_j}=\delta^\ell\alpha_x(a).
\end{aligned}$$

Since $|\ell|=N_1$, we have $\partial^\ell(\phi_k)\in S^{M_2-N_1}$ and $\delta^\ell\rho\in S^{M_1}$. We get the boundedness of $\mu^{N_1-M_1-M_2}(\xi)\int_{\mathbb{R}^n}\int_{\mathbb{R}^n}e^{-ix\cdot y}R_{N_1}(y,\xi)\alpha_x(\rho(\xi))\,\mathrm{d}x\,\mathrm{d}y$ since $\delta^\ell\rho\in S^{M_1}$ and $\partial^\ell R_{N_1}(y,\xi)$ is the rest of index N_1 in Taylor's expansion of $\partial^\ell\phi_k(\xi+\gamma y)$ for which on has $\partial^\ell\phi_k\in S^{M_2-N_1}$. □

5. The pseudo-differential calculus on finitely generated projective modules over the noncommutative n torus

We can generalize these results to arbitrary finitely generated projective right modules over the noncommutative n torus following p. 553 of [2], which considers finitely generated projective modules over an arbitrary unital $*$-algebra. Let E be a finitely generated projective right A_θ^∞-module. Since E is a finitely generated projective right A_θ^∞-module, we can write E as a direct summand $E=(A_\theta^\infty)^r e$ of a free module $(A_\theta^\infty)^r$ with direct complement $F=(A_\theta^\infty)^r(\mathrm{id}-e)$, where the idempotent $e\in M_r(A_\theta^\infty)$ is self-adjoint. Consider an $r\times r$ matrix valued symbol $\rho=(\rho_{j,k})$ where $\rho_{j,k}:\mathbb{R}^n\to A_\theta^\infty$ are scalar symbols and $\rho_{j,k}\in S^d$. Define the operator $P_\rho:E\to E$ as follows: $P_\rho(\vec{a}):=(2\pi)^{-n}\int_{\mathbb{R}^n}\int_{\mathbb{R}^n}e^{-is\cdot\xi}\rho(\xi)\alpha_s(\vec{a})\,\mathrm{d}s\,\mathrm{d}\xi$. Define the inner product $\langle\vec{a},\vec{b}\rangle:E\times E\to\mathbb{C}$ sending $(\vec{a},\vec{b})\mapsto\tau(\vec{b}^*\vec{a})$. Since $P_\rho(\vec{a})_j=(2\pi)^{-n}\int_{\mathbb{R}^n}\int_{\mathbb{R}^n}e^{-is\cdot\xi}\sum_{k=1}^r\rho_{j,k}(\xi)\alpha_s(a_k)\,\mathrm{d}s\,\mathrm{d}\xi$, Lemma 2.5 generalizes to E as follows after applying it to each component: $P_\rho(\vec{a})=\sum_m\rho(m)\vec{a}_m\prod_{j=1}^n U_j^{m_j}$. Theorems 3.1 and 4.1 generalize as follows.

Theorem 5.1. *(a) For a pseudo-differential operator P with $r\times r$ matrix valued symbol $\sigma(P)=\rho=\rho(\xi)$, the symbol of the adjoint P^* satisfies*

$$\sigma(P^*)\sim\sum_{(\ell_1,\ldots,\ell_n)\in(\mathbb{Z}_{\geq0})^n}\frac{\partial_1^{\ell_1}\cdots\partial_n^{\ell_n}\delta_1^{\ell_1}\cdots\delta_n^{\ell_n}(\rho(\xi))^*}{\ell_1!\cdots\ell_n!}.$$

(b) If Q is a pseudo-differential operator with $r\times r$ matrix valued symbol $\sigma(Q)=\rho'=\rho'(\xi)$, then the product PQ is also a pseudo-differential operator and has symbol

$$\sigma(PQ)\sim\sum_{(\ell_1,\ldots,\ell_n)\in(\mathbb{Z}_{\geq0})^n}\frac{\partial_1^{\ell_1}\cdots\partial_n^{\ell_n}(\rho(\xi))\delta_1^{\ell_1}\cdots\delta_n^{\ell_n}(\rho'(\xi))}{\ell_1!\cdots\ell_n!}.$$

Proof. First let's prove part (a). Let ρ be an $r \times r$ matrix valued symbol of order M and $\vec{a}, \vec{b} \in E$. We have

$$\begin{aligned}\langle P_\rho(\vec{a}), \vec{b}\rangle &= \tau\left(\vec{b}^* \frac{1}{(2\pi)^n}\int_{\mathbb{R}^n}\int_{\mathbb{R}^n} e^{-is\cdot\xi}\rho(\xi)\alpha_s(\vec{a})\,\mathrm{d}s\,\mathrm{d}\xi\right)\\ &= \tau\left(\left(\frac{1}{(2\pi)^n}\int_{\mathbb{R}^n}\int_{\mathbb{R}^n} e^{+is\cdot\xi}\alpha_{-s}(\rho(\xi)^*\vec{b})\,\mathrm{d}s\,\mathrm{d}\xi\right)^* \vec{a}\right) = \langle \vec{a}, P_\rho^*(\vec{b})\rangle\end{aligned}$$

where

$$\begin{aligned}P_\rho^*(\vec{b})_j &= \frac{1}{(2\pi)^n}\int_{\mathbb{R}^n}\int_{\mathbb{R}^n} e^{+is\cdot\xi}\sum_{k=1}^{r}\alpha_{-s}(\rho(\xi)^*_{j,k}b_k)\,\mathrm{d}s\,\mathrm{d}\xi\\ &= \frac{1}{(2\pi)^n}\int_{\mathbb{R}^n}\int_{\mathbb{R}^n} e^{+is\cdot\xi}\sum_{k=1}^{r}\alpha_{-s}(\rho(\xi)^*_{j,k})\alpha_{-s}(b_k)\,\mathrm{d}s\,\mathrm{d}\xi\\ &= \sum_{m,p}\left[\rho_m(p-m)\prod_{h=1}^{n}U_h^{m_h}\right]^*\left(b_p\prod_{h=1}^{n}U_h^{p_h}\right)\end{aligned}$$

so

$$\begin{aligned}\sigma(P_\rho^*)(\xi) &= \left[\sum_m \rho_m(\xi-m)\prod_{h=1}^{n}U_h^{m_h}\right]^*\\ &= \left[\frac{1}{(2\pi)^n}\int_{\mathbb{R}^n}\int_{\mathbb{R}^n} e^{-ix\cdot y}\sum_m \rho_m(\xi-y)\alpha_x\left(\prod_{h=1}^{n}U_h^{m_h}\right)\mathrm{d}x\,\mathrm{d}y\right]^*\\ &= \left[\frac{1}{(2\pi)^n}\int_{\mathbb{R}^n}\int_{\mathbb{R}^n} e^{-ix\cdot y}\alpha_x(\rho(\xi-y))\,\mathrm{d}x\,\mathrm{d}y\right]^*.\end{aligned}$$

The rest of the proof reduces to the $r = 1$ case, applying it to each entry in $\rho = (\rho_{j,k})$.

We proceed to part (b). Let ρ be an $r \times r$ matrix valued symbol of order m_1 and ϕ be an $r \times r$ matrix valued symbol of order m_2. Let $\{\varphi_k\}$ be a partition of unity and define $\phi_k(\xi) := \phi(\xi)\varphi_k(\xi)$. Let $\vec{a} \in E$. We have

$$\begin{aligned}P_{\phi_k}(P_\rho(\vec{a})) &= \frac{1}{(2\pi)^n}\int_{\mathbb{R}^n}\int_{\mathbb{R}^n} e^{-is\cdot\xi}\phi_k(\xi)\alpha_s(P_\rho(\vec{a}))\,\mathrm{d}s\,\mathrm{d}\xi\\ &= \frac{1}{(2\pi)^n}\int_{\mathbb{R}^n}\int_{\mathbb{R}^n} e^{-is\cdot\xi}\phi_k(\xi)\alpha_s\left(\frac{1}{(2\pi)^n}\int_{\mathbb{R}^n}\int_{\mathbb{R}^n} e^{-it\cdot\eta}\rho(\eta)\alpha_t(\vec{a})\,\mathrm{d}t\,\mathrm{d}\eta\right)\mathrm{d}s\,\mathrm{d}\xi\\ &= \frac{1}{(2\pi)^{2n}}\int_{\mathbb{R}^n}\int_{\mathbb{R}^n}\left\{\int_{\mathbb{R}^n}\int_{\mathbb{R}^n} e^{-is\cdot\xi-it\cdot\eta}\phi_k(\xi)\alpha_s(\rho(\eta))\alpha_{s+t}(\vec{a})\,\mathrm{d}t\,\mathrm{d}\eta\right\}\mathrm{d}s\,\mathrm{d}\xi\\ &= \frac{1}{(2\pi)^{2n}}\int_{\mathbb{R}^n}\int_{\mathbb{R}^n}\left\{\int_{\mathbb{R}^n}\int_{\mathbb{R}^n} e^{-ix\cdot\xi-i(y-x)\cdot\eta}\phi_k(\xi)\alpha_x(\rho(\eta))\alpha_y(\vec{a})\,\mathrm{d}y\,\mathrm{d}\eta\right\}\mathrm{d}x\,\mathrm{d}\xi\\ &= \frac{1}{(2\pi)^{2n}}\int_{\mathbb{R}^n}\int_{\mathbb{R}^n}\left\{\int_{\mathbb{R}^n}\int_{\mathbb{R}^n} e^{-ix\cdot(\xi-\eta)-iy\cdot\eta}\phi_k(\xi)\alpha_x(\rho(\eta))\alpha_y(\vec{a})\,\mathrm{d}y\,\mathrm{d}\eta\right\}\mathrm{d}x\,\mathrm{d}\xi\\ &= \frac{1}{(2\pi)^{2n}}\int_{\mathbb{R}^n}\int_{\mathbb{R}^n}\left\{\int_{\mathbb{R}^n}\int_{\mathbb{R}^n} e^{-ix\cdot\sigma-iy\cdot\eta}\phi_k(\sigma+\tau)\alpha_x(\rho(\tau))\alpha_y(\vec{a})\,\mathrm{d}y\,\mathrm{d}\tau\right\}\mathrm{d}x\,\mathrm{d}\sigma\\ &= \frac{1}{(2\pi)^{2n}}\int_{\mathbb{R}^n}\int_{\mathbb{R}^n} e^{-iy\cdot\tau}\left\{\int_{\mathbb{R}^n}\int_{\mathbb{R}^n} e^{-ix\cdot\sigma}\phi_k(\sigma+\tau)\alpha_x(\rho(\tau))\,\mathrm{d}x\,\mathrm{d}\sigma\right\}\alpha_y(\vec{a})\,\mathrm{d}y\,\mathrm{d}\tau\\ &= \frac{1}{(2\pi)^n}\int_{\mathbb{R}^n}\int_{\mathbb{R}^n} e^{-iy\cdot\tau}\lambda_k(\tau)\alpha_y(\vec{a})\,\mathrm{d}y\,\mathrm{d}\tau\end{aligned}$$

where $\lambda_k(\tau) = \frac{1}{(2\pi)^n}\int_{\mathbb{R}^n}\int_{\mathbb{R}^n} e^{-ix\cdot\sigma}\phi_k(\sigma+\tau)\alpha_x(\rho(\tau))\,\mathrm{d}x\,\mathrm{d}\sigma$ so

$$P_\phi(P_\rho(\vec{a})) = \frac{1}{(2\pi)^n}\int_{\mathbb{R}^n}\int_{\mathbb{R}^n} e^{-iy\cdot\tau}\lambda(\tau)\alpha_y(\vec{a})\,\mathrm{d}y\,\mathrm{d}\tau$$

where $\lambda(\tau) = \sum_{k=0}^{\infty}\lambda_k(\tau)$. Let $\lambda_k(\xi) := \frac{1}{(2\pi)^n}\int_{\mathbb{R}^n}\int_{\mathbb{R}^n} e^{-ix\cdot y}\phi_k(\xi+y)\alpha_x(\rho(\xi))\,\mathrm{d}x\,\mathrm{d}y$. Since

$$\begin{aligned}\lambda_k(\xi)_{\alpha,\gamma} &= \frac{1}{(2\pi)^n}\int_{\mathbb{R}^n}\int_{\mathbb{R}^n} e^{-ix\cdot y}\sum_{\beta=1}^{r}\phi_k(\xi+y)_{\alpha,\beta}\alpha_x(\rho(\xi)_{\beta,\gamma})\,\mathrm{d}x\,\mathrm{d}y\\ &= \sum_{\beta=1}^{r}\frac{1}{(2\pi)^n}\int_{\mathbb{R}^n}\int_{\mathbb{R}^n} e^{-ix\cdot y}\phi_k(\xi+y)_{\alpha,\beta}\alpha_x(\rho(\xi)_{\beta,\gamma})\,\mathrm{d}x\,\mathrm{d}y,\end{aligned}$$

the rest of the proof reduces to the $r = 1$ case, applying it to each summand in the above sum. □

6. Sobolev spaces on finitely generated projective modules over the noncommutative n torus

Let $\lambda(\xi) = (1+\xi_1^2+\cdots+\xi_n^2)^{1/2}\mathrm{id}_E$. Consider the following inner product on E.

Definition 6.1. Define the Sobolev inner product $\langle\cdot,\cdot\rangle_s : E\times E\to\mathbb{C}$ by $\langle\vec{a},\vec{b}\rangle_s := \langle P_{\lambda^s}(\vec{a}), P_{\lambda^s}(\vec{b})\rangle = \sum_{j,m}(1+|m_1|^2+\cdots+|m_n|^2)^s\overline{b_{j,m}}a_{j,m}$.

Note that for $s = 0$ this agrees with $\langle\cdot,\cdot\rangle$. This inner product induces the following norm.

Definition 6.2. Define the Sobolev norm $||\cdot||_s : E\to\mathbb{R}_{\geq 0}$ by $||\vec{a}||_s^2 := \langle P_{\lambda^s}(\vec{a}), P_{\lambda^s}(\vec{a})\rangle = \sum_{j,m}(1+|m_1|^2+\cdots+|m_n|^2)^s|a_{j,m}|^2$.

Using this norm, we can define the analog of Sobolev spaces on E.

Definition 6.3. Define the Sobolev space H^s to be the completion of E with respect to $||\cdot||_s$.

We can prove that a pseudo-differential operator of order $d\in\mathbb{R}$ continuously maps H^s into H^{s-d}. However we must first prove the case where $s = d$.

Theorem 6.4. *Suppose ρ is a matrix valued symbol of order d. Then, for any $\vec{a}\in E$, $||P_\rho(\vec{a})||_0 \leq C||\vec{a}||_d$ for some constant $C > 0$ and P_ρ defines a bounded operator $P_\rho : H^d\to H^0$.*

Proof. Let F be an orthogonal eigenbasis of e normalized with respect to $\langle\cdot,\cdot\rangle$, and let $F_1 := \{f_h : 1\leq h\leq r_1\}$ be the subset of eigenvectors with eigenvalue 1. Note that $\{\prod_g U_g^{m_g}f_j : m\in\mathbb{Z}^n, 1\leq j\leq r_1\}$ is an orthogonal basis of E considered as a $\mathbb{C}$-vector space, with respect to $\langle\cdot,\cdot\rangle_s$. We have $||\prod_g U_g^{m_g}f_j||_0^2 = 1$, $||\rho_m(\xi)_{h,j}\prod_g U_g^{m_g}f_j||_0^2 = |\rho_m(\xi)_{h,j}|^2$, and $||\rho(\xi)_{h,j}f_j||_0^2 = \sum_m|\rho_m(\xi)_{h,j}|^2$. Since $\rho_{h,j}$ is of order d, we have $||\rho(\xi)_{h,j}||_0 \leq C_\rho(1+|\xi|)^d$, and since $(1-|\xi|)^2\geq 0$ gives us $(1+|\xi|)^2\leq 2(1+|\xi|^2)$, we have $||\rho(\xi)_{h,j}||_0^2 \leq C_\rho^2(1+|\xi|)^{2d}\leq C_\rho^2 2^d(1+|\xi|^2)^d$. Let $k_\rho := C_\rho^2 2^d$. Then we have $\sum_m|\rho_m(\xi)_{h,j}|^2\leq k_\rho(1+|\xi|^2)^d$. Let $e_{s,m} := (1+|m_1|^2+\cdots+|m_n|^2)^{-s/2}\prod_g U_g^{m_g}$ and $E_s := \{e_{s,m}\mid m\in\mathbb{Z}\}$. By definition we have E_sF_1 orthonormal with respect to $\langle\cdot,\cdot\rangle_s$. It suffices to prove this theorem for the case $\vec{a} = e_{d,m}f_j$ by the orthonormality of E_dF_1 since $||P_\rho(\vec{a})||_0^2 = \sum_{j,m}|a_{j,m}|^2||P_\rho(e_{d,m}f_j)||_0^2$ and $||\vec{a}||_d^2 = \sum_{j,m}|a_{j,m}|^2||e_{d,m}f_j||_d^2$. Since $||e_{d,m}f_j||_d^2 = 1$, it suffices to show that $||P_\rho(e_{d,m}f_j)||_0^2\leq K$ for some constant $K > 0$. We have

$$\begin{aligned}||P_\rho(e_{d,m}f_j)||_0^2 = ||\rho(m)e_{d,m}f_j||_0^2 &= \left|\left|\rho(m)(1+|m_1|^2+\cdots+|m_n|^2)^{-d/2}\prod_g U_g^{m_g}f_j\right|\right|_0^2\\ &= \left|\left|\sum_k\rho_k(m)\prod_g U_g^{k_g}(1+|m_1|^2+\cdots+|m_n|^2)^{-d/2}\prod_g U_g^{m_g}f_j\right|\right|_0^2\end{aligned}$$

$$
\begin{aligned}
&= (1+|m_1|^2+\cdots+|m_n|^2)^{-d}\left\|\sum_k \rho_k(m)\prod_g U_g^{k_g}\prod_g U_g^{m_g} f_j\right\|_0^2 \\
&= (1+|m_1|^2+\cdots+|m_n|^2)^{-d}\left\|\sum_k \rho_k(m)w(m,k)\prod_g U_g^{k_g+m_g} f_j\right\|_0^2 \\
&= (1+|m_1|^2+\cdots+|m_n|^2)^{-d}\left\|\sum_k \rho_{k-m}(m)w(m,k-m)\prod_g U_g^{k_g} f_j\right\|_0^2 \\
&= (1+|m_1|^2+\cdots+|m_n|^2)^{-d}\left\|\sum_{h,k} \rho_{k-m}(m)_{h,j}w(m,k-m)\prod_g U_g^{k_g} f_j\right\|_0^2 \\
&= (1+|m_1|^2+\cdots+|m_n|^2)^{-d}\sum_{h,k} |\rho_{k-m}(m)_{h,j}|^2 \\
&= (1+|m_1|^2+\cdots+|m_n|^2)^{-d}\sum_{h,k} |\rho_{k}(m)_{h,j}|^2 \\
&\le (1+|m_1|^2+\cdots+|m_n|^2)^{-d} r_1 k_\rho (1+|m_1|^2+\cdots+|m_n|^2)^{d} = r_1 k_\rho
\end{aligned}
$$

where $w(k,m) := \prod_{j=1}^n U_j^{m_j}\prod_{j=1}^n U_j^{k_j}\left(\prod_{j=1}^n U_j^{m_j+k_j}\right)^{-1} \in S^1 \subset \mathbb{C}$ so our desired constant is $K = r_1 k_\rho = r_1 C_\rho^2 2^d$ and we are done. □

For the general case $s \neq d$ we need to prove a lemma saying that $||\cdot||_s = ||\cdot||_{s-t} \circ P_{\lambda^t}$.

Lemma 6.5. *For any $\vec{a} \in E$ and $s,t \in \mathbb{R}$, $\vec{a} \in H^s$ if and only if $P_{\lambda^t}(\vec{a}) \in H^{s-t}$ with $||\vec{a}||_s = ||P_{\lambda^t}(\vec{a})||_{s-t}$.*

Proof. Suppose that $\vec{a} \in H^s$ or $P_{\lambda^t}(\vec{a}) \in H^{s-t}$. Then

$$
\begin{aligned}
||P_{\lambda^t}(\vec{a})||_{s-t}^2 &= \sum_{j,m}(1+|m_1|^2+\cdots+|m_n|^2)^{s-t}\lambda^{2t}(m)|a_{j,m}|^2 \\
&= \sum_{j,m}(1+|m_1|^2+\cdots+|m_n|^2)^{s-t}(1+|m_1|^2+\cdots+|m_n|^2)^{t}|a_{j,m}|^2 \\
&= \sum_{j,m}(1+|m_1|^2+\cdots+|m_n|^2)^{s}|a_{j,m}|^2 = ||\vec{a}||_s
\end{aligned}
$$

so we know that $\vec{a} \in H^s$ and $P_{\lambda^t}(\vec{a}) \in H^{s-t}$. □

Then the general case follows quite easily.

Corollary 6.6. *Suppose ρ is a matrix valued symbol of order d. Then $||P_\rho(\vec{a})||_{s-d} \le C||\vec{a}||_s$ for some constant $C > 0$ and P_ρ defines a bounded operator $P_\rho : H^s \to H^{s-d}$.*

Proof. By Lemma 6.5, we have $||P_\rho(\vec{a})||_{s-d} = ||P_{\lambda^{s-d}}(P_\rho(\vec{a}))||_0$. By Proposition 5.1(b), the matrix valued symbol $\sigma(P_{\lambda^{s-d}} \circ P_\rho)$ is of order $d + (s-d) = s$, so Theorem 6.4 gives us $||P_{\lambda^{s-d}}(P_\rho(\vec{a}))||_0 \le C||\vec{a}||_s$ for some constant $C > 0$. □

We can also define an analog of the C^k norm on E.

Definition 6.7. Define the C^k norm $||\cdot||_{\infty,k} : E \to \mathbb{R}_{\ge 0}$ as follows: $||\vec{a}||_{\infty,k} := \sum_{|\ell|\le k} ||\delta^\ell(\vec{a})||_{C^*}$ where the C^* norm $||\cdot||_{C^*}$ is given by $||\vec{a}||_{C^*}^2 := \sup\{|\lambda| : \vec{a}^*\vec{a} - \lambda\cdot 1 \text{ not invertible}\}$.

Since, for arbitrary $\vec{a} = \sum_{j,m} a_{j,m} \prod_g U_g^{m_g} f_j$,

$$D_s^\ell \alpha_s(\vec{a}) = (-i\partial_s)^\ell \sum_{j,m} e^{is\cdot m} a_{j,m} \prod_{g=1}^n U_g^{m_g} f_j = \sum_{j,m} m^\ell e^{is\cdot m} a_{j,m} \prod_{g=1}^n U_g^{m_g} f_j$$
$$= \sum_{j,m} \delta^\ell e^{is\cdot m} a_{j,m} \prod_{g=1}^n U_g^{m_g} f_j = \delta^\ell \alpha_s(\vec{a})$$

we have $(A_\theta^k)^r e = C^k$. We can easily prove an analog of the Sobolev lemma on E as follows.

Theorem 6.8. *For $s > k+1$, $H^s \subseteq (A_\theta^k)^r e$.*

Proof. First consider the case $k = 0$. Note that $||\cdot||_{\infty,0} = ||\cdot||_{C^*}$ so for arbitrary $a_{j,m} \prod_{g=1}^n U_g^{m_g} f_j$ we have $||a_{j,m} \prod_{g=1}^n U_g^{m_g} f_j||_{\infty,0}^2 = \sup\{|\lambda| : |a_{j,m}|^2 - \lambda \cdot 1 \text{ not invertible}\} = |a_{j,m}|^2$ and for arbitrary $\vec{a} = \sum_{j,m} a_{j,m} \prod_{g=1}^n U_g^{m_g} f_j$ we have $||\vec{a}||_{\infty,0}^2 \leq \sum_{j,m} ||a_{j,m} \prod_{g=1}^n U_g^{m_g} f_j||_{\infty,0}^2 = \sum_{j,m} |a_{j,m}|^2 = ||\vec{a}||_0^2$ by the triangle inequality. We have $\vec{a} = \sum_{j,m} a_{j,m} \lambda^s(m) \lambda^{-s}(m) \prod_{g=1}^n U_g^{m_g} f_j$ so by the Cauchy-Schwarz inequality we get $||\vec{a}||_0^2 \leq ||\vec{a}||_s^2 \sum_m (1 + |m_1|^2 + \cdots + |m_n|^2)^{-s}$. Since $2s > 2$, $(1+|m_1|^2+\cdots+|m_n|^2)^{-s}$ is summable over $m \in \mathbb{Z}^n$ and $j \in \{1, \ldots, r\}$ so $||\vec{a}||_0^2 \leq C||\vec{a}||_s$. Thus we get $||\vec{a}||_{\infty,0} \leq ||\vec{a}||_0 \leq C||\vec{a}||_s$ and $H^s \subseteq (A_\theta^0)^r e$.

Now suppose $k > 0$. Using what we've proven for the previous case, we have

$$||\delta^\ell(\vec{a})||_{\infty,0} \leq C||\delta^\ell(\vec{a})||_{s-|\ell|} = C \left|\left| \sum_{j,m} m^\ell a_{j,m} \prod_{g=1}^n U_g^{m_g} f_j \right|\right|_{s-|\ell|}$$
$$< C \left|\left| \sum_{j,m} (1 + |m_1|^2 + \cdots + |m_n|^2)^{|\ell|} a_{j,m} \prod_{g=1}^n U_g^{m_g} f_j \right|\right|_{s-|\ell|}$$
$$= C||P_{\lambda^{|\ell|}}(\vec{a})||_{s-|\ell|} = C||\vec{a}||_s$$

for $|\ell| \leq k$ since $s - |\ell| \geq s - k > 1$. Therefore, $||\vec{a}||_{\infty,k} = \sum_{|\ell|\leq k} ||\delta^\ell(\vec{a})||_{\infty,0} \leq \sum_{|\ell|\leq k} C||\vec{a}||_s \leq C||\vec{a}||_s(k+1)(k+2)/2$ and we get $H^s \subseteq (A_\theta^k)^r e$. □

We get the following corollary.

Corollary 6.9. $\bigcap_{s\in\mathbb{R}} H^s = (A_\theta^\infty)^r e$.

Proof. Suppose $a \in \bigcap_{s\in\mathbb{R}} H^s$. Then for any $k \in \mathbb{Z}_{\geq 0}$, $\vec{a} \in H^{k+2}$, so by the theorem we just proved, $\vec{a} \in (A_\theta^k)^r e$. Consequently $\vec{a} \in (A_\theta^\infty)^r e$, so $\bigcap_{s\in\mathbb{R}} H^s \subseteq (A_\theta^\infty)^r e$.

Suppose $a \in (A_\theta^\infty)^r e$. Then since H^s is the completion of $(A_\theta^\infty)^r e$ with respect to $||\cdot||_s$, $(A_\theta^\infty)^r e \subseteq H^s$ for all $s \in \mathbb{R}$, and $(A_\theta^\infty)^r e \subseteq \bigcap_{s\in\mathbb{R}} H^s$. □

We can also prove an analog of the Rellich lemma on E.

Theorem 6.10. *Let $\{\vec{a}_N\} \in (A_\theta^\infty)^r e$ be a sequence. Suppose that there is a constant C so that $||\vec{a}_N||_s \leq C$ for all N. Let $s > t$. Then there is a subsequence $\{\vec{a}_{N_j}\}$ that converges in H^t.*

Proof. Let F be an orthogonal eigenbasis of e normalized with respect to $\langle\cdot,\cdot\rangle$, and let $F_1 := \{f_h : 1 \leq h \leq r_1\}$ be the subset of eigenvectors with eigenvalue 1. Let $e_{s,m} := (1 + |m_1|^2 + \cdots + |m_n|^2)^{-s/2} \prod_g U_g^{m_g}$ and $E_s := \{e_{s,m} \mid m \in \mathbb{Z}\}$. $E_s F_1$ is an orthogonal basis of E considered as a $\mathbb{C}$-vector space normalized with respect to $\langle\cdot,\cdot\rangle_s$, so we can write $\vec{a}_N := \sum_{h,k} a_{N,h,k} e_{s,k} f_h$. Then $|a_{N,h,k}|^2 \leq \sum_{h,k} |a_{N,h,k}|^2 \leq C^2$ and $|a_{N,h,k}| \leq C$. Applying the Arzela-Ascoli theorem to $\{a_{N,h,k}\}$ for some fixed (h,k), we can get a subsequence $\{a_{N_j,h,k}\}$ of

$\{a_{N,h,k}\}$ such that for any $\epsilon > 0$ there exists $M(\epsilon) \in \mathbb{N}$ such that $|a_{N_i,h,k} - a_{N_j,h,k}| < \epsilon$ whenever $i, j \geq M(\epsilon)$. Do this for all $1 \leq h \leq r_1$ and $|k_1|^2 + \cdots + |k_n|^2 \leq R^2$, replacing $\{a_N\}$ with $\{a_{N_j}\}$ each time. Then we get a subsequence $\{a_{N_j}\}$ of $\{a_N\}$ such that for any $\epsilon > 0$ there exists $M(\epsilon) \in \mathbb{N}$ such that, for all $1 \leq h \leq r$ and $|k_1|^2 + \cdots + |k_n|^2 \leq R^2$, $|a_{N_i,h,k} - a_{N_j,h,k}| < \epsilon$ whenever $i, j \geq M(\epsilon)$. Now consider the sum $||a_{N_i} - a_{N_j}||_t^2 = \sum_{h,k} |a_{N_i,h,k} - a_{N_j,h,k}|^2 (1 + |k_1|^2 + \cdots + |k_n|^2)^{t-s}$. Decompose it into two parts: one where $|k_1|^2 + \cdots + |k_n|^2 > R^2$ and one where $|k_1|^2 + \cdots + |k_n|^2 \leq R^2$. On $|k_1|^2 + \cdots + |k_n|^2 > R^2$ we estimate $(1 + |k_1|^2 + \cdots + |k_n|^2)^{t-s} < (1 + R^2)^{t-s}$ so that $\sum_h \sum_{|k_1|^2 + \cdots + |k_n|^2 \geq R^2} |a_{N_i,h,k} - a_{N_j,h,k}|^2 (1 + |k_1|^2 + \cdots + |k_n|^2)^{t-s} < (1 + R^2)^{t-s} \sum_{h,k} |a_{N_i,h,k} - a_{N_j,h,k}|^2 \leq 2r_1 C^2 (1 + R^2)^{t-s}$. If $\epsilon > 0$ is given, we choose R so that $2r_1 C^2 (1 + R^2)^{t-s} < \epsilon$. The remaining part of the sum is over $|k_1|^2 + \cdots + |k_n|^2 \leq R^2$ and can be bounded above by $\epsilon' := \epsilon - 2r_1 C^2 (1 + R^2)^{t-s}$ if $i, j \geq M(\sqrt{\epsilon'/[r_1(2R+1)^n]})$ because an n-ball of radius R centered at the origin is contained in an n-cube of side length $2R$ that has $(2R+1)^n$ lattice points. Then the total sum is bounded above by ϵ, and we are done. □

Acknowledgement
I would like to thank Farzad Fathizadeh for suggesting the problem of filling in the details in the pseudo-differential calculus on the noncommutative n-torus. I would also like thank Matilde Marcolli for helping me make plans to take my candidacy exam. I would like to thank Vlad Markovic and Eric Rains for agreeing to be on my candidacy exam committee.

References

[1] Connes A 1985 *Inst. Hautes Études Sci. Publ. Math.* **62** 257–360
[2] Connes A 1994 *Noncommutative Geometry* (San Diego, CA: Academic Press)
[3] Connes A and Moscovici H 1995 *Geom. Funct. Anal.* **5** 174–243
[4] Connes A and Marcolli M 2008 *Noncommutative Geometry, Quantum Fields and Motives* (American Mathematical Society Colloquium Publications vol 55) (Providence, RI: American Mathematical Society)
[5] Connes A 1980 *C. R. Acad. Sci. Paris Sér. A-B* **290** A599–A604
[6] Gilkey P B 1974 *The Index Theorem and the Heat Equation* (Mathematics Lecture Series vol 4) (Boston, MA: Publish or Perish)
[7] Raymond X S 1991 *Elementary Introduction to the Theory of Pseudodifferential Operators* (Boston: CRC Press)
[8] Wong M W 2014 *An Introduction to Pseudo-differential Operators* 3rd ed (Series on Analysis, Applications and Computation vol 6) (Hackensack, NJ: World Scientific Publishing)
[9] Connes A and Tretkoff P 2011 *Noncommutative Geometry, Arithmetic, and Related Topics* (Baltimore, MD: Johns Hopkins Univ. Press) pp 141–58
[10] Connes A and Moscovici H 2014 *J. Amer. Math. Soc.* **27** 639–84
[11] Fathizadeh F and Khalkhali M 2012 *J. Noncommut. Geom.* **6** 457–80
[12] Fathizadeh F and Khalkhali M 2013 *J. Noncommut. Geom.* **7** 1145–83
[13] Fathizadeh F and Khalkhali M 2015 *J. Noncommut. Geom.* **9** 473–503
[14] Fathizadeh F and Khalkhali M 2013 *Lett. Math. Phys.* **103** 1–18

IOP Conf. Series: Journal of Physics: Conf. Series **965** (2018) 012043 doi:10.1088/1742-6596/965/1/012043

A Theory of Gravity and General Relativity based on Quantum Electromagnetism

J.X. Zheng-Johansson

Institute of Fundamental Physics Research, 611 93 Nyköping, Sweden

Abstract.

Based on first principles solutions in a unified framework of quantum mechanics and electromagnetism we predict the presence of a universal attractive depolarisation radiation (DR) Lorentz force (F) between quantum entities, each being either an IED matter particle or light quantum, in a polarisable dielectric vacuum. Given two quantum entities $i = 1, 2$ of either kind, of characteristic frequencies ν_i^0, masses $m_i^0 = h\nu_i^0/c^2$ and separated at a distance r^0, the solution for F is $F = -\mathscr{G} m_1^0 m_2^0/(r^0)^2$, where $\mathscr{G} = \chi_0^2 e^4/12\pi^2\epsilon_0^2\rho_\lambda$; χ_0 is the susceptibility and ρ_λ is the reduced linear mass density of the vacuum. This force F resembles in all respects Newton's gravity and is accurate at the weak F limit; hence $\mathscr{G}$ equals the gravitational constant G. The DR wave fields and hence the gravity are each propagated in the dielectric vacuum at the speed of light c; these can not be shielded by matter. A test particle μ of mass m^0 therefore interacts gravitationally with all of the building particles of a given large mass M at r^0 apart, by a total gravitational force $F = -GMm^0/(r^0)^2$ and potential $V = -\partial F/\partial r^0$. For a finite V and hence a total Hamiltonian $H = m^0c^2 + V$, solution for the eigenvalue equation of μ presents a red-shift in the eigen frequency $\nu = \nu^0(1 - GM/r^0c^2)$ and hence in other wave variables. The quantum solutions combined with the wave nature of the gravity further lead to dilated gravito optical distance $r = r^0/(1 - GM/r^0c^2)$ and time $t = t^0/(1 - GM/r^0c^2)$, and modified Newton's gravity and Einstein's mass energy relation. Applications of these give predictions of the general relativistic effects manifested in the four classical test experiments of Einstein's general relativity (GR), in direct agreement with the experiments and the predictions given based on GR.

1. Introduction

It has long been established in experiment that Newton's inverse square law of gravity, proposed by I Newton in his *Principia* (1686), is accurate for macroscopic objects and in weak gravitational (g) field. It has been further established in experiment since around the turn of the 20th century that Newton's law of gravity is deviated in strong g fields for matter objects and light, whereas light having no rest mass is implicitly subject to Newton's law only. The deviations mainly exhibit as general relativistic (g-r) effects manifested in the three or four classical-test experiments [1, 2, 3, 4, 5] of A Einstein's general theory of relativity (GR). Einstein's GR predicts all the phenomena successfully [6, 7, 8, 5]. GR is an advanced topic in modern theoretical physics; many monograph introductions have been written on it, including by the very eminent physicists [9, 10, 11, 12, 13]. The cause of gravity yet remains unresolved up to the present, with or without a g-r effect. I. Newton proposed his inverse square law essentially on phenomenological basis. A. Einstein assumed in his GR that a geometric curvature is being produced in the empty-space and time about a material object and is the cause of gravity. GR remains a phenomenological theory.

IOP Conf. Series: Journal of Physics: Conf. Series **965** (2018) 012043 doi:10.1088/1742-6596/965/1/012043

The geometric description of GR is incompatible with the particle and field based description of quantum mechanics (QM), electromagnetism, and the other three fundamental forces.

QM and GR, along with Newton's gravity in the weak g limit, have been experimentally corroborated to great extents. The two remain yet un-united [14, 15, 16]. A unification of the two, under quantum gravity (QG), has been the holly grill of modern physics. Superstring theory is a much acknowledged contender for QG [17, 18]. Yet the super-partner particles of the theory remain to be observed in experiment; the strings and the six or so extra dimensions remain to be demonstrated in experiment, and supplied with physics foundations. As one of other aspects, the tiny strings $\sim 10^{-35}$ m are just points to the particle waves; they do not ultimately resolve the outstanding difficulty related to wave-particle duality of QM. Loop quantum gravity is focused on quantising the space, rather than unifying gravity with the other fundamental forces. QG is desired in black hole and big bang researches.

An Internally Electrodynamic (IED) model of particles, along with a polarisable vacuuonic dielectric vacuum, has been developed by the author since 2000 using overall experimental observations for particles and vacuum as input information. Based on first principles solutions for the IED particles in a unified framework of classical, quantum and relativistic mechanics, it has been possible to predict a range of the well-established basic properties and relations of particles (see [19] for a review and the original papers). One of the solutions is a depolarisation radiation (DR) Lorentz force acting between IED matter particles in a polarisable dielectric vacuum [20], which resembles in all respects Newton's gravity. In this paper, we extend the solution of gravity to including light quanta (Secs 2, 3), and the effect of gravity on the dynamical variables of the interacting entities (Sec 4), within the same framework of quantum electromagnetism. We thereby obtain a generalised quantum-electromagnetic theory of gravity and general relativity. The solutions of the theory are then applied to predict the g-r effects manifested in the four classical test experiments of GR (Sec 5). The solutions include quantum gravitational waves (composed of DR fields) from individual matter particles or light quanta; the natural extension of this to an accelerating macroscopic object is a macroscopic gravitational wave, which formal solution we shall describe in a separate paper. This meaningful research would not have been carried out in a foreseeable future, if having not been motivated by an open remark by Professor G 't Hooft, concerning any new gravity theory that can not be a complete work without including a representation of general relativity.

2. Quantum electromagnetic descriptions of matter particles and light quanta. Dielectric theory of the vacuum

We shall work in a three dimensional flat space; this space has a one to one correspondence to a polarisable dielectric vacuum. Let there be in this space a charge q, of charge density ρ_q, at the origin of a co-ordinate system x, y, z fixed therein. q is in accelerated motion, forming a current density $\mathbf{j}_q$, and generates thereby electromagnetic radiation fields $\mathbf{E}, \mathbf{B}$ according to the Maxwell equations

$$\nabla \cdot \mathbf{E} = \rho_q/\epsilon_0, \quad \nabla \times \mathbf{B} - (1/c^2)\partial_t \mathbf{E} = \mu_0 \mathbf{j}_q, \quad \nabla \times \mathbf{E} + \partial_t \mathbf{B} = 0, \quad \nabla \cdot \mathbf{B} = 0, \tag{1}$$

where $c(= 1/\sqrt{\epsilon_0\mu_0})$ is the speed of light in the vacuum; ϵ_0 is the permittivity and μ_0 permeability of the vacuum. In the empty space ($\emptyset$) left over after removal of the dielectric vacuum, we directly measure the applied radiation fields $\mathbf{E}_\emptyset$ and $\mathbf{B}_\emptyset$. An applied $\mathbf{E}_\emptyset$ at a location $\mathbf{r}$ will polarise the vacuum in the vicinity, and induce a vacuum polarisation $\mathbf{P}$ and a depolarisation (radiation) field $\mathbf{E}_p$ relative to $\emptyset$. Applying the usual dielectric theory [21, 22] here, these are given in terms of the macroscopic fields $\mathbf{E}$, $\mathbf{B}$ measured in the vacuum as

$$\mathbf{E}_\emptyset = \mathbf{E} - \mathbf{E}_p = \kappa_0 \mathbf{E}, \quad \mathbf{E}_p = -\frac{\mathbf{P}}{\epsilon_\emptyset} = -\chi_0 \mathbf{E}, \quad \chi_0 = \kappa_0 - 1;$$

$$\mathbf{B}_\emptyset = \mathbf{B} - \mathbf{B}_p = \kappa_0 \mathbf{B}, \quad \mathbf{B}_p = -\frac{\mathbf{E}_p \times \mathbf{c}}{c^2} = -\chi_0 \mathbf{B}. \tag{2}$$

$\kappa_0(> 1)$ is the dielectric constant and $\chi_0(> 0)$ is the susceptibility of the dielectric vacuum against $\emptyset$; $\epsilon_0 = \kappa_0\epsilon_\emptyset$, and $\epsilon_\emptyset$ is the permittivity of the empty space $\emptyset$. We have implicitly assumed no significant magnetisation of the vacuum by $\mathbf{B}_\emptyset$. $\mathbf{B}_p$ is thus a magnetic field purely induced by the time rate of $\mathbf{E}_p$ (Maxwell-Ampere's law). In terms of the vacuuonic vacuum model[1] [19, 22, 23], (a) the vacuum is filled of densely packed, electrically neutral but polarisable entities called vacuuons separated at a mean spacing b; $b \sim 10^{-18}$ m estimated based on the observational shortest wavelength ($1 \sim 2 \times 10^{-17}$ m) of electromagnetic radiation; and (b) each vacuuon is composed of a pair of spinning entities called p- and n- vaculeons of a charge $+e$ and $-e$, that are located at the core and on a spherical shell about the core, and are strongly bound electromagnetically.

Assuming q $(= +e$ or $-e)$ is oscillating sinusoidally about $\mathbf{r} = 0$ along z direction, with a frequency ν and amplitude $\mathcal{A}$, Eqs (1) have the usual solutions for $\mathbf{E}$ at a distance $\mathbf{r}(r, \theta, \phi)$ sufficiently far from q, which combined with (2) further yield solutions for $\mathbf{E}_p$, given as

$$\mathbf{E}(\mathbf{r}, t) = -\frac{aE_0 \sin\theta\varphi(\mathbf{r}, t)}{r}\hat{\theta}, \quad E_0 = \frac{e\mathcal{A}\omega^2}{4\pi\epsilon_0 c^2 a}, \quad \varphi(\mathbf{r}, t) = \cos(\mathbf{k}\cdot\mathbf{r} - \omega t + \alpha_0), \tag{3}$$

$$\mathbf{E}_p(\mathbf{r}, t; -\tfrac{\pi}{2}) = -\chi_0 \mathbf{E}(\mathbf{r}, t) = \frac{E_{p0} a \sin\theta\psi_p}{r}\hat{\theta}, \quad E_{p0} = -\chi_0 E_0, \quad \psi_p(\mathbf{r}, t) = -\varphi(\mathbf{r}, t; -\frac{\pi}{2}); \tag{4}$$

and $\mathbf{B} = -\frac{\mathbf{E}\times\mathbf{c}}{c^2}$, $\mathbf{B}_p = -\frac{\mathbf{E}_p\times\mathbf{c}}{c^2}$. Here $\omega = 2\pi\nu$, $k = \frac{2\pi}{\lambda}$, $\lambda = c/\nu$, α_0 is an initial phase, and a is a linear dimension defined after (5) below. q may implicitly also be in linear motion, at velocity v in $+x$ direction here. The fields (φ's) generated oppositely say in $+x, -x$ directions, as $\varphi^\dagger, \varphi^\ddagger$, are thus in general Doppler effected, differently. For the DR force in question in Sec 2, v affects E_0 through ω; $\omega = \sqrt{\omega^\dagger\omega^\ddagger} = \gamma\Omega$, where $\gamma = (1 - v^2/c^2)^{-1/2}$ and $\Omega = \lim_{v^2/c^2\to 0}\omega$ from IED solution (see [19]). v does not affect the integration of φ^j's which are written as φ in (3c).

The energy density of the $\mathbf{E}, \mathbf{B}$ fields, as $\mathbf{E}^{(c)}, \mathbf{B}^{(c)}$ in complex forms, is $\rho_\varepsilon = \epsilon_0(\mathbf{E}^{(c)} \times \mathbf{B}^{(c)})\cdot\mathbf{c}$. In the dielectric vacuum of an induced elasticity, $\mathbf{E}, \mathbf{B}$ correspond to a transverse elastic wave, of a transverse (tensile) displacement $\mathbf{u} = (\mathcal{A}/E_0)\mathbf{E}$ and amplitude $\mathcal{A}$. So accordingly $\rho_\varepsilon = \frac{1}{2}\rho_l\omega^2|u^{(c)}|^2$, where ρ_l is the linear (sub-vacuum) mass density of the perturbed vacuum. Suppose that the $\mathbf{E}, \mathbf{B}$ comprising one energy quantum $h\nu$ extend through a spherical volume V of radius r_0; $\ell = 2r_0 = \mathcal{N}_1\lambda = n_1 b$ contains $\mathcal{N}_1$ wavelengths λ's and n_1 vacuuon spacings b's. The one energy quantum, or the Hamiltonian of the one quantum, is thus equal to the integration of ρ_ε in V, with $dV = r^2 dr d\Omega$, $d\Omega = \sin\theta d\theta d\phi$,

$$\varepsilon = \int_V \epsilon_0(\mathbf{E}^{(c)} \times \mathbf{B}^{(c)})\cdot\mathbf{c}dV = \ell\sigma\epsilon_0 E_0^2 = \int_V \frac{1}{2}\rho_l\omega^2 {u^{(c)}}^2 dV = \frac{1}{2}\lambda\rho_\lambda\omega^2\mathcal{A}^2 = h\nu = mc^2. \tag{5}$$

$\sigma = \frac{4}{3}\pi a^2$ represents a cross sectional area through which ε is transported. $\rho_\lambda = \ell\rho_l/\lambda = \mathcal{N}_1\lambda\rho_l/\lambda = \mathcal{N}_1\rho_l$ is the *reduced* linear mass density of the vacuum.

$\mathbf{E}, \mathbf{B}$ may be the internal components of a simple, single charged matter particle such as a proton or an electron, generated by its charge $q = +e$ or $-e$; q is of zero rest particle mass but is oscillating with a frequency ν specified by (5) (the IED model). Its ε is in general conveyed a fraction a_q by its oscillatory charge, and a_r by its radiation field, with $0 \le a_q, a_r \le 1$. So $a_q\varepsilon + a_r\varepsilon = a_r\varepsilon = \varepsilon(= h\nu = mc^2)$, given for the extreme case $a_r = 1$ and $a_q = 0$ in (5), is the

[1] It can be shown that, combined Lorentz transformations for both the sources and the interferometer arms as of the Michelson-Morley and Kennedy-Thorndike experiments, render null and constant fringe shifts of light in two-way trips against a substantial vacuum (Zheng-Johansson J X 2005 *Inference of Basic Laws of Classical, Quantum and Relativistic Mechanics based on First-Principles of a Minimal Set* 254 pages, unpublished).

total (rest) mass energy and m is the (rest) mass of the resultant IED particle (at rest). Or, $\mathbf{E}, \mathbf{B}$ may comprise a light quantum, (originally) emitted by a matter particle of charge $q = +e$ or $-e$ here. m then represents the dynamical mass, or simply mass, of the light quantum. (5) and (3b) give a few further relations to be used:

$$\rho_l = \frac{e^2 m^2}{6\pi\epsilon_0 \hbar^2}, \quad \rho_\lambda = \mathscr{N}_1 \rho_l = \frac{\mathscr{N}_1 e^2 m^2}{6\pi\epsilon_0 \hbar^2} = \frac{\hbar}{\pi c A^2}. \tag{6}$$

3. Newtonian gravity between light quanta and matter particles

We shall in this section derive a general equation of Newton's gravity between a light quantum μ and an IED matter particle i (Secs 3.1-2) or j (Sec 3.3), their mean mutual gravity (Sec 3.4), and finally between μ (designating thereof a light quantum or IED matter particle) and an object composed of many IED matter particles (Sec 3.5). We shall disregard a g-r effect until Sec 4.

3.1 Action of μ on i μ and i are specified as follows: (a) μ is one of N_μ light quanta that are being constantly (re-) generated by an array of point sources in the vicinity of the origin on the yz plane, at $x = 0$, in the co-ordinate system x, y, z. Its source has an oscillation frequency ν_μ and amplitude $\boldsymbol{A}_\mu$. Let specifically the sources be pairs of dipole charges $q_\mu = \pm e$ of the vacuuons comprising the vacuum here, referred to as *virtual* sources. They are induced by the plane wave $\mathbf{E}_\mu$ fields propagated to $x = 0$ at time $t = 0$, that were generated say at $-x_1$ at time $-t_1$ earlier by some natural sources of unit charges $\pm e$. (b) i is one of N_i IED matter particles located at a mean position $x = r$ on the x axis. i has a mass m_i, and charge $q_i = +e$ or $-e$ oscillating with a frequency $\nu_i = m_i c^2/h$ (Eq 5) and amplitude $\boldsymbol{A}_i$. (c) $\boldsymbol{A}_\mu$ and $\boldsymbol{A}_i$ are along the z direction during a brief time under immediate consideration. So both sources $\eta = i$ and μ generate fields $\mathbf{E}_\eta, \mathbf{B}_\eta$ and $\mathbf{E}_{p_\eta}, \mathbf{B}_{p_\eta}$ according to (3)-(4) in their local co-ordinate systems, x, y, z for μ, and a parallel set of axes x_i, y_i, z_i with an origin fixed at $x = r$ for i. $\mathbf{E}_\mu, \mathbf{B}_\mu$ ($\mathbf{E}_i, \mathbf{B}_i$) are assumed shielded, not reaching i (μ). And, (d) i and μ are in dynamical equilibrium.

The fields $\mathbf{E}_{p_\mu} = \mathbf{E}_\mu - \mathbf{E}_{\emptyset_\mu}$ and $\mathbf{B}_{p_\mu} = \mathbf{B}_\mu - \mathbf{B}_{\emptyset_\mu}$ (Eq 4) generated by q_μ at $x = 0$ at time $t = 0$ will propagate to q_i at $x = r$ at time t, and act on q_i a (momentary) Lorentz force $\mathbf{F}_{tot_{\mu i}} = q_i \mathbf{E}_{p_\mu} + \mathbf{F}_{\mu i}$. $q_i \mathbf{E}_{p_\mu}$ drives q_i into a transverse motion at velocity $\boldsymbol{v}_{p_i}$ given as

$$\boldsymbol{v}_{p_i} = \boldsymbol{v}_i - \boldsymbol{v}_{\emptyset_i} = \frac{\tau_\mu q_i \mathbf{E}'_{p_\mu}}{2\pi m_i}, \quad \boldsymbol{v}_i = \int \frac{q_i \mathbf{E}_\mu}{m_i} dt = \frac{\tau_\mu q_i \mathbf{E}'_\mu}{2\pi m_i}, \quad \boldsymbol{v}_{\emptyset_i} = \frac{\tau_\mu q_i \mathbf{E}'_{\emptyset_\mu}}{2\pi m_i}, \tag{7}$$

where $\tau_\mu = 1/\nu_\mu$, $\mathbf{E}'_\mu(\mathbf{r}, t) = \frac{a_\mu E_{0_\mu} \sin\theta}{r} \cos(\alpha_\mu + \frac{\pi}{2})\hat{z} \equiv \mathbf{E}_\mu(\mathbf{r}, t; +\frac{\pi}{2})$ given after (3a), etc.; $\alpha_\mu = \mathbf{k}_\mu \cdot \mathbf{r} - \omega_\mu t$; we have set $\alpha_{0_\mu} = 0$ for the initial phase.

$$\mathbf{F}_{\mu i} = q_i(\boldsymbol{v}_i \times \mathbf{B}_\mu - \boldsymbol{v}_{\emptyset_i} \times \mathbf{B}_{\emptyset_\mu}) = q_i(\frac{\tau_\mu q_i}{2\pi m_i})[\mathbf{E}'_\mu \times \mathbf{B}_\mu - (\mathbf{E}'_\mu - \mathbf{E}'_{p_\mu}) \times (\mathbf{B}_\mu - \mathbf{B}_{p_\mu})]$$

is the difference between a Lorentz magnetic force acted by $\mathbf{B}_\mu$ in the vacuum, and one by $\mathbf{B}_{\emptyset_\mu}$ in $\emptyset$, which we know to each obey the right hand rule. Setting $\mathbf{E}'_\mu = 0$, $\mathbf{B}_\mu = 0$, we obtain the final Lorentz magnetic force acted by $\mathbf{E}'_{p_\mu}, \mathbf{B}_{p_\mu}$ on q_i, referred to as DR (Lorentz) force,

$$\mathbf{F}_{\mu i} = -q_i \boldsymbol{v}_{p_i} \times \mathbf{B}_{p_\mu} = -\frac{\tau_\mu q_i^2 \mathbf{E}'_{p_\mu} \times \mathbf{B}_{p_\mu}}{2\pi m_i} = -\frac{\tau_\mu q_i^2 \chi_0^2 \mathbf{E}'_\mu \times \mathbf{B}_\mu}{2\pi m_i} = -\frac{\mathcal{K}_{\mu i} f(\alpha_\mu)}{r^2}\hat{x},$$

$$\mathcal{K}_{\mu i} = \frac{\tau_\mu q_i^2 \chi_0^2 a_\mu^2 E_{0_\mu}^2 \sin^2\theta}{2\pi m_i c} = \frac{\tau_\mu q_i^2 \chi_0^2 a_\mu^2 \left(\frac{q_\mu A_\mu \omega_\mu^2}{4\pi\epsilon_0 c^2 a_\mu}\right)^2 \cdot (\frac{1}{2}\lambda_\mu \rho_{\lambda_\mu} h^2)}{2\pi m_i c \cdot (\frac{1}{2}\lambda_\mu \rho_{\lambda_\mu} h^2)} = \frac{\chi_0^2 e^4 m_\mu^3}{4\pi\epsilon_0^2 h^2 \rho_{\lambda_\mu} m_i}, \tag{8}$$

$f(\alpha_\mu) = |\sin(\alpha_\mu)\cos(\alpha_\mu)|$. (7a) for $\boldsymbol{v}_{p_i}$, (3b) for E_{0_μ}, $\theta = \frac{\pi}{2}$ (i is essentially a point on the x axis), $q_i = q_\mu = \pm e$, and the identity relations (5) have been used. Based on (8), $\mathbf{F}_{\mu i}$ is pointed from i to μ, hence an attraction, irrespective of the signs of q_i, q_μ; and $\mathbf{F}_{\mu i}$ in relation to $\boldsymbol{v}_{p_i}$ and $\mathbf{B}_{p_\mu}$ obeys a *left hand rule*.

For a large N_μ number of μ's being constantly (re-)generated at $x = 0$ over a long time $\Delta t(>> \mathcal{N}_{1\mu}/\nu_\mu, \mathcal{N}_{1i}/\nu_i)$, firstly the α_{0_μ}'s of different μ's are independent random variables. These reduce their relative radiation intensities by a factor $f_0 = \cos^2 \alpha_{0_\mu}$, rather than producing coherent waves. Secondly, under the influence of random environmental fields, the amplitude of each original natural source, $\boldsymbol{A}_{\mu 1}$ at $-x_1$, hence $\mathcal{A}_\mu = \frac{a}{|x_1|}|\boldsymbol{A}_{\mu 1}|$ at $x = 0$, would explore in Δt all possible orientations $\Omega(\theta, \phi)$. Its projection in the yz plane, becoming $\mathcal{A}_{\mu_{yz}} = \mathcal{A}_\mu(\sin^2\theta \sin^2\phi + \cos^2\theta)^{1/2}$ at $x = 0$, is now responsible for producing a DR force given otherwise by (8). Including the two features, $F_{\mu i}$ is written as $F_{\mu i(N_\mu)} = -(\mathcal{A}^2_{\mu_{yz}}/\mathcal{A}^2_\mu)\mathcal{K}_{\mu i} f_0 f(\alpha_\mu)/r^2$. Its average over Ω in (0,4π), and $\alpha_{0_\mu}, \alpha_\mu$ in $(0, \pi)$ is

$$\langle F_{\mu i}\rangle = \frac{1}{\pi^2}\int_0^\pi \int_0^\pi \Big(\frac{1}{4\pi}\int_\Omega F_{\mu i(N_\mu)} d\Omega\Big) d\alpha_{0_\mu} d\alpha_\mu = -\frac{\mathcal{K}_{\mu i}\mathcal{P}_{yz}\bar{f}_0\bar{f}}{r^2} = -\frac{\mathcal{K}_{\mu i}}{3\pi r^2}, \tag{9}$$

where $\mathcal{P}_{yz} = \frac{1}{4\pi}\int_\Omega (\mathcal{A}^2_{\mu_{yz}}/\mathcal{A}^2_\mu) d\Omega = \frac{2}{3}$, $\bar{f}_0 = \frac{1}{\pi}\int_0^\pi f_0 d\alpha_{0_\mu} = \frac{1}{\pi}\int_0^\pi \cos^2\alpha_{0_\mu} d\alpha_{0_\mu} = \frac{1}{2}$, and $\bar{f} = \frac{1}{\pi}\int_0^\pi f(\alpha_\mu)d\alpha_\mu = \frac{1}{\pi}\int_0^\pi |\sin\alpha_\mu \cos\alpha_\mu| d\alpha_\mu = \frac{1}{\pi}$.

3.2 Action of i on μ The vacuum about $x = 0$ is electrically and dynamically characterised by the μ-disturbance as follows: (i) $\mathbf{E}_\mu$ polarises the vacuuons in the volume $V_\mu = n_1 b\sigma = n_1 b \cdot (n_y b) \cdot (n_z b)$ occupied by the μ (wave) train. We find $n_1 n_y$ polarised vacuuons, and hence $n_1 n_y$ units of polarisation charges $+e$'s (or $-e$'s) on each enclosing Gauss surface perpendicular to $\mathbf{E}_\mu$ along the z axis. And (ii) the $n_1 n_y$ effective oscillators are driven by $\mathbf{E}_\mu$ into oscillations at the frequency ν_μ of $\mathbf{E}_\mu$, and hence imparted with a dynamical mass $\Delta m_\mu = \frac{m_\mu}{n_1 n_y}$ each; $m_\mu = h\nu/c^2$ (Eq 5). $(n_1 n_y e) \times \Delta m_\mu = e m_\mu (\propto \sqrt{\rho_{\lambda_\mu}})$ returns correctly the e and m_μ of the original natural source. The last identity relation renders an equivalent single vacuuon (virtual source) v_μ representation to be used: v_μ is instantaneously located at $x = 0$; it carries all the mass m_μ and polarisation (dipole) charges $(q_\mu =) q_+, q_- = +e, -e$, oriented along $\mathbf{E}_\mu$ direction.

$\mathbf{E}_{p_i}, \mathbf{B}_{p_i}$ generated at $x_i = 0$ $(x = r)$ and $t = 0$, will propagate to μ at $x_i = -r$ $(x = 0)$ at time t, acting on either charge q_+ or q_- of v_μ a Lorentz force $\mathbf{F}_{tot,i\mu\pm} = q_\pm \frac{1}{2}\mathbf{E}_{p_i} + \mathbf{F}_{i\mu_\pm}$. $q_\pm \frac{1}{2}\mathbf{E}_{p_i}$ drives $q_\pm$ into motion as: $m_\mu d\boldsymbol{v}_{p_{\mu\pm}}/dt = \pm e\frac{1}{2}\mathbf{E}_{p_i}$. The integrated velocities are $\boldsymbol{v}_{p_{\mu\pm}} = \int \frac{q_\pm \frac{1}{2}\mathbf{E}_{p_i}}{m_\mu} dt = \frac{\tau_i q_\pm \frac{1}{2}\mathbf{E}'_{p_i}}{2\pi m_\mu}$, where $\tau_i = 1/\nu_i$, $\mathbf{E}'_{p_i}(0,t) = \mathbf{E}_{p_i}(0, t; \frac{\pi}{2})$. $\mathbf{F}_{i\mu_\pm}$ is a DR Lorentz force acted by i on q_+ or q_- of v_μ. This force on both charges, hence on v_μ, is given as, applying directly the left hand rule (Sec 3.1),

$$\mathbf{F}_{i\mu} = \mathbf{F}_{i\mu_+} + \mathbf{F}_{i\mu_-} = -(q_+\boldsymbol{v}_{p_{\mu+}} + q_-\boldsymbol{v}_{p_{\mu-}}) \times \mathbf{B}_{p_i} = -\frac{\tau_i(q_+^2 + q_-^2)\frac{1}{2}\mathbf{E}'_{p_i} \times \mathbf{B}_{p_i}}{2\pi m_\mu} = \frac{\mathcal{K}_{i\mu} f(\alpha_i)}{r^2}\hat{x},$$

$$\mathcal{K}_{i\mu} = \frac{\tau_i \chi_0^2 q_\mu^2 a_i^2 E_{0_i}^2 \sin^2\theta_i}{2\pi m_\mu c} = \frac{\tau_i \chi_0^2 e^2 a_i^2 (\frac{e\mathcal{A}_i\omega_i^2}{4\pi\epsilon_0 c^2 a_i})^2 \times \frac{1}{2}\lambda_i \rho_{\lambda_i} h^2}{2\pi m_\mu c \times \frac{1}{2}\lambda_i \rho_{\lambda_i} h^2} = \frac{\chi_0^2 e^4 m_i^3}{4\pi\epsilon_0^2 h^2 \rho_{\lambda_i} m_\mu}, \tag{10}$$

and $f(\alpha_i) = |\sin(\alpha_i)\cos(\alpha_i)|$, where $\alpha_i = \mathbf{k}_i \cdot \mathbf{r}_i - 2\pi\nu_i t$ given for $\alpha_{0_i} = 0$. And $r_i = r + \delta x \doteq r$ in the denominator for $r >> \delta x = c\mathcal{N}_{1i}/\nu_i$, (3b) for E_{0_i} $\theta_i = \frac{\pi}{2}$, and (5) for m_i, m_μ have been used. For the N_i IED particles measured over long time Δt, with $\boldsymbol{A}_i$ and α_{0_i} being random variables similarly as of μ, the average DR Lorentz force acted by i on μ is similarly given as

$$\langle F_{i\mu}\rangle = \frac{1}{\pi^2}\int_0^\pi \int_0^\pi \Big(\frac{1}{4\pi}\int_\Omega \frac{\mathcal{A}^2_{i_{yz}}}{\mathcal{A}^2_i} f_0 F_{i\mu} d\Omega\Big) d\alpha_{0_i} d\alpha_i = \frac{\mathcal{K}_{i\mu}\mathcal{P}_{yz}\bar{f}_0\bar{f}}{r^2} = \frac{\mathcal{K}_{i\mu}}{3\pi r^2}. \tag{11}$$

3.3 Action of μ on j j is identical to i but is located at $y = r$ on the y axis perpendicular to the μ path along the x axis. Each point on the μ train (of length ℓ_μ) serves as a new virtual source re-generating (spherical) radiation fields; a point x on it (say in $[-\frac{1}{2}\ell_\mu, \frac{1}{2}\ell_\mu]$) is at a distance $\mathbf{r}' = x\hat{x} + r\hat{y} + \delta z\hat{z}$ to j. Here the $\mathbf{E}'_{p_\mu}$ and $\mathbf{B}_{p_\mu}$ along the z and x directions (the ψ_{p_μ} waves) regenerated at time $t = 0$ will propagate to j at $y = r$, across a distance $\mathbf{r}'$ at time $t' = |\mathbf{r}'|/c$, and act on j a DR Lorentz force given similar to (8) as, with $m_j = m_i$, $q_j^2 = q_i^2 = e^2$,

$$\mathbf{F}_{\mu j} = -q_j \boldsymbol{v}_{p_j} \times \mathbf{B}_{p_\mu} = -\frac{\tau_\mu q_j^2(-E'_{p_\mu}\hat{z}) \times (-E_{p_\mu}\hat{x})}{2\pi m_j c} \doteq -\frac{\mathcal{K}_{ui} f(\alpha'_\mu)}{r^2}\hat{y}, \tag{12}$$

where $\boldsymbol{v}_{p_j} = \frac{\tau_\mu q_j \mathbf{E}'_{p_\mu}}{2\pi m_j}$, $\mathbf{E}'_{p_\mu} = -E'_{p_\mu}\hat{z} = -\frac{aE_{0_\mu}\sin\alpha'_\mu}{r'}\hat{z}$, $\mathbf{B}_{p_\mu} = -\frac{E_{p_\mu}}{c}\hat{x} = -\frac{aE_{0_\mu}\cos\alpha'_\mu}{r'}\hat{x}$; $\mathcal{K}_{\mu i}$ is as given by (8); $f(\alpha'_\mu) = |\sin\alpha'_\mu \cos\alpha'_\mu|$, $\alpha'_\mu = \mathbf{k}_\mu \cdot \mathbf{r}' - \omega_\mu t' + \alpha'_{0_\mu}$, and $\alpha'_{0_\mu} = k_\mu x - \omega_\mu \cdot 0 + \alpha_{0_\mu}$. The last of Eqs (12) is given for $r >> \ell_\mu, \delta z$, so $\mathbf{r}' \doteq r\hat{y}$ and all the ψ_{p_μ} wave fields from different points x on each μ train (μ quantum) will arrive at j at $y = r$ at essentially the same time $t = r/c$. α'_{0_μ} is thus inconsequential and may be set to $\alpha'_{0_\mu} = 0$.

For the $\mathbf{E}_\mu, \mathbf{B}_\mu$ originally emitted from a (large) N_μ natural sources (at $-x_1$), the projection of $\mathcal{A}_\mu, \mathcal{A}_{\mu_{yz}}$, hence $\mathbf{E}_\mu$ in the yz plane about $x = 0$ here is randomly oriented at an angle ϑ (and $\mathbf{B}_\mu$ at $\frac{\pi}{2} - \vartheta$) to the z axis; and so are $\mathbf{E}'_\mu$ and $\mathbf{E}'_{p_\mu}$. The z-projection of the $\mathbf{E}'_{p_\mu}$, $\mathbf{E}'_{p_\mu,z} = -E'_{p_\mu}\cos\vartheta\,\hat{z}$, and the induction magnetic field along the x direction, $\mathbf{B}_{p_\mu,x} = -\frac{|\mathbf{E}_{p_\mu,z}|}{c}\hat{x} = \frac{|\mathbf{E}_{p_\mu}|\cos\vartheta}{c}\hat{x}$, are directly responsible for producing a DR force on j at time t:

$$\mathbf{F}_{\mu j(N_\mu)} = -q_j \boldsymbol{v}_{p_j,z} \times \mathbf{B}_{p_\mu,x} = -\frac{\mathcal{A}^2_{\mu_{yz}}}{\mathcal{A}^2_\mu}|\mathbf{F}_{\mu j}|\cos^2\vartheta\,\hat{y} = -\frac{(\mathcal{A}^2_{\mu_{yz}}/\mathcal{A}^2_\mu)\mathcal{K}_{\mu i} f(\alpha'_\mu) f_\vartheta}{r^2}\hat{y}, \tag{13}$$

where $f_\vartheta = \cos^2\vartheta$. For a large N_μ plane-waves propagated to $x = 0$ at time $t = 0$, ϑ may assume all possible values in $(0, 2\pi)$. The average of $F_{\mu j}$ over N_μ quanta over long time Δt is therefore

$$\langle F_{\mu j}\rangle = -\frac{1}{\pi^2}\int_0^\pi \int_0^\pi \frac{1}{4\pi}\int_\Omega F_{\mu j(N_\mu)} d\Omega d\alpha'_\mu d\vartheta = -\frac{\mathcal{K}_{\mu i}\mathcal{P}_{yz}\bar{f}\bar{f}_\vartheta}{r^2} = -\frac{\mathcal{K}_{\mu i}}{3\pi r^2}, \tag{14}$$

where $\bar{f} = \frac{1}{\pi}$ as before; $\bar{f}_\vartheta = \frac{1}{\pi}\int_0^\pi \cos^2\vartheta d\vartheta = \frac{1}{2}$.

From (14) and (9) we have $\langle F_{\mu j}\rangle = \langle F_{\mu i}\rangle$. The same features underling the actions j on μ and i on μ inevitably yield $\langle F_{j\mu}\rangle = \langle F_{i\mu}\rangle$. We shall hereafter refer to the μ-i interaction only. If μ is also an IED particle, we could have obtained $\langle F_{i\mu}\rangle$, say, directly from interchanging the subscripts of $\langle F_{\mu i}\rangle$, to be given exactly as (11). Unless otherwise specified we shall hereafter make no distinction between a light quantum and (IED) matter particle.

3.4 Average mutual DR Lorentz force $\langle F_{\mu i}\rangle$ and $\langle F_{i\mu}\rangle$ are (attractive) action and reaction forces between μ and i. Under condition (d), the two forces must be equal in amplitude (and opposite in direction), and hence in turn equal to their mean,

$$F = -\sqrt{|\langle F_{\mu i}\rangle||\langle F_{i\mu}\rangle|} = -\frac{\sqrt{\mathcal{K}_{\mu i}\mathcal{K}_{i\mu}}}{3\pi r^2} = -\frac{\mathcal{G} m_\mu m_i}{r^2}, \quad \mathcal{G} = \frac{\chi_0^2 e^4}{12\pi^2\epsilon_0^2\hbar^2\rho_\lambda}, \tag{15}$$

$$\rho_\lambda = \sqrt{\rho_{\lambda_i}\rho_{\lambda_\mu}} = \frac{\mathcal{N}_1 e^2 m_i m_\mu}{6\pi\epsilon_0\hbar^2}, \quad \chi_0^2 = \frac{8\pi^3\epsilon_0\mathcal{N}_1\mathcal{G} m_i m_\mu}{e^2}, \quad \frac{\chi_0^2}{\rho_\lambda} = \frac{12\pi^2\epsilon_0^2\hbar^2\mathcal{G}}{e^4}, \tag{16}$$

and $\mathcal{N}_1 = \sqrt{\mathcal{N}_{1i}\mathcal{N}_{1\mu}}$; $\rho_{\lambda_i}, \rho_{\lambda_\mu}$ are as given by (6b). The negative sign indicates that F is an attraction, as $\langle F_{\mu i}\rangle$ and $\langle F_{i\mu}\rangle$ are. F given by the inverse square formula (15a) resembles directly Newton's gravitational force, acting between two quantum entities (a light quantum or IED matter particle each) μ, i here; hence $\mathcal{G}$ corresponds to the gravitational constant G, $\mathcal{G} = G$. Based on (4), $\mathbf{E}_p, \mathbf{B}_p$ are propagated at the speed of light c, and so is the gravity F.

3.5 Action on μ by a macroscopic object M M is composed of large N_i number of IED matter particles i's each as specified by (a)-(d), Sec 3.1; it has a mass $M = \sum_{i=1}^{N_i} m_i$. i are atomic electrons and protons, and (for the unit charges used here) electrons and protons comprising the atomic neutrons. Assume as in typical applications that M is uniform and spherical, of a radius R; its mass centre is at $\mathbf{r}$. μ is a light quantum or IED matter particle located at $\mathbf{r} = 0$. $\mathbf{E}_{p_\mu}$, $\mathbf{B}_{p_\mu}$ of μ are the de-polarisation and induction magnetic fields produced internal of (the polarised vacuuons comprising) the dielectric vacuum, as contrasted to density fluctuations of the vacuum; assume no work has been down by the resultant F. As such, $\mathbf{E}_{p_\mu}$, $\mathbf{B}_{p_\mu}$ can not be absorbed, hence nor be shielded by the matter particles on their path. Each particle i of M therefore sees directly μ, and *vice versa*. The total DR Lorentz force between μ and M is thus

$$F = -\sum_i \sqrt{|\langle F_{\mu i}\rangle||\langle F_{i\mu}\rangle|}\,\frac{\mathbf{r}_{\mu i}}{r_{\mu i}} = -\mathscr{G}|\sum_i \frac{m_i \mathbf{r}_{\mu i}}{r_{\mu i}^3}|m_\mu = -\frac{\mathscr{G} M m_\mu}{r^2}. \tag{17}$$

$\mathscr{G}$ for each μ, i pair involves $\rho_{\lambda(\mu i)}$, $\chi^2_{0(\mu i)}$ (Eq 15b) which are separately dependent of m_μ, m_i (Eqs 16a,b), hence indicated by $_{(\mu i)}$ here. Based on (16c), the ratio $\chi^2_{0(\mu i)}/\rho_{\lambda(\mu i)}$ is independent of the masses, and so should be that of χ_0^2, ρ_λ for M and μ: $\frac{\chi^2_{0(\mu 1)}}{\rho_{\lambda(\mu 1)}} = \ldots = \frac{\chi_0^2}{\rho_\lambda} = \frac{12\pi^2\epsilon_0^2\hbar^2\mathscr{G}}{e^4}$. For the vacuum occupied by M, ρ_λ say satisfying the equalities above may be expressed by

$$\rho_\lambda = \frac{1}{N_i}\sum_{i=1}^{N_i} \rho_{\lambda(\mu i)} = \frac{\sqrt{\mathscr{N}_{1\mu}}\langle\sqrt{\mathscr{N}_{1i}}\rangle e^2 M m_\mu}{6\pi\epsilon_0\hbar^2 N_i}, \quad \langle\sqrt{\mathscr{N}_{1i}}\rangle = \sum_{i=1}^{N_i} \sqrt{\mathscr{N}_{1i}}\frac{m_i}{M}. \tag{18}$$

4. Generalised theory of gravity and relativity

We shall in this section generalise the theory of gravity of Sec 3 by including the effect of gravity on the dynamical variables of a test particle μ (Secs 4.1-2) and test macroscopic object (Sec 4.3). We refer to this effect as the *general relativistic* (g-r) effect in this work. The g-r effect manifestly coincides with the additional content of Einstein's GR over Newton's gravity; the usual light quantum and the IED particle employed in Sec 3 are already intrinsically special relativistic and governed by the Lorentz transformations. To facilitate the discussion, we re-locate M at $r = 0$ and μ at a distance $\mathbf{r}$ from it in the co-ordinate system x, y, z. And we re-express Newton's gravity here using "proper" dynamical variables r^0, m^0 measured at the limit $g \cdot r^0 \to 0$, indicated by a superscript 0, acting along $\mathbf{r}$ direction as,

$$F(r^0) = -g(r^0)m^0 = -\frac{\partial V(r^0)}{\partial r^0} = -\frac{GMm^0}{(r^0)^2}, \quad g(r^0) = \frac{GM}{(r^0)^2}, \quad V(r^0) = -\frac{GMm^0}{r^0}. \tag{19}$$

4.1 Effect of gravity on the wave and particle-dynamics variables of a single μ μ may be an IED matter particle or light quantum, and is in stationary state in a g field as specified by (19). Let firstly be no applied non-gravitational force present. At $r^0 \to \infty$, $g(r^0) \cdot r^0 = -V/m^0 \to 0$ which is a maximum. Accordingly μ has an inertial mass $m^0 = m(\infty)$, mass energy $\varepsilon^0 = m^0c^2$, and a *capacity to work* $\varepsilon^0 = m^0c^2$, which are maximum each. When brought from infinity to a finite separation $r = r(r^0)$ under F, F has done a negative work to μ, $\Delta V = -\int_\infty^{r^0} F dr = -\frac{GMm^0}{r^0} = V(r^0)$ along r-direction. Assume that the process is (quasi) static and hence in general non-adiabatic, so no kinetic energy (T) has been gained by μ; T would be lost, to such as heat. The total mechanical energy or Hamiltonian of μ, hence also its *capacity* to do work, is thus reduced by the amount $-\Delta V = -V$ to

$$H = \varepsilon^0 + V(r^0) = m^0c^2 - \frac{GMm^0}{r^0}. \tag{20}$$

μ is dually a quantum wave. For $V = 0$, μ has a usual total eigen plane wave function $\psi(\mathbf{r}^0, t^0) = Ce^{i(\mathbf{k}_d^0\cdot\mathbf{r}^0 - 2\pi\nu t^0)}$ (the same ψ is given by the IED solution through $\psi(\mathbf{r}^0, t^0) = \sum_j[\varphi^j(\mathbf{r}^0, t^0; \frac{\pi}{2}) + i\varphi^j(\mathbf{r}^0, t^0)]$) and an eigen frequency $\nu^0 = m^0c^2/h$, where $k_d^0 = (\frac{v}{c})k^0$, v is particle speed and $k^0 = 2\pi\nu^0/c$. For a finite V, and H as given in (20), we may establish the corresponding operator H_{op}, the eigen value equation and subsequently obtain (solve for) the eigen value H, in a region where $V(r^0)$ is essentially constant,

$$H_{op}\psi = H\psi, \quad H_{op} = \varepsilon^0 + V(r^0) = m^0c^2 - \frac{GMm^0}{r^0}; \tag{21}$$

$$H = h\nu = m^0c^2 - \frac{GMm^0}{r^0} = h\nu^0 - \frac{GMh\nu^0}{r^0c^2} \quad \text{or} \quad \nu = \nu^0(1 - \frac{GM}{r^0c^2}). \tag{22}$$

Here, for V is constant, the eigen function continues to be a plane wave, $\psi(\mathbf{r}, t) = Ce^{i(\mathbf{k}_d\cdot\mathbf{r} - 2\pi\nu t)}$. So $i\hbar\partial\psi/\partial t = h\nu\psi$. Its equality with $H\psi$ of (21a) gives (22a,b). Based on (22), the eigen frequency ν of the total μ wave is red shifted. If V is produced by a large (spherical) mass such as the earth, all test particles in a region of constant r^0 are subject to the same $V(r^0)/m$. It is thus meaningful to extend (22) to define here a "general-relativistic mass" m of μ as

$$mc^2(= h\nu) = m^0c^2 - \frac{GMm^0}{r^0} \quad \text{or} \quad m = m^0(1 - \frac{GM}{r^0c^2}). \tag{23}$$

More generally, μ may be moving, at a velocity v such that $\gamma = 1/\sqrt{1 - v^2/c^2} > 1$ appreciably, and subject to an applied non-gravitational potential V_{ap}. Then H is now

$$H' = m^0c^2 + V + V_{ap} = mc^2 + V_{ap}, \quad \text{or} \quad (H' - V_{ap})^2 = m^2c^4 = m_{\text{rest}}^2c^4 + (mv)^2c^2 \tag{24}$$

where $m = \gamma m_{\text{rest}}$, $m_{\text{rest}} = \lim_{v^2/c^2\to 0} m$. The corresponding eigenvalue equation now describes the μ. In typical applications, such as those in Sec 5, V_{ap} is electromagnetic and varies over a quantum length scale, $a \sim 10^{-10}$ m or shorter. Across a, $\Delta V = |V(r) - V(r + a)| \doteq |V(r)\frac{a}{r}|$ is in general $<< |V_{ap}|^2$; so V is essentially constant. (20)-(23) for a free μ thus hold directly.

4.2 Effect of gravity on the space and time co-ordinates for the single μ Based on Sec 3, gravity F is transmitted to and from μ (at r^0) through the propagation of the ψ_p wave (Eq 4) from and to M (at $r^0 = 0$) at the constant speed c across a distance r^0 given as, for $g \cdot r^0 \to 0$,

$$r^0 = t^0c = \mathscr{N}\tau^0c = \mathscr{N}\lambda^0, \quad \lambda^0 = \tau^0c, \tag{25}$$

where $\mathscr{N}$ is the number of wavelengths contained in r^0. The relationship (25a) firstly means that if ψ_p is emitted by M at $r^0 = 0$ at time $t = 0$, it then arrives to μ at position r^0 after a time t^0. Alternatively, it also means that if the first wave front of ψ_p enters μ at position r^0 at time t_1^0, it takes a further time t^0, i.e. at an absolute time $t_1^0 + t^0$ latter, for the $\mathscr{N}$ wavelengths to pass through μ. The second meaning connects t^0, r^0 directly with the local variables τ^0, λ^0 at the location r^0, and it is the so-signified t^0, and $r^0 = ct^0$, that directly characterise the magnitude of F in (19.1). In this latter sense, $r^0 = \mathscr{N}\lambda^0$ stands for a *gravitational optical*, or simply "*gravito optical*" distance traversed by the gravity wave ψ_p; and t^0 for the time.

For a finite V, the red shifted ν from ν^0 in (22b) directly describes the particle fields φ's of μ located at r^0 and hence, based on (Eq 4), the ψ_p wave emitted by μ here. For the ψ_p wave transmitted to μ (from M), we can be led to the same red shifted ν by arguing simply based on

[2] For example, for an electron e and proton p at a separation $a \sim 1 \times 10^{-10}$ m, $V_{ap} \sim -e^2/4\pi\epsilon_0 a = -14.4$ eV. For p at the sun surface, $V = -GMm_p/R = -1992$ eV, $\Delta V \doteq V(R)\frac{a}{R} = 2.86 \times 10^{-16}$ eV $<< V_{ap}$; and on the earth, $V = -0.653$ eV, ΔV is even trivially small.

Newton's law for action and reaction, assuming M and μ are in dynamical equilibrium.[3] So, the local wave variables for characterising the magnitude of gravity F transmitted either to or from μ are the red shifted ν of (22b), and λ given in (26a) below. Accordingly, the "gravito optical" distance r and time t for the $\mathcal{N}$ number of wavelengths to pass through (either into or out of) μ are defined by the local λ and $\tau = 1/\nu$ at $r(r^0)$ given by (26b,c) below:

$$\lambda = \frac{\lambda^0}{1-\frac{GM}{r^0c^2}}, \quad r = \mathcal{N}\lambda = \frac{r^0}{1-\frac{GM}{r^0c^2}}, \quad t = \frac{r}{c} = \mathcal{N}\tau = \frac{t^0}{1-\frac{GM}{r^0c^2}}. \tag{26}$$

(26) hold irrespective of the variant $\frac{GM}{r'^0c^2}$ with $r' \approx r'^0 (< r)$ before ψ_p arrives at μ or after ψ_p has left μ. Based on (26b,c), it takes elongated gravito optical distance and time to transmit the same $\mathcal{N}$ wavelengths of gravity F. μ is thus acted by a reduced F, hence able to move further apart, and re-equilibrated at the dilated r. r gives then the observational distance.

By virtue of the underlining second meaning of (25a), the effect of gravity on the space and time co-ordinates manifests exclusively along a g field line. Given a μ at a distant point $(x, 0, 0)$ on the x axis. Then in a small local region such that every point in it is connected to M by a g line parallel with the x axis, the general relativistic transformations from the proper co-ordinates x^0, y^0, z^0, t^0 to the x, y, z, t affected by the g field, are given as

$$x = \gamma_g x^0, \quad t = \frac{x}{c} = \gamma_g t^0, \quad \gamma_g = \frac{1}{1-\frac{GM}{x^0c^2}}; \quad y = y^0; \quad z = z^0. \tag{27}$$

In spherical polar co-ordinates and for a point $\mathbf{r}$ on the x axis, these become $r = \gamma_g r^0$, $t = \gamma_g t^0$; $\phi = \phi^0$; $\theta = \theta^0$. Using the above for r, keeping t as a dependent variable, the transformation for the (invariant) squared shortest line element (or geodesic) ds of light is $(ds)^2 = -c^2(dt)^2 + \gamma_g^2[(dr^0)^2 + (r^0)^2(d\theta^0)^2 + (r^0)^2 \sin^2\theta^0 (d\phi^0)^2]$, where $dr \doteq dr^0(1-\frac{GM}{r^0c^2})^{-1}$.

Based on experiment, Newton's law (19) holds accurately in the $g\cdot r \to 0$ limit. This suggests that, by retrieving from r the $g\cdot r \to 0$ value r^0 based on (26b), and using this in (19), we can obtain the correct gravity, the ratios F/m^0 and V/m^0 here, as:

$$\frac{F(r(r^0))}{m^0(r^0)} = g(r(r^0)) = -\frac{GM}{(r^0)^2} = -\frac{GM}{r^2\left(1-\frac{GM}{r^0c^2}\right)^2}, \quad \frac{V(r(r^0))}{m^0(r^0)} = -\frac{GM}{r\left(1-\frac{GM}{r^0c^2}\right)}. \tag{28}$$

From a more basic consideration, the G value in Newton's law (19) is experimentally determined on the earth (or between two Cavendish balls) which mass to r ratio is small on an astronomical scale. This G thus represents the $g\cdot r \to 0$ value. (19) is thus expected to hold exactly if the $g\cdot r \to 0$ values of all other variables (m, r) are consistently used in it; this is done in (28). Using (28a) for F, (26b) for r^0, the Newtonian equation of motion in the $g\cdot r \to 0$ limit (which we are certain to be correct) is thus

$$\frac{d^2\mathbf{r}^0}{d(t^0)^2}\left(=\frac{\mathbf{F}(r^0)}{m^0}\right) = -\frac{GM}{(r^0)^2}\hat{r}. \quad \text{Or} \quad \frac{d^2\mathbf{r}}{d(t^0)^2} = -\frac{GM}{r^2(1-\frac{GM}{rc^2})^3}\hat{r} \tag{29}$$

expressed using the observational r; the proper t^0 is kept, assuming this may be theoretically estimated. In spherical polar co-ordinates (r^0, θ^0, ϕ^0), $\mathbf{dr}^0 = dr^0\hat{r}^0 + r^0 d\theta^0\hat{\theta}^0 + r^0\sin\theta^0 d\phi^0\hat{\phi}^0$, $\frac{d^2\mathbf{r}^0}{d(t^0)^2} = \frac{d^2r^0}{d(t^0)^2}\hat{r} + (\frac{dr^0}{dt^0} + r^0\frac{d}{dt^0})\frac{d\boldsymbol{\theta}^0}{dt^0} + (\sin\theta^0\frac{dr^0}{dt^0} + r^0\cos\theta^0\frac{d\theta^0}{dt^0} + r^0\sin\theta^0\frac{d}{dt^0})\frac{d\boldsymbol{\phi}^0}{dt^0}$.

If an applied non-gravitational force F_{ap} also presents along r^0 direction, similarly we can firstly write down the eom in the $g\cdot r \to 0$ limit, $F(r^0) + F_{ap}(r^0) = m^0\frac{d^2r^0}{d(t^0)^2}$. Using the g-r

[3] One can formally describe the (self) effect of gravity of M on its own $\psi_{pM}(r,t)$ wave, by treating the ψ_{pM} perturbed vacuuons at r as a test entity.

transformations in it then gives the eom expressed in r, etc. The $g \cdot r \to 0$ eom is a statement of the "(weak) equivalence principle". Namely, the m^0 acted by $F(r^0)$ or by $F_{ap}(r^0)$ is the same mass in nature; the accelerations produced by $F(r^0)$ and $F_{ap}(r^0)$ are accordingly equivalent.

4.3 Effect of gravity on the space and time co-ordinates of a macroscopic test object m_a m_a is composed of a (large) N simple single charged matter particles μ's of masses m_μ's; it has a mass $m_a = \sum_\mu m_\mu$. m_a is (i) spherical, of a radius R, and (ii) in internal thermal and hydrodynamic equilibrium. Within m_a, each μ, μ' act on one another a central force $\mathcal{F}_{\mu,\mu'}$. In addition, each μ is subject to the g potential $V(\mathbf{r}_\mu) = -\boldsymbol{g}(\mathbf{r}_\mu)\cdot\mathbf{r}_\mu m_\mu$ of a large mass M at $\mathbf{r} = 0$ at a separation $|\mathbf{r}|$ from m_a, where $\mathbf{r}_\mu = \mathbf{r} + \langle\boldsymbol{\zeta}_\mu\rangle$ and $\langle\boldsymbol{\zeta}_\mu\rangle(\le R)$ are the expectation values of the distances of μ to the CM's of M and m_a. No other external force presents. Assume also $r >> 2R$ (condition iii), so $\lim_{2R/r=0}\boldsymbol{g}(\mathbf{r}^0_\mu) = \boldsymbol{g} \equiv \boldsymbol{g}(\mathbf{r}^0) = \frac{GM}{(r^0)^2}\hat{r}$, i.e. all μ's are subject to the same $\boldsymbol{g}$. In terms of $\mathbf{r}_\mu$, each μ moves according to Newton's eom (the correspondence principle), $m_\mu\ddot{\mathbf{r}}_\mu = m_\mu\boldsymbol{g} + \sum_{\mu'\neq\mu}\mathcal{F}_{\mu,\mu'}$. Sum over all μ, with $m_a = \sum_\mu m_\mu$ and $\sum_\mu\sum_{\mu'\neq\mu}\mathcal{F}_{\mu,\mu'} = 0$ under condition (ii):

$$\sum_\mu m_\mu\ddot{\mathbf{r}}_\mu = m_a\boldsymbol{g} + \sum_\mu\sum_{\mu'\neq\mu}\mathcal{F}_{\mu,\mu'} = m_a\boldsymbol{g} = m_a\ddot{\mathbf{r}}, \quad \mathbf{r} = \frac{\sum_\mu m_\mu\mathbf{r}_\mu}{m_a} \tag{30}$$

Using (23b), (26b) for m_μ, r_μ in (30b), with $m_a = \sum_\mu m_\mu = \sum_\mu(1-\frac{GM}{r^0c^2})m^0_\mu = (1-\frac{GM}{r^0c^2})m^0_a$, $m^0_a = \sum_\mu m^0_\mu$, we obtain the scalar form of $\mathbf{r}$, and time $t = r/c$,

$$r = \frac{|\sum_\mu m^0_\mu(1-\frac{GM}{r^0c^2})\frac{\mathbf{r}^0_\mu}{1-\frac{GM}{r^0c^2}}|}{m_a} = \frac{m^0_a r^0}{m^0_a(1-\frac{GM}{r^0c^2})} = \frac{r^0}{1-\frac{GM}{r^0c^2}}, \quad t = \frac{t^0}{1-\frac{GM}{r^0c^2}}, \tag{31}$$

where $m^0_a r^0 = |\sum_\mu m^0_\mu\mathbf{r}^0_\mu|$, or $r^0 = |\sum_\mu\frac{m^0_\mu}{m^0_a}\mathbf{r}^0_\mu|$; $t^0 = |\sum_\mu\frac{m^0_\mu}{m^0_a}\frac{\mathbf{r}^0_\mu}{c}|$; $t^0 = r^0/c$.

5. Predictions of g-r effects in the classical test experiments of GR

We apply in this section the solutions of Sec 4 to predict the g-r effects as manifested in the four classical-test experiments of GR.

(i) Anomalous mercury perihelion precession Mercury rotates along an ellipse about the sun of mass $M(=1.99\times10^{30}$ kg) under the sun's gravity F as given by (28). Its eom is (29a) in terms of r^0, t^0. Approximating the ellipse by a circle of radius $r = a\sqrt{1-e^2}(=57.24\times10^9$ m) in the xy plane here, with the sun at $r = 0$, (29a) becomes $-\frac{r^0d^2\phi^0}{d(t^0)^2}\hat{r} = \boldsymbol{g}^0$; or $-d(\frac{d\mathbf{r}^0}{dt^0}) = \boldsymbol{g}^0dt^0$, where $|d\mathbf{r}^0| = r^0d\phi^0(=d(r^0\phi^0))$, $\frac{d\mathbf{r}^0}{dt^0} = \boldsymbol{v}^0$, and $d(\frac{d\mathbf{r}^0}{dt^0}) = d\boldsymbol{v}^0 = \boldsymbol{v}^0(t^0+dt^0) - \boldsymbol{v}^0(t^0) = d\boldsymbol{v}^0_r(\equiv dv^0_r\hat{r})$. Solving the equation using similar-triangles relation gives $\frac{g^0dt^0}{v^0} = \frac{|d\mathbf{r}^0|}{r^0}$. Integrating over one rotation period $T^0 = 2\pi r^0/v^0$ gives $\frac{g^0T^0}{v^0} = \frac{\Phi^0r^0}{r^0}$. Or, $\Phi^0 = \frac{g^0T^0}{v^0} = \frac{GM(T^0)^2}{2\pi(r^0)^3}$ for the sweep angle in T^0. Using (31.a) for r^0 in it gives the sweep angle (Φ) in one rotation period $T^{0^*} = 2\pi r/v$, in terms of the observational r and v as, with Φ^0 computed as $\Phi^{0^*} = \frac{GM(T^{0^*})^2}{2\pi r^3}$,

$$\Phi = \frac{GM(T^{0^*})^2}{2\pi r^3(1-\frac{GM}{r^0c^2})^3} \doteq \Phi^{0^*}(1+\frac{3GM}{rc^2}). \quad \text{Or} \quad \Delta\Phi = \Phi - \Phi^{0^*} = \frac{6\pi GM}{rc^2} = 42''/\text{century}, \tag{32}$$

where $\Phi^{0^*} = 2\pi$ follows from geometric definition (in flat space) for one pure rotation in T^{0^*}. $\Delta\Phi$ is an extra, anomalous angle swept by a fixed point, e.g. the perihelion (across a closed path to a precessing observer) in T^{0^*}, and is manifestly an angle of precession. The observational value is $(\Delta\Phi)^{exp} = 43''/$century [1]. Mercury anomalous precession, and gravitational red shifts [3], were the available experiments when A Einstein devised (and justified) his GR [6, 7].

(ii) Gravitational red shift A light pulse μ emitted in the g field of the sun is according to Sec 4 red shifted, from λ^0 to $\lambda(R) \doteq \lambda^0(1+\frac{GM}{Rc^2})$ (Eq 26a) for the wavelength at the sun surface $r = R(R^0)$; here $V/m_\mu = -\frac{GM}{R^0}$. After further traversing a distance $r_{se} - R \doteq r_{se} (= 1.496 \times 10^{11}$ m) in time $t_{se} \doteq \frac{r_{se}}{c}$ to reach the earth, the instantaneous virtual source of μ (Secs 3.1-2) is decelerated to velocity $v_g = g(R) \cdot 0 - g(r_{se}) \cdot t_{se} = \frac{GM}{r^0_{se} c}$. So formally $\lambda(R)$ is further Doppler red shifted to

$$\lambda(r_{se}) = \lambda(R)(1+\frac{v_g}{c}) = \lambda(R)(1+\frac{GM}{r^0_{se}c^2}) \doteq \lambda(R) \doteq \lambda^0(1+\frac{GM}{Rc^2}) \quad (\text{for } r_{se} >> R)$$

$$\text{or} \quad \frac{\lambda(r_{se}) - \lambda^0}{\lambda^0} (\doteq \frac{g(R) \cdot R}{c^2}) \doteq \frac{GM}{Rc^2} = 2.11 \times 10^{-6}. \quad (33)$$

The predicted value agrees with the experiments [2] within accepted tolerance, and the prediction of Einstein's GR [7].

(iii) Light bending Consider a light pulse μ emitted by a distant star (s) just by-passing the sun (S) limb at the radius $R(R^0)$, and arriving to the earth (e). Let s, S, e be in the xy plane, the CM of the sun be at $r = 0$, and the μ path $sS - Se$ be such that it crosses the x axis at $x = x_0 = R(R^0)$; $sS - Se$ would coincide with $x = x_0$ if under zero gravity. But the sun presents a gravity $\mathbf{F}$; for $sSe >> 2R$, $\mathbf{F}$ may be approximated as $F_x = -gm^0$ in $-x$ direction, with $g = \frac{GM}{(R^0)^2}$ everywhere in $R \geq y \geq -R$; and $F_x = 0$ elsewhere (the usual "thin lens approximation"). μ thus needs be emitted at an angle α_s from the vertical line $x = x_0$, such that the μ path firstly goes from $(x, y) = (R, R)$ to $(x_0, 0)$, where $x_0 = R + R\tan\alpha_s - \Delta x_s = R$. This requires $R\tan\alpha_s - \Delta x_s = 0$, or $\alpha_s \doteq \tan\alpha_s (= \frac{\Delta x_s}{y}) = \frac{\Delta x_s}{R}$ (for $\frac{\Delta x_s}{R} << 1$). $\Delta x_s = \Delta x'_s + \Delta x''_s$ is an extra x-component distance including two terms: (I) $\Delta x'_s = vT^0 = \frac{GM}{c^2}$ traversed by the instantaneous virtual source of μ (Sec 3), which is accelerated under F_x in $-x$ direction to velocity $v = -gT^0$ in time $T^0 = \frac{R^0}{c}$, and (II) $\Delta x''_s = c(T - T^0) \doteq \frac{GM}{c^2}$ traversed by the radiation field comprising μ at the speed c, as the result of time dilation from T^0 to $T \doteq T^0(1+\frac{GM}{R^0c^2})$ (Eq 26c) under F_x in $-x$ direction. Further from $(x, y) = (R, 0)$ to $(R - \Delta x_e, -R)$, the μ path is bent similarly, such that $\alpha_e \doteq \tan\alpha_e = \frac{\Delta x_e}{R}$; but $\Delta x_e = \Delta x_s$ and $\alpha_e = \alpha_s$ for F_x is symmetric about the x axis. The total bending angle (α_b) is the sum

$$\alpha_b = \alpha_s + \alpha_e \doteq \frac{\Delta x_s + \Delta x_e}{R} = \frac{2\Delta x_s}{R} = \frac{2(\Delta x'_s + \Delta x''_s)}{R} = \frac{4GM}{Rc^2}. \quad (34)$$

The prediction agrees with experiment [4]. Einstein's GR gives the same result [8].

(iv) Shapiro time delay Consider a radar signal μ sent from the earth (e) to a planet (p) and reflected back during a superior conjunction, i.e. viewed from e, p is just above the sun (s). Let e, s, p be in the xy plane, s be at $\mathbf{r} = 0$, and the μ path be just by-passing the sun limb at $x = R$. Under the sun's gravity $\mathbf{F}(\mathbf{r}) = -g(\mathbf{r})m_\mu\hat{r}$ (Eq 28a) the distance of μ to s is dilated from r^0 to $r \doteq r^0(1+\frac{GM}{r^0c^2})$ (Eq 26b). Omitting its negligible bending under $\mathbf{F}$ (cf Sec 4.iii), the geometric μ path is just the straight line $ep(pe)$ along the y direction; a point $\mathbf{r}(x, y)$ to $y = 0$ on it has the y co-ordinate $y = \sqrt{r^2 - R^2}$; and $x = R$ everywhere. A differential line element at $\mathbf{r}(R, y)$ on the ep line is $d\mathbf{r} \doteq dy\,\hat{y}$; $dy = \frac{dr}{\sin\phi} = \frac{rdr}{y}$. Or, with (26b) for r, and $dr \doteq dr^0(1+\frac{GM}{r^0c^2})$,

$$dy = \frac{rdr}{y} = \frac{(1+2\frac{GM}{r^0c^2}+(\frac{GM}{r^0c^2})^2)\,r^0dr^0}{y}; \quad dy - dy^0 \doteq \frac{r_sdr}{\sqrt{r^2-R^2}} + \frac{r_s^2dr}{4r\sqrt{r^2-R^2}}, \quad (35)$$

where $dy^0 \doteq \frac{dr^0}{(y/r^0)}$ is to leading order the $g \cdot r \to 0$ value of dy; and $r_s = \frac{2GM}{c^2} = 2.95 \times 10^3$ m. y is not expanded for its g-dilation is of a higher order. $dy - dy^0$ is the extra (gravito optical) distance travelled by μ in the g field, and $d\tau = \frac{dy-dy^0}{c}$ gives the extra time. The total extra time from R to r in a one-way trip is

$$\tau_1|_{R\to r} = \int_{t(R)}^{t(r)} d\tau = \frac{1}{c}\int_R^r \left(\frac{r_s}{\sqrt{r'^2-R^2}} + \frac{r_s^2}{4r'\sqrt{r'^2-R^2}}\right) dr' = \frac{r_s}{c}\ln\frac{r+\sqrt{r^2-R^2}}{R} + \mathcal{O} \quad (36)$$

where $\mathcal{O} = \frac{r_s^2}{4Rc}\sec^{-1}\left(\frac{r}{R}\right)$. The total extra time, or Shapiro time delay, for μ in a round trip between the earth (at $r_e = 1.496\times10^{11}$ m) and mars (at $r_p = 2.28\times10^{11}$ m) is, omitting $\mathcal{O}$, $\tau_{\text{round}} = 2(\tau_1|_{R\to r_e} + \tau_1|_{R\to r_p}) = (2r_s/c)\ln[(r_e+\sqrt{r_e^2-R^2})(r_p+\sqrt{r_p^2-R^2})/R^2] = 247\ \mu$s. The measured value is 250 μs [5].

The author thanks Professor C Burdik for providing the opportunity to present this research at the ISQS25 in Prague, June, 2017, where the author has enjoyed communications with Dr Dahm, Professor S Catto, and other ISQS-25 participants. Professor G 't Hooft has motivated this meaningful research and has given valuable moral support for the research. Professor B Johansson has given valuable moral support for the unification research. P-I Johansson has privately financed this research and given moral support.

[1] Le Verrier U 1859 Sur la théorie de mercure et sur le mouvement du périhélie de cette planéte *Comptes rendus* **49** 379-383
[2] Evershed J 1927 The solar rotation and the Einstein displacement derived from measures of the H and K lines in prominences *Mon. Not. R. Astr. Soc.* **88** 126
[3] Forbes E G1963 A history of the solar red shift problem *Ann. Sci.* **17** 129-164
[4] Dyson F W, Eddington A S and Davidson C 1920 A determination of the deflection of light by the sun's gravitational field, from observations made at the total eclipse of 29 May 1919 *Phil. Trans. Roy. Soc.* A **220** 291-333
[5] Shapiro I I *et al* 1968 Fourth test of general relativity: preliminary results *Phys. Rev. Lett.* **20** 1265-9
[6] Einstein A 1916 Die Grundlage der allgemeinen relativitäts theorie *Ann. Phys.* **49** 769-822
[7] Einstein A 1911 Uber den Einfluss der schwerkraft auf die austreitung der lichtes *Ann. Phys. Lpz.* **35** 905
[8] See Will C M 2006 The confrontation between general relativity and experiment *Living Rev. Relativity* **9** 39
[9] 't Hooft G 2001 *Introduction to General Relativity* (Princeton: Rinton Press)
[10] Dirac P A M 1975 *General Theory of Relativity* (New York: Wiley)
[11] Misner C W, Thorne K S and d Wheeler J A 1973 *Gravitation* (San Francisco: W. H. Freeman and Company)
[12] Weinberg S 1972 *Gravitation and Cosmology: Principles and Applications of General Theory of Relativity* (New York: Wiley)
[13] Hawking S W and Ellis G F R 1973 *The Large Scale Structure of Space Time* (Cambridge: Cambridge University Press)
[14] 't Hooft G 2016 The evolution of quantum field theory, from QED to grand unification *Advanced Series on Directions in High Energy Physics* vol 26 *The Standard Theory of Particle Physics: Essays to Celebrate CERN's 60th Anniversary* ed L Maiani and L Rolandi (Singapore: World Scientific) pp 1-27
[15] 't Hooft G 2001 Obstacles on the way towards the quantization of space, time and matter - and possible resolutions *Stud. Hist. Phil. Mod. Phys.* **32** 157-180
[16] Carlip S 2001 Quantum gravity: a progress report *Rep. Prog. Phys.* **64** 885
[17] De Haro S, Dieks D, 't Hooft G, Verlinde E 2013 Forty years of string theory. Reflecting on the foundations *Found. Phys.* **43** 1-7
[18] Green M B, Schwarz J H and Witten E 1988 *Superstring Theory* vols 1,2 (Cambridge: Cambridge University Press)
[19] Zheng-Johansson J X 2010 Internally electrodynamic particle model: its experimental basis and its predictions *Phys. Atom. Nucl.* **73** 571-581 *Proc. Int. 27th Int. Colloq. Group. Theory. in Math. Phys.* ed G Pogosyan (Ireven, 2008) (*Preprint* arxiv:0812.3951)
[20] Zheng-Johansson J X, Johansson P-I, Lundin R 2006 *Suppl. Blug. J. Phys.* **33** *Quantum Theory and Symmetries IV* vol 2 ed V K Dobrev (Sofia: Heron Press) pp 771-780
[21] See e.g. Ashcroft N W and Mermin D 1976 *Solid State Physics* (Philadelphia: Saunders College) chapter 27
[22] Zheng-Johansson J X 2006 Dielectric theory of the vacuum *Preprint* arxiv:physics/0612096
[23] Zheng-Johansson J X 2012 *J. Phys.: Conf. Series* **343** 012135

IOP Conf. Series: Journal of Physics: Conf. Series **965** (2018) 012044 doi:10.1088/1742-6596/965/1/012044

Relations for Modular Forms from Vertex Operator Algebras

Alexander Zuevsky
Institute of Mathematics, Czech Academy of Sciences, Prague, Czech Republic
E-mail: zuevsky@yahoo.com

Abstract. We will give a short reminder for vertex operator algebra notion and corresponding characters. Then we discuss algebraic methods for explicit computation of the partition and correlation functions. We then illustrate general ways to find number theory identities for related modular forms by specific examples of modular form relations arising from our construction. Finally, we present new results concerning identities for prime forms on genus g Riemann surfaces and genus two n-point functions for vertex operator algebra characters.

1. Introduction

The origin of CFT is contained in the structure of corresponding vertex operator algebra (VOA) [1]. Modes of a vertex operator algebra are non-commutative and subject to certain commutation relations. n-point functions possess modular properties which (from the information point of view) corresponds to co-invariance with respect to the modular group which acts on parameters of a Riemann surface as well as on their local coordinates. This means that no matter how we change Riemann surface by using the modular group, the data coming from correlation functions of our surface remains invariant up to computable multipliers. In order to determine the higher genus correlation functions for a particular CFT, we have to understand how information on a higher genus Riemann surface can be reconstructed from data for two lower genus Riemann surfaces. It turns out that for this purpose we have to "sew" two lower genus CFT's (considered on two lower genus Riemann surfaces). But first we have to sew two Riemann surfaces themselves. For this purpose we use the Yamada sewing procedure. Finally, the procedure for sewing lower genus information allows us to obtain the higher genus n-point correlation functions which also possess modular properties. To form a genus $g_1 + g_2$ Riemann surface we sew lower genus g_1 and g_2 surfaces. The structure of vertex operator algebras allows us to compute constructively characters (partition and n-point correlation functions) and to prove explicitly their natural modularity properties. The structure of vertex operator algebras and CFT correlation functions can be used for coding.

This paper represents the author's talk given at the XXVth International Conference on Integrable Systems and Quantum symmetries (ISQS-25), 2017.

2. Geometry: construction of a genus g Riemann surface

The simplest example of forming a higher genus Riemann surface from lower genus one is the sewing the Riemann sphere to form a torus. Sew a handle to the Riemann sphere $\hat{\mathbb{C}}$ by identifying

IOP Conf. Series: Journal of Physics: Conf. Series **965** (2018) 012044 doi:10.1088/1742-6596/965/1/012044

annular regions centred at $\mathcal{A}_{\pm 1} \in \hat{\mathbb{C}}$ via a sewing condition with complex sewing parameter ρ

$$(z - \mathcal{A}_{-1})(z' - \mathcal{A}_1) = \rho,$$

This is called the canonical Sewing. We call $\rho, \mathcal{A}_{\pm 1}$ the canonical parameters. The annuli do not intersect provided

$$|\rho| < \frac{1}{4}|\mathcal{A}_{-1} - \mathcal{A}_1|^2$$

Inequivalent tori depend only on $\chi = -\frac{\rho}{(\mathcal{A}_{-1} - \mathcal{A}_1)^2}$ for $|\chi| < \frac{1}{4}$.

We may construct a general genus g Riemann surface by identifying g pairs of annuli centred at $\mathcal{A}_{\pm i} \in \hat{\mathbb{C}}$ for $i = 1, \ldots, g$ and sewing parameters ρ_i satisfying

$$(z - \mathcal{A}_{-i})(z' - \mathcal{A}_i) = \rho_i,$$

provided no two annuli intersect.

Let $x_{-i} = x - \mathcal{A}_{-i}$ and $y_j = y - \mathcal{A}_j$ be local coordinates in the neighbourhood of canonical sewing parameters $\mathcal{A}_{-i}, \mathcal{A}_j$ for $i, j \in \{\pm 1, \ldots \pm g\}$ with $i \neq -j$. The 2-point function is expanded as

$$\frac{1}{(x-y)^2} = \sum_{k,l \geq 1} (-1)^{k+1} \frac{(k+l-1)!}{(k-1)!(l-1)!} \frac{x_{-i}^{k-1} y_j^{l-1}}{(\mathcal{A}_{-i} - \mathcal{A}_j)^{k+l}}.$$

Define the canonical moment matrix R^{C}, [13] an infinite matrix indexed by $k, l = 1, 2, \ldots$ and $i, j \in \{\pm 1, \ldots \pm g\}$ where $R^{\mathrm{C}}_{i,-i}(k,l) = 0$ and

$$R^{\mathrm{C}}_{i,j}(k,l) = \frac{(-1)^k \rho_i^{k/2} \rho_j^{l/2}}{\sqrt{kl}} \frac{(k+l-1)!}{(k-1)!(l-1)!} \frac{1}{(\mathcal{A}_{-i} - \mathcal{A}_j)^{k+l}} \text{ for } i \neq -j,$$

$(I - R^{\mathrm{C}})^{-1}$ plays a central role in computing the genus g period matrix and other structures.

3. Vertex operator algebras

A Vertex Operator Algebra (VOA) [1] $(V, Y, \mathbf{1}, \omega)$ consists of a $\mathbb{Z}$-graded complex vector space $V = \bigoplus_{n \in \mathbb{Z}} V_{(n)}$ where $\dim V_{(n)} < \infty$ for each $n \in \mathbb{Z}$, a linear map $Y : V \to End(V)[[z, z^{-1}]]$ for a formal parameter z and pair of distinguished vectors: the vacuum $\mathbf{1} \in V_{(0)}$ and the conformal vector $\omega \in V_{(2)}$. For each $v \in V$, the image under the map Y is the vertex operator

$$Y(v, z) = \sum_{n \in \mathbb{Z}} v(n) z^{-n-1},$$

with *modes* $v(n) \in End(V)$, where $Y(v,z)\mathbf{1} = v + O(z)$. Vertex operators satisfy *locality* i.e. for all $u, v \in V$ there exists an integer $k \geq 0$ such that

$$(z_1 - z_2)^k [Y(u, z_1), Y(v, z_2)] = 0.$$

The vertex operator of the conformal vector ω is

$$Y(\omega, z) = \sum_{n \in \mathbb{Z}} L(n) z^{-n-2},$$

where the modes $L(n)$ satisfy the Virasoro algebra with *central charge* c

$$[L(m), L(n)] = (m-n)L(m+n) + \frac{m^3 - m}{12} \delta_{m,-n} c \, \mathrm{Id}_V.$$

We define the homogeneous space of weight k to be

$$V_{(k)} = \{v \in V | L(0)v = kv\},$$

and we write $wt(v) = k$ for $v \in V_{(k)}$. Finally, we have a translation condition

$$Y(L(-1)u, z) = \partial_z Y(u, z).$$

4. Genus zero correlation functions

For $u_1, u_2, \ldots, u_n \in V$ define the n-point (correlation) function by

$$\langle \mathbf{1}, Y(u_1, z_1)Y(u_2, z_2) \ldots Y(u_n, z_n)\mathbf{1} \rangle,$$

where the invariant form is described in subsection 17.4 in Appendix. The locality property of vertex operators implies that this formal expression coincides with the analytic expansion of a rational function of $z_1, z_2, \ldots, z_n$ in the domain $|z_1| > |z_2| > \ldots > |z_n|$. Thus the n-point function can taken to be a rational function of $z_1, z_2, \ldots, z_n \in \hat{\mathbb{C}}$, the Riemann sphere.

Theorem 4.1 (Hurley, Tuite) , *[2] For a VOA of central charge C, the Virasoro n-point function is a β-extended permanent*

$$\langle \mathbf{1}, Y(\omega, z_1) \ldots Y(\omega, z_n)\mathbf{1} \rangle = perm_{\frac{C}{2}} B,$$

for $B_{ij} = \frac{1}{(z_i - z_j)^2}, i \neq j$ and $B_{ii} = 0$, where perm denotes the permanent of a matrix.

5. The genus g partition function - canonical parameters

We now define [13] the genus g partition function for a VOA V in the canonical sewing scheme in terms of genus zero $2g$-point correlation functions as follows:

$$Z_V^{(g)}(\rho_i, \mathcal{A}_{\pm i}) = \langle \mathbf{1}, \prod_{i=1}^{g} \sum_{n_i \geq 0} \rho_i^{n_i} \sum_{v_i \in V^{(n)}} Y(v_i, \mathcal{A}_{-i}) Y(\overline{v}_i, \mathcal{A}_i)\mathbf{1} \rangle,$$

where $\overline{v}_i$ is dual to v_i. For genus one this reverts to the standard definition [6]:

Theorem 5.1 (Mason, Tuite)

$$Z_V^{(1)}(\rho, \mathcal{A}_{\pm 1}) = \mathrm{Tr}_V(q^{L_0})$$

where $q = C(\chi)$, the Catalan series for $\chi = -\frac{\rho}{(\mathcal{A}_{-1} - \mathcal{A}_1)^2}$.

6. Partition function $Z_{M_2}^{(g)}(\rho_i, \mathcal{A}_{\pm i})$ for Heisenberg VOA M_2

Theorem 6.1 (Tuite and Z.) , *[13]*

- $$Z_{M_2}^{(g)}(\rho_i, \mathcal{A}_{\pm i}) = \frac{1}{\det(I - R^{\mathrm{C}})},$$

  where R^{C} is the canonical moment matrix.
- $\det(I - R^{\mathrm{C}})$ *is holomorphic and non-vanishing.*
- *The genus g Heisenberg generating function can be computed in terms of a permanent of genus g bilinear forms of the second kind.*

7. Genus one correlation functions and Zhu recursion

The *genus one partition function* for V is defined by the trace function [1]

$$Z_V^{(1)}(\tau) = \mathrm{Tr}_V\left(q^{L(0)-c/24}\right),$$

and the *genus one n-point correlation function* by

$$Z_V^{(1)}(v_1, z_1; \ldots; v_n, z_n; \tau) = \mathrm{Tr}_V\left(Y(q_{z_1}^{L(0)} v_1, q_{z_1}) \ldots Y(q_{z_n}^{L(0)} v_n, q_{z_n}) q^{L(0)-\frac{c}{24}}\right).$$

In particular, the genus one 1-point function for $v \in V$ is

$$Z_V^{(1)}(v;\tau) = \mathrm{Tr}_V\left(o(v) q^{L(0)-c/24}\right),$$

where, for v homogeneous, $o(v) := v(wt(v)-1) : V_{(m)} \to V_{(m)}$. Every n-point function is expressible in terms of 1-point functions

$$\begin{aligned}
&Z_V^{(1)}(v_1, z_1; \ldots; v_n, z_n; \tau)\\
&= Z_V^{(1)}(Y[v_1, z_1] \ldots Y[v_n, z_n]\mathbf{1}; \tau)\\
&= Z_V^{(1)}(Y[v_1, z_1 - z_n] \ldots Y[v_n, z_{n-1} - z_n] v_n; \tau).
\end{aligned}$$

8. Genus one correlation functions and Zhu recursion

We will make repeated Zhu recursion [14] which recursively relates n-point correlation functions to $(n-1)$-point functions

Theorem 8.1 *[Zhu Recursion] Genus one n-point correlation functions obey*

$$\begin{aligned}
&Z_V^{(1)}(v_1, z_1; \ldots; v_n, z_n; \tau)\\
&= \mathrm{Tr}_V\left(o(v_1) Y(q_{z_2}^{L(0)} v_2, q_{z_2}) \ldots Y(q_{z_n}^{L(0)} v_n, q_{z_n}) q^{L(0)-c/24}\right)\\
&+ \sum_{k=2}^{n} \sum_{j \geq 0} P_{1+j}(z_1 - z_k, \tau) Z_V^{(1)}(v_2, z_2; \ldots; v_1[j] v_k, z_k; \ldots; v_n, z_n; \tau).
\end{aligned}$$

9. Genus two surfaces formed from two tori

For the general description of the genus two Riemann surface formation from to tori see [6]. Consider an oriented torus $\mathcal{S}_a^{(1)} = \mathbb{C}/\Lambda_a$ with lattice $\Lambda_a = 2\pi i(\mathbb{Z}\tau_a \oplus \mathbb{Z})$ for $\tau_a \in \mathbb{H}_1$. For local coordinate $z_a \in \mathbb{C}/\Lambda_a$ the closed disc $|z_a| \leq r_a$ is contained in $\mathcal{S}_a^{(1)}$ provided $r_a < \frac{1}{2}D(q_a)$ where

$$D(q_a) = \min_{\lambda \in \Lambda_a, \lambda \neq 0} |\lambda|,$$

is the minimal lattice distance. One introduces a complex sewing parameter ϵ where $|\epsilon| \leq r_1 r_2 < \frac{1}{4}D(q_1)D(q_2)$ and excise the disc $\{z_a, |z_a| \leq |\epsilon|/r_{\bar{a}}\}$ centered at $z_a = 0$ to form a punctured torus

$$\widehat{\mathcal{S}}_a^{(1)} = \mathcal{S}_a^{(1)} \backslash \{z_a, |z_a| \leq |\epsilon|/r_{\bar{a}}\},$$

where we here (and below) we use the convention

$$\bar{1} = 2, \quad \bar{2} = 1.$$

Defining the annulus $\mathcal{A}_a = \{z_a, |\epsilon|/r_{\bar{a}} \leq |z_a| \leq r_a\}$ we identify $\mathcal{A}_1$ with $\mathcal{A}_2$ via the sewing relation

$$z_1 z_2 = \epsilon.$$

The genus two Riemann surface $\mathcal{S}^{(2)}$ is parameterized by the sewing domain

$$\mathcal{D}_{\text{sew}} = \left\{ (\tau_1, \tau_2, \epsilon) \in \mathbb{H}_1 \times \mathbb{H}_1 \times \mathbb{C} : |\epsilon| < \frac{1}{4} D(q_1) D(q_2) \right\}.$$

10. Genus g partition function by sewing two Riemann surfaces

We define the genus g partition function by inductively sewing together lower genus 1-point functions $Z_M^{(g_i)}(u, z_i)$, $i = 1, 2$, $z_i \in \Sigma^{(g_i)}$, u belongs to n-th grading subspace of V, for a V-module M:

$$Z_M^{(g)}(z_1, z_2, \epsilon) = \sum_{u \in V_{[n]}} Z_M^{(g_1)}(u, z_1) Z_M^{(g_2)}(\bar{u}, z_2),$$

where $g = g_1 + g_2$ and z_1, z_2 are insertion points, ϵ is the sewing parameter, and $\bar{u}$, u are related by a non-degenerate bilinear invariant form on V. For instance, with suitable local coordinates in the neighbourhood of these points we find for the Heisenberg VOA M_2 one obtains

Theorem 10.1

$$Z_{M_2}^{(g)}(v_1, v_2, \epsilon) = \frac{1}{\det(I - Q_{(g_1, g_2)})} Z_{M_2}^{(g_1)} Z_{M_2}^{(g_2)},$$

where $Z_{M_2}^{(g)}$ is a non-vanishing holomorphic function on the sewing domain and is automorphic with respect to $SL(2, \mathbb{Z}) \times \ldots \times SL(2, \mathbb{Z}) \subset Sp(2g, \mathbb{Z})$ with automorphy factor $\det(C\Omega^{(g)} + D)^{-1}$, *and a multiplier system, $Q_{(g_1, g_2)}$ is a moment matrix containing genus g sewing data, and $\Omega^{(g)}$ is genus g period matrix.*

The *genus g n-point correlation function* for $a_1, \ldots, a_L$ and $b_1, \ldots, b_R$ formally inserted at $x_1, \ldots, x_L \in \widehat{\mathcal{S}}_1$ and $y_1, \ldots, y_R \in \widehat{\mathcal{S}}_2$, respectively, by

$$Z_V^{(g)} \left(a_1, x_1; \ldots; a_L, x_L | b_1, y_1; \ldots; b_R, y_R; \Omega_1^{(g_1)}, \Omega_2^{(g_2)}, \epsilon \right)$$
$$= \sum_{u \in V} Z_V^{(g_1)} \left(Y[a_1, x_1] \ldots Y[a_L, x_L] u; \Omega_1^{(g_1)} \right) Z_V^{(g_2)} \left(Y[b_R, y_R] \ldots Y[b_1, y_1] \overline{u}; \Omega_2^{(g_2)} \right).$$

11. General ways to find number theory identities for related modular forms:

Having at hands the fundamental way allowing us to construct the higher genus characters for vertex operator algebras, we proceed to the identities resulting from this method. Here we can enumerate a few general vertex operator algebra methods to derive number theory identities for related modular forms:

- Boson-Fermion correspondence;
- Comparison of alternative representations of characters;
- Higher-Lower genus relations for characters;
- Powers of η-functions according from partition functions and determinants;

- Modular-invariance properties of characters.

Let us proceed with examples of relations for modular forms that appear in computations of partition and n-point functions for vertex operator algebras. In the following sections we illustrate the above mentioned general vertex operator algebra methods by specific examples of modular relations arising from our construction.

12. Triple Jacobi identity

As a result of the boson-fermion correspondence one obtains Jacobi triple product identity [1]:

$$q^{\frac{\kappa^2}{2}-\frac{1}{24}}\prod_{l\geq 1}(1-\theta^{-1}q^{l-\frac{1}{2}-\kappa})(1-\theta q^{l-\frac{1}{2}+\kappa}) = \frac{e^{2\pi i(\alpha+\frac{1}{2})(\beta+\frac{1}{2})}}{\eta(\tau)}\vartheta^{(1)}\begin{bmatrix}\mu_1\\ \mu_2\end{bmatrix}(0,\tau),$$

where $\mu_1 = -\beta+\frac{1}{2}$, $\mu_2 = \alpha+\frac{1}{2}$. Here $\eta(\tau)$ is the Dedekind eta-function

$$\eta(\tau) = q^{1/24}\prod_{n=1}^{\infty}(1-q^n).$$

Thus we get the identity for the first power of the η-function.

13. Genus two triple Jacobi identities

We could made use of the genus one identities to derive higher genus analogues of Jacobi product identities [10]. In particular, at genus two we get an explicit analogue of classical Jacobi triple identity

$$\frac{\Theta^{(2)}\begin{bmatrix}a\\ b\end{bmatrix}(\Omega)}{\vartheta^{(1)}\begin{bmatrix}a_1\\ b_1\end{bmatrix}(\tau_1)\,\vartheta^{(1)}\begin{bmatrix}a_2\\ b_2\end{bmatrix}(\tau_2)} = \det\left(I - A_1^{(1)}A_2^{(1)}\right)^{1/2}\det(I-Q).$$

Here $\vartheta^{(1)}\begin{bmatrix}a\\ b\end{bmatrix}(\tau)$, $\Theta^{(2)}\begin{bmatrix}a\\ b\end{bmatrix}(\Omega)$, are genus one and genus two theta-functions correspondingly, and Ω is a period matrix. In the alternative ρ-parameter self-sewing (see subsection 17.2 of Appendix) of the torus to form a genus two Riemann surface:

Theorem 13.1 *[11] The ratio of genus two and genus one Riemann theta functions on $\mathcal{D}^\rho$ is given by*

$$\frac{\vartheta^{(2)}\begin{bmatrix}\alpha\\ \beta\end{bmatrix}(\Omega^{(2)})}{\vartheta^{(1)}\begin{bmatrix}\alpha_1\\ \beta_1\end{bmatrix}(\kappa w,\tau)} = e^{2i\pi\beta_2\kappa}\left(\frac{e^{i\pi B}\rho}{K(w,\tau)^2}\right)^{\frac{1}{2}\kappa^2}\det(I-T)\det(I-R)^{\frac{1}{2}}.$$

14. Modular discriminant and generalizations of the Garvan's formula

The modular discriminant is defined by

$$\Delta(\tau) = \eta(\tau)^{24},$$

where $\eta(\tau)$ is the Dedekind eta-function

$$\eta(\tau) = q^{1/24}\prod_{n=1}^{\infty}(1-q^n).$$

The fundamental classical formula for the modular discriminant is

$$\Delta(\tau) = \frac{1}{1728} \det \begin{pmatrix} E_4(\tau) & E_6(\tau) \\ E_6(\tau) & E_8(\tau) \end{pmatrix}.$$

The following formula was first proposed by F. Garvan.

$$\Delta^2(\tau) = -\frac{691}{250\,(1728)^2} \det \begin{pmatrix} E_4(\tau) & E_6(\tau) & E_8(\tau) \\ E_6(\tau) & E_8(\tau) & E_{10}(\tau) \\ E_8(\tau) & E_{10}(\tau) & E_{12}(\tau) \end{pmatrix}.$$

Using the Fay's trisecant identity we obtain in [15]:

Theorem 14.1 *For* $(\theta, \phi) \neq (1,1)$ *one has*

$$\Delta^n(\tau) = \frac{\vartheta^{(1)}\begin{bmatrix} \frac{1}{2} \\ \frac{1}{2} \end{bmatrix}(0,\tau)\, \Theta^{(1)}_{8n,8n,(\mathbf{1},\mathbf{1})}(\mathbf{x},\mathbf{y},\tau).}{\vartheta^{(1)}\begin{bmatrix} -\beta+\frac{1}{2} \\ \alpha+\frac{1}{2} \end{bmatrix}\left(\sum_{i=1}^{8n}(x_i - y_i), \tau\right)} \det\left(\mathbf{P}_{8n}(\theta,\phi)\right).$$

For $(\theta, \phi) = (1,1)$,

$$\Delta^n(\tau) = -\frac{\Theta^{(1)}_{8n+1,8n+1,(\mathbf{1},\mathbf{1})}(\mathbf{x},\mathbf{y},\tau)}{\vartheta^{(1)}\begin{bmatrix} \frac{1}{2} \\ \frac{1}{2} \end{bmatrix}\left(\sum_{i=1}^{8n+1}(x_i - y_i), \tau\right)} \det\left(\mathbf{Q}_{8n+1}\right).$$

where

$$\Theta^{(1)}_{r,s,(\mathbf{m},\mathbf{n})}(\mathbf{x},\mathbf{y},\tau) \equiv \frac{\prod_{1\le i\le r, 1\le j\le s} \vartheta^{(1)}\begin{bmatrix} \frac{1}{2} \\ \frac{1}{2} \end{bmatrix}(x_i - y_j, \tau)^{m_i n_j}}{\prod_{1\le i<k\le r} \vartheta^{(1)}\begin{bmatrix} \frac{1}{2} \\ \frac{1}{2} \end{bmatrix}(x_i - x_k, \tau)^{m_i m_k} \prod_{1\le j<l\le s} \vartheta^{(1)}\begin{bmatrix} \frac{1}{2} \\ \frac{1}{2} \end{bmatrix}(y_j - y_l, \tau)^{n_j n_l}},$$

$$\mathbf{P}_n(\theta,\phi) = \left(P_1\begin{bmatrix} \theta \\ \phi \end{bmatrix}(x_i - y_j, \tau)\right), \quad (1 \le i, j \le n),$$

$$\mathbf{Q}_n = \begin{pmatrix} P_1(x_1 - y_1, \tau) & \ldots & P_1(x_1 - y_n, \tau) & 1 \\ \vdots & \ddots & & \vdots \\ P_1(x_n - y_1, \tau) & & P_1(x_n - y_n, \tau) & 1 \\ 1 & \ldots & 1 & 0 \end{pmatrix}.$$

Using the full version of the Fay's generalized trisecant identity we derive the following

Theorem 14.2 *For* $(\theta, \phi) \neq (1,1)$,

$$\Delta^p(\tau) = \frac{\vartheta^{(1)}\begin{bmatrix} -\beta+\frac{1}{2} \\ \alpha+\frac{1}{2} \end{bmatrix}(0,\tau)\, \Theta^{(1)}_{r,s,(\mathbf{m},\mathbf{n})}(x,y,\tau)}{\vartheta^{(1)}\begin{bmatrix} -\beta+\frac{1}{2} \\ \alpha+\frac{1}{2} \end{bmatrix}\left(\sum_{i=1}^{r} m_i x_i - \sum_{j=1}^{s} n_j y_j, \tau\right)} \det(\mathbf{M}_{r,s}),$$

where

$$p = 8\left(\sum_{1\le i\le r, 1\le k\le s} m_i n_j - \sum_{1\le i<k\le r} m_i m_k - \sum_{1\le j<l\le s} n_j n_l\right).$$

IOP Conf. Series: Journal of Physics: Conf. Series **965** (2018) 012044 doi:10.1088/1742-6596/965/1/012044

15. Formulas relating prime forms through period matrices and theta-functions

Finally, in the next two sections we present new results concerning identities for prime forms (see subsection 17.1 of Appendix) on genus g Riemann surfaces and genus two n-point functions for vertex operator algebras on Riemann surfaces.

In [9] we have derived the formula relating the genus $g+1$ and genus g prime forms $K^{(g+1)}\left(x, y, \Omega^{(g+1)}\right) = K^{(g)}\left(x, y, \Omega^{(g)}\right) e^{-\frac{1}{2} b_a X_{\bar{a}\bar{a}}(\rho) b_a^T}$, and in particular,

$$K^{(2)}\left(x, y, \Omega^{(2)}\right) e^{\frac{1}{2} b_a(x,y;k)\, X_{\bar{a}\bar{a}}(\rho)\, b_a^T(x,y;k)} = K^{(1)}(x-y, \tau).$$

Here $b_a(x,y;k) = \int_y^x a_a(\cdot, k)$, and $a_a(x) X_{\bar{a}\bar{a}}(\rho) a_a^T(y) = \sum_{k,l \geq 1} a_a(x,k) X_{\bar{a}\bar{a}}(k,l,\rho) a_a(y,l)$, with $a_a(x,k)$ a certain one-form on the initial Riemann surface $\widehat{\Sigma}^{(g_a)}$ of genus g. Here $X_{\bar{a}\bar{a}}(k,l,\rho)$ is an infinite matrix determined from genus g data (see [6] for details). In [9] we have proved that $\mathcal{S}^{(g+1)}$ is holomorphic in ρ for $|\rho| < r_1 r_2$ with $\mathcal{S}^{(g+1)}(x,y) = S_\kappa^{(g)}(x,y) + O(\rho)$, for some kernel $S_\kappa^{(g)}(x,y)$.

For the genus $g+1$ prime form $K^{(g+1)}$ and genus g prime form $K^{(g)}$ we obtain here the following:

Proposition 15.1 *Let us define* $U^{(g)}(x,y) = \frac{K^{(g)}(x,p_2)\, K^{(g)}(y,p_1)}{K^{(g)}(x,p_1)\, K^{(g)}(y,p_2)}$, *and* $z_{p_1,p_2} = \int\limits_{p_1}^{p_2} \nu^{(g)}$, *for holomorphic one-forms* $\nu^{(g)}$. *One can relate the genus* $g+1$ *and* g *prime forms* $K^{(g+1)}$ *and* $K^{(g)}$ *by means of the following formulas. For* $\kappa \neq -1/2$, *we get*

$$\begin{aligned} K^{(g)}(x,y) = {} & \Theta^{(g)} \begin{bmatrix} \alpha^{(g)} \\ \beta^{(g)} \end{bmatrix} \left(x, y; \nu^{(g)} + \kappa z_{p_1,p_2}, \kappa z_{p_1,p_2} | \Omega^{(g)}\right) \left(U^{(g)}(x,y)\right)^{\kappa} \\ & \times \left[\left(K^{(g+1)}(x,y)\right)^{-1} \Theta^{(g+1)} \begin{bmatrix} \alpha^{(g+1)} \\ \beta^{(g+1)} \end{bmatrix} \left(x, y; \nu^{(g+1)}, 0 | \Omega^{(g+1)}\right) \right. \\ & \left. - \xi h_{S_\kappa^{(g)}}(x) D^\theta \left(I - T_{S_\kappa^{(g)}}^{(g)}\right)^{-1} \bar{h}^T_{S_\kappa^{(g)}}(y) \right]^{-1}. \end{aligned} \tag{1}$$

For $\kappa = -1/2$ *one obtains:*

$$\begin{aligned} K^{(g)}(x,y) = {} & \left(\vartheta \begin{bmatrix} \alpha^{(g)} \\ \beta^{(g)} \end{bmatrix} \left(\int\limits_y^x \nu^{(g)} + \frac{1}{2} z_{p_1,p_2} | \Omega^{(g)} \right) \left(U^{(g)}(x,y)\right)^{\frac{1}{2}} \right. \\ & \left. - \theta_{g+1} \vartheta \begin{bmatrix} \alpha^{(g)} \\ \beta^{(g)} \end{bmatrix} \left(\int\limits_y^x \nu^{(g)} - \frac{1}{2} z_{p_1,p_2} | \Omega^{(g)} \right) \left(U^{(g)}(x,y)\right)^{-\frac{1}{2}} \right) \\ & \times \left(\vartheta \begin{bmatrix} \alpha^{(g)} \\ \beta^{(g)} \end{bmatrix} \left(\frac{1}{2} z_{p_1,p_2} | \Omega^{(g)} \right) - \theta_{g+1} \vartheta \begin{bmatrix} \alpha^{(g)} \\ \beta^{(g)} \end{bmatrix} \left(-\frac{1}{2} z_{p_1,p_2} | \Omega^{(g)} \right) \right)^{-1} \\ & \times \left[\Theta^{(g+1)} \begin{bmatrix} \alpha^{(g+1)} \\ \beta^{(g+1)} \end{bmatrix} \left(x, y; \nu^{(g+1)}, 0 | \Omega^{(g+1)}\right) K^{(g+1)}(x,y)^{-1} \right. \\ & \left. - \xi h_{S_\kappa^{(g)}}(x) D^\theta (I - T^{(g)}_{S_\kappa^{(g)}})^{-1} \bar{h}^T_{S_\kappa^{(g)}}(y) \right]^{-1} . \square \end{aligned} \tag{2}$$

Therefore, we derive

Lemma 15.2 *For $\kappa \neq -1/2$, we get the formula for the genus g prime form*

$$K^{(g)}\left(x, y, \Omega^{(g)}\right) =$$
$$\left[e^{\frac{1}{2} b_a X_{\bar{a}\bar{a}}(\rho) b_a^T} \Theta^{(g+1)} \begin{bmatrix} \alpha^{(g+1)} \\ \beta^{(g+1)} \end{bmatrix} \left(x, y; \nu^{(g+1)}, 0 | \Omega^{(g+1)}\right) \right.$$
$$\left. - \Theta^{(g)} \begin{bmatrix} \alpha^{(g)} \\ \beta^{(g)} \end{bmatrix} \left(x, y; \nu^{(g)} + \kappa z_{p_1,p_2}, \kappa z_{p_1,p_2} | \Omega^{(g)}\right) \left(U^{(g)}(x, y)\right)^{\kappa} \right]$$
$$\left[\xi h_{S_\kappa^{(g)}}(x) D^\theta \left(I - T^{(g)}_{S_\kappa^{(g)}}\right)^{-1} \bar{h}^T_{S_\kappa^{(g)}}(y) \right]^{-1} . \square \tag{3}$$

16. Self-sewn genus two prime form formula for characters

We are able to express the genus two character for a vertex operator algebra module (see subsection 17.3 of Appendix) via genus two prime forms by the next proposition. The formula generalizes the genus one formula for the case of genus two Riemann surface.

Proposition 16.1 *With $z_{n+1} = w$, $z_{n+2} = 0$, $u_{n+1} = \sigma f_2 \Psi_\kappa$, and $u_{n+2} = \overline{\Psi}_\kappa$,*

$$Z^{(2)}_\alpha(u_1 \otimes e^{\beta_1}, z_1; \ldots, u_n \otimes e^{\beta_n}, z_n; \tau, w, \rho)$$
$$= \frac{q^{\frac{1}{2}\alpha^2}}{\eta(\tau)} \sum_{\Psi_\kappa} Q_\alpha^{\beta_1,\ldots,\beta_{n+2}}(u_1, z_1; \ldots; u_{n+2}, z_{n+2}; q) \; \exp\left(\alpha \left(\sum_{i=1}^{n+2} \beta_i z_i\right)\right)$$
$$\cdot \prod_{1 \le r < s \le n+2} \left(K^{(2)}(z_r, z_s)\right)^{\beta_r \beta_s} e^{\frac{1}{2}\beta_r\beta_s \, b_a(z_r,z_s;k) \, X_{\bar{a}\bar{a}} \, b_a^T(z_r,z_s;k)} . \square \tag{4}$$

In [9] we established formulas relating genus one and genus two prime forms via theta-function relations originating from the Szegő kernel description. Using such formulas, we prove here the genus two version of the torus formula in the self-sewing formalism.

Proposition 16.2 *With $z_{n+1} = w$, $z_{n+2} = 0$, $u_{n+1} = \sigma f_2 \Psi_\kappa$, and $u_{n+2} = \overline{\Psi}_\kappa$,*

$$Z^{(2)}_\alpha(u_1 \otimes e^{\beta_1}, z_1; \ldots, u_n \otimes e^{\beta_n}, z_n; \tau, w, \rho)$$
$$= \frac{q^{\frac{1}{2}\alpha^2}}{\eta(\tau)} \sum_{\Psi_\kappa} Q_\alpha^{\beta_1,\ldots,\beta_{n+2}}(u_1, z_1; \ldots; u_{n+2}, z_{n+2}; q) \; \exp\left(\alpha \left(\sum_{i=1}^{n+2} \beta_i z_i\right)\right)$$
$$\cdot \prod_{1 \le r < s \le n+2} \left[\Theta^{(1)} \begin{bmatrix} \alpha^{(1)} \\ \beta^{(1)} \end{bmatrix} \left(z_r, z_s; \nu^{(1)} + \kappa z_{p_1,p_2}, \kappa z_{p_1,p_2} | \Omega^{(1)}\right) \right]^{-\beta_r\beta_s} \left(U^{(1)}(z_r, z_s)\right)^{-\kappa\beta_r\beta_s}$$
$$\times \left[\left(K^{(2)}(z_r, z_s)\right)^{-1} \Theta^{(2)} \begin{bmatrix} \alpha^{(2)} \\ \beta^{(2)} \end{bmatrix} \left(z_r, z_s; \nu^{(2)}, 0 | \Omega^{(2)}\right) \right.$$
$$\left. - \xi h_{S_\kappa^{(1)}}(z_r) D^\theta \left(I - T^{(1)}_{S_\kappa^{(1)}}\right)^{-1} \bar{h}^T_{S_\kappa^{(1)}}(z_s) \right]^{-\beta_r\beta_s} . \tag{5}$$

IOP Conf. Series: Journal of Physics: Conf. Series **965** (2018) 012044 doi:10.1088/1742-6596/965/1/012044

For $\kappa = -1/2$ *we get*

$$
\begin{aligned}
&Z^{(2)}_{\alpha}(u_1 \otimes e^{\beta_1}, z_1; \ldots, u_n \otimes e^{\beta_n}, z_n; \tau, w, \rho) \\
&= \frac{q^{\frac{1}{2}\alpha^2}}{\eta(\tau)} \sum_{\Psi_\kappa} Q^{\beta_1,\ldots,\beta_{n+2}}_{\alpha}(u_1, z_1; \ldots; u_{n+2}, z_{n+2}; q)\ \exp\left(\alpha\left(\sum_{i=1}^{n+2} \beta_i z_i\right)\right) \\
&\cdot \prod_{1\le r<s\le n+2} \left(\vartheta \begin{bmatrix} \alpha^{(1)} \\ \beta^{(1)} \end{bmatrix} \left(\int_y^x \nu^{(1)} + \frac{1}{2} z_{p_1,p_2} | \Omega^{(1)} \right) \left(U^{(1)}(x,y)\right)^{\frac{1}{2}} \right. \\
&\qquad \left. -\theta_2 \vartheta \begin{bmatrix} \alpha^{(1)} \\ \beta^{(1)} \end{bmatrix} \left(\int_y^x \nu^{(1)} - \frac{1}{2} z_{p_1,p_2} | \Omega^{(1)} \right) \left(U^{(1)}(x,y)\right)^{-\frac{1}{2}} \right)^{\beta_r\beta_s} \\
&\quad \times \left(\vartheta \begin{bmatrix} \alpha^{(1)} \\ \beta^{(1)} \end{bmatrix} \left(\frac{1}{2} z_{p_1,p_2} | \Omega^{(1)} \right) - \theta_2 \vartheta \begin{bmatrix} \alpha^{(1)} \\ \beta^{(1)} \end{bmatrix} \left(-\frac{1}{2} z_{p_1,p_2} | \Omega^{(1)} \right) \right)^{-\beta_r\beta_s} \\
&\times \left[\Theta^{(2)} \begin{bmatrix} \alpha^{(2)} \\ \beta^{(2)} \end{bmatrix} \left(x, y; \nu^{(2)}, 0 | \Omega^{(2)} \right) K^{(2)}(x,y)^{-1} \right. \\
&\qquad \left. -\xi h_{S^{(1)}_\kappa}(x) D^\theta (I - T^{(1)}_{S^{(1)}_\kappa})^{-1} \bar{h}^T_{S^{(1)}_\kappa}(y) \right]^{-\beta_r\beta_s}.
\end{aligned} \tag{6}
$$

Proof: We consider the case with $\kappa \neq -1/2$,

$$
\begin{aligned}
K^{(1)}(x-y,\tau) &= \Theta^{(1)} \begin{bmatrix} \alpha^{(1)} \\ \beta^{(1)} \end{bmatrix} \left(x, y; \nu^{(1)} + \kappa z_{p_1,p_2}, \kappa z_{p_1,p_2} | \Omega^{(1)} \right) \left(U^{(1)}(x,y) \right)^{\kappa} \\
&\cdot \left[\left(K^{(2)}\left(x, y, \Omega^{(2)}\right) \right)^{-1} \Theta^{(2)} \begin{bmatrix} \alpha^{(2)} \\ \beta^{(2)} \end{bmatrix} \left(x, y; \nu^{(2)}, 0 | \Omega^{(2)} \right) \right. \\
&\qquad \left. -\xi h_{S^{(1)}_\kappa}(x) D^\theta \left(I - T^{(1)}_{S^{(1)}_\kappa} \right)^{-1} \bar{h}^T_{S^{(1)}_\kappa}(y) \right]^{-1}.
\end{aligned} \tag{7}
$$

Let us take $u_{n+1} \otimes e^{\beta_{n+1}} = \sigma f_2 \Psi_\kappa$, and $u_{n+2} \otimes e^{\beta_{n+2}} = \overline{\Psi}_\kappa$. Then we obtain

$$
\begin{aligned}
& Z^{(2)}_\alpha(u_1 \otimes e^{\beta_1}, z_1; \ldots, u_n \otimes e^{\beta_n}, z_n; \tau, w, \rho) \\
&= \sum_{\Psi_\kappa} Z^{(1)}_\alpha \left(u_1 \otimes e^{\beta_1}, z_1; \ldots, u_n \otimes e^{\beta_n}, z_n; \sigma f_2 \Psi_\kappa, w; \overline{\Psi}_\kappa, 0; q \right) \\
&= \sum_{\Psi_\kappa} Z^{(1)}_\alpha \left(u_1 \otimes e^{\beta_1}, z_1; \ldots, u_n \otimes e^{\beta_n}, z_n; u_{n+1} \otimes e^{\beta_{n+1}} w; u_{n+2} \otimes e^{\beta_{n+2}}, 0; q \right) \\
&= \sum_{\Psi_\kappa} Q^{\beta_1, \ldots, \beta_n}_\alpha (u_1, z_1; \ldots u_n, z_n; q) \; Z^{(1)}_\alpha \left(e^{\beta_1}, z_1; \ldots, e^{m_n}, z_n; e^{\beta_{n+1}} w; e^{\beta_{n+2}}, 0; q \right) \\
&= \sum_{\Psi_\kappa} \frac{q^{\frac{1}{2}\alpha^2}}{\eta(\tau)} Q^{\beta_1, \ldots, \beta_n}_\alpha (u_1, z_1; \ldots u_n, z_n; q) \; \exp\left(\alpha \left(\sum_{i=1}^{n+2} \beta_i z_i \right) \right) \\
&\quad \cdot \prod_{1 \le r < s \le n+2} \left(K^{(1)}(z_{rs}, \tau) \right)^{\beta_r \beta_s} \\
&= \sum_{\Psi_\kappa} \frac{q^{\frac{1}{2}\alpha^2}}{\eta(\tau)} Q^{\beta_1, \ldots, \beta_n}_\alpha (u_1, z_1; \ldots u_n, z_n; q) \exp\left(\alpha \left(\sum_{i=1}^{n+2} \beta_i z_i \right) \right) \\
&\quad \cdot \prod_{1 \le r < s \le n+2} \left[\Theta^{(1)} \begin{bmatrix} \alpha^{(1)} \\ \beta^{(1)} \end{bmatrix} \left(z_r, z_s; \nu^{(1)} + \kappa z_{p_1,p_2}, \kappa z_{p_1,p_2} | \Omega^{(1)} \right) \right]^{-\beta_r \beta_s} \\
&\left(U^{(1)}(z_r, z_s) \right)^{-\kappa \beta_r \beta_s} \left[\left(K^{(2)} \left(z_r, z_s \Omega^{(2)} \right) \right)^{-1} \Theta^{(2)} \begin{bmatrix} \alpha^{(2)} \\ \beta^{(2)} \end{bmatrix} \left(z_r, z_s; \nu^{(2)}, 0 | \Omega^{(2)} \right) \right. \\
&\quad \left. - \xi h_{S^{(1)}_\kappa}(z_r) D^\theta \left(I - T^{(1)}_{S^{(1)}_\kappa} \right)^{-1} \bar{h}^T_{S^{(1)}_\kappa}(z_s) \right]^{-\beta_r \beta_s}.
\end{aligned}
$$

Thus we arrive at (6).

□

Note that corresponding generalizations are possible at higher genuses.

Acknowledgments
We would like to thank the Organizers of the XXVth International Conference on Integrable Systems and Quantum symmetries (ISQS-25), and in particular, Prof. C. Burdik.

17. Appendix:

17.1. The prime form

Let us recall the definition and properties of the prime form on a genus g Riemann surfaces [7]. Consider a compact Riemann surface $\Sigma^{(g)}$ of genus g with canonical homology cycle basis $a_1, \ldots, a_g, b_1, \ldots, b_g$. In general there exists g holomorphic 1-forms ν_i, $i = 1, \ldots, g$ which we may normalize by

$$\oint_{a_i} \nu_j^{(g)} = 2\pi i \delta_{ij}. \tag{8}$$

The genus g period matrix $\Omega^{(g)}$ is defined by

$$\Omega^{(g)}_{ij} = \frac{1}{2\pi i} \oint_{b_i} \nu_j^{(g)}, \tag{9}$$

IOP Conf. Series: Journal of Physics: Conf. Series **965** (2018) 012044 doi:10.1088/1742-6596/965/1/012044

for $i,j=1,\ldots,g$. $\Omega^{(g)}$ is symmetric with positive imaginary part, i.e., $\Omega^{(g)}\in\mathbb{H}_g$, the Siegel upper half plane. The canonical intersection form on cycles is preserved under the action of the symplectic group $Sp(2g,\mathbb{Z})$ where

$$\begin{pmatrix} b \\ a \end{pmatrix} \to \begin{pmatrix} \tilde{b} \\ \tilde{a} \end{pmatrix} = \begin{pmatrix} A & B \\ C & D \end{pmatrix}\begin{pmatrix} b \\ a \end{pmatrix}, \quad \begin{pmatrix} A & B \\ C & D \end{pmatrix} \in Sp(2g,\mathbb{Z}). \tag{10}$$

This induces the modular action on $\mathbb{H}_g$

$$\Omega^{(g)} \to \tilde{\Omega}^{(g)} = \left(A\Omega^{(g)}+B\right)\left(C\Omega^{(g)}+D\right)^{-1}. \tag{11}$$

One introduces the normalized differential of the second kind defined by

$$\omega^{(g)}(x,y) \sim \frac{dxdy}{(x-y)^2}, \tag{12}$$

for $x\sim y$, for local coordinates x,y, with normalization $\int_{a_i}\omega^{(g)}(x,\cdot)=0$ for $i=1,\ldots,g$. Using the Riemann bilinear relations, one finds that $\nu_i^{(g)}(x)=\oint_{b_i}\omega^{(g)}(x,\cdot)$. We recall also the definition of the theta function with real characteristics

$$\vartheta\begin{bmatrix}\alpha \\ \beta\end{bmatrix}\left(z|\Omega^{(g)}\right) = \sum_{m\in\mathbb{Z}^g} \exp\left(i\pi(m+\alpha).\Omega^{(g)}.(m+\alpha)+(m+\alpha).(z+2\pi i\beta)\right), \tag{13}$$

for $\alpha=(\alpha_i),\beta=(\beta_i)\in\mathbb{R}^g$, $z=(z_i)\in\mathbb{C}^g$ and $i=1,\ldots,g$ with

$$\begin{aligned} \vartheta\begin{bmatrix}\alpha \\ \beta\end{bmatrix}\left(z+2\pi i(\Omega^{(g)}.r+s)|\Omega^{(g)}\right) &= e^{2\pi i\alpha.s}e^{-2\pi i\beta.r}e^{-i\pi r.\Omega^{(g)}.r-r.z}\vartheta\begin{bmatrix}\alpha \\ \beta\end{bmatrix}\left(z|\Omega^{(g)}\right), \\ \vartheta\begin{bmatrix}\alpha+r \\ \beta+s\end{bmatrix}\left(z|\Omega^{(g)}\right) &= e^{2\pi i\alpha.s}\vartheta\begin{bmatrix}\alpha \\ \beta\end{bmatrix}\left(z|\Omega^{(g)}\right), \end{aligned} \tag{14}$$

for $r,s\in\mathbb{Z}^g$.

There exists a (nonsingular and odd) character $\left[\begin{smallmatrix}\gamma \\ \delta\end{smallmatrix}\right]$ such that [7]

$$\vartheta\begin{bmatrix}\gamma \\ \delta\end{bmatrix}(0|\Omega^{(g)})=0, \qquad \partial_{z_i}\vartheta\begin{bmatrix}\gamma \\ \delta\end{bmatrix}(0|\Omega^{(g)})\neq 0. \tag{15}$$

Let

$$\zeta^{(g)}(x)=\sum_{i=1}^{g}\partial_{z_i}\vartheta\begin{bmatrix}\gamma \\ \delta\end{bmatrix}(0|\Omega^{(g)})\nu_i^{(g)}(x), \tag{16}$$

a holomorphic one-form, and let $\zeta^{(g)}(x)^{\frac{1}{2}}$ denote the form of weight $\frac{1}{2}$ on the double cover $\widetilde{\Sigma^{(g)}}$ of $\Sigma^{(g)}$. We also refer to $\zeta(x)^{\frac{1}{2}}$ as a (double-valued) $\frac{1}{2}$-form on Σ. We define the prime form $E(x,y)$ by

$$E^{(g)}(x,y)=\frac{\vartheta\left[\begin{smallmatrix}\gamma \\ \delta\end{smallmatrix}\right]\left(\int_y^x\nu^{(g)}|\Omega^{(g)}\right)}{\zeta^{(g)}(x)^{\frac{1}{2}}\zeta^{(g)}(y)^{\frac{1}{2}}} \sim (x-y)dx^{-\frac{1}{2}}dy^{-\frac{1}{2}} \quad \text{for } x\sim y, \tag{17}$$

where $\int_y^x\nu^{(g)}=\left(\int_y^x\nu_i^{(g)}\right)\in\mathbb{C}^g$. $E^{(g)}(x,y)=-E^{(g)}(y,x)$ is a holomorphic differential form of weight $(-\frac{1}{2},-\frac{1}{2})$ on $\widetilde{\Sigma^{(g)}}\times\widetilde{\Sigma^{(g)}}$. We denote by $K^{(g)}$ the functional part of $E^{(g)}$.

17.2. Self-sewing of a Riemann surface

Consider the construction a Riemann surface $\Sigma^{(g+1)}$ formed by self-sewing a handle to a Riemann surface $\Sigma^{(g)}$ of genus g. We review the Yamada formalism [5] in this scheme which we call the ρ-formalism. Consider a Riemann surface $\Sigma^{(g)}$ of genus g and let z_1, z_2 be local coordinates in the neighbourhood of two separated points p_1 and p_2. Consider two disks $|z_a| \le r_a$, for $r_a > 0$ and $a = 1, 2$. Note that r_1, r_2 must be sufficiently small to ensure that the disks do not intersect. Introduce a complex parameter ρ where $|\rho| \le r_1 r_2$ and excise the disks

$$\{z_a : |z_a| < |\rho| r_{\bar{a}}^{-1}\} \subset \Sigma^{(g)},$$

to form a twice-punctured surface

$$\widehat{\Sigma}^{(g)} = \Sigma^{(g)} \backslash \bigcup_{a=1,2} \{z_a : |z_a| < |\rho| r_{\bar{a}}^{-1}\}.$$

As before, we use the convention $\bar{1} = 2$, $\bar{2} = 1$. We define annular regions $\mathcal{A}_a \subset \widehat{\Sigma}^{(g)}$ with $\mathcal{A}_a = \{z_a : |\rho| r_{\bar{a}}^{-1} \le |z_a| \le r_a\}$ and identify them as a single region $\mathcal{A} = \mathcal{A}_1 \simeq \mathcal{A}_2$ via the sewing relation

$$z_1 z_2 = \rho, \tag{18}$$

to form a compact Riemann surface $\Sigma^{(g+1)} = \widehat{\Sigma}^{(g)} \backslash \{\mathcal{A}_1 \cup \mathcal{A}_2\} \cup \mathcal{A}$ of genus $g+1$. The sewing relation (18) can be considered to be a parameterization of a cylinder connecting the punctured Riemann surface to itself. In the ρ-formalism we define a standard basis of cycles $\{a_1, b_1, \ldots, a_{g+1}, b_{g+1}\}$ on $\Sigma^{(g+1)}$ where the set $\{a_1, b_1, \ldots, a_g, b_g\}$ is the original basis on $\Sigma^{(g)}$. Let $\mathcal{C}_a(z_a) \subset \mathcal{A}_a$ denote a closed anti-clockwise contour parameterized by z_a surrounding the puncture at $z_a = 0$. Clearly $\mathcal{C}_2(z_2) \sim -\mathcal{C}_1(z_1)$ on applying the sewing relation (18). We then define the cycle a_{g+1} to be $\mathcal{C}_2(z_2)$ and define the cycle b_{g+1} to be a path chosen in $\widehat{\Sigma}^{(g)}$ between identified points $z_1 = z_0$ and $z_2 = \rho/z_0$ on the sewn surface.

17.3. The free fermion vertex operator superalgebra (VOSA)

Consider the rank two free fermion VOSA $V(H, \mathbb{Z} + \frac{1}{2})^{\otimes 2}$ of central charge 1, [1]. The weight $\frac{1}{2}$ space $V_{\frac{1}{2}}$ is spanned by ψ^+, ψ^- with vertex operators

$$Y(\psi^\pm, z) = \sum_{n \in \mathbb{Z}} \psi^\pm(n) z^{-n-1},$$

whose modes satisfy the anti-commutation relations

$$[\psi^+(m), \psi^-(n)] = \delta_{m,-n-1}, \quad [\psi^+(m), \psi^+(n)] = [\psi^-(m), \psi^-(n)] = 0. \tag{19}$$

The VOSA is generated by $\psi^\pm$ with V spanned by Fock vectors of the form

$$\Psi(\mathbf{k}, \mathbf{l}) \equiv \psi^+(-k_1) \ldots \psi^+(-k_s)\, \psi^-(-l_1) \ldots \psi^-(-l_t)\mathbf{1}, \tag{20}$$

for distinct $0 < k_1 < \ldots < k_s$ and $0 < l_1 < \ldots < l_t$. The Virasoro vector

$$\omega^{(g)} = \frac{1}{2}(\psi^+(-2)\psi^-(-1) + \psi^-(-2)\psi^+(-1))\mathbf{1}.$$

generates a Virasoro algebra with central charge $c = 1$ for which the Fock vectors have weight

$$(\Psi(\mathbf{k}, \mathbf{l})) = \sum_{i=1}^{s} \left(k_i - \frac{1}{2}\right) + \sum_{j=1}^{t} \left(l_j - \frac{1}{2}\right). \tag{21}$$

The mode $a(0)$ generates continuous $V_{\mathbb{Z}}$-automorphisms $g = e^{-2\pi i\alpha a(0)}$ for all $\alpha \in \mathbb{C}$. In particular, we define the fermion number involution

$$\sigma = e^{\pi i a(0)},$$

where $\sigma u = (-1)^{p(u)}u$ for u of parity $p(u)$.

17.4. The invariant form on $\mathcal{M}$

It is convenient to define [10] for formal parameter z and $\chi \in \mathbb{C}$

$$(-z)^{\chi} = e^{i\pi B\chi} z^{\chi}, \tag{22}$$

where B is an *odd integer* parametrizing the formal branch cut. In [10] we introduced an invariant bilinear form $\langle \cdot, \cdot \rangle$ on $\mathcal{M}$ associated with the Möbius map

$$\begin{pmatrix} 0 & \lambda \\ -e^{i\pi B}\lambda^{-1} & 0 \end{pmatrix} : z \mapsto -\frac{\lambda^2}{e^{i\pi B} z}, \tag{23}$$

for $\lambda \neq 0$. We are particularly interested in the Möbius map $z \mapsto \rho/z$ associated with the sewing condition (18) so that we will choose

$$\lambda = e^{\frac{1}{2}i\pi B}\rho^{\frac{1}{2}}, \tag{24}$$

for the odd integer B of (22). Thus we reformulate the sewing relationship (18) as $z_1 = -\frac{\lambda^2}{z_2}$ so that $dz_1^{\frac{1}{2}} = \xi\rho^{\frac{1}{2}}/z_2\, dz_2^{\frac{1}{2}}$ for $\xi = e^{\frac{1}{2}i\pi B}$.

Define the adjoint vertex operator

$$\mathcal{Y}^{\dagger}\left(u \otimes e^{\alpha}, z\right) = \mathcal{Y}\left(e^{-z\lambda^{-2}L(1)} \left(\frac{\lambda}{e^{i\pi B} z}\right)^{2L(0)} (u \otimes e^{\alpha}), \frac{\lambda^2}{e^{i\pi B} z}\right). \tag{25}$$

A bilinear form $\langle \cdot, \cdot \rangle$ on $\mathcal{M}$ is said to be invariant if for all $u \otimes e^{\alpha}$, $v \otimes e^{\beta}$, $w \otimes e^{\gamma} \in \mathcal{M}$ we have

$$\langle \mathcal{Y}(u \otimes e^{\alpha}, z)\, (v \otimes e^{\beta}),\ w \otimes e^{\gamma} \rangle = e^{-i\pi B\alpha\beta} \langle v \otimes e^{\beta},\ \mathcal{Y}^{\dagger}(u \otimes e^{\alpha}, z)\, w \otimes e^{\gamma} \rangle. \tag{26}$$

(25) reduces to the usual definition for a VOSA when $\alpha, \beta, \gamma \in \mathbb{Z}$ [10]. Choosing the normalization $\langle \mathbf{1}, \mathbf{1} \rangle = 1$ then $\langle \cdot, \cdot \rangle$ on $\mathcal{M}$ is symmetric, unique and invertible with [10]

$$\langle u \otimes e^{\alpha}, v \otimes e^{\beta} \rangle = \lambda^{-\alpha^2} \delta_{\alpha,-\beta} \langle u \otimes e^{0}, v \otimes e^{0} \rangle. \tag{27}$$

Thus the dual of the Fock vector $\Psi = \Psi(\mathbf{k}, \mathbf{l})$ is

$$\overline{\Psi}(\mathbf{k}, \mathbf{l}) = (-1)^{st + \lfloor(\Psi)\rfloor}\, \lambda^{2(\Psi)} \Psi(\mathbf{l}, \mathbf{k}),$$

where $\lfloor x \rfloor$ denotes the integer part of x [10]. Applying (24) and (27) it follows that $\Psi_{\alpha} = \Psi_{\alpha}(\mathbf{k}, \mathbf{l})$ has dual

$$\begin{aligned} \overline{\Psi}_{\alpha}(\mathbf{k}, \mathbf{l}) &= (-1)^{st + \lfloor(\Psi)\rfloor}\, \lambda^{2(\Psi_{\alpha})} \Psi_{-\alpha}(\mathbf{l}, \mathbf{k}) \\ &= (-1)^{st + \lfloor(\Psi)\rfloor}\, e^{i\pi B(\Psi_{\alpha})}\, \rho^{(\Psi_{\alpha})} \Psi_{-\alpha}(\mathbf{l}, \mathbf{k}). \end{aligned} \tag{28}$$

IOP Conf. Series: Journal of Physics: Conf. Series **965** (2018) 012044 doi:10.1088/1742-6596/965/1/012044

References

[1] Kac V 1998 *Vertex Operator Algebras for Beginners* (Providence: American Mathematical Society)
[2] Hurley D and Tuite M P 2013 *Int. J. Math.* **24** 3050056
[3] Mason G, Tuite M P and Zuevsky A 2008 *Commun. Math. Phys.* **283** 305
[4] Mumford D 1983 *Tata Lectures on Theta I and II* (Boston: Birkhäuser)
[5] Yamada A 1980 *Kodai Math. J.* **3** 114
[6] Mason G and Tuite M P 2007 *Commun. Math. Phys.* **270** 587
[7] Fay J D 1973 *Theta Functions on Riemann surfaces* (Berlin: Springer-Verlag)
[8] Dong C, Li H and Mason G 2000 *Commun. Math. Phys.* **214** 1
[9] Tuite M P and Zuevsky A 2011 *Commun. Math. Phys.* **306** 617
[10] Tuite M P and Zuevsky A 2011 *Commun. Math. Phys.* **306** 419
[11] Tuite M P and Zuevsky A 2013 Genus two partition and correlation functions for fermionic vertex operator superalgebras II *Preprint* arXiv:1308.2441
[12] Gilroy T and Tuite M P 2015 Genus Two Zhu Theory for Vertex Operator Algebras *Preprint* arXiv:1511.07664
[13] Tuite M P and Zuevsky A 2011 *RIMS* **1756** 81
[14] Zhu Y 1996 *J. Amer. Math. Soc.* **9** 237
[15] Zuevsky A 2016 *J. of Combinatorics and Number Theory* **8** 225
[16] Zuevsky A 2017 to appear in *Int. J. of Number Theory*

IOP Conf. Series: Journal of Physics: Conf. Series **965** (2018) 012045 doi:10.1088/1742-6596/965/1/012045

Component reduction and the superconformal gravity invariants[1]

J Novak[1] and G Tartaglino-Mazzucchelli[2]
[1]Max-Planck-Institut für Gravitationsphysik, Albert-Einstein-Institut, Am Mühlenberg 1, D-14476 Golm, Germany
[2]Instituut voor Theoretische Fysica, KU Leuven, Celestijnenlaan 200D, B-3001 Leuven, Belgium

E-mail: joseph.novak@aei.mpg.de, gabriele.tartaglino-mazzucchelli@kuleuven.be

Abstract.
We present the component structure of the superconformal gravity invariants in six dimensions, which was recently elaborated in arXiv:1701.08163.

1. Introduction

Conformal gravity invariants appear in the study of conformal field theories on an arbitrary curved background in even spacetime dimensions, where they correspond to certain conformal anomalies. The conformal anomalies express the tracelessness of the energy-monentum tensor in correlation functions and have been classified by Deser and Schwimmer long ago [1]. It is known that there are two main types: type A and type B. There is always one type A anomaly corresponding to the topological Euler invariant but an increasing number of type B anomalies in increasing spacetime dimension. For instance, there is one type B anomaly in four dimensions and three type B anomalies in six dimensions (6D). The type B anomalies correspond to the conformal gravity invariants, being Weyl invariant expressions constructed from the curvature tensors and its covariant derivatives.

In six dimensions, the conformal gravity invariants may be written as spacetime integrals of the form

$$I_{\rm i} = \int {\rm d}^6 x\, e\, \mathcal{L}_{\rm i} \ , \quad {\rm i} = 1, 2, 3 \ , \tag{1.1}$$

where

$$\mathcal{L}_1 = C_{abcd} C^{aefd} C_e{}^{bc}{}_f \ , \tag{1.2a}$$

$$\mathcal{L}_2 = C_{abcd} C^{cdef} C_{ef}{}^{ab} \ , \tag{1.2b}$$

$$\mathcal{L}_3 = C_{abcd} (\delta^a_e \Box - 4 R_e{}^a + \frac{6}{5} \delta^a_e R) C^{ebcd} + \text{total derivative} \ , \quad \Box := \mathcal{D}^a \mathcal{D}_a \ . \tag{1.2c}$$

Here $\mathcal{D}_a$ is the covariant derivative, C_{abcd} is the Weyl tensor satisfying the properties

$$C_{abcd} = C_{[ab][cd]} \ , \quad C_{[abc]d} = 0 \ , \quad \mathcal{D}_{[a} C_{bc]}{}^{de} = -\frac{2}{3} \mathcal{D}_f C_{[ab}{}^{f[d} \delta_{c]}^{e]} \ , \tag{1.3}$$

[1] Based on the talk presented by JN at ISQS25 (Prague, Czech Republic, 6 – 10 June, 2017).

and R_{ab} and R are the Ricci curvature and scalar curvature, respectively. The Lagrangian $\mathcal{L}_3$ can be completed by adding a total derivative (which we suppress for simplicity) in such a way that it transforms homogeneously under Weyl transformations as $\mathcal{L}_1$ and $\mathcal{L}_2$. The properties of the curvature tensor implies that all conformal gravity invariants are described by the linear combination

$$I = c_1 I_1 + c_2 I_2 + c_3 I_3 \ , \tag{1.4}$$

where c_i are arbitrary coefficients.

In the context of superconformal field theories, which exhibit invariance under the supersymmetric extension of conformal symmetry, conformal field theories in six dimensions become more intriguing. Remarkably, the structure of the superconformal algebra implies that 6D is the highest spacetime dimension in which superconformal field theories can exist [2]. On the other hand, the only known non-trivial unitary conformal field theories in 6D are supersymmetric and arise in string theory, realizing either $\mathcal{N} = (1, 0)$ or $\mathcal{N} = (2, 0)$ superconformal symmetry. These properties place 6D superconformal field theories in a special class in the space of conformal field theories. Their type B anomalies should be supersymmetric and correspond to superconformal gravity invariants. Therefore, the construction of superconformal gravity invariants helps one elaborate on the anomaly structure of the 6D superconformal field theories.

Supersymmetry imposes constraints on the structure of the superconformal invariants, which now involves additional fields such as the gravitini. Thus one typically expects the number of independent superconformal invariants to reduce with more supersymmetry. The corresponding superconformal algebras exist for $\mathcal{N} = (n, 0)$ but are usually limited to the $n \leq 2$ cases to ensure that superconformal gravity does not contain higher spin fields. In regards to the supersymmetric type B anomalies, there are two obvious points that should be addressed for the corresponding invariants. These include the number of independent conformal gravity invariants that one can extend to the supersymmetric case and the other is their full supersymmetric forms when coupled to the rest of the Weyl multiplet of superconformal gravity whose structure in components was first described in [3]. In principle, these points can be addressed indirectly by computing the conformal anomaly of various (1, 0) or (2, 0) matter multiplets coupled to (super)gravity. Such computations have been limited to only the purely gravitational parts, see eg. [4, 5, 6], from which one can deduce the following constraints on the coefficients c_i for the $(1, 0)$ and $(2, 0)$ cases, respectively:

$$c_1 - 2c_2 - 6c_3 = 0 \qquad \mathcal{N} = (1, 0) \ , \tag{1.5a}$$
$$c_2 - 3c_3 = 0 \qquad \mathcal{N} = (2, 0) \ . \tag{1.5b}$$

The above constraints tell us that we expect two independent invariants in the $(1, 0)$ case and only one independent invariant in the $(2, 0)$ case.[2] Their purely gravitational form is given by eq. (1.4) with coefficients restricted by the constraints (1.5). Despite this information, very little is known from the computation of the purely gravitational parts of these invariants about their supersymmetric completions.

In this paper we present the results of [7] in which the full component structure of 6D conformal supergravity invariants were given. Our goals will be the following: (i) to present the direct path of [7] to constructing the component structure of 6D $\mathcal{N} = (1, 0)$ superconformal gravity from superspace; (ii) to use the invariants to deduce information about the structure of the (2, 0) invariant; and (iii) to verify the known constraints on the coefficients c_i.

This paper is organized as follows. In section 2 we give a brief review of the superspace formulation of 6D superconformal gravity in [8] and present the procedure of [7] in extracting component expressions. In section 3 we present the $\mathcal{N} = (1, 0)$ superconformal gravity invariants

[2] The $(1, 0)$ constraint also holds in the $(2, 0)$ case.

in component form. Remarkably, as will be shown in section 4, the component bosonic action of the $\mathcal{N} = (2,0)$ superconformal gravity invariant can be almost completely recovered by considering its truncation onto a $\mathcal{N} = (1,0)$ invariant. Finally, conclusions are presented in section 5.

2. From superspace to components

The $\mathcal{N} = (1,0)$ superconformal tensor calculus was developed over thirty years ago in [3] for the construction of general supergravity-matter systems. The formulation met with some success in describing certain higher-derivative invariants, such as the supersymmetric Riemann curvature squared term [9]. However, it has given little insight into the construction of further higher derivative invariants, in particular the conformal supergravity invariants, which is the focus of our presentation.

Apart from superconformal tensor calculus, there is another approach to $\mathcal{N}$-extended conformal supergravity for dimension $D \leq 6$ based on using a curved $\mathcal{N}$-extended superspace $\mathcal{M}^{D|\delta}$, where δ is the number of fermionic dimensions. Within the superspace setting, one can choose to gauge only part of the superconformal algebra and rely on super-Weyl transformations to realise the conformal structure, which leads to the conventional superspace approach. It was developed for the 4D $\mathcal{N} = 1$ and $\mathcal{N} = 2$ cases in [10, 11] and presented for the 5D $\mathcal{N} = 1$ case in [12] which was extended to the 6D $\mathcal{N} = (1,0)$ case in [13]. However, for applications involving component reduction it is technically easier to make use of another superspace formulation, known as conformal superspace [14, 15, 16, 17, 8]. It is based on gauging the entire superconformal algebra and consequently makes more direct the connection with the component formulation, superconformal tensor calculus.[3] Conformal superspace has already proved useful in the construction of higher derivative invariants, e.g. [17, 18, 19, 20], and for this reason it provides an ideal formulation for the description of the higher derivative superconformal gravity invariants.

In this section we briefly present the 6D $\mathcal{N} = (1,0)$ conformal superspace developed in [8] and demonstrate how the component fields within superconformal tensor calculus originate within it following [7]. For the sake of brevity we refer the reader to [8, 7] for our conventions and the superconformal algebra, which contains the supertranslations $P_A = (P_a, Q_\alpha^i)$, the Lorentz generators M_{ab}, the dilatations $\mathbb{D}$, the $\mathrm{SU}(2)_\mathrm{R}$ transformations J_{ij} and the special conformal transformations $K^A = (K^a, S_i^\alpha)$.

2.1. *Gauging the superconformal algebra in superspace*

To gauge the superconformal algebra one takes a $\mathcal{N} = (1,0)$ curved superspace $\mathcal{M}^{6|8}$ locally parametrised by coordinates $z^M = (x^m, \theta^\mu_{\mathtt{i}})$, where $m = 0,1,2,3,4,5$, $\mu = 1,2,3,4$ and $\mathtt{i} = \underline{1},\underline{2}$. One now associates with each generator $X_{\underline{a}} = (M_{ab}, J_{ij}, \mathbb{D}, S^\gamma_k, K^c)$ a connection one-form $\omega^{\underline{a}} = (\Omega^{ab}, \Phi^{ij}, B, \mathfrak{F}^k_\gamma, \mathfrak{F}_c) = \mathrm{d}z^M \omega_M{}^{\underline{a}}$ and with P_A the vielbein $E^A = (E^\alpha_i, E^a)$. They may be used to construct the covariant derivatives

$$\nabla_A = E_A{}^M \partial_M - \frac{1}{2}\Omega_A{}^{ab} M_{ab} - \Phi_A{}^{ij} J_{ij} - B_A \mathbb{D} - \mathfrak{F}_A{}^j_\beta S^\beta_j - \mathfrak{F}_{Ab} K^b \ . \tag{2.1}$$

It is important to note that the connections can be defined such that the generators appearing in the structure group act on the covariant derivatives in the same way as they do on P_A except with P_A replaced by ∇_A.

The supergravity gauge transformations of the covariant derivatives may be summarised as

$$\delta_\mathcal{K} \nabla_A = [\mathcal{K}, \nabla_A] \ , \quad \mathcal{K} := \xi^A \nabla_A + \frac{1}{2}\Lambda^{bc} M_{bc} + \Lambda^{ij} J_{ij} + \sigma \mathbb{D} + \Lambda^i_\alpha S^\alpha_i + \Lambda_a K^a \ , \tag{2.2}$$

[3] The conventional formulations can be obtained by partially fixing the gauge freedom.

where the gauge parameters associated with $\mathcal{K}$ satisfy natural reality properties. It should be mentioned that the supergravity transformations act on a tensor superfield U as $\delta_{\mathcal{K}}U = \mathcal{K}U$. The superfield U is said to be *primary* and of dimension Δ if $K_A U = 0$ and $\mathbb{D}U = \Delta U$.

The above geometric setup contains too many fields to describe conformal supergravity alone. In order to describe conformal supergravity it is necessary to impose constraints on the covariant derivative algebra. These are algebraic constraints on the superspace curvatures and their corresponding consequences due to the Bianchi identities. As in lower dimensions, see [14, 15, 16, 17], it is natural to constrain the covariant derivative algebra to resemble those of supersymmetric Yang-Mills. However, this will lead to the corresponding composite component connections having non-trivial dependence on the covariant fields of the Weyl multiplet, which unnecessarily complicates the corresponding component results. To remedy this, one can redefine the vector covariant derivative $\nabla_a \to \hat{\nabla}_a$ by modifying the Lorentz and K_A connections to remove such dependence and result in the "traceless frame" of [7]. We present the resulting covariant derivative algebra below.

The torsion and curvatures appear in the the (anti)commutation relations

$$[\hat{\nabla}_A, \hat{\nabla}_B\} \equiv -\mathcal{F}_{AB} = -\hat{T}_{AB}{}^C \hat{\nabla}_C - \frac{1}{2}\hat{R}(M)_{AB}{}^{cd} M_{cd} - \hat{R}(J)_{AB}{}^{kl} J_{kl} - \hat{R}(\mathbb{D})_{AB}\mathbb{D} - \hat{R}(S)_{AB}{}^k_\gamma S^\gamma_k - \hat{R}(K)_{ABc}K^c \, , \quad (2.3)$$

where $\hat{\nabla}_A = (\hat{\nabla}_a, \nabla^i_\alpha)$, and the torsion and curvatures are constrained to satisfy the Bianchi identities $[\hat{\nabla}_{[A}, \mathcal{F}_{BC\}}\} = 0$. The covariant derivative algebra is constrained in such a way that all curvatures are expressed in terms of the super-Weyl tensor $W^{\alpha\beta}$ and its covariant derivatives. The super-Weyl tensor [13] is a symmetric primary superfield of dimension 1,

$$W^{\alpha\beta} = W^{\beta\alpha} \, , \quad K^A W^{\beta\gamma} = 0 \, , \quad \mathbb{D}W^{\alpha\beta} = W^{\alpha\beta} \, , \quad (2.4)$$

and satisfies the following superspace differential constraints[4]

$$\nabla^{(i}_\alpha \nabla^{j)}_\beta W^{\gamma\delta} = -\delta^{(\gamma}_{[\alpha} \nabla^{(i}_{\beta]} \nabla^{j)}_\rho W^{\delta)\rho} \, , \quad (2.5a)$$

$$\nabla^k_\alpha \nabla_{\gamma k} W^{\beta\gamma} - \frac{1}{4}\delta^\beta_\alpha \nabla^k_\gamma \nabla_{\delta k} W^{\gamma\delta} = 8\mathrm{i}\nabla_{\alpha\gamma} W^{\gamma\beta} \, . \quad (2.5b)$$

The covariant derivative algebra is given by

$$\{\nabla^i_\alpha, \nabla^j_\beta\} = -2\mathrm{i}\varepsilon^{ij}(\gamma^c)_{\alpha\beta}\hat{\nabla}_c - 2\mathrm{i}\varepsilon^{ij}(\gamma_a)_{\alpha\beta}W^{abc}M_{bc} - \frac{3}{2}\varepsilon^{ij}\varepsilon_{\alpha\beta\gamma\delta}X^{\gamma k}S^\delta_k -\mathrm{i}\varepsilon^{ij}(\gamma^a)_{\alpha\beta}\left(\frac{1}{4}\eta_{ac}Y - \hat{\nabla}^b W_{abc} + W_a{}^{ef}W_{cef}\right)K^c \, . \quad (2.6)$$

The nontrivial torsion and curvature components in the commutator $[\hat{\nabla}_a, \hat{\nabla}^j_\beta]$ are given by

$$\hat{T}_{a\beta k}{}^{j\gamma} = -\frac{1}{2}(\gamma_a)_{\beta\delta}W^{\delta\gamma}\delta^j_k \, , \quad (2.7a)$$

$$\hat{R}(\mathbb{D})_a{}^j_\beta = -\frac{\mathrm{i}}{2}(\gamma_a)_{\beta\gamma}X^{\gamma j} \, , \quad (2.7b)$$

$$\hat{R}(M)_a{}^j_\beta{}^{cd} = \mathrm{i}\delta^{[c}_a(\gamma^{d]})_{\beta\gamma}X^{\gamma j} - \mathrm{i}(\gamma_a{}^{cd})_{\gamma\delta}X^{j\gamma\delta}_\beta + 2\mathrm{i}(\gamma_a)_{\beta\gamma}(\gamma^{cd})_\delta{}^\rho X^{j\gamma\delta}_\rho \, , \quad (2.7c)$$

$$\hat{R}(J)_a{}^j_\beta{}^{kl} = 2\mathrm{i}(\gamma_a)_{\beta\gamma}X^{\gamma(k}\varepsilon^{l)j} \, , \quad (2.7d)$$

[4] The superfield $W^{\alpha\beta}$ is equivalent to an anti-self-dual superfield $W_{abc} = \frac{1}{8}(\gamma_{abc})_{\alpha\beta}W^{\alpha\beta}$.

$$\begin{aligned}\hat{R}(S)_{a\beta\gamma}^{\ j\,k} &= -\frac{\mathrm{i}}{4}(\gamma_a)_{\beta\delta}\,Y_\gamma{}^{\delta jk} + \frac{3\mathrm{i}}{20}(\gamma_a)_{\gamma\delta}Y_\beta{}^{\delta jk} \\ &\quad -\frac{\mathrm{i}}{8}(\gamma_a)_{\beta\delta}\hat{\nabla}_{\gamma\rho}W^{\delta\rho}\varepsilon^{jk} + \frac{\mathrm{i}}{40}(\gamma_a)_{\gamma\delta}\hat{\nabla}_{\beta\rho}W^{\delta\rho}\varepsilon^{jk} \\ &\quad -\frac{\mathrm{i}}{8}(\gamma_a)_{\delta\epsilon}\,\varepsilon_{\beta\rho\tau\gamma}\,W^{\delta\rho}W^{\epsilon\tau}\varepsilon^{jk}\ , \end{aligned} \tag{2.7e}$$

$$\begin{aligned}\hat{R}(K)_{a\beta c}^{\ j} &= \frac{\mathrm{i}}{4}(\gamma_c)_{\beta\gamma}\hat{\nabla}_a X^{\gamma j} - \frac{\mathrm{i}}{4}(\gamma_{acd})_{\gamma\delta}\hat{\nabla}^d X_\beta^{j\gamma\delta} + \frac{\mathrm{i}}{3}(\gamma_a)_{\beta\delta}(\gamma_{cd})_\rho{}^\gamma\hat{\nabla}^d X_\gamma^{j\delta\rho} \\ &\quad -\frac{\mathrm{i}}{8}(\gamma_a)_{\beta\gamma}(\gamma_c)_{\delta\rho}W^{\gamma\delta}X^{\rho j} + \frac{5\mathrm{i}}{12}(\gamma_a)_{\beta\rho}(\gamma_c)_{\gamma\epsilon}W^{\gamma\delta}X_\delta^{j\rho\epsilon} \\ &\quad +\frac{\mathrm{i}}{4}(\gamma_a)_{\gamma\rho}(\gamma_c)_{\beta\epsilon}W^{\gamma\delta}X_\delta^{j\rho\epsilon} - \frac{\mathrm{i}}{2}(\gamma_a)_{\gamma\rho}(\gamma_c)_{\delta\epsilon}W^{\gamma\delta}X_\beta^{j\rho\epsilon}\ . \end{aligned} \tag{2.7f}$$

Here we have introduced the dimension 3/2 superfields

$$X_\gamma^{k\alpha\beta} = -\frac{\mathrm{i}}{4}\nabla_\gamma^k W^{\alpha\beta} - \delta_\gamma^{(\alpha}X^{\beta)k}\ , \quad X^{\alpha i} := -\frac{\mathrm{i}}{10}\nabla_\beta^i W^{\alpha\beta} \tag{2.8}$$

together with the dimension 2 descendant superfields:

$$Y_\alpha{}^{\beta ij} := -\frac{5}{2}\Big(\nabla_\alpha^{(i}X^{\beta j)} - \frac{1}{4}\delta_\alpha^\beta\nabla_\gamma^{(i}X^{\gamma j)}\Big) = -\frac{5}{2}\nabla_\alpha^{(i}X^{\beta j)}\ , \tag{2.9a}$$

$$Y := \frac{1}{4}\nabla_\gamma^k X_k^\gamma\ , \tag{2.9b}$$

$$Y_{\alpha\beta}{}^{\gamma\delta} := \nabla_{(\alpha}^k X_{\beta)k}{}^{\gamma\delta} - \frac{1}{6}\delta_\beta^{(\gamma}\nabla_\rho^k X_{\alpha k}{}^{\delta)\rho} - \frac{1}{6}\delta_\alpha^{(\gamma}\nabla_\rho^k X_{\beta k}{}^{\delta)\rho}\ . \tag{2.9c}$$

The action of the spinor covariant derivatives, as well as the S-supersymmetry generators, on the defined superfields may be found in [8]. The remaining commutator $[\hat{\nabla}_a, \hat{\nabla}_b]$ in the covariant derivative algebra follows from the superspace Bianchi identities for the covariant derivatives. Finally, the action of the generators $X_{\underline{a}}$ on the covariant derivatives is the same as their action on P_A with P_A replaced with $\hat{\nabla}_A$ with one exception. The exception is $[S_i^\alpha, \hat{\nabla}_a]$ which now obtains explicit dependence on the super-Weyl tensor,

$$\begin{aligned}[S_i^\alpha, \hat{\nabla}_b] &= -\mathrm{i}(\tilde{\gamma}_b)^{\alpha\beta}\nabla_{\beta i} + \frac{1}{10}W_{bcd}(\tilde{\gamma}^{cd})^\alpha{}_\gamma S_i^\gamma - \frac{1}{4}X_i^\alpha K_b \\ &\quad +\Big[\frac{1}{4}(\tilde{\gamma}_{bc})^\alpha{}_\beta X_i^\beta + \frac{1}{2}(\gamma_{bc})_\beta{}^\gamma X_{\gamma i}{}^{\beta\alpha}\Big]K^c\ . \end{aligned} \tag{2.10}$$

It is important to stress that the entire covariant derivative algebra is expressed directly in terms of the super-Weyl tensor and its descendents. In particular, it is trivial to see that when $W^{\alpha\beta}$ vanishes the supergeometry is superconformally flat. Furthermore, the standard Weyl multiplet of 6D $\mathcal{N} = (1,0)$ conformal supergravity is encoded in the superspace geometry. The component fields can be readily identified as certain $\theta = 0$ parts of the superspace gauge one-forms and descendants of $W^{\alpha\beta}$, which we now turn to demonstrating.

2.2. Identifying the component structure

In contrast with conformal superspace, the superconformal tensor calculus framework [3] involves gauging the superconformal algebra in spacetime. Associated respectively with local translations, Q-supersymmetry, SU(2) R-symmetry, and dilatations are the vielbein $e_m{}^a$, the gravitino $\psi_m{}_i^\alpha$, the SU(2) gauge field $\mathcal{V}_m{}^{ij}$, and a dilatation gauge field b_m. The remaining gauge symmetries are gauged by the spin connection $\omega_m{}^{cd}$, the S-supersymmetry connection $\phi_m{}_\alpha^i$, and the special conformal connection $\mathfrak{f}_{ma}$. These are composite fields, meaning that they are not independent and are expressed in terms of other fields of the Weyl multiplet. To ensure that the last three connections are composite, one imposes conventional constraints on the vielbein curvature $R(P)_{mn}{}^a$, the gravitino curvature $R(Q)_{mn}{}_i^\alpha$, and the conformal Lorentz curvature $R(M)_{mn}{}^{ab}$. Upon doing this it is necessary to introduce additional degrees of freedom since the bosonic and fermionic degrees of freedom do not match. An off-shell representation is achieved by introducing three covariant fields: a real anti-self-dual tensor T^-_{abc}, a chiral fermion $\chi^{\alpha i}$, and a real scalar field D which modify the supersymmetry algebra, the curvatures and the constraints imposed on the curvatures in a consistent way [3]. The component fields of the 6D $\mathcal{N} = (1,0)$ Weyl multiplet [3] can be identified within the geometry of conformal superspace, which begins with a manifest off-shell gauging of the superconformal algebra from the very beginning. We will elaborate further on this point below.

The identification of the component one-forms is straightforward. We start with the vielbein ($e_m{}^a$) and gravitini ($\psi_m{}_i^\alpha$). These appear as the $\theta = 0$ projections of the coefficients of $\mathrm{d}x^m$ in the supervielbein E^A,

$$e^a = \mathrm{d}x^m e_m{}^a = E^a\| \ , \qquad \psi_i^\alpha = \mathrm{d}x^m \psi_m{}_i^\alpha = 2E_i^\alpha\| \ , \tag{2.11}$$

where the double bar denotes setting $\theta = \mathrm{d}\theta = 0$.[5] The remaining fundamental and composite one-forms correspond to double-bar projections of superspace one-forms,

$$\mathcal{V}^{kl} := \Phi^{kl}\| \ , \quad b := B\| \ , \quad \hat{\omega}^{cd} := \hat{\Omega}^{cd}\| \ , \quad \hat{\phi}_\gamma^k := 2\,\hat{\mathfrak{F}}_\gamma^k\| \ , \quad \hat{\mathfrak{f}}_c := \hat{\mathfrak{F}}_c\| \ . \tag{2.12}$$

The covariant matter fields are contained within the super-Weyl tensor W_{abc} and its independent descendants as follows:

$$T^-_{abc} := -2W_{abc}| \ , \tag{2.13a}$$

$$\chi^{\alpha i} := \frac{15}{2}X^{\alpha i}| = -\frac{3\mathrm{i}}{4}\nabla_\beta^i W^{\alpha\beta}| \ , \tag{2.13b}$$

$$D := \frac{15}{2}Y| = -\frac{3\mathrm{i}}{16}\nabla_\alpha^k \nabla_{\beta k} W^{\alpha\beta}| \ . \tag{2.13c}$$

The differential constraints on the superfield $W^{\alpha\beta}$ do not allow for any other independent descendents apart from those that correspond to curvatures but these are composites of the already defined component fields.

By taking the double bar projection of $\hat{\nabla}$, we define the component vector covariant derivative $\hat{\nabla}_a$ to coincide with the projection of the superspace derivative $\hat{\nabla}_a|$,

$$e_m{}^a\hat{\nabla}_a = \partial_m - \frac{1}{2}\psi_m{}_i^\alpha \nabla_\alpha^i| - \frac{1}{2}\hat{\omega}_m{}^{cd}M_{cd} - b_m\mathbb{D} - \mathcal{V}_m{}^{kl}J_{kl} - \frac{1}{2}\hat{\phi}_m{}_\alpha^i S_i^\alpha - \hat{\mathfrak{f}}_{ma}K^a \ . \tag{2.14}$$

The projected spinor covariant derivative $\nabla_\alpha^i|$ corresponds to the generator of Q-supersymmetry, and is defined so that if $\mathcal{U} = U|$, then $Q_\alpha^i\mathcal{U} = \nabla_\alpha^i|\mathcal{U} := (\nabla_\alpha^i U)|$. Note that there is no ambiguity for the other generators as e.g. $M_{cd}\mathcal{U} = (M_{cd}U)|$, and so local diffeomorphisms, Q-supersymmetry transformations, and so forth descend naturally from their corresponding rule in superspace.

The component supercovariant curvature tensors are defined by the expressions $\hat{R}(P)_{ab}{}^c = \hat{T}_{ab}{}^c|$ and $\hat{R}(Q)_{ab}{}_k^\gamma = \hat{T}_{ab}{}_k^\gamma|$, and with $\hat{R}(M)_{ab}{}^{cd}$, $\hat{R}(J)_{ab}{}^{kl}$, $\hat{R}(\mathbb{D})_{ab}$, $\hat{R}(S)_{ab}{}_\gamma^k$ and $\hat{R}(K)_{abc}$ coinciding with the lowest components of the corresponding superspace curvatures. The constraints on the superspace curvatures determine how to supercovariantize a given component

[5] In what follows a single line next to a superfield denotes setting $\theta = 0$.

curvature. One first need to simply take the double bar projection of the superspace torsion and each of the superspace curvatures. Upon doing so one finds [7]

$$\begin{aligned}
\hat{R}(P)_{ab}{}^c &= 0\,, && (2.15a)\\
\hat{R}(Q)_{abk} &= \frac{1}{2}\hat{\Psi}_{abk} + \mathrm{i}\tilde{\gamma}_{[a}\hat{\phi}_{b]k} + \frac{1}{24}T^{-}_{cde}\tilde{\gamma}^{cde}\gamma_{[a}\psi_{b]k}\,, && (2.15b)\\
\hat{R}(\mathbb{D})_{ab} &= 2e_a{}^m e_b{}^n\partial_{[m}b_{n]} + 4\hat{\mathfrak{f}}_{[ab]} + \psi_{[a}{}^i\hat{\phi}_{b]i} + \frac{\mathrm{i}}{15}\psi_{[a}{}^j\gamma_{b]}\chi_j\,, && (2.15c)\\
\hat{R}(M)_{ab}{}^{cd} &= \mathcal{R}_{ab}{}^{cd}(\hat{\omega}) + 8\delta^{[c}_{[a}\hat{\mathfrak{f}}_{b]}{}^{d]} + \mathrm{i}\psi_{[aj}\gamma_{b]}\hat{R}(Q)^{cdj} + 2\mathrm{i}\psi_{[aj}\gamma^{[c}\hat{R}(Q)_{b]}{}^{d]j}\\
&\quad -\psi_{[aj}\gamma^{cd}\hat{\phi}_{b]}{}^j - \frac{2\mathrm{i}}{15}\delta^{[c}_{[a}\psi_{b]j}\gamma^{d]}\chi^j + \frac{\mathrm{i}}{2}\psi_{[a}{}^j\gamma^e\psi_{b]j}\,T^{-cd}_{e}\,, && (2.15d)\\
\hat{R}(J)_{ab}{}^{kl} &= \mathcal{R}_{ab}{}^{kl}(\mathcal{V}) + 4\psi_{[a}{}^{(k}\hat{\phi}_{b]}{}^{l)} + \frac{4\mathrm{i}}{15}\psi_{[a}{}^{(k}\gamma_{b]}\chi^{l)}\,, && (2.15e)
\end{aligned}$$

where we have introduced the derivatives

$$\hat{\mathcal{D}}_m = \partial_m - \frac{1}{2}\hat{\omega}_m{}^{bc}M_{bc} - b_m\mathbb{D} - \mathcal{V}_m{}^{ij}J_{ij}\,, \quad \hat{\mathcal{D}}_a = e_a{}^m\hat{\mathcal{D}}_m\,, \tag{2.16}$$

together with

$$\begin{aligned}
\hat{\Psi}_{ab}{}^{\gamma}_{k} &:= 2e_a{}^m e_b{}^n\hat{\mathcal{D}}_{[m}\psi_{n]}{}^{\gamma}_{k}\,, && (2.17a)\\
\hat{\mathcal{R}}_{ab}{}^{cd} &:= \hat{\mathcal{R}}_{ab}{}^{cd}(\omega) = e_a{}^m e_b{}^n\Big(2\partial_{[m}\omega_{n]}{}^{cd} - 2\omega_{[m}{}^{ce}\omega_{n]e}{}^d\Big)\,, && (2.17b)\\
\hat{\mathcal{R}}_{ab}{}^{kl} &:= \hat{\mathcal{R}}_{ab}{}^{kl}(\mathcal{V}) = e_a{}^m e_b{}^n\Big(2\partial_{[m}\mathcal{V}_{n]}{}^{kl} + 2\mathcal{V}_{[m}{}^{p(k}\mathcal{V}_{n]p}{}^{l)}\Big)\,. && (2.17c)
\end{aligned}$$

On the other hand, the superspace curvatures are constrained to imply

$$\begin{aligned}
\hat{R}(P)_{ab}{}^c &= 0\,, && (2.18a)\\
\gamma^b\hat{R}(Q)_{abk} &= 0\,, && (2.18b)\\
\hat{R}(M)_{ac}{}^{bc} &= 0\,, && (2.18c)
\end{aligned}$$

which explains why we called the choice of vector covariant derivatives the "traceless" frame since these conditions are modified using a different choice of vector covariant derivative. The conditions (2.18) allow one to solve for the composite connections as follows:

$$\begin{aligned}
\hat{\omega}_{abc} &= \omega(e)_{abc} - 2\eta_{a[b}b_{c]} - \frac{\mathrm{i}}{4}\psi_b{}^k\gamma_a\psi_{ck} - \frac{\mathrm{i}}{2}\psi_a{}^k\gamma_{[b}\psi_{c]k}\,, && (2.19a)\\
\hat{\phi}_m{}^k &= \frac{\mathrm{i}}{16}\Big(\gamma^{bc}\gamma_m - \frac{3}{5}\gamma_m\tilde{\gamma}^{bc}\Big)\Big(\hat{\Psi}_{bc}{}^k + \frac{1}{12}T^{-}_{def}\tilde{\gamma}^{def}\gamma_{[b}\psi_{c]}{}^k\Big)\,, && (2.19b)\\
\hat{\mathfrak{f}}_a{}^b &= -\frac{1}{8}\mathcal{R}_a{}^b(\hat{\omega}) + \frac{1}{80}\delta^b_a\mathcal{R}(\hat{\omega}) + \frac{1}{8}\psi_{[aj}\gamma^{bc}\hat{\phi}_{c]}{}^j\\
&\quad -\frac{1}{80}\delta^b_a\psi_{cj}\gamma^{cd}\hat{\phi}_d{}^j + \frac{\mathrm{i}}{16}\psi_{cj}\gamma_a\hat{R}(Q)^{bcj} + \frac{\mathrm{i}}{8}\psi_{cj}\gamma^{[b}\hat{R}(Q)_a{}^{c]j} + \frac{\mathrm{i}}{60}\psi_{aj}\gamma^b\chi^j\\
&\quad +\frac{\mathrm{i}}{16}\psi_a{}^j\gamma_c\psi_{dj}\,T^{-bcd} - \frac{\mathrm{i}}{160}\delta^b_a\psi_c{}^j\gamma_d\psi_{ej}\,T^{-cde}\,. && (2.19c)
\end{aligned}$$

One could now plug the above results into eqs. (2.15) to arrive at expressions for the curvatures. Here we will simply refer the reader to [7] for the details. We also refer the reader there for supersymmetry transformations of the independent component fields and expressions for the curvatures $\hat{R}(S)$ and $\hat{R}(K)$. The above results represent the main results necessary for the component reduction of the conformal supergravity invariants from their superspace description.

IOP Conf. Series: Journal of Physics: Conf. Series **965** (2018) 012045 doi:10.1088/1742-6596/965/1/012045

3. The $\mathcal{N} = (1,0)$ superconformal gravity invariants

It turns out there are exactly two independent $\mathcal{N} = (1,0)$ superconformal gravity invariants that were constructed in superspace in [8]. One of these invariants contains a certain combination of C^3 terms at the component level, while the other contains a $C\Box C$ term at the component level. Their construction is based on the use of the superform approach to constructing supersymmetric invariants [22, 23, 24].

The superform approach to constructing supersymmetric invariants is a general approach based on the idea that a closed super D-form automatically leads to a supersymmetric invariant in D dimensions. In six-dimensions, the approach starts with a six-form $J = \frac{1}{6!}\mathrm{d}z^{M_6} \wedge \cdots \wedge \mathrm{d}z^{M_1} J_{M_1\cdots M_6}$ that is required to be closed $\mathrm{d}J = 0$. The supersymmetric invariant is given by

$$S = \int_{\mathcal{M}^6} i^* J = \int \mathrm{d}^6 x \, e \, {}^*J|_{\theta=0} \, , \qquad {}^*J := \frac{1}{6!}\varepsilon^{mnpqrs} J_{mnpqrs} \, , \tag{3.1}$$

where $i : \mathcal{M}^6 \to \mathcal{M}^{6|8}$ is the inclusion map and i^* is its pullback, the effect of which is to project $\theta^\mu_i = \mathrm{d}\theta^\mu_i = 0$. The closure of J, $\mathrm{d}J = 0$, ensures that the action (3.1) is invariant under super-diffeomorphisms. It is also required that J transform by (at most) an exact form under all gauge transformations. The component action may then be obtained as an expansion in higher products of the gravitino fields by expressing the action in terms of the tangent frame[6]

$$\begin{aligned} S &= \int \mathrm{d}^6 x \, \frac{1}{6!}\varepsilon^{m_1\cdots m_6} E_{m_6}{}^{A_6} \cdots E_{m_1}{}^{A_1} J_{A_1\cdots A_6}|_{\theta=0} \, , \\ &= \frac{1}{6!}\int \mathrm{d}^6 x \, e \, \varepsilon^{a_1\cdots a_6}\Big[J_{a_1\cdots a_6} + 3\psi_{a_1}{}^\alpha_i J^i_{\alpha a_2\cdots a_6} + \frac{15}{4}\psi_{a_2}{}^\beta_j \psi_{a_1}{}^\alpha_i J^{ij}_{\alpha\beta a_3\cdots a_6} \\ &\quad + \frac{5}{2}\psi_{a_3}{}^\gamma_k \psi_{a_2}{}^\beta_j \psi_{a_1}{}^\alpha_i J^{ijk}_{\alpha\beta\gamma a_4 a_5 a_6} + \mathcal{O}(\psi^4)\Big]|_{\theta=0} \, . \end{aligned} \tag{3.2}$$

Appropriate closed superforms describing the supersymmetric C^3 and $C\Box C$ invariants were given in [8]. They were entirely built out of primary composite superfields that were in turn built out the the super-Weyl tensor and their covariant derivatives. The correspondence outlined in the previous section from superspace to components as well as eq. (3.2) allows one to convert these superforms into their corresponding component actions. We do not provide the tedious details of this procedure here and instead simply present a large sector of the bosonic terms. We refer the reader to [7] for the details, including the full component actions.

3.1. The supersymmetric C^3 invariant

Below we present the supersymmetric C^3 invariant. For the sake of brevity we only give the bosonic sector and suppress terms involving the covariant field T^-_{abc}. The component Lagrangian is given by

$$\begin{aligned} \mathcal{L}_{C^3} &= C_{abcd}C^{abef}C^{cd}{}_{ef} + 2\,C_{abcd}C^{aefc}C^b{}_e{}^d{}_f - \frac{3}{4}C_{abcd}\mathcal{R}^{ab\,ij}\mathcal{R}^{cd}{}_{ij} + \frac{3}{2}\mathcal{R}_{ab}{}^{ij}\mathcal{R}^{ac}{}_i{}^k\mathcal{R}^b{}_{c\,jk} \\ &\quad + \frac{3}{16}\varepsilon^{abcdef}\mathcal{R}_{abi}{}^j\mathcal{R}_{cdj}{}^k\mathcal{R}_{efk}{}^i - \frac{4}{75}D^3 - \frac{1}{10}DC_{abcd}C^{abcd} + \frac{8}{5}D\mathcal{R}_{ab}{}^{ij}\mathcal{R}^{ab}{}_{ij} \\ &\quad + \mathcal{O}(T^-) + \text{fermion terms} \, , \end{aligned} \tag{3.3}$$

where $\mathcal{O}(T^-)$ denotes the terms involving the covariant field T^-_{abc}. We have included the SU(2) Chern-Simons form as the first term on line 2 since in this form the Lagrangian is completely superconformally invariant.

[6] The expansion should be slightly modified if J is not invariant under additional gauge transformations. We refer the reader to [8] for the details.

3.2. The supersymmetric $C\Box C$ invariant

The $C\Box C$ invariant can also be given and it is described by the following component Lagrangian:

$$\begin{aligned}\mathcal{L}_{C\Box C} &= C_{abcd}(\delta^a_e\Box - 4\mathcal{R}_e{}^a + \frac{6}{5}\delta^a_e\mathcal{R})C^{ebcd} + 4C_{abcd}C^{aefc}C^b{}_e{}^d{}_f - C_{ab}{}^{cd}C_{cd}{}^{ef}C_{ef}{}^{ab}\\ &\quad -3\mathcal{R}_{abij}\Box\mathcal{R}^{abij} + 3\mathcal{R}_a{}^b\mathcal{R}^{acij}\mathcal{R}_{bcij} - \frac{3}{2}\mathcal{R}\mathcal{R}^{abij}\mathcal{R}_{abij} + 6\,\mathcal{R}_a{}^b{}_i{}^j\mathcal{R}_b{}^c{}_j{}^k\mathcal{R}_c{}^a{}_k{}^i\\ &\quad +6\,C^{abcd}\,\mathcal{R}_{ab}{}^{ij}\mathcal{R}_{cd\,ij} + \tfrac{4}{15}\,D\Box D + \frac{4}{75}\mathcal{R}D^2 + \tfrac{8}{75}\,D^3 + \tfrac{2}{5}D\,C_{abcd}C^{abcd}\\ &\quad -\tfrac{14}{5}D\,\mathcal{R}_{ab}{}^{ij}\mathcal{R}^{ab}{}_{ij} + \mathcal{O}(T^-) + \text{fermion terms} + \text{total derivative}\,, \end{aligned} \tag{3.4}$$

where we have introduced the definition $\Box := \hat{\mathcal{D}}^a\hat{\mathcal{D}}_a$ and for simplicity we have suppressed a total derivative.

4. The $\mathcal{N} = (2,0)$ superconformal gravity invariant

The $\mathcal{N} = (2,0)$ Weyl multiplet of conformal supergravity was constructed in [21]. Many of the formulae there could actually be fixed by considering their truncations from $(2,0)$ to $(1,0)$. In a similar way, almost all of the bosonic terms of a $(2,0)$ conformal supergravity invariant can be obtained. In this section, we show that there is at most one $(2,0)$ conformal supergravity invariant and present a significant portion of its bosonic terms following [7].

We begin by recalling the component structure of the Weyl multiplet of $\mathcal{N} = (2,0)$ conformal supergravity [21]. In the off-shell gauging of the $(2,0)$ superconformal algebra, one associates the following independent fields with the local translations, Q-supersymmetry, USp(4) R-symmetry, and the dilatations: the vielbein $e_m{}^a$, the gravitino $\psi_m{}^i$, the USp(4) gauge field $\mathcal{V}_m{}^{ij}$, and the dilatation gauge field b_m. The remaining gauge symmetries are associated with composite connections, which include the spin connection $\omega_m{}^{cd}$, the S-supersymmetry connection $\phi_m{}^i$ and the special conformal connection $\mathfrak{f}_{ma}$. An off-shell representation of the conformal supersymmetry algebra is achieved by introducing three covariant matter fields: $T_{abc}{}^{ij} = T_{[abc]}{}^{[ij]}$, $\chi^{i,jk} = \chi^{i,[jk]}$ and $D^{ij}{}_{kl} = D^{[ij]}{}_{[kl]} = D_{kl}{}^{ij}$. Here $T_{abc}{}^{ij}$ is anti-self-dual with respect to its Lorentz vector indices and all covariant matter fields of the Weyl multiplet are traceless with respect to the invariant antisymmetric tensor Ω^{ij} of USp(4). These covariant fields are used to build the full covariant curvatures given in [21].

We now turn to outlining how to perform the truncation from $(2,0)$ to $(1,0)$. The $(2,0)$ Weyl multiplet decomposes into a number of $(1,0)$ multiplets, including a Weyl multiplet on which half of the supersymmetry is manifest, two gravitini multiplets associated with the extra spin-3/2 gauge fields, and a Yang-Mills multiplet associated with the extra R-symmetry connections. To obtain only the fields of the $(1,0)$ Weyl multiplet we truncate away the additional gravitino multiplets and Yang-Mills multiplet as in [21]. We split the USp(4) indices $i = 1,\cdots,4$ to $(i = 1,2, i' = 1,2)$ and switch off the third and fourth gravitini $\psi_m{}^{i'} = 0$. We also identify $\mathcal{V}_m{}^{ij} = \hat{\mathcal{V}}_m{}^{ij}$ and switch off the other components of the USp(4) connection $\mathcal{V}_m{}^{i'j'} = \mathcal{V}_m{}^{ij'} = 0$. These constraints induce restrictions on the covariant fields of the Weyl multiplet so that the Q- and S-supersymmetry transformations are consistent. These constraints are given by

$$T_{abc}{}^{ij} = \varepsilon^{ij}T^-_{abc}\,, \qquad T_{abc}{}^{i'j'} = -\varepsilon^{i'j'}T^-_{abc}\,, \tag{4.1a}$$

$$\chi_i{}^{jk} = \varepsilon^{jk}\chi_i\,, \quad \chi_i{}^{j'k'} = -\varepsilon^{j'k'}\chi_i\,, \quad \chi_{i'}{}^{j'k} = \frac{1}{2}\delta^{j'}_{i'}\chi^k\,, \tag{4.1b}$$

$$D^{ij}{}_{kl} = -\varepsilon^{ij}\varepsilon_{kl}D\,, \quad D^{ij}{}_{k'l'} = \varepsilon^{ij}\varepsilon_{k'l'}D\,, \quad D^{i'j'}{}_{k'l'} = -\varepsilon^{i'j'}\varepsilon_{k'l'}D\,, \tag{4.1c}$$

$$D^{ij'}{}_{kl'} = -\frac{1}{2}\delta^i_k\delta^{j'}_{l'}D\,. \tag{4.1d}$$

Now we are in a position to consider the $(2,0)$ uplift of our $(1,0)$ conformal supergravity invariants. However, it becomes immediately apparent that each of the two $(1,0)$ invariants can not be uplifted due to the presence of the term $DC^{abcd}C_{abcd}$ in their Lagrangians. The problem is that there is no scalar built out of the fields of the $(2,0)$ Weyl multiplet which will truncate to this $(1,0)$ term. However, there is precisely one linear combination for which all such problematic terms cancel and it corresponds to

$$I_{C\Box C} + 4I_{C^3} , \tag{4.2}$$

which tells us that the $(2,0)$ combination must be unique. The $(2,0)$ uplift of the invariant fixes almost all the bosonic terms except for a few involving the covariant field $T_{abc}{}^{ij}$. For our purposes here, we will ignore all terms explicitly involving this covariant field. The result of the uplift gives the following bosonic terms for the $(2,0)$ invariant:

$$\begin{aligned}\mathcal{L}_{(2,0)} = \; & C_{abcd}(\delta^a_e\Box - 4\mathcal{R}_e{}^a + \frac{6}{5}\delta^a_e\mathcal{R})C^{ebcd} + 12C_{abcd}C^{aefc}C^b{}_e{}^d{}_f + 3C_{ab}{}^{cd}C_{cd}{}^{ef}C_{ef}{}^{ab} \\ & -3\mathcal{R}_{ab}{}^{ij}\Box\mathcal{R}^{ab}{}_{ij} + 3\mathcal{R}_a{}^b\mathcal{R}^{acij}\mathcal{R}_{bcij} - \frac{3}{2}\mathcal{R}\mathcal{R}^{abij}\mathcal{R}_{abij} + 6\,\mathcal{R}_a{}^b{}_i{}^j\mathcal{R}_b{}^c{}_j{}^k\mathcal{R}_c{}^a{}_k{}^i \\ & +3\,C^{abcd}\mathcal{R}_{ab}{}^{ij}\mathcal{R}_{cd\,ij} + \frac{1}{75}D^{ij}{}_{kl}\Box D^{kl}{}_{ij} + \frac{1}{375}\mathcal{R}D^2 - \frac{2}{1125}D^{ij}{}_{kl}D^{kl}{}_{pq}D^{pq}{}_{ij} \\ & -\frac{2}{5}D^{ij}{}_{kl}R(V)_{abi}{}^kR(V)^{ab}{}_j{}^l + \mathcal{O}(T) + \text{fermion terms} + \text{total derivative} ,\end{aligned} \tag{4.3}$$

where $\mathcal{O}(T)$ represents terms involving $T_{abc}{}^{ij}$.

5. Conclusion

In conclusion, we presented the 6D $\mathcal{N} = (1,0)$ conformal superspace formulation of conformal supergravity constructed in [8] and showed how it encodes the component fields of the Weyl multiplet. We presented the linearly independent component invariants for $\mathcal{N} = (1,0)$ conformal supergravity constructed in [7], which to date are probably the most complicated supergravity actions in six dimensions. Remarkably, these results allowed us to obtain a great deal of information about the component action of the $\mathcal{N} = (2,0)$ conformal supergravity invariant for which one can recover almost all its bosonic terms.

Using the results presented here, we can provide an independent derivation to the ones in the literature of the relations (1.5). The argument goes as follows. A general conformal gravity invariant may be written in terms of the purely gravitational parts of the $(1,0)$ and $(2,0)$ invariants as

$$c_1\mathcal{L}_1 + c_2\mathcal{L}_2 + c_3\mathcal{L}_3 = (c_1 - 2c_2 - 6c_3)\mathcal{L}_1 + (c_2 - 3c_3)\mathcal{L}_{C^3} + c_3\mathcal{L}_{(2,0)} . \tag{5.1}$$

For the $\mathcal{N} = (1,0)$ case, the first term on the right hand side must vanish since such a combination cannot be supersymmetrised, which requires the first condition of (1.5). For the $\mathcal{N} = (2,0)$ case, the second term on the right hand side must vanish since $\mathcal{L}_{C^3}$ cannot be completed alone into a (2, 0) invariant, which gives the second constraint of (1.5). Thus the constraints (1.5) follow directly from a completely supersymmetric argument.

Acknowledgments

We are grateful to Daniel Butter for collaboration on related projects. JN is supported by a Humboldt research fellowship of the Alexander von Humboldt Foundation. The work of GT-M was supported in part by the Interuniversity Attraction Poles Programme initiated by the Belgian Science Policy (P7/37) and in part by the KU Leuven C1 grant ZKD1118 C16/16/005.

IOP Conf. Series: Journal of Physics: Conf. Series **965** (2018) 012045 doi:10.1088/1742-6596/965/1/012045

References

[1] Deser S and Schwimmer A 1993 *Phys. Lett.* B **309** 279
[2] Nahm W 1978 *Nucl. Phys.* B **135** 149
[3] Bergshoeff E, Sezgin E and Van Proeyen A 1986 *Nucl. Phys.* B **264** 653
[4] Bastianelli F, Frolov S and Tseytlin A A 2000 *JHEP* **0002** 013
[5] Kulaxizi M and Parnachev A 2010 *Phys. Rev.* D **82** 066001
[6] Beccaria M and Tseytlin A A 2016 *JHEP* **1601** 001
[7] Butter D, Novak J and Tartaglino-Mazzucchelli G 2017 *JHEP* **1705** 133
[8] Butter D, Kuzenko S M, Novak J and Theisen S 2016 *JHEP* **1612** 072
[9] Bergshoeff E, Salam A and Sezgin E 1986 *Phys. Lett.* B **173** 73
[10] Howe P S 1981 *Phys. Lett.* B **100** 389
[11] Howe P S 1982 *Nucl. Phys.* B **199** 309
[12] Kuzenko S M and Tartaglino-Mazzucchelli G 2008 *JHEP* **0804** 032
[13] Linch III W D and Tartaglino-Mazzucchelli G 2012 *JHEP* **1208** 075
[14] Butter D 2010 *Annals Phys.* **325** 1026
[15] Butter D 2011 *JHEP* **1110** 030
[16] Butter D, Kuzenko S M, Novak J and Tartaglino-Mazzucchelli G 2013 *JHEP* **1309** 072
[17] Butter D, Kuzenko S M, Novak J and Tartaglino-Mazzucchelli G 2015 *JHEP* **1502**, 111
[18] Butter D, Kuzenko S M, Novak J and Tartaglino-Mazzucchelli G 2013 *JHEP* **1310**, 073
[19] Kuzenko S M, Novak J and Tartaglino-Mazzucchelli G 2014 *JHEP* **1401**, 121
[20] Kuzenko S M, Novak J and Tartaglino-Mazzucchelli G 2015 *JHEP* **1509**, 081
[21] Bergshoeff E, Sezgin E and Van Proeyen A 1999 *Class. Quant. Grav.* **16** 3193
[22] Hasler M F 1998 *Eur. Phys. J.* C **1** 729
[23] Gates Jr S J 1999 *Supersymmetries and Quantum Symmetries* ed J Wess and E A Ivanov (Berlin: Springer) p 46
[24] Gates Jr S J 1999 *Nucl. Phys.* B **541** 615

Institute of Physics
Dirac House, Temple Back
Bristol BS1 6BE UK

ISSN: 1742-6588
ISBN 978-1-5108-5852-7
9 781510 858527

CURRAN ASSOCIATES INC.
proceedings.com